Safety Symbols

Safety symbols in the following table are used in the lab activities to indicate possible hazards. Learn the meaning of each symbol. **It is recommended that you wear safety goggles and apron at all times in the lab. This might be required in your school district.**

Safety Symbols	Hazard	Examples	Precaution	Remedy
Disposal	Special disposal procedures need to be followed.	certain chemicals, living organisms	Do not dispose of these materials in the sink or trash can.	Dispose of wastes as directed by your teacher.
Biological	Organisms or other biological materials that might be harmful to humans	bacteria, fungi, blood, unpreserved tissues, plant materials	Avoid skin contact with these materials. Wear mask or gloves.	Notify your teacher if you suspect contact with material. Wash hands thoroughly.
Extreme Temperature	Objects that can burn skin by being too cold or too hot	boiling liquids, hot plates, dry ice, liquid nitrogen	Use proper protection when handling.	Go to your teacher for first aid.
Sharp Object	Use of tools or glassware that can easily puncture or slice skin	razor blades, pins, scalpels, pointed tools, dissecting probes, broken glass	Practice common-sense behavior and follow guidelines for use of the tool.	Go to your teacher for first aid.
Fume	Possible danger to respiratory tract from fumes	ammonia, acetone, nail polish remover, heated sulfur, moth balls	Be sure there is good ventilation. Never smell fumes directly. Wear a mask.	Leave foul area and notify your teacher immediately.
Electrical	Possible danger from electrical shock or burn	improper grounding, liquid spills, short circuits, exposed wires	Double-check setup with teacher. Check condition of wires and apparatus. Use GFI-protected outlets.	Do not attempt to fix electrical problems. Notify your teacher immediately.
Irritant	Substances that can irritate the skin or mucous membranes of the respiratory tract	pollen, moth balls, steel wool, fiberglass, potassium permanganate	Wear dust mask and gloves. Practice extra care when handling these materials.	Go to your teacher for first aid.
Chemical	Chemicals that can react with and destroy tissue and other materials	bleaches such as hydrogen peroxide; acids such as sulfuric acid, hydrochloric acid; bases such as ammonia, sodium hydroxide	Wear goggles, gloves, and an apron.	Immediately flush the affected area with water and notify your teacher.
Toxic	Substance may be poisonous if touched, inhaled, or swallowed.	mercury, many metal compounds, iodine, poinsettia plant parts	Follow your teacher's instructions.	Always wash hands thoroughly after use. Go to your teacher for first aid.
Flammable	Flammable chemicals may be ignited by open flame, spark, or exposed heat.	alcohol, kerosene, potassium permanganate	Avoid open flames and heat when using flammable chemicals.	Notify your teacher immediately. Use fire safety equipment if applicable.
Open Flame	Open flame in use, may cause fire.	hair, clothing, paper, synthetic materials	Tie back hair and loose clothing. Follow teacher's instruction on lighting and extinguishing flames.	Notify your teacher immediately. Use fire safety equipment if applicable.

Eye Safety
Proper eye protection should be worn at all times by anyone performing or observing science activities.

Clothing Protection
This symbol appears when substances could stain or burn clothing.

Animal Safety
This symbol appears when safety of animals and students must be ensured.

Radioactivity
This symbol appears when radioactive materials are used.

Handwashing
After the lab, wash hands with soap and water before removing goggles.

Inspire
Physical Science

McGraw Hill

Phenomenon: Refraction and Reflection

Light bends when it travels through the crystal ball, showing an inverted image of the coast.

Fun Fact

The distance between the coast and the crystal ball determines the size of the image as well as whether or not the image is flipped.

FRONT COVER: Delphotos/Alamy Stock Photo. **BACK COVER:** Delphotos/Alamy Stock Photo.

mheducation.com/prek-12

Mc Graw Hill

Send all inquiries to:
McGraw-Hill Education
STEM Learning Solutions Center
8787 Orion Place
Columbus, OH 43240

ISBN: 978-0-07-668304-8
MHID: 0-07-668304-4

Printed in the United States of America.

8 9 10 LWI 25 24 23 22

STEM

McGraw-Hill is committed to providing instructional materials in Science, Technology, Engineering, and Mathematics (STEM) that give all students a solid foundation, one that prepares them for college and careers in the 21st century.

Welcome to

Inspire Physical Science

Explore Our Phenomenal World

The Inspire High School Series brings phenomena to the forefront of learning to engage and inspire students to investigate key science concepts through their three-dimensional learning experience.

Start exploring now!

Inspire Curiosity • Inspire Investigation • Inspire Innovation

WELCOME TO *INSPIRE PHYSICAL SCIENCE*

Owning Your Learning

1 Encounter the Phenomenon

Every day, you are surrounded by natural phenomena that makes you wonder.

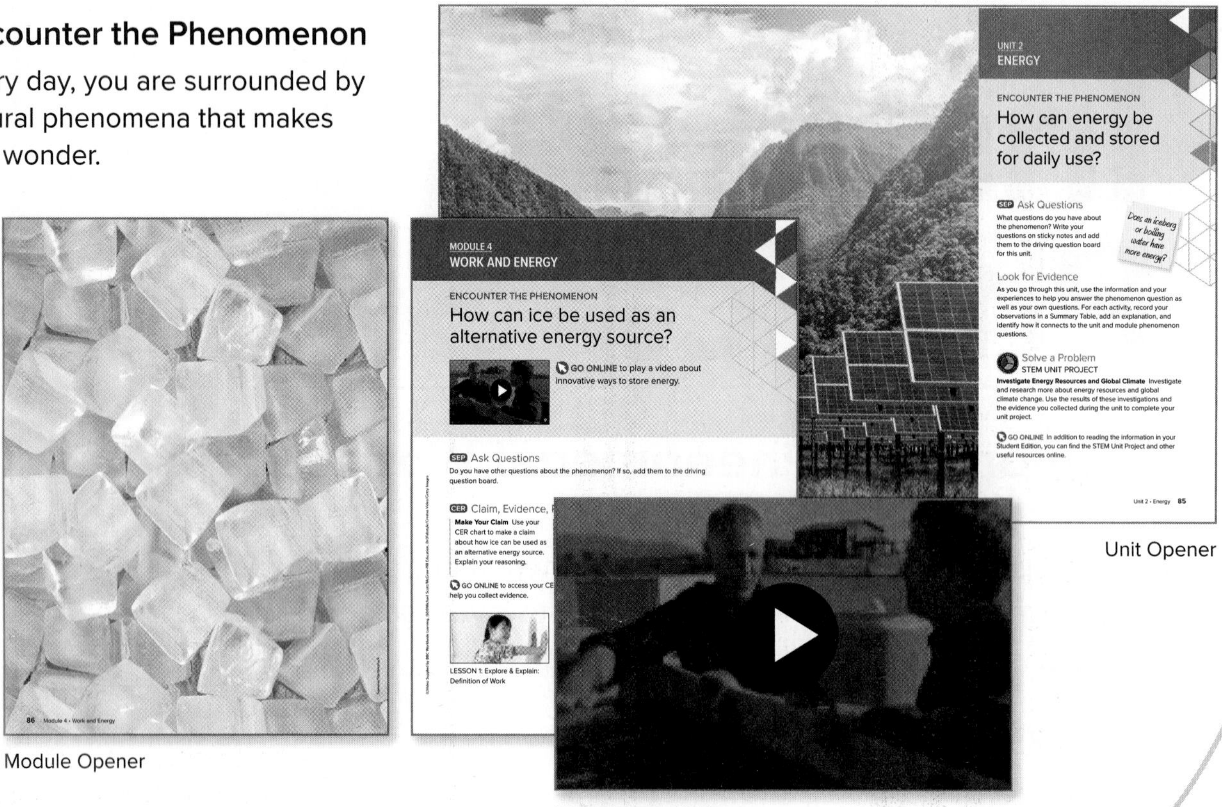

UNIT 2
ENERGY
ENCOUNTER THE PHENOMENON
How can energy be collected and stored for daily use?
Ask Questions
Does an iceberg or boiling water have more energy?
Look for Evidence
Solve a Problem
STEM UNIT PROJECT
Unit 2 • Energy 85
86 Module 4 • Work and Energy
MODULE 4
WORK AND ENERGY
ENCOUNTER THE PHENOMENON
How can ice be used as an alternative energy source?
GO ONLINE to play a video about innovative ways to store energy.
Ask Questions
Do you have other questions about the phenomenon? If so, add them to the driving question board.
Claim, Evidence,
Make Your Claim Use your CER chart to make a claim about how ice can be used as an alternative energy source. Explain your reasoning.
LESSON 1: Explore & Explain: Definition of Work

Unit Opener

Module Opener

Phenomenon Video

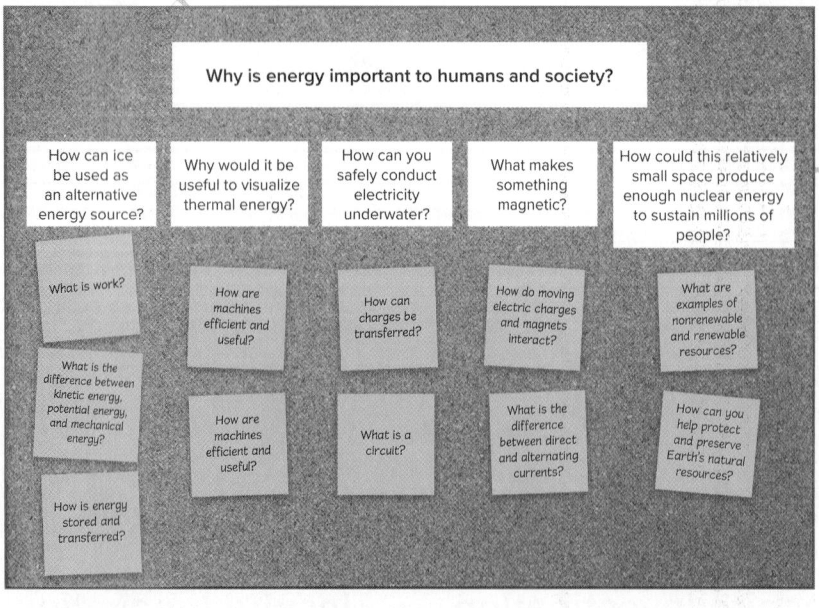

Why is energy important to humans and society?
How can ice be used as an alternative energy source?
Why would it be useful to visualize thermal energy?
How can you safely conduct electricity underwater?
What makes something magnetic?
How could this relatively small space produce enough nuclear energy to sustain millions of people?
What is work?
What is the difference between kinetic energy, potential energy, and mechanical energy?
How is energy stored and transferred?
How are machines efficient and useful?
How are machines efficient and useful?
How can charges be transferred?
What is a circuit?
How do moving electric charges and magnets interact?
What is the difference between direct and alternating currents?
What are examples of nonrenewable and renewable resources?
How can you help protect and preserve Earth's natural resources?

2 Ask Questions

At the beginning of each unit and module, make a list of the questions you have about the phenomenon. Share your questions with your classmates.

(t)Video Supplied by BBC Worldwide Learning, (b)moi/amanaimagesRF/Getty Images

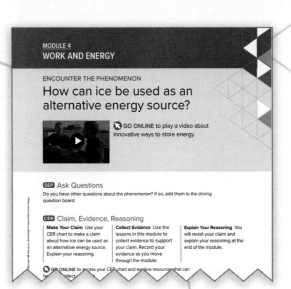

MODULE 4
WORK AND ENERGY

ENCOUNTER THE PHENOMENON

How can ice be used as an alternative energy source?

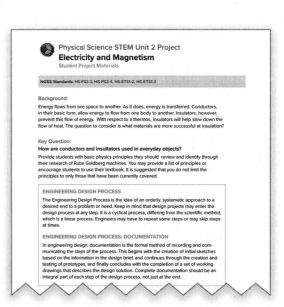

GO ONLINE to play a video about innovative ways to store energy.

SEP Ask Questions

Do you have other questions about the phenomenon? If so, add them to the driving question board.

CER Claim, Evidence, Reasoning

Make Your Claim Use your CER chart to make a claim about how ice can be used as an alternative energy source. Explain your reasoning.

Collect Evidence Use the lessons in this module to collect evidence to support your claim. Record your evidence as you move through the module.

Explain Your Reasoning You will revisit your claim and explain your reasoning at the end of the module.

GO ONLINE to access your CER chart and explore resources that can he... collect...

3 Claim, Evidence, Reasoning

As you investigate each phenomenon, you will write your claim, gather evidence by performing labs and completing reading assignments and Applying Practices, and explain your reasoning to answer the unit and module phenomena.

SUMMARY TABLE

Activity Model	Observation Evidence	Explanation Reasoning	Connection to Phenom	Questions Answered	New Questions
Applying Practices: Modeling Changes in Energy	Energy within a system can change forms and be transferred between parts of the system.	The kinetic energy of the molecules in a cup of hot water can be transferred to ice cubes, which causes them to melt.	Unit: Electrical energy can be collected and stored in batteries. Module: Ice can be used as a heat sink to convert heat into mechanical energy.	How is energy stored and transferred?	What other types of energy can be transformed into usable forms?

4 Summarize Your Work

When you collect evidence, you can record your data in a summary table and use the data to collaborate with others to answer the questions you had.

5 Apply Your Evidence and Reasoning

At the end of the unit, modules, and lessons, you can use all of the data you collected to help complete your STEM Unit Project.

Physical Science STEM Unit 2 Project
Electricity and Magnetism
Student Project Materials

NGSS Standards: HS-PS3-3, HS-PS3-4, HS-ETS1-2, HS-ETS1-3

Background:

Energy flows from one space to another. As it does, energy is transferred. Conductors, in their basic form, allow energy to flow from one body to another. Insulators, however, prevent this flow of energy. With respect to a thermos, insulators will help slow down the flow of heat. The question to consider is what materials are more successful at insulation?

Key Question:

How are conductors and insultators used in everyday objects?

Provide students with basic physics principles they should review and identify through their research of Rube Goldberg machines. You may provide a list of principles or encourage students to use their textbook. It is suggested that you do not limit the principles to only those that have been currently covered.

ENGINEERING DESIGN PROCESS

The Engineering Design Process is the idea of an orderly, systematic approach to a desired end to a problem or need. Keep in mind that design projects may enter the design process at any step. It is a cyclical process, differing from the scientific method, which is a linear process. Engineers may have to repeat some steps or may skip steps at times.

ENGINEERING DESIGN PROCESS: DOCUMENTATION

In engineering design, documentation is the formal method of recording and communicating the steps of the process. This begins with the creation of initial sketches based on the information in the design brief, and continues through the creation and testing of prototypes, and finally concludes with the completion of a set of working drawings that describes the design solution. Complete documentation should be an integral part of each step of the design process, not just at the end.

AUTHORS, ADVISORS, AND CONSULTANTS

High School Reviewers

Each teacher reviewed selected chapters of *Inspire Physical Science* and provided feedback and suggestions regarding the effectiveness of the instruction.

Danielle Chirip
University High School
Orlando, FL

Dwight Dutton
East Chapel Hill High School
Chapel Hill, NC

Candace Hebert
Alfred M. Barbe High School
Lake Charles, LA

Dr. Carol Jones
Science Consultant
Macomb Intermediate School District
Clinton Township, MI

Christina McCray
Leesville Road High School
Raleigh, NC

Shelli Pace
Captain Shreve High School
Shreveport, LA

Balmatie Sagramsingh
Cypress Creek High School
Orlando, FL

Tiffany Tammasini
Cypress Creek High School
Orlando, FL

Josephine K. Thirunayagam
Cypress Creek High School
Orlando, FL

Sara Tondra
Marysville High School
Marysville, OH

Elizabeth Trageser
Bentworth School District
Bentleyville, PA

Authors

Charles William McLaughlin, PhD
Teaching Professor, Chemistry
Montana State University
Bozeman, MT

Marilyn Thompson, PhD
Professor and Associate Director,
Measurement and Statistical Analysis
Arizona State
Tempe, AZ

Dinah Zike
Founder and president,
Dinah-Might Adventures

Content Consultants

Content consultants each reviewed selected chapters of *Inspire Physical Science* for content accuracy and clarity.

David G. Haase
Professor of Physics
North Carolina State University
Raleigh, NC

Michael O. Hurst
Associate Professor of Biochemistry
Georgia Southern University
Statesboro, GA

Sally Koutsoliotas
Associate Professor of Physics
Bucknell University
Lewisburg, PA

Dr. Maria Pacheco
Associate Professor of Chemistry
Buffalo State College
Buffalo, NY

Contributing Writers

Contributing writers helped develop chapter features.

Andrew Schroeder
Columbus, OH

Karen Sottosanti
Pickerington, OH

Stephen Whitt
Columbus, OH

Jenipher Willoughby
Forest, VA

Safety Consultant

The safety consultant reviewed lab and lab materials for safety and implementation.

Kenneth R. Roy, PhD
Director of Environmental Health and Safety
Glastonbury Public Schools
Glastonbury, CT

KEY PARTNERS

Smithsonian

Smithsonian

Following the mission of its founder James Smithson for "an establishment for the increase and diffusion of knowledge," the Smithsonian Institution today is the world's largest museum, education, and research complex. To further their vision of shaping the future, a wealth of Smithsonian online resources are integrated within this program.

SPONGELAB
A GLOBAL SCIENCE COMMUNITY

SpongeLab Interactives

SpongeLab Interactives is a learning technology company that inspires learning and engagement by creating gamified environments that encourage students to interact with digital learning experiences.

Students participate in inquiry activities and problem-solving to explore a variety of topics using games, interactives, and video while teachers take advantage of formative, summative, or performance-based assessment information that is gathered through the learning management systems.

PhET
INTERACTIVE SIMULATIONS
University of Colorado Boulder

PhET Interactive Simulations

The PhET Interactive Simulations project at the University of Colorado Boulder provides teacher and students with interactive science and math simulations. Based on extensive education research, PhET sims engage students through an intuitive, game-like environment where students learn through exploration and discovery.

UNIT 1

MOTION AND FORCES

ENCOUNTER THE PHENOMENON

Why did the person jump backward over the bar?

INTRODUCTION TO PHYSICAL SCIENCE

This module introduces the nature of science, what physical science is, and provides tools for the study of science.

MODULE 1: THE NATURE OF SCIENCE

MODULE 2: MOTION

MODULE 3: FORCES AND NEWTON'S LAWS

Robert Daly/OJO Images/Age Fotostock

Soonthorn Wongsaita/Shutterstock

UNIT 2

ENERGY

ENCOUNTER THE PHENOMENON

How can energy be collected and stored for daily use?

UNIT 3
WAVES

ENCOUNTER THE PHENOMENON

How do waves interact with our senses?

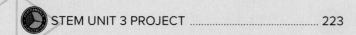

NASA images/Shutterstock

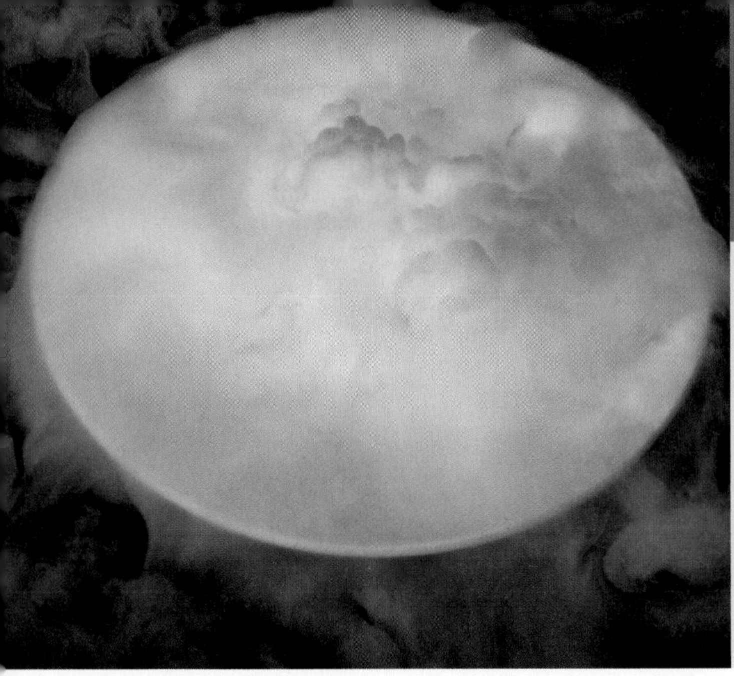

UNIT 4

MATTER

ENCOUNTER THE PHENOMENON

Why can dry ice go directly from a solid to a gas?

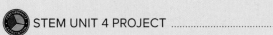

MODULE 14: SOLIDS, LIQUIDS, AND GASES

MODULE 15: CLASSIFICATION OF MATTER

MODULE 16: PROPERTIES OF ATOMS AND THE PERIODIC TABLE

MODULE 17: ELEMENTS AND THEIR PROPERTIES

UNIT 5
REACTIONS

ENCOUNTER THE PHENOMENON
Why are the jellyfish glowing?

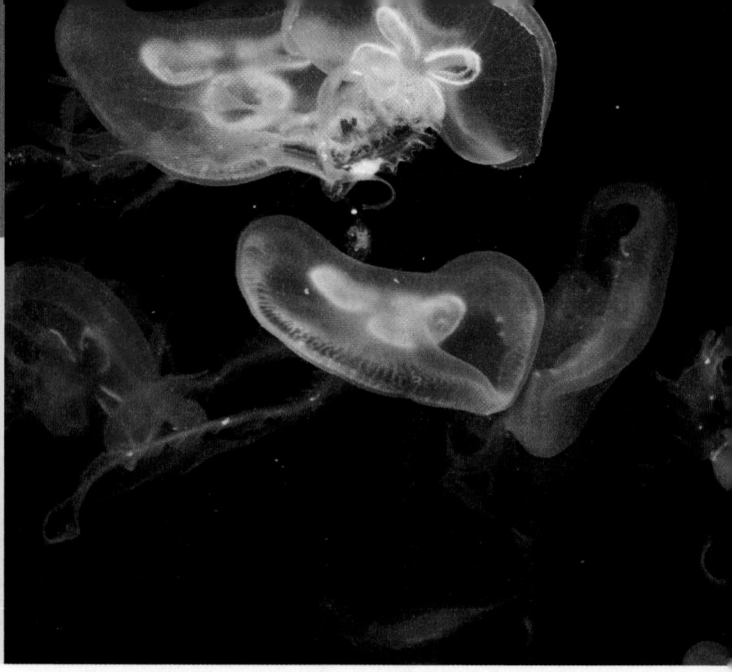

Emil O/Shutterstock

MODULE 18: CHEMICAL BONDS
ENCOUNTER THE PHENOMENON

MODULE 19: CHEMICAL REACTIONS
ENCOUNTER THE PHENOMENON

MODULE 20: RADIOACTIVITY AND NUCLEAR REACTIONS
ENCOUNTER THE PHENOMENON

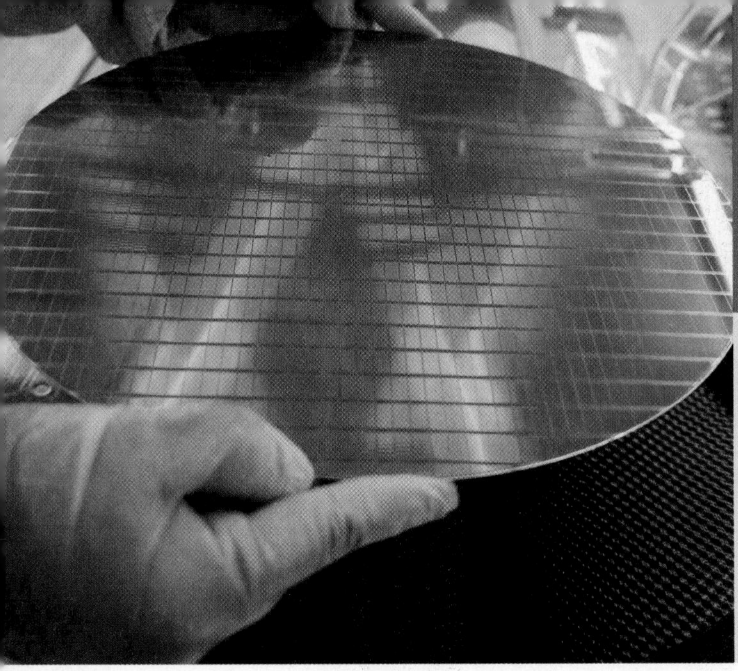

UNIT 6

APPLICATIONS OF CHEMISTRY

ENCOUNTER THE PHENOMENON

How are advancements in chemistry related to technology?

MODULE 21: SOLUTIONS

MODULE 22: ACIDS, BASES, AND SALTS

MODULE 23: ORGANIC COMPOUNDS

MODULE 24: NEW MATERIALS THROUGH CHEMISTRY

WeatherVideoHD.TV

MODULE 1

THE NATURE OF SCIENCE

ENCOUNTER THE PHENOMENON

How do we know how rainbows form?

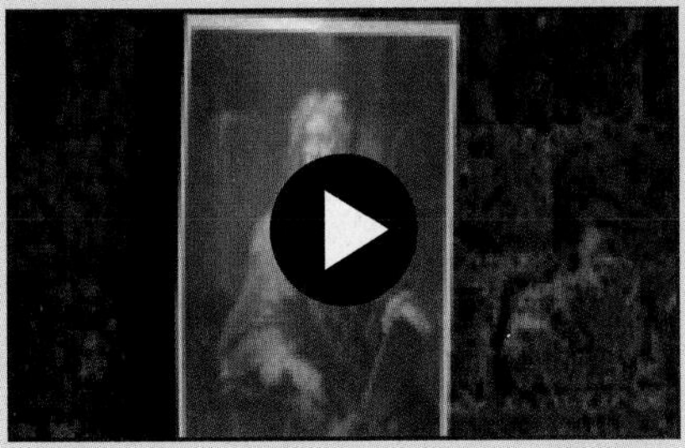

GO ONLINE to play a video about the way Newton studied light.

SEP Ask Questions

Do you have other questions about the phenomenon? If so, add them to the driving question board.

CER Claim, Evidence, Reasoning

Make Your Claim Use your CER chart to make a claim how rainbows form. Explain your reasoning.

Collect Evidence Use the lessons in this module to collect evidence to support your claim. Record your evidence as you move through the module.

Explain Your Reasoning You will revisit your claim and explain your reasoning at the end of the module.

GO ONLINE to access your CER chart and explore resources that can help you collect evidence.

LESSON 1: Explore & Explain: What is science?

LESSON 4: Explore & Explain: Global Technological Needs

Smithsonian

Additional Resources

LESSON 1

THE METHODS OF SCIENCE

FOCUS QUESTION

What are the steps of the methods of science?

What is science?

Science is not just a subject in school. It is a method for studying the natural world. After all, science comes from the Latin word *scientia,* which means "knowledge." Science is a process based on inquiry that helps develop explanations about events in nature.

Nature follows a set of rules. Many rules, such as those concerning how the human body works, are complex. Other rules, such as the fact that Earth rotates about once every 24 hours, are much simpler. Scientists, such as the one shown in **Figure 1,** ask questions and make observations to learn about the rules that govern the natural world.

Major categories of science

Science covers many different topics that can be classified according to three main categories: (1) Life science deals with living things. (2) Earth science investigates Earth and space. (3) Physical science studies matter and energy. Sometimes, though, a scientific study will overlap the categories. One scientist, for example, might study how to build better artificial limbs. Is this scientist studying energy and matter or how muscles operate? She is studying both life science and physical science.

Figure 1 This scientist is monitoring water quality from a solar-powered field laboratory.

Observe *In the photograph, what evidence do you see of the three main branches of science?*

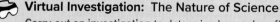

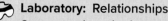

Tdub303/E+/Getty Images

3D THINKING DCI Disciplinary Core Ideas CCC Crosscutting Concepts SEP Science & Engineering Practices

COLLECT EVIDENCE

Use your Science Journal to record the evidence you collect as you complete the readings and activities in this lesson.

INVESTIGATE

GO ONLINE to find these activities and more resources.

Virtual Investigation: The Nature of Science
Carry out an investigation to determine how color affects heat absorption.

Laboratory: Relationships
Carry out an investigation to determine the patterns that exist between variables in an experiment.

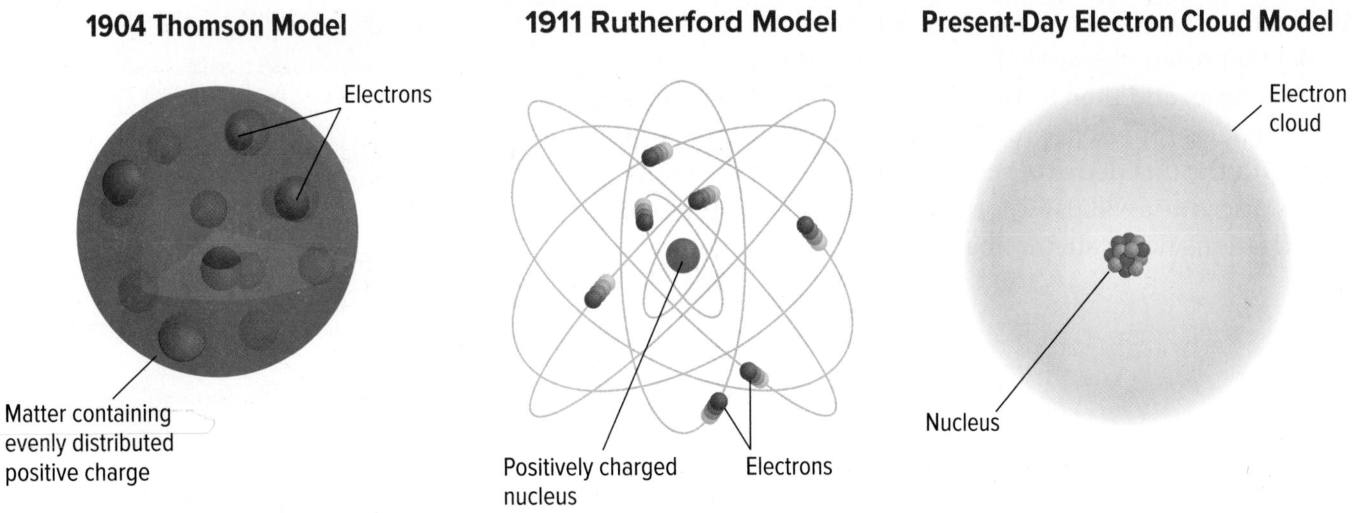

Figure 2 Our understanding of the atom has changed over time.

Science changes

Scientific explanations help us understand the natural world. Sometimes these explanations must be modified. As more is learned, earlier explanations might be found to be incomplete, or new technology might provide more accurate answers.

For example, scientists have been studying the atom for more than two centuries. Throughout this time, they have revised their thinking on what atoms might look like, how they interact, and even how they combine to form other substances.

In the early 1900s, British physicist J.J. Thomson created a model of the atom that consisted of electrons embedded in a ball of positive charge. Several years later, physicist Ernest Rutherford created a model of the atom based on new research. As shown in **Figure 2,** his model was different from Thomson's model. Instead of a solid ball, the atom consisted of a nucleus with electrons orbiting it like the planets orbit the Sun.

Later in the 20th century, scientists discovered the nucleus is not a solid ball but is made of protons and neutrons. This improved our understanding of the atom and its behavior. The present-day model of the atom is a nucleus made of protons and neutrons surrounded by an electron cloud. The electron cloud represents the space containing rapidly moving electrons.

The electron cloud model is the result of scientists pulling together evidence from many investigations. They came to an agreement that it is the best model for the information available at the time. Because it is the nature of science to be open to change, investigations into the model of the atom continue today.

Investigations

Scientists learn new information about the natural world by performing investigations. Some investigations involve simply observing something that occurs and recording the observations. Other investigations involve setting up experiments with a control to test the effect of one thing on another.

Modeling Sometimes an investigation involves building a model that resembles something, such as a model of a new space vehicle, and then testing the model to see how it acts. Other models represent processes or objects that cannot be seen with the unaided eye, such as the models of the atom. Often, a scientist will use information from several types of investigations when attempting to learn about the natural world.

Scientific Methods

Although scientists do not always follow a rigid set of steps, investigations often follow a general pattern. The pattern of investigation procedures is called the **scientific methods.** Six common steps found in the scientific methods are shown in **Figure 3.** A scientist might add new steps, repeat some steps many times, or skip steps altogether.

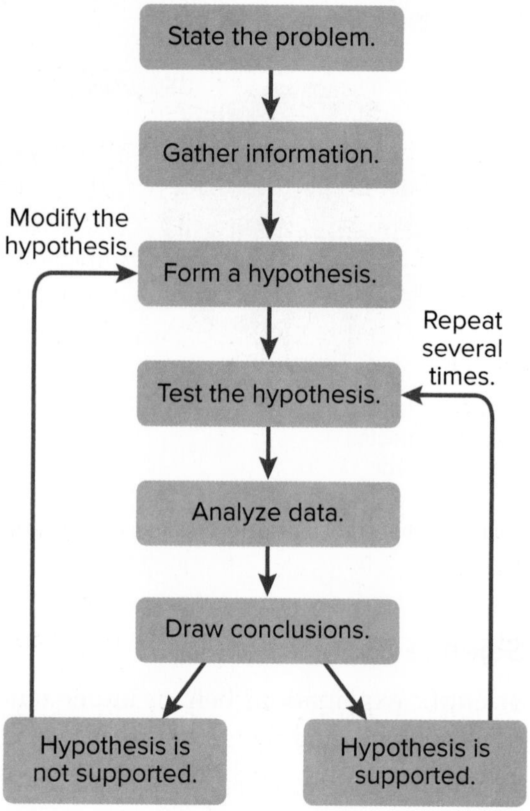

Figure 3 The series of procedures shown here is one way to use scientific methods to solve a problem.

State the problem

To begin the process, a scientist must state what he or she is going to investigate. Many investigations begin when someone observes an event in nature and wonders why or how it occurs. The question of "why" or "how" is the problem.

Scientists once posed questions about why objects fall to Earth, what causes day and night, and how to generate electricity for daily use. Many times, a statement of a problem arises when an investigation is complete and its results lead to new questions. For example, once scientists understood why we experience day and night, they wanted to know why Earth rotates.

Sometimes a new question is posed when an investigation runs into trouble. For example, some early work on guided missiles found the instruments in the nose cone did not always work properly. The original problem statement involved how to guide missiles during flight. The new statement involved how to protect the instruments in the nose cone.

Get It?

Identify What is the first step in a scientific investigation, and what form does it usually take?

Research and gather information

Before beginning an investigation, scientists research what is already known about the problem. They gather and examine observations and interpretations from reliable sources. This background helps scientists fine-tune their question and form a hypothesis.

Form a hypothesis

A **hypothesis** is a possible answer to a question or a possible solution to a problem based on what you know and what you observe. When trying to find a better material to protect the space shuttle, NASA scientists looked to other materials that were used in similar situations. Scientists knew that a ceramic coating had been found to solve the guided missile problem. They hypothesized that a ceramic material might work on the space shuttle also.

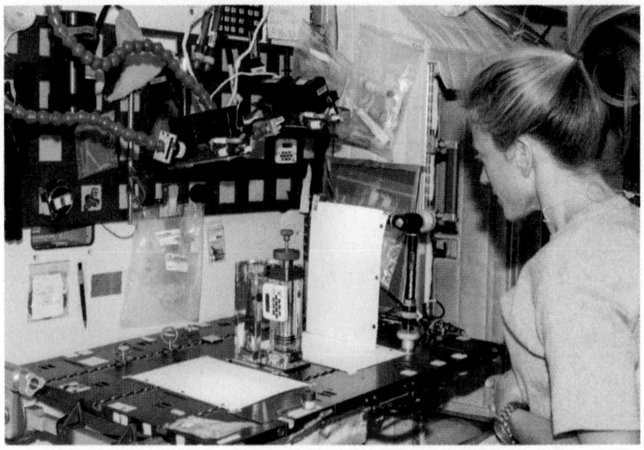

Figure 4 An astronaut aboard the *International Space Station* conducts an experiment session with the Capillary Flow Experiment (CFE). CFE observes the flow of fluid, in particular capillary phenomena, in microgravity.

Test a hypothesis

Some hypotheses can be tested by making observations. Others can be tested by building a model and relating it to real-life situations. One common way to test a hypothesis is to perform an experiment. An **experiment** tests the effect of one thing on another using a control.

Variables An experiment usually contains at least two variables. A **variable** is a quantity that can have more than a single value. **Table 1** summarizes the types of variables. For example, numerous experiments aboard space shuttles and the *International Space Station* (ISS) have studied the effects of microgravity on plants. Before these experiments could begin, scientists had to think of every factor that might affect plant growth. Each of these factors is a variable.

Independent and dependent variables In the microgravity experiment, plant growth is the **dependent variable** because its value changes according to the changes in the other variables. The variable that is changed to see how it will affect the dependent variable is called the **independent variable.** The microgravity is the independent variable. Scientists on the ISS are using microgravity as an independent variable in other experiments as well, as shown in **Figure 4.**

Constants To be sure they were testing to see how microgravity affects growth, mission specialists kept the other possible factors the same. A factor that does not change is called a **constant.** The microgravity experiments used the same soil and type of plant. Additionally, each plant was given the same amount of light and water and was kept at the same temperature. Type of soil, type of plant, amount of light, amount of water, and temperature were constants for this experiment.

Table 1 Types of Variables

Dependent Variable	changes according to the changes of the independent variable
Independent Variable	the variable that is changed to test the effect on the dependent variable
Constant	a factor that does not change when other variables change
Control	the standard by which the test results can be compared

Controls A **control** is the standard by which test results can be compared. After the mission specialists gathered their data on the plants grown in microgravity, they compared their results with the same types of plants grown on Earth's surface with the same constants. This comparison allowed them to analyze the data and form a conclusion about whether microgravity has an effect on plant growth.

Get It?
Identify What is the purpose of a control in an experiment?

Analyze the data

An important part of every investigation includes recording observations and organizing the test data into easy-to-read tables and graphs. Later in this module, you will study ways to display data. When you are making and recording observations, you should include all results, even unexpected ones. Many important discoveries have been made from unexpected results.

Scientific inferences are based on observations made using scientific methods. All possible scientific explanations must be considered. If the data are not organized in a logical manner, wrong conclusions can be drawn. When a scientist communicates and shares data, other scientists will examine that data, consider how it is analyzed, and compare it to the work of others. Scientists share their data through reports and conferences. In **Figure 5,** a scientist is presenting his data.

Draw conclusions

Based on the analysis of the data, the next step is to decide whether the hypothesis is supported. For the hypothesis to be considered valid and widely accepted, the experiment must result in the exact same data every time it is repeated. If the experiment does not support the hypothesis, the hypothesis must be reconsidered. Perhaps the hypothesis needs to be revised, or maybe the experiment's procedure needs to be refined.

Figure 5 An exciting and important part of an investigation is sharing your ideas with others.

ACADEMIC VOCABULARY
Infer
to come to a logical conclusion based on observations and evidence
After observing a trail of ants in his kitchen, Joe inferred that he had spilled some sugar.

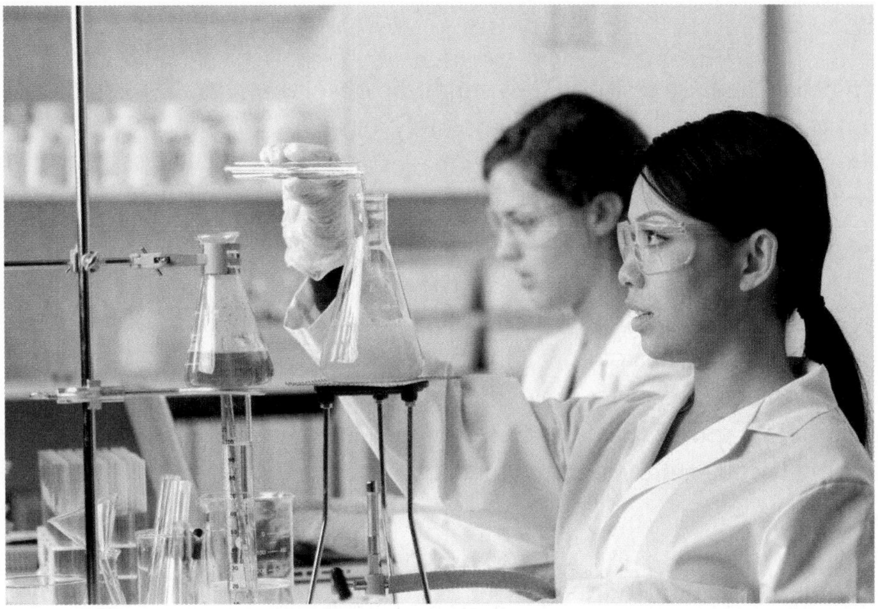

Figure 6 In order for medicine to be approved for use on humans, scientists have to run multiple trials to prove the results are objective.

Peer review

Before it is made public, science-based information is reviewed by scientists' peers—scientists who are in the same field of study. Peer review is a process by which the procedures and results of an experiment are evaluated by other scientists who are in the same field as those who are conducting similar research. Reviewing other scientists' work is a responsibility that many scientists have.

Being objective

Scientists also have a responsibility to minimize bias in their investigations. **Bias** occurs when a scientist's expectations change how the results are analyzed or conclusions are formed. Bias might cause a scientist to select a result from one trial over those from other trials. An example of bias would be presenting positive test results to promote a product and withholding unfavorable results.

Scientists can reduce bias by running as many trials as possible and by keeping accurate notes of each observation made. **Figure 6** shows a scientist who is researching a medication's effectiveness. One way to reduce bias in this case would be to conduct the experiment so that the researchers didn't know which group was the study group and which group was the control. This is called a blind experiment.

Valid experiments must also have data that are measurable. For example, a scientist performing a global warming study must base his or her data on accurate measures of global temperature. This allows others to compare the results to data they obtain from similar experiments. Most importantly, the experiment must be repeatable. Findings are supportable when other scientists around the world perform the same experiment and get the same results.

Get It?

Define What is bias in science?

Visualizing with Models

Sometimes scientists cannot see everything that they are testing. They might be observing something that is too large or small, takes too much time to see completely, or is hazardous. In these cases, scientists use models. A **model** represents an idea, event, or object to help people better understand it.

Models in history

Models have been used throughout history. Lord Kelvin, a scientist who lived in England in the 1800s, was famous for making models. To model his idea of how light moves through space, he put balls into a bowl of jelly and encouraged people to move the balls around with their hands. Recall the models of the atom in **Figure 2**. Scientists used models of atoms to represent their current understanding because of the atom's small size.

High-tech models

Scientific models don't always have to be something you can touch. Another type of model is a computer simulation, like the one shown in **Figure 7**. A computer simulation uses a computer to test a process or procedure and to collect data. Computer software is designed to safely and conveniently mimic the processes under study. Today, many scientists use computers to build models.

Computer simulations, such as the one shown in **Figure 7**, enable pilots to practice all aspects of flight without ever leaving the ground. In addition, the computer simulation can simulate harsh weather conditions or other in-flight challenges that pilots might face.

Figure 7 This is a computer simulation of an aircraft landing on a runway. The image on the screen in front of the pilots mimics what they would see if they were landing a real plane.

Identify *models in your classroom.*

Santofilme/iStock/Getty Images

Scientific Theories and Laws

A scientific **theory** is an explanation based on knowledge gained from many observations and investigations. It is not a guess. If scientists repeat an investigation and the results always support the hypothesis, the hypothesis can be called a theory. As new information becomes available, theories can be modified.

A **scientific law** is a statement about what happens in nature and that seems to be true all the time. Laws describe specific relationships under given conditions. They don't explain why or how something happens. Gravity is an example of a scientific law. The law of gravity states that any one mass will attract another mass.

A theory can be used to explain a law, but theories do not become laws. For example, many theories have been proposed to explain how the law of gravity works. Even so, there are few accepted theories in science and even fewer laws.

Figure 8 Science can't answer all questions, like questions about opinions and values. This piece of art might look very beautiful to one person but not to another.

Discuss *Can anyone prove that a piece of art is beautiful? Explain.*

The Limitations of Science

Science can help you explain many things, but science cannot explain or solve every question. It is scientists' job to develop hypotheses that can be tested and verified. Questions that cannot be tested and verified, such as those about opinions and values, are not scientific. You might take a survey to gather opinions about a piece of art, such as the painting in **Figure 8**, but it would not prove the opinions to be true or false.

Check Your Progress

Summary

- Scientists ask questions and perform investigations to learn more about the natural world.
- Scientists use scientific methods to test their hypotheses.
- Models help scientists visualize concepts.
- A theory is a possible explanation for observations, while a scientific law describes a pattern but does not explain why things happen.

Demonstrate Understanding

1. **Define** Summarize the steps you might use to carry out an investigation using scientific methods.
2. **Explain** what a law is, what a theory is, and why a theory cannot become a law.

Explain Your Thinking

3. **Analyze** What is the dependent variable in an experiment that shows how the volume of a gas changes with changes in temperature?
4. **MATH Connection** An experiment to determine how many breaths a squirrel takes per minute yields this data: minute 1: 65 breaths; minute 2: 73 breaths; minute 3: 67 breaths; minute 4: 71 breaths; minute 5: 62 breaths. Calculate the average number of breaths per minute.

LEARNSMART® Go online to follow your personalized learning path to review, practice, and reinforce your understanding.

LESSON 2

STANDARDS OF MEASUREMENT

FOCUS QUESTION

Which units are used when measuring length, volume, mass, electricity, and temperature?

Units and Standards

Suppose you and a friend want to find out whether a desk will fit through a doorway. You have no ruler, so you decide to use your hands as measuring tools. Using the width of his hands, your friend measures the doorway and says it is eight hands wide. Using the width of your hands, you measure the desk and find it is $7\frac{3}{4}$ hands wide. Will the desk fit through the doorway? You can't be sure. Even though you both used hands to measure, you didn't check to see whether your hands were the same width as your friend's hands. In other words, because you didn't use a measurement standard, you can't compare the measurements. A **standard** is an exact quantity that people agree to use to compare measurements.

Measurement Systems

Suppose the label on a ball of string says that the length of the string is 1. Is the length 1 meter (m), 1 foot (ft), or 1 centimeter (cm)? For a measurement to make sense, it must include a number and a unit, as shown in **Figure 9.**

Your family might buy lumber by the foot, milk by the gallon, and potatoes by the pound. These units are part of the U.S. customary system of measurement, which comes from units used in the former British Empire. Most countries other than the United States now use the metric system, which is based on multiples of ten.

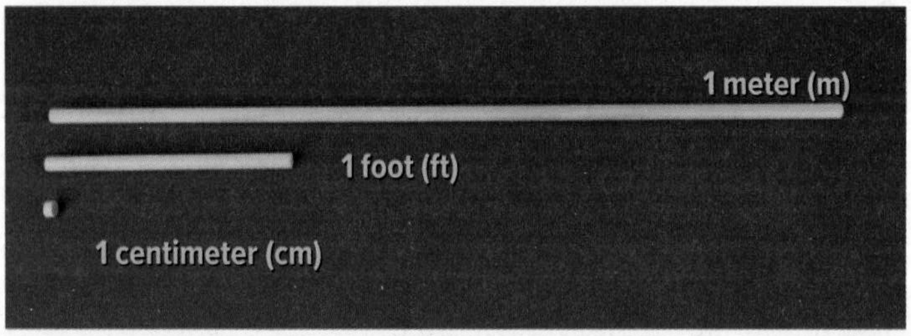

Figure 9 For data collected in an investigation or experiment, you need appropriate and consistent standards of measurement. It is important that you always use a number and a unit when describing length.

3D THINKING | DCI Disciplinary Core Ideas | CCC Crosscutting Concepts | SEP Science & Engineering Practices

COLLECT EVIDENCE

Use your Science Journal to record the evidence you collect as you complete the readings and activities in this lesson.

INVESTIGATE

GO ONLINE to find these activities and more resources.

Quick Investigation: Determine the Density of a Pencil
Carry out an investigation to determine the density of a pencil based on its physical properties.

CCC **Identify Crosscutting Concepts**
Create a table of the crosscutting concepts and fill in examples you find as you read.

International System of Units

In 1960, an improved version of the metric system was devised. Known as the International System of Units, this system is often abbreviated **SI,** from the French *Le Systeme Internationale d'Unites*. All SI standards are universally accepted and understood by scientists throughout the world.

All of the units in SI are based on fundamental physical constants that are the same throughout the universe. For example, the standard meter equals the exact distance that light travels through a vacuum in 1/299,792,458 seconds.

Table 2 SI Base Units

Quantity Measured	Unit	Symbol
Length	meter	m
Mass	kilogram	kg
Time	second	s
Electric current	ampere	A
Temperature	kelvin	K
Amount of substance	mole	mol
Intensity of light	candela	cd

A base unit in SI is one that is based on a universal physical constant. There are seven base units in SI. The names and symbols for the seven base units are shown in **Table 2.** All other SI units are derived from these seven units. The base unit of mass, the kilogram, was recently redefined to be based on a constant known as the Planck constant.

SI prefixes

The SI system is easy to use because it is based on multiples of ten. Prefixes are used with the names of the units to indicate what multiple of ten should be used with the units. For example, the prefix *kilo-* means "1000," which means that one kilometer equals 1000 meters. Likewise, one kilogram equals 1000 grams. Because *deci-* means "one-tenth," one decimeter equals one-tenth of a meter. A decigram equals one-tenth of a gram. The most frequently used prefixes are shown in **Table 3.**

Table 3 Common SI Prefixes

Prefix	Symbol	Multiplying Factor
Kilo-	k	1,000
Deci-	d	0.1
Centi-	c	0.01
Milli-	m	0.001
Micro-	μ	0.000 001
Nano-	n	0.000 000 001

Get It?

Calculate How many meters is 1 km? How many grams is 1 dg?

Converting between SI units

Sometimes quantities are measured using different units. A conversion factor is a ratio that is equal to 1. It is used to change one unit to another. For example, there are 1000 mL in 1 L, so 1000 mL = 1 L. If both sides in this equation are divided by 1 L, the equation becomes:

$$\frac{1000 \text{ mL}}{1 \text{ L}} = 1$$

To convert units, multiply by the appropriate conversion factor. For example, to convert 1.255 L to mL, multiply 1.255 L by a conversion factor. Use the conversion factor with new units (mL) in the numerator and the old units (L) in the denominator.

$$1.255 \text{ L} \times \frac{1000 \text{ mL}}{1 \text{ L}} = 1255 \text{ mL}$$

EXAMPLE Problem 1

CONVERT UNITS How long, in centimeters, is a 3075-mm rope?

Identify the Unknown: rope length in cm

List the Knowns: rope length in mm = 3075 mm

1 m = 100 cm = 1000 mm

Set Up the Problem: $\text{length in cm} = \text{length in mm} \times \frac{100\ \text{cm}}{1000\ \text{mm}}$

Solve the Problem: $\text{length in cm} = 3075\ \text{mm} \times \frac{100\ \text{cm}}{1000\ \text{mm}} = 307.5\ \text{cm}$

Check the Answer: Millimeters are smaller than centimeters, so make sure your answer in mm is greater than the measurement in cm. Because SI is based on tens, the answer in mm should differ from the length in cm by a factor of ten.

PRACTICE Problems

ADDITIONAL PRACTICE

5. If your pencil is 11 cm long, how long is it in millimeters?

6. CHALLENGE Some birds migrate 20,000 miles. If 1 mile equals 1.6 kilometers, calculate the distance these birds fly in kilometers.

Measuring Length

The word *length* is used in many ways. For example, the length of a novel is the number of pages or words it contains. In scientific measurement, however, length is the distance between two points. That distance might be the diameter of a hair or the distance from Earth to the Moon. The SI base unit of length is the meter (m). A baseball bat is about 1 m long. Metric rulers and metersticks are used to measure length. **Figure 10** compares a meter and a yard.

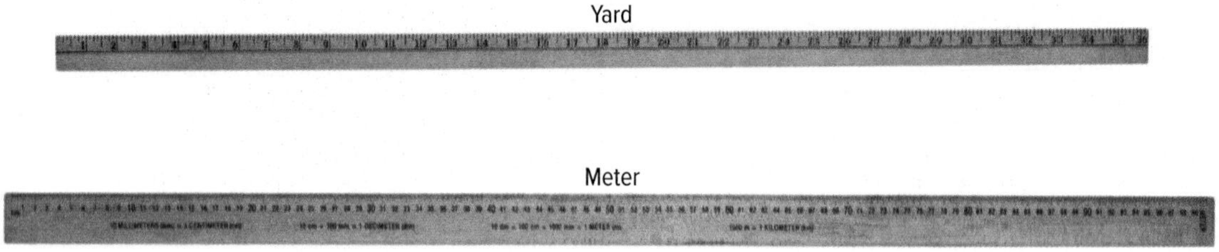

Figure 10 One meter is slightly longer than 1 yard, and 100 m is slightly longer than a football field.

Predict *whether your time for a 100-m dash would be slightly more or less than your time for a 100-y dash.*

STEM CAREER Connection

Computer Systems Analyst

Computer systems analysts study an organization's current computer systems and procedures and design solutions to help the organization operate more efficiently and effectively.

Figure 11 The size of the object being measured determines the appropriate unit to use. A tape measure measures in meters. A micrometer measures very small lengths.

Choosing a unit of length

As shown in **Figure 11,** the unit with which you measure will depend on the size of the object being measured. For example, the diameter of a shirt button is about 1 cm. You would most likely measure the length of your pencil in centimeters but the length of your classroom in meters. What unit would you use to measure the distance from your home to school? You would probably want to use a unit larger than a meter. The kilometer (km), which is 1000 m, is used to measure these kinds of distances.

By choosing an appropriate unit, you avoid large-digit numbers and numbers with many decimal places. It is easier to read and calculate 21 km than 21,000 m. And 13 mm is easier to use than 0.013 m.

Measuring Volume

The amount of space occupied by an object is called its **volume.** If you want to know the volume of a solid rectangle, such as a brick, you measure its length, width, and height and multiply the three numbers and their units together: $V = l \times w \times h$. For a brick, your length measurements would most likely be in centimeters. The volume would then be expressed in cubic centimeters (cm^3) because when you multiply, you add the exponents. To find out how large of a load a moving van can carry, your length measurements would probably be in meters, and the volume would be expressed in cubic meters (m^3).

Sometimes, liquid volumes, such as doses of medicine, are expressed in cubic centimeters. One cubic centimeter and one milliliter are the same volume.

$$1 \text{ mL} = 1 \text{ cm}^3$$

Suppose you wanted to convert a measurement in liters to cubic centimeters. You use conversion factors to convert L to mL and then mL to cm^3.

$$1.5 \text{ L} \times \frac{1000 \text{ mL}}{1 \text{ L}} \times \frac{1 \text{ cm}^3}{1 \text{ mL}} = 1500 \text{ cm}^3$$

SCIENCE USAGE v. COMMON USAGE

volume

Science usage: the amount of space occupied by an object
To measure the volume of the cube, Raul submerged it in a beaker of water.

Common usage: the degree of loudness
Alisa couldn't hear the radio, so she turned up the volume.

Table 4 Densities of Some Materials at 20°C

Material	Density (g/cm^3)
Hydrogen	0.00009
Oxygen	0.0014
Water	1.0
Aluminum	2.7
Iron	7.9
Gold	19.3

Measuring Mass and Density

Matter is anything that takes up space and has mass. A table-tennis ball and a golf ball have about the same volume. If you pick them up, you notice a difference. The golf ball has more mass. **Mass** is a measurement of the quantity of matter in an object. The mass of a golf ball is about 45 g. It is almost 18 times the mass of a table-tennis ball, which is about 2.5 g. A bowling ball has a mass of about 5000 g. This makes its mass roughly 100 times greater than the mass of the golf ball and 2000 times greater than the table-tennis ball's mass.

Density

A cube of polished aluminum and a cube of silver that are the same size look similar and have the same volume, but they have different masses. The mass and volume of an object can be used to find the density of the material of which the object is made. **Density** is the mass per unit volume of a material. You find density by dividing an object's mass by the object's volume. For example, the density of an object having a mass of 10 g and a volume of 2 cm^3 is 5 g/cm^3. **Table 4** lists the densities of some familiar materials.

Derived units

The measurement unit for density, g/cm^3, is a combination of SI units. A unit obtained by combining different SI units is called a derived unit. An SI unit multiplied by itself also is a derived unit. Thus, the liter, which is based on the cubic decimeter, is a derived unit. A cubic meter, m^3, is another example of a derived unit.

Measuring Time and Temperature

It is often necessary to keep track of how long it takes for something to happen. Time is the interval between two events. The SI unit for time is the second. In the laboratory, you will use a stopwatch or a clock with a second hand to measure time.

Another type of measurement common to science is temperature. You will learn the scientific meaning of the word *temperature* in a later module, but for now, think of temperature as a measure of how hot or how cold something is. **Table 5**, on the next page, summarizes SI measurements, including time and temperature.

Table 5 SI Dimensions

Unit	Example	
Millimeters		A dime is about 1 mm thick.
Meters	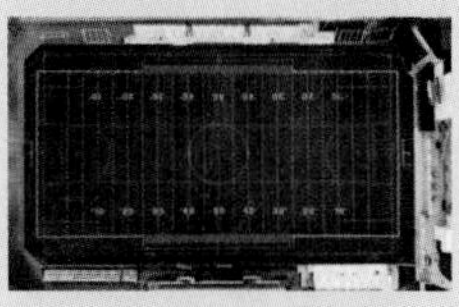	A football field is about 91 m long.
Kilometers		The distance from your house to the store can be measured in kilometers.
Milliliters		A teaspoonful of medicine is about 5 mL.
Liters		This carton holds 1.89 L of milk.
Grams/cm^3		This stone sinks because it is denser—has more grams per cubic centimeter—than water.
Meters/second		The speed of a roller-coaster car can be measured in meters per second.
Kelvin		Water boils at 373 K and freezes at 273 K.
Grams	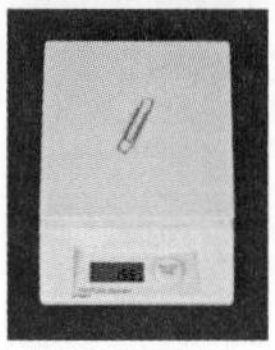	The mass of a paper clip can be measured in grams.

(t to b)Sven Kahns/McGraw-Hill Education, Ahturner/Shutterstock, Alita Xander/123RF, Brazzo/iStock/Getty Images, Ingram Publishing, Matt Meadows/McGraw-Hill Education, Rarpia/iStock/Getty Images, Matt Meadows/McGraw-Hill Education, Matt Meadows/McGraw-Hill Education, Matt Meadows/McGraw-Hill Education

Comparing Temperature Scales

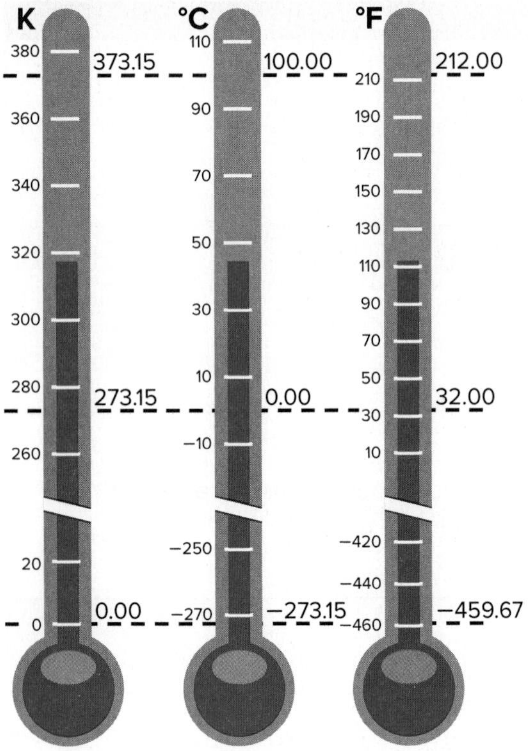

Figure 12 These three thermometers illustrate the three most common temperature scales. The dotted lines show absolute zero, the freezing point of water, and the boiling point of water.

State *the boiling point of water on the three scales.*

Celsius

Look at **Figure 12.** For much scientific work, temperature is measured on the Celsius (C) scale. On this scale, the freezing point of water is 0°C, and the boiling point of water is 100°C. Between these points, the scale is divided into 100 equal divisions. Each one represents 1°C. On the Celsius scale, average human body temperature is 37°C, and a typical room temperature is between 20°C and 25°C.

Kelvin and Fahrenheit

The SI unit of temperature is the kelvin (K). Zero on the Kelvin scale (0 K) is the coldest possible temperature, also known as absolute zero. Absolute zero is roughly equal to −273°C, which is 273°C below the freezing point of water.

Most laboratory thermometers are marked only with the Celsius scale. Because the divisions on the Celsius and Kelvin scales are the same size, the Kelvin temperature can be found by adding 273 to the Celsius reading. So, on the Kelvin scale, water freezes at 273 K and boils at 373 K. Notice that degree symbols are not used with the Kelvin scale.

The temperature measurement with which you are probably most familiar is the Fahrenheit scale. On the Fahrenheit scale, the freezing point of water is 32°F, and the boiling point is 212°F. A temperature difference of 1° on the Fahrenheit scale is 5/9° on the Celsius scale.

Check Your Progress

Summary

- The International System of Units, or SI, was established to provide a standard of measurement and to reduce confusion.
- Conversion factors are used to change one unit to another and involve using a ratio equal to 1.
- The size of an object determines which unit you will use to measure it.

Demonstrate Understanding

7. **Explain** why it is important to have exact standards of measurement.
8. **Make a Table** Organize the following measurements from smallest to largest and include the multiplying factor for each: kilometer, nanometer, centimeter, meter, and micrometer.

Explain Your Thinking

9. **Explain** why density is a derived unit.
10. **MATH Connection** Make the following conversions: 27°C to kelvins, 20 dg to milligrams, and 3 m to decimeters.
11. **MATH Connection** What is the density of an unknown metal that has a mass of 158 g and a volume of 20 mL? Use **Table 4** to identify this metal.

LEARNSMART® Go online to follow your personalized learning path to review, practice, and reinforce your understanding.

LESSON 3

COMMUNICATING WITH GRAPHS

FOCUS QUESTION

When would you use a bar graph instead of a line graph?

A Visual Display

Scientists often graph the results of their experiments to make it easier to detect patterns in the data. A **graph** is a visual display of information or data. **Figure 13** is a graph that shows the time and distance from home as a girl walked her dog. The horizontal axis, called the *x*-axis, measures time. Time is the independent variable, because as it changes, it affects the measure of another variable. The distance from home that the girl and the dog walk is the other variable. It is the dependent variable and is measured on the vertical axis, called the *y*-axis.

Graphs are useful for displaying numerical information in business, science, sports, advertising, and many everyday situations. Graphs make it easier to understand patterns by displaying data in a visual manner.

Scientists often graph their data to detect patterns that would not have been evident in a table. Businesspeople may graph sales dollars to determine trends. Different kinds of graphs—line, bar, and circle—are appropriate for displaying different types of information. Additional information on making and using graphs can be found in the Math Skill Handbook in the back of this book.

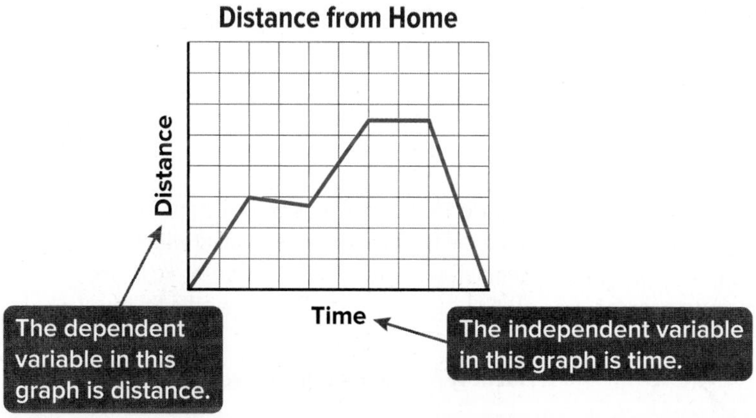

Distance from Home
Distance
Time
The dependent variable in this graph is distance.
The independent variable in this graph is time.

Figure 13 This graph tells the story of the motion that takes place when a girl takes her dog for a 10-minute walk. The dependent variable is on the *y*-axis, and the independent variable is on the *x*-axis.

3D THINKING | DCI Disciplinary Core Ideas | CCC Crosscutting Concepts | SEP Science & Engineering Practices

COLLECT EVIDENCE

Use your Science Journal to record the evidence you collect as you complete the readings and activities in this lesson.

INVESTIGATE

GO ONLINE to find these activities and more resources.

Lab: Who contributes carbon dioxide?
Carry out an investigation to calculate the quantity of carbon dioxide produced per person in different countries.

Teaching Activity: Reading Graphs
Analyze a graph to determine how they make it easier to understand and detect patterns within data.

Line Graphs

A line graph can show any relationship in which the dependent variable changes due to a change in the independent variable. Line graphs often show how one or more variables change over time. In this case, time is the independent variable. You can use a line graph to track many things, such as how certain stocks perform or how the population changes over any period of time—a month, a week, or a year.

Displaying data on line graphs

You can show more than one event on the same graph as long as the relationship between the variables is identical. Suppose a builder had to choose one thermostat from among three different kinds for a new school. He tested them to find out which was the best brand to install throughout the building. He installed different thermostats in classrooms A, B, and C. He set each thermostat at 20°C. He turned on the furnace and checked the temperatures in the three rooms every 5 min for 25 min. He recorded his data in **Table 6.**

The builder then plotted the data on the graph in **Figure 14.** He could see from the table that the data did not vary much for the three classrooms. So, he chose small intervals for the y-axis and left out part of the scale (the part between 0°C and 15°C). This allowed him to spread out the area on the graph where the data points lie.

You can easily see the contrast in the colors of the three lines and their relationship to the black horizontal line. The black line represents the thermostat setting and is the control. The control is what the resulting room temperature of the classrooms should be if the thermostats are working efficiently.

Table 6 Room Temperature

Time*(min)	Classroom Temperature (°C)		
	A	B	C
0	16	16	16
5	17	17	16.5
10	19	19	17
15	20	21	17.5
20	20	23	18
25	20	25	18.5

*minutes after turning on heat

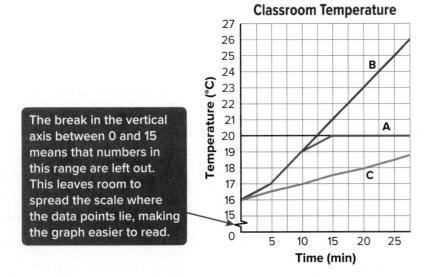

The break in the vertical axis between 0 and 15 means that numbers in this range are left out. This leaves room to spread the scale where the data points lie, making the graph easier to read.

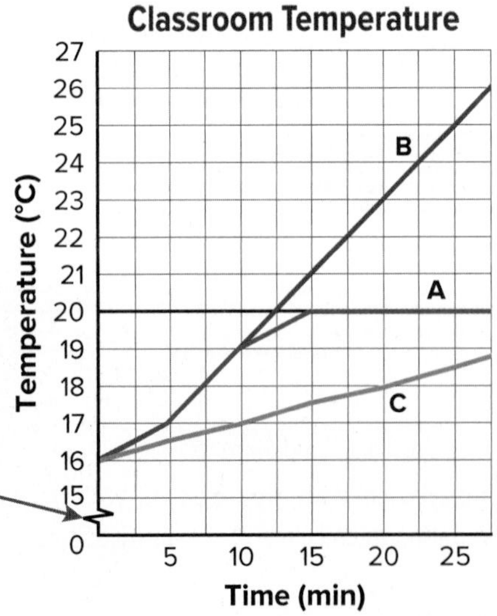

Figure 14 The builder's graph compares the time the furnace has been running with the temperature in each of three rooms, A, B, and C. The black line at $y = 20$°C shows the thermostat setting.

Identify *the thermostat that reached 20°C first.*

Constructing line graphs

In addition to choosing a scale that makes a graph readable, other factors are involved in constructing useful graphs. The most important factor in making a line graph is always using the *x*-axis for the independent variable. The *y*-axis is always used for the dependent variable. Recall that the dependent variable changes in response to the changes that you make to the independent variable, and the independent variable is the variable that you change to see how it will affect the dependent variable.

Another factor in constructing a graph involves units of measurement. You must use consistent units when graphing data. For example, you might use a Celsius thermometer for one part of your experiment and a Fahrenheit thermometer for another. But you must first convert your temperature readings to the same unit of measurement before you make your graph.

Once the data is plotted as points, a straight line or a curve is drawn based on those points. This should not be done like a connect-the-dots game. Instead, a best-fit line or a most-probable smooth curve is placed among the data points, as shown in **Figure 15.**

In the past, graphs had to be made by hand, with each point plotted individually. Today, scientists, mathematicians, and students use a variety of tools, such as computer programs and graphing calculators, to help them draw and interpret graphs.

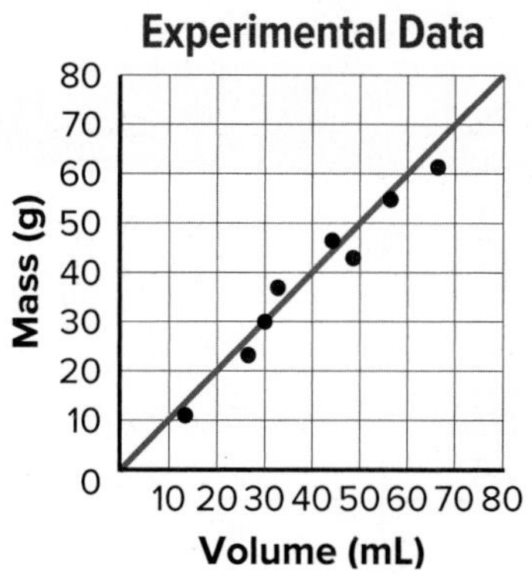

Figure 15 Generally, the line or curve that you draw will not intersect all of your data points.

APPLY SCIENCE

Make and Use Graphs

Line graphs are useful tools for showing the relationship between an independent and a dependent variable. In an experiment, you checked the air temperature at certain hours of the day and recorded it in the data table shown here.

Time	Temperature
8:00 A.M.	27°C
12:00 P.M.	32°C
4:00 P.M.	30°C

Identify the Problem

time = independent variable

temperature = dependent variable

Temperature is the dependent variable because it varies with time. Graph time on the *x*-axis and temperature on the *y*-axis. Mark equal increments on the graph and include all measurements. Plot each point on the graph by finding the time on the *x*-axis and moving up until you find the recorded temperature on the *y*-axis. Continue placing points on the graph. Then, connect the points from left to right.

Solve the Problem

1. Based on your graph, what was the temperature at 10:00 A.M.? What was the temperature at 2:00 P.M.?
2. What is the relationship between time and temperature?
3. Why is a line graph a useful tool for viewing this data?
4. For what other types of data might a line graph be useful?

Table 7 Classroom Size

Number of Students	Number of Classrooms
20	1
21	3
22	3
23	2
24	3
25	5
26	5
27	3

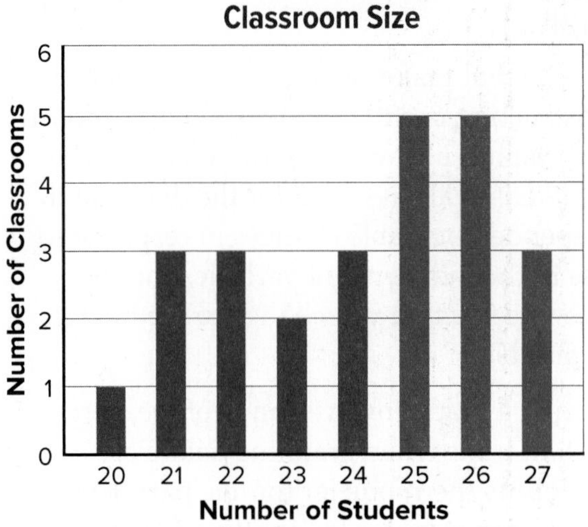

Figure 16 The height of each bar corresponds to the number of classrooms having a particular number of students.

Bar Graphs

A bar graph is useful for comparing information or displaying data that do not change continuously. Suppose you counted the number of students in every classroom in your school and organized your data in **Table 7.** You could show these data in a bar graph like the one in **Figure 16.** Notice you can easily determine which classrooms have the greatest and least numbers of students. You can also easily see that there are the same numbers of classrooms with 21 and 22 students and with 25 and 26 students.

Bar graphs can be used to compare oil or crop production, to compare costs of different products, or as data in promotional materials. Similar to a line graph, the independent variable is plotted on the *x*-axis, and the dependent variable is plotted on the *y*-axis.

Recall that you might need to place a break in the scale of the graph to better illustrate your results. For example, if your data set included the points 1002, 1010, 1030, and 1040 and the intervals on the scale were every 100 units, you might not be able to see the difference from one bar to another. If you had a break in the scale and started your data range at 1000 with intervals of ten units, you could make a more accurate comparison.

Get It?

Describe possible data for which using a bar graph would be better than using a line graph.

Circle Graphs

A circle graph, sometimes called a pie chart, is used to show how some fixed quantity is broken into parts. The circular pie represents the total. The slices represent the parts and usually are represented as percentages of the total.

Figure 17 illustrates how a circle graph could be used to show the percentage of buildings in a neighborhood using each of a variety of heating fuels. You easily can see that more buildings use gas heat than any other kind of system. What other information does the graph provide?

Sources of Heat

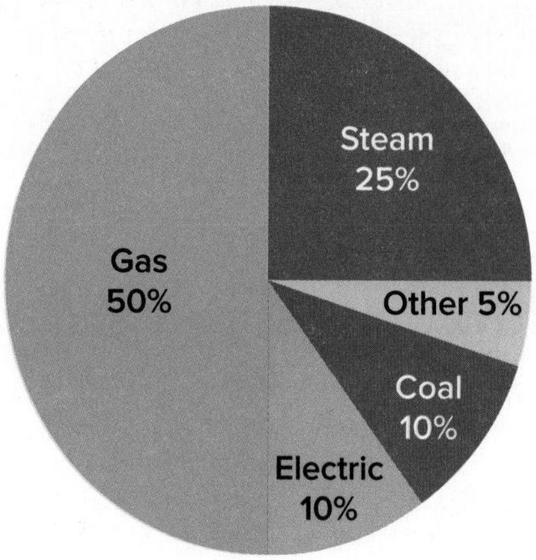

Figure 17 A circle graph shows the different parts of a whole quantity.

To create a circle graph, start with the total of what you are analyzing. Suppose the survey of heating fuels counted 72 buildings in the neighborhood. For each type of heating fuel, you divide the number of buildings using each type of fuel by the total (72). Then multiply that decimal by 360° to determine the angle that the decimal makes in the circle. For example, 18 buildings use steam. Therefore, $18 \div 72 = 0.25$, and $0.25 \times 360° = 90°$ on the circle graph. You then would measure 90° on the circle with your protractor. You can also calculate that if this graph shows 50 percent of the buildings use gas, then 36 of the buildings use gas ($0.50 \times 72 = 36$).

When you create a graph, think carefully about which type of graph you will use and how you will present your data. In addition, consider the conclusions you may draw from your graph. Make sure your conclusions are based on sound information and that you present your information clearly.

Check Your Progress

Summary

- Graphs are a visual representation of data.
- Scientists often graph their data to detect patterns.
- A line graph shows a relationship between an independent and a dependent variable.
- Bar graphs are best used to compare information collected by counting.
- A circle graph shows how a fixed quantity is broken down into parts.

Demonstrate Understanding

12. **Identify** the kind of graph that would best show the results of a survey of 144 people, of which 75 ride a bus, 45 drive cars, 15 carpool, and 9 walk to work.
13. **State** which type of variable is plotted on the *x*-axis and which type is plotted on the *y*-axis.
14. **Compare and Contrast** How are line, bar, and circle graphs similar? How are they different?

Explain Your Thinking

15. **Explain** why the points in a line graph can be connected.
16. **MATH Connection** In a survey, it was reported that 56 out of 245 people would rather drink orange juice than coffee in the morning. Calculate the percentage of a circle graph that orange-juice drinkers would occupy.

LEARNSMART® Go online to follow your personalized learning path to review, practice, and reinforce your understanding.

LESSON 4
SCIENCE AND TECHNOLOGY

FOCUS QUESTION

How does society affect the technology that we use?

What is technology?

The terms *science* and *technology* often are used interchangeably. However, these terms have very different definitions. Science is an exploration process. Scientific processes are used to gain knowledge to explain and predict natural occurrences. Scientists often pursue scientific knowledge for the sake of learning new information. There may or may not be a plan to use the knowledge.

When scientific knowledge is used to solve a human need or problem, as shown in **Figure 18,** the result is referred to as technology. **Technology** is the application of scientific knowledge to benefit people. Given this definition, is an aspirin tablet technology? Is a car technology? What about the national highway system? Although these examples appear to be very different, they all represent examples of technology. Technology can be

- any human-made object (such as a radio, computer, or pen),
- methods or techniques for making any object or tool (such as the process for making glass or ceramics),
- knowledge or skills needed to operate a human-made object (such as the skills needed to pilot an airplane), or
- a system of people and objects used to do a particular task (such as the Internet, a system for sharing information).

Figure 18 The prosthetic leg in this image is a technological object. The knowledge needed to interpret this image is also technology.

3D THINKING DCI Disciplinary Core Ideas CCC Crosscutting Concepts SEP Science & Engineering Practices

COLLECT EVIDENCE

Use your Science Journal to record the evidence you collect as you complete the readings and activities in this lesson.

INVESTIGATE

GO ONLINE to find these activities and more resources.

Quick Investigation: Research the Past

Carry out an investigation to ask a senior citizen what types of technologies existed when they were teenagers.

LabA: Care Package

Carry out an investigation to model which materials work best in the packaging of foods, and construct an explanation for how the packaging design may reduce consumer costs.

sportpoint/Shutterstock

Technological objects

The value of a technological object changes through time, as shown in **Figure 19.** What is considered new technology today might be considered an antique tomorrow. For example, special feathers called quills were used long ago to write with ink. The quill pen was the height of technology for over 1000 years. Then, in the middle of the nineteenth century, metallic pens and writing points came into use. The modern ballpoint pen was not widely used until the 1940s.

Technological methods or techniques

Just as writing instruments have changed over time, so have techniques for performing various tasks. Long ago, people would sit for hours and copy each page of a book by hand. Books were expensive and could be bought by only the very rich. Today, books can be created in different ways. They can be made on a computer, printed with a computer printer, and bound with a simple machine. Modern printing presses, like those shown in **Figure 20,** are used to produce the majority of books used today, including your textbook.

The methods used for printing books have changed over time. Each technique has its own technology. Other techniques that characterize technology include using a compact disc to store information, using a refrigerator to preserve food, and using e-mail to correspond with friends.

Technological knowledge or skills

Technology also can be the knowledge or skills needed to perform a task. For example, computer skills are needed to run the software that is used to make books and other documents. Printing press operators must use their skills to print books successfully. Any time a complex machine is used to perform a task, technological skills must be used by the operator.

Figure 19 Technology changes through time. A sundial was once the common method used to tell time. Today, we often rely on wristwatches or digital clocks.

Figure 20 Hundreds of years ago, books had to be written and published by hand. Now, printing presses can generate thousands of pages an hour.

Technological systems

A network of people and objects that work together to perform a task is also technology. One example of this technology is the Internet. The Internet is a collection of computers and software that is used to exchange information. A technological system is a collection of various types of technology that are combined to perform a specific function. The airline industry is an example of a technological system. This industry is a collection of objects, methods, systems, knowledge, and procedures. The airports, pilots, fuel, and ticketing process form a technological system that is used to move people and goods.

Get It?

Identify another example of a technological system. Explain why it fits into this category.

Global Technological Needs

The value of technology may vary for different people and at different times. The technology that is needed in the United States is not necessarily needed in other parts of the world. Developing countries have different needs for technology than do industrialized countries.

Developing countries

The people of some countries work hard to meet basic needs such as food, shelter, clothing, safe drinking water, and health care. For example, the family shown in **Figure 21** lives without electricity and running water in their home. Instead of going to the grocery store for the food their family eats, they might grow most of it themselves. Children might walk to school instead of taking the school bus.

Figure 21 The technological needs of this family might be different from those of a family in another location. All people need safe water, food supplies, health care, education, and a safe place to work, but the technology involved in meeting these needs can differ from location to location.

Compare and contrast *these needs with the needs of your family.*

Brittak/E+/Getty Images

Meeting basic needs Technological solutions in developing countries are often limited to supplying a family's basic needs. Technology that would supply adequate and safe drinking water and food supplies would be valued before technology such as access to the Internet. Increasing the accessibility of basic health care would improve the quality of life and increase the life expectancy in developing countries. The technology valued by rural people in developing countries contrasts with the technology valued by people in those countries' industrialized cities.

Figure 22 The basic needs for survival are available to most people in industrialized countries. Value is placed on technology that improves the quality of life, such as devices that make tasks easier and devices that provide entertainment.

Describe *three technological objects that you value that would likely be of less value to a family in a developing country.*

Industrialized countries

The United States is considered an industrialized country. Technology gives people the ability to clean polluted waters. Improving the quality of the food supply is also valued. But the level of urgency and focus differs from those in developing countries.

Most areas of the United States have adequate and safe water and food supplies. Because the needs for survival are met in industrialized nations, money often is spent on different types of technology. The technology used in industrialized countries is designed to improve people's quality of life.

Look at the home in **Figure 22,** and compare it to the home on the previous page. Most homes in the United States have electricity and running water. Quality health care is available to many people. The life expectancy for Americans is in the late-seventies, generally higher than life expectancy in developing countries. Most homes in the United States contain many different types of technology, including computers, telephones, and televisions. Money is spent on such medical procedures as cosmetic surgery to remove wrinkles and eye surgery to eliminate the need for wearing glasses.

Contrasting needs As you can see, the human needs in developing countries and industrialized countries are very different. Both developing and industrialized countries value technologies that meet basic human needs, but the actual technology required in each type of country may be very different.

Get It?

Compare and contrast the technological needs of developing and industrialized countries.

Figure 23 Consumers often decide which technologies will be developed. If consumers do not purchase a product, additional money usually will not be spent on the production or improvement of the product.

Social Forces That Shape Technology

Science and society are closely connected. **Society** is a group of people that share similar values and beliefs. Discoveries in science and technology bring about changes in society. In turn, society affects how new technologies develop. The development of technology is affected by society and its changing values, politics, and economics.

In the past 100 years, attitudes about automobiles have changed in the United States. Many people became able to own cars due to the changes in technology and manufacturing. As car ownership increased, so did fossil fuel consumption. With rising gasoline prices, some consumers began buying more fuel-efficient cars. The automotive industry has researched and developed technologies that make cars more fuel-efficient. Today, hybrid cars use both gasoline and electricity.

Personal values

People will support the development of technologies that agree with their personal values, directly and indirectly. Purchasing technology is a direct way in which people support the development of technology. For example, if consumers continue to purchase fuel-efficient cars, like the one shown in **Figure 23,** additional money will be spent on improving the technology. If consumers fail to buy a product, companies usually will not spend additional money on that type of technology. People also support the development of technology directly when they give their money to organizations that are committed to a specific project, such as cancer research.

There are also many ways to indirectly support technologies that agree with one's personal values. For example, people vote for a congressional candidate based on the candidate's views on various issues. This is an indirect way in which people's personal values influence whether technology projects receive funding and support.

Robert Churchill/E+/Getty Images

Economic Forces That Shape Technology

Many factors influence how much money is spent on technology. Before funding is given for a project, several questions should be answered. What is the benefit of this product? What is the cost? Who will buy this product? All of these questions should be answered before money is invested in a project. Various methods exist to fund new and existing technology.

Federal government

One way in which funds are allocated for research and development of technology is through the federal government. Every year, Congress and the president place large amounts of money in the federal budget for scientific research and development. These funds are reserved for specific types of research, such as agriculture, defense, energy, and transportation. This money is given to companies and institutions in the form of contracts and grants to do specific types of research. Citizens can affect how the government funds technology by voting, as shown in **Figure 24**, and other political activities.

Private foundations

Some scientific research is funded using money from private foundations. Funds are raised for various types of disease research, such as breast cancer and muscular dystrophy, through events such as races and telethons. Many private foundations focus on research for a specific cause.

Private industries

Research and development is also funded by private industries. Industries budget a portion of their profits for research and development. Investing in research and development can make money for the company in the long-term. Bringing new products to the marketplace is one way companies make profits.

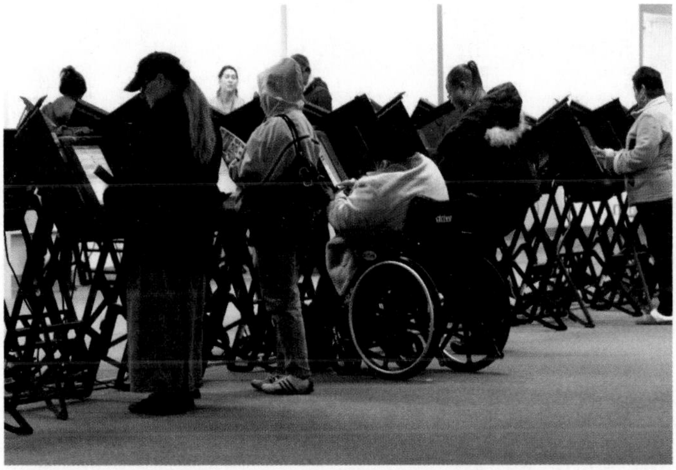

Figure 24 Participating in the political process is one way that people can influence which technologies are developed and which are not.

Explain *how voting for members of Congress influences which technologies will be developed.*

WORD ORIGINS

technology

comes from the Greek word *technologia*, which means "an ordered treatment of an art"

Technology helps our everyday lives.

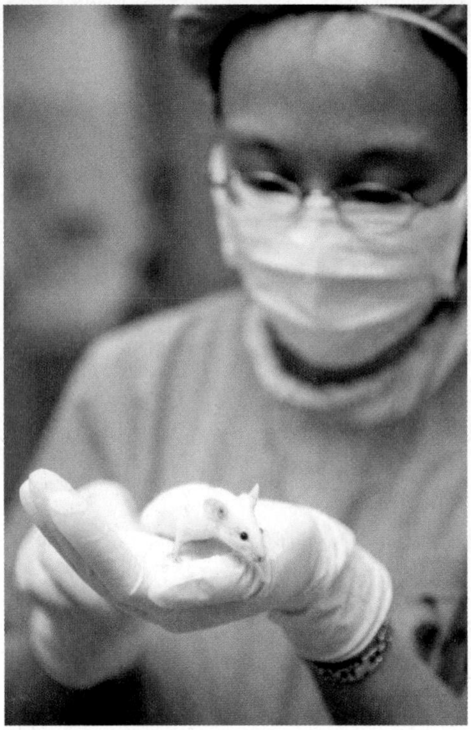

Figure 25 Scientists must be aware of ethical issues when testing on animal subjects. Although animals are still used in some scientific testing, today there are laws and guidelines to regulate this testing.

Moral and Ethical Issues

When people need to distinguish between right and wrong, what is fair, and what is in the best interest of all people, moral and ethical issues are raised. Ethics help scientists establish standards that they agree to follow when they collect, analyze, and report data. Scientists are expected to conduct investigations honestly and openly.

Some ethical questions in science concern the use of animals, such as the mouse shown in **Figure 25,** as well as humans. The inhumane treatment of humans and animals in past experiments has led to public outcry. For example, human test subjects have been put through experiments against their will or without being informed of the risks associated with the research or the true nature of the experiments. Ethical questions about these practices helped to create laws and guidelines to prevent unethical treatment of both humans and animals in scientific research.

Biotechnology

Any technological application using living things or living systems is called biotechnology. Breeding animals for certain traits, using fermentation to obtain cheese or wine, and baking with yeast are biotechnologies that humans have used for many years. Today, biotechnology includes diverse areas of research, such as stem cells and genetically engineered crops. Some of these research areas are controversial because they challenge society's values and beliefs. All areas must be examined for their impact on individuals, society, and the environment.

Check Your Progress

Summary

- Technology can be an object, a technique, a skill, or a system.
- Societal and economic forces influence which technologies will be developed and used around the world.
- The development of new technology is influenced by voting and buying habits.
- The federal government, private foundations, and private industries fund the research and development of technology.

Demonstrate Understanding

17. **Classify** the types of technology, and give at least two examples of each type.
18. **Explain** why the types of technology that are valued can vary.
19. **Describe** how private citizens have a voice in which projects the federal government will fund.

Explain Your Thinking

20. Evaluate Would cell-phone technology be of use in a developing country? Explain your answer.
21. **MATH Connection** In 2010, the Department of Defense's overall budget was approximately $534 billion. In the same year, the Department of Defense budgeted $79.1 billion for research, development, tests, and evaluations. What percentage of the budget does this represent?

LEARNSMART® Go online to follow your personalized learning path to review, practice, and reinforce your understanding.

Imagemore/Getty Images

Scientific Methods

Abu Ali al-Hasan ibn al-Haytham is sometimes known as the father of the scientific methods, the first true scientist, and the founder of optics. He was born in a.d. 965 in what is now Iraq, but he spent most of his life in Cairo, Egypt. As a young man, he read the works of Aristotle, the third-century b.c. Greek philosopher. He approved of Aristotle's reliance on experience, rather than untested ideas, in pursuit of truth.

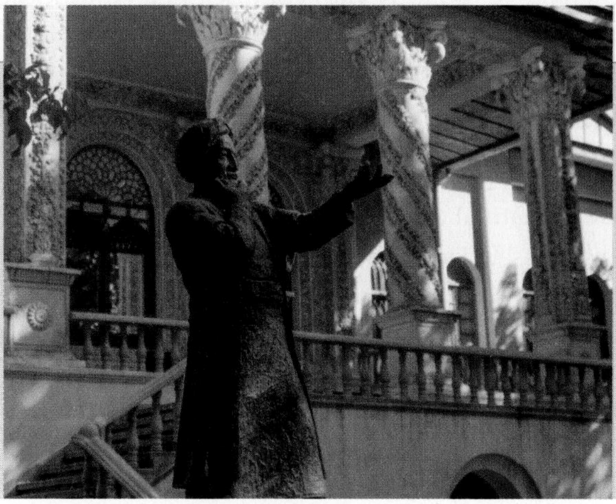

Ibn al-Haytham is often regarded as the founder of scientific methods.

In one of his essays, Ibn al-Haytham, whose statue is shown in the photo at right, wrote that "the seeker after truth is . . . the one who submits to argument and demonstration and not to the sayings of a human being whose nature is fraught with all kinds of imperfection and deficiency." Ibn al-Haytham studied and wrote about many scientific disciplines, including mathematics (especially geometry), astronomy, optics, physics, and medicine. He became one of the first scientists to perform experiments to test his hypotheses. In addition, he contributed to the fields of philosophy and psychology.

Experiments with optics

The study of optics is the branch of physics that studies light and vision. Earlier thinkers, such as the second-century a.d. Greek scientist Ptolemy, thought that people could see because rays of light came out of their eyes and fell on the objects they were observing. Ibn al-Haytham performed experiments to test the behavior of light and the workings of the human eye. His observations led him to conclude that people could see because rays of light entered, not exited, their eyes.

A lasting influence

Ibn al-Haytham's insistence on experimentation and on making quantifiable observations helped develop scientific methods as we know them today. Centuries after his death, Ibn al-Haytham's most important work, *Kitab al-Manazir* (*Optics*), influenced European scientists such as Roger Bacon, a 13th-century philosopher who is sometimes credited with laying the foundation for the use of scientific methods.

Because of Ibn al-Haytham's lifelong quest to develop a rational way of exploring his ideas, today's scientists and students of science have scientific methods, a powerful tool for testing hypotheses.

SEP DCI CCC

APPLY SCIENTIFIC REASONING AND EVIDENCE

Describe and explain what characterizes science and its methods, what resembles science but does not meet its criteria (pseudoscience), and what is clearly not science. Identify a question you think can be answered through science and a question that is outside its boundaries, and explain your reasoning.

MODULE 1

STUDY GUIDE

GO ONLINE to study with your Science Notebook.

Lesson 1 THE METHODS OF SCIENCE

- Scientists ask questions and perform investigations to learn more about the natural world.
- Scientists use scientific methods to test their hypotheses.
- Models help scientists visualize concepts.
- A theory is a possible explanation for observations, while a scientific law describes a pattern but does not explain why things happen.

- scientific methods
- hypothesis
- experiment
- variable
- dependent variable
- independent variable
- constant
- control
- bias
- model
- theory
- scientific law

Lesson 2 STANDARDS OF MEASUREMENT

- The International System of Units, or SI, was established to provide a standard of measurement and to reduce confusion.
- Conversion factors are used to change one unit to another and involve using a ratio equal to 1.
- The size of an object determines which unit you will use to measure it.

- standard
- SI
- volume
- matter
- mass
- density

Lesson 3 COMMUNICATING WITH GRAPHS

- Graphs are a visual representation of data.
- Scientists often graph their data to detect patterns.
- A line graph shows a relationship between an independent variable and a dependent variable.
- Bar graphs are best used to compare information collected by counting.
- A circle graph shows how a fixed quantity is broken down into parts.

- graph

Lesson 4 SCIENCE AND TECHNOLOGY

- Technology can be an object, a technique, a skill, or a system.
- Societal and economic forces influence which technologies will be developed and used around the world.
- The development of new technology is influenced by voting and buying habits.
- The federal government, private foundations, and private industries fund the research and development of technology.

- technology
- society

THREE-DIMENSIONAL THINKING

Module Wrap-Up

REVISIT THE PHENOMENON

How do we know how rainbows form?

CER Claim, Evidence, Reasoning

Explain Your Reasoning Revisit the claim you made when you encountered the phenomenon. Summarize the evidence you gathered from your investigations and research and finalize your Summary Table. Does your evidence support your claim? If not, revise your claim. Explain why your evidence supports your claim.

GO FURTHER

SEP Data Analysis Lab

Graphing Scientific Data

Interpolation is a method used to approximate values that are between points of a graph. Extrapolation is a method for approximating values that are beyond the range of the data. The data in **Table 1** were obtained from an experiment conducted to find out how the volume of a gas changes when its temperature changes. Use this data to construct and interpret a graph.

Table 1

Temperature (K)	Volume (cm^3)
0	a.
100	71
140	b.
210	155
273	c.
280	195
360	257
400	d.
600	e.

Procedure

1. Draw a graph on a piece of graph paper.
2. Mark the *x*-axis for the independent variable and the *y*-axis for the dependent variable.
3. Plot a point for each temperature/volume set of data in the table. Draw the line that best fits the data points.
4. Extend the line to include all temperatures from 0 K to 600 K.

CER Analyze and Interpret Data

1. **Claim** Use your graph to predict values for the volume of a gas at 0 K, 140 K, 273 K, 400 K, and 600 K, and place these values in the data table.
2. **Evidence and Reasoning** Suppose you had drawn the graph in a "dot-to-dot" fashion. Why would it be difficult to extrapolate from this type of graph?
3. **Claim, Reasoning** Why isn't it necessary for all of the data points to be on the drawn line of the graph?
4. **Claim** Write a sentence that describes the relationship between the temperature and the volume of a gas.

WeatherVideoHD.TV

UNIT 1

MOTION AND FORCES

ENCOUNTER THE PHENOMENON

How could this athlete jump higher?

SEP Ask Questions

What questions do you have about the phenomenon? Write your questions on sticky notes and add them to the driving question board for this unit.

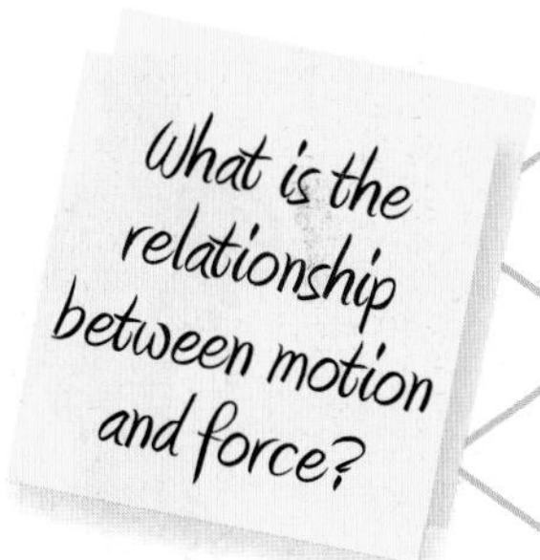

What is the relationship between motion and force?

Look for Evidence

As you go through this unit, use the information and your experiences to help you answer the phenomenon question as well as your own questions. For each activity, record your observations in a Summary Table, add an explanation, and identify how it connects to the unit and module phenomenon questions.

Solve a Problem

STEM UNIT PROJECT

Design a Safety Inspection for an Amusement Park Ride
Amusement park rides can be fun, but they must be carefully designed and maintained. Research incidents and accidents of mobile amusement park rides, as well as safety measures that can be taken to prevent them. Use the results of these investigations and the evidence you collected during the unit to model a safety inspection and the resulting safety report.

GO ONLINE In addition to reading the information in your Student Edition, you can find the STEM Unit Project and other useful resources online.

Stephen Barnes/Sport/Alamy Stock Photo

MODULE 2

MOTION

ENCOUNTER THE PHENOMENON

Why is this motor bike traveling in an arc?

GO ONLINE to play a video about the projectile motion of a fireball.

SEP Ask Questions

Do you have other questions about the phenomenon? If so, add them to the driving question board.

CER Claim, Evidence, Reasoning

Make Your Claim Use your CER chart to make a claim about why this motor bike is traveling in an arc. Explain your reasoning.

Collect Evidence Use the lessons in this module to collect evidence to support your claim. Record your evidence as you move through the module.

Explain Your Reasoning You will revisit your claim and explain your reasoning at the end of the module.

GO ONLINE to access your CER chart and explore resources that can help you collect evidence.

LESSON 2: Explore & Explain: Motion and Position

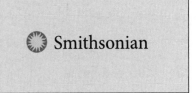

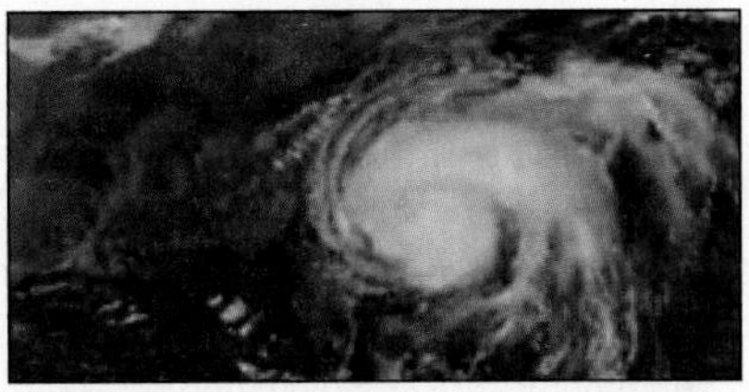

LESSON 2: Explore & Explain: Velocity

Smithsonian

Additional Resources

(t)Video Supplied by BBC Worldwide Learning, (bl)Allkindza/iStock/Getty Images, (br)GOES 13/NASA/NOAA GOES Project, Dennis Chesters

LESSON 1
DESCRIBING MOTION

FOCUS QUESTION
Which factors describe the motion of an object?

Motion and Position

You do not always need to see something move to know that motion has taken place. For example, suppose you look out a window and see a mail truck stopped next to a mailbox, as shown in **Figure 1.** One minute later, you look out again and see the same truck stopped farther down the street. Although you did not see the truck move, you know it moved because its position relative to the mailbox changed.

Reference points

A reference point is needed to determine the position of an object. In **Figure 1,** the reference point might be a mailbox. **Motion** is a change in an object's position relative to a reference point. How you describe an object's motion depends on the reference point that is chosen. For example, the description of the mail truck's motion in **Figure 1** would be different if the reference point were a tree instead of a mailbox.

After a reference point is chosen, a frame of reference can be created. A frame of reference is a coordinate system in which the position of the object is measured. The x-axis and y-axis of the reference frame are drawn so that they are perpendicular to each other and intersect the reference point.

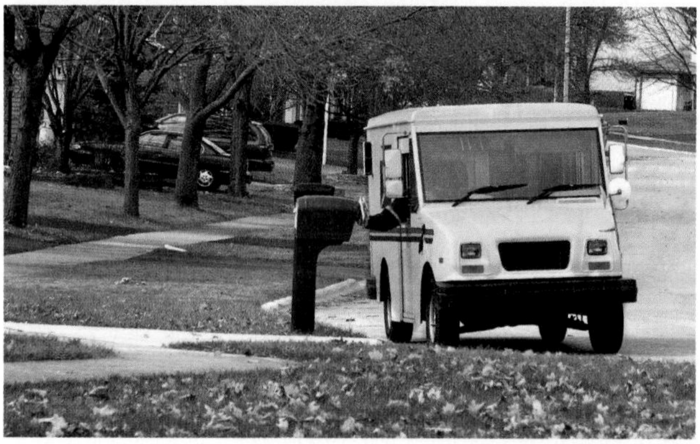

Figure 1 As the mail truck follows its route, it stops at each mailbox along the street.

Explain *How would you know the mail truck has moved?*

Allkindza/iStock/Getty Images

3D THINKING DCI Disciplinary Core Ideas CCC Crosscutting Concepts SEP Science & Engineering Practices

COLLECT EVIDENCE

Use your Science Journal to record the evidence you collect as you complete the readings and activities in this lesson.

INVESTIGATE

GO ONLINE to find these activities and more resources.

Virtual Investigation: Describing Motion
Carry out an investigation to determine the relationship between distance, average speed, and time of cars in a race, and to predict the distance the cars will travel.

Quick Investigation: Measure Average Speed
Carry out an investigation to determine the average speed of a toy car and how average speed changes when forces of different magnitude are applied to the system.

Coordinate systems

Figure 2 shows a map of the city where the mail truck is delivering mail. The map has a coordinate system drawn on it. The *x*-axis is in the east-west direction, the *y*-axis is in the north-south direction, and each division represents a city block. The post office is located at the origin. The mail truck is located at 3 blocks east ($x = 3$) and 2 blocks north ($y = 2$) of the post office.

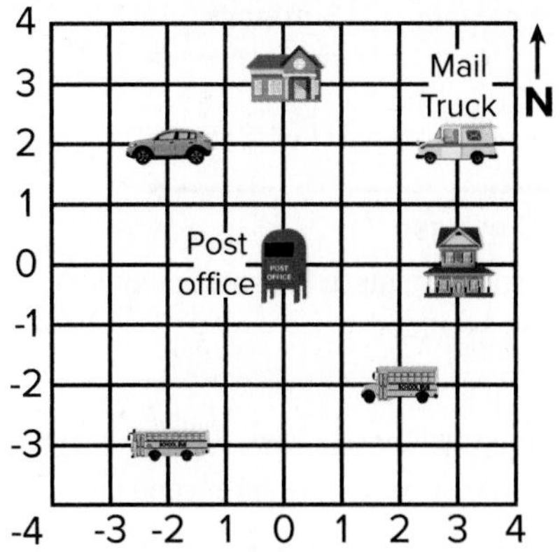

Figure 2 A coordinate system is like a map. The reference point is at the origin, and each object's position can be described with its coordinates.

Identify *the position of the orange car.*

Change in Position

Have you ever run a 50-m dash? Describing how far and in what direction you moved was an important part of describing your motion.

Distance

In a 50-m dash, each runner travels a total distance of 50 m. The SI unit of distance is the meter (m). Longer distances are measured in kilometers (km). One kilometer is equal to 1000 m. Shorter distances are measured in centimeters (cm) or millimeters (mm). One meter is equal to 100 cm and to 1000 mm.

Displacement

Suppose a runner jogs to the 50-m mark and then turns around and runs back to the 20-m mark, as shown in **Figure 3.** The runner travels 50 m in the original direction (east) plus 30 m in the opposite direction (west), so the total distance that she ran is 80 m. How far is she from the starting line? The answer is 20 m. Sometimes, you may want to know the change in an object's position relative to the starting point. An object's **displacement** is the distance and direction of the object's change in position. In **Figure 3,** the runner's displacement is 20 m east.

The length of the runner's displacement and the total distance traveled would be the same if the runner's motion were in a single direction. For example, if the runner ran east from the starting line to the finish line without changing direction, then the distance traveled would be 50 m, and the displacement would be 50 m east.

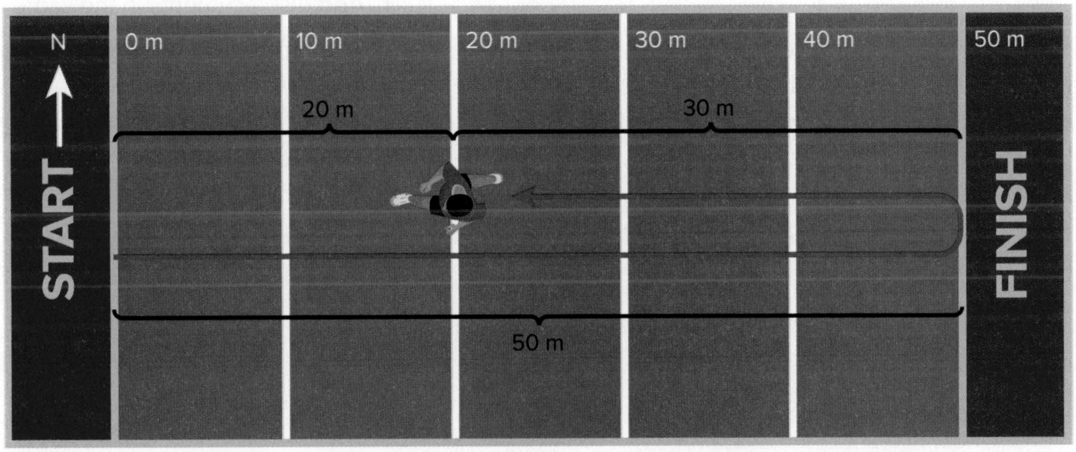

Figure 3 An object's displacement is not the same as the total distance that the object traveled. The runner's displacement is 20 m east of the starting line. However, the total distance the runner traveled is 80 m.

Describe *the difference between the total distance traveled and the displacement.*

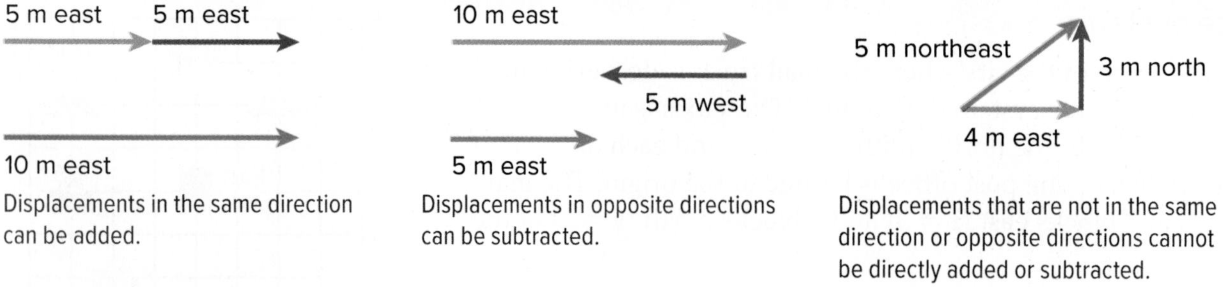
5 m east
5 m east
10 m east
10 m east
5 m west
5 m east
5 m northeast
3 m north
4 m east
Displacements in the same direction can be added.
Displacements in opposite directions can be subtracted.
Displacements that are not in the same direction or opposite directions cannot be directly added or subtracted.

Figure 4 These arrows represent the students' walks. The green arrows show the first part of the walk, and the purple arrows show the second part. The orange arrows show the students' displacements.

Adding displacements

You know that you can add distances to get the total distance. For example, 2 m + 3 m = 5 m. But how would you add the displacements 5 m east and 10 m east? Directions in math problems are much like units: you can add numbers with like directions. For example, suppose a student walks 5 m east, stops at a crosswalk, and then walks another 5 m east, as shown on the left in **Figure 4.** His displacement is

5 m east + 5 m east = 10 m east

But what if the directions are not the same? Then compare the two directions. If the directions are exactly opposite, the distances can be subtracted. Suppose a student walks 10 m east, turns around, and walks 5 m west, as shown in the center of **Figure 4.** The size of the displacement would be

10 m − 5 m = 5 m

The direction of the total displacement is always the direction of the larger displacement. In this case, the larger displacement is east, so the total displacement is 5 m east.

Get It?

Determine the total displacement of a dog that runs 15 m north, 6 m south, then 8 m north.

Now suppose the two displacements are neither in the same direction nor in opposite directions, as illustrated on the right in **Figure 4.** Here, the student walks 4 m east and then 3 m north. The student walks a total distance of 7 m, but the displacement is 5 m in a roughly northeast direction. The displacements of 4 m east and 3 m north cannot be directly added or subtracted, and they should be discussed separately. The rules for adding displacements are summarized in **Table 1.**

Table 1 Rules for Adding Displacements

1. Add displacements in the same direction.
2. Subtract displacements in opposite directions.
3. Displacements that are not in the same or in opposite directions cannot be directly added together.

SCIENCE USAGE v. COMMON USAGE

position

Science usage: the location of an object in relation to a reference point
The cat's position was 3 meters west of the house.

Common usage: a point of view; a job or rank
After graduation, I accepted a teller position at the bank.

Speed

Think back to the mail truck moving down the street. You could describe the movement by the distance traveled or by the displacement. You might also want to describe how fast the truck is moving. To do this, you need to know how far it travels in a given amount of time. To describe how fast an object moves, scientists use the object's speed. **Speed** is the distance an object travels per unit of time.

Calculating speed

Any change over time is called a rate. For example, you could describe how quickly water is leaking from a tank by stating how many liters are lost each hour. This would be the rate of water leakage. If you think of distance as the change in position, then speed is the rate of change in position. Speed can be calculated from the equation shown below.

Speed Equation

$$\text{speed (in meters/second)} = \frac{\text{distance (in meters)}}{\text{time (in seconds)}}$$

$$s = \frac{d}{t}$$

In SI units, distance is measured in meters, and time is measured in seconds. Therefore, the SI unit for speed is meters per second (m/s). Sometimes, it is more convenient to express speed in other units, such as kilometers per hour (km/h). **Table 2** shows the speeds of some common objects.

Table 2 Common Speeds

Motion	Speed (m/s)
Olympic 100-m dash	10 m/s
Car on city street (35 mph)	16 m/s
Car on interstate highway (65 mph)	29 m/s
Commercial airplane	250 m/s

EXAMPLE Problem 1 ADDITIONAL PRACTICE

CALCULATE SPEED A car traveling at a constant speed covers a distance of 750 m in 25 s. What is the car's speed?

Identify the Unknown: speed: s

List the Knowns: distance: $d = 750$ m
time: $t = 25$ s

Set Up the Problem: $s = \frac{d}{t} = \frac{750 \text{ m}}{25 \text{ s}}$

Solve the Problem: $s = \frac{750 \text{ m}}{25 \text{ s}} = 30$ m/s

Check the Answer: 30 m/s is approximately the speed limit on a U.S. interstate highway, so the answer is reasonable.

PRACTICE Problems

1. A passenger elevator travels from the first floor to the 60th floor, a distance of 210 m, in 35 s. What is the elevator's speed?
2. A motorcycle is moving at a constant speed of 40 km/h. How long does it take the motorcycle to travel a distance of 10 km?
3. How far does a car travel in 0.75 h if it is moving at a constant speed of 88 km/h?
4. **CHALLENGE** A long-distance runner is running at a constant speed of 5 m/s. How long does it take the runner to travel 1 km?

Constant speed

Suppose you are in a car traveling on a nearly empty freeway. You look at the speedometer and see that the car's speed hardly changes. If the car neither slows down nor speeds up, the car is traveling at a constant speed. If you are traveling at a constant speed, you can calculate your speed by dividing any distance interval by the time it took you to travel that distance. The speed you calculate will be the same regardless of the interval you choose.

Changing speed

Usually, speed is not constant. Think about riding a bicycle for a distance of 5 km. The bicycle's speed will vary, as in **Figure 5.** As you start out, your speed increases from 0 km/h to 20 km/h. You slow down to 10 km/h as you pedal up a steep hill and speed up to 30 km/h going down the other side of the hill. You stop for a red light, speed up again, and move at a constant speed for a while. Finally, you slow down and come to a stop.

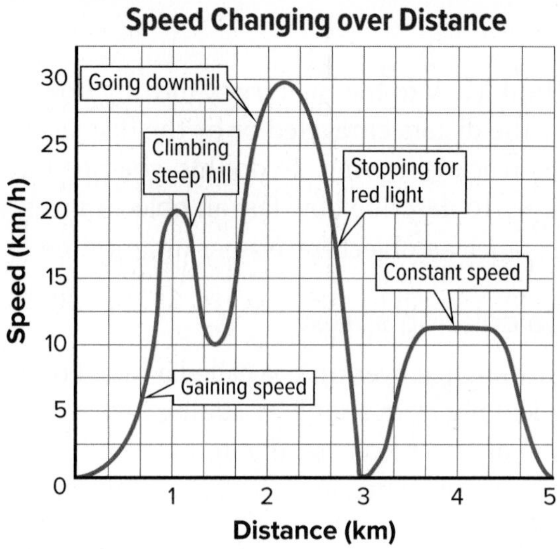

Figure 5 The cyclist's speed varies from 0 km/h to 30 km/h during this trip.

Explain *how you can describe the speed of an object when the speed is changing.*

Checking your watch, you find that the trip took 15 min. How would you express your speed on such a trip? Would you use your fastest speed, your slowest speed, or some speed between the two? Two common ways of expressing a changing speed are average speed and instantaneous speed.

Get It?

Identify two common ways of expressing a changing speed.

Average speed Average speed is one way to describe the speed of the bicycle trip. Average speed is the total distance traveled divided by the total time of travel. It can be calculated using the relationships between speed, distance, and time. For the bicycle trip just described, the total distance traveled was 5 km, and the total time was $\frac{1}{4}$ h, or 0.25 h. Therefore, the average speed was

$$s = \frac{d}{t} = \frac{5 \text{ km}}{0.25 \text{ h}} = 20 \text{ km/h}$$

Get It?

Identify how to calculate average speed.

STEM CAREER Connection

Delivery Truck Driver

With more and more online sales, companies must have a way to get the products to customers' doors. Delivery truck drivers pick up products at a distribution center and drive them to businesses and homes for delivery. Acceleration, velocity, and time are important factors for delivery truck drivers as they try to make their deliveries on a tight schedule.

Instantaneous speed Suppose you watch a car's speedometer, like the one in **Figure 6,** go from 0 km/h to 80 km/h. A speedometer shows how fast a car is going at one point in time, or at one instant. The speed shown on a speedometer is the instantaneous speed. Instantaneous speed is the speed at a given point in time. When something is speeding up or slowing down, its instantaneous speed is changing. The speed is different at different points in time. If an object is moving with constant speed, the instantaneous speed does not change. The speed is the same at every point in time.

Figure 6 A speedometer gives the car's instantaneous speed. Instantaneous speed is the speed at one instant in time.

Get It?
Identify two examples of motion in which an object's instantaneous speed changes.

Graphing Motion

The motion of an object over a period of time can be shown on a distance-time graph. For example, the graph in **Figure 7** shows the distance traveled by three swimmers during a 30-minute workout. Time is plotted along the horizontal axis of the graph, and the distance traveled is plotted along the vertical axis of the graph.

Each axis must have a scale that covers the range of numbers to be plotted. In **Figure 7,** the distance scale must range from 0 to 2400 m, and the time scale must range from 0 to 30 min. Next, the *x*-axis is divided into equal time intervals, and the *y*-axis is divided into equal distance intervals.

Once the scales for each axis are in place, the data points can be plotted. In **Figure 7,** there is a data point plotted for each swimmer every 2.5 minutes. After plotting the data points, a line is drawn to connect the points.

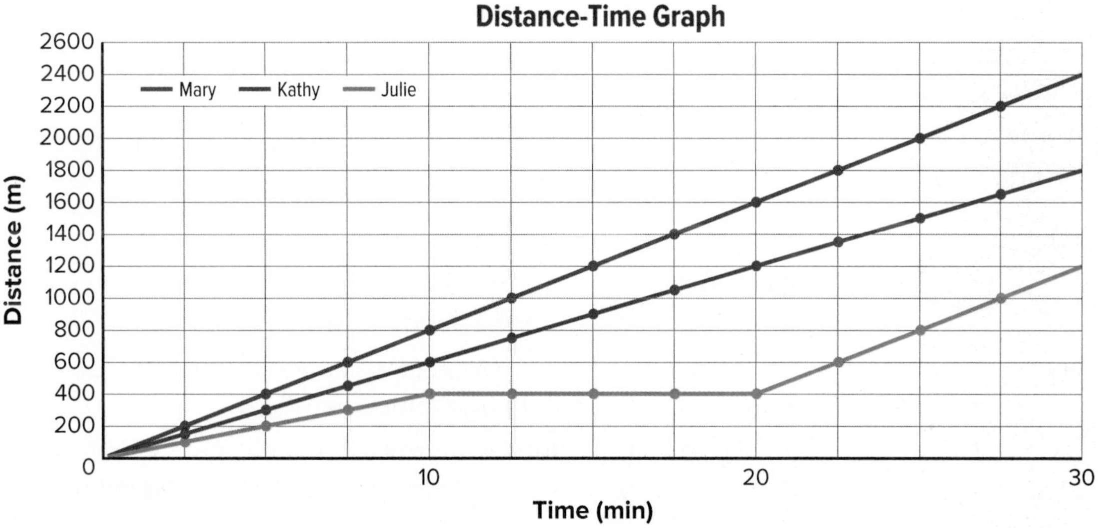

Figure 7 This graph shows how far each girl swam during a 30 minute workout. Time is divided into 2.5-minute intervals along the *x*-axis. Distance swam is divided into 200-m intervals along the *y*-axis.

Examine *the graph and determine which girl swam the farthest during the workout.*

Speed on distance-time graphs

If an object moves with constant speed, the increase in distance over equal time intervals is the same. As a result, the line representing the object's motion is a straight line. For example, look at the graph of the swimmers' workouts in **Figure 8.** The straight red line represents the motion of Mary, who swam with a constant speed of 80 m/min.

The green line represents the motion of Julie, who did not swim with a constant speed. She swam with a constant speed of 40 m/min for 10 minutes, rested for 10 minutes, and then swam with a constant speed of 80 m/min for 10 minutes.

The graph shows that the line representing the motion of the faster swimmer is steeper. The steepness of a line on a graph is the line's slope. The slope of a line on a distance-time graph equals the object's speed. Because Mary has a greater speed (80 m/min) than Kathy (60 m/min), the line representing her motion has a steeper slope.

Now look at the green line representing Julie's motion. During the time she is resting, her line is horizontal. A horizontal line on a distance-time graph has zero slope and represents an object at rest.

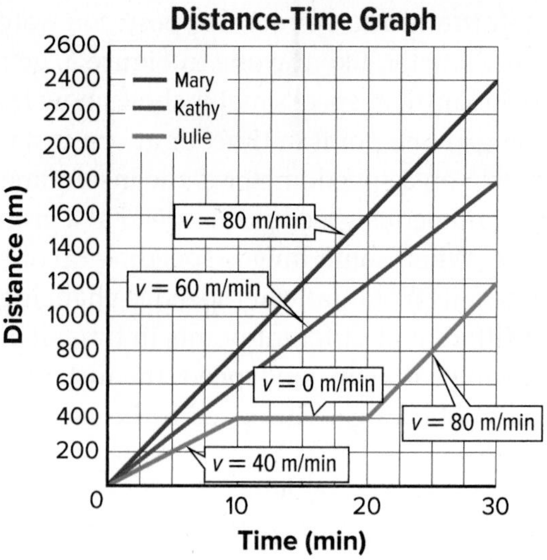

Figure 8 An object's speed is equal to the slope of the line on a distance-time graph.

Identify *the part of the graph that shows one of the swimmers resting for 10 min.*

Check Your Progress

Summary

- Motion occurs when an object changes its position relative to a reference point.
- Displacement is the distance and direction of a change in position from the starting point.
- Speed is the rate at which an object's position changes.
- On a distance-time graph, time is the *x*-axis, and distance is the *y*-axis.
- The slope of a line plotted on a distance-time graph is the speed.

Demonstrate Understanding

5. **Describe** the trip from your home to school using the words *position, distance, displacement,* and *speed.*
6. **Explain** whether an object's displacement could be greater than the distance the object travels.
7. **Describe** the motion represented by a horizontal line on a distance-time graph.
8. **Describe** the difference between average speed and constant speed.

Explain Your Thinking

9. **Explain** During a trip, can a car's instantaneous speed ever be greater than its average speed? Explain.
10. **MATH Connection** Michiko walked a distance of 1.60 km in 30 min. Find her average speed in m/s.
11. **MATH Connection** A car travels at a constant speed of 30.0 m/s for 0.80 h. Find the total distance traveled in km.

LEARNSMART® Go online to follow your personalized learning path to review, practice, and reinforce your understanding.

LESSON 2

VELOCITY AND MOMENTUM

FOCUS QUESTION

How does the velocity of an object affect its momentum?

Velocity

You turn on the radio and hear a news story about a hurricane. The storm, traveling at a speed of 20 km/h, is located 500 km east of your location. Should you worry?

Unfortunately, you do not have enough information to answer that question. Knowing only the speed of the storm is not much help. Speed describes only how fast something is moving. To decide whether you need to move to a safer area, you also need to know the direction that the storm is moving. In other words, you need to know the velocity of the storm. **Velocity** includes the speed of an object and the direction of its motion. Velocity has the same units as speed, m/s. If you had been told that the hurricane was traveling straight toward your house at 20 km/h, you would have known to evacuate.

Velocity and speed

Because velocity depends on direction as well as speed, the velocity of an object can change even if the speed of the object remains constant. For example, the race cars in **Figure 9** have constant speeds through a turn. Even though the speeds remain constant, their velocities change because they change direction throughout the turn.

Get It?

Describe how velocity and speed are different.

Figure 9 These cars travel at constant speed, but not with constant velocity. The cars' velocities change because their direction of motion changes.

3D THINKING DCI Disciplinary Core Ideas CCC Crosscutting Concepts SEP Science & Engineering Practices

COLLECT EVIDENCE

Use your Science Journal to record the evidence you collect as you complete the readings and activities in this lesson.

INVESTIGATE

GO ONLINE to find these activities and more resources.

Lab: Motion Graphs

Carry out an investigation to measure the speed of a toy car and create a distancetime graph to model at what points its speed increased, decreased, or remained constant.

Laboratory: Velocity and Momentum

Carry out an investigation to determine how the relationship between the momentum of an object and how long a force acts on it.

Same speed, different velocities It is possible for two objects to have the same speed but different velocities. For example, the two escalators pictured in **Figure 10** are moving at the same speed but in different directions. The speeds of the two sets of passengers are the same, but their velocities are different because they are moving in different directions. Cars traveling in opposite directions on a road with the same speed also have different velocities.

Figure 10 The two escalators move with a speed of 0.5 m/s. But the closer escalator's velocity is 0.5 m/s downward, and to the left, while the closer escalator's velocity is 0.5 m/s upward and to the left.

Motion of Earth's Crust

Can you think of something that is moving so slowly that you cannot detect its motion, but you can see evidence of its motion over long periods of time? As you look around the surface of Earth from year to year, its basic structure seems the same. Mountains, plains, and oceans seem to remain unchanged. Yet, if you examined geologic evidence of what Earth's surface looked like over the past 250 million years, you would see that large changes have occurred. **Figure 11** shows how, according to the theory of plate tectonics, the positions of landmasses have changed during this time. Changes in the landscape occur constantly as continents drift slowly over Earth's surface.

These moving plates cause geologic changes, such as the formation of mountain ranges, earthquakes, and volcanic eruptions. The movement of the plates changes the size of the oceans. The Pacific Ocean is getting smaller, and the Atlantic Ocean is getting larger. As they collide and spread apart, the plates' movement also changes the shape of the continents.

Plates move so slowly that their speeds are given in units of centimeters per year. Along the San Andreas Fault in California, two plates move past each other with an average speed of about 1 cm per year. The Australian Plate moves faster and pushes Australia north at an average speed of about 17 cm/y. Therefore, the velocity of the Australian plate is 17 cm/y north.

Figure 11 Geologic evidence suggests that Earth's surface is changing. The continents have moved slowly over time and are still moving today.

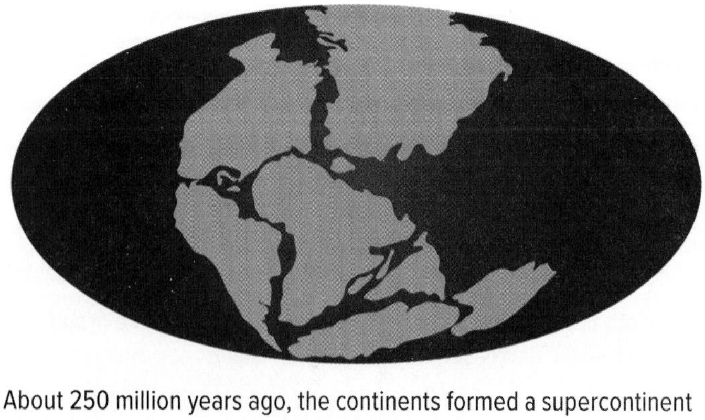

About 250 million years ago, the continents formed a supercontinent called Pangaea.

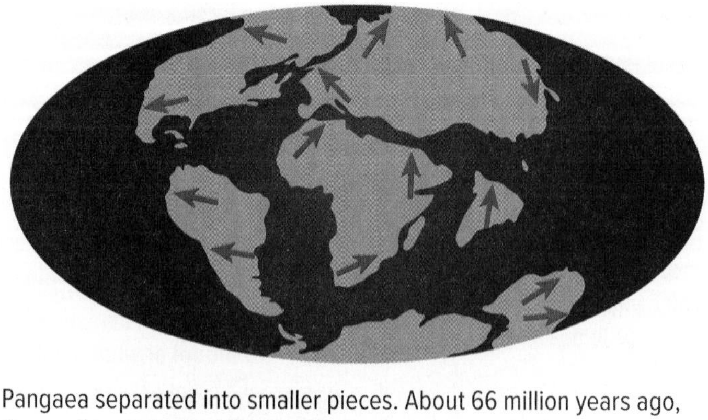

Pangaea separated into smaller pieces. About 66 million years ago, the continents looked like the figure above.

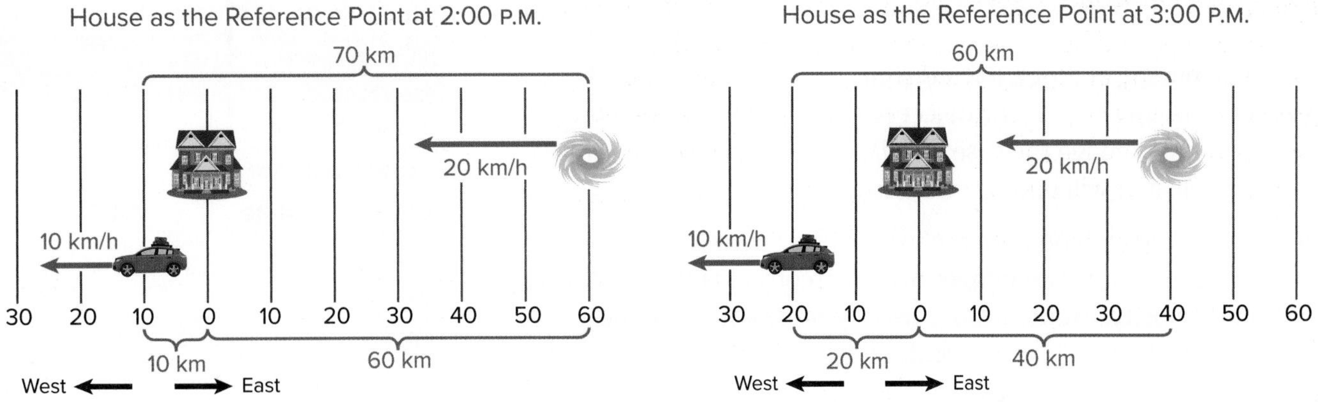

Figure 12 If the house is chosen for the reference point, the car appears to be traveling 10 km/h west, and the hurricane appears to be traveling 20 km/h west.

Relative Motion

Have you ever watched cars pass you on the highway? Cars traveling in the same direction often seem to creep by, while cars traveling in the opposite direction seem to zip by. This apparent difference in speeds is because the reference point–your vehicle–is also moving.

The choice of a moving reference point affects how you describe motion. For example, the motion of a hurricane can be described using a stationary reference point, such as a house. **Figure 12** shows the locations and velocities of a hurricane and a car relative to a house at 2:00 P.M. and 3:00 P.M. The distance between the hurricane and the house is decreasing at a rate of 20 km/h. The distance between the house and the car is increasing at a rate of 10 km/h.

How would the description of the hurricane's motion be different if the reference point were a car traveling at 10 km/h west? **Figure 13** shows the motion of the hurricane and the house relative to the car. A person in the car would say that the hurricane is approaching with a speed of 10 km/h and that the house is moving away at a speed of 10 km/h. It is important to notice that **Figure 12** and **Figure 13** show the same changes, but they use different reference points. Velocity and position always depend on the point of reference chosen.

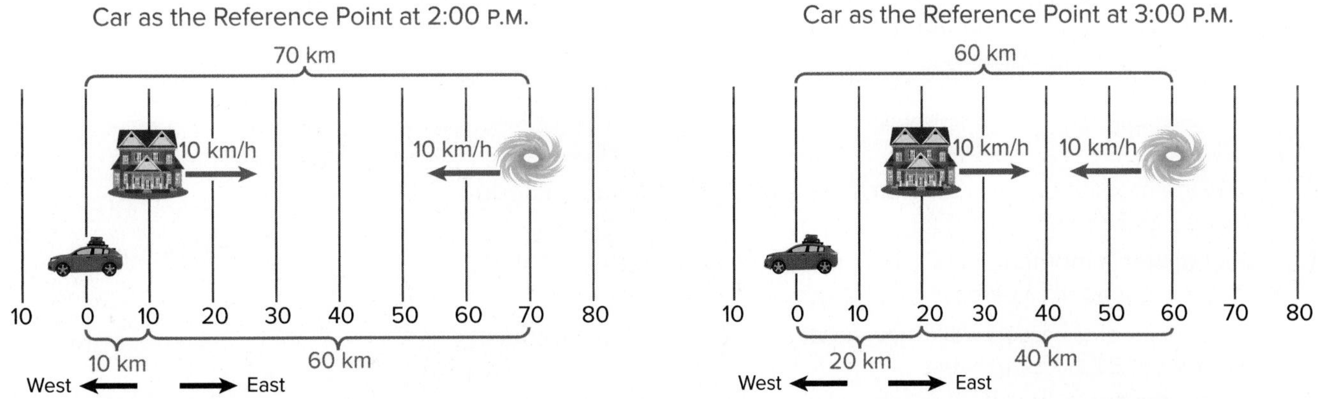

Figure 13 If the car is chosen as the reference point, the hurricane appears to be moving toward the car at 10 km/h, and the house is moving away from the car at 10 km/h.

Momentum

An object is moving at 2 m/s toward a glass vase. Will the vase be damaged in the collision? If the object has a small mass, like a bug, a collision will not damage the vase. But if the object has a larger mass, like a car, a collision will damage the vase.

A useful way of describing both the velocity and mass of an object is to state its momentum. The **momentum** of an object is the product of its mass and velocity. Momentum is usually represented by the symbol p and is defined for a particular frame of reference.

Momentum Equation

momentum (in kg·m/s) = **mass** (in kg) × **velocity** (in m/s)

$$p = mv$$

The unit for momentum is kg·m/s. Like velocity, momentum has a size and a direction. An object's momentum is always in the same direction as its velocity. **Table 3** shows the momenta of some common objects.

Table 3 Typical Momenta

Object	Momentum (kg·m/s)
Tossed baseball	0.15
Person walking	100
Car on interstate	45,000

Get It?

Explain how two objects could have the same velocity but different momentums.

EXAMPLE Problem 2

SOLVE FOR MOMENTUM At the end of a race, a sprinter with a mass of 80.0 kg has a velocity of 10.0 m/s east. What is the sprinter's momentum?

Identify the Unknown: momentum: p

List the Knowns: mass: m = 80.0 kg
velocity: v = **10.0 m/s east**

Set Up the Problem: $p = mv$ = (80.0 kg) × (**10.0 m/s**) **east**

Solve the Problem: p = (80.0 kg)(**10.0 m/s**) **east** = 800.0 kg·m/s east

Check the Answer: Our answer makes sense because it is greater than the momentum of a walking person but much less than the momentum of a car on the highway.

PRACTICE Problems ADDITIONAL PRACTICE

12. What is the momentum of a car with a mass of 1300 kg traveling north at a speed of 28 m/s?

13. A baseball has a momentum of 6.0 kg·m/s south and a mass of 0.15 kg. What is the baseball's velocity?

14. Find the mass of a person walking west at a speed of 0.8 m/s with a momentum of 52.0 kg·m/s west.

15. **CHALLENGE** The mass of a basketball is three times greater than the mass of a softball. Compare the momenta of a softball and a basketball if they are moving at the same velocity.

Figure 14 Both the car and the truck have a velocity of 30 m/s west, but the truck has a much larger momentum.

Comparing momenta

Think about the car and the truck in **Figure 14.** Which has the greater momentum? The truck does because it has more mass. When two objects travel at the same velocity, the object with more mass has a greater momentum. A difference in momenta is why a car traveling at 2 m/s might damage a porcelain vase, but an insect flying at 2 m/s will not.

Now consider two 1-mg insects. One insect flies at a speed of 2 m/s, and the other flies at a speed of 4 m/s. The second insect has a greater momentum. If two objects have the same mass, the object with the greater velocity has the greater momentum.

Check Your Progress

Summary

- The velocity of an object includes the object's speed and its direction of motion relative to a reference point.
- An object's motion is always described relative to a reference point.
- The momentum of an object is defined for a particular frame of reference and is the product of the object's mass and velocity: $p = mv$.

Demonstrate Understanding

16. **Describe** a car's velocity as it goes around a track at a constant speed.
17. **Explain** why streets and highways have speed limits rather than velocity limits.
18. **Identify** For each of the following news stories, determine whether the object's speed or velocity is given: the world record for the 100-meter dash is about 10 m/s; the wind is 30 km/h from the northwest; a 200,000 kg train was traveling north at 70 km/h when it derailed; a car was issued a ticket for traveling at 140 km/h on the interstate.

Explain Your Thinking

19. **Describe** You are walking toward the back of a bus that is moving forward with a constant velocity. Describe your motion relative to the frame of reference of the bus and relative to the frame of reference of a point on the ground.
20. **MATH Connection** What is the momentum of a 100-kg football player running north at a speed of 4 m/s?
21. **MATH Connection** Compare the momenta of a 6300-kg elephant walking 0.11 m/s and a 50-kg dolphin swimming 10.4 m/s.

LEARNSMART® Go online to follow your personalized learning path to review, practice, and reinforce your understanding.

LESSON 3
ACCELERATION

FOCUS QUESTION
How does acceleration result in projectile motion?

Velocity and Acceleration

You are in a stopped car when the light turns green. The driver steps on the gas pedal, and the car starts moving faster and faster. Just as speed is the rate of change of position, **acceleration** is the rate of change of velocity. When the velocity of an object changes, the object is accelerating. Remember that velocity includes the speed and direction of an object. Therefore, a change in velocity can be either a change in speed or a change in direction. Acceleration occurs when an object changes its speed, its direction, or both.

When you think of acceleration, you might think of speeding up. However, an object that is slowing down also is accelerating, as is an object that is changing direction. **Figure 15** shows the three ways an object can accelerate. Like velocity and momentum, acceleration has a direction. In **Figure 15,** you can see that when the car is speeding up, its acceleration and velocity are in the same direction. When the car is slowing down, its acceleration is in the opposite direction of its velocity. When the car changes direction, the acceleration is neither in the same direction nor opposite direction as the car's velocity.

Get It?
Identify three ways that an object can accelerate.

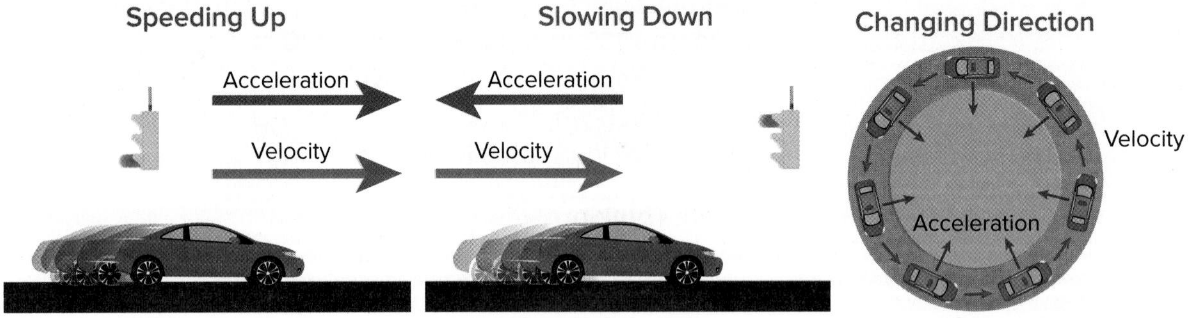

Figure 15 An object, such as this car, is accelerating if it is speeding up, slowing down, or changing direction.

3D THINKING
DCI Disciplinary Core Ideas   CCC Crosscutting Concepts   SEP Science & Engineering Practices

COLLECT EVIDENCE
Use your Science Journal to record the evidence you collect as you complete the readings and activities in this lesson.

INVESTIGATE
GO ONLINE to find these activities and more resources.

Laboratory: Projectile Motion
Carry out an investigation to analyze and use data on the projectile motion of objects to model and predict different trajectories.

LabA: The Momentum of Colliding Objects
Carry out an investigation to observe and calculate the momentum of different balls, and to compare the results of collisions involving different amounts of momentum.

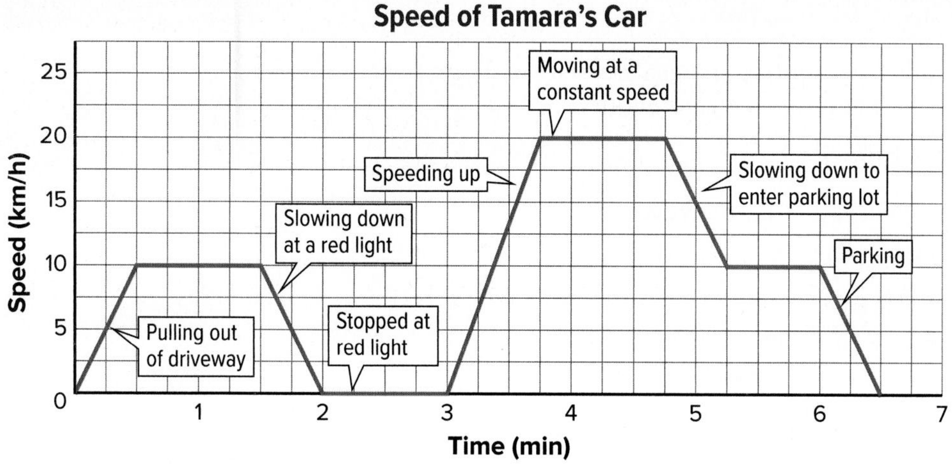

Figure 16 For objects that are speeding up and slowing down, the slope of the line on a speed-time graph is the acceleration.

Identify *the time intervals during which Tamara's car is not accelerating.*

Speed-time graphs and acceleration

When an object travels in a straight line and does not change direction, a graph of speed versus time can provide information about the object's acceleration. **Figure 16** shows the speed-time graph of Tamara's car as she drives to the store. Just as the slope of a line on a distance-time graph is the object's speed, the slope of a line on a speed-time graph is the object's acceleration. For example, when Tamara pulls out of her driveway, the car's acceleration is 0.33 km/min^2, which is equal to the slope of the line from $t = 0$ to $t = 0.5$ min.

Calculating acceleration

Acceleration is the rate of change in velocity. To calculate the acceleration of an object, the change in velocity is divided by the length of the time interval over which the change occurs. The change in velocity is the final velocity minus the initial velocity. If the direction of motion does not change and the object moves in a straight line, the size of the change in velocity can be calculated from the change in speed. Then, the acceleration of an object can be calculated from the following equation.

Acceleration Equation

$$\text{acceleration (in meters/second}^2\text{)} = \frac{\text{change in velocity (in meters/second)}}{\text{time (in seconds)}}$$

$$a = \frac{v_f - v_i}{t}$$

In SI units, velocity has units of m/s and time has units of s, so the SI unit of acceleration is m/s^2. In some cases, your calculations will result in a negative acceleration. The negative sign means *in the opposite direction*. For example, an acceleration of -10 m/s^2 north is the same as 10 m/s^2 south.

CCC CROSSCUTTING CONCEPTS

Systems and System Models Create a table to demonstrate the importance of defining the initial conditions of a system when discussing acceleration. Change the initial conditions of the system's motion to show how a given acceleration can result in different final velocities.

EXAMPLE Problem 3

CALCULATE ACCELERATION A skateboarder has an initial velocity of 3 m/s west and comes to a stop in 2 s. What is the skateboarder's acceleration?

Identify the Unknown: acceleration: a

List the Knowns: initial velocity: $v_i = \mathbf{3\ m/s\ west}$

final velocity: $v_f = \mathbf{0\ m/s\ west}$

time: $t = \mathbf{2\ s}$

Set Up the Problem: $a = \frac{(v_f - v_i)}{t} = \frac{(0\text{ m/s} - 3\text{ m/s})}{2\text{ s}}$ west

Solve the Problem: $a = \frac{(0\text{ m/s} - 3\text{ m/s})}{2\text{ s}} = -1.5\text{ m/s}^2$ west

The acceleration has a negative sign, so the direction is reversed.

$a = 1.5\text{ m/s}^2$ east

Check the Answer: The magnitude of the acceleration (1.5 m/s²) is reasonable for a skateboard that takes 2 s to slow from 3 m/s to 0 m/s. The acceleration is in the opposite direction of the velocity, so the skateboard is slowing down, as we expected.

PRACTICE Problems

ADDITIONAL PRACTICE

22. An airplane starts at rest and accelerates down the runway for 20 s. At the end of the runway, its velocity is 80 m/s north. What is its acceleration?

23. A cyclist starts at rest and accelerates at 0.5 m/s² south for 20 s. What is the cyclist's final velocity?

24. **CHALLENGE** A ball is dropped and falls with an acceleration of 9.8 m/s² downward. It hits the ground with a velocity of 49 m/s downward. How long did it take the ball to fall to the ground?

Motion in Two Dimensions

So far, we have discussed only motion in a straight line. But most objects are not restricted to moving in a straight line. Recall that we cannot add measurements that are not in the same or opposite directions. So, we will discuss motion in each direction separately. For example, suppose a student walked three blocks north and four blocks east. The trip could be described this way: the student walked north for three blocks at 1 m/s and then walked east for four blocks at 2 m/s.

Recall that objects that change direction are accelerating. For an object that is changing direction, its acceleration is not in the same or opposite direction as its velocity. This means that we cannot use the acceleration equation. Just as with displacement and velocity, accelerations that are not in the same or opposite directions cannot be directly combined.

Get It?

Explain why you cannot use the acceleration equation for an object that changes direction.

Circular motion

Think about a horse's horizontal motion on a carousel, such as the one in **Figure 17.** The horse moves in a circular path. Its speed remains constant, but it is accelerating because its direction of motion changes. The change in the direction of the horse's velocity is toward the center of the carousel. The horse's velocity is perpendicular to the inward acceleration. Acceleration toward the center of a curved or circular path is called **centripetal acceleration.** In the same way, Earth experiences centripetal acceleration as it orbits the Sun in a nearly circular path.

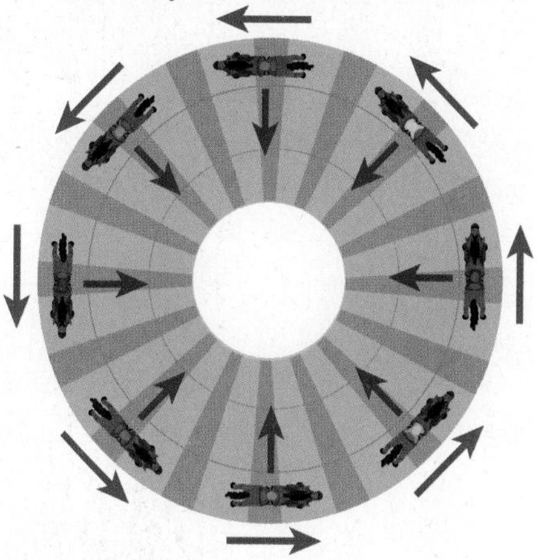

Figure 17 The horizontal speed of the horses in this carousel is constant, but the horses are accelerating because their direction is changing constantly. The acceleration of each horse is toward the center of the circular carousel.

Get It?
Define the term *centripetal acceleration.*

Projectile motion

If you have tossed a ball to someone, you have probably noticed that thrown objects do not travel in straight lines. They curve downward. That is why quarterbacks, dart players, and archers aim above their targets. Anything that is thrown or shot through the air is called a projectile. Earth's gravity causes projectiles to follow a curved path.

Horizontal and vertical motion When you throw or shoot an object, such as the rubber band in **Figure 18,** the force exerted by your hand gives the object a horizontal velocity. For example, after the rubber band is released, its horizontal velocity is constant. The rubber band does not accelerate horizontally. If there were no gravity, the rubber band would move along the straight dotted line in **Figure 18.**

However, when you release a rubber band, gravity causes it to accelerate downward. The rubber band has an increasing vertical velocity. The result of these two motions is that the rubber band travels in a curve, even though its horizontal and vertical motions are completely independent of each other.

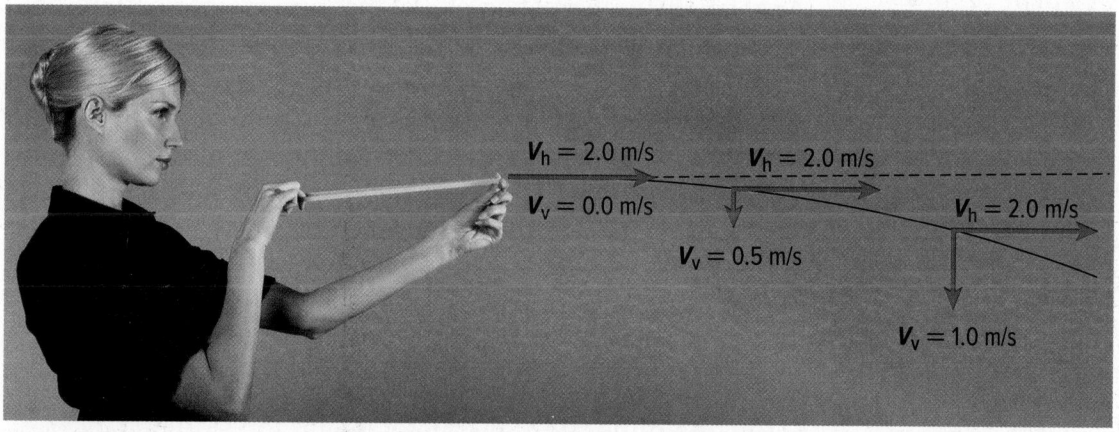

Figure 18 The woman gives the rubber band a horizontal velocity. The horizontal velocity of the rubber band remains constant, but gravity causes the rubber band to accelerate downward. The combination of these two motions causes the rubber band to move in a curved path.

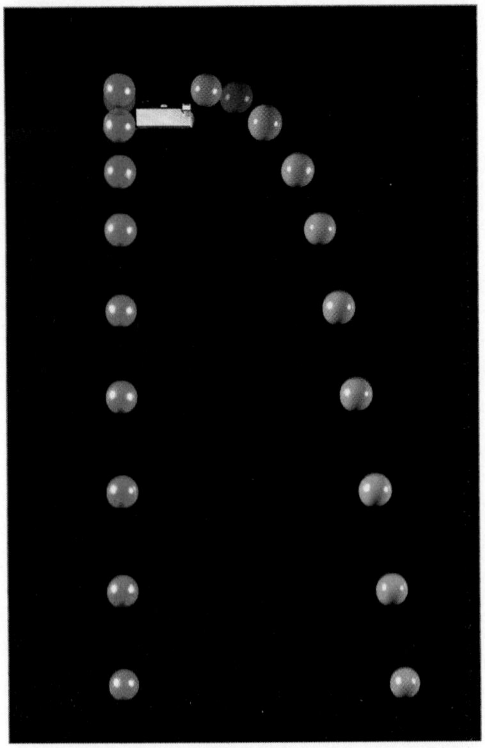

Figure 19 The dropped ball and the thrown ball in this multiflash photograph have the same downward acceleration.

Throwing and dropping If you were to throw a ball as hard as you could in a perfectly horizontal direction, would it take longer to reach the ground than if you dropped a ball from the same height? Surprisingly, it will not. A thrown ball and a dropped ball will hit the ground at the same time. Both balls in **Figure 19** travel the same vertical distance in the same amount of time. However, the ball thrown horizontally travels a greater horizontal distance than the ball that is dropped.

Amusement park acceleration

Riding roller coasters in amusement parks can give you the feeling of danger, but these rides are designed to be safe. Engineers use the laws of physics to design amusement park rides that are thrilling but harmless. Roller coasters are constructed of steel or wood. Because wood is not as strong as steel, wooden roller coasters do not have hills that are as high and as steep as those of some steel roller coasters.

The highest speeds and accelerations are usually produced on steel roller coasters. Steel roller coasters can offer multiple steep drops and inversion loops, which give the riders large accelerations. As riders move down a steep hill or an inversion loop, they will accelerate toward the ground due to gravity. When riders go around a sharp turn, they are also accelerated. This acceleration is due to a change in direction.

Check Your Progress

Summary

- Acceleration is the rate of change of velocity.
- The speed of an object increases if the acceleration is in the same direction as the velocity.
- The speed of an object decreases if the acceleration and the velocity of the object are in opposite directions.
- If an object is moving in a straight line, the change in velocity equals the final speed minus the initial speed.
- Acceleration toward the center of a curved or circular path is called centripetal acceleration.

Demonstrate Understanding

25. **Describe** the acceleration of your bicycle as you ride it from your home to the store.
26. **Determine** the change in velocity of a car that starts at rest and has a final velocity of 20 m/s north.
27. **Analyze** the motion of an object that has an acceleration of 0 m/s^2.

Explain Your Thinking

28. **Compare** Suppose a car is accelerating so that its speed is increasing. First, describe the line that you would plot on a speed-time graph for the motion of the car. Then describe the line that you would plot on a distance-time graph.
29. **MATH Connection** A ball is dropped from a cliff and has an acceleration of 9.8 m/s^2. How long will it take the ball to reach a speed of 24.5 m/s?
30. **MATH Connection** A sprinter leaves the starting blocks with an acceleration of 4.5 m/s^2. What is the sprinter's speed 2 s later?

LEARNSMART® Go online to follow your personalized learning path to review, practice, and reinforce your understanding.

Autonomous Vehicles Go Subterranean

In 2004, the Defense Advanced Research Projects Agency (DARPA) held a competition that opened the door for the development of autonomous vehicles. At the time, driverless cars were the stuff of science fiction, but the DARPA Grand Challenge helped move self-driving technology forward. Today, car companies are testing autonomous cars on rural and urban roads, and these vehicles may soon be part of our everyday lives.

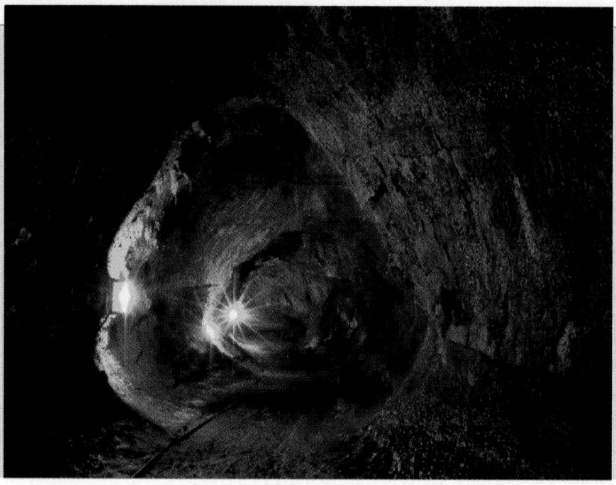

Underground tunnels provide a testing ground for advances in autonomous vehicles.

Physics at the wheel

How can a vehicle operate autonomously? To move safely, it must interpret the environment, evaluate its position relative to the destination, navigate obstacles, and control speed and direction of motion. Light beams emitted by lasers and radio waves given off by a radar unit bounce off surrounding objects and geographical features. Reflection times determine distances between a vehicle and these objects and help evaluate changes in position and terrain.

Additional position data is provided by the Global Positioning System (GPS) and sensors that measure wheel rotation and direction of motion. The vehicle's computer integrates all incoming data, compares it to a map of the route, and makes necessary adjustments in steering, throttling, and braking.

DARPA Subterranean Challenge

The DARPA Subterranean (SubT) Challenge requires competitors to map, navigate, and search underground environments such as natural cave networks and human-made tunnels. Teams of engineers, scientists, and citizens from around the world will compete in the Challenge, developing hardware and software for testing on a variety of courses to conquer tight passages, vertical shafts, underground water, unstable structures, and darkness while not being able to depend on GPS.

The Challenge will not produce the ultimate below-ground autonomous vehicle, but if it advances the technology used to work in underground environments, future vehicles could help save lives.

COMMUNICATE SCIENTIFIC INFORMATION

Brainstorm the potential benefits of using automated vehicles in underground settings. Use your ideas to design a poster advertising an automated underground vehicle service. Share your poster with the class.

MODULE 2
STUDY GUIDE

GO ONLINE to study with your Science Notebook.

Lesson 1 DESCRIBING MOTION

- Motion occurs when an object changes its position relative to a reference point.
- Displacement is the distance and direction of a change in position from the starting point.
- Speed is the rate at which an object's position changes.
- On a distance-time graph, time is the x-axis, and distance is y-axis.
- The slope of a line plotted on a distance-time graph is the speed.

- motion
- displacement
- speed

Lesson 2 VELOCITY AND MOMENTUM

- The velocity of an object includes the object's speed and its direction of motion relative to a reference point.
- An object's motion is always described relative to a reference point.
- The momentum of an object is defined for a particular frame of reference and is the product of the object's mass and velocity: $p = mv$.

- velocity
- momentum

Lesson 3 ACCELERATION

- Acceleration is the rate of change of velocity.
- The speed of an object increases if the acceleration is in the same direction as the velocity.
- The speed of an object decreases if the acceleration and the velocity of the object are in opposite directions.
- If an object is moving in a straight line, the change in velocity equals the final speed minus the initial speed.
- Acceleration toward the center of a curved or circular path is called centripetal acceleration.

- acceleration
- centripetal acceleration

THREE-DIMENSIONAL THINKING

Module Wrap-Up

REVISIT THE PHENOMENON

Why is this motor bike traveling in an arc?

CER Claim, Evidence, Reasoning

Explain Your Reasoning Revisit the claim you made when you encountered the phenomenon. Summarize the evidence you gathered from your investigations and research. Does your evidence support your claim? If not, revise your claim. Explain why your evidence supports your claim.

STEM UNIT PROJECT

Now that you've completed the module, revisit your STEM unit project. You will summarize your evidence and apply it to the project.

GO FURTHER

SEP Data Analysis Lab

The 400 Meter Dash

The 400-m sprint, or dash, is a footrace that is equal to one lap around a running track. To complete the distance of 400 m on a standard running track, the starting positions of the runners are staggered, with one runner actually starting at the finish line. At the sound of a starting pistol, the athletes take off from their fixed positions and speed up to advance beyond the other runners. Some runners have a strong "kick," or an ability to increase their velocity at the end of the race.

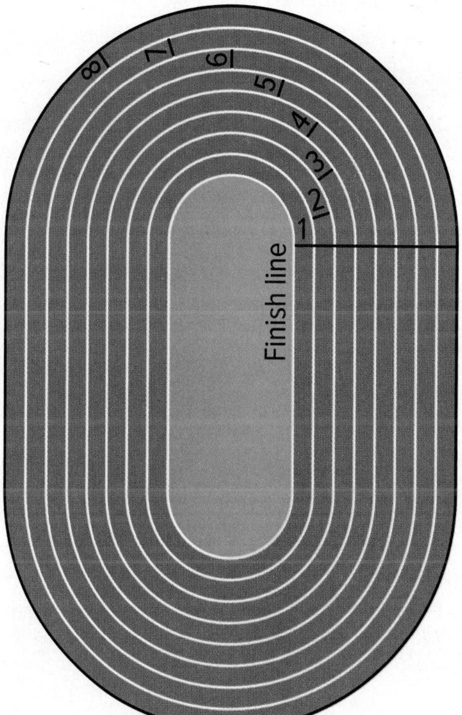

The diagram on the right shows the starting positions for eight racers in a 400-m race. Look at the diagram and answer the questions below.

CER Analyze and Interpret Data

1. **Claim and Evidence** The starting positions indicated on the diagram are typical for a 400-m dash. Why are the runners not all starting together in a straight line?
2. **Claim, Evidence, Reasoning** At the completion of a race, what is the displacement of the runner in lane 1? Is this the same for all the runners? Explain your answer.
3. **Claim, Evidence, Reasoning** If a male runner in the fourth starting position ran the 400-m race in 44.40 s, how would you calculate his average speed? Explain your answer.

Stephen Barnes/Sport/Alamy Stock Photo

format35/iStock/Getty Images

FORCES AND NEWTON'S LAWS

ENCOUNTER THE PHENOMENON

Why can a car stop faster than a train?

GO ONLINE to play a video about how leaves affect friction on the wheels of a train.

SEP Ask Questions

Do you have other questions about the phenomenon? If so, add them to the driving question board.

CER Claim, Evidence, Reasoning

Make Your Claim Use your CER chart to make a claim about why a car can stop faster than a train. Explain your reasoning.

Collect Evidence Use the lessons in this module to collect evidence to support your claim. Record your evidence as you move through the module.

Explain Your Reasoning You will revisit your claim and explain your reasoning at the end of the module.

GO ONLINE to access your CER chart and explore resources that can help you collect evidence.

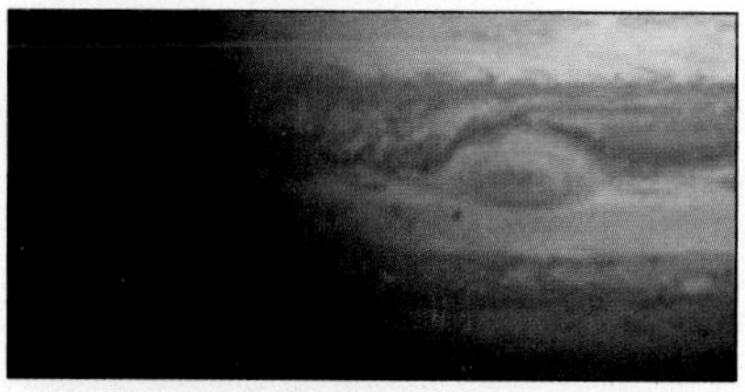

LESSON 1: Explore & Explain: Gravity

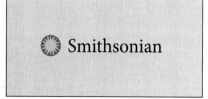

LESSON 3: Explore & Explain: What happens in a crash?

Smithsonian

Additional Resources

(t)Video Supplied by BBC Worldwide Learning, (bl)Cassini/NASA, (br)TRL Ltd./Science Source

LESSON 1
FORCES

FOCUS QUESTION

How does friction affect motion?

What is force?

Catching a basketball and hitting a baseball with a bat are examples of applying force to an object. A **force** is a push or a pull. In both examples, the applied force changes the movement of the ball. Sometimes it is obvious that a force has been applied. But other forces are not as noticeable. For instance, are you conscious of the force that the floor exerts on your feet? Can you feel the force of the atmosphere pushing against your body or gravity pulling on your body? Think about all of the forces that you exert in a day. Every push, pull, stretch, or bend is a force being applied to an object.

Changing motion

What happens to the motion of an object when you exert a force on it? A force can cause the motion of an object to change. Think of kicking a soccer ball, as shown in **Figure 1.** The player's foot strikes the ball with a force that causes the ball to stop and then move in the opposite direction. If you have played billiards, you know that you can make a ball at rest roll into a pocket by striking it with another ball. The force from the moving ball causes the ball at rest to move in the direction of the force. In each case, the velocity of the ball was changed by a force.

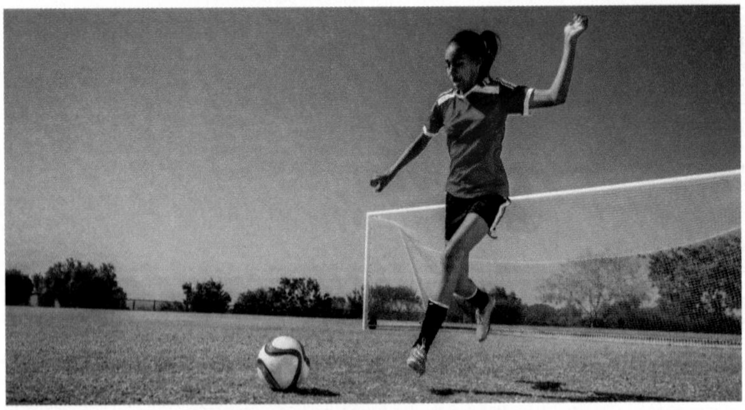

Figure 1 When the player kicks the soccer ball, she is exerting a force on the ball. This kick will cause the ball's motion to change.

Erik Isakson/Blend Images

3D THINKING DCI Disciplinary Core Ideas CCC Crosscutting Concepts SEP Science & Engineering Practices

COLLECT EVIDENCE

Use your Science Journal to record the evidence you collect as you complete the readings and activities in this lesson.

INVESTIGATE

GO ONLINE to find these activities and more resources.

Laboratory: Friction Predictions

Carry out an investigation to observe and compare the forces in a system needed to move an object over surfaces with varying degrees of friction.

Quick Investigation: Compare Friction

Carry out an investigation to observe and compare how surfaces with different levels of friction interact in a system to prevent an object from sliding.

Net force

When two or more forces act on an object at the same time, the forces combine to form the net force. The **net force** is the sum of all of the forces acting on an object. Forces have a direction, so they follow the same addition rules as displacement, as listed in **Table 1.** Forces are measured in SI units of newtons (N). A force of about 3 N is needed to lift a full can of soda at a constant speed.

Table 1 Rules for Adding Forces

1. Add forces in the same direction.
2. Subtract forces in opposite directions.
3. Forces that are neither in the same direction nor in opposite directions cannot be directly added together.

Unbalanced forces Look at **Figure 2A.** The students are each pushing on the box in the same direction. These forces are combined, or added together, because they are exerted on the box in the same direction. The students in **Figure 2B** are pushing in opposite directions. Here, the direction of the net force is the same as the direction of the greater force. In other words, the student who pushes harder causes the box to move in the direction of that push. The net force is the difference between the two forces because they are in opposite directions. In **Figure 2A** and **Figure 2B,** the net force has a value that is not zero, and the box moves. The forces that the students apply are considered unbalanced forces.

Balanced forces Now suppose that the students are pushing with the same size force but in opposite directions, as shown in **Figure 2C.** The net force on the box is zero because the two forces cancel each other. Forces that are equal in size and opposite in direction are called balanced forces. Unbalanced forces cause changes in motion. Balanced forces do not cause a change in motion.

Figure 2 Forces can be balanced or unbalanced.
Identify *another example of unbalanced forces and another example of balanced forces.*

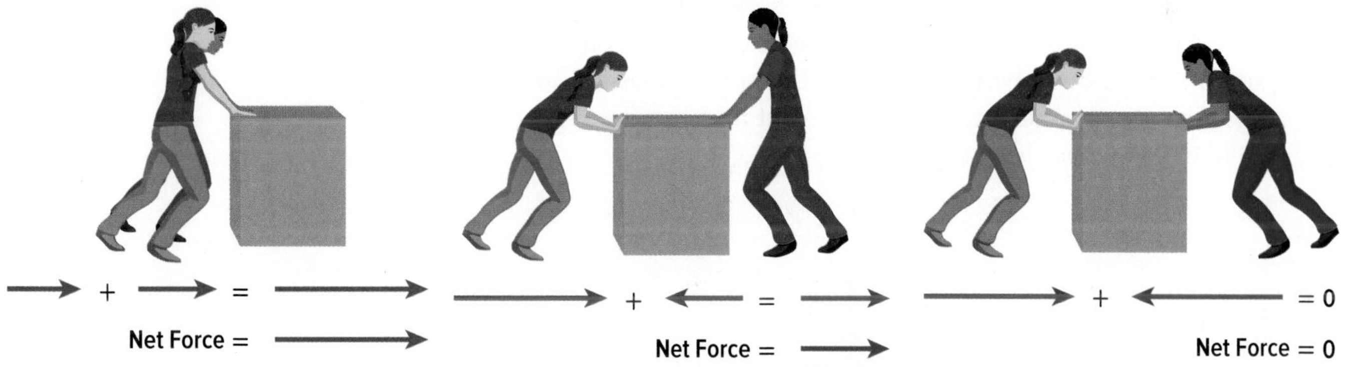

[A] Two students push on the box in the same direction. These forces are unbalanced. The net force is the sum of the two forces, and the box will move in the direction that the students push.

[B] Two students push on the box with unequal forces in opposite directions. These forces are unbalanced. The net force is the difference of the two forces, and the box will move in the direction of the greater force.

[C] Two students push on the box with equal forces but in opposite directions. These forces are balanced. The net force is zero, and the box does not move.

Friction

Suppose you give a skateboard a push with your hand. After you let go, the skateboard slows down and eventually stops. Because the skateboard's motion is changing as it slows down, there must be a force acting on it. The force that slows the skateboard is called friction. **Friction** is the force that opposes the sliding motion of two surfaces that are touching each other.

What causes friction?

Would you believe that the surface of a highly polished piece of metal is rough? Surfaces that appear smooth actually have many bumps and dips. These bumps and dips can be seen when the surface is examined under a microscope, as shown in **Figure 3.** If two surfaces are in contact, welding or sticking occurs where the bumps touch each other. These microwelds are the source of friction. To move one surface over the other, a force must be applied to break the microwelds.

Figure 3 The surface of this teapot looks and feels smooth, but it is rough at the microscopic level.

Get It?
Describe the source of friction.

The amount of friction between two surfaces depends on the kinds of surfaces and the force pressing the surfaces together. Rougher surfaces have more bumps and can form more microwelds, increasing the amount of friction. In addition, a larger force pushing the two surfaces together will cause more of the bumps to come into contact, as shown in **Figure 4.** The microwelds will be stronger, and a greater force must be applied to break them apart.

Static friction

Suppose you have a cardboard box filled with books, such as the one in **Figure 5,** and you want to move that box. The box is resting on what seems to be a smooth floor, but when you push on the box, it does not budge. The box experiences no change in motion, so the net force on the box is zero. The force of friction cancels your push.

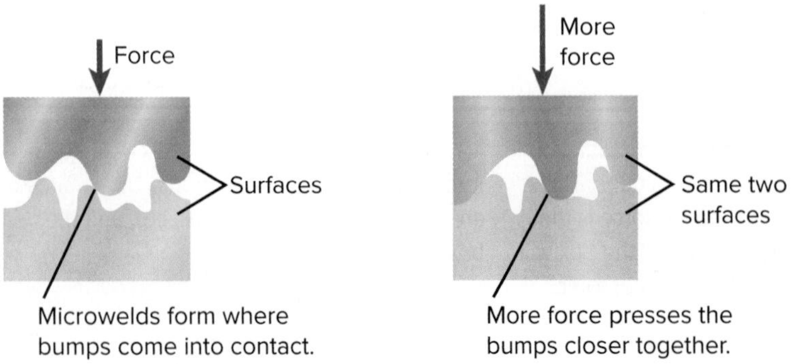
Force
Surfaces
Microwelds form where bumps come into contact.
More force
Same two surfaces
More force presses the bumps closer together.

Figure 4 Friction is caused by microwelds that form between two surfaces. Microwelds are stronger when the two surfaces are pushed together with a greater force.

Explain *how the area of contact between the surfaces changes when they are pushed together.*

Static friction balances the applied force. The box remains at rest and does not accelerate.

Sliding friction and the applied force are unbalanced. The box accelerates to the right.

Figure 5 Friction opposes the sliding motion of two surfaces that are touching each other.
Describe *the net force on each box.*

This type of friction is called static friction. Static friction prevents two surfaces from sliding past each other and is due to the microwelds that have formed between the bottom of the box and the floor. Your push is not great enough to break the microwelds, and the box does not move, as shown in **Figure 5.**

Sliding friction

If you and a friend push together, as shown on the right in **Figure 5,** the box moves. Together, you and your friend have exerted enough force to break the microwelds between the floor and the bottom of the box. But if you stop pushing, the box quickly comes to a stop. To keep the box moving, you must continually apply a force. This is because sliding friction opposes the motion of the box as the box slides across the floor. Sliding friction opposes the motion of two surfaces sliding past each other and is caused by microwelds constantly breaking and forming as the objects slide past each other. The force of sliding friction is usually less than the force of static friction.

Rolling friction

You may think of friction as a disadvantage. But wheels, like the ones shown in **Figure 6,** would not work without friction. As a wheel rolls, static friction acts over the area where the wheel and surface are in contact. This special case of static friction is sometimes called rolling friction.

You may have seen a car that was stuck in snow, ice, or mud. The driver steps on the gas, but the wheels just spin without the car moving. The force used to rotate the tires is greater than the force of static friction between the wheels and the ground, so the tires slide instead of gripping the ground. Spreading sand or gravel on the surface increases the friction until the wheels stop slipping and begin rolling. When referring to tires on vehicles, people often use the term *traction* instead of *friction*.

Figure 6 Rolling friction between the bicycle's wheels and the pavement keeps the wheels from slipping.

Gravity

At this moment, you are exerting an attractive force on everything around you—your desk, your classmates, and even the planet Jupiter, millions of kilometers away. This attractive force acts on all objects with mass and is called gravity. **Gravity** is an attractive force between any two objects that depends on the masses of the objects and the distance between them.

Gravity is one of the four basic forces called the fundamental forces. The other basic forces are the electromagnetic force, the strong nuclear force, and the weak nuclear force. Gravity acts on all objects with mass, and the electromagnetic force acts on all charged particles. Both gravity and the electromagnetic force have an infinite range. The nuclear forces affect only particles in the nuclei of atoms.

The law of universal gravitation

In the 1660s, English scientist Isaac Newton used data on the motions of the planets to find the relationship between the gravitational force between two objects, the objects' masses, and the distance between them. This relationship is called the law of universal gravitation and can be written as the following equation.

$$F = G\frac{m_1 m_2}{d^2}$$

In this equation, G is the universal gravitational constant, and d is the distance between the centers of the two masses, m_1 and m_2. The law of universal gravitation states that the gravitational force increases as the mass of either object increases and as the objects move closer, as shown in **Figure 7.** The force of gravity between any two objects can be calculated if their masses and the distance between them are known.

If the mass of either of the objects increases, the gravitational force between them increases.

If the objects are closer together, the gravitational force between them increases.

Figure 7 The law of universal gravitation states that the gravitational force between two objects depends on their masses and the distance between them.

Gravity and you The law of universal gravitation explains why you feel Earth's gravity but not the Sun's gravity or this book's gravity. While the Sun has much more mass than Earth, the Sun is too far away to exert a noticeable gravitational attraction on you. And while this book is close, it does not have enough mass to exert an attraction that you can feel. Only Earth is both close enough and massive enough that you can feel its gravitational attraction.

The range of gravity According to the law of universal gravitation, the gravitational force between two masses decreases rapidly as the distance between the masses increases. For example, if the distance between two objects increases from 1 m to 2 m, the gravitational force between them becomes one-fourth as large. If the distance increases from 1 m to 10 m, the gravitational force between the objects is one-hundredth as large. However, no matter how far apart two objects are, the gravitational force between them never completely drops to zero. Because the gravitational force between two objects never disappears, gravity is called a long-range force.

Get It?
Explain why gravity is called a long-range force.

The gravitational field Forces that act at a distance, without requiring contact between objects, can be explained using the concept of a field. A **field** is a region of space that has a physical quantity (such as a force) at every point. All objects are surrounded by a gravitational field. **Figure 8** shows that Earth's gravitational field is strongest near Earth and becomes weaker as the distance from Earth increases. The strength of the gravitational field, represented by the letter g, is measured in newtons per kilogram (N/kg).

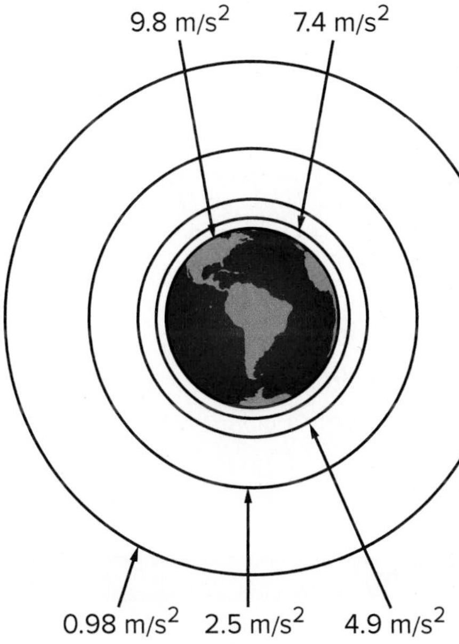

Figure 8 Earth's gravitational field exists at all points in space. It is strongest near the surface and decreases in strength as one moves away from Earth.

Weight The gravitational force exerted on an object is the object's **weight.** The universal law of gravitation can be used to calculate weight, but scientists use a simplified version of this equation that combines m_1, d^2, and G into a single number called the gravitational strength, g.

Weight Equation

weight (N) = mass (kg) × **gravitational strength** (N/kg)

$$F_g = mg$$

We use F_g for weight because weight is the force due to gravity. Weight has units of newtons (N) because it is a force. The g in the subscript stands for *gravity.* The gravitational strength, g, has the units N/kg. Recall that 1 N = 1 kg · m/s^2. So, g can also be written with units of m/s^2.

Weight and mass Weight and mass are not the same. Weight is a force, and mass is a measure of the amount of matter an object contains. But, according to the weight equation, weight and mass are related. Weight increases as mass increases.

Weight on Earth We often need to know an object's weight on Earth. At Earth's surface, m_1 is Earth's mass, and d is Earth's radius. Using those values in the law of universal gravitation shows that g = 9.8 N/kg near Earth's surface. **Table 2** lists the weights of some objects on Earth.

Table 2 Weight of Common Objects on Earth

Object	Weight
Cell phone	1 N
Backpack full of books	100 N
Jumbo jet	3.4 million N

EXAMPLE Problem 1

SOLVE FOR WEIGHT An elephant has a mass of 5000 kg. What is the elephant's weight?

Identify the Unknown: weight: F_g

List the Knowns: mass: m = **5000 kg**

gravitational strength: g = **9.8 N/kg**

Set Up the Problem: $F_g = mg$

Solve the Problem: F_g = (**5000 kg**)(**9.8 N/kg**) = **49,000 N**

Check the Answer: The gravitational strength is about 10 N/kg, so we would expect the elephant's weight to be about 50,000 N. Our answer (F_g = 49,000 N) makes sense.

PRACTICE Problems

ADDITIONAL PRACTICE

1. A squirrel has a mass of 0.5 kg. What is its weight?
2. A boy weighs 400 N. What is his mass?
3. **CHALLENGE** An astronaut has a mass of 100 kg and has a weight of 370 N on Mars. What is the gravitational strength on Mars?

Weight away from Earth An object's weight usually refers to the gravitational force between the object and Earth. But the weight of an object can change, depending on the gravitational force on the object. For example, the gravitational strength on the Moon is 1.6 N/kg, about one-sixth as large as Earth's gravitational strength. As a result, a person, such as the astronaut in **Figure 9,** would weigh only about one-sixth as much on the Moon as on Earth.

Figure 9 Although the astronaut has the same mass on the Moon, he weighs less than he does on Earth. He can take longer steps and jump higher than on Earth.

Finding other planets

Earth's motion around the Sun is affected by the gravitational pulls of the other planets in the solar system. In the same way, the motion of every planet in the solar system is affected by the gravitational pulls of all of the other planets.

In the 1840s, the most distant planet known was Uranus. The motion of Uranus calculated from the law of universal gravitation disagreed slightly with its observed motion. Some astronomers suggested that there must be an undiscovered planet affecting the motion of Uranus. Using the law of universal gravitation and the laws of motion, two astronomers independently calculated the orbit of this planet. As a result of these calculations, the planet Neptune was found in 1846.

Check Your Progress

Summary

- A force is a push or a pull on an object.
- The net force on an object is the combination of all of the forces acting on the object.
- Unbalanced forces cause the motion of objects to change.
- Friction is the force that opposes the sliding motion of two surfaces that are in contact.
- Gravity is an attractive force between all objects that have mass.

Demonstrate Understanding

4. **Describe** two forces that would change the motion of a bicycle traveling along a road.
5. **Explain** Can there be forces acting on an object if the object is at rest? Must there be an unbalanced force acting on a moving object? Explain your answers.
6. **Explain** Why does coating surfaces with oil reduce friction between the surfaces?
7. **Distinguish** between the mass of an object and the object's weight.

Explain Your Thinking

8. **Predict** Suppose Earth's mass increased but Earth's diameter did not change. Describe how the gravitational force between Earth and an object on its surface would change.
9. **MATH Connection** On Earth, what is the weight of a large-screen TV that has a mass of 75 kg?
10. **MATH Connection** Two students push on a box in the same direction, and one student pushes in the opposite direction. What is the net force on the box if each student pushes with a force of 50 N?

LEARNSMART® Go online to follow your personalized learning path to review, practice, and reinforce your understanding.

LESSON 2

NEWTON'S LAWS OF MOTION

FOCUS QUESTION

How do forces affect acceleration?

Isaac Newton and the Laws of Motion

In 1665, Cambridge University in England closed for 18 months because of the bubonic plague. Isaac Newton, a 23-year-old student, used this time to develop the law of universal gravitation, calculus, and three laws describing how forces affect the motion of objects. Newton's laws of motion apply to the motion of everyday objects, such as cars and bicycles, as well as the motion of planets and stars.

Newton's First Law of Motion

Recall that forces change an object's motion. Newton's first law explains the relationship between force and change in motion. **Newton's first law of motion** states that an object moves at a constant velocity unless an unbalanced force acts on it. This means that a moving object will continue to move in a straight line at a constant speed unless an unbalanced force acts on it. If an object is at rest, its velocity is zero, and it stays at rest unless an unbalanced force acts on it. Newton's first law is sometimes called the law of inertia. **Inertia** (ih NUR shuh) is the tendency of an object to resist any change in its motion. **Figure 10** shows a toy car hitting a block and stopping. The block on top of the car is not attached to the car and keeps traveling. The block's forward motion demonstrates the property of inertia.

Figure 10 You might have seen the law of inertia without even knowing it. For example, as the car hits the wooden block and stops, the red block continues to move forward.

Describe *another example of inertia.*

3D THINKING | DCI Disciplinary Core Ideas | CCC Crosscutting Concepts | SEP Science & Engineering Practices

COLLECT EVIDENCE

Use your Science Journal to record the evidence you collect as you complete the readings and activities in this lesson.

INVESTIGATE

GO ONLINE to find these activities and more resources.

Applying Practices: Newton's Second Law

HS-PS2-1. Analyze data to support the claim that Newton's second law of motion describes **the mathematical relationship among** the net force on a macroscopic object, its mass, and its acceleration.

Inertia and mass Does a bowling ball have the same inertia as a table-tennis ball? Why is there a difference? Swatting a ball with a table-tennis paddle would not change the motion of a bowling ball much, but it would greatly accelerate a table-tennis ball. A greater force would be needed to change the motion of a bowling ball because it has greater inertia.

Recall that mass is the amount of matter in an object. An object's inertia is related to its mass. The greater an object's mass is, the greater its inertia is. A bowling ball has more mass than a table-tennis ball, so the bowling ball has a greater inertia.

You will sometimes hear people say that when an object begins to move, inertia is overcome. This is not true. The object still has mass when it is moving, so it still has inertia. As long as the mass is the same, the object has the same inertia.

Newton's Second Law of Motion

Newton's first law of motion states that the motion of an object changes only if an unbalanced force acts on the object. Newton's second law of motion describes how the forces exerted on an object, its mass, and its acceleration are related.

Force and acceleration

How are throwing a ball as hard as you can and tossing it gently different? When you throw hard, you exert a greater force on the ball. The ball has a greater velocity when it leaves your hand. The hard-thrown ball has a greater change in velocity, and the change occurs over a shorter period of time. Recall that acceleration equals the change in velocity divided by the time it takes for the change to occur. So, a hard-thrown ball is accelerated more than a gently-thrown ball, as shown in **Figure 11.**

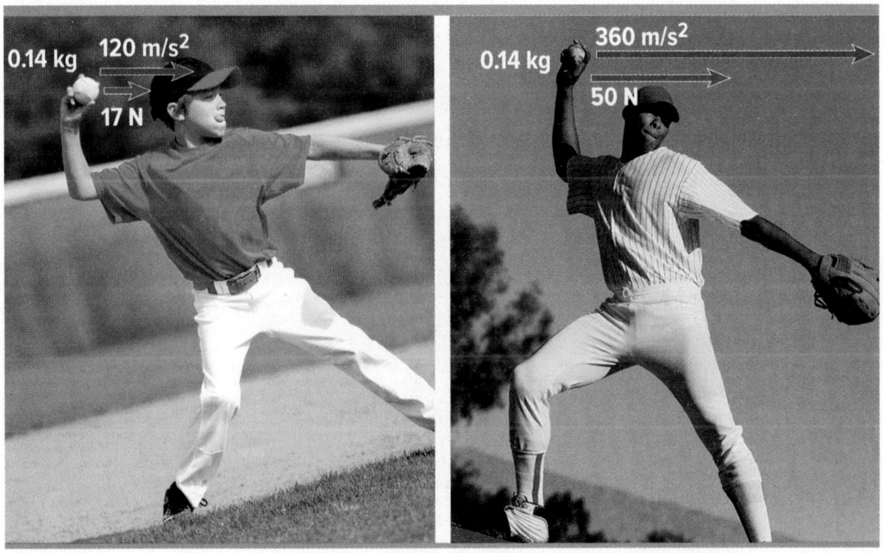

Figure 11 According to Newton's second law of motion, the harder you throw a baseball, the greater its acceleration will be.

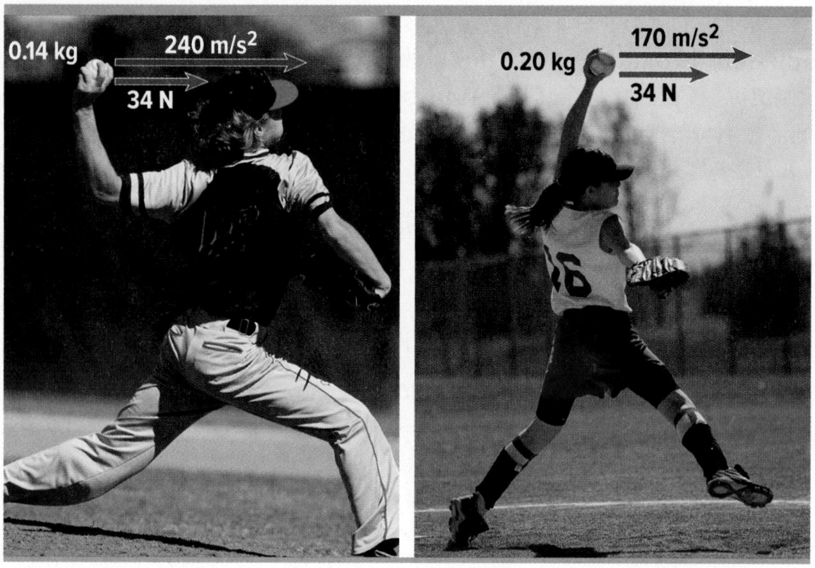

Figure 12 If both pitchers apply the same force, the baseball will experience a greater acceleration than the softball because the baseball has less mass.

Mass and acceleration

If you throw a softball and a baseball as hard as you can, as shown in **Figure 12,** why do they not have the same speed? The difference is due to their masses. A softball has a mass of about 0.20 kg, but a baseball's mass is about 0.14 kg. The softball has less velocity after it leaves your hand than the baseball does, even though you exerted the same force. The softball has a lesser final speed because it experienced a lesser acceleration. The acceleration of an object depends on its mass as well as the force exerted on it. Force, mass, and acceleration are related.

Get It?

Identify You apply a force of 2 N to a toy car and to a real car. Which car has the greater acceleration?

Relating force, mass, and acceleration

Newton's second law of motion states that an object's acceleration is in the same direction as the net force on the object and is equal to the net force exerted on it divided by its mass. Newton's second law can be written as the following equation.

The Second Law of Motion Equation

$$\textbf{acceleration}\text{ (in meters/second}^2\text{)} = \frac{\textbf{net force}\text{ (in newtons)}}{\textbf{mass}\text{ (in kilograms)}}$$

$$a = \frac{F_{net}}{m}$$

In the above equation, acceleration has units of meters per second squared (m/s^2), and mass has units of kilograms (kg). Recall that the net force is the sum of all of the forces on an object. Remember that the SI unit for force is a newton (N) and that $1\ N = 1\ kg{\cdot}m/s^2$. Just like velocity and acceleration, force has a size and a direction.

(l)tammykayphoto/123RF, (r)Carson Granci/DesignPics/SuperStock

Calculating net force Newton's second law can also be used to calculate the net force if mass and acceleration are known. To do this, the equation for Newton's second law must be solved for the net force, F_{net}. To solve for the net force, multiply both sides of the above equation by the mass.

$$m \times a = \cancel{m} \times \frac{F_{net}}{\cancel{m}}$$

The mass, *m*, on the right side cancels.

$$F_{net} = ma$$

For example, when a tennis player hits a ball, the racket and the ball might be in contact for only a few thousandths of a second. Because the ball's velocity changes over such a short period of time, the ball's acceleration could be as high as 5000 m/s². The ball's mass is 0.06 kg, so the size of the net force exerted on the ball would be 300 N.

$$F_{net} = ma = (0.06\ \text{kg})(5000\ \text{m/s}^2) = 300\ \text{kg} \cdot \text{m/s}^2 = 300\ \text{N}$$

EXAMPLE Problem 2

CALCULATE ACCELERATION You push a wagon that has a mass of 12 kg. If the net force on the wagon is 6 N south, what is the wagon's acceleration?

Identify the Unknown: acceleration: a

List the Knowns: mass: m = 12 kg net force: F_{net} = **6 N south**

Set Up the Problem: $a = \frac{F_{net}}{m}$

Solve the Problem: $a = \frac{\text{6 N south}}{\text{12 kg}} = \text{0.5 m/s}^2 \text{ south}$

Check the Answer: The value of the net force (6) is less than the value of the wagon's mass (12), so we would expect the acceleration's value to be less than 1. Our answer (0.5 m/s²) makes sense.

PRACTICE Problems

ADDITIONAL PRACTICE

11. If a helicopter's mass is 4500 kg and the net force on it is 18,000 N upward, what is its acceleration?

12. What is the net force on a dragster with a mass of 900 kg if its acceleration is 32.0 m/s² west?

13. A car pulled by a tow truck has an acceleration of 2.0 m/s² east. What is the mass of the car if the net force on the car is 3000 N east?

14. CHALLENGE What is the net force on a skydiver falling with a constant velocity of 10 m/s downward?

ACADEMIC VOCABULARY

period
a length of time

Each class period is 45 minutes.

CCC CROSSCUTTING CONCEPTS

Cause and Effect With a partner, repeat Example Problem 2 for wagons with different mass. Explain the relationship between acceleration and mass.

Newton's Third Law of Motion

What happens when you push against a wall? If the wall is sturdy, nothing happens. But if you pushed against a wall while wearing roller skates, you would go rolling backward. This is a demonstration of **Newton's third law of motion.** Newton's third law of motion states that when one object exerts a force on a second object, the second object exerts a force on the first that is equal in strength and opposite in direction.

Sometimes, Newton's third law is written as "to every action force there is an equal and opposite reaction force." However, one force is not causing the second force. They occur at the same time. It does not matter which object is labeled Object 1 and which is labeled Object 2. Think about a boat tied to a dock with a taut rope. You could say that the action force is the boat pulling on the rope and the reaction force is the rope pulling on the boat. But it would be just as correct to say that the action force is the rope pulling on the boat and the reaction force is the boat pulling on the rope.

Forces on different objects do not cancel

If these two forces are equal in size and opposite in direction, you might wonder how some things ever happen. For example, if the box in **Figure 13** pushes on the student when the student pushes on the box, why does the box move? According to the third law of motion, action and reaction forces act on different objects. Recall that the net force is the sum of the forces on a single object. The left picture in **Figure 13** shows the forces on the student. The net force is zero, and the student remains at rest. The right picture shows that there is a net force of 20 N to the right on the box, and the box accelerates to the right.

Get It?

Explain why action and reaction forces do not cancel.

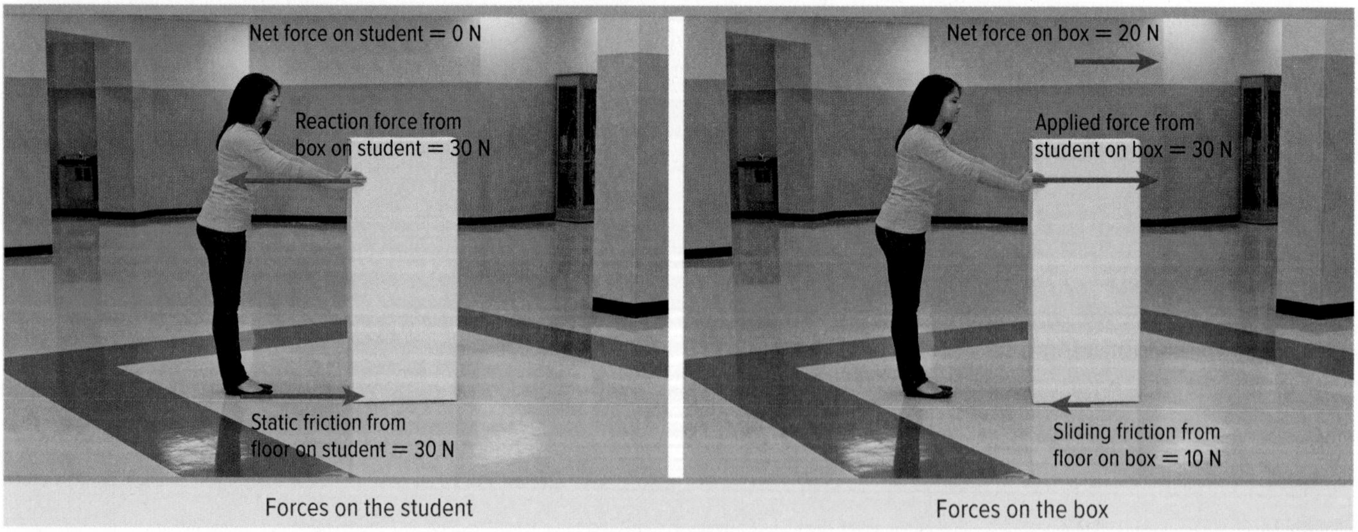

Figure 13 The student pushes on the box, and the box pushes on the student. To understand the motion of each object, the other forces on that object must be examined.

Forces are interactions

Newton's third law depends on a very important fact: forces are interactions between objects. For example, it makes no sense to say "The box has a force of 30 N." Is this force acting on the box? Is the box pushing on something? What is causing this force? However, it does make sense to say "The student applied a force of 30 N to the box."

Furthermore, both objects experience a force from the interaction. Look at the skaters in **Figure 14.** The male skater is pulling upward and to the left on the female skater, while the female skater is pulling downward and to the right on the male skater. The two forces are equal in size but opposite in direction. Both skaters feel a pull of the same size.

Even if the two objects have different masses, they will feel forces of the same size. For example, if a bug flies into the windshield of a truck, the bug and the truck exert forces on each other. According to Newton's third law, the bug and the truck experience forces of the same size even though their masses and accelerations are different.

Figure 14 The skaters exert forces on each other. According to Newton's third law of motion, the two forces are equal in size but opposite in direction. Both skaters feel the same amount of force.

Check Your Progress

Summary

- Newton's first law of motion states that the motion of an object at rest or moving with constant velocity will not change unless an unbalanced force acts on the object.
- Inertia is the tendency of an object to resist a change in motion.
- Newton's second law of motion states that the acceleration of an object depends on its mass and the net force exerted on it.
- According to Newton's third law of motion, when an object exerts a force on a second object, the second object exerts a force on the first object that is equal in strength and opposite in direction.

Demonstrate Understanding

15. **Interpret and Apply** Use Newton's laws of motion to describe what happens when you kick a soccer ball.
16. **Explain** why Newton's first law of motion is sometimes called the law of inertia.
17. **Determine** whether the inertia of an object changes as the object's velocity changes.
18. **Explain** why an object with a smaller mass has a greater acceleration than an object with a larger mass if the same force acts on each.

Explain Your Thinking

19. **Identify** You push a book across a table. The book moves at a constant speed, but you do not move. Identify all of the forces on you. Then, identify all of the forces on the book.
20. **MATH Connection** A student pushes on a 5-kg box with a force of 20 N forward. The force of sliding friction is 10 N backward. What is the acceleration of the box?
21. **MATH Connection** You push yourself on a skateboard with a force of 30 N east and accelerate at 0.5 m/s^2 east. Find the mass of the skateboard if your mass is 58 kg.

LEARNSMART® Go online to follow your personalized learning path to review, practice, and reinforce your understanding.

LESSON 3

USING NEWTON'S LAWS

FOCUS QUESTION

How do Newton's three laws explain the change of motion that occurs in a collision?

What happens in a crash?

Newton's first law of motion can explain what happens in a car crash. When a car traveling about 50 km/h collides head-on with something solid, the car crumples, slows down, and stops within approximately 0.1 s. According to Newton's first law, the passengers will continue to travel at the same velocity that the car was moving unless a force acts on them.

This means that within 0.02 s after the car stops, any unbelted passengers will slam into the windshield, dashboard, steering wheel, or the backs of the front seats, as shown on the left in **Figure 15.** These unbelted passengers are traveling at the car's original speed of 50 km/h (about 30 miles per hour).

Even in a low-speed crash, inertia causes the unrestrained dummies to slam into the windshield or seat in front of them.

This crash dummy was restrained safely with a safety belt and cushioned with an airbag in this low-speed crash.

Figure 15 The crash dummies have inertia and resist changes in motion.

3D THINKING DCI Disciplinary Core Ideas CCC Crosscutting Concepts SEP Science & Engineering Practices

COLLECT EVIDENCE

Use your Science Journal to record the evidence you collect as you complete the readings and activities in this lesson.

INVESTIGATE

GO ONLINE to find these activities and more resources.

Applying Practices: Conservation of Momentum

HS-PS2-2. Use mathematical representations to support the claim that the total momentum of a system of objects is conserved when there is no net force on the system.

Applying Practices: Egg Heads

HS-PS2-3. Apply science and engineering ideas to design, evaluate, and refine a device that minimizes the force on a macroscopic object during a collision.

Safety belts

Passengers wearing safety belts, also shown in **Figure 15,** will be slowed down by the force of the safety belt. This prevents passengers from being thrown out of their seats. Car-safety experts say that about half of the people who die in car crashes would survive if they wore safety belts. Thousands of others would suffer fewer serious injuries.

The force needed to slow a person's speed from 50 km/h to 0 in 0.1 s is equal to 14 times the force of gravity on the person. Therefore, safety belts are designed to loosen a little as they restrain passengers. The loosening increases the time it takes to slow the person down, resulting in a smaller acceleration and a smaller force on the passenger.

Airbags

Airbags also reduce injuries in car crashes by providing a cushion that reduces the acceleration of the passengers and prevents them from hitting the dashboard. When impact occurs, a chemical reaction occurs in the airbag, producing nitrogen gas. The airbag expands rapidly and then deflates just as quickly as the nitrogen gas escapes out of tiny holes in the bag. The entire process is completed in about 0.04 s.

Newton's Second Law and Gravitational Acceleration

Recall that the gravitational force exerted on an object is equal to the object's mass times the strength of gravity ($F_g = mg$). If gravity is the only force acting on an object ($F_{net} = F_g$), then Newton's second law states that the object's acceleration is the force of gravity divided by the object's mass. Notice that the mass cancels, and the object's acceleration due to gravity is equal to the gravitational strength.

$$a = \frac{F_{net}}{m} = \frac{F_g}{m} = \frac{\cancel{m}g}{\cancel{m}} = g$$

When discussing the acceleration due to gravity, g is written with the units m/s^2. On Earth, any falling object with only gravity acting on it will accelerate at $9.8\ m/s^2$ toward Earth.

Air resistance

The gravitational force causes objects to fall toward Earth. However, objects falling in the air experience air resistance. **Air resistance** is a friction-like force that opposes the motion of objects that move through the air. Air resistance acts in the direction opposite to the motion of an object moving through air. If the object is falling downward, air resistance exerts an upward force on the object.

The amount of air resistance on an object depends on the size, shape, and speed of the object, as well as the properties of the air. Air resistance, not the object's mass, is the reason feathers, leaves, and pieces of paper fall more slowly than do pennies, acorns, and billiard balls. If there were no air resistance, then all objects, including a feather and a billiard ball, would fall with the same acceleration, as shown in **Figure 16.**

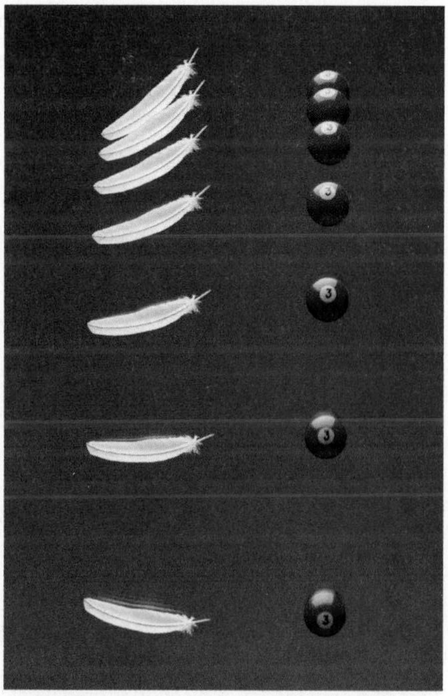

Figure 16 In a vacuum, there is no air resistance. As a result, a feather and a billiard ball fall with the same acceleration in a vacuum.

Erich Schrempp/Science Source

Get It?
Explain why some objects fall faster than others.

Size and shape The more spread out an object is, the more air resistance it will experience. Picture dropping two plastic bags, as shown in **Figure 17.** One is crumpled into a ball, and the other is spread out. When the bags are dropped, the crumpled bag falls faster than the spread-out bag. The downward force of gravity on both bags is the same, but the upward force of air resistance on the crumpled bag is less. As a result, the net downward force on the crumpled bag is greater.

Speed and terminal velocity The amount of air resistance also increases as the object's speed increases. As an object falls, gravity causes it to accelerate downward. But as an object falls faster, the upward force of air resistance increases. So, the net force on the object decreases as it falls, as shown with the skydiver in **Figure 18.**

Eventually, the upward air resistance force becomes large enough to balance the downward force of gravity, and the net force on the object is zero. Then the acceleration of the object is zero, and the object falls with a constant speed called the terminal velocity. **Terminal velocity** is the maximum speed an object will reach when falling through a substance, such as air.

A falling object's terminal velocity depends on its size, shape, and mass. For example, the air resistance on an open parachute is much greater than the air resistance on the skydiver alone. When her parachute opens, the skydiver's terminal velocity will be small enough that she can land safely.

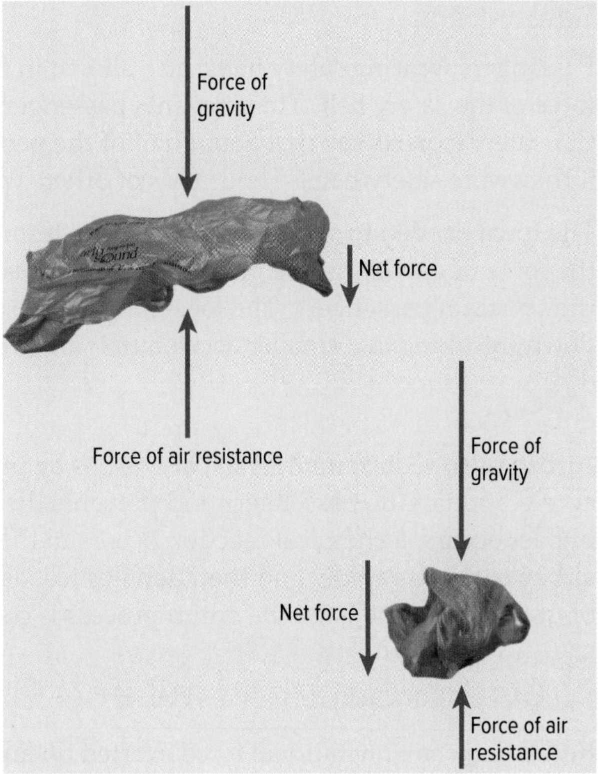

Figure 17 An object's surface size and shape determine how fast it will fall. Because of its greater surface area, the bag on the left has more air resistance acting on it as it falls.

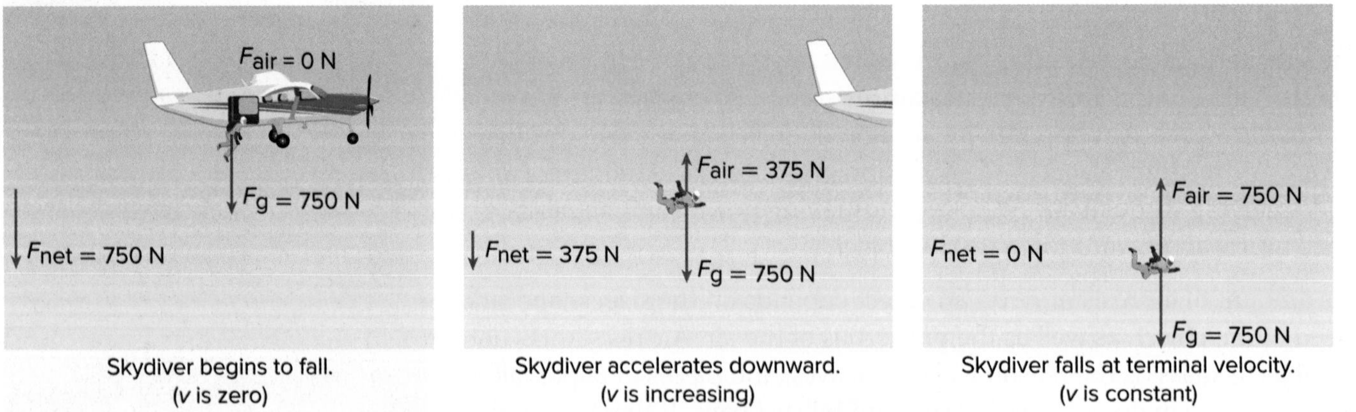

Figure 18 As the skydiver's speed increases, so does the force of air resistance acting on her. At some point, the force of the air resistance equals the force due to gravity, and the skydiver falls at a constant speed called terminal velocity.

Free fall

Suppose an object were falling and there were no air resistance. Gravity would be the only force acting on the object. If gravity is the only force acting on an object, the object is said to be in **free fall.** For example, the feather and the billiard ball falling in a vacuum, shown in **Figure 16,** are in free fall. Another example is an object in orbit. Earth, for example, is in free fall around the Sun. If Earth did not have a velocity perpendicular to the gravitational force, it would fall into the Sun. Similarly, satellites are in free fall around Earth.

Weightlessness You might have seen pictures of astronauts and equipment floating inside an orbiting spacecraft. They are said to be experiencing weightlessness. But, according to the law of universal gravitation, the strength of Earth's gravitational field at a typical orbiting altitude is about 90 percent of its strength at Earth's surface. So, an 80-kg astronaut would weigh about 700 N in orbit and would not be weightless.

What does it mean to say that something is weightless? Think about how you measure your weight. When you stand on a scale, as shown on the left in **Figure 19,** you are at rest. The net force on you is zero, according to Newton's second law of motion. The scale exerts an upward force that balances your weight. The dial on a scale shows the size of the upward force, which is the same as your weight.

Now suppose you stand on a scale in an elevator that is in free fall, as shown on the right in **Figure 19.** You would no longer push down on the scale at all. The scale dial would read zero, even though the force of gravity has not changed. An orbiting spacecraft is in free fall, and objects in it seem to float because they are all falling around Earth at the same rate.

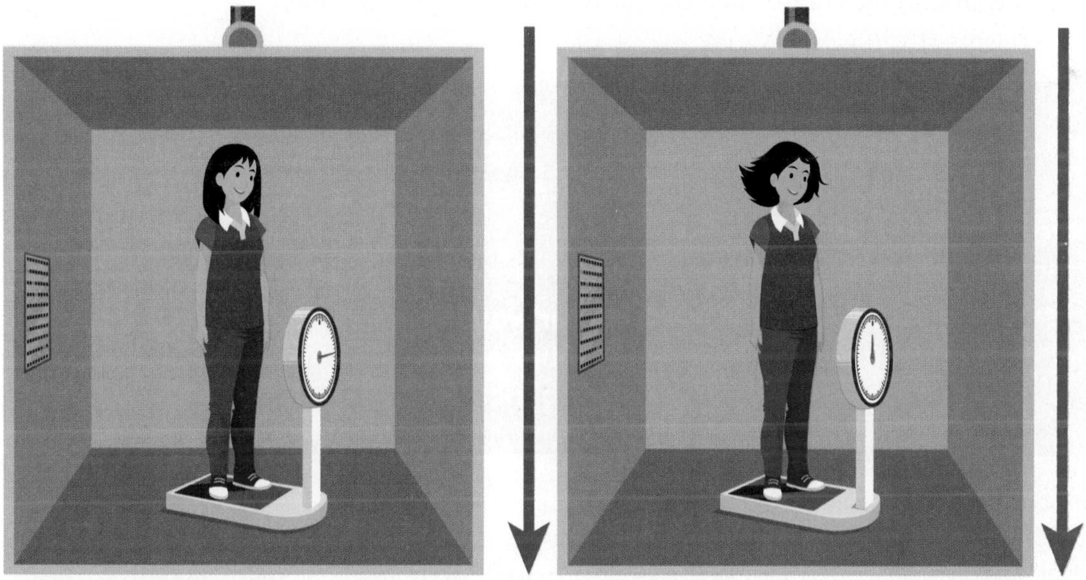

When the elevator is stationary, the scale shows the girl's weight.

If the elevator were in free fall, the scale would read zero.

Figure 19 The reading on the scale depends on the upward force the scale exerts on the girl. If both the girl and the scale are in free fall, the girl experiences the sensation of weightlessness, even though the force of gravity has not changed.

Figure 20 The riders on this amusement-park ride are traveling in a circle because of the centripetal force acting on them.

Identify *the force that acts as the centripetal force.*

Centripetal Forces

Orbiting objects, such as space shuttles, are traveling in nearly circular paths. According to Newton's first law, this change in motion is caused by a net force acting on the object. Newton's second law states that because the object's acceleration is toward the center of the curved path, the net force is also toward the center. A **centripetal force** is a force exerted toward the center of a curved path.

Anything that moves in a circle is doing so because a centripetal force is accelerating it toward the center. Many different forces can act as a centripetal force. Gravity is the centripetal force that keeps planets orbiting the Sun. On the amusement-park ride in **Figure 20,** the centripetal force is the walls pushing on the people.

When a car rounds a level curve on a highway, friction between the tires and the road acts as the centripetal force. If the road is slippery, the frictional force might not be large enough to keep the car moving around the curve. Then the car will slide in a straight line, as shown in **Figure 21.**

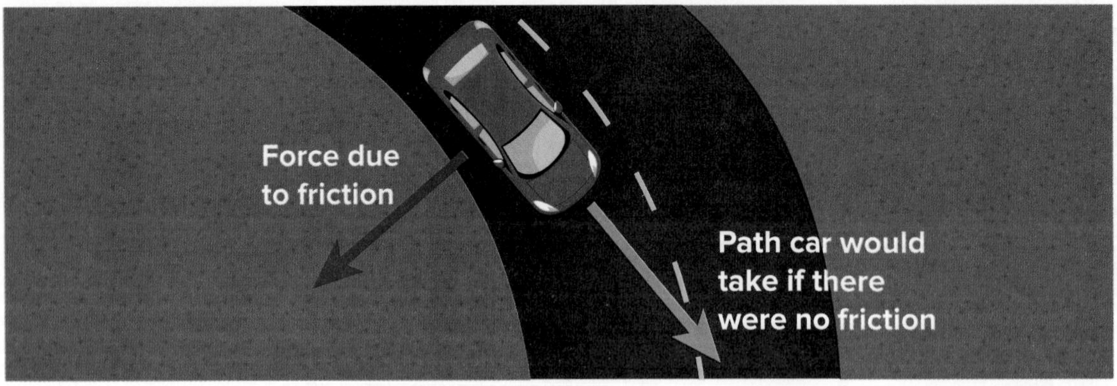

Figure 21 If the inward frictional force is too small, the car will continue in a straight line and not make it around the curve.

STEM CAREER Connection

Civil Engineer

Civil engineers conceive of and design roadways in cities. They must take into account factors such as current and future traffic flows, design of intersections and interchanges, building materials, geometric alignment, and pavement maintenance.

SCIENCE USAGE v. COMMON USAGE

gravity

Science usage: an attractive force between any two objects that depends on their masses and the distance between them *Gravity is the force that keeps Earth orbiting around the Sun.*

Common usage: seriousness

If you understood the gravity of the situation, you would not find it funny.

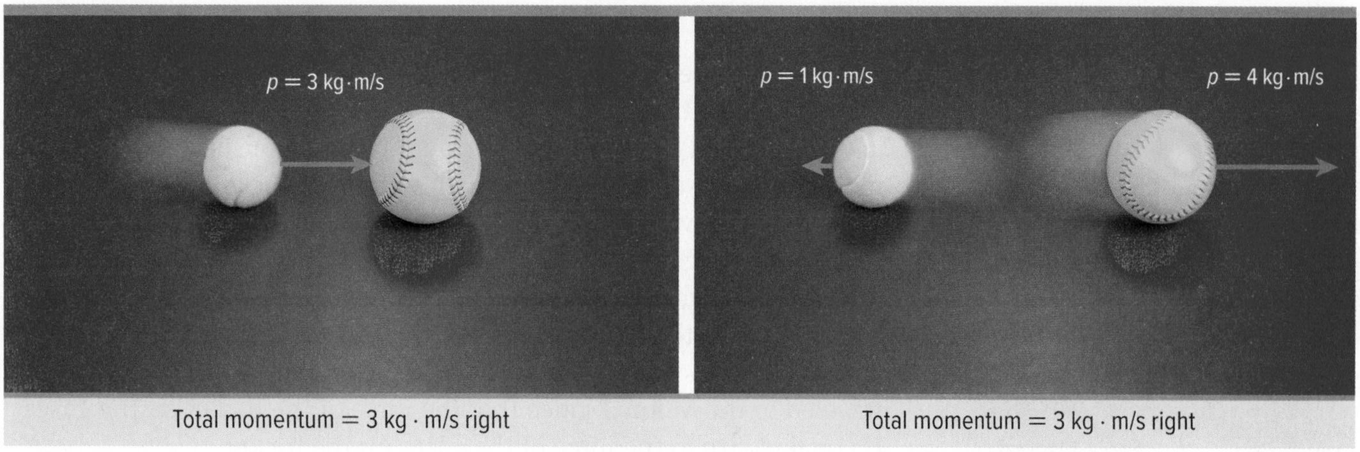

Figure 22 When two objects collide and there are no external forces acting on the objects, momentum is conserved.

Force and Momentum

Recall that acceleration is the difference between the initial and final velocities, divided by the time. Therefore, we can write Newton's second law in the following way.

$$F = ma = m \times \frac{(v_f - v_i)}{t} = \frac{(mv_f - mv_i)}{t}$$

Recall that an object's momentum equals its mass multiplied by its velocity. In the equation above, mv_f is the final momentum, and mv_i is the initial momentum. The equation states that the net force exerted on an object equals the change in its momentum divided by the time over which the change occurs. In fact, this is how Newton originally wrote the second law of motion.

Conservation of momentum

Newton's second and third laws of motion can be used to describe what happens when objects collide. For example, consider the collision of the balls in **Figure 22.** We will assume that friction is too small to cause a noticeable change in the balls' motion. In this ideal case, there are no external forces, but the balls exert forces on each other during the collision. According to Newton's third law, the forces are equal in size and opposite in direction. Therefore, the momentum lost by the first ball is gained by the second ball, and the total momentum of the two balls is the same before and after the collision. This is the **law of conservation of momentum**—if no external forces act on a group of objects, their total momentum does not change.

Collisions with multiple objects When a cue ball hits the group of motionless balls, as shown in **Figure 23,** the cue ball slows down and the rest of the balls begin to move. The momentum that the group of balls gained is equal to the momentum that the cue ball lost. Momentum is conserved.

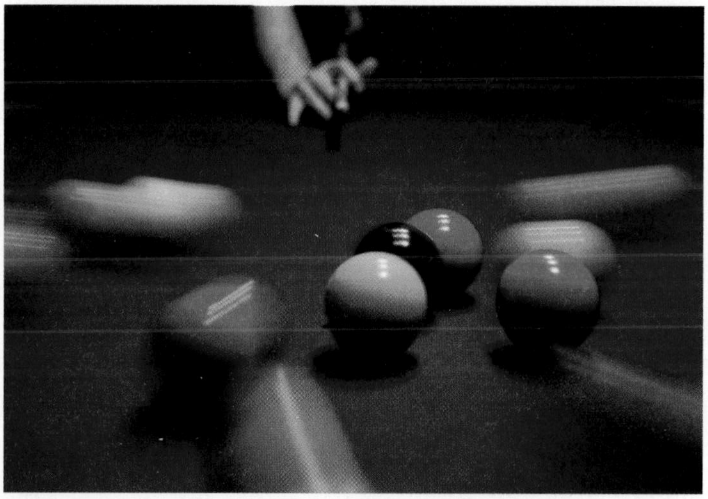

Figure 23 Momentum is conserved in collisions with more than two objects if there are no external forces acting on the objects. The momentums of the cue ball just before the collision is equal to the total of the momentums of the billiard balls (including the cue ball) just after the collision.

Figure 24 The force on a rocket depends on the amount and velocity of the gas expelled from its engine.

Rocket propulsion

Suppose you are standing on skates holding a softball. You exert a force on the softball when you throw it. According to Newton's third law, the softball exerts a force on you. This force pushes you backward in the direction opposite the softball's motion. Rockets use the same principle to move, even in the vacuum of outer space. In a rocket engine, burning fuel produces hot gases. The rocket engine exerts a force on these gases and causes them to escape out the back of the rocket. By Newton's third law, the gases exert a force on the rocket and push it forward. **Figure 24** shows one of the Apollo rockets that traveled to the Moon. Notice that the force of the rocket on the gases is equal in size to the force of the gases on the rocket.

Momentum is conserved when a rocket ejects the hot gas. If the rocket is initially at rest, then the total momentum of the rocket and the fuel is zero. After the fuel is burned and the hot gas is expelled, the gas travels backward with a momentum of $m_{gas}\ v_{gas}$ and the rocket travels forward with a momentum of $m_{rocket}\ v_{rocket}$. These momenta are equal in size, but opposite in direction. By controlling how much gas is ejected and the gas's velocity, the rocket's motion can be controlled.

Check Your Progress

Summary

- In a car crash, an unrestrained passenger will continue moving at the velocity of the car before the crash.
- Air resistance is a force that opposes an object's motion through the air.
- If gravity is the only force acting on an object, then the object is in free fall.
- Objects appear to be weightless in free fall.
- A force that causes an object to move in a circular path is called a centripetal force.
- The law of conservation of momentum states that if objects exert forces only on each other, their total momentum is conserved.

Demonstrate Understanding

22. **Describe** Use Newton's laws to describe how inertia, gravity, and air resistance affect skydivers as they fall, open their parachutes, and reach terminal velocity.
23. **Discuss** the advantages of wearing a safety belt when riding in a vehicle.
24. **Explain** why planets orbit the Sun instead of traveling off into space.
25. **Describe** what happens to the momentum of two billiard balls that collide.
26. **Explain** how a rocket can move through outer space, where there is no matter for it to push on.

Explain Your Thinking

27. **Predict** Suppose you are standing on a scale in an elevator that is accelerating upward. Will the scale read your weight as larger or smaller than the weight it reads when you are stationary? Explain.
28. **MATH Connection** A fuel-filled rocket is at rest. It burns its fuel and expels hot gas. The gas has a momentum of 1500 kg·m/s backward. What is the momentum of the rocket?

LEARNSMART® Go online to follow your personalized learning path to review, practice, and reinforce your understanding.

Extreme Altitudes

In July of 1969, *Apollo 11* rocketed toward an historic meeting with the Moon. The editors at *The New York Times* were in a distinctly uncomfortable position. Forty-nine years before, the newspaper published an editorial that ridiculed the scientist who suggested that rockets could travel in space. Three days before astronauts first walked on the Moon, editors wrote, "*The Times* regrets the error."

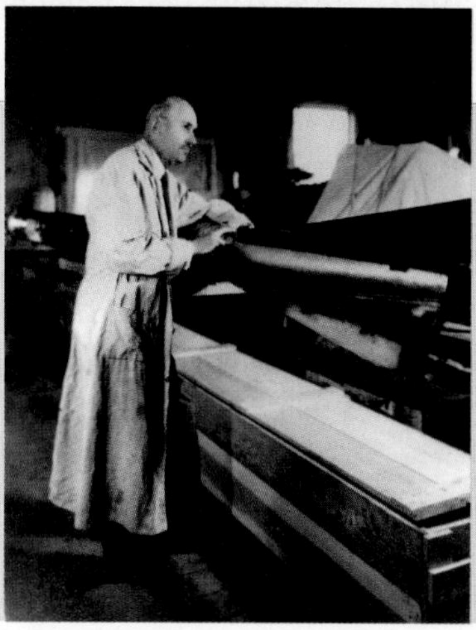

Robert Goddard works on a rocket in 1935.

Rocket man

Robert Goddard, shown in the photo, was passionate about rocketry. In 1919, Goddard wrote a report to the Smithsonian Institution entitled "A Method of Reaching Extreme Altitudes." In this paper, Goddard proposed that a large rocket propelled by powerful fuel could reach the Moon. Goddard was unprepared for the scornful reaction.

Public scorn, private determination

An editorial in *The New York Times* ridiculed Goddard and accused him of lacking "the knowledge ladled out daily in high schools." According to the writer, even students knew that a rocket could not function in space because there was nothing for the fuel exhaust to push against as it exited the rocket to create forward motion.

Goddard continued his experiments, despite the public humiliation. On March 16, 1926, his 3-m rocket, *Nell,* soared into the sky at 28 m/s. *Nell* reached a height of 12.5 m, becoming the first liquid-fueled rocket. This launch began the age of modern rocketry that led to space exploration and landing on the Moon.

Faulty physics

Why was the editorial incorrect? Consider Newton's third law of motion: For every action, there is an equal and opposite reaction. The writer believed that because the exhaust would have nothing to push against in space, there could be no opposing force pushing the rocket forward.

However, he failed to consider what happens inside the rocket. As rocket fuel is burned, a gas is generated. The gas expands rapidly, pushing against everything around it, and escapes through a nozzle at the bottom of the rocket. The escaping exhaust exerts a force on the rocket, causing the rocket to move in the opposite direction.

OBTAIN, EVALUATE, AND COMMUNICATE INFORMATION

Robert Goddard said, "Every vision is a joke until the first man accomplishes it; once realized, it becomes commonplace." Research a modern technology that might have been ridiculed had it been proposed in 1920. Use your findings to write a brief report.

STUDY GUIDE

GO ONLINE to study with your Science Notebook.

Lesson 1 FORCES

- A force is a push or a pull on an object.
- The net force on an object is the combination of all the forces acting on the object.
- Unbalanced forces cause the motion of objects to change.
- Friction is the force that opposes the sliding motion of two surfaces that are in contact.
- Gravity is an attractive force between all objects that have mass.

- force
- net force
- friction
- gravity
- field
- weight

Lesson 2 NEWTON'S LAWS OF MOTION

- Newton's first law of motion states that the motion of an object at rest or moving with constant velocity will not change unless an unbalanced force acts on the object.
- Inertia is the tendency of an object to resist a change in motion.
- Newton's second law of motion states that the acceleration of an object depends on its mass and the net force exerted on it.
- According to Newton's third law of motion, when an object exerts a force on a second object, the second object exerts a force on the first object that is equal in strength and opposite in direction.

- Newton's first law of motion
- inertia
- Newton's second law of motion
- Newton's third law of motion

Lesson 3 USING NEWTON'S LAWS

- In a car crash, an unrestrained passenger will continue moving at the velocity of the car before the crash.
- Air resistance is a force that opposes an object's motion through the air.
- If gravity is the only force acting on an object, then the object is in free fall.
- Objects appear to be weightless in free fall.
- A force that causes an object to move in a circular path is called a centripetal force.
- The law of conservation of momentum states that if objects exert forces only on each other, their total momentum is conserved.

- air resistance
- terminal velocity
- free fall
- centripetal force
- law of conservation of momentum

THREE-DIMENSIONAL THINKING

Module Wrap-Up

REVISIT THE PHENOMENON

Why can a car stop faster than a train?

CER Claim, Evidence, Reasoning

Explain Your Reasoning Revisit the claim you made when you encountered the phenomenon. Summarize the evidence you gathered from your investigations and research and finalize your Summary Table. Does your evidence support your claim? If not, revise your claim. Explain why your evidence supports your claim.

STEM UNIT PROJECT

Now that you've completed the module, revisit your STEM unit project. You will apply your evidence from this module and complete your project.

GO FURTHER

SEP Data Analysis Lab

Friction and the Curve Ball

The curve ball was invented by a young pitcher named Arthur "Candy" Cummings. Although Cummings first threw the curve ball during a game while pitching for the Brooklyn Excelsiors in 1867, he actually invented his technique many years before. Why does the ball curve? It's all about friction.

The snapping action of the pitcher's wrist puts a spin on the ball. That spin changes the friction between the air and the ball. After it's thrown, parts of the ball experience more air friction and parts of the ball experience less. A curved path results from the ball moving toward the least amount of friction.

Specifically, one movement of the pitcher's wrist when the ball is released causes a top spin, making the top of the ball move forward against the air (more friction) and the bottom move in the same direction as the air (less friction). Like any curve ball, the ball curves toward the least amount of friction—downward.

CER Analyze and Interpret Data

1. **Claim and Reasoning** What effect might the stitches on a baseball have on the baseball's path?
2. **Claim, Evidence, Reasoning** Do you think a baseball curves better at the top of a high mountain or down on a flat plain? Explain.
3. **Claim, Evidence, Reasoning** Describe how the type of spin a pitcher applies will influence the path of a baseball.

Soonthorn Wongsaita/Shutterstock

UNIT 2
ENERGY

ENCOUNTER THE PHENOMENON

How can energy be collected and stored for daily use?

SEP Ask Questions

What questions do you have about the phenomenon? Write your questions on sticky notes and add them to the driving question board for this unit.

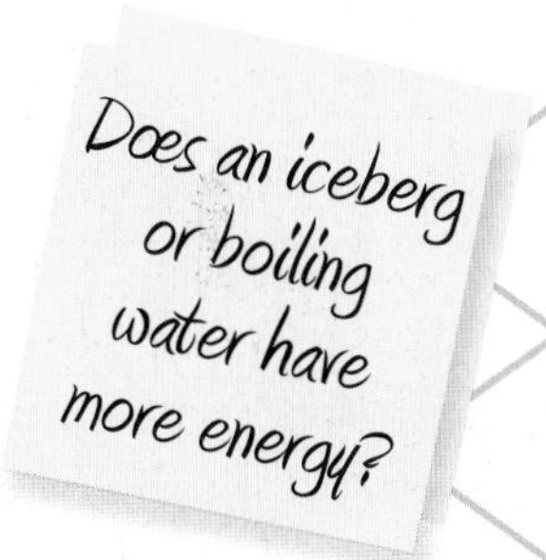

Does an iceberg or boiling water have more energy?

Look for Evidence

As you go through this unit, use the information and your experiences to help you answer the phenomenon question as well as your own questions. For each activity, record your observations in a Summary Table, add an explanation, and identify how it connects to the unit and module phenomenon questions.

Solve a Problem
STEM UNIT PROJECT

Design a Power Bank for Storing Electrical Power Solar and wind power have many benefits, but one potential drawback is that they do not always generate energy at a constant rate or at times people use the most energy. Research ways that excess electrical power can be stored for later use. Use the results of these investigations and the evidence you collected during the unit to design a power bank for an alternative energy source for use during off times when power is not sufficiently generated.

GO ONLINE In addition to reading the information in your Student Edition, you can find the STEM Unit Project and other useful resources online.

Sommai/Shutterstock

WORK AND ENERGY

ENCOUNTER THE PHENOMENON

How can ice be used as an alternative energy source?

GO ONLINE to play a video about innovative ways to store energy.

SEP Ask Questions

Do you have other questions about the phenomenon? If so, add them to the driving question board.

CER Claim, Evidence, Reasoning

Make Your Claim Use your CER chart to make a claim about how ice can be used as an alternative energy source. Explain your reasoning.

Collect Evidence Use the lessons in this module to collect evidence to support your claim. Record your evidence as you move through the module.

Explain Your Reasoning You will revisit your claim and explain your reasoning at the end of the module.

GO ONLINE to access your CER chart and explore resources that can help you collect evidence.

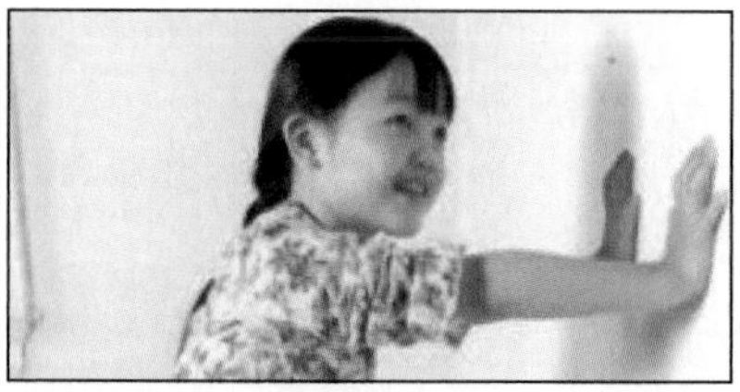

LESSON 1: Explore & Explain: Definition of Work

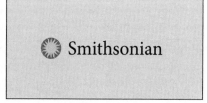

LESSON 3: Explore & Explain: The Law of Conservation of Energy

Smithsonian

Additional Resources

LESSON 1

WORK AND MACHINES

FOCUS QUESTION

How do machines utilize and change work?

Definition of Work

To many people, the word work means something that people do to earn money. In that sense, work can be anything from fixing cars to designing Web sites. The word work might also mean exerting a force with muscles. However, in science, the word work is used in a different way.

Motion and work

Press your hand against the surface of your desk as hard as you can. Have you done any work on the desk? The answer is no, no matter how tired you get from the effort. In science, **work** is force applied through a distance. If you push against the desk and it does not move, then you have not done any work on the desk because the desk has not moved.

Force and direction of motion

Imagine that you are pushing a lawn mower, as shown in **Figure 1**. You could push on this mower in many different directions. You could push it horizontally. You could also push down on the mower or push on it at an angle. Think about how the mower's motion would be different each time. The direction of the force that you apply to the lawn mower affects how much work you do on it.

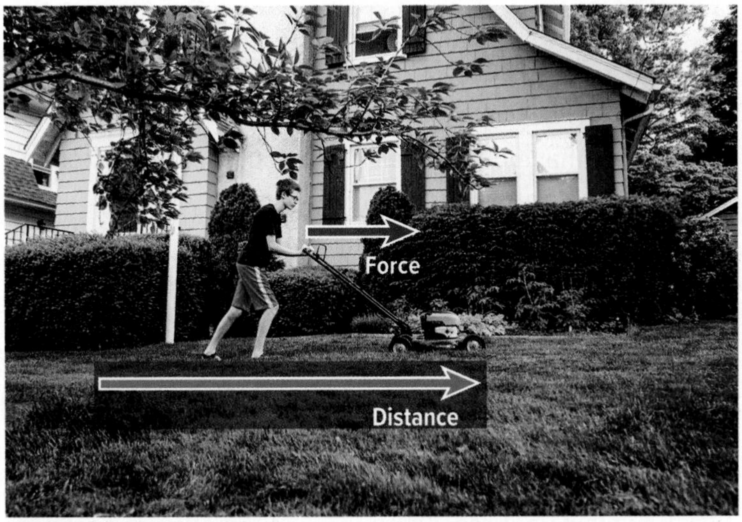

Figure 1 You apply a force through a distance when you push a lawn mower over a lawn. In other words, you do work on the lawn mower when you push it over a lawn.

Explain *whether you could do work on the mower without moving it.*

3D THINKING DCI Disciplinary Core Ideas CCC Crosscutting Concepts SEP Science & Engineering Practices

COLLECT EVIDENCE

Use your Science Journal to record the evidence you collect as you complete the readings and activities in this lesson.

INVESTIGATE

GO ONLINE to find these activities and more resources.

Lab: Mechanical Advantage and Efficiency

Develop and use a model to calculate the work needed to lift objects and the effect an inclined plane has on improving mechanical advantage and efficiency.

Laboratory: Pulleys

Develop and use a model to calculate the work needed to lift objects and the effect a single fixed pulley and a block and tackle have on improving mechanical advantage and efficiency.

Force parallel to motion Imagine that you push on the lawn mower in **Figure 1** with a force of 25 N and through a distance of 4 m. In what direction would you push to do the maximum amount of work on the mower? You do the maximum amount of work when you push the lawn mower in the same direction as it is moving. When force and motion are parallel, which means they are in the same direction, work is equal to force multiplied by distance.

Work Equation

work (in joules) = **applied force** (in newtons) × **distance** (in meters)

$$W = Fd$$

If force is measured in newtons (N) and distance is measured in meters (m), then work is measured in joules (J). You do about 1 J of work on a cell phone when you pick it up off the floor.

EXAMPLE Problem 1

SOLVE FOR WORK You push a refrigerator with a horizontal force of 100 N. If you move the refrigerator a distance of 5 m while you are pushing, how much work do you do?

Identify the Unknown: work: W

List the Knowns: applied force: $F = 100$ N distance: $d = 5$ m

Set Up the Problem: $W = Fd$

Solve the Problem: $W = (100\ \text{N})(5\ \text{m}) = 500\ \text{J}$

Check the Answer: Check to see whether the units match on both sides of the equation.

units of W = (units of F) × (units of d) = N × m = J

PRACTICE Problems ADDITIONAL PRACTICE

1. A couch is pushed with a horizontal force of 80 N and moves a distance of 5 m across the floor. How much work is done in moving the couch?
2. How much work do you do when you lift a 100-N child 0.5 m?
3. The brakes on a car do 240,000 J of work in stopping the car. If the car travels a distance of 40 m while the brakes are being applied, how large is the average force that the brakes exert on the car?
4. **CHALLENGE** The force needed to lift an object is equal in size to the gravitational force on the object. How much work is done in lifting an object that has a mass of 5 kg a vertical distance of 2 m?

Force perpendicular to motion When you carry books while walking at a constant velocity, you might think that your arms are doing work on those books. After all, you are exerting a force on the books to hold them, and the books are moving with you. Your arms might even feel tired. However, in this case, the force exerted by your arms does zero work on the books. This is because there is a 90° angle between this force on the books and the motion of the books. When a force is perpendicular to motion, the work from that force is zero.

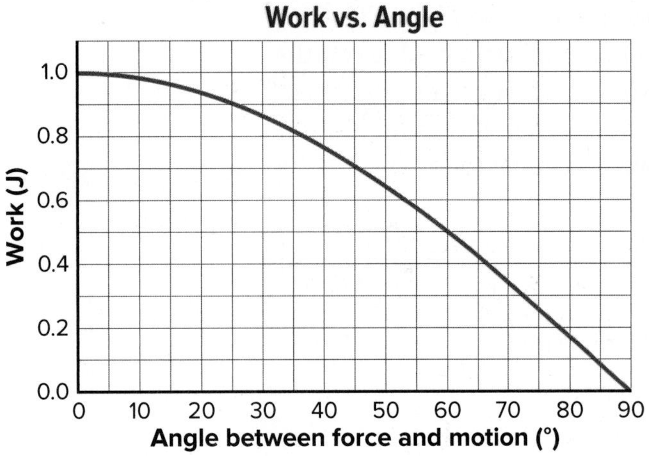

Figure 2 This graph shows the work done on an object for different angles between force and motion when 1 N is applied to the object and the object moves 1 m.

Interpret Graphs *At what angle between force and motion is work one-half as much as when force and motion are in the same direction?*

Other directions If a force on an object and that object's motion are parallel, then the work from that force equals that force's magnitude multiplied by distance. If the force on an object and that object's motion are perpendicular, then the work from that force equals zero. How much work is done when the angle between force and motion is not parallel or perpendicular? For these other directions, the work done is less than the force multiplied by the distance but more than zero. **Figure 2** shows a graph of how direction affects work.

Get It?

Describe the work done on an object when the force on that object and the motion of that object are perpendicular.

Figure 3 The batter does work on the ball only while the ball is in contact with her bat. After the ball leaves the bat, she no longer exerts any force on the ball.

Identify *a force that is doing work on a ball when that ball is falling through the air.*

When is work done?

Suppose you give a book a push; it leaves your hand and slides along a table for a distance of 1 m before it comes to a stop. The distance that you use to calculate the work you did on the book is how far the object moved while you applied a force. Even though the book moved a total of 1 m, you do work on the book only while you touch it.

You can only do work on an object while you are applying a force on that object. In **Figure 3**, the girl did work on the softball only while the bat was in contact with the softball. Even if the ball is hit hard enough to sail over the left-field fence for a home run, the only work done on the ball was while it was in contact with the bat.

ACADEMIC VOCABULARY

contact

union or junction of surfaces

Your hand must be in contact with the book to push on it.

Machines

A **machine** is a device that changes the force or increases the motion from work. Imagine trying to lift a grand piano. Without help, this would be impossible. However, you could probably push the piano up a ramp or lift the piano with a pulley system. In both cases, you use a machine to change the force on the piano.

However, you might be more interested in increasing speed than increasing force. Using only the power of your muscles, how fast could you move? You could travel much faster on a machine called a bicycle than on foot.

Types of machines

When you cut your food with a knife, use a screwdriver, or chew your food, you are using a simple machine. A **simple machine** is a machine that does work with only one movement of the machine. There are six types of simple machines: lever, pulley, wheel and axle, inclined plane, screw, and wedge. The pulley and the wheel and axle are modified levers, and the screw and the wedge are modified inclined planes. A common example of each type of simple machine is shown in **Figure 4**.

Get It?

Identify at least two simple machines that can be found at a playground.

A **compound machine** is a combination of two or more simple machines. For example, a pair of scissors is a compound machine. It combines two wedges and two levers. A bicycle is also a compound machine.

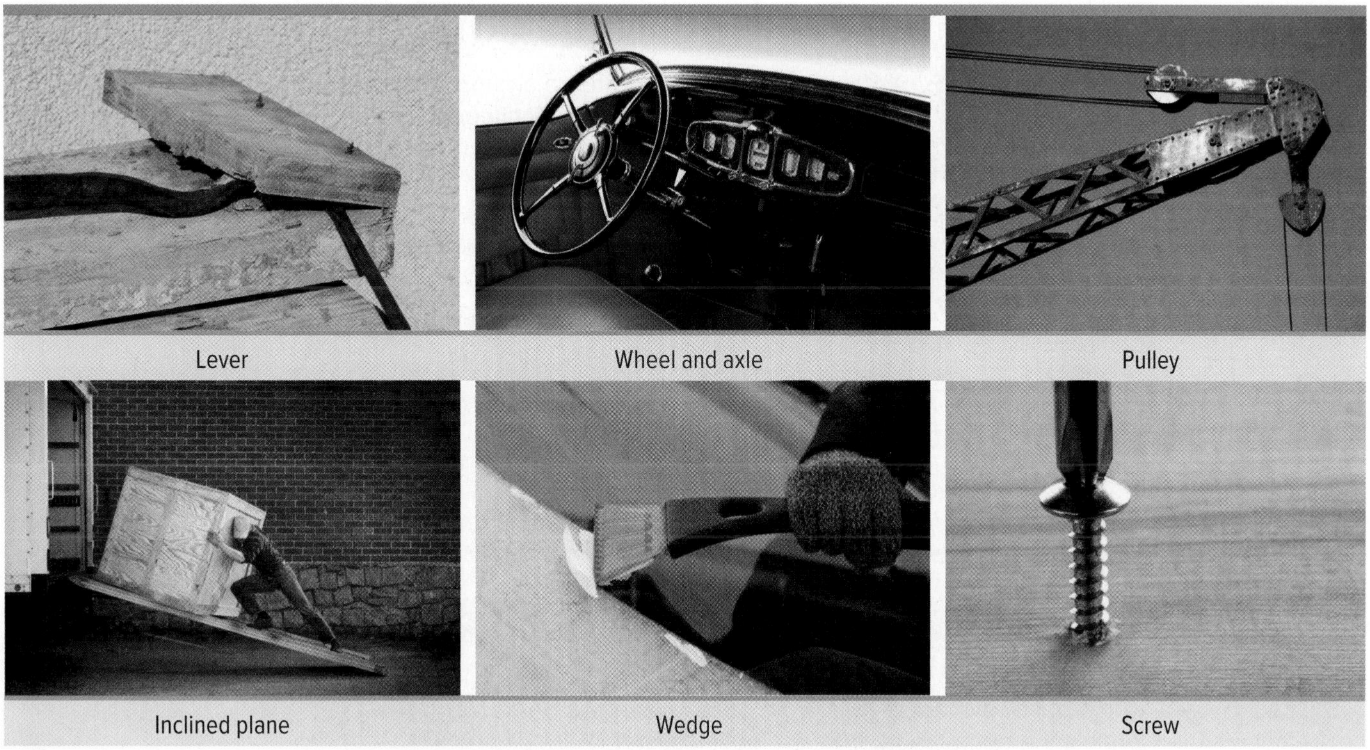

Figure 4 The six types of simple machines are levers, wheels and axles, pulleys, inclined planes, wedges, and screws. These machines can be combined to form compound machines.

Describe *a compound machine that includes at least two simple machines.*

(tl)Liane Matrisch/Hemera/Getty Images, (tc)Christian Delbert/123RF, (tr)Martina Berg/age fotostock, (bl)stevecoleimages/Photodisc/Getty Images, (bc)Hanson Ng/DesignPics/age fotostock, (br)Jason_V/iStock/Getty Images

Efficiency

Machines can increase force or increase speed. You might think this means that you get more work out of a machine than you put into a machine because work is related to force and motion. However, no machine can increase both force and speed at the same time. In fact, you always put more work into a machine than you get out of that machine. This is a fundamental scientific law that cannot be broken by building better machines.

Get It?

Compare You use a ramp (an inclined plane) to help load a heavy crate into a truck. How does the work you put into this simple machine compare with the work you get out of the machine?

Efficiency is the ratio of output work to input work. Efficiency is often measured in percent.

Efficiency Equation

$$\text{efficiency (\%)} = \frac{\textbf{output work} \text{ (in joules)}}{\textbf{input work} \text{ (in joules)}} \times 100$$

$$e = \frac{W_{out}}{W_{in}} \times 100$$

Machines can be made more efficient by reducing friction. This is usually done by adding a lubricant, such as oil or grease, to surfaces that rub together. However, all machines are less than 100 percent efficient.

EXAMPLE Problem 2

SOLVE FOR EFFICIENCY You do 20 J of work in pushing a crate up a ramp. If the output work from the inclined plane is 11 J, then what is the efficiency of the inclined plane?

Identify the Unknown: efficiency: e

List the Knowns: work in: $W_{in} = 20$ J

work out: $W_{out} = 11$ J

Set Up the Problem: $e = \frac{W_{out}}{W_{in}} \times 100$

Solve the Problem: $e = \frac{11 \text{ J}}{20 \text{ J}} \times 100$

e = 55 percent

Check the Answer: The work out is about half of the work in.
Therefore, an answer close to 50 percent is reasonable.

PRACTICE Problems

ADDITIONAL PRACTICE

5. Find the efficiency of a machine that does 800 J of work if the input work is 2000 J.
6. The input work on a pulley system is 75 J. If the pulley system is 84 percent efficient, then what is the output work from the pulley system?
7. **CHALLENGE** Workers do 8000 J of work on a 2000-N crate to push it up a ramp. If the ramp is 2 m high, then what is the efficiency of the ramp?

How are machines useful?

How are machines useful if a machine's output work is always less than its input work? Machines change the way work is done. They can increase speed, change the direction of a force, or increase force.

Increase speed Bicycles are machines that increase speed. You can travel more quickly on a bicycle than on foot. However, to increase speed, a bicycle decreases force. Look at the cyclist in the top panel of **Figure 5**. Although the cyclist can reach his destination faster by biking instead of walking, his legs must apply a larger force to the pedals to cross the distance.

Change direction of force Some machines change the direction of an applied force. The wedge-shaped blade of the ax in **Figure 5** is one example. You exert a downward force on an ax when chopping wood. The shape of the blade changes your downward force into outward forces that split the wood.

Increase force A car jack, such as the one in the bottom panel of **Figure 5**, increases force but decreases speed. The upward force exerted on the car is greater than the downward force that you exert on the handle. However, you move the car jack handle faster than the jack lifts the car.

We can describe the effectiveness of a machine at increasing force by its mechanical advantage. **Mechanical advantage** is the ratio of output force to input force.

Mechanical Advantage Equation

$$\text{mechanical advantage} = \frac{\textbf{output force} \text{ (in newtons)}}{\textbf{input force} \text{ (in newtons)}}$$

$$MA = \frac{F_{out}}{F_{in}}$$

The input force is the force that a person or a device such as a motor, applies to the machine. The output force is the force that the machine applies to another object. In the car jack example in **Figure 5**, the man applies an input force to the car jack, and the car jack applies an output force to the car. The mechanical advantage of the car jack is greater than 1 because the output force is greater than the input force.

Figure 5 A machine can change work to increase speed, change the direction of a force, or increase force.

EXAMPLE Problem 3

SOLVE FOR MECHANICAL ADVANTAGE A crate weighs 950 N. If you can use a pulley system to lift that crate with a force of only 250 N, then what is the mechanical advantage of the pulley system?

Identify the Unknown: mechanical advantage: MA

List the Knowns: output force: $F_{out} = 950\ \text{N}$

input force: $F_{in} = 250\ \text{N}$

Set Up the Problem: $MA = \frac{F_{out}}{F_{in}}$

Solve the Problem: $MA = \frac{950\ \text{N}}{250\ \text{N}}$

$MA = 3.8$

Check the Answer: The weight of the crate is very close to four times the force needed to lift the crate. Therefore, the mechanical advantage of the crate should be close to 4. Our answer is close to 4, so it is reasonable.

PRACTICE Problems

ADDITIONAL PRACTICE

8. Calculate the mechanical advantage of a hammer if the input force is 125 N and the output force is 2,000 N.

9. CHALLENGE Find the force needed to lift a 3,000-N weight using a machine with a mechanical advantage of 15.

Check Your Progress

Summary

- Work is force applied through a distance.
- A machine can increase speed, change the direction of a force, or increase force.
- The output work from a machine is never as great as the input work on that machine.
- Efficiency is the ratio of output work to input work.
- Mechanical advantage is the ratio of output force to input force.

Demonstrate Understanding

10. **Give an example** of a machine that increases speed, a machine that changes the direction of force, and a machine that increases force.
11. **Contrast** the scientific definition of *work* with its everyday meaning.
12. **Compare** the output force with the input force for a machine that has a mechanical advantage that is greater than one.

Explain Your Thinking

13. **Describe** how lubricating a machine affects the output work from that machine. How would the input and output forces be affected?
14. **MATH Connection** If you push a book 1.5 m across a table using a constant force of 10.0 N, how much work do you do on the book?
15. **MATH Connection** What is the efficiency of a ramp if the input work is 96 J and the output work is 24 J?

LEARNSMART® Go online to follow your personalized learning path to review, practice, and reinforce your understanding.

LESSON 2

DESCRIBING ENERGY

FOCUS QUESTION

How do you classify and calculate different forms of energy?

Change Requires Energy

When something is able to change its surroundings or itself, it has energy. **Energy** is the ability to cause change. The availability of energy limits what can occur in any system. Without energy, nothing would ever change. The moving tennis racket in **Figure 6** has energy. That racket causes change when it deforms the tennis ball and changes the tennis ball's motion.

Work transfers energy

The tennis racket in **Figure 6** also does work on the tennis ball, applying a force to that ball through a distance. When this happens, the racket transfers energy to the ball. Therefore, energy can also be described as the ability to do work.

Because energy can be described as the ability to do work, energy can be measured with the same units as work. Energy, like work, can be measured in joules. Imagine that the tennis racket in **Figure 6** does 250 J of work on the tennis ball. Then, 250 J of energy is transferred from the racket to the ball.

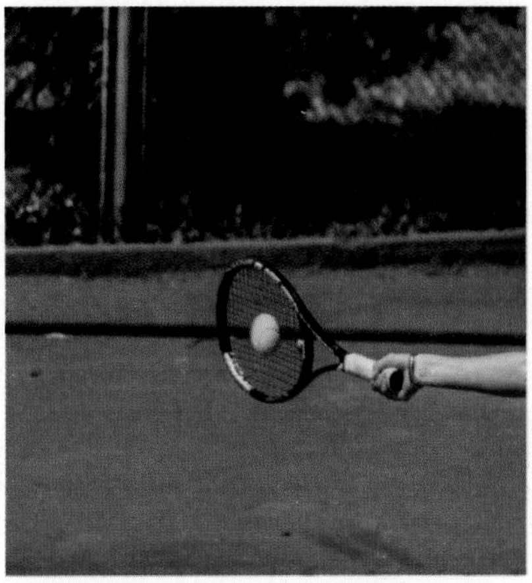

Figure 6 The tennis racket causes changes to occur when it hits the tennis ball.

Describe *the changes that are occurring.*

Get It?

Identify What limits the speed at which the ball will leave the racket?

Systems The tennis racket and the tennis ball in **Figure 6** are systems. A **system** is anything around which you can imagine a boundary. A system can be a single object, such as a tennis ball, or a group of objects, such as the solar system. Energy is a quantitative property of a system. When one system does work on a second system, energy is transferred from the first system to the second system.

3D THINKING DCI Disciplinary Core Ideas CCC Crosscutting Concepts SEP Science & Engineering Practices

COLLECT EVIDENCE

Use your Science Journal to record the evidence you collect as you complete the readings and activities in this lesson.

INVESTIGATE

GO ONLINE to find these activities and more resources.

Quick Investigation: Interpret Data from a Slingshot

Carry out an investigation to observe and compare the forces in a system needed to move an object using a slingshot with varying degrees of applied force.

Revisit the Encounter the Phenomenon Question

What information from this lesson can help you answer the Module question?

Different Forms of Energy

Turn on an electric light, and a dark room becomes bright. Turn on a portable music player, and sound comes through your headphones. In both situations, a change occurs. These changes differ from each other and from the tennis racket hitting the tennis ball in **Figure 6.** This is because energy has many different forms. These forms include mechanical energy, electrical energy, chemical energy, and radiant energy.

Figure 7 shows some everyday situations in which you might notice energy. Automobiles make use of the chemical energy of gasoline. Many household appliances require electrical energy to function. Radiant energy from the Sun warms Earth. In short, energy plays a role in every activity that you do.

Get It?
Identify three different forms of energy.

An energy analogy

Is the chemical energy of food the same as the energy that comes from the Sun or the energy that comes from gasoline? Money can be used in an analogy to help you understand energy. Money exists in a variety of forms, such as coins, dollar bills, and twenty-dollar bills. You can convert money from one form to another. For example, you could obtain four quarters for a dollar bill. Regardless of its form, money is money. The same is true for energy. Energy from the Sun that warms you and energy from the food that you eat are only different forms of the same thing.

Chemical potential energy | Electrical energy | Radiant energy

Figure 7 Energy can be stored and transferred from one place to another. For example, the energy due to the chemical bonds in gasoline is easy to transport in cars. Electrical energy is transferred from a power plant to appliances in your home. Radiant energy is transferred from the Sun to Earth.

SCIENCE USAGE v. COMMON USAGE

energy

Science usage: the ability to cause change
You transfer energy when you do work.

Common usage: the capacity of acting or being active
That soccer player had a lot of energy on the field today.

Kinetic energy

When you think of energy, you might think of objects in motion. Objects in motion can collide with other objects and cause change. Therefore, objects in motion have energy. **Kinetic energy** is energy due to motion. A car moving along a highway and a ballet dancer leaping through the air have kinetic energy. The kinetic energy from an object's motion depends on that object's mass and speed.

Kinetic Energy Equation

kinetic energy (in joules) = $\frac{1}{2}$ **mass** (in kg) × [**speed** (in m/s)]^2

$$KE = \frac{1}{2}mv^2$$

If mass is measured in kilograms (kg) and speed is measured in meters per second (m/s), then kinetic energy is measured in joules (J). If you drop a softball from just above your knee, the kinetic energy from that ball's falling motion is about 1 J just before the ball reaches the floor.

EXAMPLE Problem 4

SOLVE FOR KINETIC ENERGY A jogger with a mass of 60.0 kg is moving forward at a speed of 3.0 m/s. What is the jogger's kinetic energy from this forward motion?

Identify the Unknown: kinetic energy: KE

List the Knowns: mass: m = 60.0 kg speed: v = 3.0 m/s

Set Up the Problem: $KE = \frac{1}{2}mv^2$

Solve the Problem: $KE = \frac{1}{2}(60.0\text{ kg})(3.0\text{ m/s})^2$

$KE = \frac{1}{2}(60.0\text{ kg})(9.0\text{ m}^2/\text{s}^2)$

$KE = 270\text{ J}$

Check the Answer: Check the last step by estimating. Round 9.0 m^2/s^2 up to 10 m^2/s^2. Then, $\frac{1}{2}(60.0\text{ kg})(10\text{ m}^2/\text{s}^2) = 300\text{ J}$. This is close to 270 J, so the final calculation was reasonable.

PRACTICE Problems

ADDITIONAL PRACTICE

16. A baseball with a mass of 0.15 kg is moving at a speed of 40.0 m/s. What is the baseball's kinetic energy from this motion?

17. **CHALLENGE** A 1500-kg car doubles its speed from 50 km/h to 100 km/h. By how many times does the kinetic energy from the car's forward motion increase?

CCC CROSSCUTTING CONCEPTS

Matter and Energy Compare the kinetic energies of different objects that travel at the same speed but have different masses. Then compare objects that have the same mass but travel at different speeds. Use your results to explain whether doubling mass or doubling speed has a greater effect on kinetic energy.

Potential energy

Energy does not always involve motion. Even motionless objects can have energy. **Potential energy** is energy that is stored due to the interactions between objects. One example is the energy stored between an apple hanging on a tree and Earth. Energy is stored between the apple and Earth because of the gravitational force between the apple and Earth. Another example is the energy stored between objects that are connected by a compressed spring or a stretched rubber band.

Get It?

Explain how a book can have energy even if it is not moving.

Elastic potential energy If you stretch a rubber band and let it go, it sails across the room. As it flies through the air, it has kinetic energy due to its motion. Where did this kinetic energy come from? Just as there is potential energy due to gravitational forces, there is also potential energy due to the elastic forces between the particles that make up a stretched rubber band. The energy of a stretched rubber band or a compressed spring is called elastic potential energy. **Elastic potential energy** is energy that is stored by compressing, stretching, or bending an object.

Get It?

Describe how the elastic potential energy of a trampoline changes as a person jumps on it.

Chemical potential energy The food that you eat and the gasoline in cars also have stored energy. This stored energy is due to the chemical bonds between atoms. **Chemical potential energy** is energy that is due to chemical bonds. You might notice chemical potential energy when you burn a substance. When an object is burned, chemical potential energy becomes thermal energy and radiant energy. **Figure 8** shows the process for burning methane.

| Methane gas | + | Oxygen gas | → | Carbon dioxide gas | + | Water vapor |

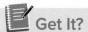

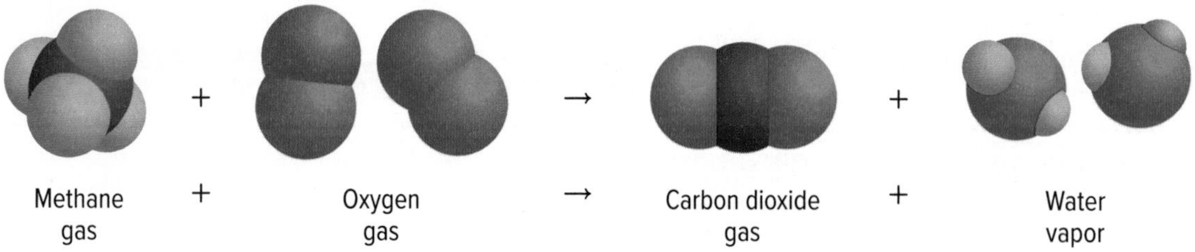

Figure 8 When methane burns, it combines with oxygen to form carbon dioxide and water vapor. In this chemical reaction, chemical potential energy is converted to other forms of energy.

WORD ORIGINS

potential

comes from the Latin word *potens*, a form of *posse*, which means *to be able*

The rock on the cliff has potential energy because it is able to cause change if it falls.

Gravitational potential energy Consider the blue vase in **Figure 9.** Together, the blue vase and Earth have potential energy. **Gravitational potential energy** is energy that is due to the gravitational forces between objects. Gravitational potential energy is often shortened to GPE.

Any system that has objects that are attracted to each other through gravity has gravitational potential energy. An apple and Earth have gravitational potential energy. The solar system also has gravitational potential energy. The gravitational potential energy of a system containing just Earth and another object depends on the object's mass, Earth's gravity, and the object's height. Recall that near Earth's surface, g is equal to 9.8 N/kg.

Gravitational Potential Energy Equation

gravitational potential energy (J)
= **mass** (kg) × **gravity** (N/kg) × **height** (m)

$GPE = mgh$

Figure 9 The gravitational potential energy of any system containing only an object on the bookcase and Earth depends on the object's mass, the strength of Earth's gravity, and the object's height. The object's height is measured relative to a reference level. The floor, the ground, the ceiling, and Earth's center are possible reference levels.

Height and gravitational potential energy Look at the bookcase in **Figure 9.** Moving a vase from one shelf to another changes its GPE because the relative position of the object in Earth's gravitational field changes. Moving the vase to a higher shelf increases GPE as more energy is stored in the vase-Earth system, and moving it to a lower shelf decreases GPE as less energy is stored in the vase-Earth system.

Now imagine that this bookcase is on the second floor of a building and that this building is at the top of a large hill. How should you measure the heights of the objects on the shelves? You could measure the heights from the floor. You could also measure the heights from the ceiling, the ground outside, the bottom of the hill, or Earth's center.

To calculate gravitational potential energy, height is measured from a reference level. This means that gravitational potential energy varies depending on the chosen reference level.

Relative to the floor, the GPE of a system containing just the blue vase and Earth is about 90 J. Relative to the ceiling, the GPE of this same system might be about –40 J. Relative to Earth's center, this system's GPE is about 300 million J. All of these statements are correct. In addition, the GPE of the blue vase-Earth system is greater than the GPE of the green vase-Earth system for every reference level. However, statements such as "The gravitational potential energy is 100 J" are meaningless unless a reference level is given.

EXAMPLE Problem 5

SOLVE FOR GRAVITATIONAL POTENTIAL ENERGY A 4.0-kg ceiling fan is placed 2.5 m above the floor. What is the gravitational potential energy of the Earth-ceiling fan system relative to the floor?

Identify the Unknown: gravitational potential energy: *GPE*

List the Knowns:
mass: $m = 4.0$ kg
gravity: $g = 9.8$ N/kg
height: $h = 2.5$ m

Set Up the Problem: $GPE = mgh$

Solve the Problem: $GPE = (4.0 \text{ kg})(9.8 \text{ N/kg})(2.5 \text{ m}) = 98 \text{ N} \cdot \text{m} = 98 \text{ J}$

Check the Answer: Round 9.8 N/kg to 10 N/kg. Then, $GPE = (4.0 \text{ kg})(10 \text{ N/kg})(2.5 \text{ m}) = 100$ J. This is very close to the answer above. Therefore, that answer is reasonable.

PRACTICE Problems

ADDITIONAL PRACTICE

18. An 8.0-kg history textbook is placed on a 1.25-m high desk. What is the gravitational potential energy of the textbook-Earth system relative to the floor?

19. CHALLENGE What is the GPE of the textbook-Earth system in problem 18, relative to the desktop?

Check Your Progress

Summary

- Forms of energy include mechanical, electrical, chemical, thermal, and radiant energy.
- Kinetic energy is the energy that a moving object has because of its motion.
- Potential energy is stored energy due to the interactions between objects.
- Different forms of potential energy include elastic potential energy, chemical potential energy, and gravitational potential energy.

Demonstrate Understanding

20. **Describe** a change caused by kinetic energy as well as a change that involves potential energy.
21. **Infer** whether a system can have kinetic energy and potential energy at the same time.
22. **Differentiate** elastic potential energy and chemical potential energy.

Explain Your Thinking

23. **Compare** The different molecules that make up the air in a room have, on average, the same kinetic energy. How does the speed of the different molecules that make up the air depend on their masses?
24. **MATH Connection** A 0.06-kg ball is moving at 5.0 m/s. What is the kinetic energy from this motion?
25. **MATH Connection** A 0.50-kg apple is 2.0 m above the reference level. What is the GPE of the apple-Earth system?

LEARNSMART® Go online to follow your personalized learning path to review, practice, and reinforce your understanding.

LESSON 3

CONSERVATION OF ENERGY

FOCUS QUESTION

How is energy transformed for a swinging object?

The Law of Conservation of Energy

Suppose you are riding on a roller coaster like the one in **Figure 10**. As your height above the ground changes, gravitational potential energy changes. As your speed changes, kinetic energy changes. Think about the motion of the roller-coaster cars. When the cars are high above the ground, GPE is large and kinetic energy is small. When the cars are low, GPE is small and kinetic energy is large. Energy is changing back and forth between GPE and kinetic energy. In addition, some kinetic energy is slowly converted into other, less useful, forms of energy during a roller-coaster ride. However, because energy cannot be destroyed, the total energy remains constant.

The **law of conservation of energy** states that energy cannot be created or destroyed. Energy can only be converted from one form to another or transferred from one place to another. Describing energy as a quantity is possible because a system's total energy is conserved even as energy transfers between objects and transforms between forms within the system.

Figure 10 Energy can be transferred or transformed, but it cannot be created or destroyed. For the roller-coaster cars, energy is converted back and forth between kinetic energy and gravitational potential energy. In addition, some kinetic energy is converted to other, less useful, forms of energy. However, the total amount of energy is constant.

Get It?
State the law of conservation of energy.

Conserving resources

You might have heard about energy conservation or have been asked to conserve energy. These ideas are related to using energy resources, such as coal and oil, wisely. However, the law of conservation of energy is a universal principle that states that total energy remains constant.

3D THINKING | DCI Disciplinary Core Ideas | CCC Crosscutting Concepts | SEP Science & Engineering Practices

COLLECT EVIDENCE

Use your Science Journal to record the evidence you collect as you complete the readings and activities in this lesson.

INVESTIGATE

GO ONLINE to find these activities and more resources.

Applying Practices: Modeling Changes in Energy
HS-PS3-1. Create a computational model to calculate the change in the energy of one component in a system when the change in energy of the other component(s) and energy flows in and out of the system are known.

Energy Transformations

You might not think that a vase on a table has any relationship with energy—until it falls. You probably associate energy more with race cars roaring by you or the Sun warming your skin on a summer day. All of these situations involve energy transformations.

Mechanical energy transformations

Bicycles, roller coasters, and swings can often be described in terms of mechanical energy. **Mechanical energy** is the sum of the kinetic energy and potential energy of the objects in a system. Mechanical energy includes the kinetic energy of objects, elastic potential energy, and gravitational potential energy. It does not include nuclear energy, thermal energy, or chemical potential energy.

Mechanical energy and total energy are not the same because there are types of energy that are not mechanical energy. As a result, mechanical energy is not necessarily conserved. However, the mechanical energy of a system often remains constant or nearly constant. When this is the case, energy is only transformed between different kinds of mechanical energy.

Falling objects Look at the apple tree in **Figure 11.** An apple-Earth system, which is a system that includes an apple and Earth, has gravitational potential energy. The apple-Earth system does not have kinetic energy while the apple is hanging from the tree because the apple is not moving.

However, when the apple falls, it gets closer to Earth, so the GPE of the apple-Earth system decreases. This potential energy is transformed into kinetic energy as the apple's speed increases.

If potential energy is being converted into kinetic energy, then the mechanical energy of the apple-Earth system does not change as the apple falls. The potential energy that the apple-Earth system loses is gained back as kinetic energy. The form of mechanical energy changes, but the total amount of mechanical energy remains the same.

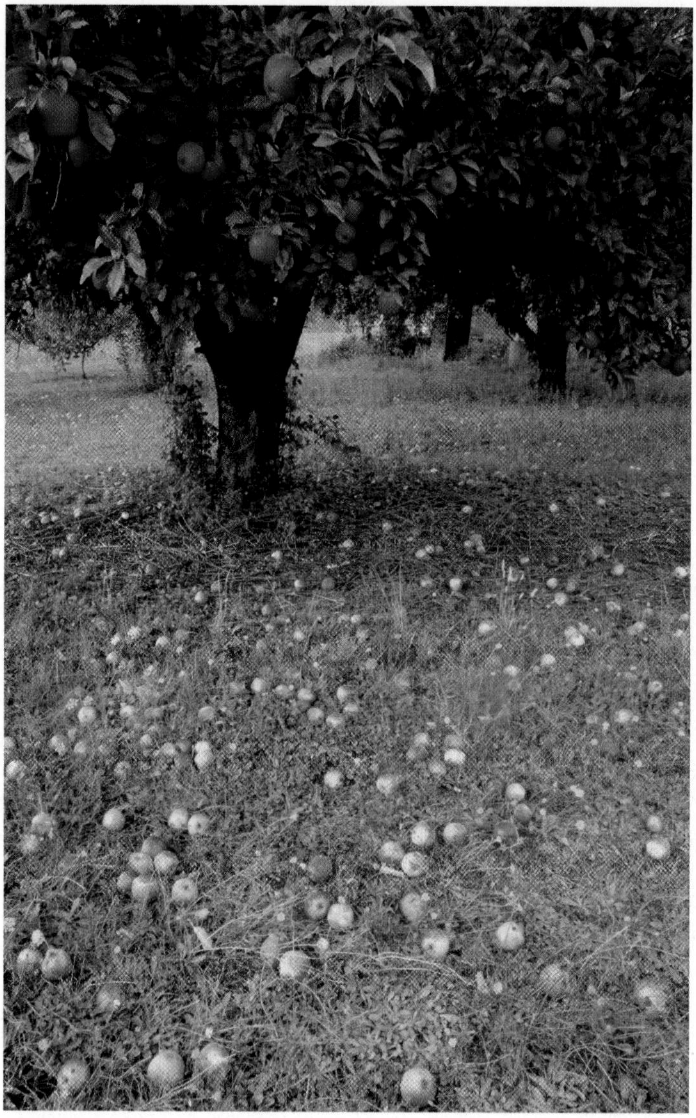

Figure 11 The gravitational potential energy of an apple-Earth system could be converted into kinetic energy. However, the mechanical energy of the apple-Earth system remains nearly constant as the apple falls.

Estimate *the gravitational potential energy of one of the apples on the tree relative to the ground.*

Get It?

Describe what happens to the mechanical energy of the apple-Earth system as the apple falls from the tree.

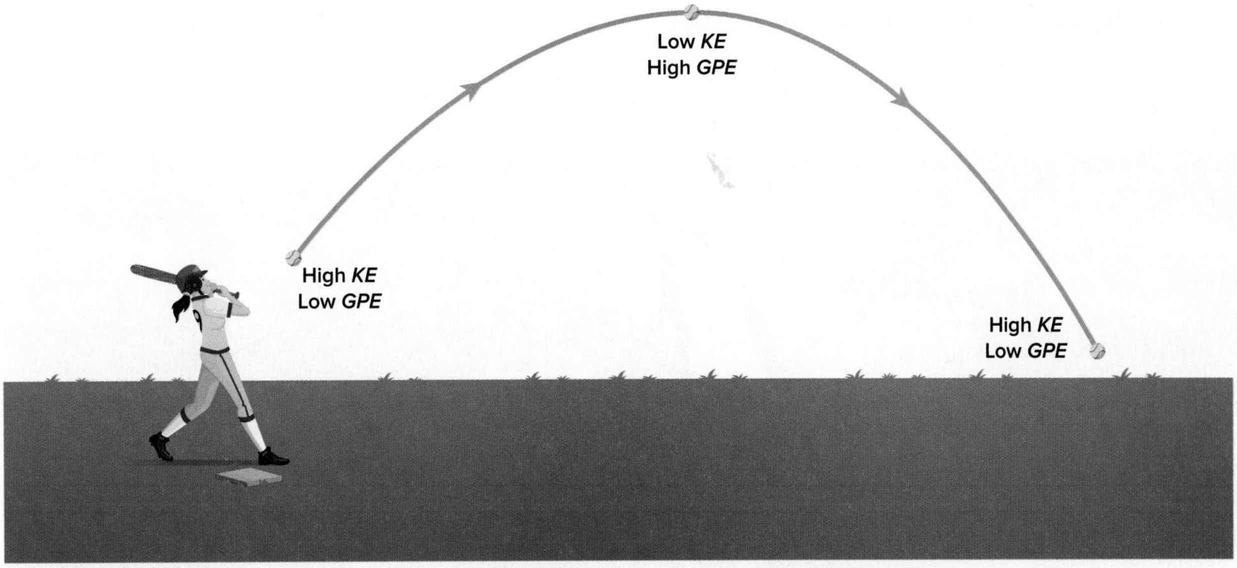

Figure 12 Kinetic energy is transformed into gravitational potential energy as the ball rises. As the ball falls, gravitational potential energy is transformed back into kinetic energy.

Predict *How large will the mechanical energy of the ball-Earth system be after the ball has reached the ground and rolled to a stop? Use the ground as the reference level.*

Projectile motion Energy transformations also occur during projectile motion when an object moves in a curved path. Look at **Figure 12** and consider the ball-Earth system. When the ball leaves the bat, the ball is moving fast, so the system's kinetic energy is relatively large.

Get It?

Explain whether the ball has potential energy when it leaves the bat.

The ball's speed decreases as it rises, so the system's kinetic energy decreases. However, the system's gravitational potential energy increases as the ball goes higher. At the top of the ball's path, the system's GPE is larger and its kinetic energy is smaller than they were when the ball left the bat. Then, as the baseball falls, the system's GPE decreases as its kinetic energy increases. However, the mechanical energy of the ball-Earth system remains constant as the ball rises and falls.

Get It?

Explain why the ball is moving faster just before it hits the ground than it was when it left the bat, but the mechanical energy of the ball-Earth system is still constant.

WORD ORIGINS

kinetic

comes from the Greek word kinetikos, which means putting in motion

A truck traveling on an interstate highway has a lot of kinetic energy.

Figure 13 Visualizing Energy Transformations

A ride on a swing illustrates how kinetic energy changes to potential energy and back to kinetic energy.

Swings The mechanical energy transformations for a swing, like the one shown in **Figure 13**, are similar to the mechanical energy transformations for a roller coaster.

Conservation of energy means that the total change of energy in any system is always equal to the total energy transferred into or out of the system. The ride starts with a push, which transfers kinetic energy to the rider. As the swing rises, the rider loses speed but gains height. In energy terms, kinetic energy changes to GPE. At the top of the rider's path, GPE is at its greatest.

Then, as the swing moves back downward, gravitational potential energy changes back to kinetic energy. At the bottom of each swing, the kinetic energy is at its maximum and the GPE is at its minimum. As the rider swings back and forth, energy is continually transformed between kinetic energy and GPE. However, the rider swings less and less on each cycle unless he or she pumps the swing or gets someone to provide a push. What is happening to the rider's mechanical energy?

Chris Frank/Moment Open/Getty Images

Other energy transformations

Consider the swing again. Think about what happens when you continue to swing without getting a push or pumping. The swing slows down and eventually comes to a stop. The mechanical energy of the swing-Earth system decreases. At first, it might appear that energy is being destroyed. However, recall that there are other forms of energy aside from mechanical energy. Energy transformations often involve these other forms.

The effect of friction If the mechanical energy of the swing-Earth system decreases, then some other forms of energy must increase by an equal amount to keep the total amount of energy the same. What could these other forms of energy be? Think about friction and air resistance. With every movement, the swing's ropes or chains rub on their hooks, and air pushes on the rider, as illustrated in **Figure 14**.

Friction and air resistance convert some of the mechanical energy into a less-useful form—thermal energy. Thermal energy is the energy of heat and hot objects. With every pass of the swing, the temperature of the hooks and the air increases slightly. Mechanical energy is not destroyed. Instead, friction and air resistance transform mechanical energy into thermal energy. This thermal energy is soon transferred to the surrounding air.

Get It?

Infer why the wheels of a car get hot when the car is driven.

To keep the swing going, you must constantly put energy into the swing-Earth system. You can do this by pumping the swing, transforming the chemical potential energy from the food that you eat into additional mechanical energy.

Figure 14 In a swing-air-Earth system, air resistance and friction transform mechanical energy into thermal energy, a less useful form of energy. To keep swinging, you need to supply more mechanical energy by pumping your legs or getting a push.

Describe *how the kinetic energy and gravitational potential energy of the swing-Earth system change with time.*

CCC CROSSCUTTING CONCEPTS

Systems and System Models Create a poster to illustrate a system and the energy transformations that occur in it, such as an object that starts on a ramp, slides down the ramp and across the floor, and eventually stops.

Transforming electrical energy Energy transformations can also involve electrical energy. Think about all the electric devices that you use every day. Electric stoves and toasters transform electrical energy into thermal energy. Televisions transform electrical energy into radiant energy and sound energy. The electric motor in a washing machine transforms electrical energy into mechanical energy. Lightbulbs transform electrical energy into radiant energy. **Figure 15** illustrates the energy change that occurs in a lightbulb.

What other devices have you used today that make use of electrical energy? You might have been awakened by an alarm clock, styled your hair, made toast, listened to music, or played a video game. What form or forms of energy is electrical energy converted to in each of these examples?

Transforming chemical potential energy Fuel stores energy in the form of chemical potential energy. For example, most cars run on gasoline, which has chemical potential energy. A car engine transforms this chemical potential energy into thermal energy and then into mechanical energy for the car's motion. A car engine also gets very hot when it is used. This is evidence that much of the thermal energy is never converted to mechanical energy.

Some energy transformations are less obvious because they do not result in visible motion, sound, heat, or light. Every green plant converts radiant energy into chemical potential energy. If you eat an ear of corn, the chemical potential energy from the corn is transferred to your body. Your body then extracts this energy for functions such as breathing, pumping blood, moving, speaking, and thinking.

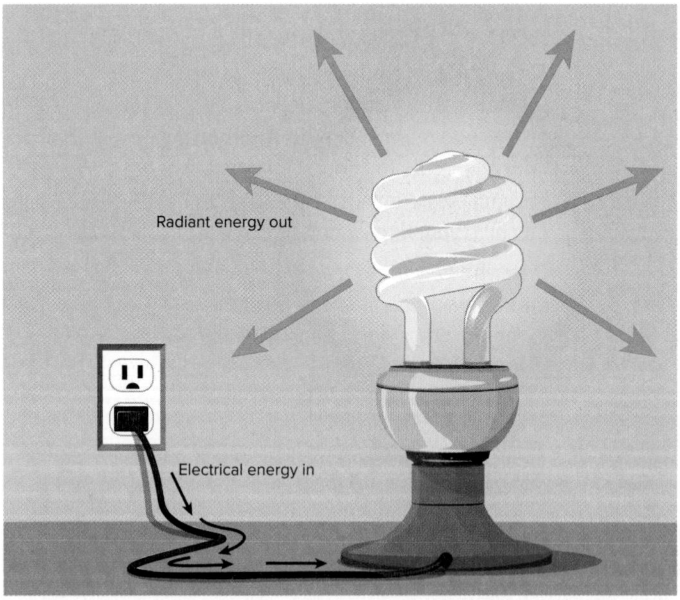

Figure 15 A lightbulb is a device that transforms electrical energy into radiant energy.

STEM CAREER Connection

Small-Engine Mechanic

If you like working with your hands and keeping things running efficiently, you might consider becoming a small-engine mechanic. Without proper care and repair, an engine can become less efficient, causing more chemical potential energy to transform into thermal energy instead of mechanical energy.

Power—how fast energy changes

Think again about the energy that your body extracts from food every day. You probably get enough energy from food in one day to jump nearly 10 km into the air. If this is true, then why can't you do this? You might have enough energy, but you don't have enough power. **Power** is the rate at which energy is converted. Power can be found using the following equation.

Power Equation

$$\text{Power (in watts)} = \frac{\text{Energy (in joules)}}{\text{time (in seconds)}}$$

$$P = \frac{E}{t}$$

Power is measured in watts; 1 watt equals 1 joule per second. A 13-W lightbulb transforms 13 J of electrical energy into radiant energy each second. A typical person can develop a power of only about 500 W for a jump. This results in a jump that is less than 1 m high for a person with average mass.

If power is the amount of energy converted per second, how can you become more powerful? One way to increase power is to increase the amount of energy converted in each second. Switching from a 13-W lightbulb to a 60-W lightbulb means you convert more than four times the energy each second.

Another way to increase power is to reduce the time needed to convert the energy. If you climb a flight of stairs in 10 seconds, you will increase your power if you climb the same stairs in less time.

EXAMPLE Problem 6

SOLVE FOR POWER You transform 950 J of chemical energy into mechanical energy to push a sofa. If it took you 5.0 s to move the sofa, what was your power?

Identify the Unknown: power: P

List the Knowns: Energy transformed: $E = 950\ \text{J}$

time: $t = 5.0\ \text{s}$

Set Up the Problem: $P = \frac{E}{t}$

Solve the Problem: $P = \frac{950\ \text{J}}{5.0\ \text{s}} = 190\ \text{W}$

Check the Answer: A typical human can develop a power of 400 W to 1000 W for short periods of time. Developing a power of 190 W would require some exertion but would not be too difficult. The answer is reasonable.

PRACTICE Problems

ADDITIONAL PRACTICE

26. If a runner's power is 400 W as she runs, how much chemical energy does she convert into other forms in 10.0 minutes?

27. CHALLENGE One horsepower is a unit of power equal to 746 W. How much energy can a 150-horsepower engine transform in 10.0 s?

Energy conversions in your body

You transfer energy from your surroundings to your body when you eat. The chemical potential energy of food supplies the cells in your body with the energy that they need to function. Energy from food is often measured in Calories (C).

You have probably seen descriptions of Calories per serving on food packages, such as the sides of cereal boxes or milk cartons. One Calorie is equal to about 4,000 J.

Every gram of fat in a food supplies a person with about 10 C (40,000 J) of energy. Carbohydrates and proteins each supply about 5 C (20,000 J) of energy per gram. Everything that your body does requires energy. The number of Calories that you need for different activities depends on your weight, your body type, and your degree of physical activity. **Table 1** shows the amount of energy needed to do various activities.

Table 1 Calories Used in 1 Hour

Type of Activity	Body Frames		
	Small	Medium	Large
Sleeping	48	56	64
Sitting	72	84	96
Eating	84	98	112
Standing	96	112	123
Walking	180	210	240
Playing tennis	380	420	460
Bicycling (fast)	500	600	700
Running	700	850	1000

Check Your Progress

Summary

- According to the law of conservation of energy, energy cannot be created or destroyed.
- Energy can be transformed from one form to another.
- Mechanical energy is the sum of all the potential energies and kinetic energies of all the objects in a system.
- Power is the rate at which energy is converted from one form to another.

Demonstrate Understanding

28. **Apply** the law of conservation of energy and describe the energy transformations that occur as you coast down a long hill on a bicycle and then apply the brakes to make the bike stop at the bottom.
29. **Identify** whether each of the following is a form of mechanical energy: elastic potential energy, chemical potential energy, gravitational potential energy.
30. **Explain** how friction affects the mechanical energy of a system.

Explain Your Thinking

31. **Compare** A roller coaster is at the top of a hill and rolls to the top of a lower hill. If mechanical energy is constant, then at the top of which hill is the kinetic energy from the roller coaster's motion greater?
32. **MATH Connection** Approximately how much electrical energy does a 5-W lightbulb convert to radiant and thermal energy in one hour?
33. **MATH Connection** The mechanical energy of a bicycle at the top of a hill is 6000 J. The bicycle stops at the bottom of the hill by applying the brakes. If the gravitational potential energy of the bicycle-Earth system is 2000 J at the bottom of the hill, how much mechanical energy was converted into thermal energy?

LEARNSMART® Go online to follow your personalized learning path to review, practice, and reinforce your understanding.

World's Fastest Bicycle

Could a person pedaling a bicycle over flat terrain move faster than a car on a highway? With a unique type of bicycle called a recumbent, the answer is yes.

The recumbent bike at bottom is in an aerodynamic shell.

Mechanical disadvantage

How could a bicycle convert an ordinary human into the fastest creature on Earth? The answer is mechanical *dis*advantage. The idea behind many simple machines is to increase force by decreasing distance.

On the other hand, a bicycle increases the effort of the rider, but the bicycle moves a greater distance. Look at the hand-powered cycle below. Riders of this type of cycle must exert great effort with their arms to accelerate their cycles forward. However, the payoff is a much higher top speed.

Recumbent bicycles, such as the one shown at top right, maximize this mechanical disadvantage. At the same time, recumbent bicycles position the rider to be able to exert maximum force on the pedals.

This hand-powered bicycle makes use of mechanical disadvantage.

Against the wind

At the World Human Powered Speed Challenge, held every year in the Nevada wilderness, the enemy is the wind. The top competitors travel faster than 130 km/h (81 mph).

At that speed, air resistance would pummel an unprotected rider, so the bikes have shells, such as the one shown above. These shells, made of Kevlar and carbon fibers for lightness and strength, deflect air around and past the rider. The shell's design allows the wheel to steer only a few degrees in either direction. These bicycles are built for straight-line speed, not maneuverability. In the early evening, when winds are calm, riders take up to 6 km to get up to speed. Timing happens at marked spots 200 meters apart along a closed highway. Riders cover this distance in less than 6 seconds.

ANALYZE DATA

Gather and graph data for the speed records of a human-powered bicycle race that date back at least 30 years, and interpret the results.

MODULE 4
STUDY GUIDE

GO ONLINE to study with your Science Notebook.

Lesson 1 WORK AND MACHINES

- Work is force applied through a distance.
- A machine can increase speed, change the direction of a force, or increase force.
- The output work from a machine is never as great as the input work on that machine.
- Efficiency is the ratio of output work to input work.
- Mechanical advantage is the ratio of output force to input force.

- work
- machine
- simple machine
- compound machine
- efficiency
- mechanical advantage

Lesson 2 DESCRIBING ENERGY

- Forms of energy include mechanical, electrical, chemical, thermal, and radiant energy.
- Kinetic energy is the energy that a moving object has because of its motion.
- Potential energy is stored energy due to the interactions between objects.
- Different forms of potential energy include elastic potential energy, chemical potential energy, and gravitational potential energy.

- energy
- system
- kinetic energy
- potential energy
- elastic potential energy
- chemical potential energy

Lesson 3 CONSERVATION OF ENERGY

- According to the law of conservation of energy, energy cannot be created or destroyed.
- Energy can be transformed from one form to another.
- Mechanical energy is the sum of all the potential energies and kinetic energies of all the objects in a system.
- Power is the rate at which energy is converted from one form to another.

- law of conservation of energy
- mechanical energy
- power

THREE-DIMENSIONAL THINKING

Module Wrap-Up

REVISIT THE PHENOMENON

How can ice be used as an alternative energy source?

CER Claim, Evidence, Reasoning

Explain Your Reasoning Revisit the claim you made when you encountered the phenomenon. Summarize the evidence you gathered from your investigations and research and finalize your Summary Table. Does your evidence support your claim? If not, revise your claim. Explain why your evidence supports your claim.

STEM UNIT PROJECT

Now that you've completed the module, revisit your STEM unit project. You will summarize your evidence and apply it to the project.

GO FURTHER

SEP Data Analysis Lab

Conservation of Energy in the Heart

It is important to eat good food for energy and health. The human body changes some of the energy from food to electrical energy. Electrical energy is used by the nervous system, which includes the brain. Electrical signals in the body travel over the nerves to every muscle in the body, telling the muscles to move.

Mechanical Energy

The brain is always sending signals to the heart. If it didn't, the heart would stop. But where does this electrical energy go after it signals the heart to pump? Conservation of energy says that the energy is not lost. Remember that the electrical energy makes the muscles around the chambers of the heart contract or squeeze together. When they do this, the electrical energy becomes mechanical energy.

Heat

Then, the energy is again transformed. This time it becomes heat. When the heart muscles contract, some of the energy they use becomes heat energy. Have you ever been cold and shivered? That is the muscles in your body using electricity to make heat.

CER Analyze and Interpret Data

1. **Claim and Evidence** What kind of energy does the heart receive from the brain?
2. **Claim, Evidence, Reasoning** How is energy conserved in the process of a heartbeat?
3. **Claim, Evidence, Reasoning** What is the source of the electrical energy in the brain?
4. While you are alive, does your brain ever stop sending electricity to your heart? Explain your answer.

PandaWild/Shutterstock

MODULE 5

THERMAL ENERGY

ENCOUNTER THE PHENOMENON

Why would it be useful to visualize thermal energy?

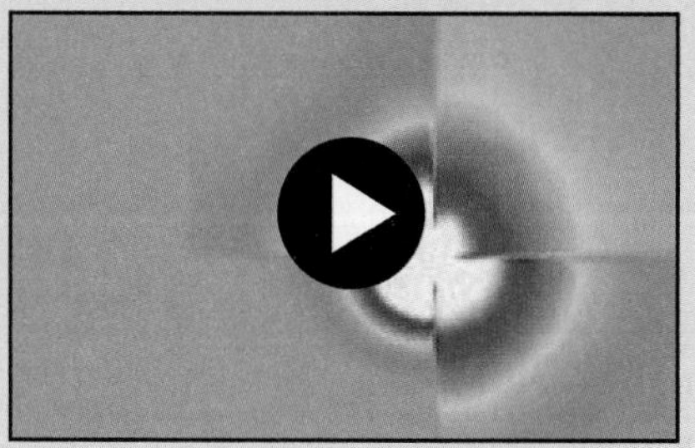

GO ONLINE to play a video about a variety of metals transferring heat from a flame at different rates.

SEP Ask Questions

Do you have other questions about the phenomenon? If so, add them to the driving question board.

CER Claim, Evidence, Reasoning

Make Your Claim Use your CER chart to make a claim about why it would be useful to visualize thermal energy. Explain your reasoning.

Collect Evidence Use the lessons in this module to collect evidence to support your claim. Record your evidence as you move through the module.

Explain Your Reasoning You will revisit your claim and explain your reasoning at the end of the module.

GO ONLINE to access your CER chart and explore resources that can help you collect evidence.

LESSON 1: Explore & Explain: Heat

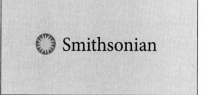

LESSON 3: Explore & Explain: Thermal Energy into Mechanical Energy

Smithsonian

Additional Resources

LESSON 1

TEMPERATURE, THERMAL ENERGY, AND HEAT

FOCUS QUESTION

Why are there different colors on a thermal image?

Temperature

You can use the words *hot* and *cold* to describe temperature. Something is hot when its temperature is high. When you heat water on a stove, its temperature increases. How are temperature and heat related?

Matter in motion

The matter around you is made of tiny particles–atoms, ions, and molecules. In all materials, these particles are in constant, random motion. They move in all directions at different speeds. These particles have kinetic energy because they are moving. The greater their speeds, the greater their kinetic energy. If the particles that make up an object have more kinetic energy, then that object feels hotter.

For example, in **Figure 1,** the particles that make up the electric stove burner on the left have more kinetic energy than the particles that make up the burner on the right. As a result, the burner on the left feels hotter than the burner on the right. Is there a more exact relationship between temperature and kinetic energy?

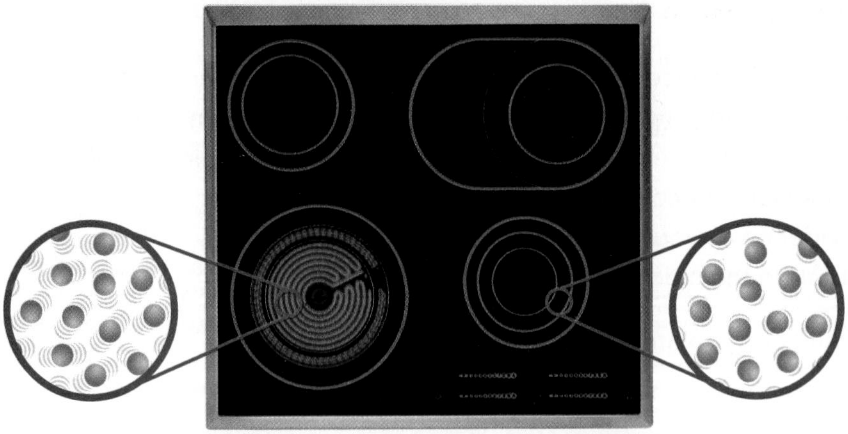

Figure 1 The particles that make up the left burner are moving faster than the particles that make up the right burner.

Predict *whether the kinetic energy of the particles that make up the water in a lake would be greater on a summer day or on a winter day.*

3D THINKING DCI Disciplinary Core Ideas CCC Crosscutting Concepts SEP Science & Engineering Practices

COLLECT EVIDENCE

Use your Science Journal to record the evidence you collect as you complete the readings and activities in this lesson.

INVESTIGATE

GO ONLINE to find these activities and more resources.

Applying Practices: Coffee Cup Calorimetry

HS-PS3-4. Plan and conduct an investigation to provide evidence that the transfer of thermal energy when two components of different temperature are combined within a closed system results in a more uniform energy distribution among the components in the system (second law of thermodynamics).

Kinetic energy and temperature

The **temperature** of an object is the measure of the average kinetic energy of the particles that make up that object. The temperature of hot tea is higher than the temperature of iced tea. Therefore, the particles that make up the hot tea have more kinetic energy on average. In SI units, scientists measure temperature in kelvins (K). However, people around the world use the Celsius scale much more often. One kelvin is the same as one degree Celsius.

Thermal Energy

If you let cold butter sit at room temperature awhile, it warms and becomes softer. The particles that make up the room-temperature air have more kinetic energy than the particles that make up the cold butter. Collisions between the air and the butter transfer energy from the faster-moving air particles to the slower-moving butter particles. The butter particles then move faster, and the temperature of the butter increases.

Recall that Earth and a ball exert gravitational forces on each other. Earth and the ball have potential energy. Similarly, the particles that make up matter exert attractive electromagnetic forces on each other and have potential energy. When you move a ball away from Earth, gravitational potential energy increases. When you move the particles that make up matter farther apart, the potential energy of that matter increases. **Figure 2** shows the kinetic and potential energies of particles.

The **thermal energy** of an object is the sum of the kinetic energy and the potential energy of all of the particles that make up that object. Thermal energy is one of the ways in which energy manifests itself at the macroscopic scale. At the microscopic scale, thermal energy can be modeled as a combination of energy associated with the motion of particles and energy associated with the relative position of particles. As the butter warmed and the kinetic energy of its particles increased, the butter's thermal energy increased.

Get It?

Explain the difference between temperature and thermal energy.

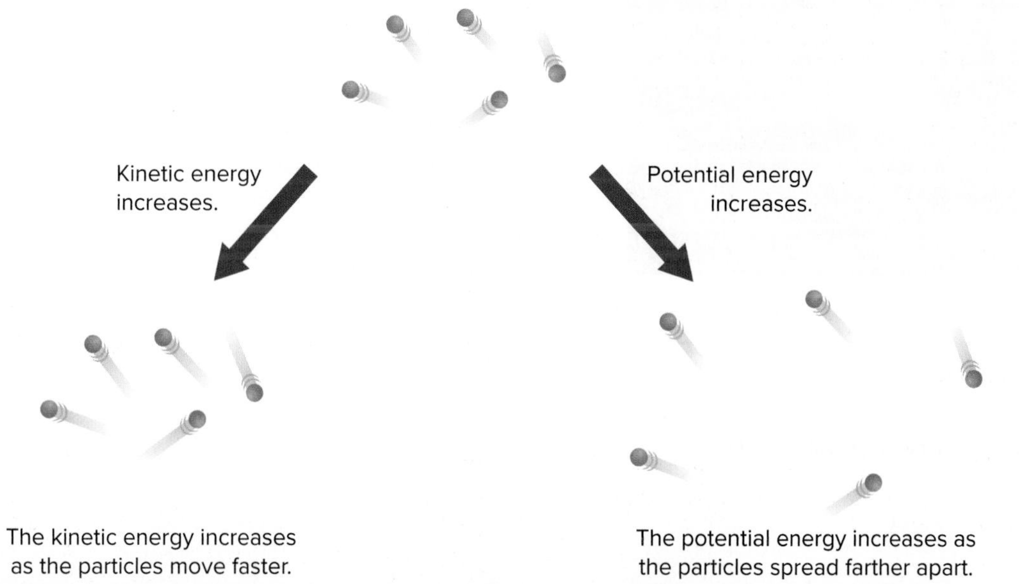

Kinetic energy increases.
Potential energy increases.
The kinetic energy increases as the particles move faster.
The potential energy increases as the particles spread farther apart.

Figure 2 The particles that make up a substance gain kinetic energy when their speeds increase. The potential energy of these particles increases when the separation between those particles increases.

Temperature and thermal energy

Thermal energy depends on temperature. The average kinetic energy of the particles that make up an object increases when the temperature of that object increases. Thermal energy is the total kinetic and potential energy of all of the particles that make up an object. The thermal energy of the object increases when the average kinetic energy of the particles that make up that object increases. Therefore, the thermal energy of an object increases as its temperature increases.

Mass and thermal energy

Thermal energy also depends on mass. If the mass of the object increases, the thermal energy of that object also increases. Suppose you have a glass of water and a beaker of water, both at the same temperature. However, the beaker contains twice as much water as the glass. The average kinetic energy of the water molecules is the same in both containers. However, there are twice as many water molecules in the beaker. Therefore, the water in the beaker has twice as much thermal energy as the water in the glass.

Heat

Suppose you place a pot of water on a hot burner, as shown in **Figure 3.** When you do this, you transfer thermal energy from the warmer stove to the cooler pot and water. **Heat** is energy that is transferred between objects due to a temperature difference between those objects. Warmer objects always heat cooler objects, but the reverse never occurs. For example, a hot stove will heat a cold pot of water. However, a cold pot of water can never heat a hot stove.

Figure 3 The warmer stove heats the cooler water.

Describe *the transfers of thermal energy in this figure.*

SCIENCE USAGE v. COMMON USAGE

heat

Science usage: to transfer thermal energy due to a temperature difference
The stove heats the pot of soup.

Common usage: a condition of being hot
The summer heat is really getting to him.

Ryerson Clark/E+/Getty Images

Specific Heat

Have you ever been to the beach during the summer? The ocean was probably cool, but the sand was probably hot. Energy from the Sun falls on the water and sand at nearly the same rate. However, the Sun's energy changes the sand's temperature more quickly than it changes the water's temperature.

A substance's temperature changes when that substance absorbs thermal energy. This temperature change depends on the amount of thermal energy that the substance absorbs and the mass of the substance. This temperature change also depends on the nature of the substance.

The **specific heat** of a material is the amount of heat needed to raise the temperature of 1 kg of that material by 1°C. Scientists measure specific heat in joules per kilogram degree Celsius [J/(kg · °C)]. **Table 1** compares the specific heats of some familiar materials.

Compare water with iron in **Table 1.** Water has a very high specific heat. Metals, such as iron, have low specific heats. To raise equal masses of water and iron 1°C, water must absorb almost 10 times more thermal energy than iron. **Figure 4** explains why this is so.

Table 1 Comparison of Specific Heats*

Substance	Specific Heat [J/(kg·°C)]
Water	4200
Wood	1700
Sand	830
Carbon (graphite)	710
Iron	450

*Values have been rounded.

Get It?

Define *specific heat.*

Water as a coolant

A coolant is a substance that can absorb a great amount of thermal energy with little change in temperature. Water is useful as a coolant because it can absorb thermal energy without a large change in temperature. For example, people use water as a coolant in automobile engines. Thermal energy transfers from the engine to the water as long as the water temperature is lower than the engine temperature.

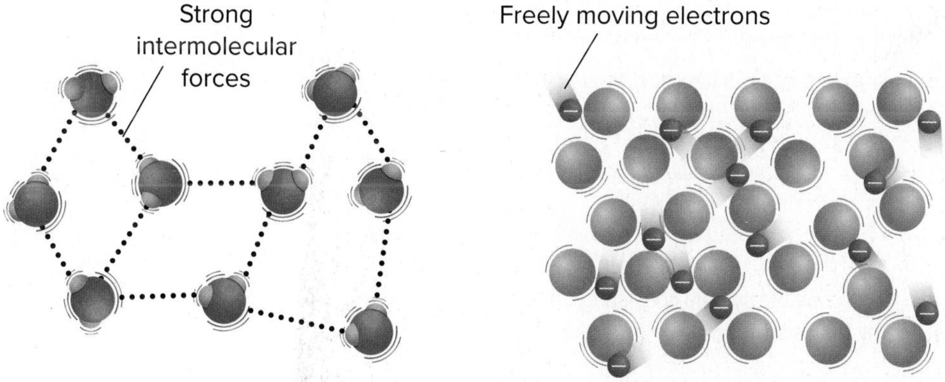

When thermal energy is added to water, some of the added thermal energy has to overcome some of the attraction between the molecules before those molecules can start moving faster.

In metals, electrons can move freely. When thermal energy is added, no strong attractions need to be overcome before the electrons can start to move faster.

Figure 4 The change in temperature due to the transfer of thermal energy depends on the nature of the substance. Water molecules have strong attractions to each other. Therefore, the specific heat of water is high. The electrons in metals move easily, transferring thermal energy throughout that metal quickly. As a result, the specific heats of metals are low.

Calculating thermal energy changes

If Q is the change in thermal energy, C is specific heat, T_f is the final temperature, T_i is the initial temperature, and $T_f - T_i$ is the temperature change, you can calculate the change in thermal energy with the following equation.

Thermal Energy Equation

change in thermal energy (J)

= mass (kg) · **temperature change** (°C) · **specific heat** $\left(\frac{J}{kg \cdot °C}\right)$

$$Q = m\left(T_f - T_i\right)C$$

EXAMPLE Problem 1

SOLVE FOR THERMAL ENERGY A wooden block has a mass of 20.0 kg and a specific heat of 1700 J/(kg·°C). Find the change in thermal energy of the block as it warms from 15.0°C to 25.0°C.

Identify the Unknown: change in thermal energy: Q

List the Knowns:
mass: m = 20.0 kg
final temperature: T_f = 25.0°C
initial temperature: T_i = 15.0°C
specific heat: C = 1700 J/(kg·°C)

Set Up the Problem: $Q = m(T_f - T_i)C$

Solve the Problem:
Q = (20.0 kg)(25.0°C − 15.0°C)(1700 J/(kg·°C))
= (20.0 kg)(10.0°C)(1700 J/(kg·°C))
= 340,000 J

Check the Answer: Do the units match on both sides of the equation? Changes in thermal energy (Q) are measured in units of joules (J). On the right side, (units of mass)(units of temperature change)(units of specific heat) = $\cancel{kg} \times \cancel{°C} \times J/(\cancel{kg} \cdot \cancel{°C})$ = J. The units on both sides of the equation match.

PRACTICE Problems ADDITIONAL PRACTICE

1. The air in a room has a mass of 50 kg and a specific heat of 1000 J/(kg·°C). What is the change in thermal energy of the air when it warms from 20°C to 30°C?
2. The temperature of a 2.0-kg block increases by 5°C when 2000 J of thermal energy are added to the block. What is the specific heat of the block?
3. **CHALLENGE** A wooden block has a mass of 0.200 kg, has a specific heat of 710 J/(kg·°C), and is at a temperature of 20.0°C. What is the block's final temperature if its thermal energy increases by 2130 J?

CCC CROSSCUTTING CONCEPTS

Energy and Matter Consider Example Problem 1. Describe the system consisting of the wooden block in terms of the flow of thermal energy that is taking place. Produce a diagram to model the energy flow.

Measuring Specific Heat

A scientist can calculate the specific heat of a material from the measurements that he or she takes using a calorimeter, such as the one shown in **Figure 5.** To determine the specific heat of a material using a calorimeter, a scientist measures the mass of a sample of the material and the mass and initial temperature of the water in the calorimeter. The scientist then heats the sample, measures the sample's temperature, and places the sample in the water in the inner chamber of the calorimeter.

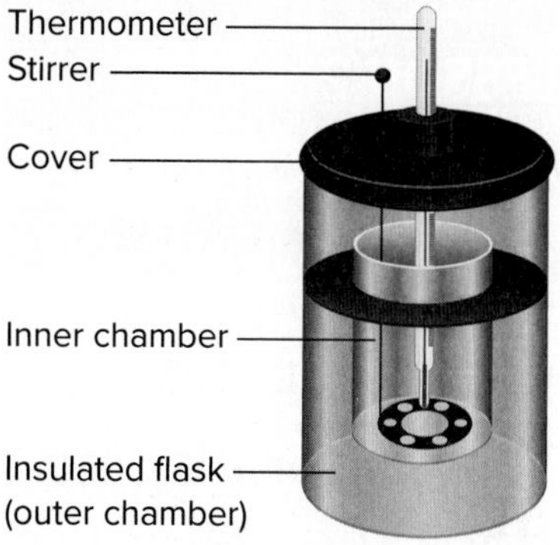

Figure 5 A calorimeter can be used to determine the specific heat of a material. The sample is placed in the inner chamber.

The sample cools as thermal energy is transferred to the water, and the temperature of the water increases. The transfer of thermal energy continues until the sample and the water are at the same temperature. Then the initial and final temperatures of the water are known, and the amount of heat gained by the water can be calculated from the water's mass, temperature change, and specific heat.

At this point, the scientist knows the mass of the substance, the thermal energy change of the substance, and the temperature change of the substance. From there, the specific heat of the unknown substance can be calculated using the thermal energy equation. With this information, the scientist might be able to accurately identify the substance.

Check Your Progress

Summary

- The temperature of an object is a measure of the average kinetic energy of the particles that make up the object.
- Thermal energy is the sum of the kinetic and potential energies of all of the particles in an object.
- Heat is a transfer of thermal energy due to a temperature difference.
- The specific heat of a material is the amount of heat needed to raise the temperature of 1 kg of the material by 1°C.

Demonstrate Understanding

4. **Describe** how the motions of the particles that make up an object change when the object's temperature increases.
5. **Describe** the energy transfer that occurs when you touch a block of ice with your hand.
6. **Infer** When one object heats another, does the temperature increase of one object always equal the temperature decrease of the other object? Explain.
7. **Explain** why water is often used as a coolant.

Explain Your Thinking

8. **Explain** whether the following statement is true: For any two objects, the one with the higher temperature always has more thermal energy.
9. **MATH Connection** Estimate the change in the thermal energy of water in a pond with a mass of 1000 kg and a specific heat of 4200 J/(kg·°C) if the water cools by 1°C.
10. **MATH Connection** Calculate the specific heat of a metal if 0.3 kg of the metal absorbs 9000 J of heat as the metal warms by 10°C.

LEARNSMART® Go online to follow your personalized learning path to review, practice, and reinforce your understanding.

LESSON 2

CONDUCTION, CONVECTION, AND RADIATION

FOCUS QUESTION

Why, after placing a pot of water on a hot stove, does the water not immediately boil?

Conduction

Energy cannot be created or destroyed, but it can be transported from one place to another and transferred between systems. Conduction, convection, and radiation are all processes that transfer energy. **Conduction** is the transfer of thermal energy by collisions between the particles that make up matter. Conduction occurs because particles that make up matter are in constant motion.

Collisions transfer thermal energy

If you leave a metal spoon in a pot of soup that is cooking on the stove, the spoon might get too hot to touch. As one end of the spoon heats up, the kinetic energy of the particles that make up that part of the spoon increases. These particles collide with neighboring particles. Conduction transfers thermal energy to the other end of the spoon as particles with more kinetic energy transfer kinetic energy to particles with less kinetic energy. Conduction transfers thermal energy without transferring matter. Conduction spreads thermal energy from warmer areas to cooler areas, as shown in **Figure 6.**

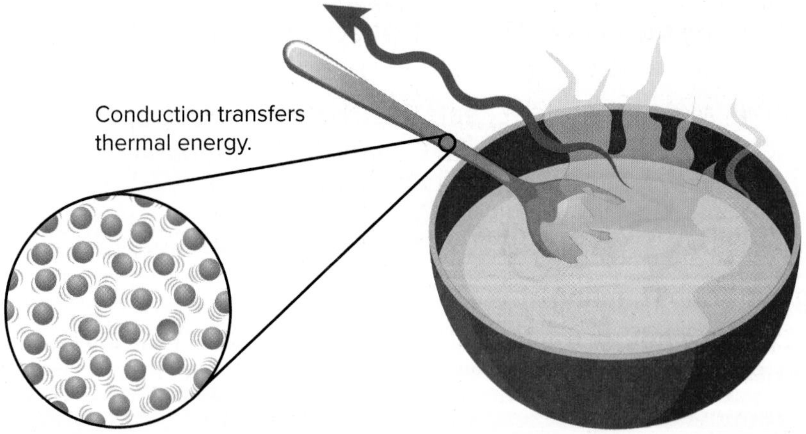
Conduction transfers
thermal energy.

Figure 6 Conduction occurs within a material as faster-moving particles collide with slower-moving particles.

State *If two objects are in contact, would conduction be from the cooler material to the warmer material or from the warmer material to the cooler material?*

3D THINKING DCI Disciplinary Core Ideas CCC Crosscutting Concepts SEP Science & Engineering Practices

COLLECT EVIDENCE

Use your Science Journal to record the evidence you collect as you complete the readings and activities in this lesson.

INVESTIGATE

GO ONLINE to find these activities and more resources.

Lab: Convection in Gases and Liquids

Carry out an investigation to demonstrate convection currents in water and air and to infer how birds utilize convection currents to conserve energy while in flight.

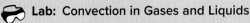

Lab: Conduction in Gases

Carry out an investigation to measure the temperature changes in the air near a heat source and to observe the conduction of thermal energy in air.

Thermal conductors

The rate at which conduction transfers thermal energy depends on the material. Conduction is faster in solids and liquids than it is in gases. In gases, particles are farther apart. Therefore, collisions among particles occur less frequently in gases.

The best conductors of thermal energy are metals. This is one reason why manufacturers often make cooking pots, like those in **Figure 7,** out of metal. In a piece of metal, some electrons are not bound to individual atoms. These electrons can move easily through the metal. Collisions between these electrons and other particles in the metal enable more rapid thermal energy transfers than in other materials. Silver, copper, and aluminum are among the best conductors of thermal energy.

Figure 7 These pots are made of metals that are excellent conductors of thermal energy.

Infer *why chefs often prefer pots that are good conductors of thermal energy.*

Convection

Unlike solids, liquids and gases are fluids that flow. In fluids, convection can transfer thermal energy. **Convection** is the transfer of thermal energy in a fluid by the movements of warmer and cooler fluid. When conduction occurs, more energetic particles collide with less energetic particles and transfer thermal energy. When convection occurs, more energetic particles move from one place to another.

Most substances expand as their temperatures increase. That is, as the particles move faster, they tend to be farther apart. Recall that density is the mass of a material divided by its volume. When a fluid expands, its volume increases, but its mass does not change. Therefore, a fluid's density decreases when that fluid is heated.

Because fluids decrease in density as they are heated, a fluid that absorbs thermal energy also decreases in density. The density of a warmer sample of a fluid is less than the density of a cooler sample of that same fluid. The same is true for the parts of a fluid. The warmer parts of a fluid are less dense than the cooler parts of a fluid.

These differences in density within a fluid drive convection. The warmer portions of the fluid rise to the top of the fluid, and the cooler portions sink to the bottom. If a fluid is heated from below, convection currents form.

Get It?

Explain how convection and density are related.

ACADEMIC VOCABULARY

transfer

to convey from one place to another

She is going to transfer her pictures from her camera to the Web site.

Convection currents

How does convection occur? Look at the lamp shown in **Figure 8.** Some of these types of lamps contain oil and alcohol. When the oil is cool, its density is greater than that of the alcohol's so it sits at the bottom of the lamp. When the light heats the two liquids in the lamp, the oil becomes less dense than the alcohol. The oil rises to the top of the lamp because it is less dense than the alcohol. As the oil rises, conduction transfers thermal energy from the warmer oil to the cooler alcohol. As a result, the oil cools as it rises.

By the time the oil reaches the top of the lamp, it cools and again becomes denser than the alcohol. As a result, the oil sinks. This rising-and-sinking action illustrates a convection current. Convection currents transfer thermal energy from warmer to cooler parts of the fluid. In a convection current, both conduction and convection transfer thermal energy.

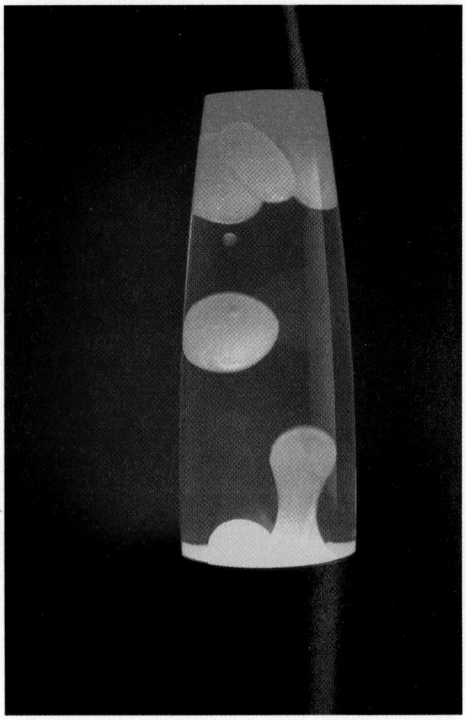

Figure 8 Heat from the light at the bottom of the lamp causes fluid to expand and rise. This creates convection currents in the lamp.

Explain *why the substances in the lamp rise and sink.*

> **Get It?**
> **Contrast** conduction with convection.

Deserts and rainforests

Earth's atmosphere is a fluid that is made of various gases. The atmosphere is warmer at the equator than it is at the North Pole and the South Pole. Also, the atmosphere is warmer at Earth's surface than it is at higher altitudes. These temperature differences produce convection currents that carry thermal energy away from the equator. Moist, warm air near the equator rises. As this air rises, it cools and loses moisture. Rain then falls over the equator.

The cooler, drier air sinks down toward the ground north and south of the equator. Desert zones form as a result. The temperature of this air increases as the air sinks. **Figure 9** shows how these convection currents result in rain forests and deserts.

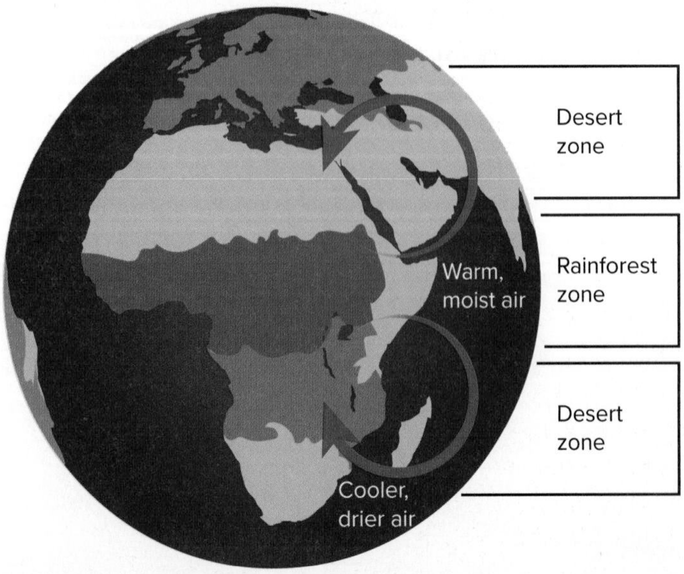

Figure 9 Sunlight on Earth is most intense at the equator. Convection currents form around Earth's equator as a result. Warm, moist air at the equator rises. As this air rises, it cools and loses its moisture in the form of rain that sustains rainforests near the equator. The convection currents then carry the dry air farther north and south. Deserts have formed where this air descends.

Incredible_movements/Shutterstock

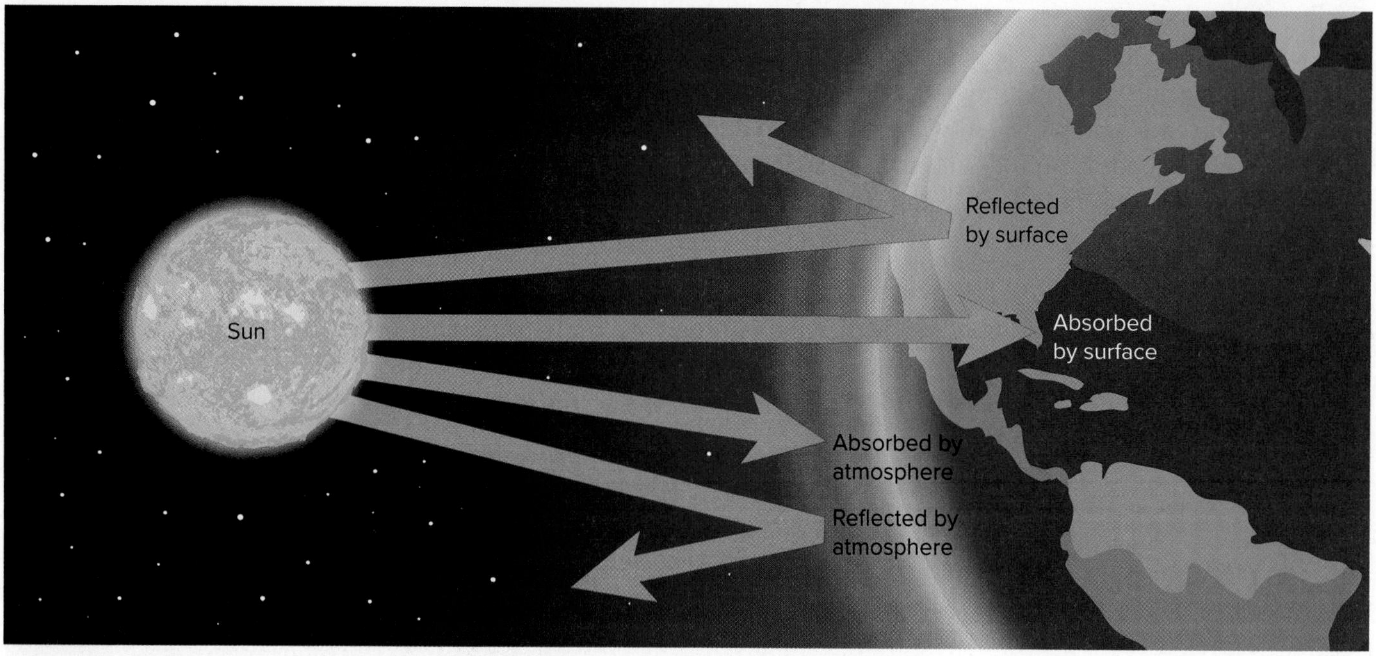

Figure 10 The arrows indicate radiation from the Sun. Not all of the Sun's radiation reaches Earth. Some of it is reflected by the atmosphere. Earth's surface also reflects some of the radiation that reaches it.

Radiation

Energy from the Sun reaches Earth, but how does that energy travel through space? Almost no matter exists in the space between Earth and the Sun, so neither conduction nor convection could warm Earth. Instead, radiation transfers energy from the Sun to Earth.

Radiation is the transfer of energy by electromagnetic waves, such as light and microwaves. These waves travel through space even when matter is not present. Energy that is transferred by radiation is often called radiant energy. When you stand near a fire, radiation transfers energy from the fire and increases the thermal energy of your body.

Get It?

Explain how energy travels through space.

Radiation and matter

When radiation strikes a material, that material absorbs, reflects, and transmits some of the energy. **Figure 10** shows what happens to radiation from the Sun as it reaches Earth. The amount of energy that a material absorbs, reflects, and transmits depends on the type of material. The thermal energy of a material increases when that material absorbs radiant energy.

Radiation in solids, liquids, and gases

In a solid, liquid, or gas, radiation travels through the space between particles. Particles can absorb and re-emit this radiation. This energy then travels through the space between particles, and other particles then absorb and re-emit the energy. Radiation usually passes more easily through gases than through solids or liquids because particles are much farther apart in gases than in solids or liquids. Thus, radiation transfers energy more rapidly and efficiently through gases than through liquids or solids.

Controlling Heat

You might not realize it, but you probably do a number of things every day to control thermal energy transfers between your body and your body's surroundings. For example, when it is cold outside, you put on a coat or a jacket before you leave your home. When you reach into an oven to pull out a hot dish, you might put a thick, protective mitten over your hand.

In both cases, you have used various materials to help control transfers of thermal energy. Your jacket decreases how much thermal energy your body transfers to the surrounding air, keeping you from getting cold. The oven mitten decreases how much thermal energy the hot dish transfers to your hand, preventing that hot dish from burning your hand.

Animals and heat

The animals shown in **Figure 11** have special features that help them control thermal energy transfers between their bodies and their bodies' surroundings. For example, the Antarctic fur seal's thick coat and the emperor penguin's thick layer of fat help to keep them from transferring thermal energy to their surroundings. This helps them survive in a climate where the temperature is often below freezing. In the desert, however, the scaly skin of the desert spiny lizard has just the opposite effect. It reflects the Sun's rays and keeps the animal from becoming too hot.

An animal's color also can play a role in keeping it warm or cool. The black feathers on the penguin's back, for example, allow it to absorb radiant energy.

Get It?
Identify two different animal adaptations for controlling the transfer of thermal energy.

Antarctic fur seal | Emperor penguins | Spiny lizard

Figure 11 Animals have different adaptations that help them control heat. Both the Antarctic fur seal and the emperor penguin have a thick layer of fat that reduces transfers of thermal energy to those animals' surroundings. The scaly skin of the desert spiny lizard reflects radiation from the Sun to prevent overheating.

Infer *why some animals go through annual cycles of growing a thicker coat of fur and then later shedding it in a process called molting.*

Figure 12 Tiny pockets of air help reduce the transfer of the people's thermal energy to the colder, outside air.

Thermal insulators

A **thermal insulator** is a material through which thermal energy moves slowly. Wood, fiberglass, and air are all good thermal insulators. Metals and other good conductors are poor thermal insulators. In conductors, thermal energy moves more rapidly from one place to another.

Insulated clothing Gases, such as air, are usually much better thermal insulators than solids or liquids. Some thermal insulators contain many pockets of trapped air. These air pockets conduct thermal energy poorly and also keep convection currents from forming. Winter jackets, like the one worn by the woman in **Figure 12,** work in this way. When you put on a jacket like this one, the fibers in the jacket trap the air and hold it next to you. The air slows the transfer of your body's thermal energy into its surroundings. Under the jacket, a blanket of warm air covers you.

Get It?

Explain how trapped air makes a material, such as a winter jacket, a good thermal insulator.

Insulated buildings

Insulation, or materials that are thermal insulators, helps prevent thermal energy transfers out of buildings in cold weather and thermal energy transfers into buildings in warm weather. Manufacturers usually make building insulation from a fluffy material, such as fiberglass, that contains pockets of trapped air. Builders pack insulation into a structure's outer walls and attic, where it reduces thermal energy transfers between the structure and the surrounding air.

Insulation helps furnaces and air conditioners work more efficiently. In the United States, about 50 percent of the energy used in homes is for climate control. Repairing and improving insulation can dramatically reduce heating and cooling costs.

Get It?

Explain why insulation is just as important in hot climates as it is in cold climates.

Thermoses You might have used a thermos bottle, like the one shown in **Figure 13,** to carry hot soup or iced tea. A thermos bottle reduces energy transfers between the bottle's contents and its surroundings. As a result, the temperature of the contents changes very little, even over many hours.

In order to minimize thermal energy transfers from conduction and convection, a thermos bottle has two glass walls with very little air between those walls. Scientists call any region with very low gas density a vacuum. Glass is a good thermal insulator, and a vacuum is an extremely good thermal insulator.

To reduce energy transfers from radiation, manufacturers often coat a thermos bottle's inside and outside glass surfaces with aluminum. This makes each surface highly reflective. The inner reflective surface prevents radiation from transferring energy out of the liquid. The outer reflective surface prevents radiation from heating the liquid.

A thermos bottle keeps the liquid inside it warm or cool. Think about the things that you do to stay warm or cool. Sitting in the shade reduces energy transfers from radiation. Opening or closing windows affects thermal energy transfers from convection. Putting on a jacket reduces thermal energy transfers from conduction. In what other ways do you control heat?

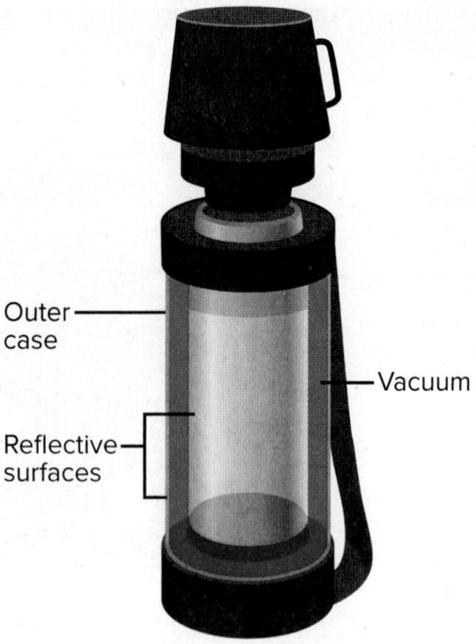

Figure 13 There is very little air between the two surfaces of a thermos bottle. This minimizes transfers of thermal energy by conduction and convection. Thermos bottles also often have reflective surfaces, which minimize energy transfers by radiation.

Check Your Progress

Summary

- Conduction is the transfer of thermal energy by collisions between more energetic and less energetic particles.
- Convection is the transfer of thermal energy by the movement of warmer and cooler materials.
- Radiation is the transfer of energy by electromagnetic waves.
- Thermal insulators are used to reduce the transfer of thermal energy from one place to another.

Demonstrate Understanding

11. **Identify** which method of thermal energy transfer would be fastest through a vacuum, which would be fastest through a gas, and which would be fastest through a solid.
12. **Explain** why the air temperature near the ceiling of a room tends to be warmer than the air temperature near the floor.
13. **Predict** whether plastic foam, which contains pockets of air, would be a good thermal conductor or a good thermal insulator.

Explain Your Thinking

14. **Infer** Several days after a snowfall, the roofs of some homes on a street have almost no snow on them, while the roofs of other homes are still snow-covered. Give one reason, related to home insulation, that might cause this difference.
15. **MATH Connection** Averaged over a year in the central United States, radiation from the Sun transfers about 200 W to each square meter of Earth's surface. If a house is 10 m long by 10 m wide, how much solar energy falls on the house each second?

LEARNSMART® Go online to follow your personalized learning path to review, practice, and reinforce your understanding.

LESSON 3

USING THERMAL ENERGY

FOCUS QUESTION

What do a refrigerator and a car engine have in common?

Heating Systems

Almost everywhere in the United States, air temperatures at some time become cold enough that people seek out sources of warmth in addition to the Sun. As a result, most modern homes and public buildings contain some type of heating system. The best heating system for any building depends on the local climate and how the building is constructed.

The simplest and earliest heating system was probably a campfire. Later, people began burning wood and coal in stoves and furnaces to keep warm. Burning a material transforms chemical energy into thermal energy. Conduction, convection, and radiation then transfer thermal energy from a stove to the surrounding air. One disadvantage of this system is that energy transfers from the room in which the stove is located to other rooms in the building can be slow.

Forced-air systems

The most common type of heating system that people in the United States use today is the forced-air system, as shown in **Figure 14.** In this system, a furnace burns fuel and heats a volume of air. A fan then blows the warm air through a series of large pipes called ducts. The ducts lead to vents in each room. Cool air returns to the furnace through additional vents, and the furnace reheats it.

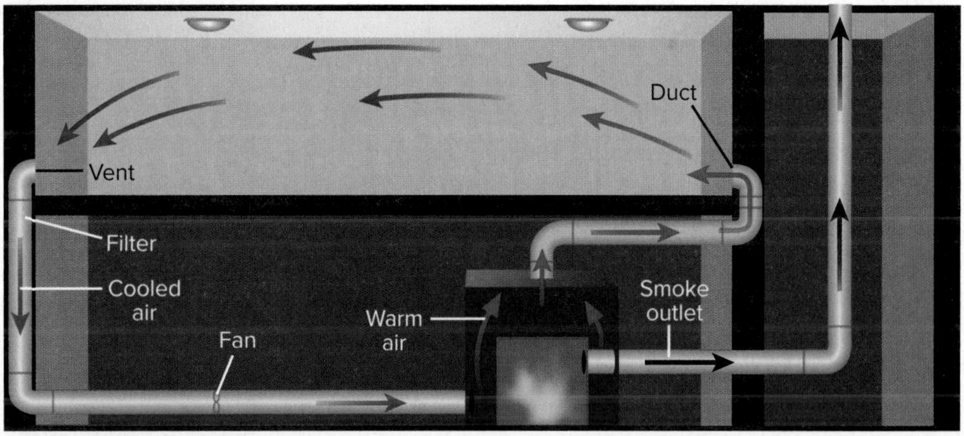

Figure 14 In forced-air systems, air heated by the furnace gets blown through ducts that usually lead to every room.

3D THINKING | DCI Disciplinary Core Ideas | CCC Crosscutting Concepts | SEP Science & Engineering Practices

COLLECT EVIDENCE

Use your Science Journal to record the evidence you collect as you complete the readings and activities in this lesson.

INVESTIGATE

GO ONLINE to find these activities and more resources.

Applying Practices: Modeling Energy at Different Scales

HS-PS3-2. Develop and use models to illustrate that energy at the macroscopic scale can be accounted for as a combination of energy associated with the motion of particles (objects) and energy associated with the relative positions of particles (objects).

Radiator systems

Before forced-air systems were widely used, radiators heated many homes and buildings. A radiator is a closed metal container that contains hot water or steam. In radiator heating systems, a central furnace heats a tank of water. A system of pipes carries the hot water or steam to radiators in other rooms.

Contrary to the name, radiators do not only transfer energy through radiation. Instead, conduction transfers the thermal energy in the hot water or steam to the metal of the radiator and then to the surrounding air. Convection helps spread this energy throughout the room.

Electric heating systems

An electric heating system has no central furnace. Instead, electric heating coils transform electrical energy into thermal energy. Portable space heaters contain such coils. Conduction transfers thermal energy from the heating coils to the surrounding air, and convection distributes thermal energy throughout the room.

Get It?
Identify the energy transformation that occurs in an electric heating system.

Solar heating

The Sun emits an enormous amount of radiant energy that strikes Earth every day. This energy can be used to help heat homes and other buildings through both passive solar heating and active solar heating.

Passive solar heating In passive solar heating systems, materials inside a building absorb radiant energy from the Sun and heat up during the day. At night, when the building begins to cool, thermal energy absorbed by these materials helps keep the rooms warm.

Active solar heating In active solar heating, a solar collector is used. A **solar collector** is a device that transforms radiant energy from the Sun into thermal energy. Radiation from the Sun heats air or water in the solar collector. A pump circulates the hot fluid to radiators in rooms of the house. Both passive solar heating and active solar heating are shown in **Figure 15.**

Get It?
Describe the function of a solar collector.

Passive solar heating system

Active solar heating system

Figure 15 In the passive solar heating system shown, the windows allow the Sun to transfer thermal energy into the room through radiation. Materials, such as wood, absorb and store some of this thermal energy. At night, these materials transfer this stored energy back to the room. In an active solar heating system, water or air is heated in a solar collector and then pumped throughout the building.

Compare and contrast *passive solar heating with active solar heating.*

(t)Jean-Yves Bruel/Photographer's Choice/Getty Images, (b)Bruce Hands/The Image Bank/Getty Images

Thermodynamics

There is another way to increase the thermal energy of an object besides heating it. Have you ever rubbed your hands together to warm them on a cold day? Your hands get warmer and their temperature increases, even though you are not heating them near a fire or a stove. You do work to increase your hands' thermal energy when you rub them together. Thermal energy, heat, and work are related. **Thermodynamics** is the study of the relationships among thermal energy, heat, and work.

Heat and work increase thermal energy

You can warm your hands by placing them near a fire because the fire heats your hands by radiation. If you rub your hands and hold them near a fire, the increase in your hands' thermal energy is even greater. Both the work you do and the heat from the fire increase your hands' thermal energy.

In the example above, your hands can be considered a system. Recall that a system is anything around which you can draw a boundary. A system can be a group of objects, such as a galaxy, a car's engine, or something as simple as a ball. An example of a system is shown in **Figure 16.**

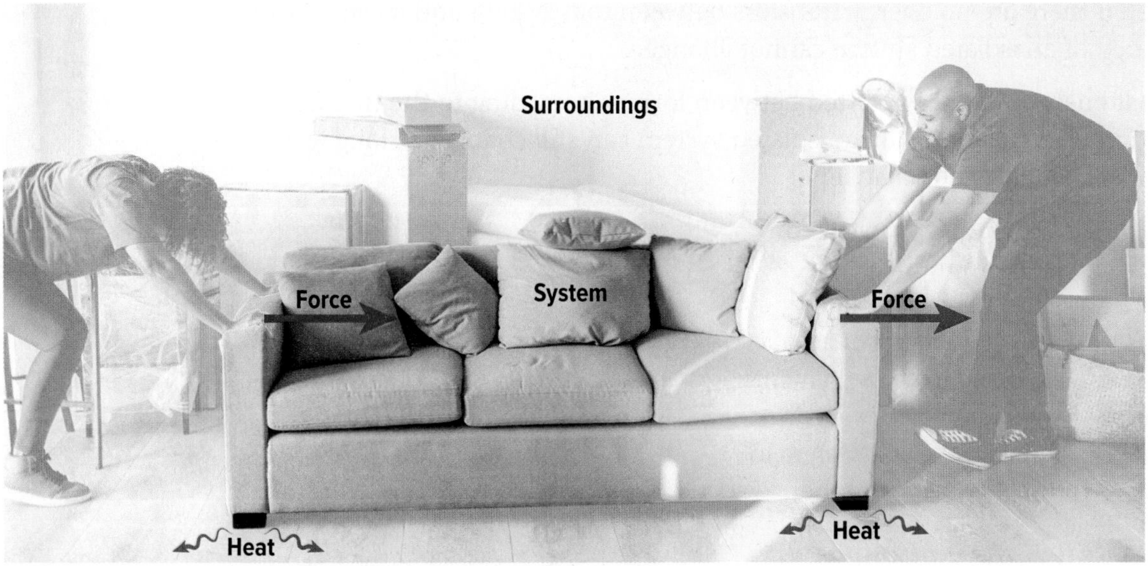

Figure 16 A couch can be a system. If you define the couch in this figure as the system, then everything else is the surroundings. Work is done on this system by pushing and pulling the couch along the floor. As the couch slides along the floor, it heats the floor slightly through friction. The work done on this system is about equal to the heat from this system. The total energy of this system is nearly constant.

Describe *a system in which thermal energy is being transferred into that system.*

CCC CROSSCUTTING CONCEPTS

Systems and System Models Make a short video in which you use a small object such as a block to model the system shown in **Figure 16**. Narrate your video, describing the inputs and outputs of energy.

A system can interact with its surroundings in many ways. You increase the energy of a system whenever you do work on that system or heat that system. The work done on a system is the work done by something outside the system's boundary on something inside the system's boundary. A system can also heat its surroundings or do work on its surroundings. When a system does work on its surroundings or heats its surroundings, the total energy of the system decreases. The total energy of the surroundings increases by the same amount.

The first law of thermodynamics

The **first law of thermodynamics** states that if the mechanical energy of a system is constant, the increase in thermal energy of that system equals the sum of the thermal energy transfers into that system and the work done on that system. This means that there are two ways to increase the temperature of a system. One way is to heat that system. Another way is to do work on that system. For example, you could increase the temperature of your hands both by warming them near a fire and by rubbing them together.

Get It?

Identify two ways to increase the temperature of a system.

Isolated and non-isolated systems

A system is isolated if there are no energy transfers between that system and its surroundings. The total energy of an isolated system cannot change.

However, recall that energy can be converted between forms. According to the first law of thermodynamics, the thermal energy of an isolated system can still change, as long as the total energy of that system does not change.

A system is non-isolated if energy is transferred between the system and its surroundings. For example, a pan on a hot stove is a non-isolated system. This means that the total energy of a non-isolated system can change. However, remember that energy cannot be created or destroyed. Energy can only be transferred from one system to another or converted from one form to another.

The second law of thermodynamics

Energy spontaneously transfers from the warm radiator to the cat in **Figure 17.** Could the reverse ever happen? Could energy spontaneously transfer from the cat to the warmer radiator?

Figure 17 Thermal energy could never spontaneously spread from this cat to the warmer radiator.

WORD ORIGIN

thermodynamics

comes from the Greek word *thermos*, which means *hot*, and the Greek word *dynamis*, which means *power*

The people who design engines must have a good understanding of thermodynamics.

Energy could never spontaneously transfer from the cat to the radiator. This is due to the second law of thermodynamics. The **second law of thermodynamics** states that energy spontaneously spreads from regions of higher concentration to regions of lower concentration. More generally, uncontrolled systems always evolve toward more stable states—that is, toward more uniform energy distribution. It is possible to transfer energy from regions of lower concentration to regions of higher concentration, but this process does not happen spontaneously.

Converting Thermal Energy into Mechanical Energy

If you push a book sitting on a table, the book will slide and then stop. Friction between the book and table converted the book's mechanical energy into thermal energy.

In the example above, mechanical energy was converted completely into thermal energy. In mechanical systems designed to do work, the conversion of mechanical energy into thermal energy in the surrounding environment means that useful energy is being converted into a less useful form. Is it possible to do the reverse and convert thermal energy completely into mechanical energy? Remember that a system's mechanical energy is related to the motion of the objects in that system as well as the interactions between the objects of that system. A system's thermal energy is related to the motions and interactions of all of the particles in that system.

Because there are far more particles than objects in a system, thermal energy is spread among more particles than mechanical energy is spread among objects. As a result, mechanical energy tends to transform into thermal energy. Therefore, it is not possible to completely convert thermal energy into mechanical energy.

Heat engines

A **heat engine** is a device that converts some thermal energy into mechanical energy. A car's engine is an example of a heat engine. A car's engine converts the chemical energy in gasoline into thermal energy. The engine then transforms some of this thermal energy into mechanical energy and rotates the car's wheels, as shown in **Figure 18.** However, only about 25 percent of the thermal energy released by the burning gasoline is converted into mechanical energy.

Figure 18 Burning fuel in the engine's cylinders transforms chemical energy into thermal energy that is then converted into mechanical energy as the pistons move up and down. The crankshaft, transmission, and differential convert the up-and-down motion of the pistons into rotation of the wheels.

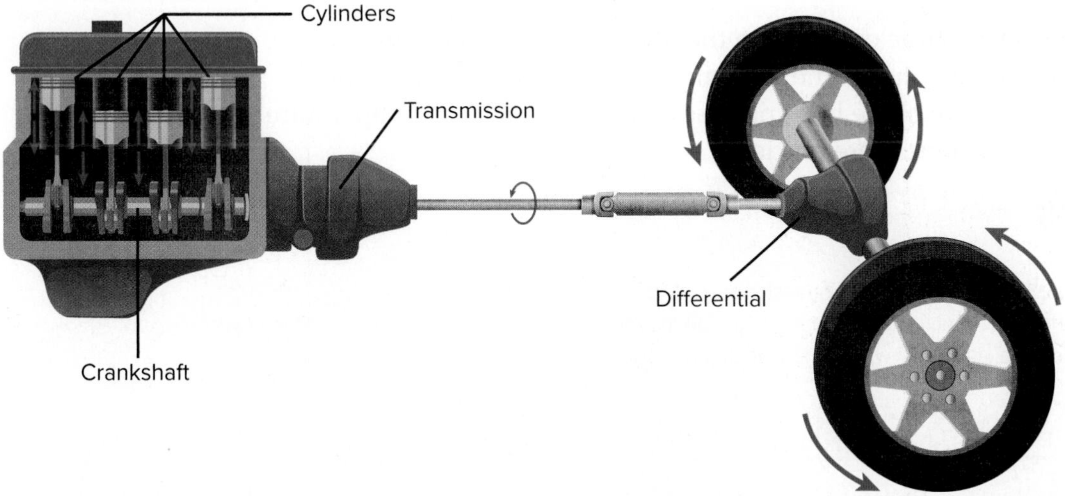

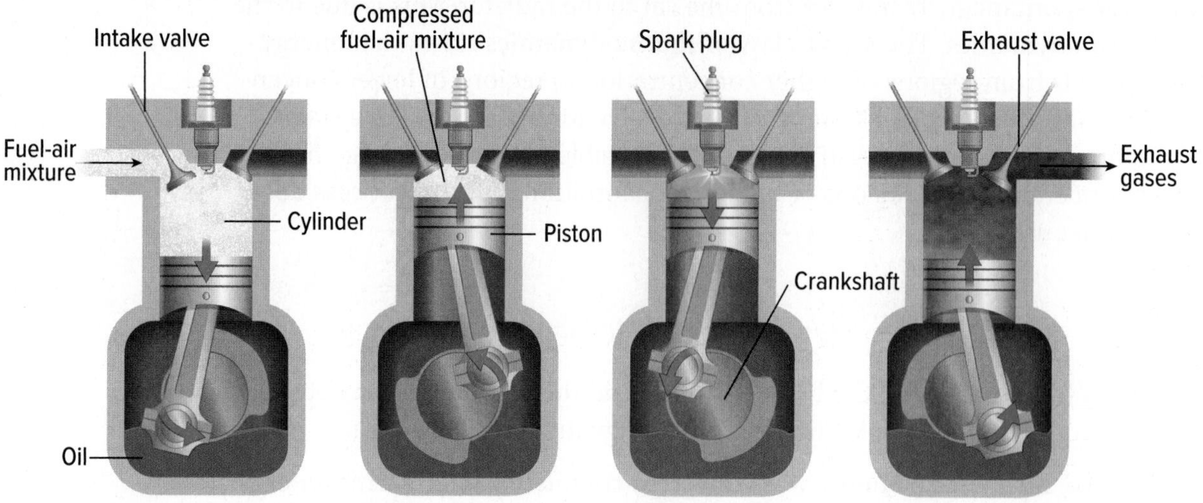
Intake valve
Compressed fuel-air mixture
Spark plug
Exhaust valve
Fuel-air mixture
Exhaust gases
Cylinder
Piston
Crankshaft
Oil
Intake The intake valve opens as the piston moves downward, drawing a mixture of gasoline and air into the cylinder.
Compression The intake valve closes as the piston moves upward, compressing the fuel-air mixture.
Power A spark plug ignites the fuel-air mixture. As the mixture burns, hot gases expand, pushing the piston down.
Exhaust As the piston moves up, the exhaust valve opens, and the hot gases are pushed out of the cylinder.

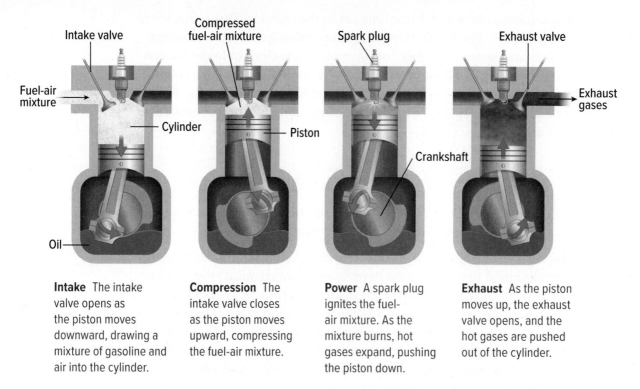

Figure 19 The up-and-down movement of a piston in an automobile engine consists of four separate strokes. These four strokes form a cycle that is repeated many times per second by each piston.

Identify *the stroke during which the spark plug ignites.*

Internal combustion engines Almost all cars are powered by internal combustion engines. An **internal combustion engine** is a heat engine that burns fuel inside a set of cylinders. Each cylinder contains a piston that moves up and down. Each up or down movement of the piston is called a stroke. Automobile and diesel engines have four strokes per cycle. **Figure 19** shows the four-stroke cycle in an automobile engine.

Another type of heat engine is an external combustion engine. The cylinders in external combustion engines are heated by burning fuel outside the cylinders. Old-fashioned steam engines are external combustion engines.

Efficiency of heat engines Approximately three-fourths of the chemical energy that is transformed into thermal energy in a car engine is never converted into mechanical energy. The inefficiency of heat engines is not due only to friction. Even if friction could be completely eliminated, a typical heat engine still would not be 100 percent efficient.

Instead, the efficiency of a heat engine depends on the difference in temperature between the burning gases in the cylinder and the temperature of the air outside the engine. Increasing the temperature of the burning gases and decreasing the temperature of the surroundings make the engine more efficient.

STEM CAREER Connection

Mechanical Engineer

Have you ever thought about how you could make a car go faster or a refrigerator work more efficiently? If so, a career as a mechanical engineer might be for you. Mechanical engineers design, develop, and test a wide variety of engines and other machines.

Doing Work to Transfer Thermal Energy

Is it possible to transfer thermal energy from a cooler area to a warmer area? You may think that the answer is no due to the second law of thermodynamics. However, thermal energy can be transferred from a cooler area to a warmer area if work is done in the process. Refrigerators and air conditioners function based on this principle.

Refrigerators

A refrigerator does work as it transfers thermal energy from inside the cool refrigerator to the warmer room. The energy to do the work comes from electrical energy the refrigerator obtains from an electrical outlet. A refrigerator makes the room that it is in warmer.

Figure 20 shows how a refrigerator operates. Liquid coolant is pumped through an expansion valve and changes into a gas. When the coolant expands as it changes into a gas, it pushes outward on its surroundings. This means that the coolant does work on its surroundings. As a result, the coolant transfers energy to its surroundings and the coolant cools. The cold gas is pumped through pipes inside the refrigerator, where it absorbs thermal energy from the area where you keep your food. When this happens, the inside of the refrigerator cools.

The gas then is pumped to a compressor that does work on the gas by compressing it. This makes the gas warmer than the temperature of the room. The warm gas is pumped through the condenser coils. Because the gas is warmer than the room, thermal energy spreads from the gas to the room. As the gas heats the room, it cools and changes back to a liquid and enters the expansion valve again. The cycle is then repeated.

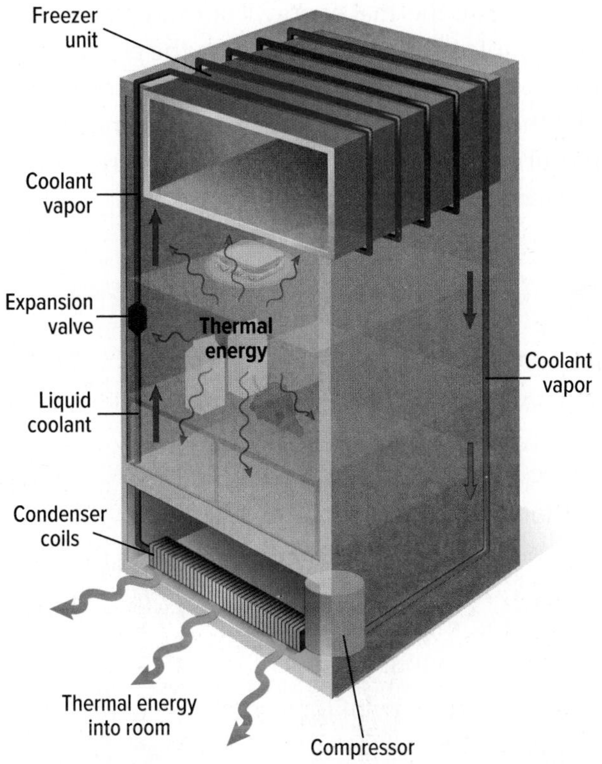

Figure 20 A refrigerator must do work on the coolant in order to transfer thermal energy from inside the refrigerator to the warmer air outside. Work is done when the compressor compresses the coolant vapor, causing its temperature to increase.

Air conditioners

An air conditioner operates like a refrigerator, except that warm air from the room is forced to pass over tubes containing the coolant. Thermal energy is transferred from the warm air to the coolant. The thermal energy that is absorbed by the coolant is then transferred to the air outdoors.

CCC CROSSCUTTING CONCEPTS

Energy and Matter With a partner, examine Figure 20 and read the text about how a refrigerator operates. Create a flow chart or other graphic organizer that describes the movement of energy and matter within a refrigerator.

Heat pumps

A heat pump is a two-way air conditioner. In warm weather, a heat pump operates as an ordinary air conditioner. It does work to transfer thermal energy from the cooler building to the warmer outdoors. This cools the building while warming the outdoors very slightly. In cold weather, a heat pump operates like an air conditioner in reverse. It does work to transfer thermal energy from the cooler outdoors to the warmer building. This warms the building while cooling the building's surroundings very slightly.

Energy transformations and thermal energy

Many energy transformations occur around you every day that convert one form of energy into a more useful form. However, when these energy transformations occur, some of the energy is usually converted into a less useful form: thermal energy. For example, friction converts mechanical energy into thermal energy when the electric generators in **Figure 21** rotate. A laptop computer converts electrical energy into thermal energy. The thermal energy from these energy transformations is no longer in a useful form and is transferred to the surroundings by conduction and convection.

Figure 21 Electric generators transform mechanical energy into electrical energy. However, some energy is always converted into thermal energy during any energy transformation.

Check Your Progress

Summary

- The first law of thermodynamics states that if the mechanical energy of a system is constant, then the increase in thermal energy of that system equals the work done on that system plus the thermal energy transferred into that system.
- The second law of thermodynamics states that energy spontaneously spreads from regions of higher concentration to regions of lower concentration.
- A heat engine transforms thermal energy into mechanical energy.
- Refrigerators, air conditioners, and heat pumps do work to transfer thermal energy from a cooler region to a warmer region.

Demonstrate Understanding

16. **Describe** a device that transforms thermal energy into another useful form.
17. **Explain** how the thermal energy of an isolated system changes with time if the mechanical energy of that system is constant.
18. **Compare and contrast** an active solar heating system with a radiator system.
19. **Predict** whether energy will ever spontaneously transfer from a cold pot of water to a hot stove.
20. **Diagram** how the thermal energy of the coolant changes as the coolant flows through a refrigerator.

Explain Your Thinking

21. **Predict** Suppose you vigorously shake a bottle of fruit juice. Predict how the temperature of the juice will change. Explain your reasoning.
22. **MATH Connection** Suppose you push down on the handle of a bicycle pump with a force of 20 N. The handle moves 0.3 m, and there is no heat between the pump and its surroundings. What is the change in thermal energy of the bicycle pump?

LEARNSMART® Go online to follow your personalized learning path to review, practice, and reinforce your understanding.

"20 Degrees Cooler Inside!"

Imagine the splashy movie theater signs all across the country in the summer months of the 1920s. Many were festooned with pictures of polar bears and false icicles. The idea of cool relief from the sweltering summer heat caught on quickly, and theater owners played up the attraction that air-conditioning held for their hot and weary summertime patrons.

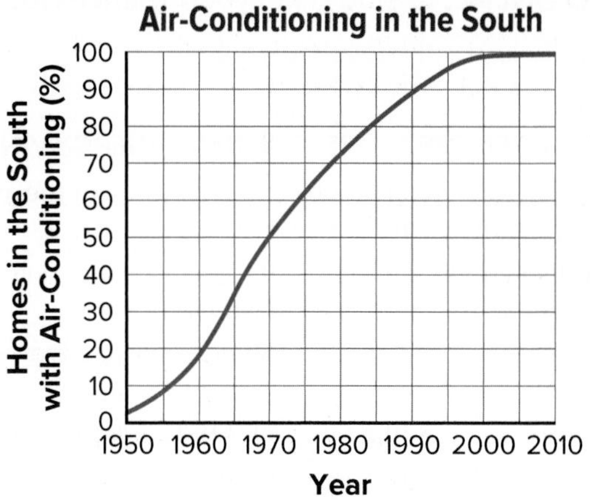

The graph shows how home air-conditioning grew in the South after World War II.

A simple idea

The concept behind air-conditioning is as simple as stepping out of a swimming pool or a shower. When you leave the water, the chill that you feel is not necessarily from cold air. When liquids evaporate, they absorb thermal energy. Thermal energy transfers to the water from your body, through your skin, so you feel cooler.

Air-conditioning systems do not use water. Instead, air conditioners circulate a substance with a very low boiling point. As the substance changes from a liquid to a gas, it absorbs thermal energy and cools the surrounding air. That cool air is pumped throughout the building, whether it is in a movie theater, a shopping mall, or your own home.

The modern air conditioner was invented in 1902 by Willis Carrier. Carrier designed his air conditioner to cool machinery at a printing plant.

Theaters caught on to the air-conditioning trend in the 1920s, but it was not until the 1950s that home air-conditioning caught on with America's middle class. According to some historians, this change altered architecture as well as where and how people live.

Architecture and air-conditioning

Air-conditioning had the greatest impact in areas with warmer climates. Over time, more and more houses in these regions were built with air-conditioning in mind. As a result, interior rooms got bigger, and front porches began to disappear.

Where people live

The existence of air-conditioning also encouraged people to move to warmer climates. Between 1910 and 1950, the South lost more than 10 million people due to migration to cooler climates. That trend reversed after 1950, as in-home air-conditioning became more common. In the 1970s alone, over 3 million more people moved to the South as moved away. The graph above shows how the percentage of homes in the South using air-conditioning changed with time.

DEVELOP A MODEL TO ILLUSTRATE

Research ways architects and engineers increase the efficiency of climate control systems. Focus on building design, materials used, and advances in climate-control technology. Develop a simple model of a new home designed to have an efficient climate-control system.

MODULE 5
STUDY GUIDE

GO ONLINE to study with your Science Notebook.

Lesson 1 TEMPERATURE, THERMAL ENERGY, AND HEAT

- The temperature of an object is a measure of the average kinetic energy of the particles that make up the object.
- Thermal energy is the sum of the kinetic and potential energies of all of the particles in an object.
- Heat is a transfer of thermal energy due to a temperature difference.
- The specific heat of a material is the amount of heat needed to raise the temperature of 1 kg of the material by 1°C.

- temperature
- thermal energy
- heat
- specific heat

Lesson 2 CONDUCTION, CONVECTION, AND RADIATION

- Conduction is the transfer of thermal energy by collisions between more energetic and less energetic particles.
- Convection is the transfer of thermal energy by the movement of warmer and cooler materials.
- Radiation is the transfer of energy by electromagnetic waves.
- Thermal insulators are used to reduce the transfer of thermal energy from one place to another.

- conduction
- convection
- radiation
- thermal insulator

Lesson 3 USING THERMAL ENERGY

- The first law of thermodynamics states that if the mechanical energy of a system is constant, then the increase in thermal energy of that system equals the work done on that system plus the thermal energy transferred into that system.
- The second law of thermodynamics states that energy spontaneously spreads from regions of higher concentration to regions of lower concentration.
- A heat engine transforms thermal energy into mechanical energy.
- Refrigerators, air conditioners, and heat pumps do work to transfer thermal energy from a cooler region to a warmer region.

- solar collector
- thermodynamics
- first law of thermodynamics
- second law of thermodynamics
- heat engine
- internal combustion engine

THREE-DIMENSIONAL THINKING

Module Wrap-Up

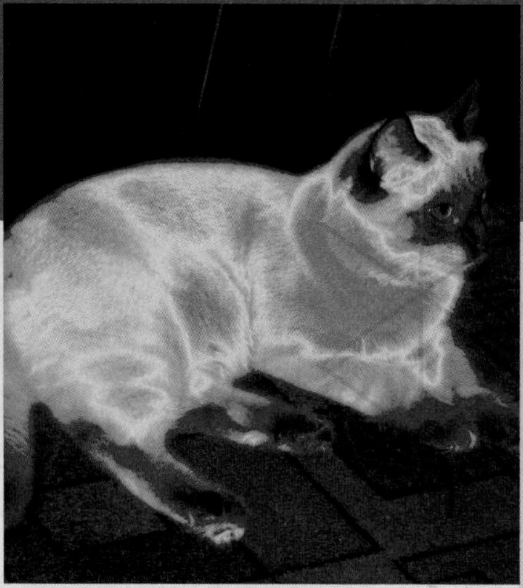

REVISIT THE PHENOMENON

Why would it be useful to visualize thermal energy?

CER Claim, Evidence, Reasoning

Explain Your Reasoning Revisit the claim you made when you encountered the phenomenon. Summarize the evidence you gathered from your investigations and research and finalize your Summary Table. Does your evidence support your claim? If not, revise your claim. Explain why your evidence supports your claim.

STEM UNIT PROJECT

Now that you've completed the module, revisit your STEM unit project. You will summarize your evidence and apply it to the project.

GO FURTHER

SEP Data Analysis Lab

Home Heating

Analyze the heating system in your home. Find someone who knows the heating system well. Maybe a parent or a brother or sister could help you. Other sources of help are a heating and cooling service person, representative of a utility company, and literature on energy conservation measures. Once you have found someone, work with your source to answer the following questions.

CER Analyze and Interpret Data

1. **Identify** What type of heating system does your home have?
2. **Identify** What type of fuel is used to heat your home?
3. **Identify** How much does it cost to heat your home per year?
4. **Claim, Reasoning** What kind of low-cost actions could you take to help conserve thermal energy in your home?
5. **Identify** Suppose you had to replace your current heating system. What options are available to you?
6. **Claim, Reasoning** Suppose you already have a highly efficient heating system. What kind of major improvements could make your home even more heating efficient?
7. **Evidence, Reasoning** Your heating system should have maintenance done on an annual basis. Why is this important?

PandaWild/Shutterstock

Stefan Sauer/dpa/AFP/Getty Images

MODULE 6

ELECTRICITY

ENCOUNTER THE PHENOMENON

How can you safely conduct electricity underwater?

GO ONLINE to play a video about the first trans-Atlantic cable laid in 1858.

SEP Ask Questions

Do you have other questions about the phenomenon? If so, add them to the driving question board.

CER Claim, Evidence, Reasoning

Make Your Claim Use your CER chart to make a claim about how you can safely conduct electricity under water. Explain your reasoning.

Collect Evidence Use the lessons in this module to collect evidence to support your claim. Record your evidence as you move through the module.

Explain Your Reasoning You will revisit your claim and explain your reasoning at the end of the module.

GO ONLINE to access your CER chart and explore resources that can help you collect evidence.

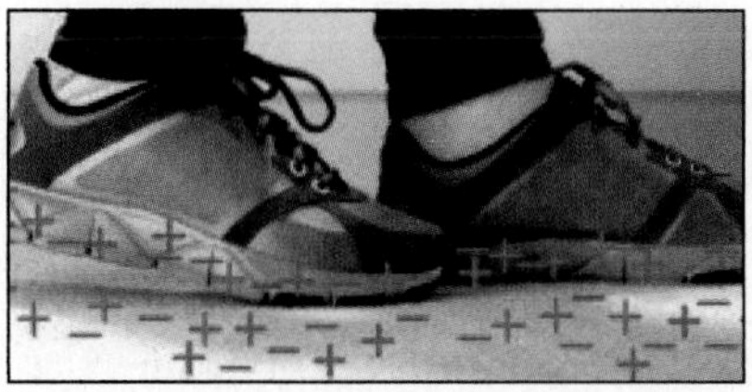

LESSON 2: Explore & Explain: Positive and Negative Charges

LESSON 2: Explore & Explain: Resistance

Smithsonian

Additional Resources

(t)Video Supplied by BBC Worldwide Learning, (bl)Geoff Butler/McGraw-Hill Education, (br)Arthur S Aubry/Stockbyte/Getty Images

LESSON 1
ELECTRIC CHARGE

FOCUS QUESTION

What is the difference between charging by contact and charging by induction?

Positive and Negative Charges

Matter is composed of atoms. These atoms are composed of protons, neutrons, and electrons. Protons have positive electric charge, and electrons have negative electric charge. Neutrons have no electric charge.

The amount of positive charge on a proton equals the amount of negative charge on an electron. An atom has equal numbers of protons and electrons, so the positive and negative charges cancel out, and an atom has no net electric charge. Objects with no net charge are said to be electrically neutral. The amount of electric charge is measured in coulombs (C). There are 6250 million billion protons in 1 C of electric charge and 6250 million billion electrons in −1 C of electric charge.

Transferring charge

Electrons are bound more tightly to some atoms and molecules. For example, compared to the electrons in the atoms that make up a carpet, electrons are bound more tightly to the atoms that make up the soles of your shoes. **Figure 1** shows that when you walk on the carpet, electrons transfer from the carpet to the soles of your shoes.

Before the shoes scuff against the carpet, both the soles of the shoes and the carpet are electrically neutral.

As the shoes scuff against the carpet, electrons transfer from the carpet to the soles of the shoes.

Figure 1 Particles in the shoes' soles hold electrons more tightly than particles in the carpet hold electrons.
Compare *the total number of charges in the left panel with the total number of charges in the right panel.*

3D THINKING | DCI Disciplinary Core Ideas | CCC Crosscutting Concepts | SEP Science & Engineering Practices

COLLECT EVIDENCE

Use your Science Journal to record the evidence you collect as you complete the readings and activities in this lesson.

INVESTIGATE

GO ONLINE to find these activities and more resources.

Applying Practices: Gravitational and Electrostatic Forces
HS-PS2-4. Use mathematical representations of Newton's Law of Gravitation and Coulomb's Law to describe and predict the gravitational and electrostatic forces between objects.

As a result, the soles of your shoes gain an excess of electrons and become negatively charged. The carpet has lost electrons and has an excess of positive charge. **Static electricity** is the accumulation of excess electric charge on an object.

Conservation of charge

When your shoe soles become charged, they illustrate the law of conservation of charge. The **law of conservation of charge** states that charge can be transferred from object to object, but it cannot be created or destroyed. In **Figure 1,** electrons are transferred from the carpet to the shoes. However, the total charge does not change. Usually, it is electrons, not protons, that transfer from one object to another.

Get It?
Compare the number of electrons and the number of protons in a charged object.

Charges Exert Forces

Have you noticed how clothes sometimes cling together when removed from the dryer? The clothes cling together because of the forces that electric charges exert on each other.

As laundry tumbles in a dryer, electrons are transferred from some objects onto other objects. The atoms that make up some clothes gain electrons and become negatively charged. The atoms that make up other clothes lose electrons and become positively charged. Items that are oppositely charged attract each other and stick together, as shown in **Figure 2.**

Figure 3 shows that unlike charges attract each other, and like charges repel each other. Charged particles do not attract or repel particles with no charge, such as neutrons, through the electric force. The force between electric charges depends on the amount of charge as well as on the distance between charges. The force increases as the amount of charge increases and as the charges get closer together.

Coulomb's Law

In the late 1770's Charles Coulomb investigated the force between electric charges. He found that the force depends on the amount of charge as well as on the distance between the charges. The force between charged particles or objects decreases as the objects get farther apart. The force increases as the amount of charge on either object increases.

For particles that are very small (like electrons), this relationship can be written as the following equation. It is known as Coulomb's law.

$$F = \frac{kq_1q_2}{d^2}$$

In this equation, k is Coulomb's constant and d is the distance between the centers of the two charges q_1 and q_2. The value of k is approximately 9.0×10^9 N·m²/C².

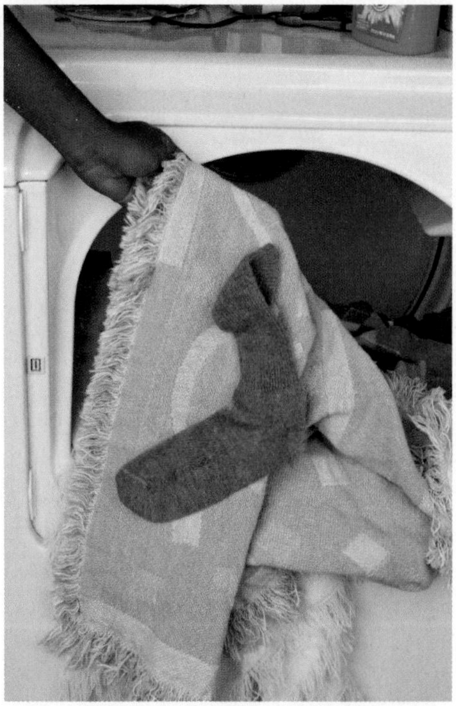

Figure 2 The items in this dryer are oppositely charged, causing them to stick together.

Predict *what would happen if the items in the dryer were identically charged.*

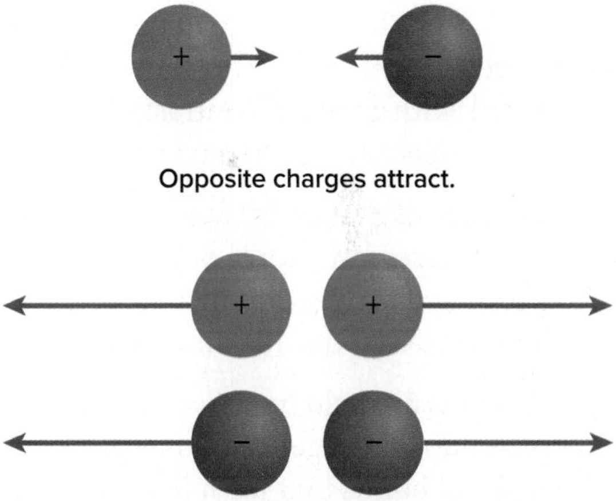
+
−
Opposite charges attract.
+
+
−
−
Like charges repel.

Figure 3 Positive charges and negative charges exert forces on each other.

Describe *what would happen if two protons were placed near each other.*

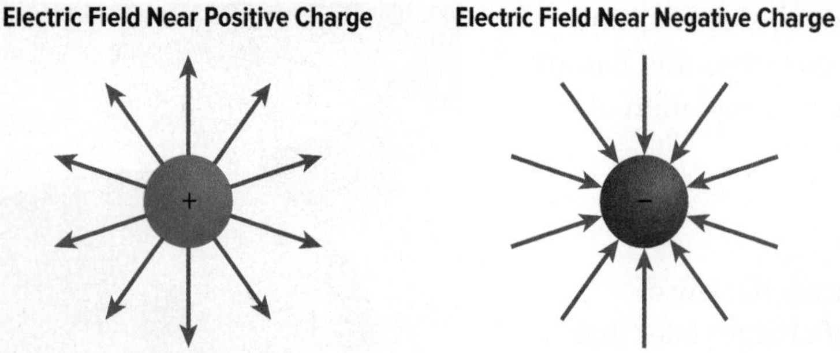

Figure 4 Surrounding every electric charge is an electric field that exerts forces on other electric charges. The arrows represent electric fields near positive and negative charges. They point in the direction a positive charge would move.

Electric fields

Electric charges do not need to be touching to exert forces on each other. An **electric field** surrounds every electric charge and exerts the force that causes other electric charges to be attracted or repelled. Any charge that is placed in an electric field will be pushed or pulled by the field. Electric fields are usually represented by arrows that indicate how the electric field would make a positive charge move, as shown in **Figure 4.** Electric field arrows point away from positive charges and toward negative charges.

Get It?

Explain why the arrows point away from the positive charge and toward the negative charge in **Figure 4.**

Comparing electric forces and gravitational forces

You might have noticed similarities between the electric force and the gravitational force. Both forces are non-contact forces that act through fields. Each force also has associated law stating that the force depends on the size of the charge or mass, the distance between them, and a constant. However, gravity is always attractive and acts between objects with mass. In contrast, the electric force can be repulsive or attractive and acts only on electrically charged objects. The forces also differ in strength.

The force of gravity between you and Earth seems to be strong. Yet, compared with electric forces, the force of gravity is very weak. For example, the electric force between a proton and an electron in a hydrogen atom is about a thousand trillion trillion trillion times larger (about 10^{39} times larger) than the gravitational force between them.

All atoms are held together by electric forces between protons and electrons that are tremendously larger than the gravitational forces between the same particles. The chemical bonds that form between atoms in molecules are also due to the electric forces between the atoms. These electric forces are much larger than the gravitational forces between the atoms.

Get It?

Compare the strength of electric forces between protons and electrons to the strength of gravitational forces between them.

However, the electric forces between the objects around you are much less than the gravitational forces between them. Most objects that you see are nearly electrically neutral and have almost no net electric charge. There is usually no noticeable electric force between these objects. But even if a small amount of charge is transferred between objects, the electric force between those objects can be noticeable.

Conductors and Insulators

If you reach for a metal doorknob after walking across a carpet, you might see a spark. The spark is caused by electrons moving from your hand to the doorknob, as shown in **Figure 5.** Recall that electrons were transferred from the carpet to your shoes. How did these electrons move from your shoes to your hand?

Conductors

A **conductor** is a material through which electrons move easily. Electrons on your shoes repel each other, and some are pushed onto your skin. Because your skin is an effective conductor, the electrons spread over your skin, including your hand.

The best electrical conductors are metals. The atoms that make up metals have electrons that are able to move easily through the material. In **Figure 5,** the electrons move from your hand to the doorknob because the doorknob is made of metal. What would happen if the doorknob were made of glass?

Figure 5 As you walk across a carpeted floor, excess electrons can accumulate on your body. When you reach for a metal doorknob, electrons flow from your hand to the doorknob, and you feel a shock. If the lights are out, you might even see a spark.

Insulators

Glass is an insulator. An **insulator** is a material in which electrons are not able to move easily. Electrons are held tightly to the atoms that make up insulators. Most plastics are insulators. The plastic coating around electric wires, such as the one shown in **Figure 6,** prevents dangerous electric shocks. If the doorknob in **Figure 5** were made of glass, you would not experience an electric shock from moving charges.

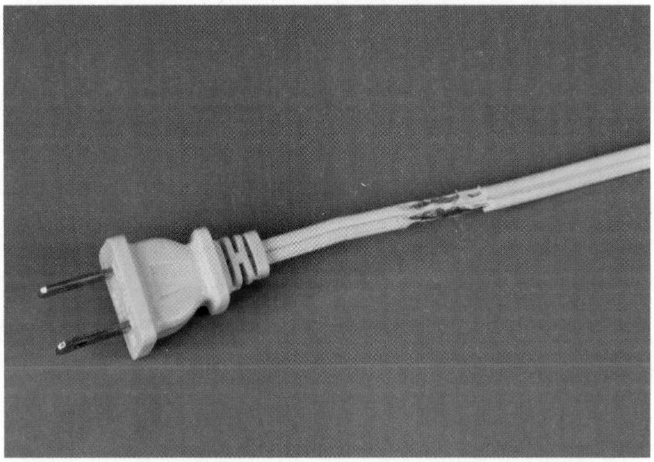

Figure 6 The plastic coating around wires is an insulator. An electrical cord is hazardous when the conducting wire is exposed.

Get It?
Contrast conductors and insulators.

SCIENCE USAGE v. COMMON USAGE

conductor

Science usage: a material in which electrons are able to move easily

Copper is used in wiring because it is a good conductor.

Common usage: the leader of a musical ensemble

The conductor bowed to the audience after the orchestra's performance.

Figure 7 The balloon on the left is neutral. The balloon on the right is negatively charged and produces a positively charged area on the sleeve by repelling electrons.

Determine *the direction of the electric force acting on the balloon on the right.*

Charging by Contact

Every example of charging that has been discussed so far is an example of charging by contact. **Charging by contact** is the process of transferring charge by touching or rubbing. Rubbing two materials together can result in a transfer of electrons. One material is left with a positive charge, and the other is left with an equal amount of negative charge.

Charging by Induction

Because electric forces act at a distance, charged objects brought near a neutral object will cause electrons to rearrange their positions on the neutral object. Suppose you charge a balloon by rubbing it with a cloth. If you bring the negatively charged balloon near your sleeve, the extra electrons on the balloon repel the electrons in the sleeve.

The electrons near the sleeve's surface move away from the balloon, leaving a positively charged area on the surface of the sleeve, as shown in **Figure 7.** When a nearby charged object causes the rearrangement of electrons on a neutral object, it is called **charging by induction.**

Get It?

Contrast charging by contact with charging by induction.

Lightning

How is getting shocked when you touch a metal doorknob similar to lightning? Both are static discharges. A static discharge is a transfer of charge between two objects because of a buildup of static electricity.

A thundercloud is a mighty static electricity generator. As air masses move and swirl in the cloud, areas of positive and negative charge build up. Eventually, enough charge builds up to cause a static discharge between the cloud and the ground. As the electric charges move through air, they collide with atoms and molecules. These collisions cause the atoms and molecules in the air to emit light. You see this light as a flash of lightning, as shown in **Figure 8** on the next page.

Get It?

Describe What is lightning?

Figure 8 Visualizing Lightning

When humid, sun-warmed air rises to meet a colder air layer, storm clouds can form. Lightning strikes occur when positive charges at the ground attract negative charges at the bottom of a storm cloud.

Through processes not yet well understood, many of the particles in a storm cloud become charged. Convection currents in the storm cloud cause charge separation. The bottom of the cloud becomes negatively charged, and the top of the cloud becomes positively charged.

The negative charges at the bottom of the cloud repel the negative charges in the ground, tree, and house, inducing a positive charge on the ground below the cloud. A person standing beneath the cloud might feel a tingling sensation at this point.

The positive charges at the ground attract the negative charges in the cloud. The negative charges in the cloud move toward the ground once the bottom of the cloud has accumulated enough negative charges.

As the negative charges approach the ground, positively charged ions surge upward and meet those negative charges. The resulting connection is the spark that you see as a lightning flash.

Thunder

Not only does lightning produce a brilliant flash of light, it also generates powerful sound waves. The electrical energy in a lightning bolt rips electrons off atoms in the atmosphere and produces great amounts of heat.

The surrounding air temperature can rise to about 30,000°C, several times hotter than the Sun's surface. The heat causes air in the lightning bolt's path to expand rapidly, producing sound waves that you hear as thunder.

Grounding

Lightning strikes can cause power outages, fires, injury, and loss of life. The sensitive electronics in a computer can be harmed by large static discharges. A discharge can occur any time that charge builds up in one area.

Providing a path for charge to reach Earth prevents any charge from building up. Earth is a large, neutral object that is also a conductor of charge. Any object connected to Earth by a good conductor will transfer any excess electric charge to Earth. Connecting an object to Earth with a conductor is called grounding.

Get It?
Explain the purpose of grounding.

Buildings often have a metal lightning rod that provides a conducting path from the highest point on the building to the ground to prevent damage by lightning, as shown in **Figure 9.** Plumbing fixtures, such as metal faucets, sinks, and pipes, often provide a convenient ground connection. Look around. Do you see anything that might act as a path to the ground?

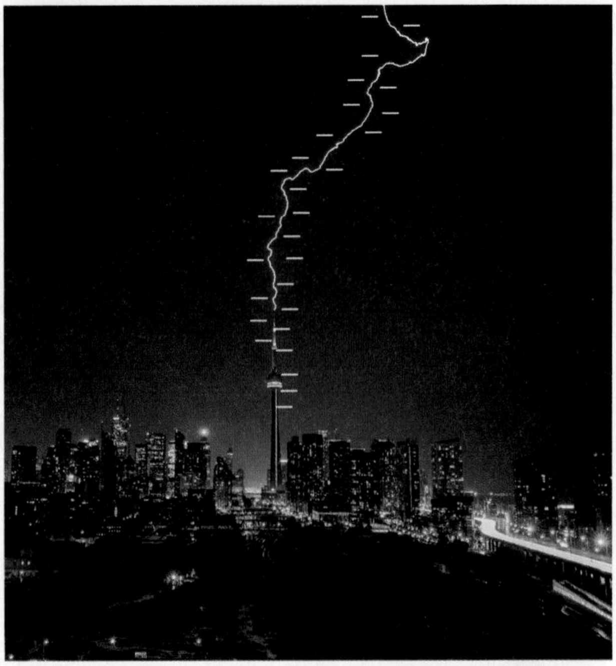

Figure 9 Due to its shape, size, and position, a lightning rod attracts electric charges from storm clouds. A lightning rod guides the charge from a lightning bolt safely to the ground.

Electroscopes

An **electroscope** is a device that can detect electric charge. One kind of electroscope is shown in **Figure 10.** This electroscope is made of two thin, metal leaves attached to a metal rod with a metal knob at the top. The leaves are allowed to hang freely from the metal rod. When the device is not charged, the leaves hang straight down, as shown in **Figure 10.** When the device is charged, electric forces push the leaves apart so they are separated.

Get It?
Identify the purpose of an electroscope.

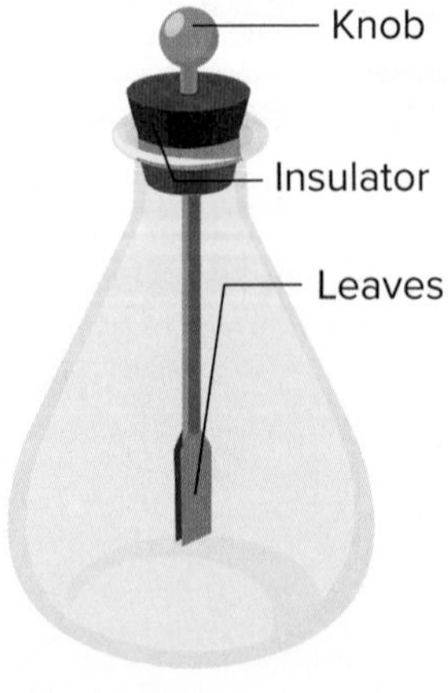

Figure 10 The electroscope's metal leaves hang freely below. This indicates that the electroscope is not charged.

Charging an electroscope

Suppose a negatively charged object touches the knob. Because the metal is a good conductor, electrons travel down the rod and into the leaves. Both leaves become negatively charged as they gain electrons, as shown in the middle of **Figure 11.** Because the leaves have similar charges, they repel each other.

When a positively charged object touches the knob of an uncharged electroscope, electrons flow out of the metal leaves and onto the rod. Each leaf then becomes positively charged, as shown on the right side of **Figure 11.** Once again, the leaves repel each other.

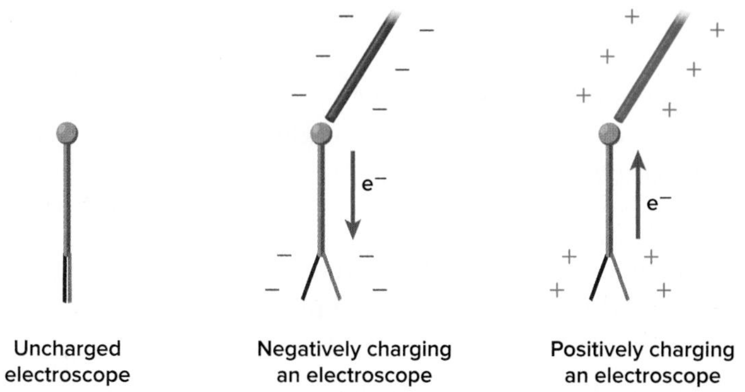

Figure 11 Notice the position of the leaves on the electroscope when they are negatively charged or positively charged.

Predict *Imagine bringing a negatively charged rod close to a positively charged electroscope. In what way will the leaves of the electroscope move?*

Check Your Progress

Summary

- There are two types of electric charge: positive charge and negative charge.
- Electric charge can be transferred between objects but cannot be created or destroyed.
- An electric charge is surrounded by an electric field that exerts forces on other charges.
- Charging by contact is the transfer of charge between two objects that are touching.
- Charging by induction is the rearrangement of charge in an object due to the influence of another charged object nearby.

Demonstrate Understanding

1. **Predict** what would happen if you touched the knob of a positively charged electroscope with a negatively charged object. Explain your prediction.
2. **Compare and contrast** electric force with gravitational force.
3. **Differentiate** between conductors and insulators.
4. **Explain** how electrically neutral objects can become charged even though charge cannot be created or destroyed.

Explain Your Thinking

5. **Infer** Humid air is a better electrical conductor than dry air. Explain why you are more likely to receive a shock after walking across a carpet when the air is dry than when the air is humid.
6. **MATH Connection** A 0.020-kg balloon is charged by rubbing and then stuck to the ceiling. If the strength of Earth's gravity on an object is 9.8 N/kg, what is the electric force on the balloon?

LEARNSMART® Go online to follow your personalized learning path to review, practice, and reinforce your understanding.

LESSON 2

ELECTRIC CURRENT

FOCUS QUESTION

How can the flow of electricity be compared to that of water?

Current and Voltage Difference

Electrical devices, such as stereos, lights, and toasters, work when there is an electric current through them. **Electric current** is the net movement of electric charges in a single direction. Electric current is measured in amperes (A). One ampere is equal to one coulomb of electric charge flowing past a point every second.

In a metal wire without an electric current, electrons are in constant motion in all directions. As a result, there is no net movement of electrons in one direction. However, when there is an electric current in the wire, electrons continue their random movements, but they also drift in the direction of the current. The movement of an electron in an electric current is similar to a ball bouncing down a flight of stairs. Even though the ball changes direction when it strikes a stair, the net motion of the ball is downward.

Voltage difference

In some ways, the electric force that causes charges to flow is similar to the force acting on the water in a pipe. Water flows from higher pressure to lower pressure, as shown in **Figure 12.** In a similar way, electric current is from higher voltage to lower voltage. A **voltage difference** is related to the force that causes electric charges to flow. Voltage difference is measured in volts (V).

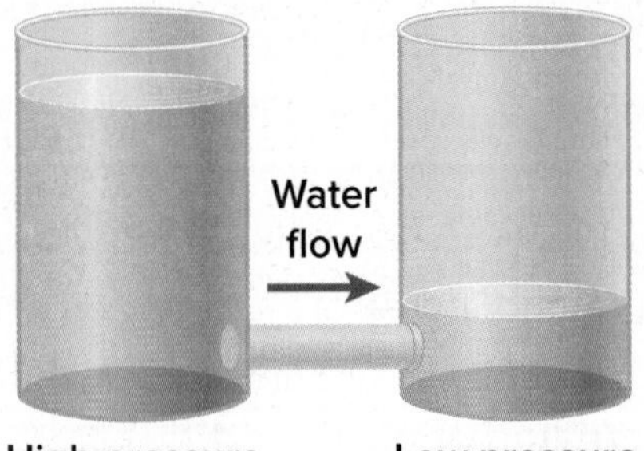

The force that causes water to flow is related to a pressure difference.

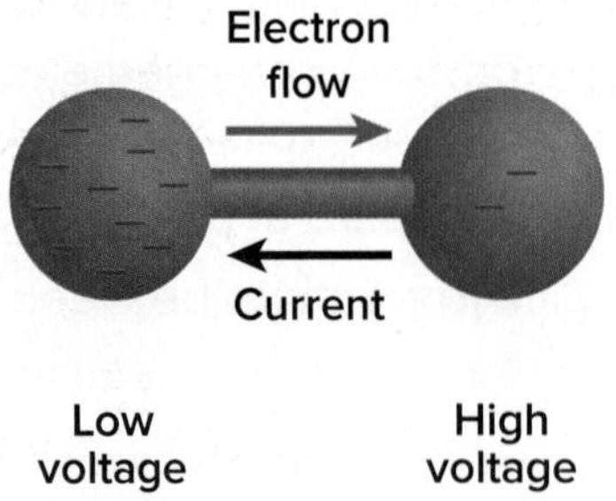

The force that causes a current is related to a voltage difference.

Figure 12 A voltage difference in a conductor produces an electric current, just as a pressure difference in the pipe produces a water current. Water flows from higher-pressure regions to lower-pressure regions. Electric current is from higher-voltage regions to lower-voltage regions. However, as you will read on the next page, electron flow is actually from lower voltage to higher voltage.

Compare *an electric current and a water current.*

3D THINKING DCI Disciplinary Core Ideas CCC Crosscutting Concepts SEP Science & Engineering Practices

COLLECT EVIDENCE

Use your Science Journal to record the evidence you collect as you complete the readings and activities in this lesson.

INVESTIGATE

GO ONLINE to find these activities and more resources.

Laboratory: Wet Cell Battery

Develop and use a model wet cell battery to determine how the voltage difference of the cell depends on the electrode materials.

Quick Investigation: Investigate Battery Addition

Carry out an investigation to measure the effect of adding lightbulbs to a circuit and to infer a relationship between voltage difference and current.

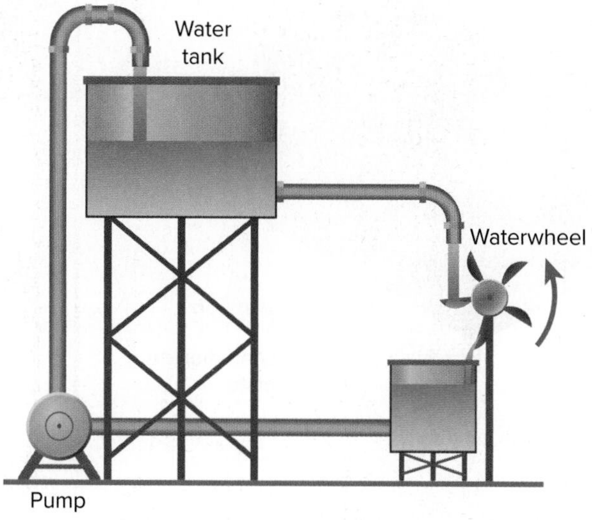

A pump provides the pressure difference that keeps water flowing.

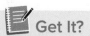

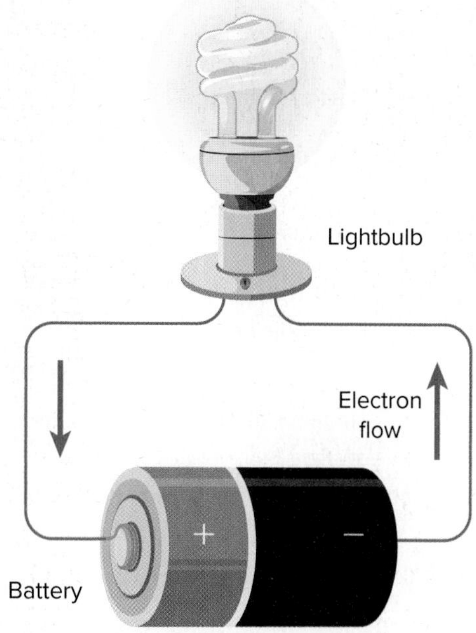

A battery provides the voltage difference that keeps electric charge flowing.

Figure 13 Water or electric charge will flow continually only through a closed loop. If any part of the loop is broken or disconnected, the flow stops.

Electric circuits

A water circuit is shown in **Figure 13.** Water flows out of the tank and falls on a waterwheel, causing it to rotate. A pump then lifts the water back up into the tank. The constant flow of water would stop if the pump stopped working. The flow of water would also stop if one of the pipes broke. Then water could no longer flow in a closed loop, and the waterwheel would stop rotating.

Figure 13 also shows an electric circuit. Just as the water current stops if there is no longer a closed loop, the electric current stops if there is no longer a closed path to follow. An **electric circuit** is a closed path that electric current follows. If the battery, the lightbulb, or one of the wires is removed from the path in **Figure 13,** the electric circuit is broken, and there will be no current.

Current and electron flow

For historical reasons, we think of current as being in the direction that positive charge flows. However, in almost all circuits, positive charge does not flow. Instead, it is the negatively charged electrons that actually move through the circuit. Because current is in the direction of positive charge flow, the flow of electrons and the current are in opposite directions. The direction of the electric current is always from higher voltage to lower voltage, but the electrons in a circuit actually flow from lower voltage to higher voltage.

Get It?

Contrast the direction of current with the direction of electron flow.

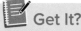

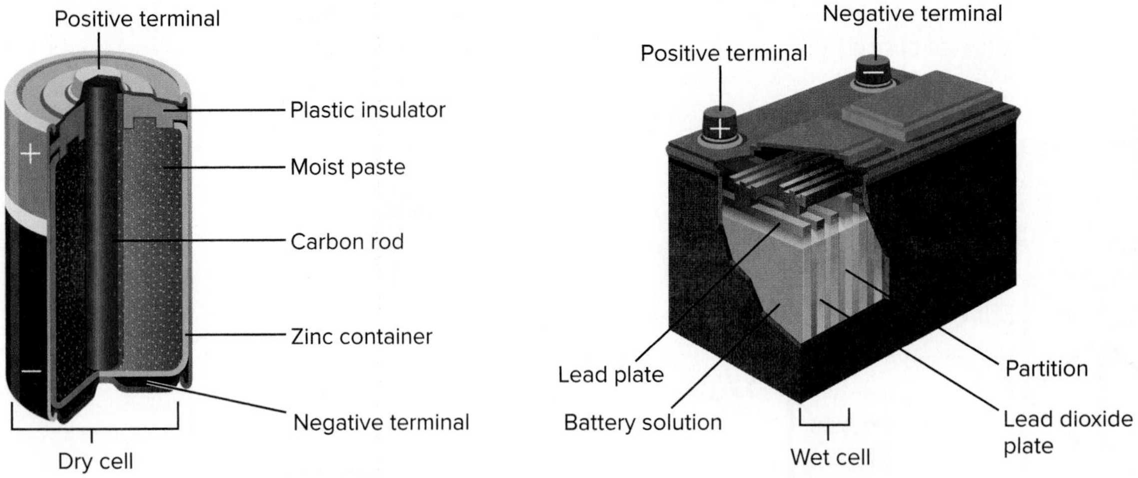

Figure 14 Chemical reactions in batteries produce a voltage difference between the positive and negative terminals.

Infer *when these chemical reactions start and stop.*

Batteries

In order to keep water flowing continually in a water circuit, a pump is used to provide a pressure difference. In a similar way, to keep electric charges continually flowing in an electric circuit, a voltage difference needs to be maintained in the circuit. Power supplies, such as the batteries shown in **Figure 14,** can provide this voltage difference.

Dry-cell batteries

You are probably most familiar with dry-cell batteries. The cylindrical batteries that are placed in flashlights are dry-cell batteries. A cell consists of two electrodes surrounded by a material called an electrolyte. The electrolyte enables charges to move from one electrode to the other.

Look at the dry cell shown in **Figure 14.** One electrode is the carbon rod, and the other is the zinc container. The electrolyte is a moist paste containing several chemicals. When the two terminals of a dry-cell battery are connected in a circuit, a chemical reaction occurs. Electrons are transferred between some of the compounds in this chemical reaction.

As a result, the carbon rod becomes positive, forming the positive terminal. Electrons accumulate on the zinc, making it the negative terminal. The voltage difference between these two terminals causes a current through a closed circuit.

Wet-cell batteries

A wet-cell battery is another common type of battery. A wet cell, like the one shown in **Figure 14,** contains two connected plates made of different metals or metallic compounds in an electrolyte. Unlike in the dry cell, the electrolyte in a wet cell is a conducting liquid solution. A wet-cell battery contains several wet cells connected together. The most common type of wet-cell battery in use today is a car battery.

Get It?

Contrast a dry-cell battery with a wet-cell battery.

Lead-acid batteries Most car batteries are lead-acid batteries, like the wet-cell battery in **Figure 14.** A typical lead-acid battery contains a series of six wet cells composed of lead and lead dioxide plates in a sulfuric acid solution. The chemical reaction in each cell provides a voltage difference of about 2 V, giving a total voltage difference of 12 V.

Electrical outlets

A voltage difference is also provided at electrical outlets, such as a wall socket. This voltage difference is usually higher than the voltage difference provided by batteries. Most types of household devices are designed to use the voltage difference supplied by a wall socket. In the United States, the average voltage difference across the two holes in a wall socket is usually 120 V. Certain devices, such as clothes dryers, plug into special 240-V electrical outlets. In other areas of the world, the average voltage difference across the two holes of an electric outlet may be 110 V, 240 V, or another value.

Resistance

If you look inside a computer, a radio, or a telephone, you might find striped objects like those in **Figure 15.** These objects are called resistors, and they are designed to resist the flow of electrons. Electrical engineers use resistors to reduce the current through all or part of a circuit. Resistors help protect more delicate electronic components. These components can melt or break if too much current is sent through them.

Current and resistance

When electrons move through a resistor, some energy is transferred to that resistor. **Resistance,** which is measured in ohms (Ω), is the tendency for a material to resist the flow of electrons and to convert electrical energy into other forms of energy, such as thermal energy. This is why mobile phones can get hot during a long phone call. Almost all materials have some resistance. For example, even copper wires have some resistance. The resistances of long copper wires can noticeably affect the currents through a circuit.

Figure 15 The striped objects in this photograph are resistors. The different-colored stripes indicate the resistance of each resistor. Resistors are designed to reduce the current through all or part of a circuit. In sensitive electronic devices, it is important to prevent the current from being too high.

Explain *why desktop computers have resistors.*

Ohm's Law

The voltage difference, current, and resistance in a circuit are related. The relationship among voltage difference, current, and resistance in a circuit is known as Ohm's law. If I stands for electric current, Ohm's law can be written as the following equation.

Ohm's Law

$$\text{current (amperes)} = \frac{\text{voltage difference (volts)}}{\text{resistance (ohms)}}$$

$$I = \frac{V}{R}$$

According to **Ohm's law,** the current in a circuit equals the voltage difference divided by the resistance. If voltage is measured in volts (V) and resistance is measured in ohms (Ω), then current is measured in amperes (A).

EXAMPLE Problem 1

SOLVE FOR CURRENT The voltage difference in a graphing calculator is 6 V, and the resistance is 1200 Ω. What is the current through the batteries of the graphing calculator?

Identify the Unknown: current: I

List the Knowns: voltage difference: $V = 6$ V resistance: $R = 1200\ \Omega$

Set Up the Problem:

$$I = \frac{V}{R}$$

$$I = \frac{6\ \text{V}}{1200\ \Omega}$$

Solve the Problem:

$$I = 0.005 \frac{\text{V}}{\Omega} = 0.005\ \text{A}$$

The current through the batteries of the graphing calculator is 0.005 A.

Check the Answer: The current in a graphing calculator is probably very small, and the answer for current is very small. Therefore, the answer is reasonable.

PRACTICE Problems ADDITIONAL PRACTICE

7. You measure the voltage difference of a circuit to be 5 V and the resistance to be 1000 Ω. What is the current in the circuit?
8. The current through a circuit is 0.0030 A. What is the resistance of this circuit if the voltage difference across the circuit is 12 V?
9. **CHALLENGE** The current in a lamp is 0.5 A when plugged into a standard wall outlet. What is the resistance of the lamp when it is plugged into a standard wall outlet?

WORD ORIGIN

circuit

comes from the Latin word *circuitus,* which means "a going around"

The electrons go around the electric circuit.

Alternating Current and Direct Current

Have you ever seen an AC/DC adapter like the one in **Figure 16?** This is a special piece of equipment that allows you to plug a battery-powered device into a wall socket. Most chargers for portable music players, mobile phones, and laptops are AC/DC adapters.

Alternating current

The *AC* in AC/DC stands for *alternating current.* This is the kind of current from household electrical outlets. In the United States, electric power grids are built so that alternating current changes direction 120 times each second. Household appliances, such as toasters and hair dryers, are built to use alternating current.

Direct current

The *DC* in AC/DC stands for *direct current.* Battery-powered devices, such as flashlights, use direct current. Unlike alternating current, direct current never changes direction. An AC/DC adapter is a device that converts the alternating current of an electrical outlet into direct current. With an AC/DC adapter, you can charge a phone battery with the current from an electrical outlet.

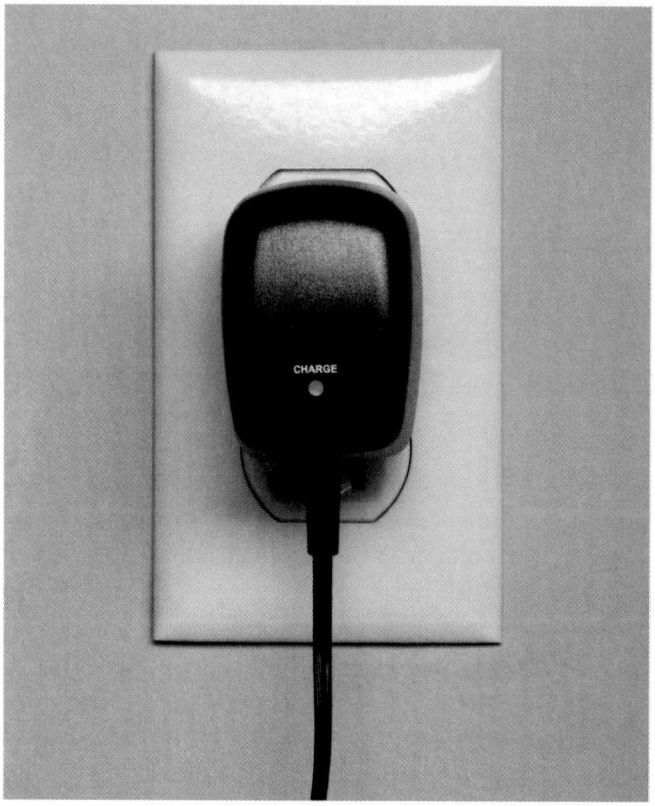

Figure 16 An AC/DC adapter converts the alternating current from an electrical outlet into direct current.

Check Your Progress

Summary

- Electric current is the net movement of electric charge in a single direction.
- A voltage difference is related to the force that causes charges to flow.
- A circuit is a closed conducting path.
- Resistance is the tendency of a material to resist the flow of electric charge.
- Ohm's law relates the current, resistance, and voltage difference in an electric circuit.

Demonstrate Understanding

10. **Compare and contrast** a current through a circuit with a static discharge.
11. **Compare and contrast** the cause of a flow of water in a pipe and the cause of a flow of electrons in a wire.
12. **Explain** how a carbon-zinc dry cell produces a voltage difference between the positive and negative terminals.
13. **Identify** two ways to increase the current in a simple circuit.

Explain Your Thinking

14. **Explain** how the resistance of the heating element in an electric heater changes as it gets hotter, transferring thermal energy to the heater's surroundings at a faster and faster rate.
15. **MATH Connection** Calculate the voltage difference in a circuit that has a resistance of 24 Ω if the current is 0.50 A.

LEARNSMART® Go online to follow your personalized learning path to review, practice, and reinforce your understanding.

LESSON 3
MORE COMPLEX CIRCUITS

FOCUS QUESTION

What are the different ways to connect an electrical circuit?

Series and Parallel Circuits

If you look inside any electrical device, you can see the components of an electric circuit. You might see wires, resistors, a power supply, an electric motor, or other components. Most real circuits are not simple circuits with only one power source, one device, and one path for the current. Multiple power sources, multiple devices, and multiple paths for the current are very common. Before investigating a device's circuits, always unplug the device or remove the batteries, and check the instruction manual for safety warnings.

Series circuits

The two main types of more complex circuits are series circuits and parallel circuits. A **series circuit** is an electric circuit with only one branch, as shown in **Figure 17.** Series circuits are used in flashlights. Because the parts of a series circuit are wired one after another, the amount of current is the same through every part. When any part of a series circuit is disconnected, there is no current through the circuit.

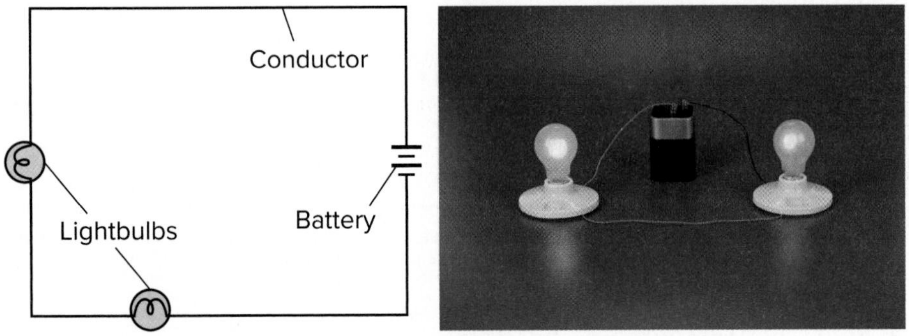

Figure 17 A series circuit provides only one path for the current to follow.
Infer *What happens to the brightness of each bulb as more bulbs are added?*

3D THINKING DCI Disciplinary Core Ideas CCC Crosscutting Concepts SEP Science & Engineering Practices

COLLECT EVIDENCE
Use your Science Journal to record the evidence you collect as you complete the readings and activities in this lesson.

INVESTIGATE
GO ONLINE to find these activities and more resources.

Lab: Compare Series and Parallel Circuits
Develop and use models of series and parallel circuits to compare and constrast their structure and functions.

Review the News
Obtain information from a current news story about electrical energy or electrical power. Evaluate your source and communicate your findings to your class.

Open circuits

A typical remote control requires two AAA batteries to operate, as shown in **Figure 18.** If you remove one of the batteries, the remote control will not work. Why does removing just one battery prevent the remote control from working at all?

You might think that the other battery does not provide enough electrical energy to power the remote control. However, there is another reason. The batteries in a remote control are connected in series. The arrows in **Figure 18** show the path of electric current through the batteries.

Because the parts of a series circuit are wired one after another, the amount of current is the same through every part. When any part of a series circuit is disconnected, there is no current through the circuit. This is called an open circuit. Removing a battery from a remote control results in an open circuit.

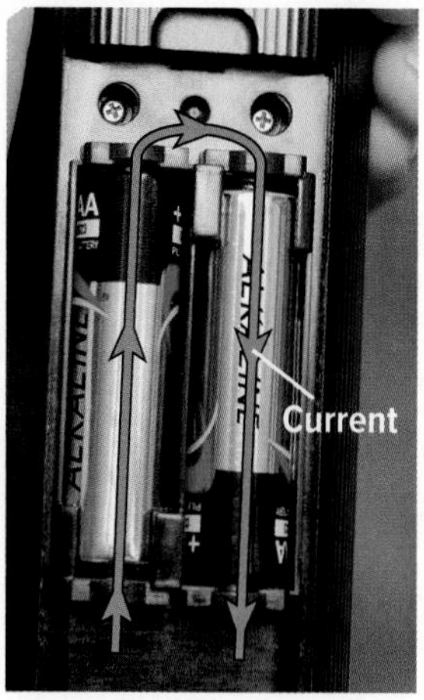

Figure 18 These two batteries in a remote control are connected in series. The arrows show the path of current through the batteries. If one battery is removed, then the circuit is broken and there is no current.

Parallel circuits

What would happen if your home were wired in a series circuit and you turned off one light? All the other lights and appliances in your home would go out too. This is why houses are wired with parallel circuits. **Parallel circuits** contain two or more branches for current. Devices on each branch can be turned on or off separately.

Look at the parallel circuit in **Figure 19.** There can be a current through both or either of the branches. Because all branches connect the same two points of the circuit, the voltage difference is the same in each branch. Then, according to Ohm's law, the current is greater through the branches that have lower resistance.

When one branch of a parallel circuit is opened, such as when you turn off a light, the current continues through the other branches. Homes use parallel wiring so that individual devices can be turned off without affecting the entire circuit.

Get It?

Explain what happens when you open one branch of a parallel circuit.

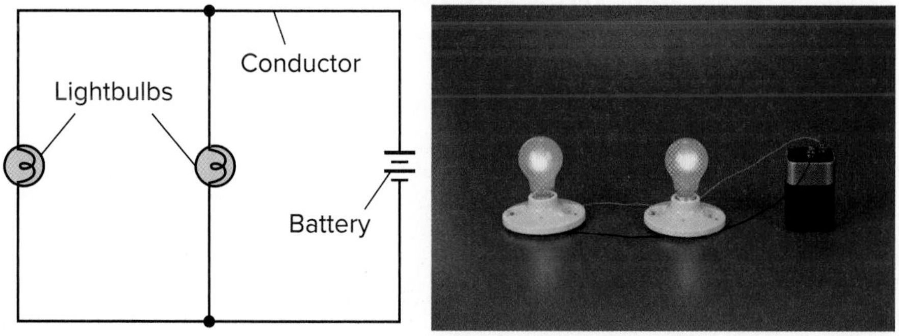

Figure 19 In parallel circuits, the current follows more than one path.

Compare *the voltage differences for each of the branches.*

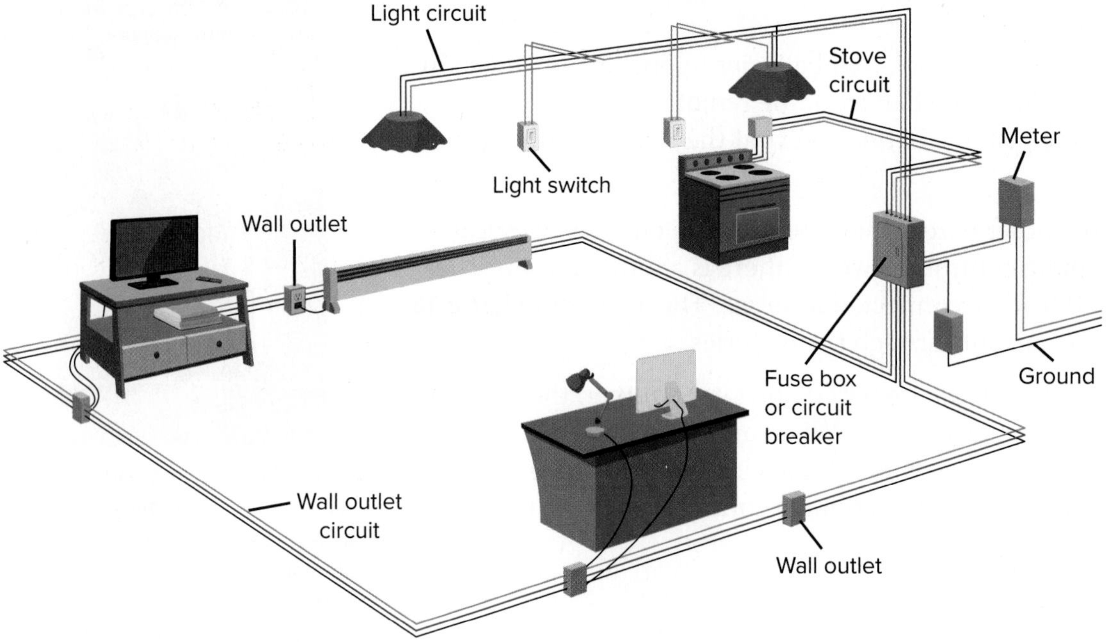

Figure 20 The wiring in a house must allow for the individual use of various appliances and fixtures.
Identify *the type of circuit that is most common in household wiring.*

Household Circuits

Consider how many different things in your home require electrical energy. You do not see the wires because most of them are hidden behind the walls, ceilings, and floors. **Figure 20** shows part of a home without the walls, ceilings, or floors so that you can see these wires. This wiring is mostly a combination of parallel circuits.

In the United States, the voltage difference in most of the branches is 120 V. In some branches that are used for electric stoves or electric clothes dryers, the voltage difference is 240 V. The main switch and circuit breaker or fuse box serve as an electrical headquarters for your home. Parallel circuits branch out from the breaker or fuse box to wall sockets, major appliances, and lights.

Safety devices

In a home, many appliances draw current from the same circuit. If more appliances are connected, there will be more current through the wires. As the amount of current increases, so does the amount of heat produced in the wires. If the wires get too hot, the insulation can melt, and the bare wires can cause a fire. To protect against overheating of the wires, all household circuits should contain either a fuse or a circuit breaker.

STEM CAREER Connection

Electrician

Do you like thinking logically and solving problems? Do you also like working with your hands? If so, you might like working as an electrician. Electricians can be independent contractors or work for small or large companies. They can work indoors or outdoors, in homes or in industry. No matter what the environment, electricians have to understand how electric circuits and their components work as they install and repair electrical systems.

Fuse Circuit breaker

Figure 21 Fuses and circuit breakers are designed to stop the current when a device is in danger of overheating.

Evaluate *which device, a fuse or a circuit breaker, would be more convenient to have in the home.*

Fuses An electrical fuse is shown on the left in **Figure 21.** An electrical fuse contains a small piece of metal that melts if the current becomes too high. When this piece of metal melts, it causes a break in the circuit, stopping the current. To enable charge to flow again in the circuit, the fuse must be replaced.

Circuit breakers A circuit breaker is another device that prevents a circuit from overheating and causing a fire. In a circuit breaker, a switch is automatically flipped when the current becomes too great. Flipping the switch opens the circuit and stops the current. Circuit breakers usually can be reset by pushing the switch back to its "on" position. Many residences in the United States contain a box of circuit breakers similar to the one shown on the right in **Figure 21.**

Get It?
Identify the purpose of fuses and circuit breakers in household circuits.

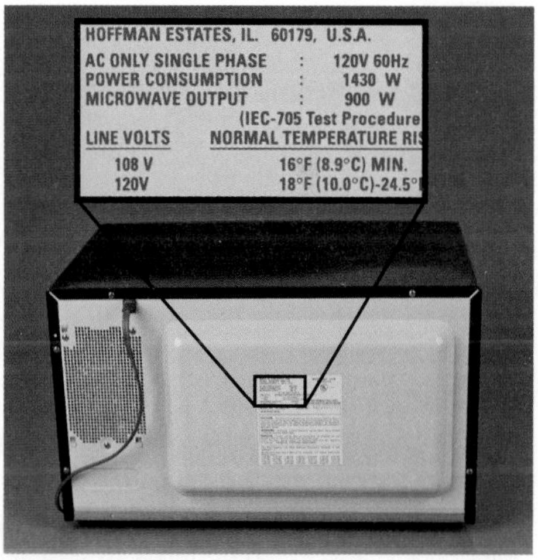

Figure 22 Many appliances are labeled with a power rating.

Electrical Power and Energy

The reason that electricity is so useful is that electrical energy can easily be converted to other types of energy. For example, electrical energy is converted to mechanical energy as the blades of a fan rotate to cool you. An electric heater converts electrical energy into thermal energy.

ACADEMIC VOCABULARY

device
a piece of equipment designed to perform a special function

A toaster is an electrical device.

CCC CROSSCUTTING CONCEPTS

Cause and Effect Suppose the radio in your car stops working. You look in the fuse box under the dashboard and find several fuses like the one shown in **Figure 21.** The glass on one of the fuses appears discolored, and the fuse element inside is broken. However, this does not necessarily mean that this fuse breaking caused the radio to stop working. What further empirical evidence would confirm that this broken fuse is the reason the radio is no longer working?

(tl)Steve Cole/Photodisc/Getty Images, (tr)Andersen Ross/Photodisc/Getty Images, (b)Matt Meadows/McGraw-Hill Education

Electrical power is the rate at which electrical energy is converted to another form of energy. The electrical power used by appliances varies. In general, heating appliances, such as electric stoves, use much more electrical power than electronic devices, such as computers. Appliances, like the microwave oven in **Figure 22,** often are labeled with a power rating that describes how much power the appliance uses.

Calculate electrical power

A device's electrical power depends on the voltage difference across that device and the current through that device. Electrical power can be calculated from the following equation.

Electrical Power Equation

electrical power (in watts) =

current (in amperes) × **voltage difference** (in volts)

$$P = IV$$

If current is measured in amperes and voltage difference is measured in volts, then electrical power is calculated in watts (W). Because the watt is a small unit of power, electrical power is often expressed in kilowatts (kW). One kilowatt equals 1000 watts.

EXAMPLE Problem 2

SOLVE FOR ELECTRICAL POWER The current in a clothes dryer is 15 A when it is plugged into a 240-volt outlet. What is the power of the clothes dryer?

Identify the Unknown: power: P

List the Knowns: current: $I = 15$ A voltage difference: $V = 240$ V

Set Up the Problem: $P = IV$

Solve the Problem: $P = (15 \text{ A})(240 \text{ V}) = 3600 \text{ W}$

The clothes dryer has a power of 3600 W.

Check the Answer: You can find the power rating for a clothes dryer by looking at the instruction manual, the manufacturer's Web site, or the clothes dryer itself. The power rating on a typical clothes dryer is about 4000 W. Therefore, an answer of 3600 W seems reasonable.

PRACTICE Problems

ADDITIONAL PRACTICE

16. A toaster oven is plugged into an outlet in which the voltage difference is 120 V. How much power does the toaster oven use if the current in the oven is 10 A?
17. A video-disc player that is not playing still uses 6.0 W of power. What is the current into the video-disc player if it is plugged into a standard 120-V outlet?
18. A flashlight bulb uses 2.4 W of power when the current in the bulb is 0.8 A. What is the voltage difference supplied by the batteries?
19. **CHALLENGE** A hair dryer is rated at 1.2 kW. When it is plugged into a standard outlet and turned on, what is the current in the hair dryer?

Calculate electrical energy

Electric companies provide the electrical energy that you can then transform into other forms of energy, such as mechanical energy or thermal energy. The electrical energy that electric companies provide can be calculated from this equation.

Electrical Energy

electrical energy (in kWh) =

electrical power (in kW) × time (in h)

$$E = Pt$$

Electrical energy is usually measured in units of kilowatt hours (kWh). One kilowatt hour is equal to 3.6 million joules and is enough energy to lift an average-sized car more than 200 meters.

EXAMPLE Problem 3

SOLVE FOR ELECTRICAL ENERGY A microwave oven with a power rating of 1200 W is used for 0.25 h. How much electrical energy did the power company provide for the microwave?

Identify the Unknown: electrical energy provided: E

List the Knowns: electrical power used: $P = 1200\ \text{W} = 1.2\ \text{kW}$

time: $t = 0.25\ \text{h}$

Set Up the Problem: $E = Pt$

Solve the Problem: $E = (1.2\ \text{kW})(0.25\ \text{h}) = 0.30\ \text{kWh}$

The power company provided 0.30 kWh of electrical energy for the microwave.

Check the Answer: Use rounding to estimate the answer. The microwave oven's power is very close to 1000 W, which is the same as 1 kW. Therefore, using the microwave for 0.25 h should require approximately 0.25 kWh of energy. Our answer, 0.30 kWh, is very close to 0.25 kWh. Therefore, our answer is reasonable.

PRACTICE Problems

ADDITIONAL PRACTICE

20. A refrigerator operates on average for 10.0 h a day. If the power rating of the refrigerator is 700 W, how much electrical energy does the refrigerator use in one day?

21. A TV with a power rating of 200 W uses 0.8 kWh of electrical energy in one day. For how many hours was the TV on during this day?

22. A hair dryer has a power of 1200 W. How much electrical energy does the hair dryer use in 3 minutes?

23. **CHALLENGE** An electric light is plugged into a 120-V outlet. If the current in the bulb is 0.50 A, how much electrical energy does the bulb use in 15 minutes?

The cost of electrical energy

The electrical energy that power companies provide costs money. The cost of using an appliance can be computed by multiplying the electrical energy used by the amount that the power company charges for each kWh. For example, if a 100-W lightbulb is left on for 5 h, the amount of electrical energy provided by the power company is

$$E = Pt = (0.1 \text{ kW})\ (5 \text{ h}) = 0.5 \text{ kWh}$$

If the power company charges $0.15 per kWh, the cost of using the bulb for 5 h is

$$\begin{aligned} \text{cost} &= (\text{kWh provided})\ (\text{cost per kWh}) \\ &= (0.5 \text{ kWh})\ (\$0.15/\text{kWh}) \\ &= \$0.08 \end{aligned}$$

The cost of using some sample household appliances is given in **Table 1,** where the cost per kWh is assumed to be $0.15/kWh.

Table 1 Sample Costs of Using Home Appliances

Appliance	Hair Dryer	Stereo	Color Television
Power rating (W)	1000	100	200
Hours used daily	0.25	2.0	4.0
kWh used monthly	7.5	6.0	24.0
Cost per kWh	$0.15	$0.15	$0.15
Monthly cost	$1.13	$0.90	$3.60

Check Your Progress

Summary

- Two types of circuits are series circuits and parallel circuits.
- Fuses and circuit breakers are used to prevent wires from overheating.
- Electrical power is the rate at which electrical energy is converted into other forms of energy.

Demonstrate Understanding

24. **Compare** series circuits with parallel circuits.
25. **Explain** what determines the current in each branch of a parallel circuit.
26. **Infer** whether a circuit breaker should be connected in parallel to the circuit that it is protecting.

Explain Your Thinking

27. **Relate** A parallel circuit with four branches is connected to a battery. Explain how the amount of current from the battery is related to the amount of current in the branches of the circuit.
28. **MATH Connection** Calculate the current into a desktop computer plugged into a 120-V outlet if the power used is 180 W.
29. **MATH Connection** A circuit breaker trips when the current in the circuit reaches 15 A. If the voltage is 120 V, how much power is being used when the breaker is tripped?
30. **MATH Connection** Estimate the monthly cost of using a 700-W refrigerator that runs for 10 h a day if the cost per kWh is $0.20.

LEARNSMART® Go online to follow your personalized learning path to review, practice, and reinforce your understanding.

War of the Currents

Look on the bottom or side of almost any electrical appliance. You will probably see writing that says something like "120 V, 60 Hz." This strange writing is the result of a dispute over one hundred years ago between inventors Thomas Alva Edison and Nikola Tesla. Historians call this battle the War of the Currents.

These modern electric generators in Boulder City, Nevada, transform the energy from Hoover Dam into electrical energy.

Edison's direct current

Edison favored direct current (DC). But there were big problems with Edison's direct-current scheme. As electric current passes through a wire, electrical energy is transformed into thermal energy. As the amount of current increases and the distance that current travels increases, more electrical energy is transformed into waste thermal energy. In Edison's system, power stations would be needed on nearly every block of a city.

Tesla's idea

Nikola Tesla devised another option. By using alternating current (AC), Tesla was able to send electrical energy at low current, but very high voltage, across great distances. Lower current meant less electrical energy transformed into wasted thermal energy. With AC, power stations could be located many miles from towns and cities.

Before AC entered a home, it would be transformed into lower voltage (and higher current) electrical energy using a device called a transformer. Transformers depend on rapidly changing currents and do not work with DC. Fellow inventor George Westinghouse worked with Tesla to develop Tesla's idea, seeing in it a way to beat Edison in the race to electrify the world.

Tesla's triumph

In 1893, Tesla's AC generators powered the Chicago World's Fair. Then, in 1895, AC had its greatest triumph when Tesla's system was used for the Edward Dean Adams Power Plant. The Adams Power Plant transformed the energy of Niagara Falls into electrical energy and was used to light up the city of Buffalo, New York. By the end of the nineteenth century, it was clear that AC had become the winner in the War of the Currents.

The photo above shows electric generators at Hoover Dam. Electrical energy from these generators provides power to homes throughout Nevada, Arizona, and Southern California.

SEP DCI CCC

EVALUATE TECHNOLOGICAL SOLUTIONS

Research the devices credited to Edison and Tesla. Compare these two inventors' influences on today's technology and possible future technologies.

MODULE 6
STUDY GUIDE

GO ONLINE to study with your Science Notebook.

Lesson 1 ELECTRIC CHARGE

- There are two types of electric charge: positive charge and negative charge.
- Electric charge can be transferred between objects but cannot be created or destroyed.
- An electric charge is surrounded by an electric field that exerts forces on other charges.
- Charging by contact is the transfer of charge between two objects that are touching.
- Charging by induction is the rearrangement of charge in an object due to the influence of another charged object nearby.

- static electricity
- law of conservation of charge
- electric field
- conductor
- insulator
- charging by contact
- charging by induction
- electroscope

Lesson 2 ELECTRIC CURRENT

- Electric current is the net movement of electric charge in a single direction.
- A voltage difference is related to the force that causes charges to flow.
- A circuit is a closed conducting path.
- Chemical reactions in a battery produce a voltage difference between the positive and negative battery terminals.
- Resistance is the tendency of a material to resist the flow of electric charge.
- Ohm's law relates the current, resistance, and voltage difference in an electric circuit.

- electric current
- voltage difference
- electric circuit
- resistance
- Ohm's law

Lesson 3 MORE COMPLEX CIRCUITS

- Two types of circuits are series circuits and parallel circuits.
- Fuses and circuit breakers are used to prevent wires from overheating.
- Electrical power is the rate at which electrical energy is converted into other forms of energy.

- series circuit
- parallel circuit
- electrical power

THREE-DIMENSIONAL THINKING

Module Wrap-Up

REVISIT THE PHENOMENON

How can you safely conduct electricity underwater?

CER Claim, Evidence, Reasoning

Explain Your Reasoning Revisit the claim you made when you encountered the phenomenon. Summarize the evidence you gathered from your investigations and research and finalize your Summary Table. Does your evidence support your claim? If not, revise your claim. Explain why your evidence supports your claim.

STEM UNIT PROJECT

Now that you've completed the module, revisit your STEM unit project. You will summarize your evidence and apply it to the project.

GO FURTHER

SEP Data Analysis Lab

Electricity in Everyday Life

For each of the following activities, predict what you will observe. Perform each activity and write down your actual observation. Give an explanation for your results.

CER Analyze and Interpret Data

1. **Claim, Evidence, Reasoning** Run a plastic comb through your hair several times. Hold the comb next to a stream of running water.
2. **Claim, Evidence, Reasoning** Tie two inflated balloons together with a string. Hold the balloons next to each other, and rub both with a piece of wool.
3. **Claim, Evidence, Reasoning** Hold a piece of newspaper flat against the wall. Stroke across the surface of the newspaper with your hand.
4. **Claim, Evidence, Reasoning** Stretch a piece of clear plastic food wrap over a glass jar and fold down the sides.
5. **Claim, Evidence, Reasoning** Add an antistatic dryer sheet to some clothes in a dryer. Turn the dryer on for 15 minutes.

Stefan Sauer/dpa/AFP/Getty Images

Paul Ridsdale pictures/Alamy Stock Photo

MAGNETISM AND ITS USES

ENCOUNTER THE PHENOMENON

What makes something magnetic?

GO ONLINE to play a video about electromagnets and magnetic fields.

SEP Ask Questions

Do you have other questions about the phenomenon? If so, add them to the driving question board.

CER Claim, Evidence, Reasoning

Make Your Claim Use your CER chart to make a claim about what makes something magnetic. Explain your reasoning.

Collect Evidence Use the lessons in this module to collect evidence to support your claim. Record your evidence as you move through the module.

Explain Your Reasoning You will revisit your claim and explain your reasoning at the end of the module.

GO ONLINE to access your CER chart and explore resources that can help you collect evidence.

LESSON 1: Explore & Explain: Magnetic Fields

LESSON 3: Explore & Explain: Generators

Smithsonian

Additional Resources

(t)Video Supplied by BBC Worldwide Learning, (bl)©Alchemy/Alamy Stock Photo, (br)Jordan Ash/McGraw-Hill Education

LESSON 1
MAGNETISM

FOCUS QUESTION
How do magnets interact with other materials?

Magnets

More than 2000 years ago, the Greeks discovered deposits of a mineral that was a natural magnet. They noticed that chunks of this mineral could attract pieces of iron. This mineral was found in a region of Turkey that was known as Magnesia, so the Greeks called the mineral magnetic stone. This mineral is now called magnetite.

Since the discovery of magnetite, many devices have been developed that rely on magnets. In the twelfth century, Chinese sailors used magnetite to make compasses that improved navigation. **Figure 1** shows common magnets and familiar devices that use magnets today. In science, **magnetism** refers to the properties and interactions of magnets.

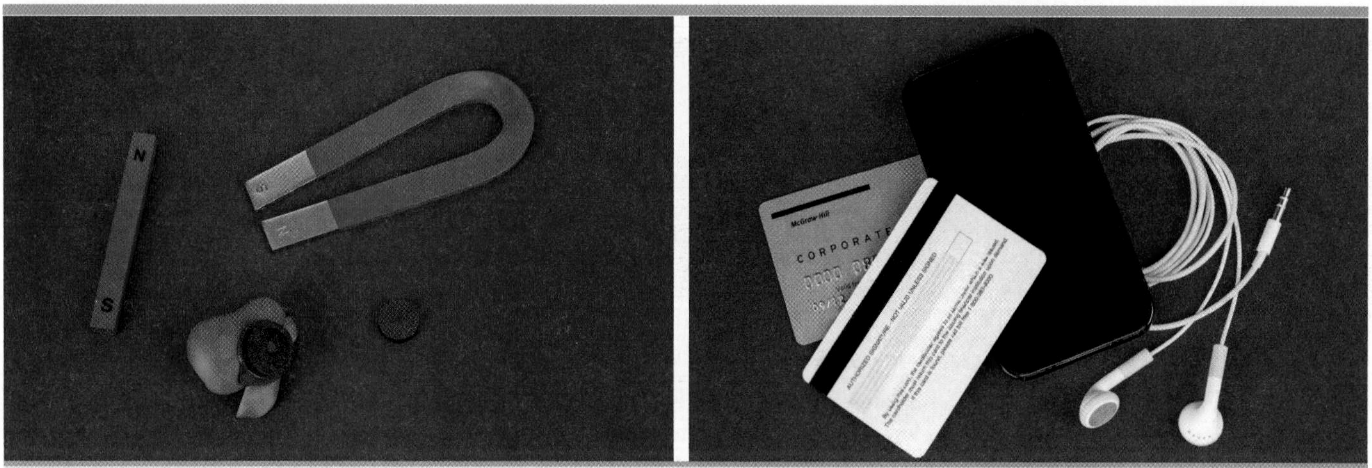

Common magnets

Devices that use magnets

Figure 1 Magnets can be found on your refrigerator door and in many devices that you use every day, such as TVs, headphones, video game systems, and telephones.

(l)Matt Meadows/McGraw-Hill Education, (r)©Holly Hildreth/McGraw-Hill Education

3D THINKING DCI Disciplinary Core Ideas CCC Crosscutting Concepts SEP Science & Engineering Practices

COLLECT EVIDENCE

Use your Science Journal to record the evidence you collect as you complete the readings and activities in this lesson.

INVESTIGATE

GO ONLINE to find these activities and more resources.

Applying Practices: Modeling Magnetic Fields
HS-PS3-5. Develop and use a model of two objects interacting through electric or magnetic fields to illustrate the forces between objects and the changes in energy of the objects due to the interaction.

Magnetic force

You probably have played with magnets and might have noticed that two magnets exert forces on each other. Depending on which ends of the magnets are close together, the magnets either repel or attract each other. You might have noticed that the interaction between two magnets can be felt even before the magnets touch. The strength of the force between two magnets increases as the magnets move closer together and decreases as the magnets move farther apart.

Magnetic strength

You might also have noticed that some magnets are stronger than others. For example, a magnet from your refrigerator can be used to pick up paper clips, but a much stronger magnet would be needed to lift a car.

Magnetic Fields

A magnet can exert a force on a distant object because of its magnetic field. A **magnetic field** is a region of space that surrounds a magnet and exerts a force on other magnets and objects made of magnetic materials. The magnetic field and the magnetic force are related. A stronger magnetic field means that a magnetic object placed in the field will experience a stronger magnetic force. In other words, stronger magnets have stronger magnetic fields.

The strength of the magnetic field affects the use of the magnet. For example, the magnetic field used for a type of medical test called magnetic resonance imaging (MRI) is about 200 times stronger than the magnetic field of a refrigerator magnet. This is why one must remove all magnetic metal objects before entering an MRI room.

Magnetic field lines

The magnetic field can be represented by lines of force called magnetic field lines. **Figure 2** shows iron filings lined up along the magnetic field lines surrounding a bar magnet. The closer together the magnetic field lines are, the stronger the magnetic field is at that point. Field lines are closer together near the magnet, as shown in **Figure 2.** This agrees with our observation that the magnetic force and the magnetic field are strongest close to the magnet. **Figure 3** shows the magnetic fields around a horseshoe magnet and a disk magnet.

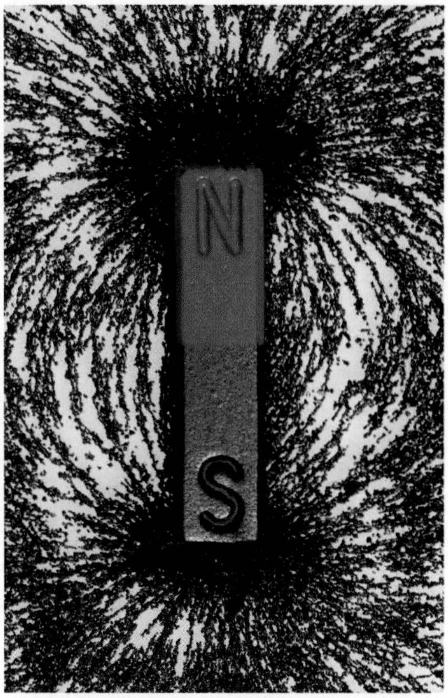

Iron fillings sprinkled around a magnet line up along the magnetic field lines.

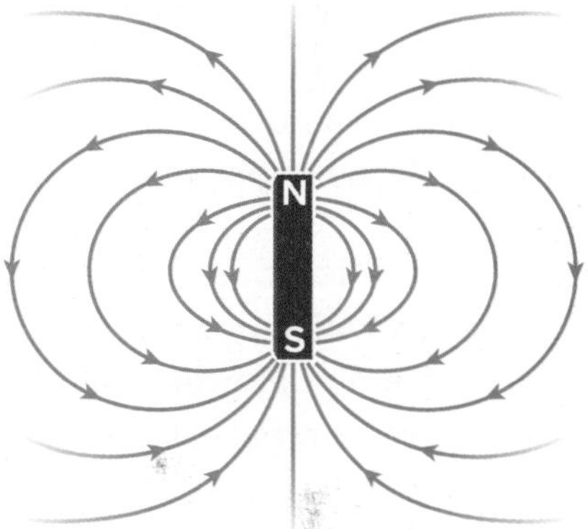

A magnetic field is represented by magnetic field lines.

Figure 2 A magnet is surrounded by a magnetic field. Field lines that are closer together indicate a stronger magnetic field.

Identify *where the magnetic field is strongest.*

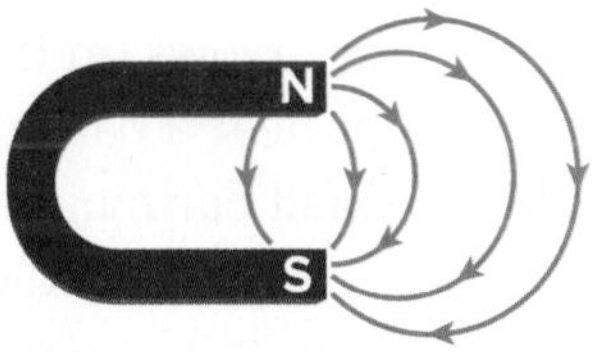

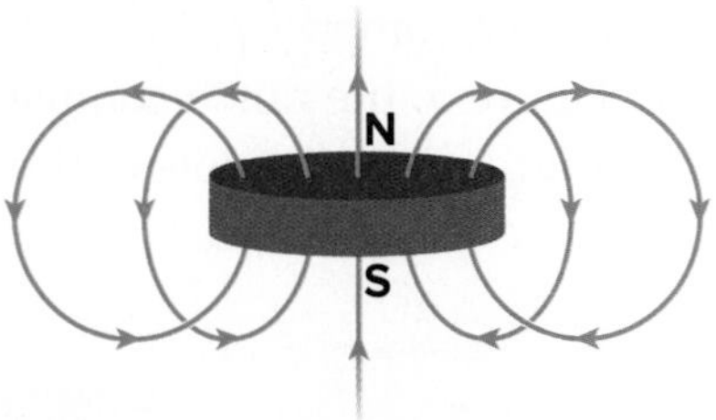

Figure 3 Horseshoe magnets and disk magnets are also surrounded by magnetic fields.

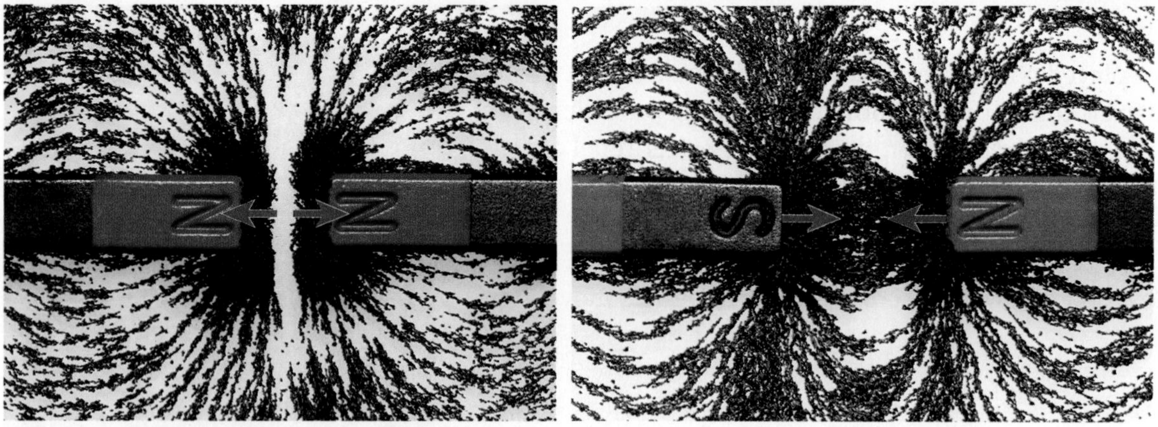

Like poles repel. Notice that in the area between the magnets, the field lines from the north pole of one magnet bend away from the field lines from the north pole of the other magnet.

Unlike poles attract. Notice that in the area between the magnets, the field lines connect the north pole of one magnet to the south pole of the other magnet.

Figure 4 When two magnetic poles are placed near each other, their magnetic fields combine. The magnets will either attract or repel each other, depending on which poles are closest together.

Magnetic poles

You might have also noticed in **Figure 2** that the magnetic field lines are closest together at the ends of the bar magnet. This means the magnetic field is strongest at the ends. The regions of a magnet that exert the strongest force are called **magnetic poles.** All magnets have a north pole and a south pole. For a bar magnet, the north and south poles are at the opposite ends.

How magnetic poles interact Whether two magnets attract or repel each other depends on which poles are brought close together. Two north poles or two south poles repel each other. However, a north pole and a south pole always attract each other. Like magnetic poles repel each other, and unlike poles attract each other. **Figure 4** illustrates these interactions.

When two magnets are brought close to each other, their magnetic fields combine to produce a new magnetic field. If you look again at **Figure 4,** you will see iron filings illustrating the magnetic fields that result when like poles and unlike poles of bar magnets are brought close to each other.

Magnetic field direction

A magnetic field also has a direction. The magnetic field always points away from north magnetic poles and toward south magnetic poles. The direction of the magnetic field around a bar magnet is shown by the arrows on the magnetic field lines in **Figure 5.** When a compass is brought near a bar magnet, the compass needle rotates. The compass needle is a small bar magnet with a north pole and a south pole. The force exerted on the compass needle by the magnetic field causes the needle to rotate until it lines up with the magnetic field lines, as shown in **Figure 5.** The north pole of a compass points in the direction of the magnetic field. Notice that in **Figure 5,** the north compass needles point along the field lines toward the south magnetic pole of the magnet.

Figure 5 Compass needles placed around a bar magnet line up along magnetic field lines. The north poles of the compass needles are shaded red.

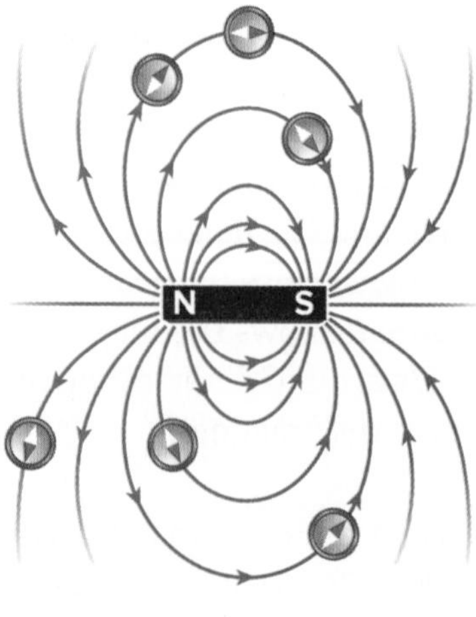

Earth's magnetic field

A compass can help determine direction because the needle's north pole points to a location near Earth's geographic north pole. Earth's geographic north pole is the northernmost point on Earth. Earth acts like a giant bar magnet with a magnetic field that extends into space. A compass needle will align with Earth's magnetic field lines, as shown in **Figure 6.** Earth's magnetic field is not very strong in comparison to common magnets. The magnetic field of a refrigerator magnet is about 100 times stronger than Earth's magnetic field.

Earth's magnetic poles The north pole of a magnet is defined as the end of the magnet that points toward Earth's geographic north pole. Sometimes the north pole and south pole of magnets are called the north-seeking pole and the south-seeking pole.

Opposite magnetic poles attract, so the north pole of a compass needle is attracted to a south magnetic pole. Therefore, Earth's south magnetic pole is near its geographic north pole, as seen in **Figure 6.** However, Earth's magnetic poles move slowly with time and sometimes switch places. Measurements of magnetism in rocks show that Earth's magnetic poles have changed places more than 150 times in the past 70 million years.

The source of Earth's magnetic field No one is certain what produces Earth's magnetic field, but many scientists think that it is generated by Earth's core. Earth's core is thought to be made of iron and nickel. The solid inner core is surrounded by a liquid outer core. The circulation of the molten iron and nickel in Earth's outer core is believed to produce Earth's magnetic field.

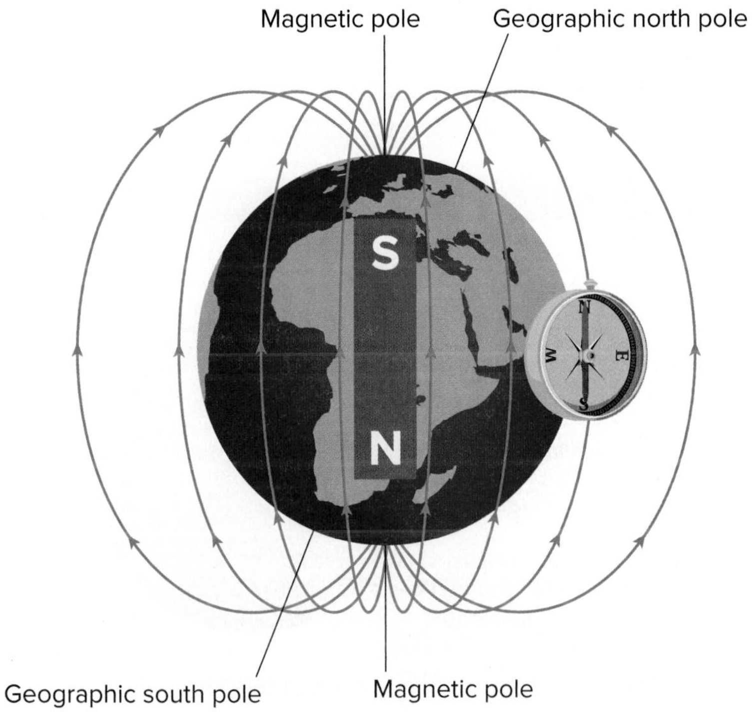

Figure 6 Currently, Earth's south magnetic pole is located about 1000 km from the geographic north pole.

Predict *which way a compass needle would point if Earth's magnetic poles switched places.*

Magnetic Materials

You might have noticed that a magnet will not attract all metal objects. For example, a magnet will not attract pieces of aluminum foil. Only a few metals, such as iron, cobalt, and nickel, are attracted by magnets or can be made into permanent magnets.

What makes these elements magnetic? Remember that every atom contains electrons. Electrons have magnetic properties. In the atoms of most elements, the magnetic properties of the electrons cancel out. But in the atoms of magnetic elements, these magnetic properties do not cancel out, and each atom behaves like a small magnet with its own magnetic field and north and south poles.

Get It?

Explain why the atoms of magnetic materials behave like small magnets.

Although the atoms have their own magnetic fields, objects made from these metals are not always magnets. For example, if you hold an iron nail close to a refrigerator door and let go, the nail falls to the floor. However, you can make the nail behave like a magnet by placing it in a magnetic field. A magnetized nail would stick to the refrigerator door.

Magnetic domains—a model for magnetism

In magnetic materials such as iron, the magnetic field created by each atom exerts a force on other nearby atoms. These forces cause large groups of neighboring atoms to align. This means that almost all north magnetic poles in the group point in the same direction. These groups of atoms with aligned magnetic poles are called **magnetic domains.** Because the magnetic poles of the individual atoms are aligned, the domain itself behaves like a larger magnet with a north pole and a south pole.

APPLY SCIENCE

How can magnetic parts of a car be recovered?

Every year, over 10 million cars are scrapped. A junk car is fed into a shredder, and then large magnets are used to separate many of its metal parts from its nonmetal parts. The retrieved materials can be reused, saving both natural resources and energy. How much of a car does a magnet actually help separate? Find the answer by interpreting the circle graph.

Identify the Problem

The circle graph shows the average percent by mass of the different materials in a car. According to the graph, how much of a car can a magnet separate for recycling?

Solve the Problem

1. Cars are made of several metals, including iron, steel (a mixture of iron and carbon), aluminum, copper, and zinc. Which of these metals would the magnet attract?

Percent Mass of Materials in a Car

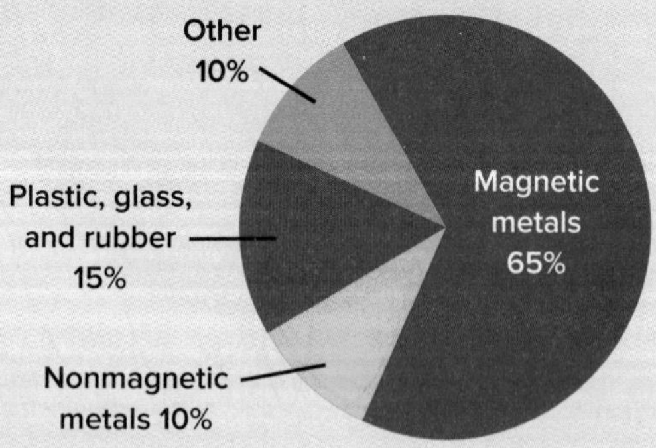

2. If a scrapped car has a mass of 1500 kg, what is the mass of the materials that can be recovered using a magnet?
3. If the average mass of a scrapped car is 1500 kg and 10 million cars are scrapped each year, what is the total mass of magnetic metals that could be recovered from scrapped cars each year?

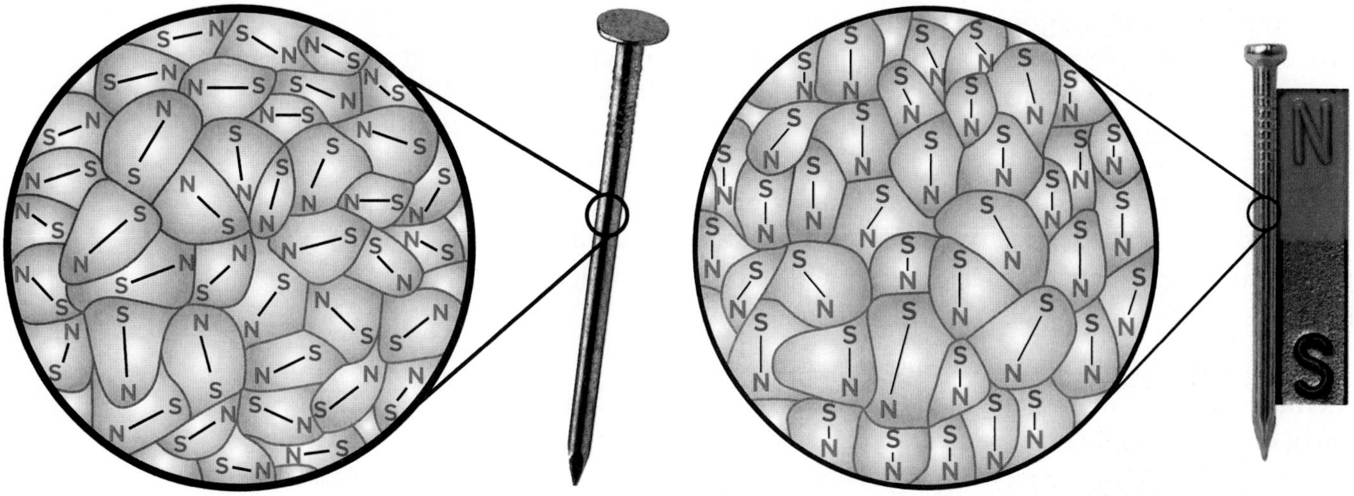

An unmagnetized iron nail is made up of microscopic domains that point in different directions.

The domains of a magnetic material will tend to align themselves along the magnetic field lines of a nearby magnet.

Figure 7 The randomly arranged magnetic domains in an iron nail will align when placed in a magnetic field. With the domains aligned, the nail is a magnet.

Random arrangement of domains An iron nail contains an enormous number of these magnetic domains. So, why does the nail not behave like a magnet? Even though each domain behaves like a magnet, the poles of the domains are arranged randomly and point in different directions, as shown on the left in **Figure 7.** As a result, the magnetic fields from the domains cancel each other, and the nail is not magnetic.

Lining up domains If you place a nail in a magnetic field, many of the domains align in the direction of the magnetic field, as shown on the right in **Figure 7.** The like poles of many of the domains point in the same direction and no longer cancel each other out. The nail itself now acts as a magnet.

Permanent magnets

When domains align, their magnetic fields add together and create a magnetic field inside the material. In a permanent magnet, this field is strong enough to prevent the constant motion of the atoms from bumping the domains out of alignment. A permanent magnet is any magnet whose domains remain aligned without an external field.

SCIENCE USAGE v. COMMON USAGE

field

Science usage: a region of space with a physical property (such as a force) that has a definite value at every point

The magnetic field is strongest close to the ends of the bar magnet.

Common usage: an area of land used for planting crops; an area of interest

From the front porch, the farmer could see the tractor and the cornfield.

Making permanent magnets Permanent magnets can be made by stroking a magnetic material with another permanent magnet. They can also be made by heating magnetic material and letting it cool in a magnetic field.

How long does a magnet last? The domains in permanent magnets do not have to remain aligned forever. However, they are still considered permanent magnets because they remained magnetized when there was no external magnetic field.

How do magnetic materials become unmagnetized? When the external magnetic field is removed, the constant motion and vibration of the atoms bump the magnetic domains out of alignment, and the domains return to a random arrangement.

How quickly this happens depends on the magnet's material and its environment. For example, heating a permanent magnet causes its atoms to move faster. If the magnet is heated enough, its atoms move fast enough to bump the domains out of alignment.

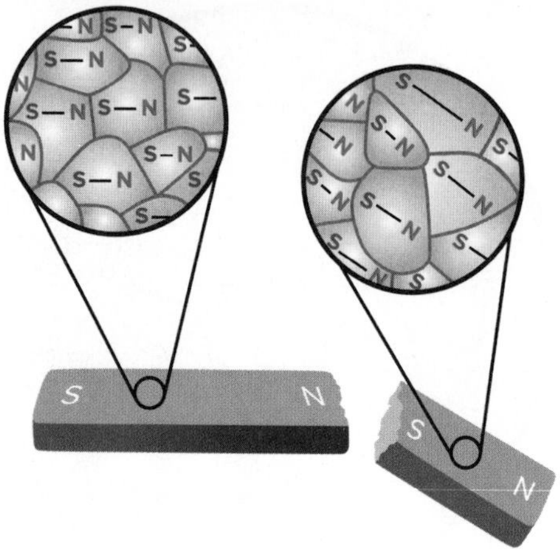

Figure 8 No matter how many pieces you break a magnet into, each piece still has a north pole and a south pole.

Can a pole be isolated?

After breaking a magnet, is one piece a north pole and one piece a south pole? Look at the domain model of the broken magnet in **Figure 8.** Recall that even single atoms of magnetic materials act as tiny magnets. Because every magnet is made of many aligned smaller magnets, even small pieces have both a north pole and a south pole.

Check Your Progress

Summary

- A magnet exerts a force on other magnets.
- A magnet has a north pole and a south pole. Like magnetic poles repel, and unlike poles attract.
- Atoms in magnetic materials behave like magnets.
- Magnetic domains are regions in a magnetic material where the magnetic poles of the atoms are aligned.
- In a permanent magnet, the magnetic domains remain aligned without an external magnetic field.

Demonstrate Understanding

1. **Explain** Why does a magnet exert a force on another magnet when the two magnets are not in contact?
2. **Describe** the magnetic field when two unlike magnetic poles are close together. Draw a diagram to illustrate your answer.
3. **Describe** how a compass needle moves when it is placed in a magnetic field.
4. **Explain** why only certain materials are magnetic.
5. **Explain** how heating a bar magnet would change its magnetic field.

Explain Your Thinking

6. **Explain** Use the magnetic domain model to explain why a magnet sticks to a refrigerator door.
7. **MATH Connection** Magnetic domains have an average volume of 0.0001 mm^3. If a magnet has dimensions of 50 mm by 10 mm by 4 mm, how many domains does the magnet contain?

LEARNSMART® Go online to follow your personalized learning path to review, practice, and reinforce your understanding.

LESSON 2

ELECTRICITY AND MAGNETISM

FOCUS QUESTION

How do electromagnets function in electrical devices?

Electric Current and Magnetism

In 1820, Danish physics teacher Hans Christian Oersted found that electricity and magnetism were related. While doing a demonstration involving electric current, he happened to have a compass near an electric circuit. He noticed that the electric current affected the direction of the compass needle. Oersted hypothesized that the electric current must produce a magnetic field around the wire and that the direction of the field depends on the direction of the current.

Oersted's hypothesis that an electric current creates a magnetic field was correct. It is now known that all moving charges, like those in an electric current, produce magnetic fields. Around a current-carrying wire, the magnetic field lines form circles, as shown in **Figure 9.** The direction of the magnetic field around the wire reverses when the direction of the current in the wire reverses. The strength of the magnetic field can be increased by increasing the current in the wire. Also, as you move farther away from the wire, the strength of the magnetic field decreases.

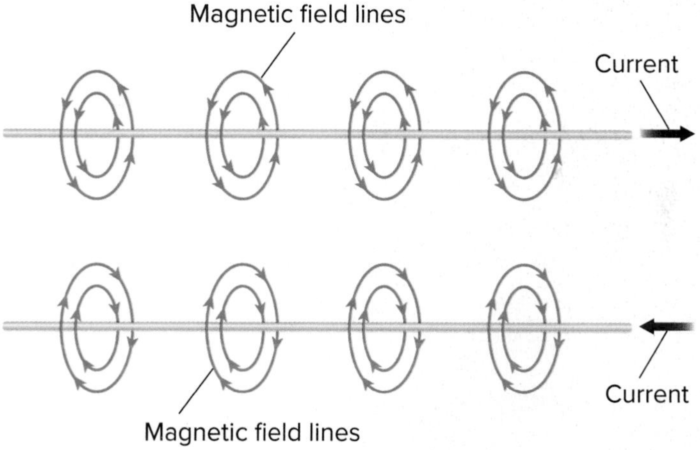

Figure 9 When there is electric current through a wire, a magnetic field forms around the wire. The direction of the magnetic field depends on the direction of the current in the wire.

Get It?
Explain how the strength of the magnetic field around a wire can be increased.

3D THINKING DCI Disciplinary Core Ideas CCC Crosscutting Concepts SEP Science & Engineering Practices

COLLECT EVIDENCE

Use your Science Journal to record the evidence you collect as you complete the readings and activities in this lesson.

INVESTIGATE

GO ONLINE to find these activities and more resources.

Laboratory: Creating Electromagnets
Develop and use a model of several electromagnets to compare the scale, proportion, and quantity of a magnetic field produced by an electric current.

Electromagnetism

Moving electric charges are surrounded by magnetic fields. The resulting force between the moving electric charge and magnets led scientists to the realization that the electric force and the magnetic force are parts of the same force, called the electromagnetic force. The **electromagnetic force** is the attractive or repulsive force between electric charges and magnets. Like gravity, the electromagnetic force is one of the fundamental forces.

The interaction between electric charges and magnets is called **electromagnetism.** Electromagnetism is what makes magnets so useful. Electromagnetism is essential in producing, transmitting, and using electricity.

Electromagnets

The magnetic field around a current-carrying wire can be made stronger by changing the shape of the wire. When there is a current in a wire loop, such as the one shown on the left in **Figure 10,** the magnetic field inside the loop is stronger than the field around a straight wire. A single wire wrapped into a cylindrical wire coil is called a **solenoid.** The magnetic field inside a solenoid is stronger than the field in a single loop. The magnetic field around each loop in the solenoid adds together to form the field shown in the center in **Figure 10.**

An **electromagnet** is a temporary magnet created when there is a current in a wire coil. Often, the current-carrying coil is wrapped around an iron core, as shown on the right in **Figure 10.** The coil's magnetic field temporarily magnetizes the iron core. As a result, the electromagnet's field can be more than 1000 times greater than the field of the solenoid.

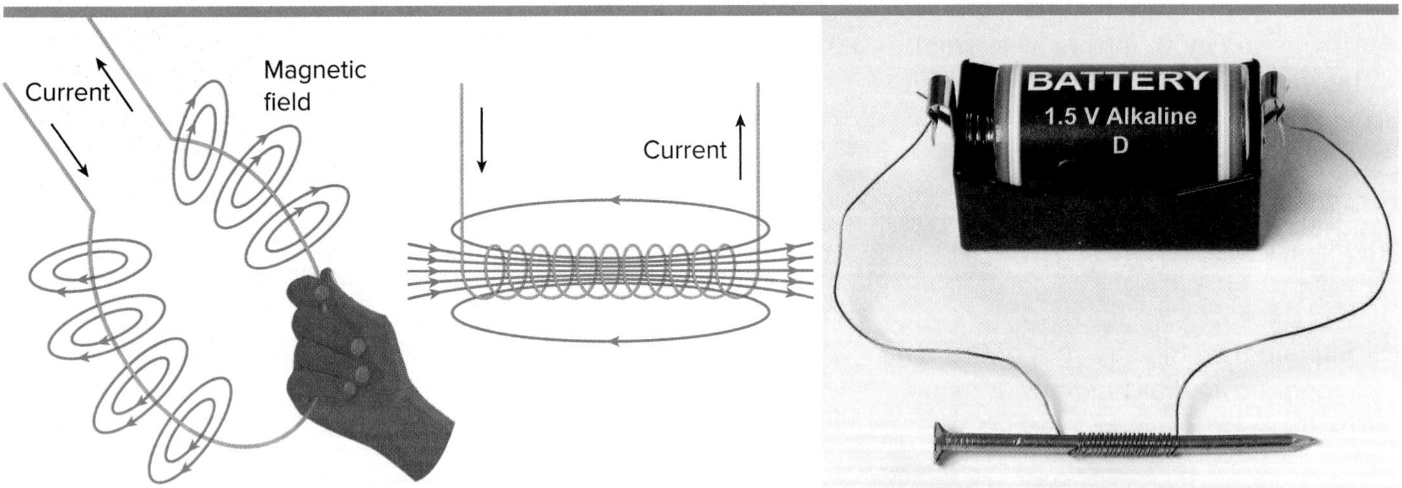

A wire loop has a magnetic field inside the loop. | Winding the wire into a coil forms a solenoid. | A solenoid wrapped around an iron core forms an electromagnet.

Figure 10 Coiling a current-carrying wire increases the strength of the magnetic field. Adding an iron core further increases the strength of the field.

WORD ORIGINS

solenoid

comes from the Greek words *solen–*, meaning "channel" or "pipe," and **–oid,** meaning "like" or "similar to"

Solenoids are used in electromagnets.

CCC CROSSCUTTING CONCEPTS

Cause and Effect With a partner, review the information about electromagnets and solenoids on this page and in **Figure 10.** Write the experimental procedure you would use to test how the number of loops in a solenoid affects the strength of an electromagnet.

Properties of electromagnets

An electromagnet behaves like any other magnet when there is current in the solenoid. One end of the electromagnet is a north pole, and the other end is a south pole. If placed in a magnetic field, an electromagnet will align itself along the magnetic field lines. An electromagnet will also exert a magnetic force on magnetic materials and other magnets.

Electromagnets are useful because their magnetic properties can be controlled by the user. The strength of the magnetic field can be increased by adding more turns of wire to the solenoid or by increasing the current in the wire. An electromagnet can even be turned off by shutting off the current to the wire.

Get It?

Compare and contrast permanent magnets and electromagnets.

Making an electromagnet rotate

The forces exerted on an electromagnet by another magnet can be used to make the electromagnet rotate. **Figure 11** shows an electromagnet suspended between the poles of a permanent magnet. The poles of the electromagnet are repelled by the like poles and attracted by the unlike poles of the permanent magnet.

When the electromagnet is in the position shown on the left side of **Figure 11,** there is a downward force on the left side and an upward force on the right side of the electromagnet. These forces cause the electromagnet to rotate as shown.

The electromagnet continues to rotate until its poles are next to the opposite poles of the permanent magnet, as shown on the right of **Figure 11.** Like a compass needle, the electromagnet will come to rest so that it is aligned in the magnetic field.

Figure 11 An electromagnet can be made to rotate in a magnetic field.

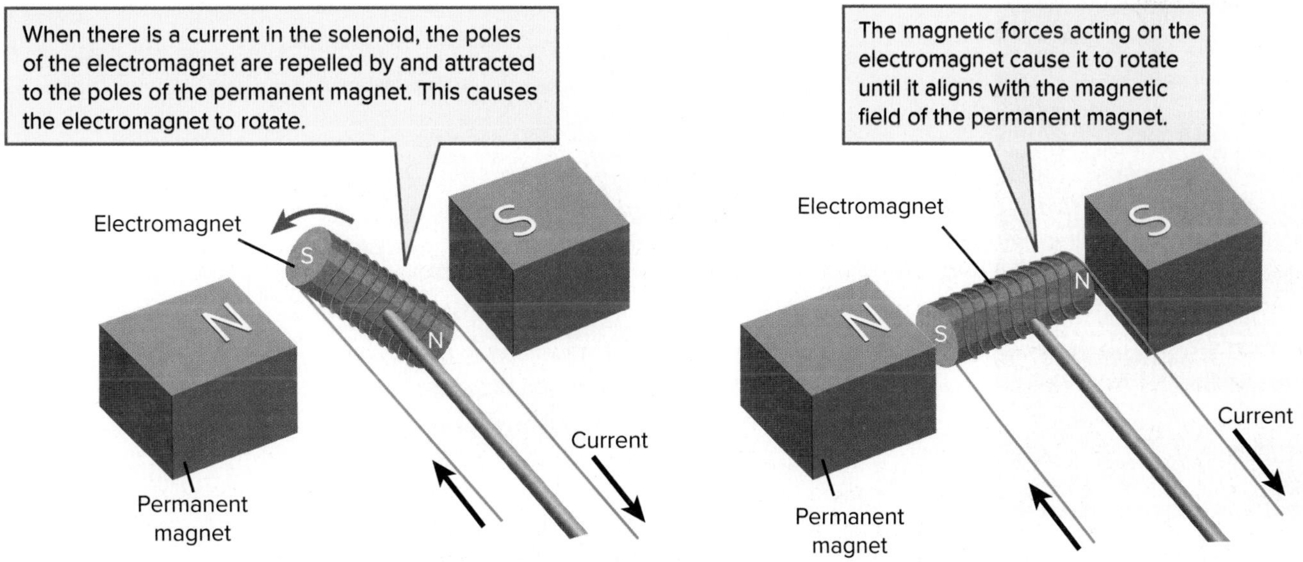
When there is a current in the solenoid, the poles of the electromagnet are repelled by and attracted to the poles of the permanent magnet. This causes the electromagnet to rotate.
Electromagnet
S
N
S
N
Current
Permanent magnet
The magnetic forces acting on the electromagnet cause it to rotate until it aligns with the magnetic field of the permanent magnet.
Electromagnet
S
N
S
N
Current
Permanent magnet

Using Electromagnets

When an electromagnet rotates, electrical energy is converted into mechanical energy to do work. Electromagnets do work in various devices, such as stereo speakers and electric motors.

Speakers

How does musical information played on an electronic device become sound that you can hear? The sound is produced by a loudspeaker, which is an electromagnet connected to a flexible speaker cone. The electromagnet changes electrical energy into the mechanical energy that vibrates the speaker cone to produce the sounds that you hear. **Figure 12** shows the parts of a loudspeaker.

When you listen to music, the electronic device produces a voltage that changes according to the musical information in the music file. This varying voltage causes a varying electric current in the electromagnet. Both the size and the direction of the electric current change, depending on the information in the music file.

The varying electric current causes both the strength and the direction of the magnetic field in the electromagnet to change. This changing magnetic field direction causes the electromagnet to be attracted to or repelled by the surrounding permanent magnet. The permanent magnet is fixed, so the changing magnetic force makes the electromagnet move back and forth. This causes the speaker cone to vibrate and makes the surrounding air vibrate to create sound waves.

Get It?
Summarize how a loudspeaker uses an electromagnet to produce sound.

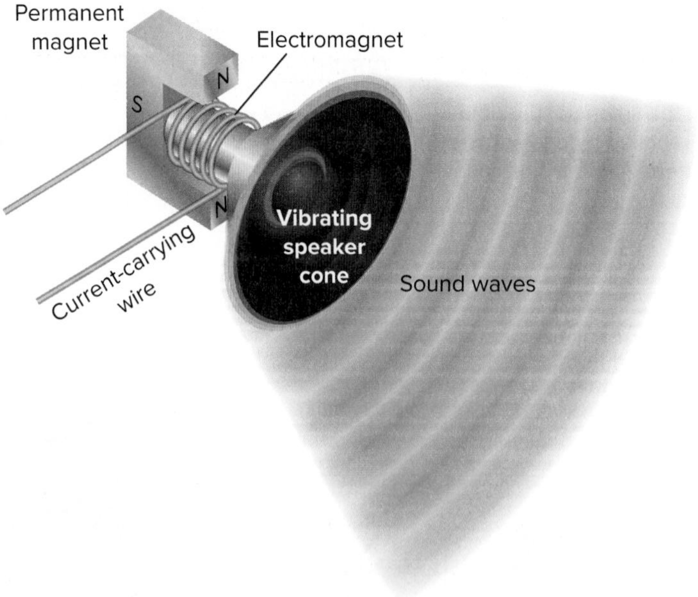

Figure 12 Changes in the electric current cause the electromagnet to move back and forth. The attached speaker cone vibrates and produces sound.

Explain *why a speaker needs a permanent magnet to produce sound.*

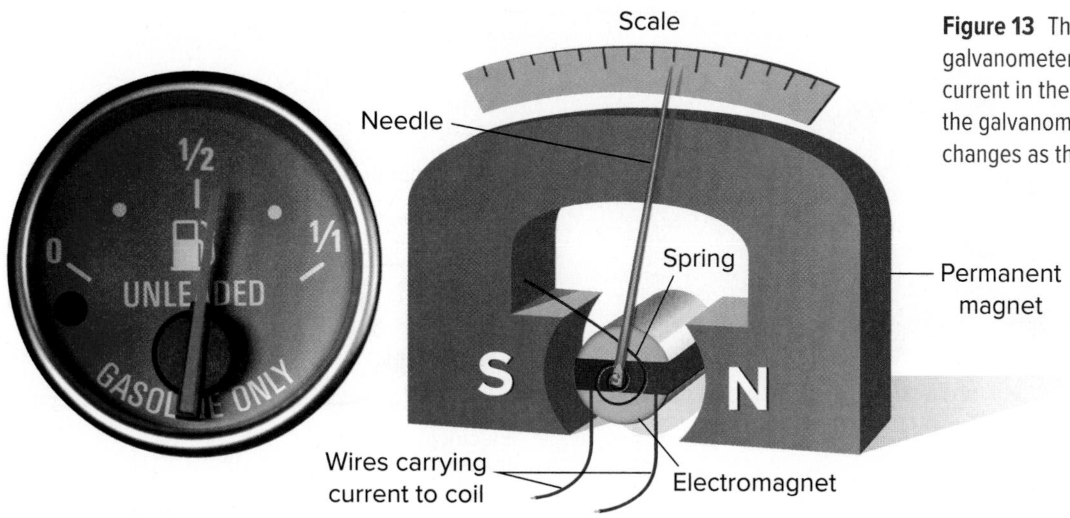

Figure 13 The rotation of the needle in a galvanometer depends on the amount of current in the electromagnet. The current in the galvanometer in a car's fuel gauge changes as the amount of fuel changes.

Galvanometers

You have probably noticed the gauges in the dashboard of a car. One gauge shows the amount of gasoline in the tank. How does a change in the amount of gasoline in a tank make a needle move in a gauge on the dashboard? The fuel gauge is a galvanometer. A **galvanometer** is a device that uses an electromagnet to measure electric current.

A diagram of a galvanometer is shown in **Figure 13.** In a galvanometer, the electromagnet is connected to a small spring. When there is a current in the electromagnet, it will rotate until the force exerted by the spring is balanced by the magnetic forces on the electromagnet. Changing the amount of current in the electromagnet changes the strength of the force between the electromagnet and the permanent magnet. Therefore, the amount that the needle rotates is related to the amount of current in the electromagnet. If the galvanometer is marked with a calibrated scale, it can be used to measure the current in a circuit.

In a car, a float in the fuel tank is attached to a sensor. The sensor sends a current to the fuel gauge galvanometer. As the level of the float in the tank changes, the amount of current sent by the sensor changes, and the rotation of the needle changes. The gauge is calibrated so that the current sent when the tank is full causes the needle to rotate to the full mark on the scale.

Electric Motors

Do you ever use an electric fan, like the one in **Figure 14,** to keep cool on hot summer days? A fan uses an electric motor to turn the blades. An **electric motor** is a device that changes electrical energy into mechanical energy. Electric motors are used in all types of industry, agriculture, and transportation. If you were to look carefully, you might find electric motors in every room of your house. For example, electric motors are used in vacuum cleaners, computers, and hair dryers.

Figure 14 This electric fan uses an electric motor to transform electrical energy into mechanical energy that turns the blades.

List *three additional devices that contain electric motors.*

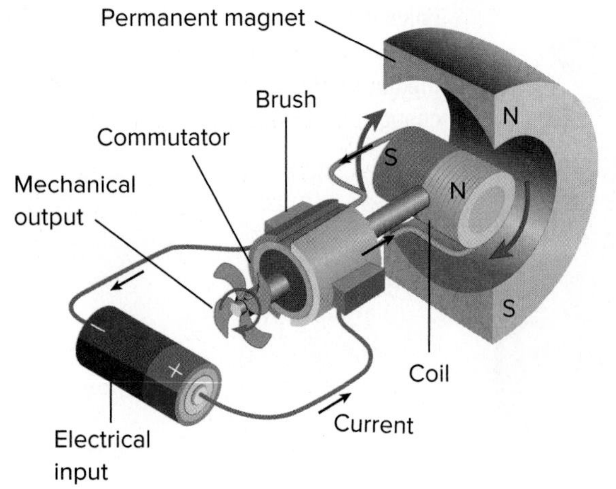

Step 1 When there is a current in the coil, the magnetic force between the permanent magnet and the coil causes the coil to rotate.

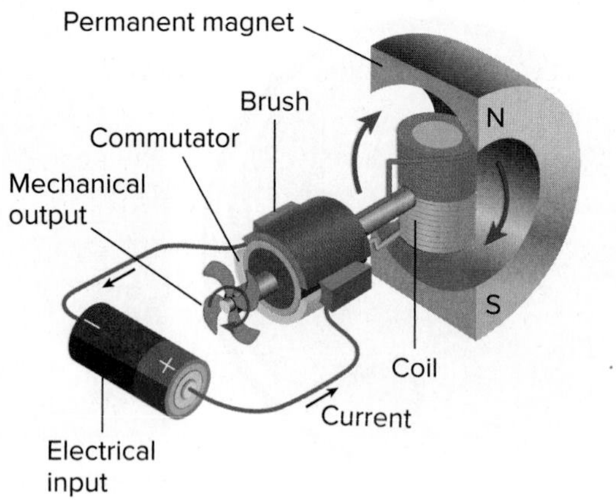

Step 2 In this position, the brushes are not in contact with the commutator, and there is no current in the coil. The inertia of the coil keeps it rotating.

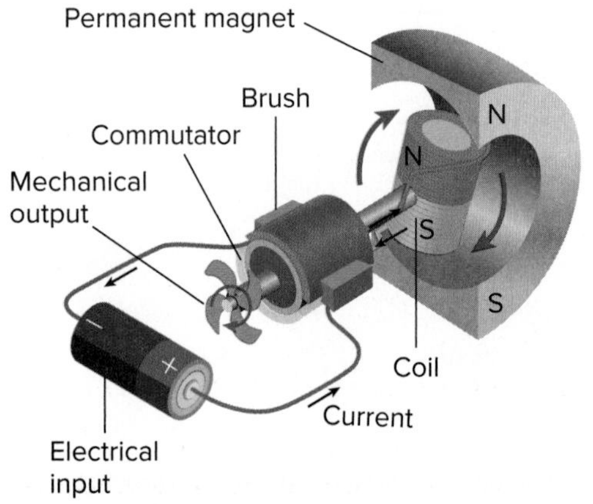

Step 3 The commutator reverses the direction of the current in the coil. This flips the north and south poles of the magnetic field around the coil.

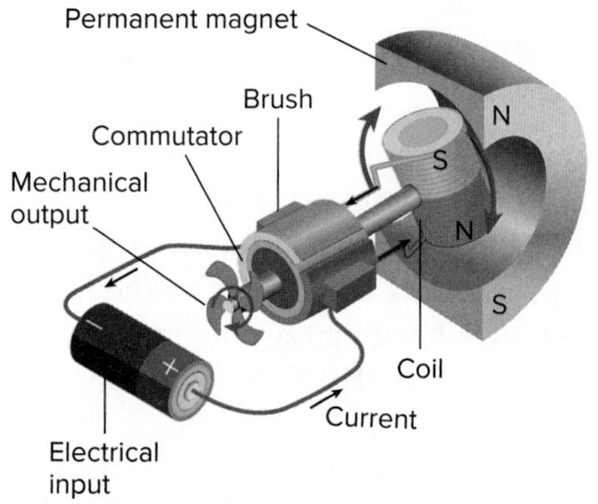

Step 4 The coil rotates until its poles are opposite the poles of the permanent magnet. The commutator reverses the current, and the coil keeps rotating.

Figure 15 In a simple electric motor, a coil rotates between the poles of a permanent magnet. To keep the coil rotating, the current must change direction twice during each rotation.

A simple electric motor

A diagram of the simplest type of electric motor is shown in **Figure 15.** The main parts of a simple electric motor include a wire coil, a permanent magnet, and a source of electric current, such as a battery. The electric current makes the coil an electromagnet.

A simple electric motor also includes components called brushes and a commutator. The brushes are conducting pads connected to the battery. The brushes make contact with the commutator, which is a conducting metal ring that is split. Each half of the commutator is connected to one end of the coil so that the commutator rotates with the coil. The brushes and the commutator form a closed electric circuit between the battery and the coil.

Making the motor spin

When there is a current in the coil, the forces between the coil and the permanent magnet cause the coil to rotate, as shown in step 1 of **Figure 15.** The coil rotates until it reaches the position shown in step 2. The brushes line up with the gaps in the commutator and no longer make contact with it. As a result, there is no current in the coil, and no magnetic forces are exerted on the coil. However, the inertia of the coil causes it to continue rotating.

In step 3, the coil has rotated so that the brushes are again in contact with the commutator. However, the halves of the commutator that are in contact with the positive and negative battery terminals have switched. The current in the coil has reversed direction. Now the top of the electromagnet is a north magnetic pole, and the bottom is a south pole. These poles are repelled by the nearby like poles of the permanent magnet, causing the magnet to continue to rotate in the same direction.

In step 4, the coil rotates until its poles are next to the opposite poles of the permanent magnet. It will continue to rotate until the brushes are not in contact with the commutator. Inertia keeps the electromagnet rotating, and it returns to the position of step 1. The cycle repeats. The coil will rotate as long as the battery remains connected. **Figure 16** shows a simple motor.

Figure 16 The coil in this simple motor will continue to rotate unless the battery is removed or its energy source is depleted.

Check Your Progress

Summary

- Moving electric charges, such as electric currents, are surrounded by magnetic fields.
- The electromagnetic force is one of the fundamental forces.
- An electromagnet is a temporary magnet consisting of a current-carrying wire wrapped around an iron core.
- The magnetic field strength of an electromagnet can be controlled by changing the number of wire loops and by changing the current in the wire.
- In a simple electric motor, an electromagnet rotates between the poles of a permanent magnet.

Demonstrate Understanding

8. **Explain** what happens when a magnet is placed near a current-carrying wire.
9. **Infer** A bar magnet is repelled when an electromagnet is brought close to it. Describe how the bar magnet would have moved if the current in the electromagnet had been reversed.
10. **Define** the term *electromagnetic force*. Give an example of a machine that uses this force.
11. **Describe** two ways that you could increase the strength of the magnetic field produced by an electromagnet.
12. **Explain** how a simple electric motor transforms electrical energy into mechanical energy.

Explain Your Thinking

13. **Predict** How would an electromagnet's magnetic field change if the iron core were replaced by an aluminum core?
14. **MATH Connection** The magnetic field around a current-carrying wire at a distance of 1 cm is twice as strong as at 2 cm. How does the field strength at 0.5 cm compare to the field strength at 1 cm?

LEARNSMART® Go online to follow your personalized learning path to review, practice, and reinforce your understanding.

PRODUCING ELECTRIC CURRENT

FOCUS QUESTION

How can magnetism be used to produce electric current?

Electromagnetic Induction

Working independently in 1831, Michael Faraday in Britain and Joseph Henry in the United States both found that moving a loop of wire through a magnetic field caused an electric current in the wire. They also found that moving a magnet through a loop of wire produces a current, as shown in **Figure 17.** In both cases, the mechanical energy associated with the motion of the wire loop or the magnet is converted into electrical energy associated with the current in the wire.

When the magnet and wire loop are moving relative to each other, the magnetic field inside the loop changes with time and induces, or generates, an electric current in the wire coil. The generation of an electric current by a changing magnetic field is **electromagnetic induction**.

Get It?
Define the term *electromagnetic induction.*

Electromagnetic induction can occur in other ways, too. For example, if the current in a wire changes with time, then the magnetic field around the wire is also changing. This changing magnetic field can induce a current in a nearby coil. Both types of electromagnetic induction are important to generating and transmitting the electrical energy that you use.

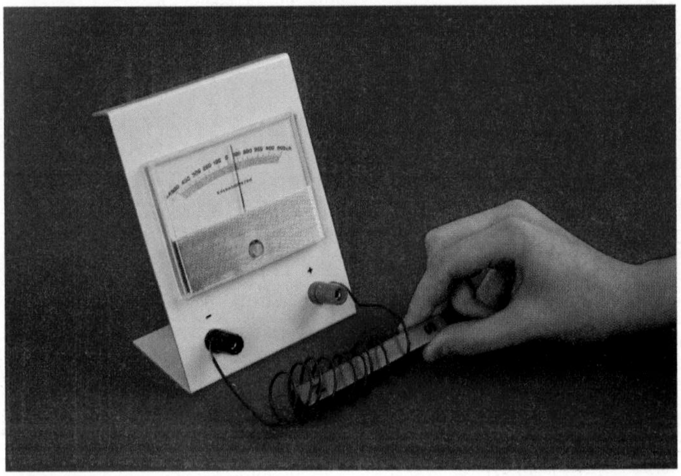

Figure 17 When the magnet is moved inside the solenoid, it induces a current in the solenoid. This solenoid is attached to a galvanometer that measures the current in microamperes.

3D THINKING DCI Disciplinary Core Ideas CCC Crosscutting Concepts SEP Science & Engineering Practices

COLLECT EVIDENCE

Use your Science Journal to record the evidence you collect as you complete the readings and activities in this lesson.

INVESTIGATE

GO ONLINE to find these activities and more resources.

Applying Practices: Investigate Electromagnetism
HS-PS2-5. Plan and conduct an investigation to provide evidence that an electric current can produce a magnetic field and that a changing magnetic field can produce an electric current.

Generators

Most of the electrical energy that you use every day is provided by generators. A **generator** uses electromagnetic induction to transform mechanical energy into electrical energy. **Figure 18** shows a hand-operated generator being used to power a flashlight. The mechanical energy is provided by turning the handle on the generator. The generator transforms the mechanical energy into electrical energy, which lights the bulb.

Figure 18 The coil in a generator is rotated by an external source of mechanical energy. Here, the student supplies the mechanical energy by turning the handle.

Simple generators

A diagram of a simple generator is shown in **Figure 19.** In this type of generator, a current is produced in the coil as the coil rotates between the poles of a permanent magnet. As the generator's wire coil rotates through the magnetic field of the permanent magnet, a current is induced in the coil. After the wire coil makes one-half of a revolution, the ends of the coil are moving past the opposite poles of the permanent magnet. This causes the current to change direction.

As the coil keeps rotating, the current that is produced continues to change direction. The direction of the current in the coil changes twice with each revolution. The frequency with which the current changes direction can be controlled by regulating the rotation rate of the generator. In the United States, electrical energy for residential use is produced by generators that rotate 60 times a second (60 Hz). Therefore, the current changes direction 120 times per second.

Using electric generators

Generators similar to the one in **Figure 19** are used in cars, where they are called alternators. The alternator provides electrical energy to operate lights and other accessories. Spark plugs in the car's engine also use this electrical energy to ignite the fuel in the cylinders of the engine. Once the engine is running, it provides the mechanical energy that is used to turn the coil in the alternator.

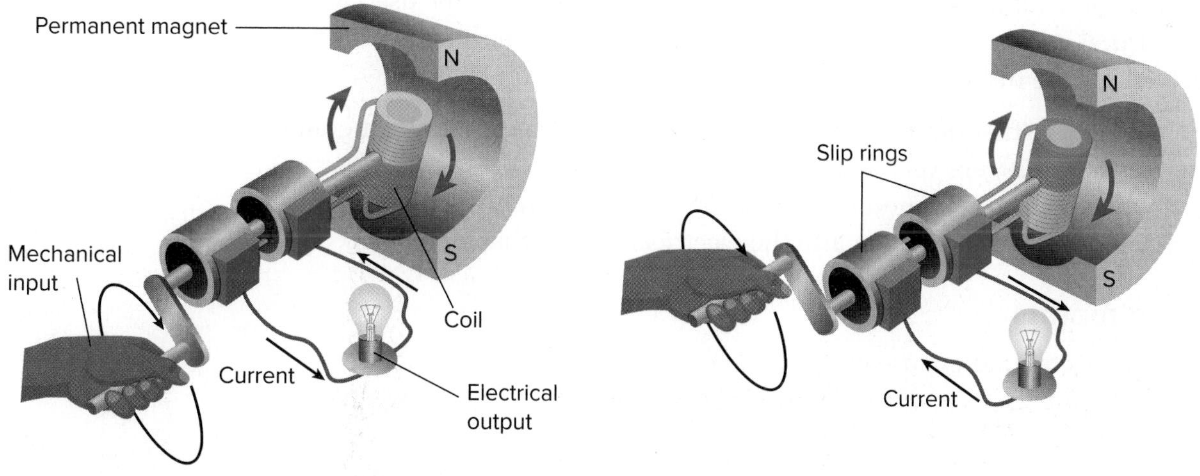

Figure 19 Moving the coil in the magnetic field of the permanent magnet induces a current in the coil. The current in the coil changes direction each time the ends of the coil move past the poles of the permanent magnet.

These wind turbines capture mechanical energy from wind to generate electricity. When the wind blows, the spinning turbine rotates a permanent magnet in the attached electric generator at the top of the tower.

Moving water provides the mechanical energy to rotate the large magnets in Hoover Dam's generators. Each can generate more than 100,000 kW of electrical power.

Figure 20 At a power plant, an energy source, such as wind or water, is used to rotate a turbine. The rotating turbine is attached to a permanent magnet, which induces a current in the coil.

Rotating the permanent magnet

Suppose that the coil in a generator is fixed and the permanent magnet rotates. The current generated would be the same as when the coil rotates and the magnet does not move. The huge generators used in electrical power plants are made this way. Mechanical energy is used to rotate the magnet, and the current is induced in the stationary coil.

Generating electricity for your home

Unless you have a generator in your home, the electrical energy needed to watch television or to wash your clothes comes from a power plant. Each power plant must have an energy source.

For example, some power plants burn fossil fuels or use nuclear reactions to produce thermal energy. This thermal energy is used to heat water and produce steam. Thermal energy is then converted into mechanical energy as the steam pushes the turbine blades. A **turbine** (TUR bine) is a large wheel that rotates when pushed by water, wind, or steam. In some areas, fields of wind turbines, like those shown in **Figure 20,** can be used to capture the mechanical energy in wind. Other power plants use the mechanical energy in falling water to drive the turbine.

The turbine is connected to the rotating magnet in the generator. Power plant generators, like those shown in **Figure 20,** are very large. The electromagnets in these generators have many coils of wire wrapped around huge iron cores. The generator changes the mechanical energy of the rotating turbine into the electrical energy that you use.

Generators and motors

Both generators and electric motors use electromagnets to convert electrical and mechanical energy. **Figure 21** summarizes the differences and similarities between electric motors and generators.

FIGURE 21 Visualizing Motors and Generators

Motors convert electrical energy into mechanical energy. In contrast, generators convert mechanical energy into electrical energy.

<table>
<tr><th>Motor</th><th>Generator</th></tr>
<tr><td colspan="2">Input</td></tr>
<tr><td>

Electrical Energy

An outside source provides electrical energy. This provides electric current which changes the motor's wire coil into an electromagnet.

</td><td>

Mechanical Energy

An outside source provides mechanical energy. This mechanical energy causes the generator's turbine to rotate.

</td></tr>
<tr><td colspan="2">Output</td></tr>
<tr><td>

Mechanical Energy

The mechanical energy from the motor is used to turn the fan's blades.

</td><td>

Electrical Energy

The electrical energy from the generator is transmitted and used to light the bulb.

</td></tr>
<tr><td colspan="2">Diagram</td></tr>
<tr><td>

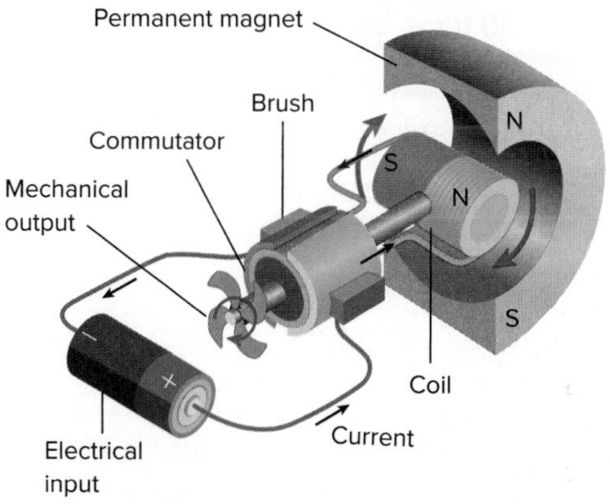

The attractive and repulsive forces between the electromagnet and the permanent magnet cause the electromagnet to rotate. Its rotation causes the shaft to rotate, providing a mechanical output.

</td><td>

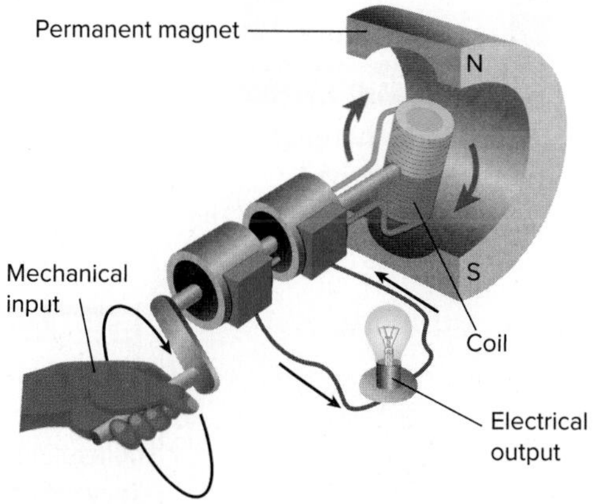

The mechanical input rotates the coil or permanent magnet. The changing magnetic field induces a current in the coil, providing an electrical output.

</td></tr>
</table>

Direct and Alternating Currents

Because power outages can occur, some electric devices, such as alarm clocks, use batteries as a backup source of electrical energy. However, the current produced by a battery is different from the current produced by an electric generator. A battery produces a direct current. **Direct current** (DC) is electric current that is always in one direction through a wire.

When you plug an appliance into a wall outlet, it is receiving an alternating current. **Alternating current** (AC) is electric current that reverses the direction in a regular pattern. In North America, the alternating current in a wall socket has a frequency of 60 Hz and an average voltage of 120 V. Electronic devices that use backup batteries usually require direct current to operate. The device's electronic components reduce the voltage of the alternating current and convert it to direct current.

Transformers

To reduce the voltage without changing the amount of electrical energy, many devices use transformers. A **transformer** is a device that increases or decreases the voltage of an alternating current. A transformer is made of a primary coil and a secondary coil wrapped around the same iron core. A transformer that increases the voltage is called a step-up transformer, and a transformer that decreases the voltage is called a step-down transformer. Both types of transformers are shown in **Figure 22.**

How a transformer works

As an alternating current passes through the primary coil, the coil's magnetic field magnetizes the iron core. The primary coil's magnetic field changes direction at the same frequency as the current in the primary coil. The magnetic field in the iron core also changes direction at that frequency. The changing magnetic field in the iron core induces an alternating current in the secondary coil. The input current and the output current alternate at the same frequency.

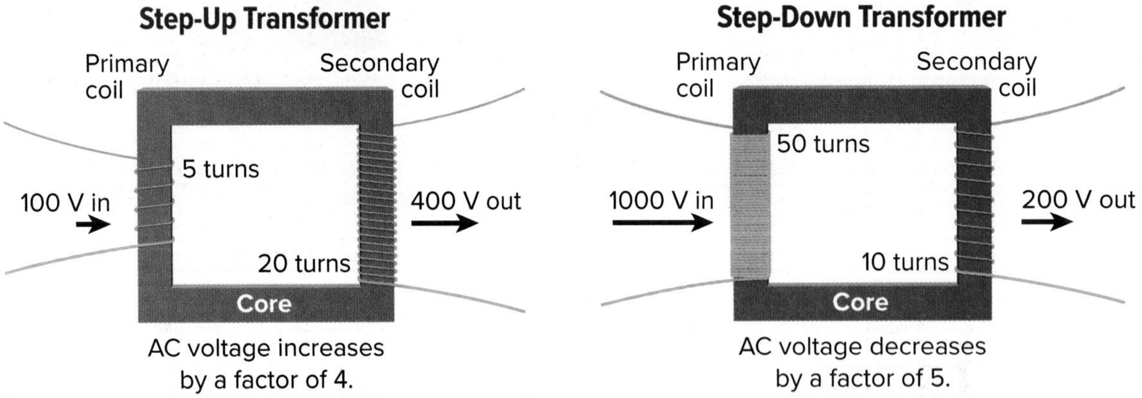

A step-up transformer increases voltage. The secondary coil has more turns than the primary coil.

A step-down transformer decreases voltage. The secondary coil has fewer turns than the primary coil.

Figure 22 Transformers increase or decrease voltage, depending on the ratio of turns in the primary coil to turns in the secondary coil.

Infer *the output voltage of the step-up transformer if the input voltage were 120 V.*

ACADEMIC VOCABULARY

primary
first in order of time or importance

Elementary school is sometimes called primary school.

STEM CAREER Connection

Electrical Engineer
Have you thought about ways to make better electric cars or develop more efficient power generation equipment? Electrical engineers design, develop, and test electrical equipment. They typically have at least a bachelor's degree, but research jobs require a master's degree or Ph.D.

Calculate output voltage

The output voltage of a transformer depends on the input voltage and the number of turns in the primary and secondary coils. This relationship can be expressed in an equation.

Transformer Voltage Equation

$$\frac{\text{output voltage (in volts)}}{\text{input voltage (in volts)}} = \frac{\text{turns in secondary coil}}{\text{turns in primary coil}}$$

$$\frac{V_{out}}{V_{in}} = \frac{N_2}{N_1}$$

In this equation, *N* stands for the number of turns in the coil. Because *N* is just a number, it has no units. Consider the step-up transformer in **Figure 22.** The secondary coil has four times as many turns as the primary coil has. So, the ratio of the output voltage to the input voltage is four to one. For this transformer, the output voltage is four times greater than the input voltage.

EXAMPLE Problem 1

SOLVE FOR OUTPUT VOLTAGE A transformer has 150 turns in its primary coil and 50 turns in its secondary coil. If the input voltage is 12 V, what is the output voltage?

Identify the Unknown: output voltage: V_{out}

List the Knowns: input voltage: $V_{in} = 12$ V

turns in the primary coil: $N_1 = 150$

turns in the secondary coil: $N_2 = 50$

Set Up the Problem: Use the equation: $\frac{V_{out}}{V_{in}} = \frac{N_2}{N_1}$

Solve for $V_{out} = \frac{V_{in}N_2}{N_1}$

Solve the Problem: $V_{out} = \frac{12 \text{ V} \times 50}{150} = 4 \text{ V}$

Check the Answer: The number of turns in the secondary coil is one-third the number of turns in the primary coil. An output voltage (4 V) that is one-third the input voltage (12 V) makes sense.

PRACTICE Problems

ADDITIONAL PRACTICE

15. A transformer has 90 turns in the primary coil and 9 turns in the secondary coil. If the output voltage is 6 V, what is the input voltage?
16. You have a transformer with 50 turns in the primary coil. If you want to increase the voltage from 5 V to 6 V, how many turns should be in the secondary coil?
17. **CHALLENGE** High voltage power lines transmit electrical energy at 240,000 V. Before the current reaches your home, the voltage is decreased to the standard voltage for a wall socket. If a single step-down transformer is used, what is the ratio of turns in the primary coil to turns in the secondary coil?

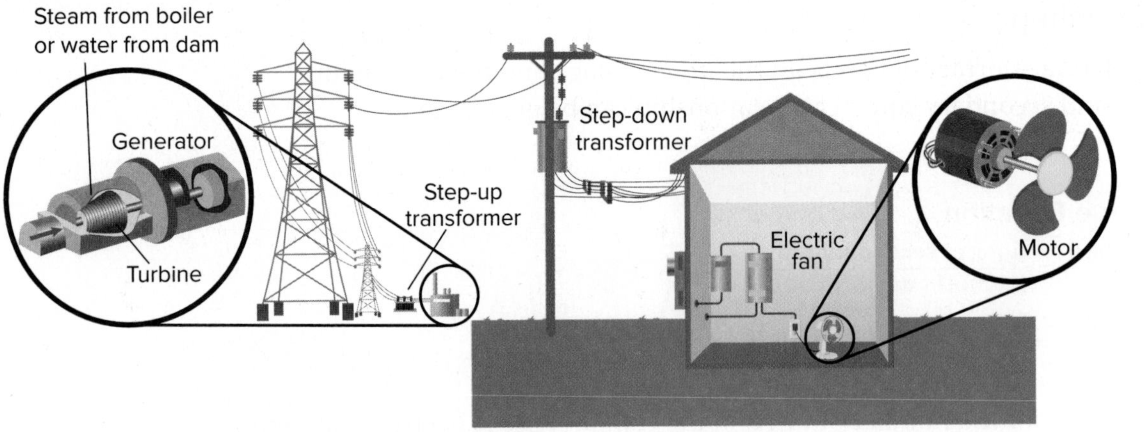

Figure 23 Many steps are involved in the production and transportation of electric current to your home.
Identify *the steps that involve electromagnetic induction.*

Transmitting electrical energy

Electrical energy is often transmitted along wires known as power lines, as shown in **Figure 23.** During transmission, some of the electrical energy is converted into thermal energy, due to the resistance of the wires, and cannot be used to operate electrical devices. Electric resistance increases as the wires get longer. Power lines can be many kilometers long. As a result, large amounts of electrical energy transform into thermal energy that heats the surroundings.

To reduce this wasteful conversion, a step-up transformer increases the voltage to more than 100,000 V before transmission. After transmission, a step-down transformer decreases the voltage to 120 V for consumer use. This process is shown in **Figure 23.**

Check Your Progress

Summary

- A changing magnetic field can induce a current in a wire.
- A generator transforms mechanical energy into electrical energy.
- Direct current is current in one direction. Alternating current changes direction in a regular pattern.
- A transformer increases or decreases the voltage of an alternating current.

Demonstrate Understanding

18. **Explain** why a magnet sitting next to a wire does not induce a current in the wire.
19. **Define** the term *electromagnetic induction,* and explain how a generator uses electromagnetic induction.
20. **Describe** the difference between the current from a battery and the current from an electric socket.
21. **Summarize** the steps involved when a transformer changes the voltage of an alternating current.

Explain Your Thinking

22. **Explain** Why is the output voltage from a transformer zero if the current in the primary coil is a direct current?
23. **MATH Connection** A transformer has 1,000 turns of wire in the primary coil and 50 turns in the secondary coil. If the input voltage is 2,400 V, what is the output voltage?

LEARNSMART® Go online to follow your personalized learning path to review, practice, and reinforce your understanding.

STEM AT WORK

Locating Land Mines

The end of a war does not always make civilians safe from weapons. In nearly 70 countries, millions of buried explosive devices called land mines pose a silent, long-lasting threat. When a person steps on a land mine, the mine explodes and can cause blindness, burns, loss of limbs, and death. Humanitarian groups make neutralizing land mines a high priority.

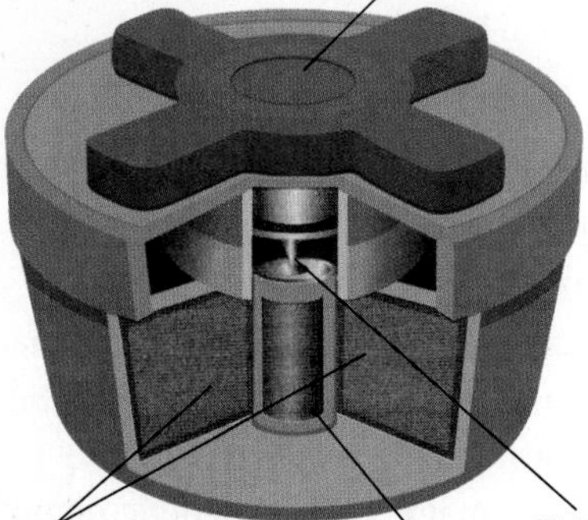

Antipersonnel land mines contain a firing mechanism, a detonator, and an explosive material encased in plastic.

Anatomy of a land mine

Antipersonnel land mines target humans rather than heavy vehicles. They are relatively small, as shown in the figure below, and are triggered by only a few kilograms of pressure. Locating them is difficult, but their components, shown in the figure on the right, provide clues. Some of a mine's components, such as the firing pin, are often made of metal and have magnetic properties, so metal detectors are often used to find land mines.

Land mines are often difficult to find because of their small size.

How do metal detectors work?

An alternating electric current is sent through a wire coil at the bottom of the detector. The coil is an electromagnet that creates an alternating magnetic field.

Some buried metallic objects respond to this pulsating magnetic field with weak magnetic fields. A receiver coil detects this field and converts it to an audible signal, pinpointing the object's location. Metal detection is dangerous and inefficient. Signals from buried cans, rusty nails, and land mines must all be investigated, slowing the process. In addition, manufacturers deliberately limit the metal in their mines, making them difficult to locate.

A different approach

Using the magnetic properties of the explosive material in land mines could be a better way to locate them. Researchers discovered that some atomic nuclei in trinitrotoluene (TNT) produce a detectable signal in the presence of a magnetic field. An instrument that uses this new technique has the potential to save lives worldwide.

RESEARCH, COMPARE, AND ANALYZE

Research other innovative methods for locating land mines. Create a chart to compare these methods. Choose one method and prepare a presentation about why you believe it is a safe, effective way to detect buried mines.

Dmitry Melnikov/123RF

MODULE 7

STUDY GUIDE

GO ONLINE to study with your Science Notebook.

Lesson 1 MAGNETISM

- A magnet exerts a force on other magnets.
- A magnet has a north pole and a south pole. Like magnetic poles repel, and unlike poles attract.
- Atoms in magnetic materials behave like magnets.
- Magnetic domains are regions in a magnetic material where the magnetic poles of the atoms are aligned.
- In a permanent magnet, the magnetic domains remain aligned without an external magnetic field.

- magnetism
- magnetic field
- magnetic pole
- magnetic domain

Lesson 2 ELECTRICITY AND MAGNETISM

- Moving electric charges, such as currents, are surrounded by magnetic fields.
- The electromagnetic force is one of the fundamental forces.
- An electromagnet is a temporary magnet consisting of a current-carrying wire wrapped around an iron core.
- The magnetic field strength of an electromagnet can be controlled by changing the number of wire loops and by changing the current in the wire.
- In a simple electric motor, an electromagnet rotates between the poles of a permanent magnet.

- electromagnetic force
- electromagnetism
- solenoid
- electromagnet
- galvanometer
- electric motor

Lesson 3 PRODUCING ELECTRIC CURRENT

- A changing magnetic field can induce a current in a wire.
- A generator transforms mechanical energy into electrical energy.
- Direct current is current in one direction. Alternating current changes direction in a regular pattern.
- A transformer increases or decreases the voltage of an alternating current.

- electromagnetic induction
- generator
- turbine
- direct current
- alternating current
- transformer

THREE-DIMENSIONAL THINKING

Module Wrap-Up

REVISIT THE PHENOMENON

What makes something magnetic?

CER Claim, Evidence, Reasoning

Explain Your Reasoning Revisit the claim you made when you encountered the phenomenon. Summarize the evidence you gathered from your investigations and research and finalize your Summary Table. Does your evidence support your claim? If not, revise your claim. Explain why your evidence supports your claim.

STEM UNIT PROJECT

Now that you've completed the module, revisit your STEM unit project. You will summarize your evidence and apply it to the project.

GO FURTHER

SEP Data Analysis Lab

The New Wave in Electric Currents

A turbine is an engine with rotators, or blades, that spin. When the turbine rotator spins, electricity is produced. Until recently, the sources of energy that turn the turbine rotators have been steam, fossil fuels, or river water. However, many of these sources are being used up, or their use is polluting the environment.

Ocean waves and tides are environmentally friendly forms of energy. Waves represent constant energy movement that could provide electricity for large regions of the planet. A problem with using waves is that wave energy produces a back-and-forth motion, and turbines are built to work continuously in one direction. One experimental turbine uses the changes in air pressure produced by waves flowing into and out of an enclosed space. The following diagrams show how this kind of turbine works.

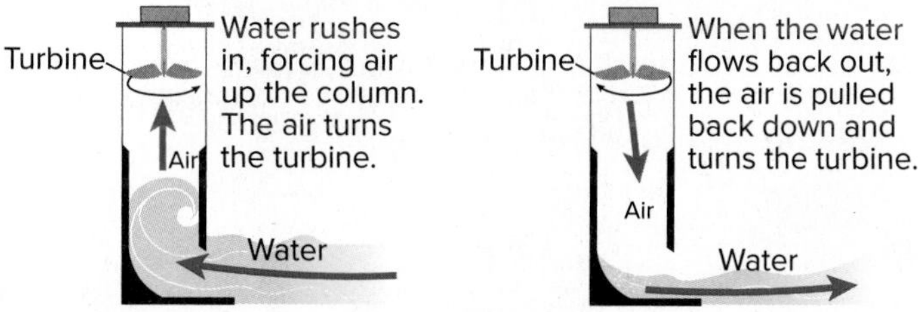

Turbine
Water rushes in, forcing air up the column. The air turns the turbine.
Air
Water
Turbine
When the water flows back out, the air is pulled back down and turns the turbine.
Air
Water

CER Analyze and Interpret Data

1. **Claim, Evidence, Reasoning** Why are ocean waves a promising form of energy to harvest?
2. **Claim, Evidence, Reasoning** Why are people searching for new energy sources?
3. **Claim, Evidence, Reasoning** State one problem with using wave energy as a source of electric production, and explain how the problem has been solved by the prototype discussed here.

Paul Ridsdale pictures/Alamy Stock Photo

Maximilian Stock Ltd/Science Source

MODULE 8

ENERGY SOURCES AND THE ENVIRONMENT

ENCOUNTER THE PHENOMENON

How could this relatively small space produce enough nuclear energy to sustain millions of people?

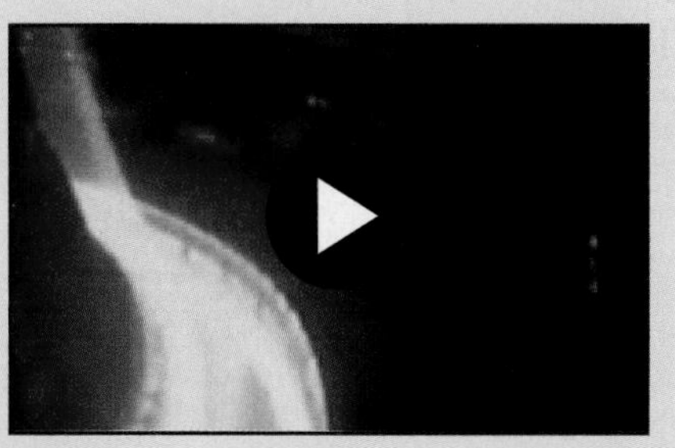

GO ONLINE to play a video about tokamak reactors.

SEP Ask Questions

Do you have other questions about the phenomenon? If so, add them to the driving question board.

CER Claim, Evidence, Reasoning

Make Your Claim Use your CER chart to make a claim about how this relatively small space can produce enough energy to sustain millions of people. Explain your reasoning.

Collect Evidence Use the lessons in this module to collect evidence to support your claim. Record your evidence as you move through the module.

Explain Your Reasoning You will revisit your claim and explain your reasoning at the end of the module.

GO ONLINE to access your CER chart and explore resources that can help you collect evidence.

LESSON 3: Explore & Explain: Energy from the Sun

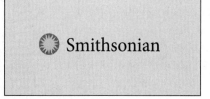

LESSON 4: Explore & Explain: Impact on Water

Smithsonian

Additional Resources

LESSON 1

FOSSIL FUELS

FOCUS QUESTION

How many ways have you relied on energy resources today?

Energy Resources

How many different ways have you relied on energy resources today? You can see energy being used in many ways, such as those shown in **Figure 1.** Furnaces and stoves use thermal energy to heat buildings and to cook food, respectively. Air conditioners use electrical energy to cool homes. Cars and other vehicles use mechanical energy to transport people and materials from one area to another.

Energy transformation

According to the law of conservation of energy, energy cannot be created or destroyed. Energy can only be transformed from one form to another. To use energy means to transform energy from one form into another. For example, you use energy when the chemical potential energy from coal, oil, or natural gas is transformed into thermal energy that is used to heat your home. The chemical potential energy from coal, oil, or natural gas can also be transformed into electrical energy that is delivered to your home and used to power a variety of devices.

Although energy cannot be destroyed, it can be converted to less useful forms. For example, when an electric current travels through power lines, about 10 percent of the electrical energy is converted to thermal energy. This reduces the amount of useful electrical energy that is delivered to homes, schools, and businesses.

Figure 1 Energy is used in many ways. Automobiles use energy from the burning of gasoline.

Identify *other processes in this photo that require energy resources.*

3D THINKING DCI Disciplinary Core Ideas CCC Crosscutting Concepts SEP Science & Engineering Practices

COLLECT EVIDENCE

Use your Science Journal to record the evidence you collect as you complete the readings and activities in this lesson.

INVESTIGATE

GO ONLINE to find these activities and more resources.

Review the News

Obtain information from a current news story about anthropogenic impacts from fossil fuel usage. **Evaluate** your source and **communicate** your findings to your class.

manttek/Shutterstock

Energy use in the United States

In 2017, the United States accounted for 16.5 percent of the total energy consumption in the world, using more energy than any other country aside from China. The top chart in **Figure 2** shows energy consumption in the United States in 2017 by sector. About 20 percent of the energy was used in homes for heating and cooling, to run appliances, to provide lighting, and for other household needs. About 29 percent was used for transportation, powering vehicles such as cars and planes. Another 18 percent was used by businesses to heat, cool, and light shops and buildings. About 32 percent of the energy was used by industry and agriculture for manufacturing and food production.

The bottom chart in **Figure 2** shows that about 80 percent of the energy used in the United States in 2017 came from burning fossil fuels. Nuclear power plants provided 9 percent, and alternative energies supplied 11 percent.

U.S. Energy Consumption by Sector

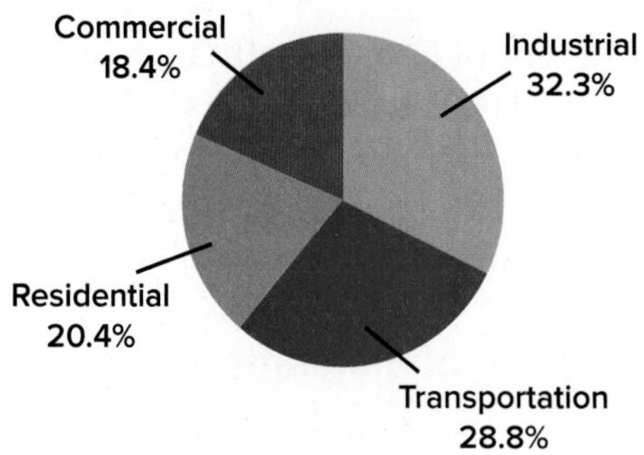

U.S. Energy Consumption by Source

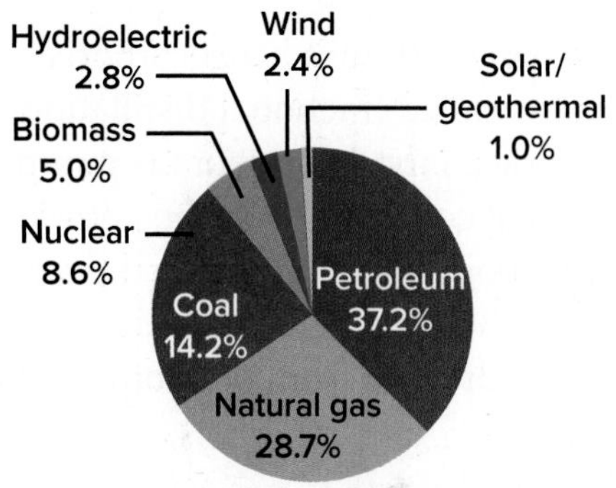

Figure 2 These circle graphs break down energy consumption in the United States in 2017. The top graph shows the percentage of energy used by different sectors. The bottom graph shows the percentage of energy obtained from different sources.

Interpret *Which energy source supplies the greatest amount of energy in the United States?*

Fossil Fuel Formation

In one hour of driving, a car might use two or three gallons of gasoline. It might be hard to believe that it took millions of years to make the fuels that are used to power your car, to produce electricity, and to heat your home. Coal, natural gas, and petroleum, also called crude oil, are **fossil fuels** because they form from the remains of ancient plants and animals that were buried and altered over millions of years.

Combustion reactions

When fossil fuels are burned, a combustion reaction occurs. During this reaction, carbon and hydrogen atoms combine with oxygen in the air to form carbon dioxide and water. This process converts the chemical potential energy that is stored in the bonds between atoms into thermal energy and light. Compared to wood, the energy stored in fossil fuels is much more concentrated. In fact, burning 1 kg of coal releases two to three times more energy than burning 1 kg of wood. **Figure 3** shows the energy content of different fuels.

Energy Content of Fuels

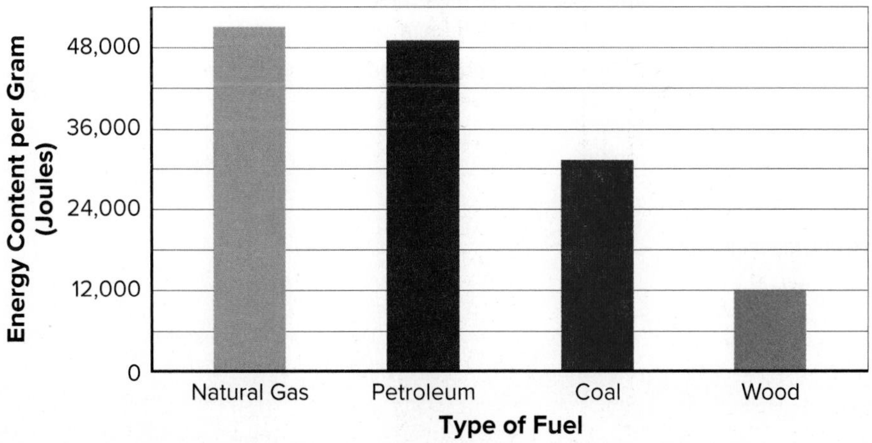

Figure 3 The fuel with the greatest chemical potential energy per gram releases the greatest amount of energy.

Petroleum

Millions of liters of petroleum, a fossil fuel, are pumped from wells within Earth's crust every day. **Petroleum** is a flammable liquid formed from the decay of ancient organisms, such as microscopic plankton and algae. It is a mixture of thousands of chemical compounds. Most of these compounds are hydrocarbons, which means that their molecules are made of different arrangements of carbon and hydrogen atoms.

Fractional distillation

Hydrocarbon compounds found in petroleum differ based on the number and arrangement of carbon and hydrogen atoms. The composition and structure of a hydrocarbon determines its chemical and physical properties.

The many different hydrocarbon compounds found in petroleum can be separated in a process called fractional distillation. This separation occurs in distillation towers at oil refineries. First, petroleum is pumped into the bottom of the tower and heated. The chemical compounds in the petroleum boil at different temperatures. Materials with the lowest boiling points rise to the top of the tower as vapor and are collected from the tower. Hydrocarbons with high boiling points, such as asphalt and waxes, remain liquid and are drained from the bottom of the tower.

Get It?

Explain how chemical compounds in petroleum are separated.

Petroleum use

Petroleum supplies around 37 percent of all energy generated in the United States each year. However, about 15 percent of petroleum-based materials in the United States are not used for fuel. Look at the materials in your home or classroom. Do you see any plastics? In addition to being a source of fuel, petroleum is used to make plastics, synthetic fabrics, cosmetics, and medicines, such as those shown in **Figure 4.** Also, lubricants such as grease and motor oil, as well as wax-based products and asphalt, are made from petroleum.

Get It?

Identify four specific non-fuel items that you use every day that are made of petroleum-based materials.

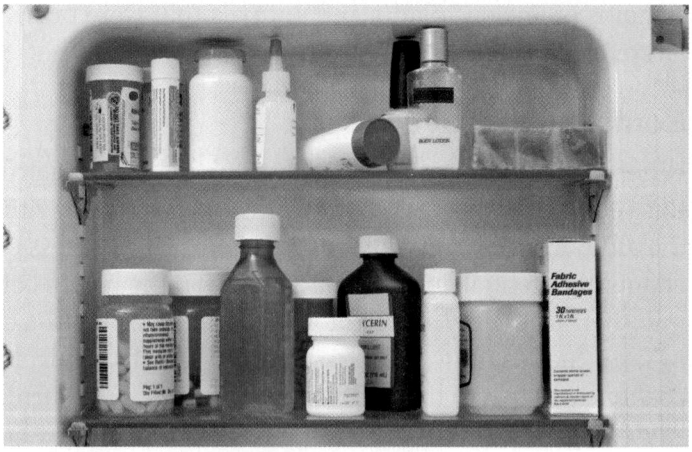

Figure 4 Petrolatum, also known as petroleum jelly, is blended with paraffin wax to make medicines, moisturizers, and other toiletries that you might find in a bathroom cabinet.

Identify *objects in your classroom that are petroleum-based products.*

WORD ORIGIN

petroleum

from the Middle English *petra*, meaning "rock", and *oleum*, meaning "oil"

Petroleum is a nonrenewable resource that is limited in supply.

Natural Gas

The chemical processes that produced petroleum when ancient organisms decayed and were buried in the seafloor also formed natural gas. Due to differences in density, lightweight natural gas compounds are found trapped on top of petroleum deposits. Natural gas is a fossil fuel composed mostly of methane, but it also contains other gaseous hydrocarbon compounds, such as propane and butane.

Natural gas contains more chemical potential energy per kilogram than petroleum or coal does. Additionally, natural gas burns cleaner than other fossil fuels, produces fewer pollutants, and leaves no ash residue. Natural gas is burned to provide energy for cooking, heating, and manufacturing. About 29 percent of the energy used in the United States comes from the combustion of natural gas. There is a good chance that your home has a stove, a furnace, a hot-water heater, or a clothes dryer that is powered by natural gas. Some cars and buses also are powered by natural gas.

Get It?

Identify advantages to using natural gas instead of petroleum or coal for energy.

Coal

Coal is a solid fossil fuel that can be found in mines, such as the one shown in **Figure 5.** During the first half of the twentieth century, most houses in the United States were heated by the combustion of coal. In fact, during this time, coal provided more than half of the energy used in the United States. Now, almost two-thirds of the energy used comes from petroleum and natural gas, and only about one-seventh comes from coal. About 90 percent of all the coal that is used in the United States is burned by power plants to produce electricity.

Figure 5 Coal mines are common in Pennsylvania, West Virginia, and Ohio. This region of the United States was home to ancient swamps hundreds of millions of years ago. Coal formed from the remains of plants that lived in these swamps.

Origin of coal

Coal mines were once the sites of ancient swamps. Coal formed as swampy plant material, buried beneath sediments, decayed and compacted into peat. Over millions of years, heat and pressure converted the peat into coal.

Coal is a mixture of hydrocarbons and other chemical compounds. Compared to petroleum and natural gas, coal contains more chemical impurities, such as sulfur and nitrogen-based compounds. As a result, more pollutants, including sulfur dioxide and nitrogen oxides, are produced when coal is burned.

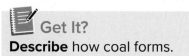

Get It?
Describe how coal forms.

Coal use

Coal is the most abundant fossil fuel in the world. The amount of coal available is estimated to last between 200 and 250 years at our current rate of consumption. Because of its supply, scientists are looking for ways to make coal a cleaner energy source. For example, filters on smoke stacks and stricter government standards have reduced harmful particulates released into the atmosphere when coal is burned.

U.S. Electricity Generation by Source

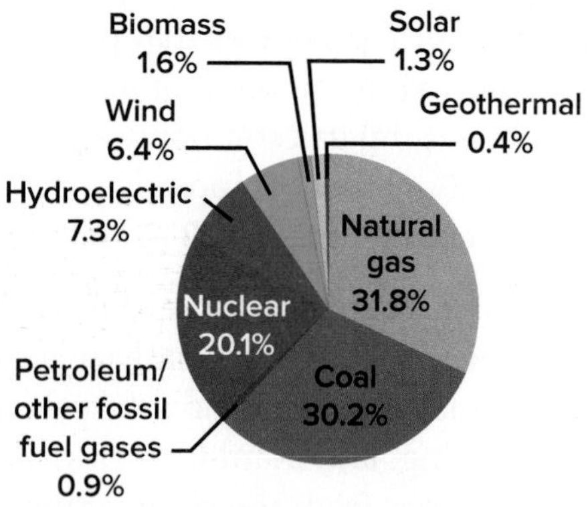

Figure 6 This circle graph shows the percentage of electrical energy that came from different energy sources used in the United States in 2017.

Electricity

Figure 6 shows that 63 percent of the electrical energy used in the United States in 2017 was produced by burning fossil fuels, such as natural gas and coal. How is the chemical potential energy stored in fossil fuels converted to electrical energy in a power plant? The process of energy conversion is shown in **Figure 7.**

Figure 7 Power plant efficiency describes how much energy is available to do work and produce electricity.
Determine *which stage in this process is the most inefficient.*

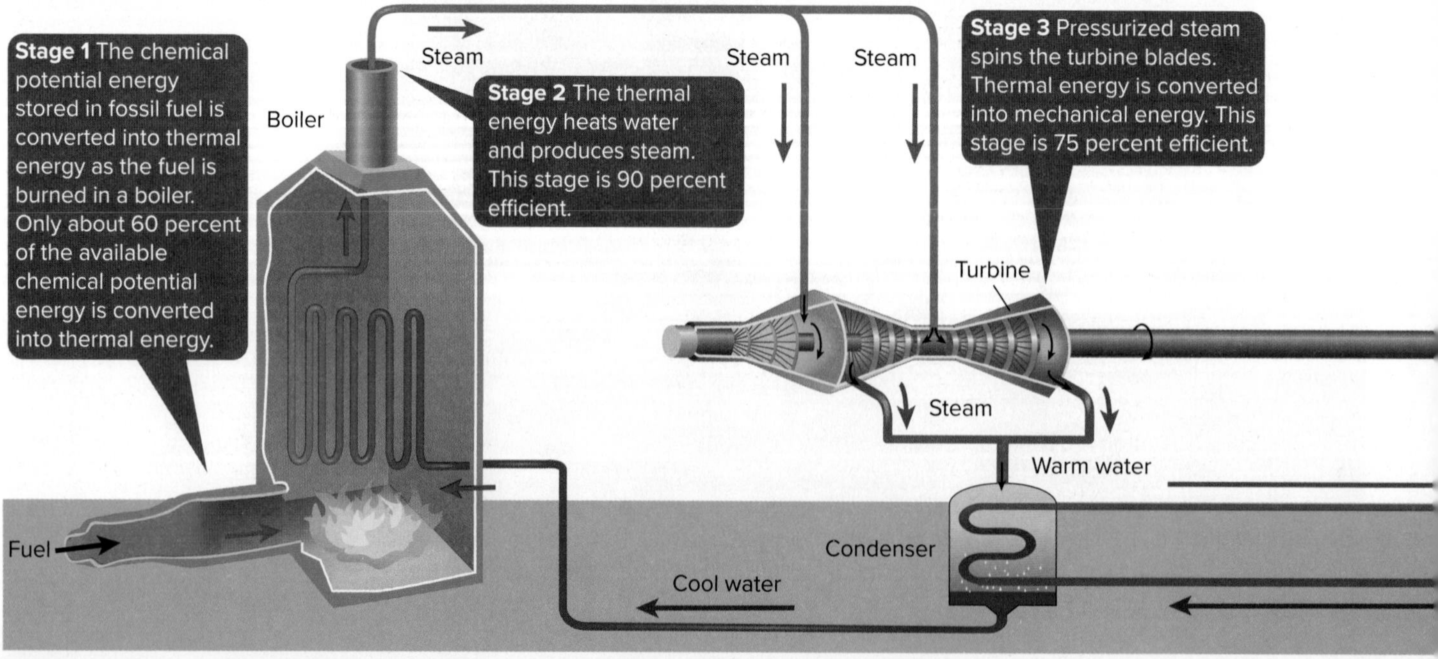

Stage 1 The chemical potential energy stored in fossil fuel is converted into thermal energy as the fuel is burned in a boiler. Only about 60 percent of the available chemical potential energy is converted into thermal energy.
Boiler
Steam
Stage 2 The thermal energy heats water and produces steam. This stage is 90 percent efficient.
Steam
Steam
Stage 3 Pressurized steam spins the turbine blades. Thermal energy is converted into mechanical energy. This stage is 75 percent efficient.
Turbine
Steam
Warm water
Condenser
Cool water
Fuel

Fuel burned in a boiler or combustion chamber converts chemical potential energy into thermal energy, which heats water and produces pressurized steam. This steam strikes the blades of a turbine, causing it to spin, converting thermal energy into mechanical energy. The shaft of the turbine connects to an electric generator, which converts mechanical energy into electrical energy. The electrical energy is then transmitted to homes, schools, and businesses through power lines.

Power plant efficiency

In the power plant, not all of the chemical potential energy stored in fuel is converted into electrical energy. Some energy is converted into thermal energy. As a result, no stage of the process of electricity production is 100 percent efficient.

The overall efficiency of a fossil fuel-burning power plant is roughly 35 percent. This means that only 35 percent of the energy stored in fossil fuels is transported to homes, schools, and businesses as electrical energy. The remaining 65 percent is converted into thermal energy. Often, this heat is released into the environment.

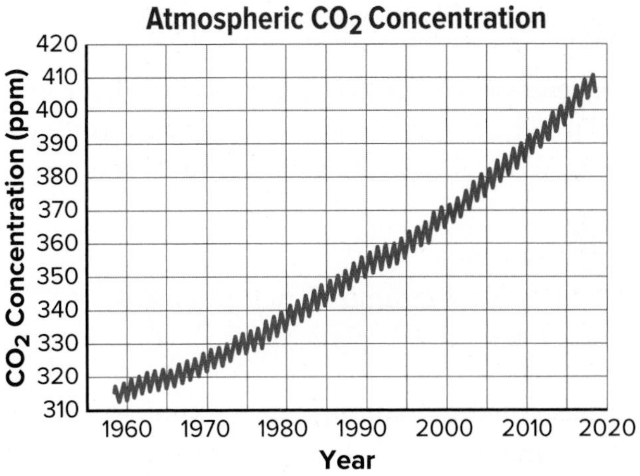

Atmospheric CO2 Concentration
CO2 Concentration (ppm)
420
410
400
390
380
370
360
350
340
330
320
310
1960
1970
1980
1990
2000
2010
2020
Year

Figure 8 The carbon dioxide concentration in Earth's atmosphere has been measured at Mauna Loa Observatory in Hawaii. From 1958 to 2018, the carbon dioxide concentration has increased by 1.56 parts per million (ppm) per year.

Predict *how the concentration of carbon dioxide will change in the next several decades based on the graph trend.*

The Cost of Fossil Fuels

Although fossil fuels are common energy resources, their uses have associated costs and risks. Burning fossil fuels releases small particulates into the atmosphere, which can cause breathing problems. Fossil fuels also release carbon dioxide (CO_2) when they are burned. **Figure 8** shows how the CO_2 concentration in the atmosphere has increased from 1958 to 2018. This increase in atmospheric CO_2 concentration is the chief contributor to global climate change.

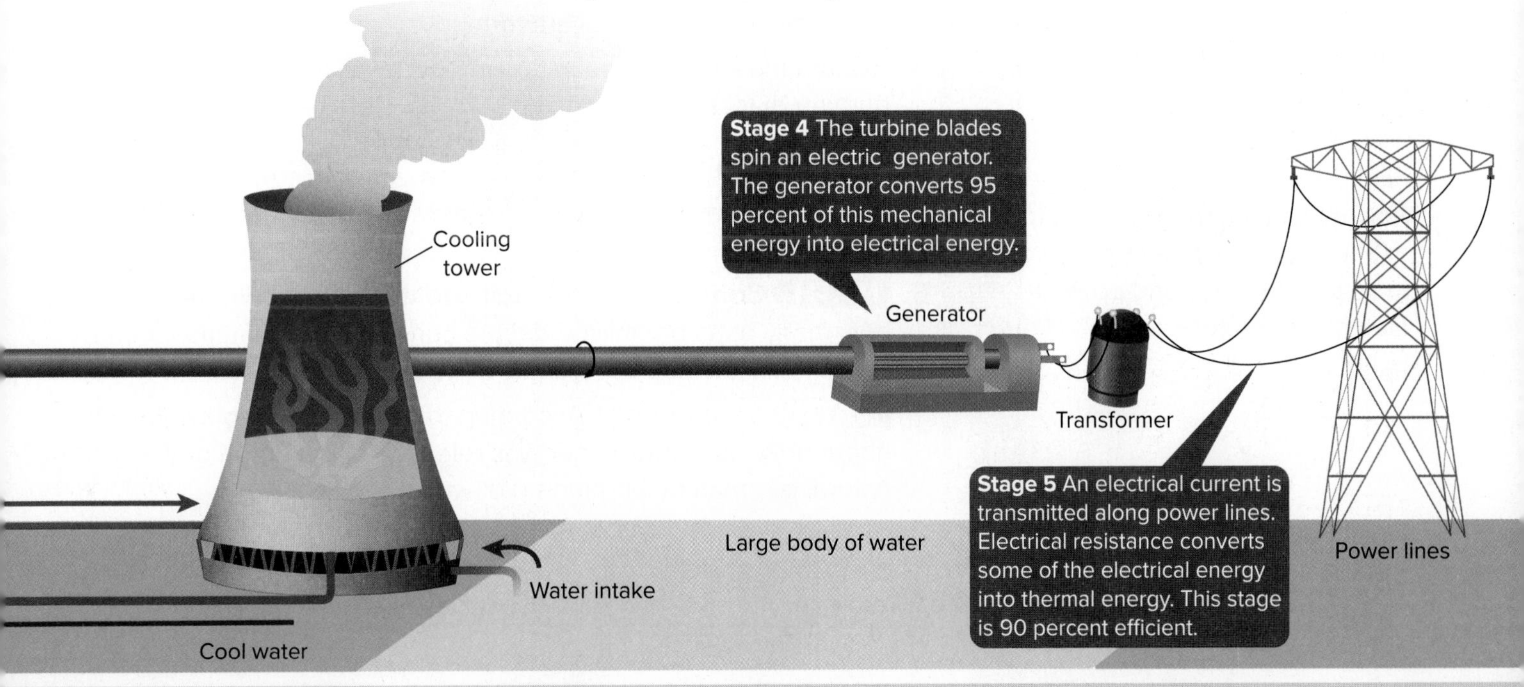

Stage 4 The turbine blades spin an electric generator. The generator converts 95 percent of this mechanical energy into electrical energy.
Cooling tower
Generator
Transformer
Stage 5 An electrical current is transmitted along power lines. Electrical resistance converts some of the electrical energy into thermal energy. This stage is 90 percent efficient.
Large body of water
Power lines
Water intake
Cool water

Nonrenewable Resources

Nonrenewable resources are resources that cannot be replaced by natural processes as quickly as they are used. All fossil fuels are nonrenewable resources.

Get It?

Identify three examples of nonrenewable resources.

Because they are nonrenewable resources, fossil fuels are decreasing in supply. As supplies of fossil fuels run out, fossil fuels will become more difficult to obtain. This will cause fuel to become more costly than it is today.

Even as fossil fuel supplies decrease, the demand for energy continues to increase. One way to meet these energy demands is to search for energy alternatives. Scientists have discovered numerous oil shale reserves in the United States, as shown in **Figure 9.** When oil shale is heated to extremely high temperatures, it releases an organic chemical compound called kerogen. Kerogen is a petroleum-like substance that has potential for meeting increasing energy demands as fossil fuel resources are consumed.

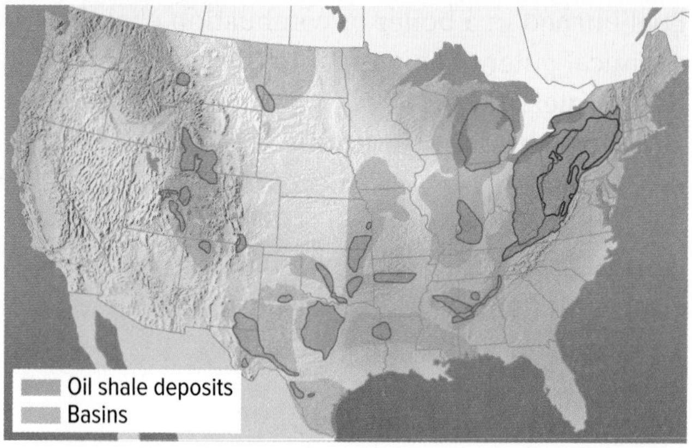

Figure 9 As population increases and fossil fuel resources decrease, scientists search for new solutions to the energy crisis. Energy alternatives, such as oil shale deposits found in basins in the central and midwestern United States, might help meet these energy demands.

Check Your Progress

Summary

- Energy cannot be created or destroyed; it can only be transformed from one form to another.
- Petroleum, natural gas, and coal are fossil fuels.
- Petroleum is a mixture of hydrocarbons.
- Power plants burn fossil fuels to extract chemical potential energy that spins turbines and powers electric generators.
- Fossil fuels are nonrenewable resources.

Demonstrate Understanding

1. **Describe** the advantages and disadvantages of using fossil fuels to generate electricity.
2. **Explain** how you use energy resources daily.
3. **Describe** how fossil fuels are formed.
4. **Explain** the law of conservation of energy as it applies to the burning of fossil fuels.

Explain Your Thinking

5. **Explain** Why are fossil fuels considered to be a nonrenewable resource?
6. **MATH Connection** According to the graph in **Figure 8,** by how many parts per million did the concentration of atmospheric carbon dioxide increase from 1958 to 2018?
7. **MATH Connection** According to the graph in **Figure 3,** about how much more energy is released by burning 1 g of natural gas than by burning 1 g of wood?

LEARNSMART® Go online to follow your personalized learning path to review, practice, and reinforce your understanding.

LESSON 2
NUCLEAR ENERGY

FOCUS QUESTION
What happens to nuclear waste?

Fusion

The Sun is a giant nuclear reactor in the sky. It transforms energy through a process called fusion. **Fusion** occurs when atomic nuclei combine at very high temperatures. In this process, a small amount of mass is transformed into a tremendous amount of thermal energy.

Fusion-based power plants are not practical. One problem with fusion is that it occurs at millions of degrees Celsius. Under these conditions, reactors use a great deal of energy. Another problem is containment–what kind of chamber can hold a reaction under these extreme conditions?

Fission

Energy is released when the nucleus of an atom splits apart in a process called **fission.** During fission, a small amount of mass is converted into a tremendous amount of thermal energy. Unlike fusion, fission-based power plants are practical. Sixty-five power plants in the United States, including the one shown in **Figure 10,** transform energy by fission reactions. These plants convert nuclear energy into electrical energy and produce 9 percent of the energy used in the United States.

Figure 10 A nuclear power plant generates electricity using the thermal energy released in fission. This concrete tower is a cooling tower that releases waste heat, a product of the fission reaction.

3D THINKING DCI Disciplinary Core Ideas CCC Crosscutting Concepts SEP Science & Engineering Practices

COLLECT EVIDENCE

Use your Science Journal to record the evidence you collect as you complete the readings and activities in this lesson.

INVESTIGATE

GO ONLINE to find these activities and more resources.

Revisit the Encounter the Phenomenon Question
What information from this lesson can help you asnwer the Unit and Module questions?

CCC **Identify Crosscutting Concepts**
Create a table of the crosscutting concepts and fill in examples you find as your read.

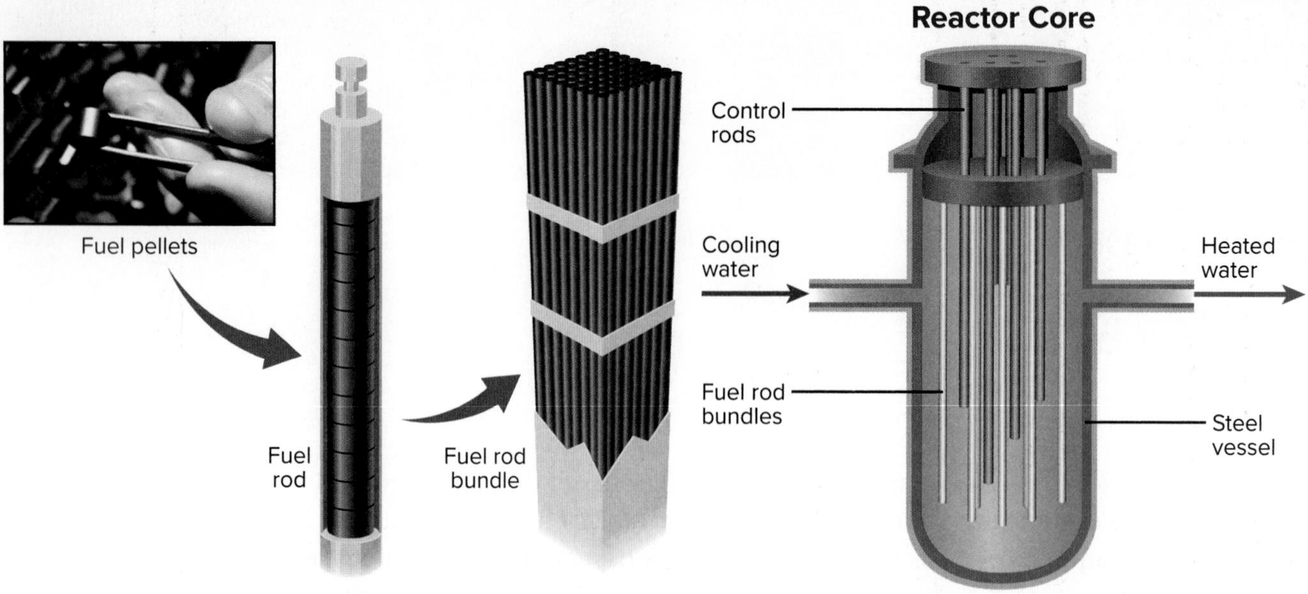

Nuclear fuel pellets are stacked together to form fuel rods. The fuel rods are bundled together and covered with a metal alloy.

The bundles are inserted into the reactor, core where a coolant removes the heat produced by the fission reaction.

Figure 11 The core of a nuclear reactor contains the fuel rod bundles. Control rods that absorb neutrons are inserted between the fuel rod bundles.

Nuclear Reactors

A **nuclear reactor** uses energy from controlled nuclear reactions to generate electricity. Although nuclear reactors vary in design, all reactors share some similarities. They contain fuel that can undergo fission and control rods that are used to control the nuclear reactions. They have a cooling system that keeps the reactor from being damaged by the enormous amount of heat produced. The actual fission of the radioactive fuel occurs in a relatively small part of the reactor known as the core, shown in **Figure 11.**

Figure 12 The reactor core, which contains the fuel rod bundles, is submerged in a cooling chamber.

Nuclear fuel

Only certain elements have nuclei that can undergo fission. Naturally occurring uranium contains the isotope U-235, which has nuclei that can be split apart. Naturally occurring uranium typically contains 0.72 percent of the isotope U-235. The uranium used in a reactor is enriched so that it contains 3–5 percent U-235. The fuel that is used in a nuclear reactor is usually uranium dioxide.

Fuel rods

The reactor core contains uranium dioxide fuel in the form of tiny pellets like the ones shown in **Figure 11.** The pellets are about the size of a pencil eraser and are placed end to end in fuel rods. The fuel rods are then bundled and covered with a metal alloy. The core of a typical reactor, shown in **Figure 12,** has about 100,000 kg of uranium contained in fuel rods. For every kilogram of uranium that undergoes fission, 1 g of matter is converted into energy, releasing energy equal to burning more than 3 million kg of coal.

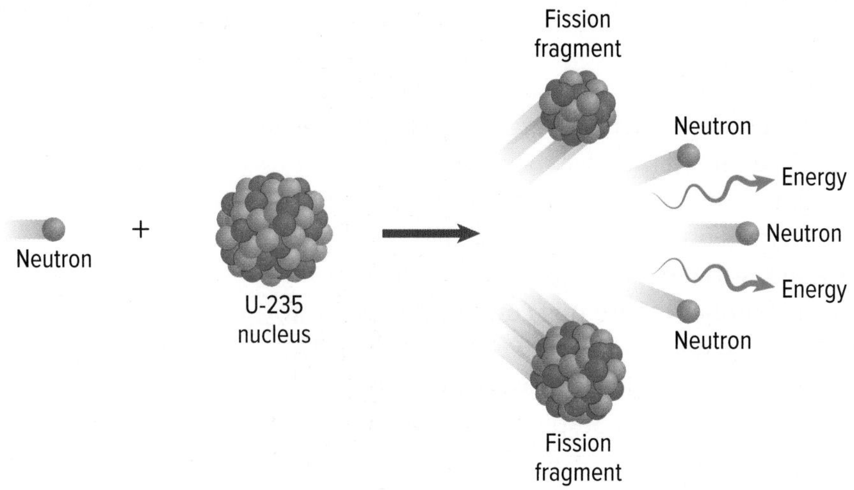

Figure 13 When a neutron strikes the nucleus of a U-235 atom, the nucleus splits apart into two smaller nuclei. In the process, two or three neutrons also are emitted. The smaller nuclei are called fission products.

Explain *what happens to the neutrons that are released in this reaction.*

The nuclear chain reaction

How does a fission reaction proceed in the reactor core? As U-235 nuclei undergo fission, neutrons are released and are absorbed by other U-235 nuclei. When a U-235 nucleus absorbs a neutron, it splits into two smaller nuclei and two or three free neutrons, as shown in **Figure 13.** These neutrons strike other U-235 nuclei, triggering the release of more neutrons, and fission continues.

Because every uranium atom that splits apart releases free neutrons that cause other uranium atoms to split apart, this process is called a nuclear chain reaction. In the chain reaction, the number of nuclei that are split can more than double at each stage of the process. As a result, an enormous number of nuclei can be split after only a small number of stages. For example, if you start with one uranium nucleus and the number of nuclei involved doubles at each stage, after only 50 stages, more than a quadrillion nuclei might be split. Nuclear chain reactions take place in a matter of milliseconds. If the process is not controlled, the chain reaction could release a tremendous amount of energy in the form of an explosion.

A constant rate To control the chain reaction, some of the neutrons that are released when U-235 splits apart must be prevented from colliding with other U-235 nuclei. These neutrons are absorbed by control rods containing boron or cadmium, which are inserted into the reactor core, as shown in **Figure 11.** Moving these control rods deeper into the reactor causes them to absorb more neutrons and to slow down the chain reaction. Eventually, only one of the neutrons released in the fission of each of the U-235 nuclei strikes another U-235 nucleus, so energy is released at a constant rate.

Get It?
Explain how a nuclear chain reaction is controlled in a nuclear reactor.

STEM CAREER CONNECTION

Nuclear Technician

If you enjoy working with computers and other equipment and are meticulous and safety-conscious, the job of nuclear technician might interest you. Nuclear technicians in the nuclear power industry monitor and operate nuclear reactors, ensuring they are functioning safely and efficiently to deliver electricity to homes, schools, and businesses.

Nuclear Power Plants

Nuclear power plants produce an electric current in a way that is similar to fossil fuel-burning power plants. As shown in **Figure 14,** the thermal energy released in fission is used to heat water and produce pressurized steam. To transfer thermal energy from the reactor core, the core contains a fluid coolant. The hot coolant is pumped into a heat exchanger. In the heat exchanger, thermal energy is transferred from the hot coolant to water, causing the water to boil and produce pressurized steam that spins a turbine. When the steam leaves the turbine, it enters a chamber where it is condensed back into liquid water. Cool water absorbs the thermal energy released during condensation. The thermal energy is then carried to the cooling tower, where it is released into the environment. The overall efficiency of nuclear power plants is about 35 percent, which is similar to that of fossil fuel-burning power plants.

Get It?

Explain how nuclear power plants produce an electric current.

Benefits and Risks of Nuclear Power

Extracting energy from atomic nuclei has its advantages. Nuclear power plants do not produce the air pollutants that fossil fuel-burning power plants release into the atmosphere. Also, nuclear power plants do not release carbon dioxide into the atmosphere.

Nuclear power plants also have their disadvantages. For example, nuclear power plants are very expensive to build, and the building process can take 10 or more years to complete. Nuclear power plants also produce radioactive waste that can be harmful to living organisms and the environment.

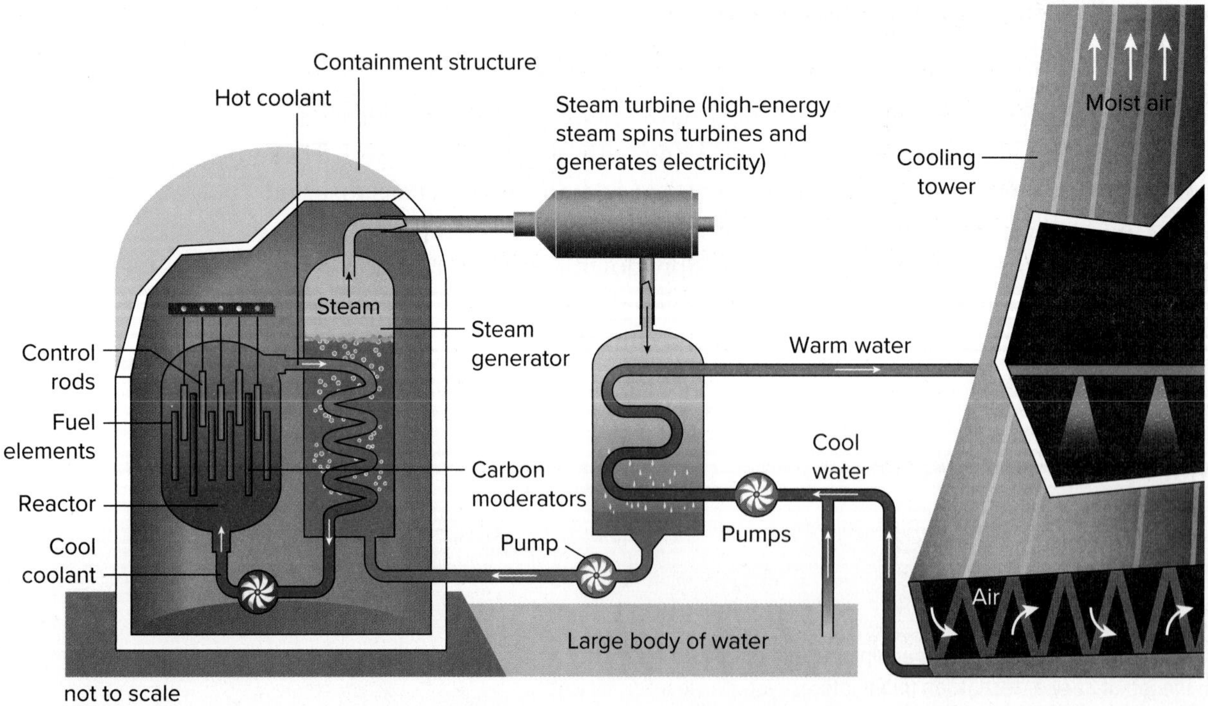

Figure 14 A nuclear power plant converts water into pressurized steam that spins a turbine and generates electricity.

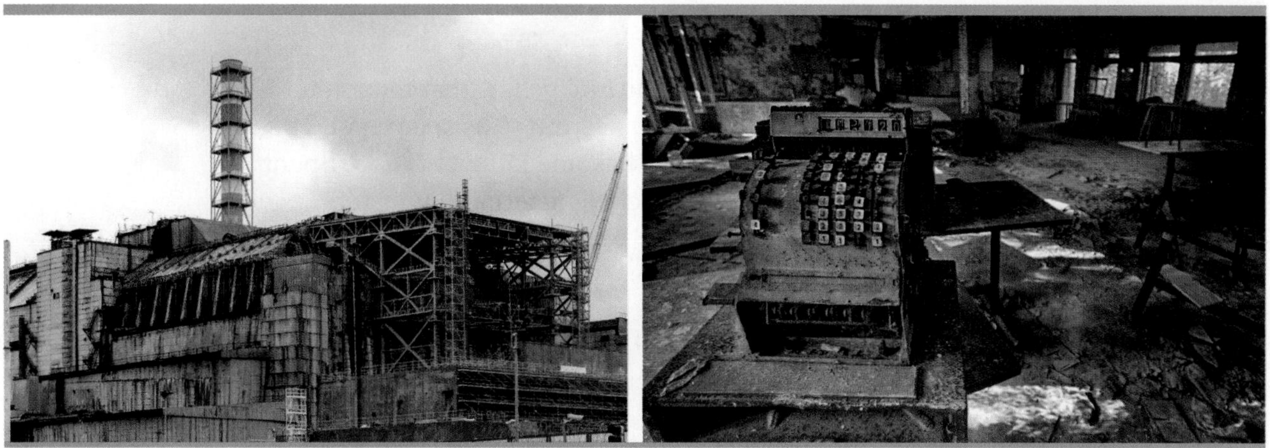

During a safety check on April 26, 1986, a nuclear reactor at the Chernobyl Nuclear Power Plant near Pripyat, Ukraine, exploded, causing the worst nuclear disaster of all time.

Pripyat was abandoned after the explosion. Though it remains uninhabited, it is possible to obtain permits to visit the area affected by the explosion.

Figure 15 A steam explosion in the nuclear reactor at Chernobyl Nuclear Power Plant melted fuel rods and ignited the reactor's graphite cover, setting the facility on fire.

The release of radioactivity

Nuclear power plants around the world operate safely every day. However, like all forms of energy production and resource extraction, nuclear power has associated economic, social, environmental, and geopolitical costs and risks. One of the serious risks of nuclear power is the release of harmful radiation from power plants. The fuel rods contain radioactive elements. Some of these radioactive elements could harm living organisms if they are released from the reactor core of a nuclear power plant.

New technologies and regulations can reduce these risks. For example, to prevent accidents, nuclear reactors have elaborate systems of safeguards, strict safety precautions, and highly trained workers. In spite of this, accidents have still occurred.

For example, an accident occurred when a reactor core at the Chernobyl Nuclear Power Plant near Pripyat, Ukraine, overheated during a routine safety test on April 26, 1986. Materials in the core caught fire and caused a chemical explosion that blew a hole in the reactor, as described in **Figure 15.** This resulted in the release of radioactive materials that were carried by winds and deposited over a large area. As a result of the accident, 50 people died from acute radiation sickness, and about 4000 cancer-related cases have been attributed to the release of radioactivity from the explosion.

The World Health Organization estimates that approximately 600,000 people were exposed to levels of radiation that continue to pose a risk to their health. Newer nuclear power plants are designed to prevent accidents like the one that occurred at Chernobyl. But there is always a possibility that an accident might occur.

SCIENCE USAGE v. COMMON USAGE

system

Science usage: the particular reaction or process being studied *The universe consists of a system and its surroundings.*

Common usage: an organized or established procedure
She designed a system in which each person would have an equal opportunity to earn a raise.

The Disposal of Nuclear Waste

After about three years of use, there is too little U-235 in the fuel pellets that remain for the chain reaction to continue. The fuel pellets left are now referred to as spent fuel. The spent fuel includes radioactive fission products in addition to some leftover U-235. Spent fuel is a form of nuclear waste. **Nuclear waste** is any radioactive material that results when radioactive materials are used.

Get It?
Describe the origin and composition of spent fuel.

While many people support the idea of using nuclear energy as an alternative to fossil fuels, they do not necessarily support the idea of nuclear waste disposal in their state. Many people refer to this antinuclear attitude as the "Not in My Backyard" syndrome. Nuclear waste disposal has been a controversial subject and continues to fuel debate about nuclear energy use.

Low-level waste

Low-level nuclear waste usually contains a small amount of radioactive material. Additionally, low-level waste usually contains radioactive materials with short half-lives. Low-level waste is a by-product of electricity generation, medical research and treatments, the pharmaceutical industry, and food preparation. Low-level wastes also include used water and air filters from nuclear power plants and discarded smoke detectors. Low-level waste is kept isolated from people and the environment. It is treated as hazardous material and is stored in spill-safe containers underground.

Get It?
Identify industries that produce low-level waste.

APPLY SCIENCE

Can a contaminated radioactive site be reclaimed?

With the discovery of radium in the early 1900s, extensive mining for the element began in the Denver, Colorado, area. Radium is a radioactive element that was used to make watch dials and instrument panels that glowed in the dark. After World War I, the radium industry collapsed. The area was left contaminated with 97,000 tons of radioactive soil and debris containing heavy metals and radium, which is now known to cause cancer. The soil was used as fill or foundation material, left in place, or mishandled.

Identify the Problem

In the 1980s, one area became known as the Denver Radium Superfund Site and was cleaned up by the Environmental Protection Agency. The land then was reclaimed by a local commercial establishment.

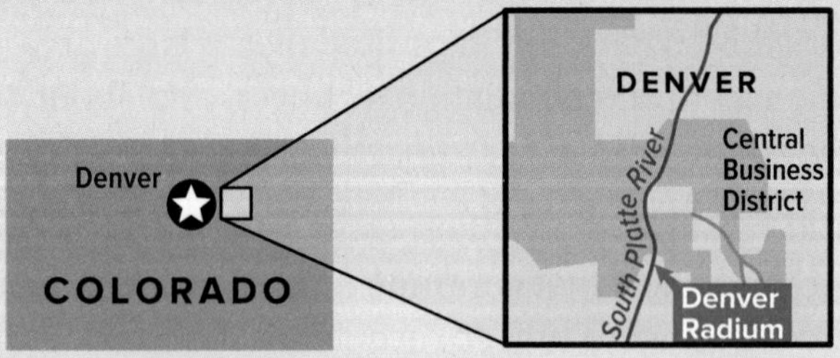

Solve the Problem

1. The contaminated soil was placed in one area, and a protective cap was placed over it. This area was restricted from being used for residential homes. Explain why it is important for the protective cap to be maintained and why homes should not be built in this area.
2. The advantages of cleaning this site are economical, environmental, and social. Give an example of each.

High-level waste

High-level nuclear waste is generated in nuclear power plants and by nuclear weapons programs. After spent fuel is removed from a reactor, it is stored in steel-lined concrete pools filled with water, as shown in **Figure 16,** or in airtight steel or concrete and steel canisters.

Many of the radioactive materials in high-level nuclear waste become nonradioactive after a relatively short amount of time. However, the spent fuel also contains materials that will remain radioactive for tens of thousands of years. For this reason, high-level waste must be disposed of in extremely durable, safe, and stable containers.

Figure 16 Spent nuclear fuel is stored in spill-safe containers at nuclear power plants and often submerged within specially designed pools.

Get It?

Describe What are the differences between low-level and high-level nuclear wastes?

One method proposed for the disposal of high-level waste is to seal the waste in ceramic glass, which is placed in protective metal containers. The containers then are buried hundreds of meters belowground in stable rock formations or in salt deposits.

Check Your Progress

Summary

- Nuclear power plants produce about 8 percent of the energy used each year in the United States.
- Nuclear reactors use the energy released in the fission of U-235 to produce electricity.
- The energy released in the fission reaction is used to make steam. The steam spins a turbine that powers an electric generator.
- Nuclear power generation produces high-level nuclear waste.

Demonstrate Understanding

8. **Compare and contrast** the advantages and disadvantages of nuclear power plants and power plants that burn fossil fuels.
9. **Describe** nuclear fission and how the chain reaction in a nuclear reactor is controlled.
10. **Describe** nuclear fusion and the problems associated with using nuclear fusion reactions as an energy source.
11. **Explain** why a chain reaction occurs when uranium-235 undergoes fission.

Explain Your Thinking

12. **Classify** A research project produced 10 g of nuclear waste with a short half-life. How would you classify this waste, and how should it be disposed?
13. **MATH Connection** Naturally occurring uranium contains 0.72 percent of the isotope uranium-235. What is the mass of uranium-235 in 2000 kg of naturally occurring uranium?

LEARNSMART® Go online to follow your personalized learning path to review, practice, and reinforce your understanding.

RENEWABLE ENERGY RESOURCES

FOCUS QUESTION

Why aren't renewable energy sources rapidly implemented to replace fossil fuels?

Energy Options

The demand for energy is increasing every day as Earth's population increases. As demand increases, our supply of nonrenewable energy resources decreases. Use of nuclear energy produces high-level waste that has to be disposed of safely. As a result, alternative energy sources are being developed to meet increasing energy demands. Some alternative energy sources are renewable resources. A **renewable resource** is an energy source that is replaced by natural processes faster than humans can consume the resource.

Energy from the Sun

The average amount of solar energy that shines down on the United States in one year is 1000 times greater than the total energy used in one year. Because the Sun is expected to continue producing energy for billions of years, solar energy is inexhaustible in our lifetime. Solar energy is a renewable resource.

Despite being renewable, less than 1 percent of the energy in the United States is produced using solar power. There are several ways to produce solar power. One way is to use a photovoltaic cell, shown in **Figure 17.** A **photovoltaic cell** converts radiant energy directly into electrical energy. Photovoltaic cells are also called solar cells.

Figure 17 Photovoltaic cells convert radiant energy into electrical energy. Some vehicles have optional photovoltaic panels made of solar cells that are used to cool the car without the use of the engine.

3D THINKING DCI Disciplinary Core Ideas CCC Crosscutting Concepts SEP Science & Engineering Practices

COLLECT EVIDENCE

Use your Science Journal to record the evidence you collect as you complete the readings and activities in this lesson.

INVESTIGATE

GO ONLINE to find these activities and more resources.

Laboratory: Using the Sun's Energy

Develop and use a model of a solar water heater to evaluate a technological solution that reduces impacts of human activities on natural systems.

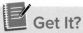

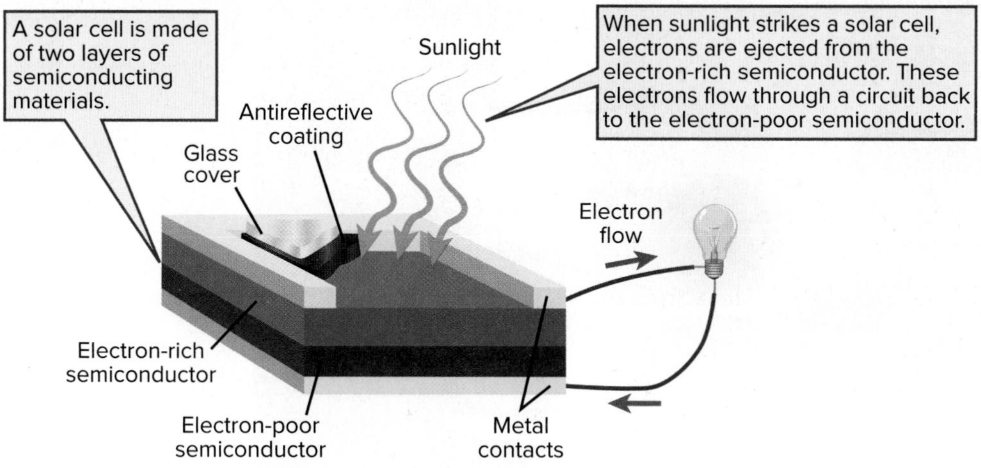
A solar cell is made of two layers of semiconducting materials.
Sunlight
When sunlight strikes a solar cell, electrons are ejected from the electron-rich semiconductor. These electrons flow through a circuit back to the electron-poor semiconductor.
Antireflective coating
Glass cover
Electron flow
Electron-rich semiconductor
Electron-poor semiconductor
Metal contacts

Figure 18 Radiant energy from sunlight strikes the surface of a solar cell, exciting electrons and causing them to flow through an electric circuit.

Identify *two devices that use solar cells for power.*

How solar cells work Solar cells are made of two layers of semiconducting material sandwiched between two layers of conducting metal, as shown in **Figure 18.** One layer of the semiconducting material is rich in electrons, and the other layer is electron poor. When sunlight strikes the surface of the solar cell, electrons flow through an electric circuit from the electron-rich material to the electron-poor material. This process of converting radiant energy from the Sun directly to electrical energy is only about 7–11 percent efficient.

The transformation of radiant energy into electrical energy using solar cells is more expensive than the transformation of thermal energy into electrical energy by combustion. However, in remote areas where power lines are not available, solar cells are a practical energy source.

Get It?

Explain how the structure and materials of solar cells enables them to transform radiant energy into electrical energy.

Parabolic troughs Other promising solar technologies concentrate solar power into a receiver. One such system is called the parabolic trough. The trough focuses the sunlight on a tube that contains a heat-absorbing fluid, such as synthetic oil or liquid salt. Sunlight heats the fluid, which circulates through a boiler, where it turns water to steam that spins a turbine to generate an electric current.

One of the world's largest concentrating solar power plants is located in the Mojave Desert in California. This facility consists of nine units that generate more than 350 megawatts of power. These nine units can generate enough electricity to meet the demands of approximately 500,000 people. The units also use natural gas as a backup power source for generating an electric current at night and on cloudy days, when solar energy is unavailable.

Get It?

State why backup power sources are needed for solar power plants.

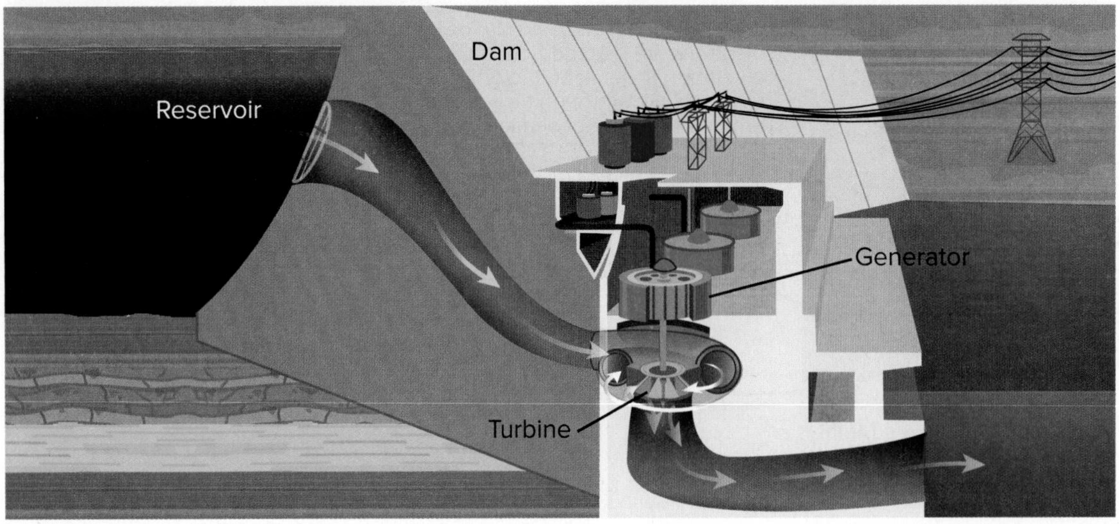

Figure 19 The gravitational potential energy of water behind the dam is converted to electrical energy in a hydroelectric power plant.

Explain *the energy conversions that occur as a hydroelectric dam produces electrical energy.*

Energy from water

Rapidly moving water can spin a turbine and power an electric generator. The gravitational potential energy of water is great when the water is retained by a dam. This energy is released when the water flows through tunnels near the base of the dam, which is called a hydroelectric dam. **Figure 19** shows how rushing water spins a turbine, converting gravitational potential energy to mechanical energy and then to electrical energy.

Hydroelectricity Electric current produced from the energy of moving water is called **hydroelectricity.** About 7 percent of the electrical energy used in the United States comes from hydroelectric power plants. Hydroelectric power plants convert mechanical energy into electrical energy with almost no pollution. They are almost twice as efficient as fossil fuel-burning or nuclear power plants. Another advantage of hydroelectric power is that the bodies of water held back by dams can form lakes that provide water for drinking, crop irrigation, and recreation. After initial construction costs, hydroelectric power plants are more cost-efficient than other energy resources.

However, dams and hydroelectric power plants can disturb the balance of natural ecosystems. Some species of ocean fish migrate back to the rivers in which they were hatched to breed. This migration can be blocked by dams, causing a decline in the fish population. Fish ladders, such as those shown in **Figure 20,** enable fish to migrate upstream, past some dams. Also, hydroelectric power plants can change the temperature of the water and cause river sediments to build up, both of which affect plant and animal habitats.

Figure 20 A fish ladder is like a series of steps that allows fish to swim past a dam.

ACADEMIC VOCABULARY

Efficient

productive without waste

Efficient heating and cooling systems help conserve fossil fuels.

CCC CROSSCUTTING CONCEPTS

Energy and Matter Review the information about hydroelectric power plants on this page and in **Figure 19.** Write a paragraph, citing evidence from the text and figure, to describe the energy flow within this system.

Energy from the oceans

The gravitational pull of the Moon and the Sun on Earth's oceans causes tides. Hydroelectric power can be produced by these tides. As the tide rises, water spins a turbine, which transforms mechanical energy into electrical energy. The water is then trapped behind a dam. At low tide, the water behind the dam is released to flow back out to sea, converting even more energy to electricity.

Hydroelectric power can also be produced by waves. There are several new technologies that harness wave energy. One type focuses wave energy into a channel. As waves enter a channel, they spin turbines, converting mechanical energy into electrical energy. Plans are also in place to harness mechanical energy from ocean currents, as shown in **Figure 21.**

Energy from the ocean is nearly pollution free, and the efficiency of tidal and wave power plants is similar to that of hydroelectric power plants. However, only a few places on Earth have large enough differences between high tide and low tide for oceans to be a useful energy source.

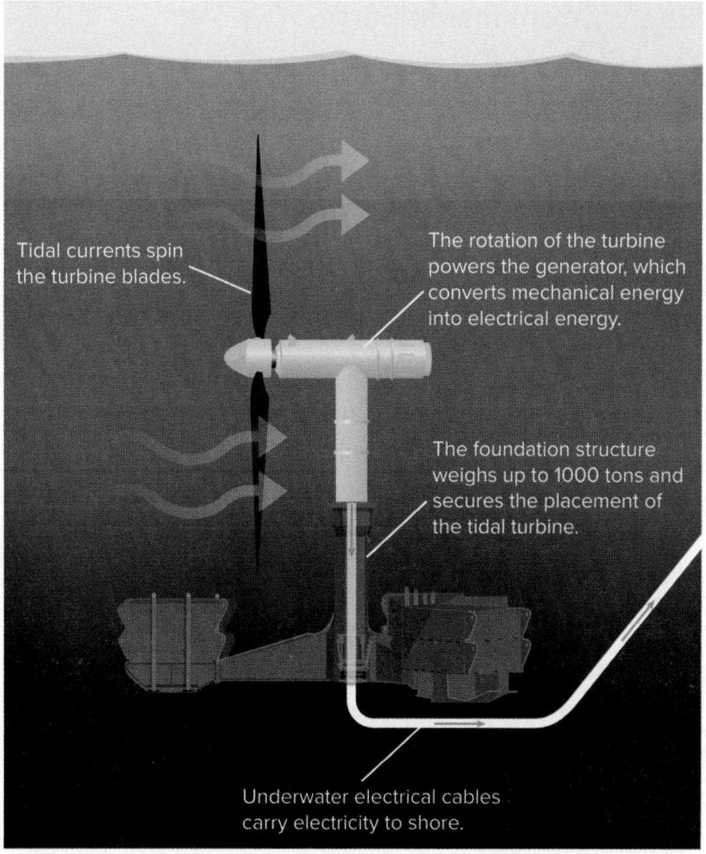

Tidal currents spin the turbine blades.
The rotation of the turbine powers the generator, which converts mechanical energy into electrical energy.
The foundation structure weighs up to 1000 tons and secures the placement of the tidal turbine.
Underwater electrical cables carry electricity to shore.

Figure 21 In Florida, researchers estimate that underwater turbines spun by Gulf Stream currents could generate the equivalent of ten nuclear power plants' worth of electricity.

Wind energy

Windmills can convert wind energy into electrical energy. As the wind blows, it spins a propeller that is connected to an electric generator. The greater the wind speed and the longer the wind blows, the greater the amount of wind energy that is converted into electrical energy. Windmill farms, like the one shown in **Figure 22,** can contain several hundred windmills.

One disadvantage of wind energy is that only a few places on Earth have enough wind to meet our energy needs. Even then, wind energy cannot be stored without the use of batteries. Windmills can be noisy and change the appearance of a landscape. They also can disrupt the migration patterns of some birds.

The advantages of using wind energy are that wind generators do not consume nonrenewable resources, and they do not pollute air or water. Research is underway to improve the design of wind generators and to increase their efficiency. By 2030, the U.S. Department of Energy wants to increase our use of wind energy so it provides 20 percent of our total electrical power.

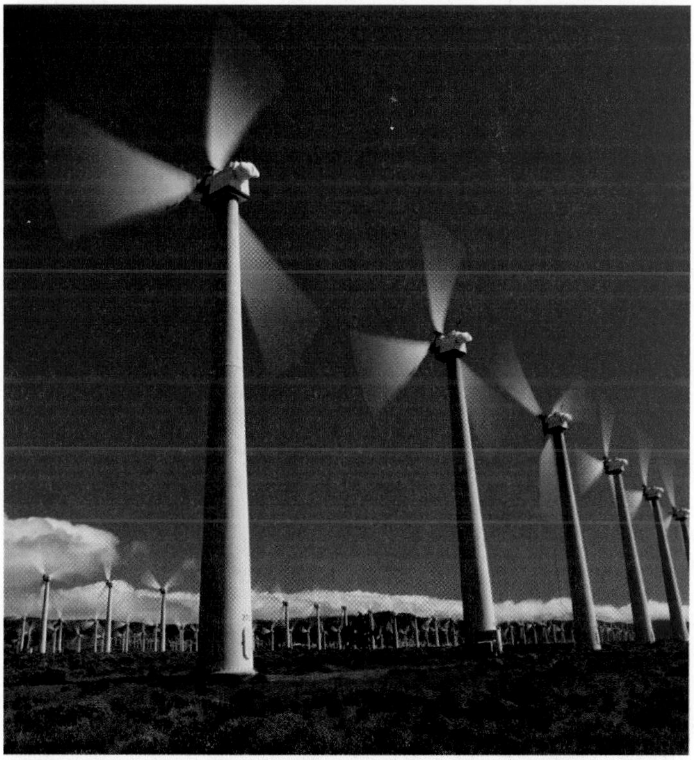

Figure 22 Wind energy is converted to electrical energy as a spinning propeller turns a generator.

Energy from inside Earth

Unstable radioactive elements in Earth's core transform nuclear energy into thermal energy. As these unstable elements decay, thermal energy is transferred from the core into Earth's mantle and crust. This is called geothermal heat. Geothermal heat can cause the rock beneath Earth's crust to melt. Molten rock beneath Earth's surface is called magma. The thermal energy that is contained in and around magma is called **geothermal energy.**

Get It?

Identify the process that transforms energy inside Earth into thermal energy.

In some areas, Earth's crust has cracks or areas of weakness that allow magma to rise toward the surface. Active volcanoes, for example, permit hot gases and magma from deep within Earth to escape. Perhaps you have seen a geyser, such as Old Faithful in Yellowstone National Park, spewing hot groundwater and steam. The groundwater erupting from the geyser is heated by magma close to Earth's surface. In some areas, hot groundwater is pumped directly into homes to provide warmth.

Geothermal power plants Geothermal energy can be converted into electrical energy, as shown in **Figure 23.** Where magma is close to the surface, the surrounding rocks are hot. Water is pumped into the ground through a well, where it comes into contact with hot rock and changes into steam. The steam then returns to the surface, where it spins a turbine that powers an electric generator.

The efficiency of geothermal power plants is about 16 percent. Although geothermal power plants can produce sulfur-based compounds, pumping the water that condenses from steam back into the ground reduces this pollution. This makes geothermal power plants a source of clean energy. However, one disadvantage is that the use of geothermal energy is limited to volcanically active areas where magma is close to the surface. Seven states in the United States have geothermal power plants—California, Nevada, Utah, Hawaii, Oregon, Idaho, and New Mexico.

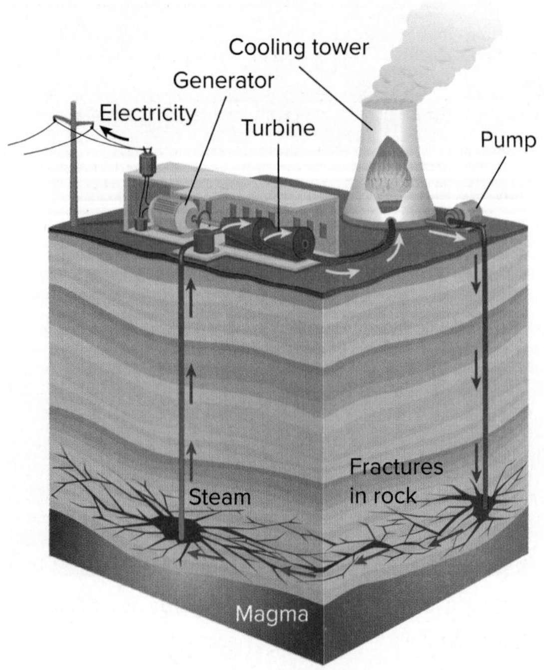

Figure 23 Geothermal power plants convert geothermal energy to electrical energy. Water is changed to steam by the hot rock. The steam is pumped to the surface where it drives a turbine attached to an electric generator.

Alternative Fuels

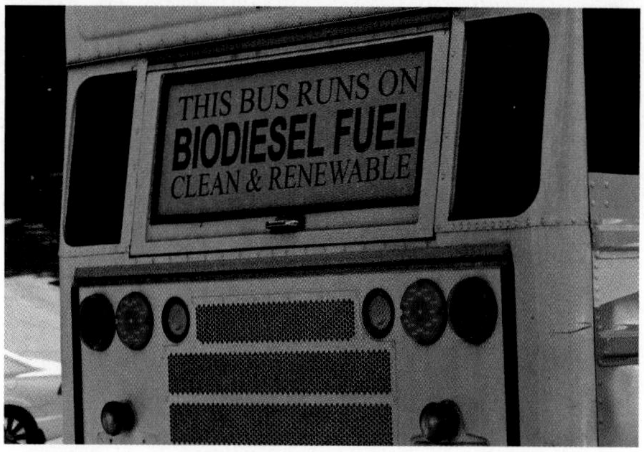

Figure 24 Vegetable oils, animal fats, and recycled cooking oils can be used to make alternative fuels such as biodiesel for transportation.

The use of fossil fuels would be greatly reduced if cars could run on alternative energy resources alone. For example, some cars use electrical energy supplied by batteries as their primary power source. Hybrid cars use electric motors and gasoline engines.

Hydrogen

Hydrogen fuel cells are another possible alternative energy resource. A fuel cell behaves like a battery. It combines hydrogen with oxygen in air to generate electrical energy, water, and heat. There are several problems with using hydrogen fuel as an alternative energy resource, however. First, obtaining hydrogen requires more energy than the energy that is released by the fuel-cell reaction. Second, hydrogen fuel cells are built from expensive platinum parts. And third, there is a lack of hydrogen fueling stations, as storing hydrogen is considered to be dangerous and difficult.

Biomass

Biomass is one of the oldest energy sources. **Biomass** is renewable organic matter, such as wood, soy, corn, sugarcane fibers, rice hulls, and animal manure. It can be burned in the presence of oxygen, which converts the stored chemical potential energy to thermal energy. **Figure 24** shows a bus powered by biodiesel derived from biomass.

Check Your Progress

Summary

- Solar cells convert radiant energy into electrical energy.
- Hydroelectric power plants convert gravitational potential energy into electrical energy.
- Wind energy is converted into electrical energy using a propeller attached to an electric generator.
- Alternative energy sources, such as the Sun, water, wind, and Earth's internal heat, can help reduce human dependence on fossil fuels.

Demonstrate Understanding

14. **Explain** the need to develop and use alternative energy sources.
15. **Describe** three ways that solar energy can be used.
16. **Explain** the similarities among electricity generation by hydroelectric, tidal, and wind sources.
17. **Infer** why geothermal energy is unlikely to become a major energy source.

Explain Your Thinking

18. **Analyze** On what single energy source do most energy alternatives depend, either directly or indirectly?
19. **MATH Connection** A house uses solar cells that generate 1.5 kW of electrical power to supply some of its energy needs. If the solar panels supply the house with 40 percent of the power it needs, how much total power does the house use?

LEARNSMART® Go online to follow your personalized learning path to review, practice, and reinforce your understanding.

LESSON 4

ENVIRONMENTAL IMPACTS

FOCUS QUESTION

How might irreversible climate change impact your daily life?

Population and Carrying Capacity

A **population** includes all the individuals of one species living in a particular area. You can see in **Figure 25** that it took thousands of years for the human population to reach 1 billion people. In the mid-1800s, human population began to increase at a rapid rate because of advances in modern medicine and the availability of clean water and better nutrition. People began to live longer. In addition, the number of births increased because more people survived to a child-bearing age.

Carrying capacity

Every person alive today uses and is dependent on Earth's natural resources. However, Earth has a carrying capacity. **Carrying capacity** is the greatest number of individuals of a particular species that the environment can support, given the natural resources available. If natural resources are consumed too quickly or the environment becomes threatened, then populations suffer. Unless Earth's natural resources are treated with care, the human population could reach its carrying capacity.

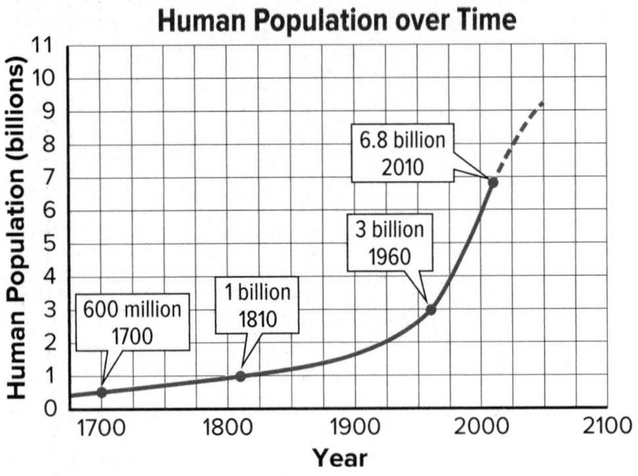

Figure 25 Human population growth rates remained fairly steady until the middle of the nineteenth century. The growth rate then began to increase rapidly.

Estimate *What will the human population be in the year 2030?*

3D THINKING DCI Disciplinary Core Ideas CCC Crosscutting Concepts SEP Science & Engineering Practices

COLLECT EVIDENCE

Use your Science Journal to record the evidence you collect as you complete the readings and activities in this lesson.

INVESTIGATE

GO ONLINE to find these activities and more resources.

Applying Practices: Environmental Consulting: Finding Solutions

HS-PS2-1. Evaluate competing design solutions for developing, managing, and utilizing energy and mineral resources **based on cost-benefit ratios.**

Figure 26 At a water treatment plant, pollutants are removed from wastewater to help keep our waterways clean.

People and the Environment

You have an impact on the environment every day. The electrical energy that you use most likely comes from burning fossil fuels. The cars and buses you use for transportation burn fossil fuel. Fossil fuels are mined from Earth and have an impact on the air that you breathe. The water that you use must be treated, as shown in **Figure 26,** to remove as many pollutants as possible before it is recycled back into waterways. **Pollutants** include any substance that contaminates the environment.

You also use plastics and paper every day. Plastics are petroleum-based products. When petroleum is refined, it produces pollutants. In the process of harvesting trees to make paper, trees are cut down. They are transported using fossil fuels, and water and air can be polluted in the paper-making process.

Impact on Land

Land is affected when resources such as fossil fuels, water, soil, or trees are extracted from Earth. You might not think of land as a natural resource, but it is as important as fossil fuels, clean water, and clean air. We use land for agriculture, forests, urban development, and even waste management. These uses impact the land and the natural resources it provides.

Agriculture

The pears and apples that you purchase at the grocery store were grown on farms, which cover 16 million km^2 of Earth's total land area. To feed the world's growing population, some farmers are planting higher-yielding seeds and using stronger nitrate- and phosphate-based fertilizers. Herbicides and pesticides are also used for weed and pest control. These methods increase the amount of food grown, but, if not managed properly, they can have a negative impact by possibly polluting soil and water and endangering animals.

CCC CROSSCUTTING CONCEPTS

Stability and Change Write a paragraph to describe how the growing human population can destabilize Earth's systems. Cite evidence from the text on this page and the previous page. Then, suggest three steps you can take that will act to stabilize one or more of Earth's systems.

Organic farms Organic farming methods, as shown in **Figure 27,** use natural fertilizers, crop rotation, and biological pest controls. These methods help reduce pollution and other negative impacts on land. However, organic farming methods cannot currently produce the food that is necessary to feed the world's growing population.

Figure 27 Organic farms can reduce the environmental impact of fertilizers, pesticides, and herbicides on land.

Deforestation

Approximately 25 percent of Earth's total land area is covered by forests. Whether you are writing on paper with a pencil, sitting in a wooden chair, or wiping your face with a napkin, you are using products derived from wood. This wood comes from forests worldwide.

Deforestation is the clearing of forest land for agriculture, grazing, urban development, or logging. It is estimated that the amount of forested land decreases by 94,000 km^2 each year. Many of these forests are home to diverse populations of plants and animals. Cutting down trees could lead to the extinction of some of these organisms. In addition, plants remove carbon dioxide from the atmosphere. Deforestation increases the concentration of carbon dioxide in the atmosphere. Scientists believe that an increase in carbon dioxide has contributed to an increase in atmospheric temperatures worldwide.

Urban development

With a growing population, the percentage of land area devoted to urban development has increased. Highways, office buildings, stores, housing developments, and parking lots are under construction every day. This development can lead to negative impacts on land. For example, paving land prevents water from soaking into the soil. Instead, water runs off into sewers or streams, increasing stream discharge and the threat of flooding. Because water is unable to seep through pavement, this also decreases the amount of water that seeps into the ground.

Some communities, businesses, and private organizations preserve areas rather than pave them. As the population grows, more urban areas have been set aside for recreation and preservation for future generations to enjoy. Some urban areas have been designated as historic sites, parks, and monuments by the federal, state, and local governments, such as Central Park in New York City, shown in **Figure 28.**

Figure 28 Some land in urban areas, such as New York City's Central Park, is preserved for recreation.

Waste

Whether or not you realize it, you impact land when you throw garbage into a trash can. About 55 percent of our garbage is disposed of in sanitary landfills. The rest is recycled or burned. Some of the wastes release substances, such as lead from batteries, that are harmful to humans and animals. Wastes that are poisonous, that cause cancer, or that can catch fire are classified as **hazardous wastes.**

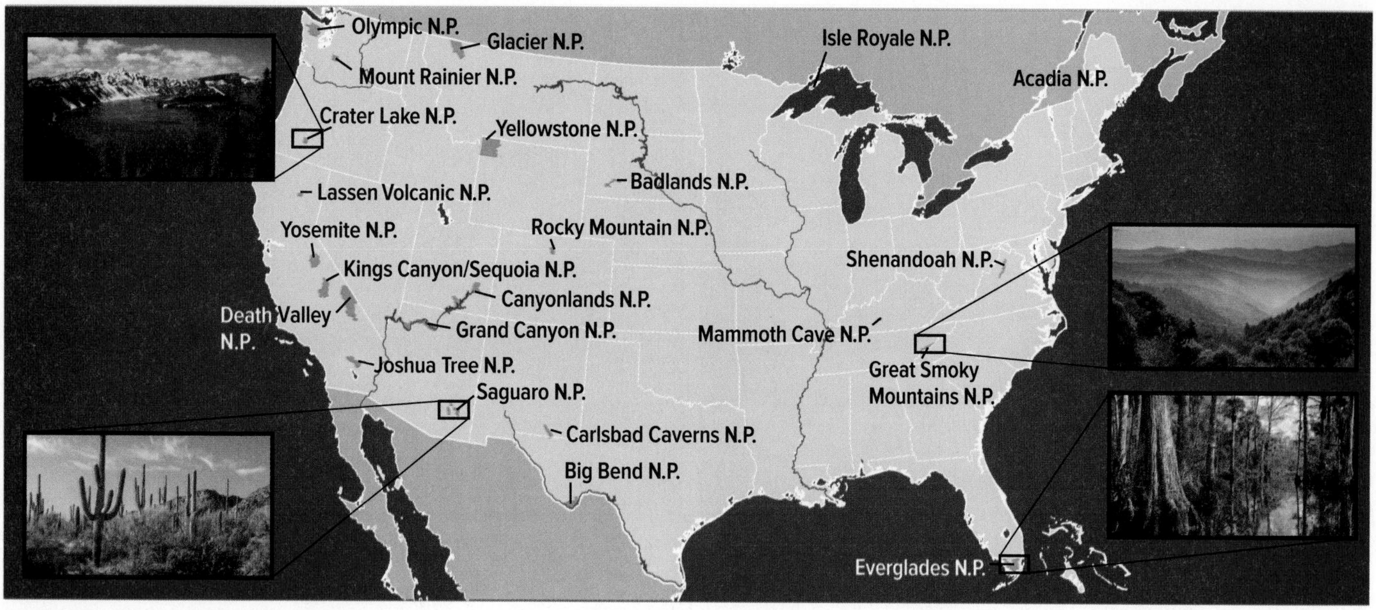

Figure 29 The U.S. National Park Service preserves and protects nearly 400 natural, cultural, and recreational areas across the country.

Identify *any national parks located close to where you live.*

National and state parks

National and state parks are areas of land, like those shown in **Figure 29,** that are preserved and protected by the U.S. government. These forests, wetlands, grasslands, and parks in the United States are safe from urban development, waste disposal, and extensive deforestation. Parks are home to plants, animals, and waterways. Millions of people visit parks, such as Grand Canyon National Park, each year.

Many countries throughout the world also set aside land for protection and preservation. As the world population grows, the impact on land may worsen. Preserving this land in its natural state will benefit generations to come.

Impact on Water

Life on Earth would not be possible without water. Plants need water to convert radiant energy into food energy. Some animals, such as fish, frogs, and whales, make bodies of water their homes. Approximately 60 percent of the human body is composed of water. How are living things affected when water becomes polluted?

Sources of water pollution

Many streams and lakes in the United States are polluted. Polluted water contains harmful chemicals and sometimes organisms that cause disease. Water can also be polluted with sediments, such as silt and clay. Sediment from runoff makes water cloudy and can limit the sunlight and oxygen supply, which then affects fish and wildlife.

Industry Mining can release metals into water. Metals such as mercury, lead, nickel, and cadmium are poisonous. However, environmental laws limit the amount of these harmful chemicals that can be released into the environment, and they protect natural resources and the people who depend upon them.

(tl)Robert Glusic/Photodisc/Getty Images, (tr)KenCanning/E+/Getty Images, (bl)Pete Ryan/National Geographic RF/Getty Images, (br)RicciPhotos/iStock/Getty Images

Prince William Sound, Alaska

Gulf of Mexico, USA

Oil and gas Oil and gas can run off roads and parking lots into lakes and rivers when it rains. It can also leak from oil tankers or pipelines associated with offshore drilling, as shown in **Figure 30.** Oil and gas are pollutants that can cause cancer. Today, environmental laws require that all new gasoline storage tanks have a double layer of steel or fiberglass to prevent spills. These laws help protect soil and water from oil spills.

Figure 30 An oil tanker accident near Prince William Sound, Alaska, on March 24, 1989, resulted in nearly two decades of environmental clean up. A deadly explosion in the Gulf of Mexico on April 20, 2010, resulted in the release of millions of gallons of oil. Recovery efforts to save and rehabilitate marine habitats, fish, and wildlife will likely continue here for decades.

Human waste When you flush a toilet or take a shower, you create wastewater. Wastewater, also called sewage, contains human waste, household detergents, and soaps. Sewage contains harmful organisms that can make people ill.

In most cities in the United States, underground pipes route water from homes, schools, and businesses to sewage treatment plants. Sewage treatment plants remove pollutants in a series of steps. These steps purify the water by removing solid materials from the sewage, killing harmful microorganisms, and reducing the amount of nitrogen and phosphorous in the water. The water is then recycled back into the environment.

Impact on Air

Like water, air is essential for all life on Earth. Air pollution can affect human health and threaten plants and animals. Air pollution comes from natural and manufactured sources. For example, cars, buses, and trucks burn fuel for energy and release exhaust into the atmosphere. Factories and power plants emit pollutants during production, as shown in **Figure 31.** Dust from farms and construction sites also contributes to air pollution. Natural sources of pollution include particles and gases emitted into air from erupting volcanoes and forest fires.

Figure 31 Significant sources of air pollution include cars and factories.

 Get It?

List the sources of land, water, and air pollution.

(t)USCG, (tr)Tigergallery/Shutterstock, (b)Anton Papulov/123RF

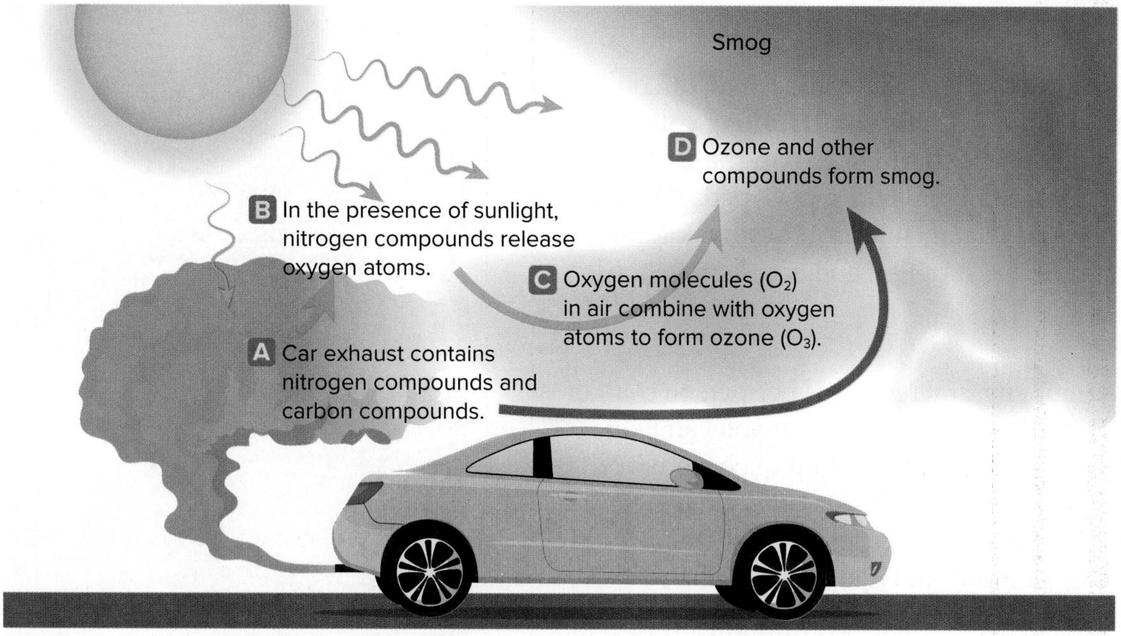
Smog
D Ozone and other compounds form smog.
B In the presence of sunlight, nitrogen compounds release oxygen atoms.
C Oxygen molecules (O_2) in air combine with oxygen atoms to form ozone (O_3).
A Car exhaust contains nitrogen compounds and carbon compounds.

Figure 32 Exhaust from cars contributes to smog formation. Sunlight helps fuel reactions that form smog compounds. These compounds include nitrogen-based compounds and ozone.

Types of air pollution

Have you ever observed a thick, brown haze on the horizon? The brown haze that you see forms from vehicle exhaust and factory and power plant pollution. This haze is often referred to as photochemical smog. **Photochemical smog** is a term used to describe the pollution that results from the reaction between sunlight and vehicle or factory exhaust.

Smog Major sources of photochemical smog include cars, factories, and power plants. Pollutants are released into the air when fossil fuels, such as gasoline are burned, as shown in **Figure 32,** emitting sulfur-, nitrogen-, and carbon-based compounds. These compounds react with oxygen in the presence of sunlight. One of the products of this reaction is ozone (O_3). Ozone that forms high in the atmosphere protects you from ultraviolet (UV) radiation from the Sun. Ozone near Earth's surface, however, can cause breathing problems.

CFCs The protective ozone high in the atmosphere is concentrated in a layer roughly 20 km above Earth's surface. This layer is called the ozone layer, and it is at risk of being destroyed. Chlorofluorocarbons (CFCs) are compounds that leak from old air conditioners and refrigerators and react with ozone. This reaction destroys ozone molecules. Even though the use of CFCs has been declining due to environmental laws, these compounds can remain in the atmosphere for decades.

Acid precipitation When sulfur-, nitrogen-, and carbon-based compounds from vehicles and factories react with moisture in the air, they form acids. When acidic moisture falls from the sky as precipitation, it is called **acid precipitation.** Acid precipitation can corrode metals and cause harm to plants and animals.

CCC CROSSCUTTING CONCEPTS

Stability and Change Read about the types of air pollution listed on this page. How do the factors described destabilize the system of Earth's atmosphere? Create a list of factors that destabilize Earth's atmosphere. Include a sentence or two describing the effects of each factor.

Reducing Pollution

Pollution is often difficult to contain. Even if one state or country reduces air pollution, pollutants from another state or country can blow across borders. For example, burning coal in the midwestern United States might cause acid precipitation in Canada. Water pollution can enter a river or stream and travel several kilometers downstream, into groundwater supplies, and across state borders.

How can you help?

The United States uses more natural resources per person than most countries in the world. There are many ways that you can help conserve resources. You can reduce the amount of consumable materials you use. You can compost some yard and kitchen waste rather than throwing it in the trash. You can also reuse and recycle many different materials, as shown in **Figure 33.**

Figure 33 Many communities offer recycling programs that allow paper, plastic, and glass to be reused rather than taken to landfills.

Scientists and engineers can make major contributions by developing technologies that produce less pollution and waste. You can help by using these technologies. For example, energy-efficient appliances can help your family reduce its energy use. Low-flush toilets, leak-free faucets, and dishwashers and washing machines that run on less water will help you reduce your water use. Driving fuel-efficient vehicles or using alternate modes of transportation, such as a bicycle or a bus, will help lessen your impact on air.

Check Your Progress

Summary

- Land resources are threatened by agriculture, deforestation, industry, and waste.
- All life on Earth requires clean air and water.
- Water can be contaminated with sediment, industrial pollutants, and human waste.
- Smog, acid precipitation, and CFCs cause air pollution.

Demonstrate Understanding

20. **Discuss** what you can do to lessen your environmental impact on natural resources such as land, water, and air.
21. Describe how urban development can increase flooding.
22. **Infer** the effect of deforestation on the carrying capacity of the Amazon rain forest.
23. **Identify** three pollutants released into the air when fossil fuels are burned.

Explain Your Thinking

24. **Infer** Southern Florida is home to many dairy and sugarcane farms. Everglades National Park, including its shallow river system, is also located there. What kinds of pollutants might affect plants and animals in the Everglades?
25. **MATH Connection** A decrease of one unit on the pH scale means a solution is ten times more acidic. A decrease of two units means the solution is 100 times more acidic. How much more acidic is acid precipitation (pH = 4.0) than pure water (pH = 7.0)?

LEARNSMART® Go online to follow your personalized learning path to review, practice, and reinforce your understanding.

Earth Day, 1970

Environmental crises captured national attention in the late 1960s. In 1966, more than 160 deaths were caused by chemical smog that settled in New York City for three days. In 1969, an oil well exploded and spread crude oil across 55 kilometers of coastline near Santa Barbara, California. During that same year, the contaminated Cuyahoga River in Ohio caught fire. These disasters inspired an environmental movement that would eventually lead to new laws that protect and preserve our natural resources.

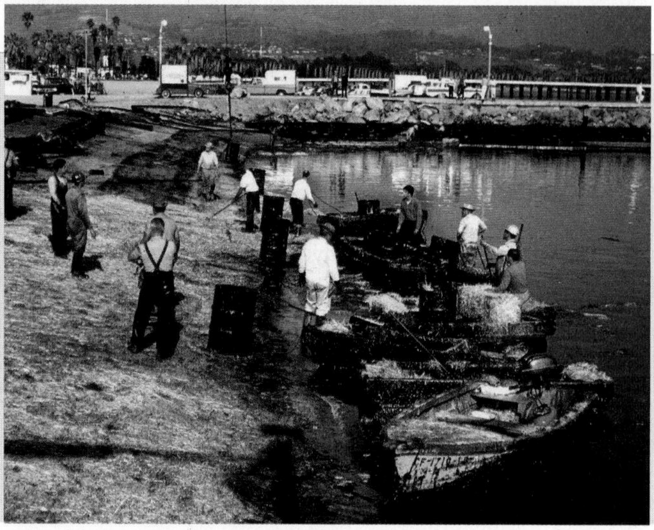

Volunteers clean the Santa Barbara coastline after the devastating 1969 oil spill.

Terry Herbig/Zumapress/Newscom

Change in the air

An environmental movement swept across the United States in the 1970s. Influential books like *Silent Spring*, written by biologist Rachel Carson, painted a bleak picture of a polluted world. The oil spill in Santa Barbara, California, shown in the photo, and nationally broadcast images of volunteers rescuing oil-covered seals and seabirds was also a catalyst for change. In 1970, one year after the catastrophe, 20 million Americans participated in the first Earth Day.

Safeguarding the environment

Earth Day led the U.S. government to respond to calls for environmental reform. In 1970, President Nixon created the Environmental Protection Agency (EPA). The EPA was responsible for conducting research and proposing new laws to protect the environment. The agency also had the ability to enforce these laws and to hold individuals and corporations accountable for meeting new environmental standards.

The Clean Air and Water Acts

One of the EPA's first laws was the Clean Air Act of 1970. The law created air pollution standards, limiting the amount of carbon monoxide, nitrogen dioxide, and ozone released into the environment. It also required the eventual phase-out of lead as an ingredient in gasoline. The Clean Water Act followed in 1972, limiting the amount of pollutants that could be released into rivers, lakes, and streams.

Looking forward

Protecting natural resources has become a worldwide effort. More than 200 countries signed the Montreal Protocol in 1987 to end the production of ozone-destroying chemicals. More recently, 195 members of the United Nations Framework Convention on Climate Change signed the Paris Agreement, committing to take action on mitigating climate change. Many hope a global effort will create a healthier planet for the future.

ENGAGE IN ARGUMENT FROM EVIDENCE

Construct an argument about the importance of the EPA to society as well as to the environment. Cite evidence from your reading.

MODULE 8
STUDY GUIDE

GO ONLINE to study with your Science Notebook.

Lesson 1 FOSSIL FUELS

- Energy cannot be created or destroyed; it can only be transformed from one form to another.
- Petroleum, natural gas, and coal are fossil fuels.
- Petroleum is a mixture of hydrocarbons.
- Power plants burn fossil fuels to extract chemical potential energy that spins turbines and powers electric generators.
- Fossil fuels are nonrenewable resources.

- fossil fuels
- petroleum
- nonrenewable resource

Lesson 2 NUCLEAR ENERGY

- Nuclear power plants produce about 8 percent of the energy used each year in the United States.
- Nuclear reactors use the energy released in the fission of U-235 to produce electricity.
- The energy released in the fission reaction is used to make steam. The steam spins a turbine that powers an electric generator.
- Nuclear power generation produces high-level nuclear waste.

- fusion
- fission
- nuclear reactor
- nuclear waste

Lesson 3 RENEWABLE ENERGY RESOURCES

- Solar cells convert radiant energy into electrical energy.
- Hydroelectric power plants convert gravitational potential energy into electrical energy.
- Wind energy is converted into electrical energy using a propeller attached to an electric generator.
- Alternative energy sources, such as the Sun, water, wind, and Earth's internal heat, can help reduce human dependence on fossil fuels.

- renewable resource
- photovoltaic cell
- hydroelectricity
- geothermal energy
- biomass

Lesson 4 ENVIRONMENTAL IMPACTS

- Land resources are threatened by agriculture, deforestation, industry, and waste.
- All life on Earth requires clean air and water.
- Water can be contaminated with sediment, industrial pollutants, and human waste.
- Smog, acid precipitation, and CFCs cause air pollution.

- population
- carrying capacity
- pollutant
- hazardous waste
- photochemical smog
- acid precipitation

THREE-DIMENSIONAL THINKING

Module Wrap-Up

REVISIT THE PHENOMENON

How could this relatively small space produce enough nuclear energy to sustain millions of people?

CER Claim, Evidence, Reasoning

Explain Your Reasoning Revisit the claim you made when you encountered the phenomenon. Summarize the evidence you gathered from your investigations and research and finalize your Summary Table. Does your evidence support your claim? If not, revise your claim. Explain why your evidence supports your claim.

STEM UNIT PROJECT

Now that you've completed the module, revisit your STEM unit project. You will apply your evidence from this module and complete your project.

GO FURTHER

SEP Data Analysis Lab

Oil from the Arctic

Oil is the leading source of energy in the United States. It supplies about 40 percent of our total energy needs. One important domestic source of crude oil, the Prudhoe Bay Oil Field, lies under the permafrost of an icy, frigid area of Alaska called the North Slope. To transport oil from the North Slope to the rest of the United States, it must first flow south through the 1300-km Trans-Alaskan Pipeline System. Research the Alaska pipeline and answer the following questions.

CER Analyze and Interpret Data

1. **Claim** Look at a map of Alaska. Find Prudhoe Bay and Valdez. What type of terrain does the Trans-Alaskan Pipeline travel through?
2. **Claim, Evidence** Many people feared that the Trans-Alaskan Pipeline would damage the environment that it passed through. What precautions have been taken to protect the environment along its route?
3. **Claim, Evidence, Reasoning** Do you think that the costs of building the Trans-Alaskan Pipeline were worth the final product—domestic oil? Explain your answer.

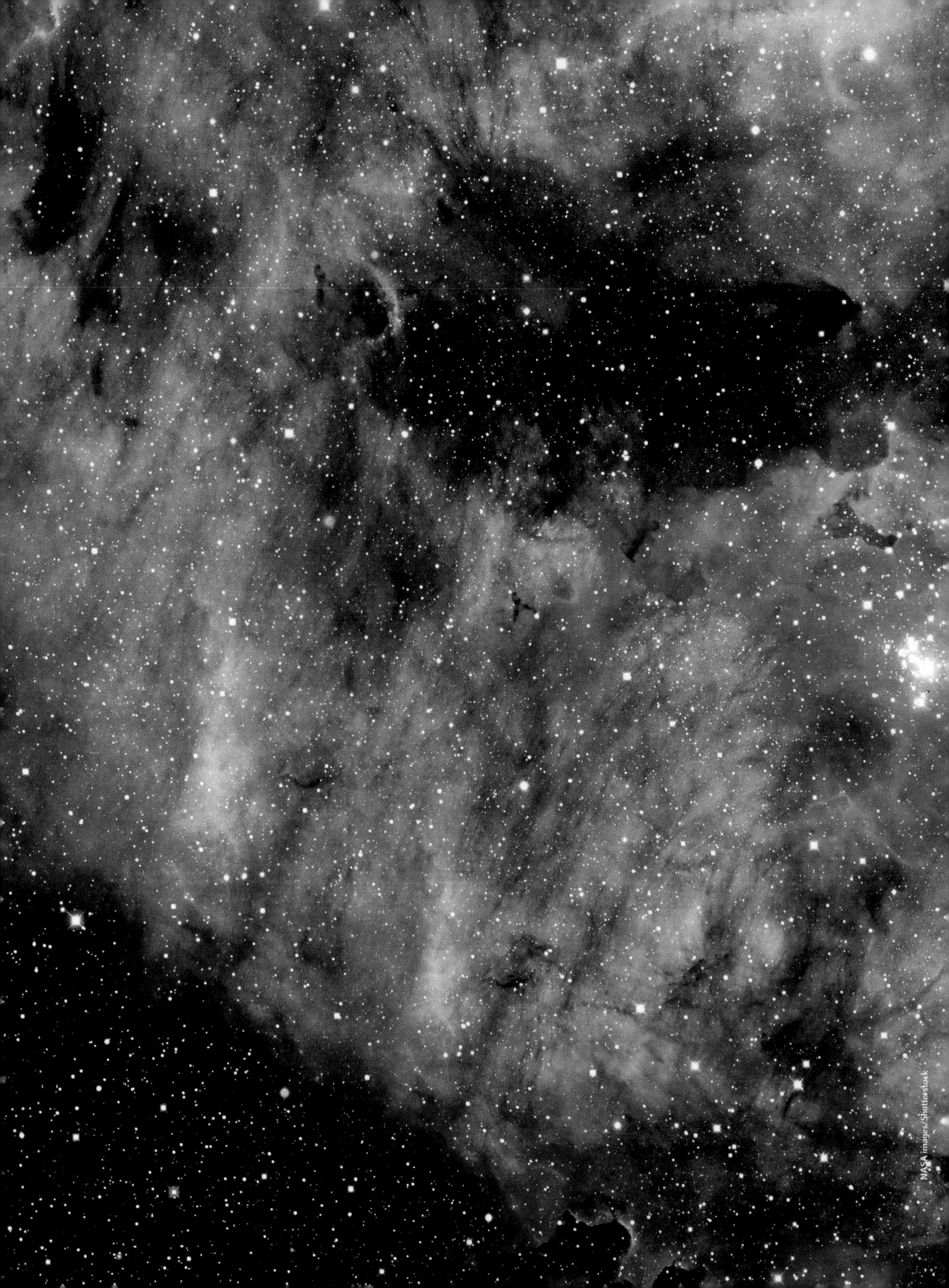

UNIT 3

WAVES

ENCOUNTER THE PHENOMENON

How do waves interact with our senses?

SEP Ask Questions

What questions do you have about the phenomenon? Write your questions on sticky notes and add them to the driving question board for this unit.

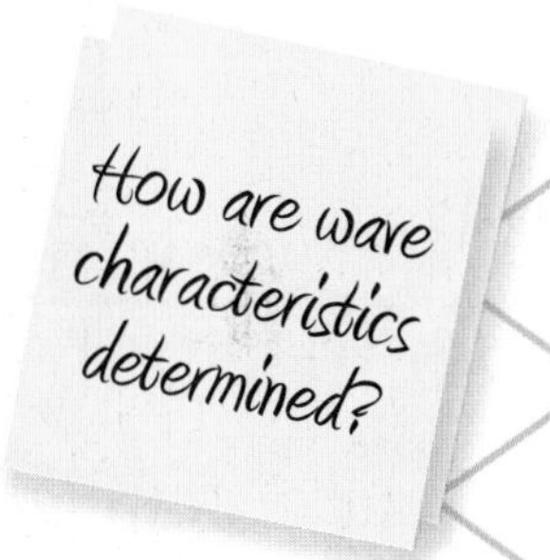
How are wave characteristics determined?

Look for Evidence

As you go through this unit, use the information and your experiences to help you answer the phenomenon question as well as your own questions. For each activity, record your observations in a Summary Table, add an explanation, and identify how it connects to the unit and module phenomenon questions.

Solve a Problem

STEM UNIT PROJECT

Design a Sustainable Electrical Power Grid When a hurricane decimates an area, such as the Caribbean islands, traditional power grids are also destroyed. Research how solar power and different types of power grids are being used to avoid these problems if another hurricane hits. Use the results of these investigations and the evidence you collected during the unit to design a sustainable power grid that can be incorporated as a community rebuilds following a hurricane.

GO ONLINE In addition to reading the information in your Student Edition, you can find the STEM Unit Project and other useful resources online.

Trevor Mogg/Alamy Stock Photo

MODULE 9
INTRODUCTION TO WAVES

ENCOUNTER THE PHENOMENON

How do buildings move during an earthquake?

GO ONLINE to play a video how buildings absorb the shock of an earthquake.

SEP Ask Questions

Do you have other questions about the phenomenon? If so, add them to the driving question board.

CER Claim, Evidence, Reasoning

Make Your Claim Use your CER chart to make a claim about how buildings move during an earthquake. Explain your reasoning.

Collect Evidence Use the lessons in this module to collect evidence to support your claim. Record your evidence as you move through the module.

Explain Your Reasoning You will revisit your claim and explain your reasoning at the end of the module.

GO ONLINE to access your CER chart and explore resources that can help you collect evidence.

LESSON 1: Explore & Explain: Waves Defined

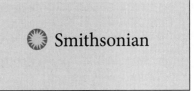

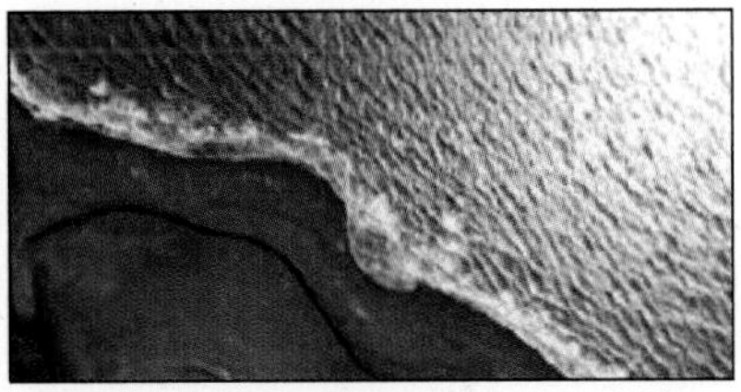

LESSON 3: Explore & Explain: Diffraction

Smithsonian

Additional Resources

LESSON 1

THE NATURE OF WAVES

FOCUS QUESTION

How would you describe the motion of a stadium-style wave?

Waves Defined

Figure 1 shows a disturbance from a pebble splashing into a pond. This disturbance travels outward from the spot where the pebble splashed. The pebble produced a small wave in the pond. A **wave** is a repeating disturbance that transfers energy through matter or space. Other examples of waves include ocean waves, sound waves, seismic waves, and light waves. Do these and other types of waves have anything in common with one another?

Waves and Energy

Because a falling pebble is moving, the falling pebble has kinetic energy. As the falling pebble splashes into the pond in **Figure 1,** the pebble transfers some of its energy to nearby water molecules, causing them to move. Those molecules then pass the energy along to neighboring water molecules, which, in turn, transfer it to their neighbors. The energy travels farther and farther from the source of the disturbance. However, the water itself does not move outward. What you see is energy traveling in the form of a wave on the surface of the water.

Figure 1 Falling pebbles transfer their kinetic energy to the molecules of water in a pond, forming waves. The waves travel outward in all directions from the source of the splash.

3D THINKING DCI Disciplinary Core Ideas CCC Crosscutting Concepts SEP Science & Engineering Practices

COLLECT EVIDENCE

Use your Science Journal to record the evidence you collect as you complete the readings and activities in this lesson.

INVESTIGATE

GO ONLINE to find these activities and more resources.

Probeware Lab: Measuring Earthquakes

Develop and use a model of a seismograph to simulate the causes and effects of the movement of continental tectonic plates.

Waves and matter

Imagine that you are in a boat on a lake. Approaching waves bump against your boat, but they do not carry it along with them as they pass. The boat does move up and down, and maybe even a short distance back and forth, because the waves transfer some of their energy to the boat.

But after the waves have passed, the boat is still in nearly the same place. The waves do not even carry the water along with them. The water that is around the boat after the wave passes is the same water as was there before the wave passed. Only the energy travels along with the waves.

All waves, whether they are water waves, sound waves, light waves or earthquake waves, have this property: they carry energy without transporting matter from place to place.

Get It?
Identify what waves carry.

Making waves

A wave will travel only as long as it has energy to carry. When you drop a pebble into a pond, small waves form, as shown in **Figure 1.** These waves carry energy. However, this energy is gradually transferred to the surrounding water and air. At the same time, the remaining energy in the waves spreads out as the waves spread out. As the energy spreads out and is transferred away from the waves, those waves shrink and disappear.

Get It?
Describe what happens to the energy carried by a water wave.

Suppose you are holding a rope at one end, and you give it a shake. You would create a pulse that would travel along the rope to the other end, and then the rope would be still again, as shown in **Figure 2.**

Now suppose you shake your end of the rope up and down for a while. You would make waves that travel along the rope. When you stop shaking your hand up and down, the rope will be still again. It is the up-and-down motion of your hand that creates the wave.

Anything that moves up and down or back and forth in a rhythmic way is vibrating. The vibrating movement of your hand at the end of the rope created the wave. In fact, all waves are produced by something that vibrates.

Get It?
Identify what produces waves.

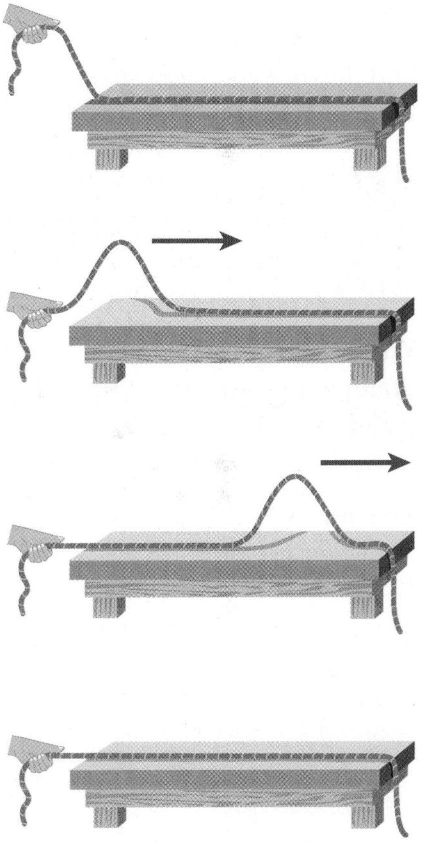

Figure 2 The hand's up and down motion produces a wave.

Explain *what happened to the energy that was carried by the wave in this rope.*

Mechanical Waves

Sound waves travel through air to reach your ears. Ocean waves travel through water to reach the shore. A **medium** is matter through which a wave travels. The medium can be a solid, a liquid, a gas, or a combination of these. Not all waves need a medium. Some waves, such as light and radio waves, can travel through a vacuum. **Mechanical waves,** such as sound waves, are waves that can travel only through matter. Mechanical waves can be either transverse waves or longitudinal waves.

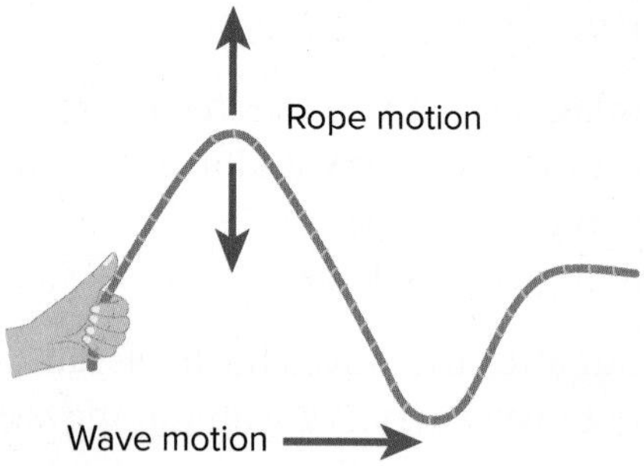

Figure 3 The wave along the rope travels horizontally, but the rope itself moves vertically, up and down.

Get It?
Describe the connection between a medium and a mechanical wave.

Transverse waves

In a **transverse wave,** particles in the medium move back and forth at right angles to the direction that the wave travels. For example, **Figure 3** shows how a wave along a rope travels horizontally, but the portions of the rope that the wave passes through move up and down. When you shake one end of a rope while your friend holds the other end, you are making transverse waves.

Get It?
Compare the direction that a transverse wave travels with the direction that matter in that wave vibrates.

Water waves When wind blows across the surface of the ocean, water waves form. Water waves are often thought of as transverse waves, but this is not entirely correct. The water in water waves does move up and down as the waves go by. But the water also moves short distances back and forth along the direction that the wave is traveling.

This movement happens because the low part of the wave can be formed only by pushing water forward or backward toward the high part of the wave, as shown on the left in **Figure 4.** This is much like a child pushing sand into a pile. Sand must be pushed in from the sides to make the pile. As the wave passes, the water that was pushed aside moves back to its initial position, as shown on the right in **Figure 4.**

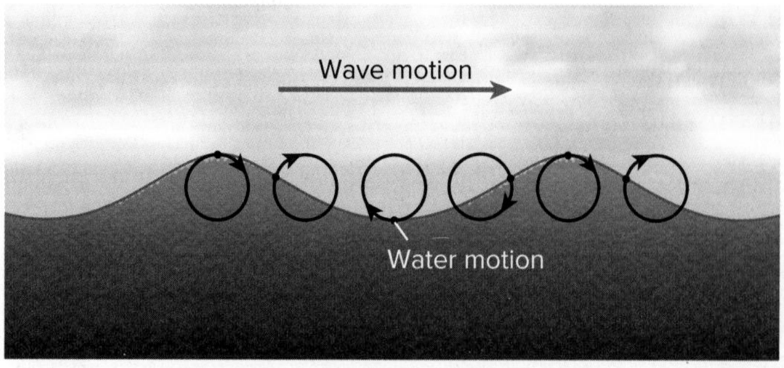

The low point of a water wave is formed when water is pushed aside and up to the high point of the wave.

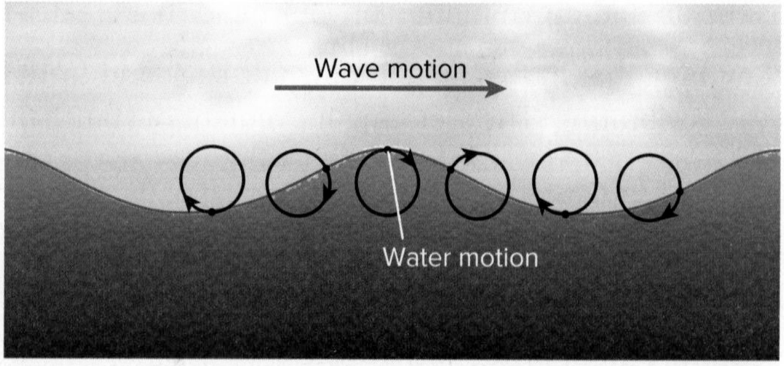

The water that is pushed aside returns to its initial position.

Figure 4 A water wave causes water to move back and forth, as well as up and down. Water is pushed back and forth to form the high and low points on the wave.

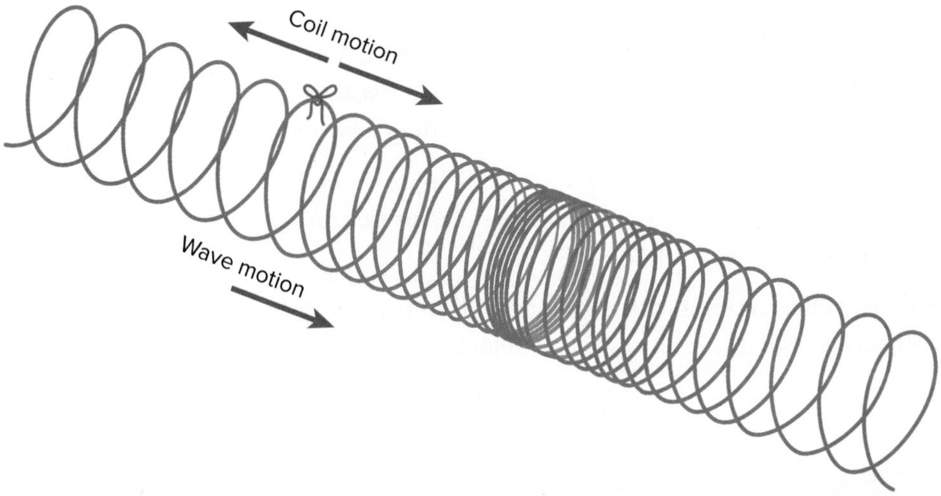

Figure 5 The longitudinal wave through this spring toy moves in only one direction, but the coils of this spring toy move back and forth.

Describe *another example of a longitudinal wave.*

Longitudinal waves

In a **longitudinal wave,** matter in the medium moves back and forth along the same direction that the wave travels. You can model longitudinal waves with a coiled spring toy, as shown in **Figure 5.** Squeeze several coils together at one end of the spring. Then let go of the coils, still holding onto coils at both ends of the spring. A wave will travel along the spring. As the wave moves, it looks as if the whole spring is moving toward one end.

Suppose you watched the coil with yarn tied to it, as in **Figure 5.** You would see that the yarn moves back and forth as the wave passes and then stops moving after the wave has passed. The wave carries energy, but not matter, forward along the spring. Longitudinal waves are sometimes called compressional waves.

Sound waves Sound waves are longitudinal waves. When a noise is made, such as when a locker door slams shut, nearby molecules in the air are pushed together by the vibrations caused by the slamming door. The molecules in the air are squeezed together, like the coils of the spring toy in **Figure 5.** These compressions travel through the air to make a sound wave. The sizes of a sound wave's compressions, as well as the distances between those compressions, determine the nature of that sound.

Sound waves in liquids and solids Sound waves can also travel through liquids and solids, such as water and wood. Particles in these mediums are pushed together and move apart as the sound waves travel through them.

Get It?

Describe how sound waves travel through solids.

SCIENCE USAGE v. COMMON USAGE

medium

Science usage: the matter through which waves travel
Air is often the medium through which a sound wave travels.

Common usage: something in a middle position
The person's uniform is a size medium.

Seismic waves

Forces within Earth's interior can cause regions of Earth's crust to move, bend, or even break. Movement in the crust, which occurs along faults, can result in a rapid release of energy. This energy travels away from the fault in the form of seismic (SIZE mihk) waves, as shown in **Figure 6.**

Seismic waves can be longitudinal waves or transverse waves. Scientists have found out much about Earth's interior by studying these seismic waves.

Seismic waves can travel through Earth as well as along Earth's surface. When the energy from seismic waves is transferred to objects on Earth's surface, those objects move and shake.

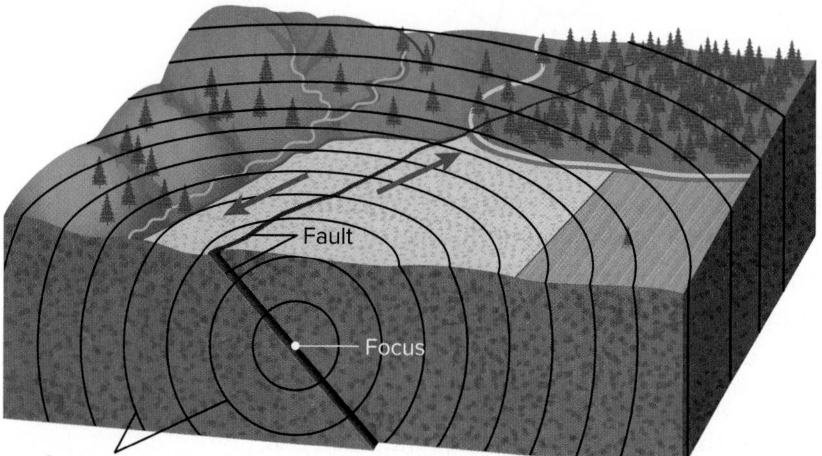

Figure 6 A fault is a break in Earth's crust. The red arrows show the direction that Earth's crust is moving at a fault. When Earth's crust shifts or breaks, the energy that is released is transmitted outward, causing an earthquake. The point where the earthquake originates is called the focus.

Explain *how earthquake waves and sound waves are similar.*

Check Your Progress

Summary

- A wave is a repeating disturbance that transfers energy through matter or space.
- Waves carry energy without transporting matter.
- In a transverse wave, matter in the medium moves at right angles to the direction that the wave travels.
- In a longitudinal wave, matter in the medium moves back and forth along the direction that the wave travels.

Demonstrate Understanding

1. **Describe** the motion of an unanchored rowboat when a water wave passes. Does the wave move the boat forward?
2. **Contrast** how you would move a spring to make a transverse wave with how you would move a spring to make a longitudinal wave.
3. **Identify** evidence that seismic waves transfer energy without transferring matter.
4. **Identify** a mechanical wave that is also a longitudinal wave.

Explain Your Thinking

5. **Describe** how the world would be different if all waves were mechanical waves.
6. **MATH Connection** The average speed of sound in water is 1500 m/s. How long would it take a sound wave to travel 9000 m in water?

LEARNSMART® Go online to follow your personalized learning path to review, practice, and reinforce your understanding.

LESSON 2

WAVE PROPERTIES

FOCUS QUESTION

What are the different ways to measure one wavelength?

The Parts of a Wave

What makes sound waves, water waves, and seismic waves different from each other? Waves can differ in how much energy they carry and in how fast they travel. Waves also have other characteristics that make them different from each other.

Suppose you shake the end of a rope and make a transverse wave. The transverse wave in **Figure 7** has alternating high points and low points. **Crests** are the high points of a transverse wave. **Troughs** are the low points of a transverse wave. The imaginary line that is half the vertical distance between a crest and a trough is called the rest position.

On the other hand, a longitudinal wave has no crests and troughs. When a longitudinal wave passes through a medium, it creates regions where the medium becomes crowded together and more dense, as shown in **Figure 7.** These regions are compressions. A **compression** is the more dense region of a longitudinal wave. **Figure 7** also shows that the coils in the region next to a compression are spread apart, or less dense. The less-dense region of a longitudinal wave is called a **rarefaction.**

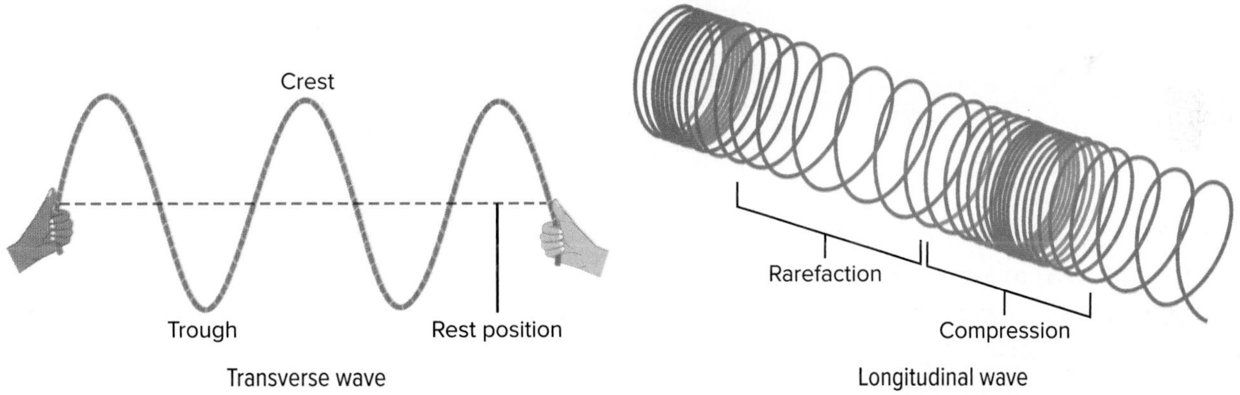

Figure 7 Different types of waves are described in different ways. Transverse waves have crests and troughs, but longitudinal waves have compressions and rarefactions.

3D THINKING DCI Disciplinary Core Ideas CCC Crosscutting Concepts SEP Science & Engineering Practices

COLLECT EVIDENCE

Use your Science Journal to record the evidence you collect as you complete the readings and activities in this lesson.

INVESTIGATE

GO ONLINE to find these activities and more resources.

Applying Practices: Wave Characterisitcs

HS-PS4-1. Use mathematical representations to support a claim regarding relationships among the frequency, wavelength, and speed of waves traveling in various media.

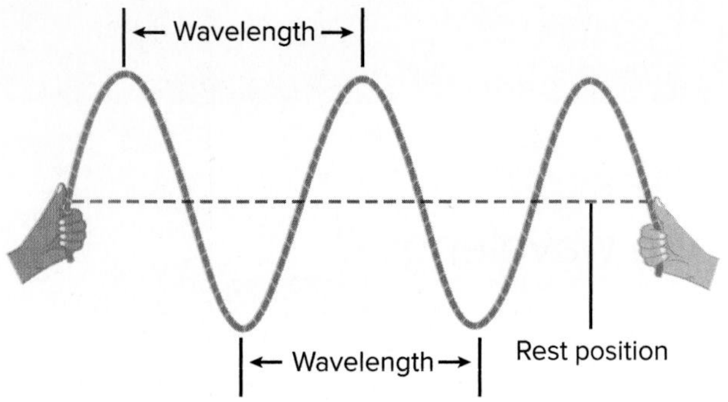

For transverse waves, a wavelength can be measured from crest to crest or trough to trough.

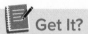

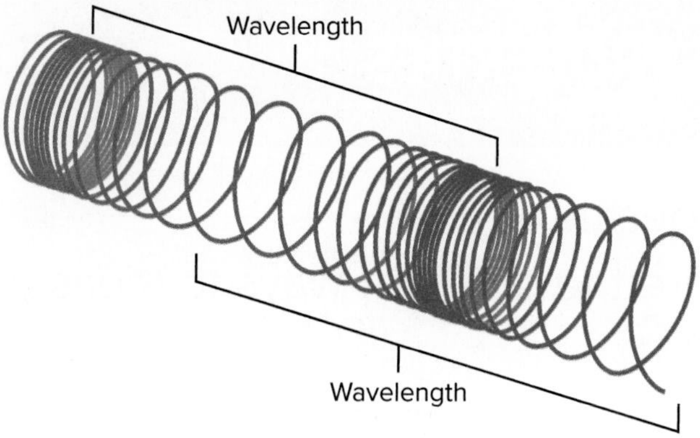

The wavelength of a longitudinal wave can be measured from compression to compression or from rarefaction to rarefaction.

Figure 8 One wavelength is the distance from one point on a wave to the nearest point just like it.

Wavelength

All waves have wavelength. **Wavelength** is the distance between one point on a wave and the nearest point just like it. **Figure 8** shows that for transverse waves, the wavelength is the distance from crest to crest or trough to trough. These two distances are equal on a transverse wave.

A wavelength of a longitudinal wave is the distance from the middle of one compression to the middle of the next compression, as shown in **Figure 8.** The wavelength of a longitudinal wave is also the distance from the middle of one rarefaction to the middle of the next rarefaction. These two distances are equal on a longitudinal wave.

Get It?

Describe how wavelength is defined for transverse waves and for longitudinal waves.

Frequency and Period

When you tune your radio to a station, you are choosing radio waves of a certain frequency. The **frequency** of a wave is the number of wavelengths that pass a fixed point each second. Frequency is also the number of times that a point on a wave moves up and down or back and forth each second.

You can find the frequency of a transverse wave by counting the number of crests that pass a point each second. You can find the frequency of a longitudinal wave by counting the number of compressions that pass a point each second.

Frequency is expressed in hertz (Hz). A frequency of 1 Hz means that one wavelength passes by in 1 s. In SI units, 1 Hz is the same as 1/s.

The **period** of a wave is the amount of time it takes one wavelength to pass a point. As the frequency of a wave increases, the period decreases. In SI units, period has units of seconds.

Wavelength is related to frequency

If you make transverse waves with a rope, you increase the frequency by moving the rope up and down faster. Moving the rope faster also makes the wavelength shorter. This relationship is always true—as frequency increases, wavelength decreases.

If you double the frequency of a wave, you halve the wavelength of that wave. If you double the wavelength of a wave, you halve that wave's frequency. **Figure 9** compares the wavelengths and frequencies of two different waves.

The frequency of a wave is always equal to the rate of vibration of the source that creates it. If you move the rope up, down, and back up in 1 s, the frequency of the wave that you generate is 1 Hz. If you move the rope up, down, and back up five times in 1 s, the resulting wave has a frequency of 5 Hz.

Get It?
Describe how the wavelength and the frequency of a wave are related.

Wave Speed

Suppose you are at a large stadium watching a baseball game, but you are high up in the bleachers, far from the action. The batter swings, and you see the ball rise. An instant later, you hear the crack of the bat hitting the ball. You see the impact before you hear it because light waves travel much faster than sound waves. Therefore, the light waves reflected from the flying ball reach your eyes before the sound waves created by the crack of the bat reach your ears.

The speed of a wave depends on the medium through which it is traveling. Sound waves usually travel faster in liquids and solids than they do in gases. However, light waves travel more slowly in liquids and solids than they do in gases or in a vacuum.

The speed of a wave also depends on the temperature of the medium through which it is traveling. Sound waves usually travel faster in a material if the temperature of the material is increased. For example, sound waves travel faster in air at 20°C than in air at 0°C.

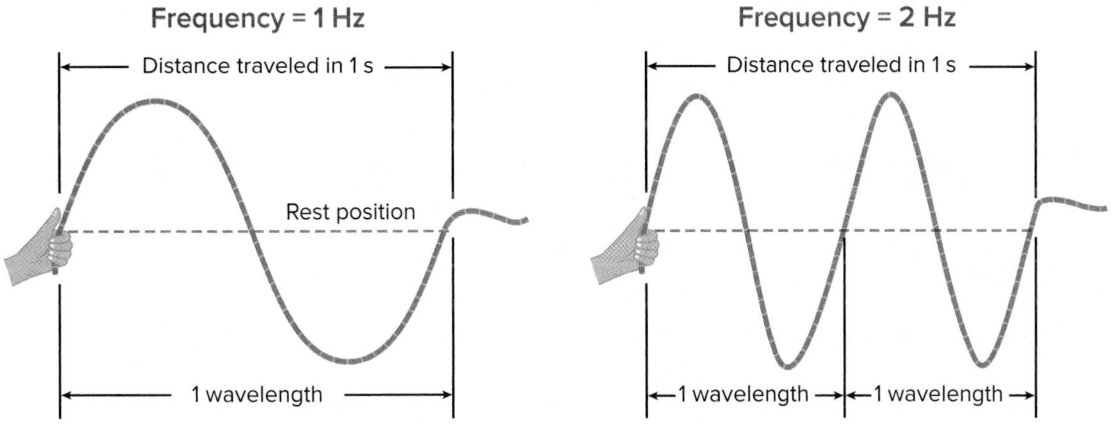

The rope is moved down, up, and down again one time in 1 s. The wave has a frequency of 1 Hz.

The rope is shaken down, up, and down again twice in 1 s. The wave has a frequency of 2 Hz.

Figure 9 When frequency increases, the number of wavelengths that pass in one second increases.

Identify *which of the waves in this figure has the higher frequency.*

Calculating wave speed

The speed of a wave depends on the medium in which the wave travels. However, the wave speed, wave frequency, and wavelength are related. The speed of a wave can be calculated from the following equation.

Wave Speed Equation

speed (in m/s) = **frequency** (in Hz) × **wavelength** (in m)

$$v = f\lambda$$

In this equation, v represents the wave speed, f is the frequency, and the Greek letter λ (lambda) represents the wavelength.

Get It?

Identify Based on the equation, how would the wavelength of a wave be affected if the speed of the wave doubles but the frequency of the wave stays the same?

Why does multiplying the frequency unit Hz by the distance unit m give the unit for speed m/s? Recall that the SI unit Hz is the same as 1/s. Therefore, m × Hz is the same as m × 1/s. Both are equivalent to m/s.

EXAMPLE Problem 1

SOLVE FOR WAVE SPEED What is the speed of a sound wave that has a wavelength of 2.00 m and a frequency of 170.5 Hz?

Identify the Unknown: wave speed: v

List the Knowns: frequency: f = 170.5 Hz wavelength: λ = 2.00 m

Set Up the Problem: $v = f\lambda$

Solve the Problem: v = (170.5 Hz)(2.00 m)

= (170.5 waves/s)(2.00 m)

= 341 m/s

The speed of the sound wave is 341 m/s.

Check the Answer: The speed of sound through air varies but is typically close to 340 m/s. Therefore, this answer seems reasonable.

PRACTICE Problems ADDITIONAL PRACTICE

7. A wave traveling in water has a frequency of 250 Hz and a wavelength of 6.0 m. What is the speed of the wave?
8. The lowest-pitched sounds humans can generally hear have a frequency of roughly 20 Hz. What is the approximate wavelength of these sound waves if their wave speed is 340 m/s?
9. A particular radio station broadcasts radio waves at 100 MHz (100 million Hz). If radio waves travel at the speed of light (300 million m/s), then what is the wavelength of the radio waves that the station is broadcasting?
10. **CHALLENGE** A sound wave with a frequency of 100.0 Hz travels in water with a speed of 1500 m/s and then travels in air with a speed of 340 m/s. Approximately how many times larger is the wavelength in water than in air?

Amplitude

Why do some earthquakes cause terrible damage, while others are hardly felt? This is because the disturbance from a wave can vary. **Amplitude** is a measure of the size of the disturbance from a wave. If the wave's amplitude is greater, then the disturbance from that wave is also greater. Amplitude is measured differently for longitudinal waves and transverse waves.

Amplitude of longitudinal waves

The amplitude of a longitudinal wave is related to how tightly the medium is pushed together at the compressions and how much the medium is pulled apart at the rarefactions. The more tightly pushed together the medium is at the compressions, the denser the medium. The denser the medium is at the compressions, the larger the wave's amplitude is and the greater the disturbance from the wave. The denser the medium is at the compressions, the less dense the medium is in the rarefactions. Therefore, another indication of high amplitude is whether the medium is stretched out more in the rarefactions. **Figure 10** shows longitudinal waves with different amplitudes.

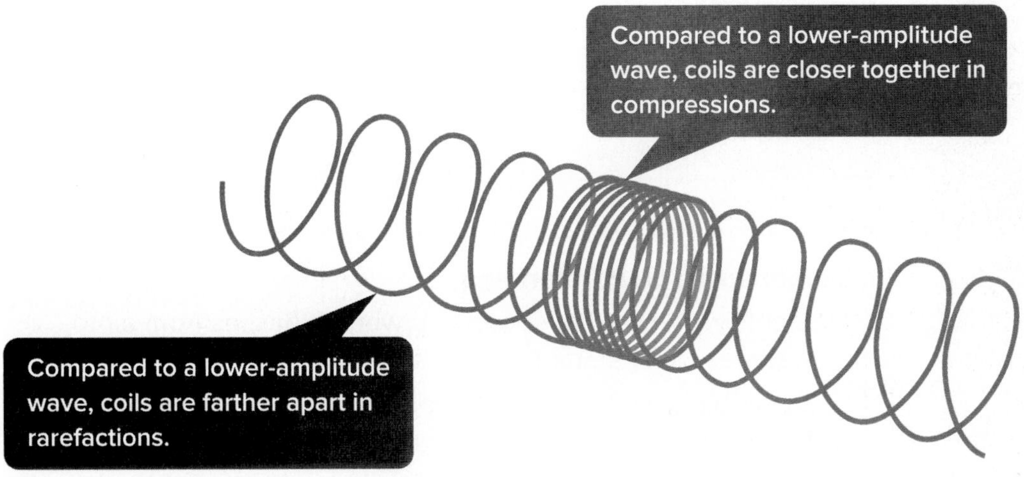

Compared to a lower-amplitude wave, coils are closer together in compressions.
Compared to a lower-amplitude wave, coils are farther apart in rarefactions.

Higher-amplitude wave

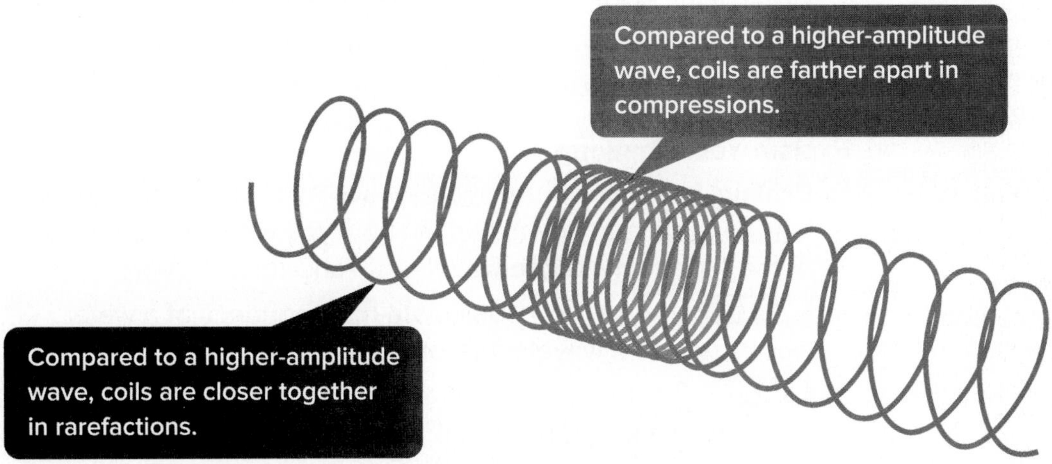

Compared to a higher-amplitude wave, coils are farther apart in compressions.
Compared to a higher-amplitude wave, coils are closer together in rarefactions.

Lower-amplitude wave

Figure 10 The coils in the high-amplitude wave's compression are closer together than the coils in the low-amplitude wave's compression.

Amplitude of transverse waves

If you have ever been knocked over by an ocean wave, you know that the higher the wave, the greater the disturbance from that wave. Remember that the amplitude of a wave increases as the disturbance from that wave increases. So, a tall ocean wave has a greater amplitude than a short ocean wave does.

The amplitude of any transverse wave is the vertical distance from the crest or trough of the wave to the rest position of the medium, as shown in **Figure 11.** Tall waves have a large amplitude, and short waves have a small amplitude. The amplitude of any transverse wave is also half the vertical distance from crest to trough.

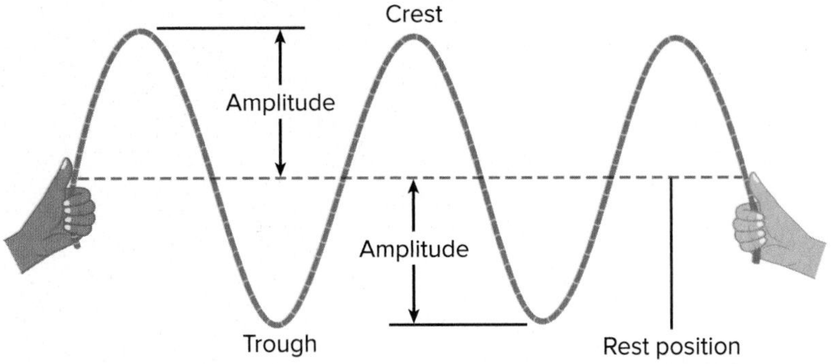

Figure 11 The amplitude of a transverse wave is half the vertical distance between the crests and the troughs.
Describe *how you could create waves with different amplitudes in a piece of rope.*

Check Your Progress

Summary

- Wavelength is the distance between a point on a wave and the nearest point just like it.
- Wave frequency is the number of wavelengths passing a fixed point each second.
- Wave period is the amount of time it takes one wavelength to pass a fixed point.
- The speed of a wave is the product of its frequency and its wavelength.
- As the amplitude of a wave increases, the disturbance from that wave increases.

Demonstrate Understanding

11. **Identify** a wave that speeds up when it passes from air to water as well as one that slows down.
12. **Describe** the difference between a longitudinal wave with a large amplitude and one with a small amplitude.
13. **Describe** how the wavelength of a wave changes if the wave slows down but its frequency does not change.
14. **Explain** how the frequency of a wave changes when the period of the wave increases.

Explain Your Thinking

15. **Explain** You make a transverse wave by shaking the end of a long rope up and down. Explain how you would shake the end of the rope to make the wavelength shorter.
16. **MATH Connection** Calculate the frequency of a water wave that has a wavelength of 0.5 m and a speed of 4 m/s.
17. **MATH Connection** An FM radio station broadcasts radio waves with a frequency of 96,000,000 Hz. What is the speed of these radio waves if they have a wavelength of 3.1 m?

LEARNSMART® Go online to follow your personalized learning path to review, practice, and reinforce your understanding.

LESSON 3
THE BEHAVIOR OF WAVES

FOCUS QUESTION
How does light behave when it travels from air to water?

Reflection

If you are one of the last people to leave your school building at the end of the day, you will probably find that the hallways are quiet and empty. When you close your locker door, the sound echoes down the empty hall. Your footsteps also make a hollow sound. Thinking you are all alone, you might be startled by your own reflection in a classroom window. The echoes and your image looking back at you from the window are caused by wave reflection. Without wave reflection, you could not even see the lockers in your school's hallway.

Reflection occurs when a wave strikes an object and bounces off it. All types of waves—including sound, water, and light waves—can be reflected. How does the reflection of light allow the girl in **Figure 12** to see herself in the mirror? It happens in two steps. First, light strikes her face and bounces off her face. Then, the light reflected off her face strikes the mirror and is reflected into her eyes.

Figure 12 Some of the light that strikes this girl's face is reflected into the mirror. Some of that light then reflects off the mirror into her eyes.

Describe *how the path of the light reflected from the girl's face would be different if the mirror were not present.*

3D THINKING — DCI Disciplinary Core Ideas — CCC Crosscutting Concepts — SEP Science & Engineering Practices

COLLECT EVIDENCE
Use your Science Journal to record the evidence you collect as you complete the readings and activities in this lesson.

INVESTIGATE
GO ONLINE to find these activities and more resources.

Laboratory: Wave in Motion
Develop and use a model of water ripples to support a claim regarding relationships among the frequency, wavelength, and speed of waves traveling through various media.

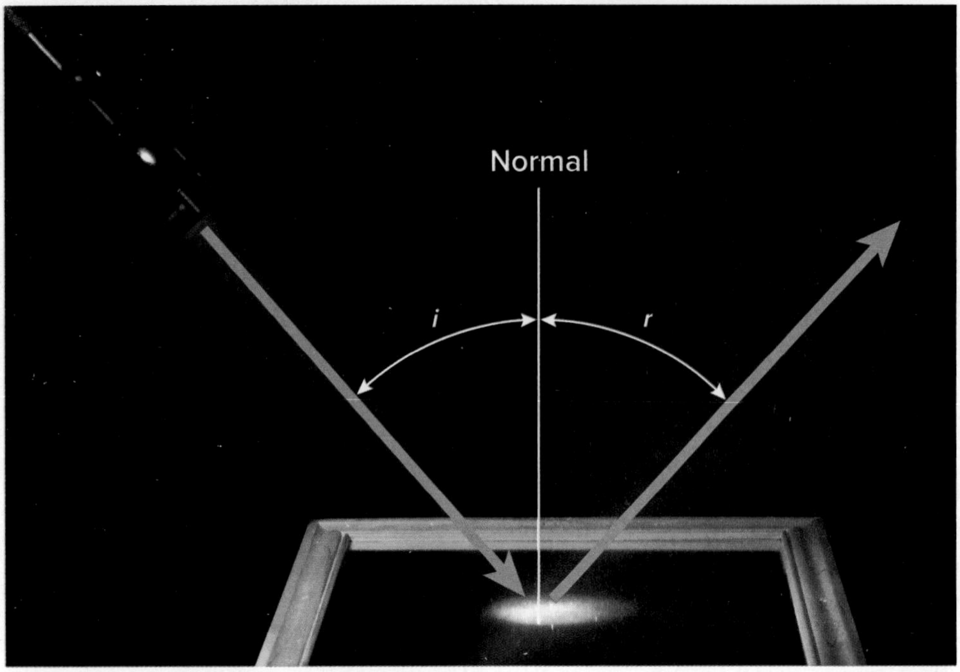

Figure 13 The angle between the incident light beam and the normal is equal to the angle between the reflected light beam and the normal.

Identify *If the angle of incidence is 40°, what is the angle of reflection?*

Echoes

Surfaces often reflect sound waves as well as light waves. Echoes are a result of reflecting sound waves. Sound waves form when your foot hits the floor, and the waves travel through the air to both your ears and other objects. When sound waves reach another object, such as a row of lockers, they reflect off that object. Sometimes, those reflected waves travel back to your ears.

Some animals use echoes to learn about their surroundings. For example, dolphins make clicking sounds and listen to the echoes. These echoes enable the dolphin to locate objects.

The law of reflection

Look at the two light beams in **Figure 13.** The beam striking the mirror is called the incident beam. The beam that bounces off the mirror is called the reflected beam. The line drawn perpendicular to the surface of the mirror is called the normal. The angle formed by the incident beam and the normal is the angle of incidence, labeled *i*. The angle formed by the reflected beam and the normal is the angle of reflection, labeled *r*.

According to the law of reflection, the angle of incidence is always equal to the angle of reflection. All reflected waves, including light waves, sound waves, and water waves, obey this law. Objects that bounce from a surface sometimes behave like waves that are reflected from a surface. For example, suppose you throw a bounce pass while playing basketball. The angle between the ball's direction and the normal to the floor is the same before and after it bounces.

STEM CAREER Connection

Acoustician

Have you enjoyed the thrilling sound of a movie soundtrack or the witty banter of actors in a play in a darkened theater? The design and planning that resulted in a good audio experience might be the work of an acoustician. Acousticians are experts in acoustics. Their work helps to control echoes within a space to make every seat the best seat in the house.

Refraction

Do you notice anything unusual in **Figure 14?** The straw looks as if it is broken into two pieces. But if you pulled the straw out of the water, you would see that it is unbroken. This illusion is caused by refraction. How does refraction work?

Remember that a wave's speed depends on the medium through which it is traveling. When a wave passes from one medium to another, such as when a light wave passes from air to water, it changes speed. If the wave is traveling at an angle when it passes from one medium to another, it changes direction, or bends, as it changes speed.

Refraction is the bending of a wave caused by a change in its speed as it travels from one medium to another. The greater the change in speed, the more the wave bends. The top panel in **Figure 15** shows how a wave refracts when it passes into a material in which that wave slows down. The wave is refracted (bent) toward the normal. The bottom panel in **Figure 15** shows how a wave refracts when it passes into a medium in which it speeds up. Then, the wave is refracted away from the normal.

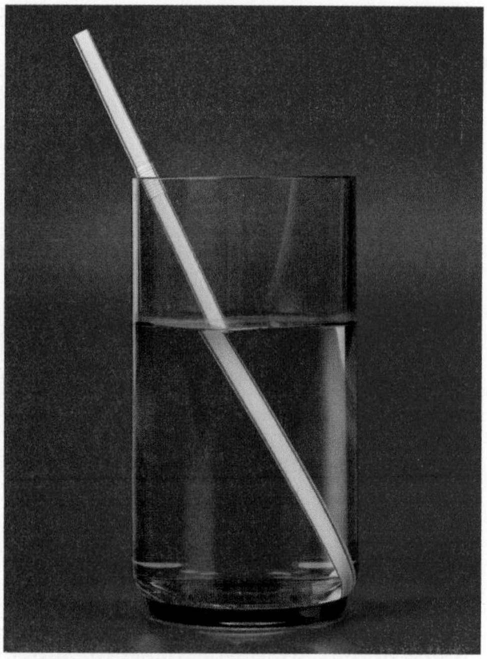

Figure 14 Light waves refract as they enter and leave the water. This results in the illusion of the broken straw.

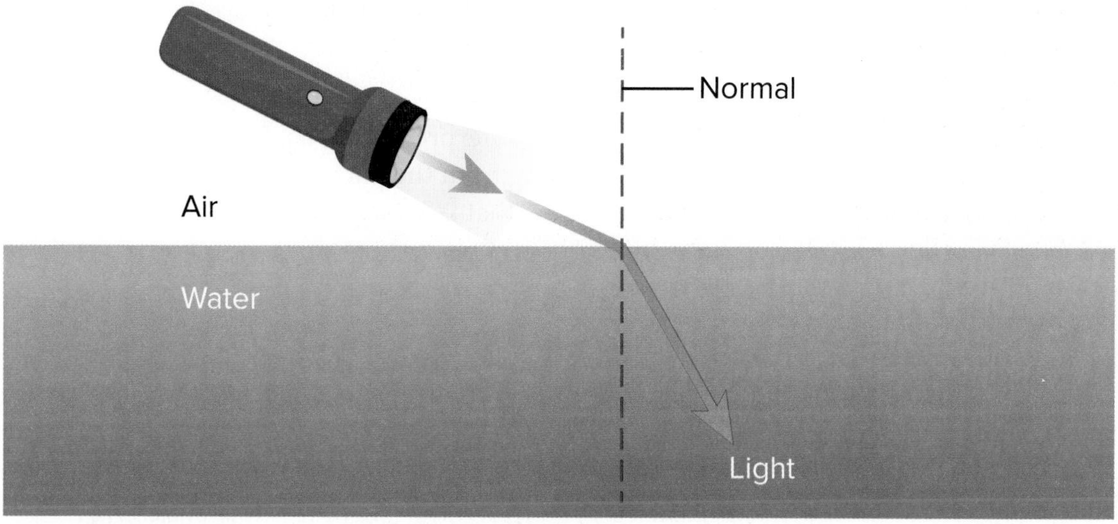

When light waves travel from air to water, they slow down and bend toward the normal.

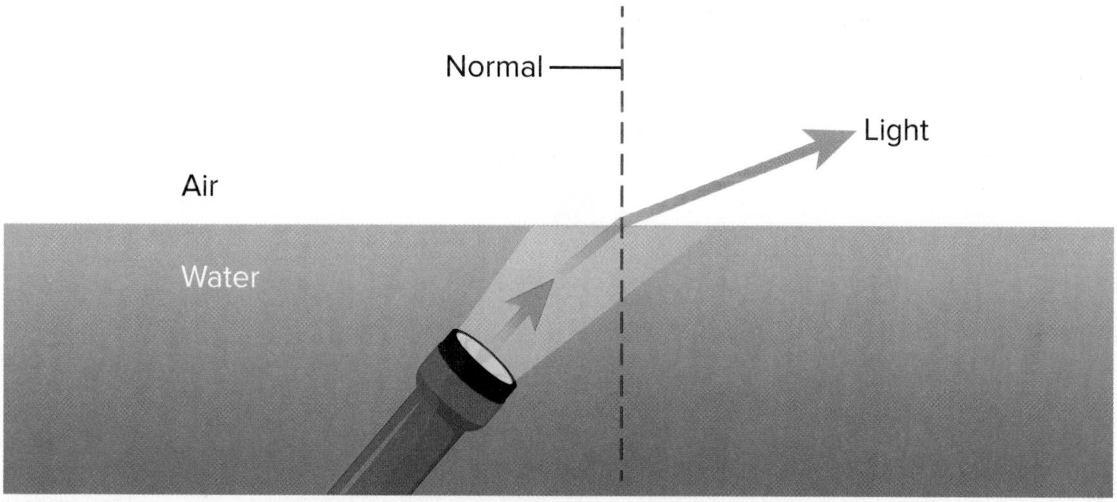

When light waves travel from water to air, they speed up and bend away from the normal.

Figure 15 Light waves travel more slowly in water than in air. This causes light waves to refract when they move from water to air or air to water.

Predict *how the beam would bend if the speed of light were the same in both air and water.*

Figure 16 The fish in this photo are farther from the surface of the water than they appear. Refraction causes this pond to appear to be much shallower than it actually is.

Refraction of light in water

Have you ever gazed at fish in a pond, such as those in **Figure 16?** You may have noticed that objects that are underwater seem closer to the surface than they really are. **Figure 17** shows how refraction causes this illusion.

In **Figure 17,** the light waves reflected from the swimmer's foot are refracted away from the normal and enter the girl's eyes. However, the brain assumes that all light waves have traveled in a straight line. The light waves that enter the girl's eyes seem to have come from a foot that was higher in the water.

This is also why the straw in **Figure 14** seems to be broken. The light waves coming from the part of the straw that is underwater are refracted, but your brain interprets them as if they have traveled in a straight line. However, the light waves coming from the part of the straw above the water are not refracted. So, the part of the straw that is underwater looks as if it has shifted.

Get It?

Explain Refraction is responsible for many optical illusions. How is your brain fooled into seeing a broken straw?

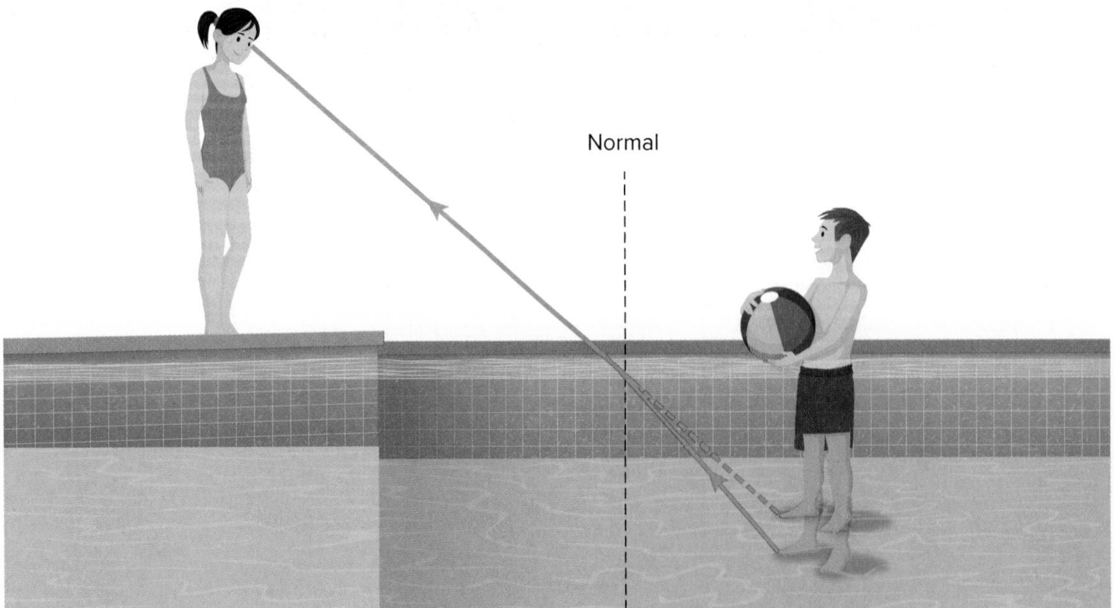

Figure 17 From the girl's perspective, refraction causes the boy to appear shorter than he actually is. Light rays from the boy's foot refract downward at the water's surface.

Diffraction

When waves strike an object, several things can happen. The waves can be reflected. If the object is transparent, light waves can be refracted as they pass through it. Often, some waves are reflected and some waves are refracted. If you look into a glass window, sometimes you can see your reflection in the window as well as objects behind it. Light is passing through the window and is also being reflected at its surface.

Waves can also behave another way when they strike an object. The waves can bend around the object. **Figure 18** shows ocean waves changing direction and bending after they strike an island.

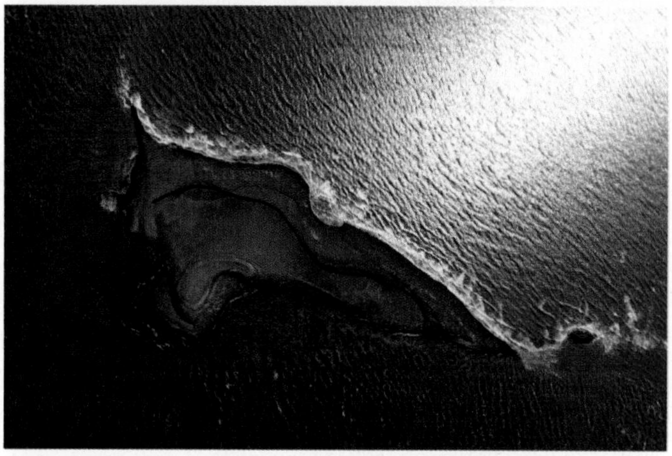

Figure 18 Diffraction can cause ocean waves to change direction as they pass around an island.

Get It?
Compare how light would respond differently when striking an object that is not transparent than when striking an object that is transparent, such as a glass window.

Diffraction is the bending of a wave around an object. Diffraction and refraction both cause waves to bend. The difference is that refraction occurs when waves pass through an object, while diffraction occurs when waves pass around an object. All waves, including water waves, sound waves, and light waves, can be diffracted.

Get It?
Contrast refraction with diffraction.

Waves also can be diffracted when they pass through a narrow opening, as shown in **Figure 19.** After they pass through the opening, the waves spread out. In this case, the waves are bending around the corners of the opening.

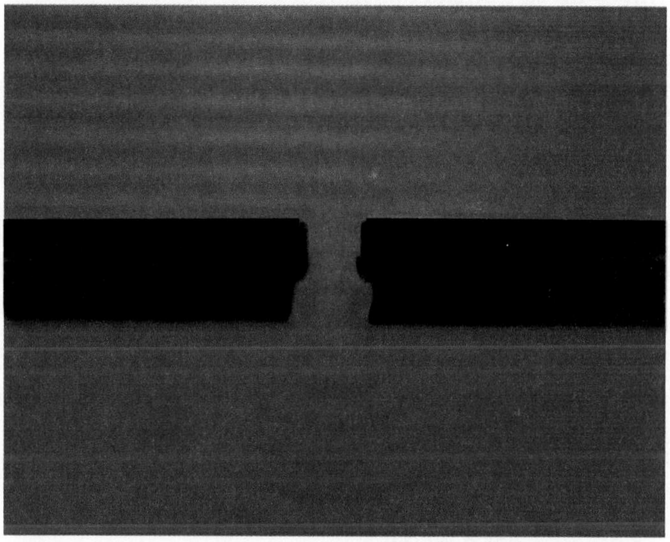

Figure 19 Waves diffract and spread out when they pass through a small opening or around a small obstacle.

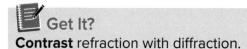

Get It?
Describe What are two situations in which a wave will diffract?

CCC CROSSCUTTING CONCEPTS

Cause and Effect So far, you have seen how waves interact with matter through reflection, refraction, and diffraction. Prepare a demonstration that provides empirical evidence of the cause and effect relationships associated with each of these interactions.

Less diffraction occurs if the wavelength is smaller than the obstacle.

More diffraction occurs if the wavelength is the same size as the obstacle.

Figure 20 The diffraction of waves around an obstacle depends on how the wavelength compares with the size of the obstacle.

Diffraction and wavelength

How much does a wave bend when it strikes an object or an opening? The amount of diffraction that occurs depends on how big the obstacle or opening is compared to the wavelength, as shown in **Figure 20.**

When an obstacle is roughly the same size as or smaller than the wavelength of a wave, the wave bends around it. But when the obstacle is much larger than the wavelength, the waves do not diffract as much. If the obstacle is much larger than the wavelength, almost no diffraction occurs. Instead, the obstacle casts a shadow.

Get It?

Describe the effect of wavelength and object size on the diffraction of a wave.

Hearing around corners Suppose you are walking down the hallway and you hear sounds coming from a classroom on the left before you reach the open classroom door. However, you cannot see into the room until you reach the doorway.

Why can you hear the sound waves but not see the light waves while you are still in the hallway? The wavelengths of sound waves are similar in size to a door opening. Sound waves diffract around the door and spread out down the hallway. Light waves have a much shorter wavelength. They are hardly diffracted at all by the door. So, you cannot see into the room until you get to the door.

Diffraction of radio waves Diffraction also affects a radio's reception. AM radio waves have longer wavelengths than FM radio waves do. Because of their longer wavelengths, AM radio waves diffract around obstacles, such as buildings and mountains.

The FM waves, with their short wavelengths, do not diffract as much. As a result, AM radio reception is often better than FM reception around tall buildings and natural barriers, such as hills.

Get It?

Compare the diffraction of FM radio waves with the diffraction of AM radio waves.

Interference

Suppose two waves travel toward each other on a long rope, as in the top panel of **Figure 21.** What happens when the two waves meet? The two waves pass through each other, and each one continues to travel in its original direction, as shown in the middle and bottom panels of **Figure 21.** However, when the waves meet in the middle panel of **Figure 21,** they form a new wave that looks different from either of the original waves.

When two waves arrive at the same place at the same time, they combine to form a new wave. **Interference** is the process of two or more waves overlapping and combining to form a new wave. This new wave exists only while the two original waves continue to overlap. Two waves can combine through either constructive interference or destructive interference.

Figure 21 When two waves are traveling on a rope, they interfere while they overlap. The two waves combine to form a new wave and then move on through each other.

Describe *the amplitude of the combined wave.*

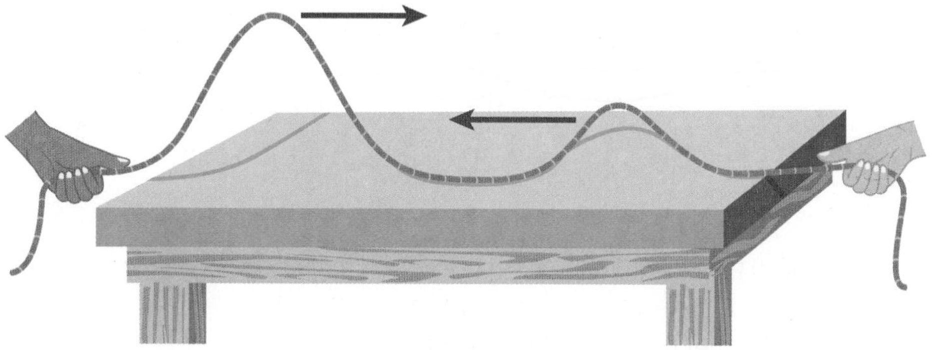

Two waves travel toward each other on a rope.

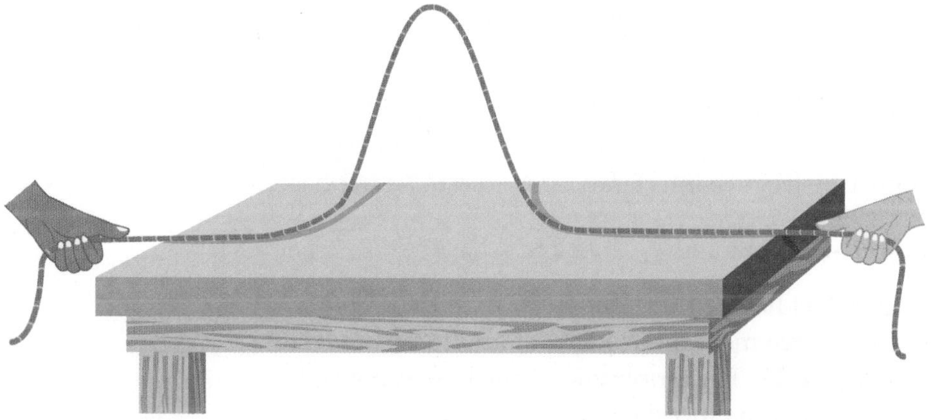

As the waves overlap, they interfere to form a new wave. While the two waves overlap, they continue to pass through each other.

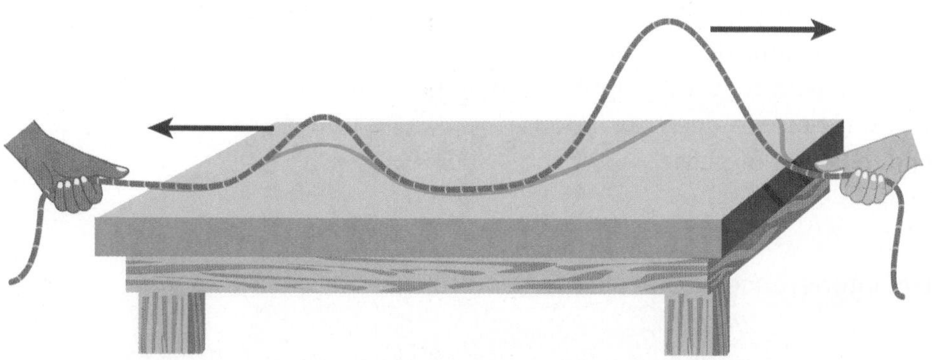

Afterward, the waves continue unchanged, as if those waves had never met.

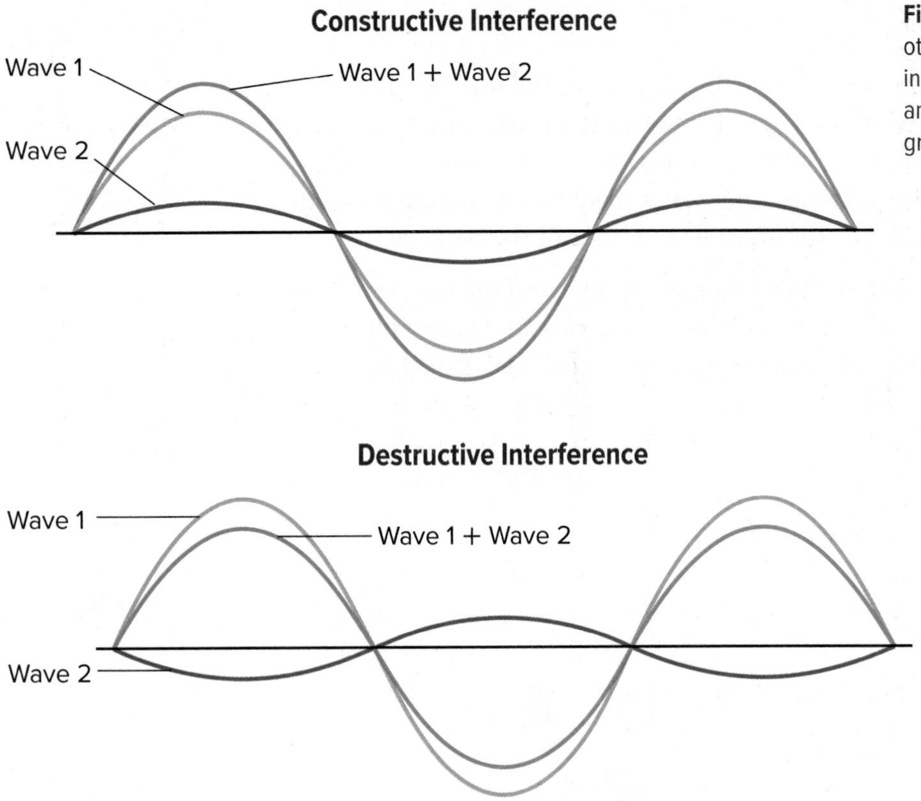

Figure 22 When waves pass through each other, constructive interference or destructive interference can occur. In both panels, the blue and yellow waves interfere to form the green wave.

Constructive interference

In constructive interference, as shown in the top panel of **Figure 22,** the waves add together. This happens when the crests of two or more transverse waves arrive at the same place at the same time and overlap. The amplitude of the new wave that forms is equal to the sum of the amplitudes of the original waves.

Constructive interference also occurs when the compressions of different longitudinal waves overlap. If the waves are sound waves, for example, constructive interference produces a louder sound. Waves undergoing constructive interference are said to be in phase.

Destructive interference

In destructive interference, the waves subtract from each other as they overlap. This happens when the crests of one transverse wave meet the troughs of another transverse wave, as shown in the bottom panel of **Figure 22.** The amplitude of the new wave is the difference between the amplitudes of the waves that overlapped.

With longitudinal waves, destructive interference occurs when the compression of one wave overlaps with the rarefaction of another wave. One way to think of this is that the compressions of one wave "fill in" the rarefactions of another wave. The compressions and rarefactions combine and form a wave with reduced amplitude. When destructive interference happens with sound waves, it causes a decrease in loudness. Waves undergoing destructive interference are said to be out of phase.

Get It?

Describe the effects of constructive interference and destructive interference on the amplitude of a wave.

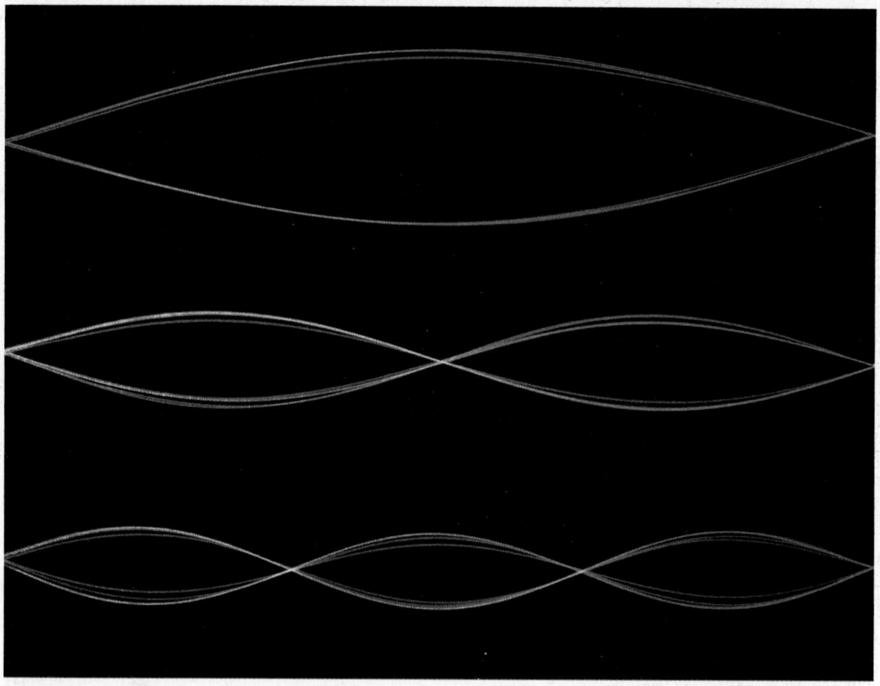

Figure 23 In each of these three wave patterns, waves are passing along the rope. However, the rope looks as though it is standing still in each case.

Explain *how nodes form in a standing wave.*

Standing waves

When you make transverse waves on a rope, you might attach one end to a fixed point, such as a doorknob, and shake the other end. The waves that you produce then reflect back from the doorknob. What happens when the incident and reflected waves meet?

As the two waves travel in opposite directions along the rope, they continually pass through each other. Interference takes place as the waves from each end overlap along the rope. At any point where a crest meets a crest, a new wave with a larger amplitude forms. But at points where crests meet troughs, the waves cancel each other and no motion occurs.

The interference of the two identical waves makes the rope vibrate in a special way, as shown in **Figure 23.** The waves create a pattern of crests and troughs that do not seem to be moving. Because the wave pattern stays in one place, it is called a standing wave. A **standing wave** is a special type of wave pattern that forms when waves equal in wavelength and amplitude but traveling in opposite directions continuously interfere with each other. Standing waves have **nodes,** which are locations where the interfering waves always cancel. The nodes always stay in the same place on the rope. Meanwhile, the wave pattern vibrates between the nodes.

Standing waves in music When the string of a violin is played with a bow, it vibrates and creates standing waves. The standing waves in the string help produce a rich, musical tone. Other instruments also rely on standing waves to produce music. Some instruments, such as flutes, create standing waves in a column of air. In other instruments, such as drums, a tightly stretched piece of material vibrates in a special way to create standing waves. As the material in a drum vibrates, nodes are created on the surface of the drum.

Resonance

When you were younger, you might have played on a swing like the one in **Figure 24.** You probably noticed that you could make the swing go higher by pumping your legs and arms. It was not necessary to pump hard, but timing was important. If you pumped in time to the swing's rhythm, you could go quite high.

Figure 24 This child times the movements of her arms and legs to make the swing go higher. Resonance is the process of increasing vibration through well-timed pushes and pulls on the object that is vibrating. This child uses resonance as she swings.

You can accomplish similar effects with sounds. Suppose you have a tuning fork that has a single natural frequency of 440 Hz, which means that the tuning fork naturally vibrates at 440 Hz when struck. Now think of a sound wave with a frequency of 440 Hz striking the tuning fork. Because the sound wave has the same frequency as the natural frequency of the tuning fork, the tuning fork will begin to vibrate. **Resonance** is the process by which an object is made to vibrate by absorbing energy at its natural frequencies.

Sometimes resonance can cause an object to absorb a large amount of energy. An object vibrates more and more strongly as it absorbs energy at its natural frequencies. If the object absorbs enough energy, it might break.

Check Your Progress

Summary

- When reflection of a wave occurs, the angle of incidence equals the angle of reflection.
- Refraction occurs when a wave changes direction as it moves from one medium to another.
- Diffraction occurs when a wave changes direction by bending around an obstacle.
- Interference occurs when two or more waves overlap and form a new wave.

Demonstrate Understanding

18. **Describe** the path that light waves take when you see your image in a mirror.
19. **Compare** the loudness of sound waves that constructively interfere with the loudness of sound waves that destructively interfere.
20. **Describe** how one tuning fork's vibrations can cause another tuning fork to vibrate.
21. **Infer** Sound waves often bend around columns in large concert halls. Is this a result of refraction or diffraction?

Explain Your Thinking

22. **Model** Suppose the speed of light was greater in water than in air. Draw a diagram to show whether an object underwater would seem deeper or closer to the surface than it actually is.
23. **MATH Connection** The angle between a flashlight beam that strikes a mirror and the reflected beam is 80 degrees. What is the angle of incidence?

LEARNSMART® Go online to follow your personalized learning path to review, practice, and reinforce your understanding.

Vanished!

Invisibility is a common theme in science fiction and fantasy. Some recent developments in technology suggest that invisibility technology may be possible.

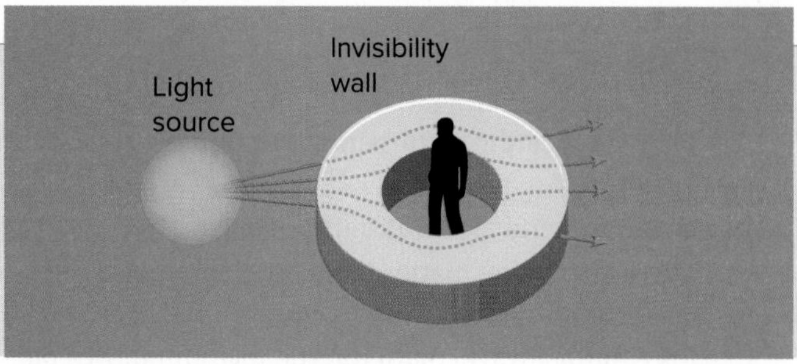

Metamaterials can cause light waves to bend around an area, preventing those waves from ever reaching the objects hidden within that area.

"Impossible" refraction

Most materials refract light waves in one direction only, from one side of the normal to the other. However, a new class of materials, called metamaterials, can refract light waves so that they do not cross the normal, as shown below.

A metamaterial first refracts a light beam like an ordinary material would, and then it refracts light in the other direction. As a result, a metamaterial can bend a light beam completely around an area, making that area invisible. The figure at right shows how a wall composed of a metamaterial might conceal a person.

An ordinary material causes light to refract across the normal. A metamaterial can act like an ordinary material, or it can refract light backward.

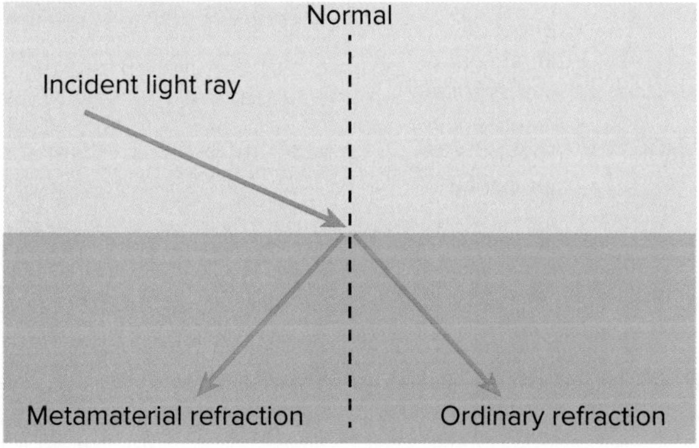

Visible light

Early metamaterials made objects invisible only to microwaves. The wavelengths of visible light waves are about 50,000 times smaller than the wavelengths of microwaves. Metamaterial components for visible light were not developed until 2010.

Other applications

While invisibility cloaks may still be in the far future, this research is leading in some surprising directions today. Tsunamis and earthquakes are powerful natural events that both involve wave motion. If metamaterials can render objects invisible to light waves, then similar materials might make buildings and even whole coastlines invisible to destructive water or seismic waves. In the future, destructive waves might be redirected harmlessly around vulnerable structures.

COMMUNICATE SCIENTIFIC INFORMATION

Write a future news article or prepare a newscast that explains the scientific basis of a use of metamaterials that is designed to protect against damage caused by destructive water or seismic waves.

MODULE 9

STUDY GUIDE

GO ONLINE to study with your Science Notebook.

Lesson 1 THE NATURE OF WAVES

- A wave is a repeating disturbance that transfers energy through matter or space.
- Waves carry energy without transporting matter.
- In a transverse wave, matter in the medium moves at right angles to the direction that the wave travels.
- In a longitudinal wave, matter in the medium moves back and forth along the direction that the wave travels.

- wave
- medium
- mechanical wave
- transverse wave
- longitudinal wave

Lesson 2 WAVE PROPERTIES

- Wavelength is the distance between a point on a wave and the nearest point just like it.
- Wave frequency is the number of wavelengths passing a fixed point each second.
- Wave period is the amount of time it takes one wavelength to pass a fixed point.
- The speed of a wave is the product of its frequency and its wavelength.
- As the amplitude of a wave increases, the disturbance from that wave increases.

- crest
- trough
- compression
- rarefaction
- wavelength
- frequency
- period
- amplitude

Lesson 3 THE BEHAVIOR OF WAVES

- When reflection of a wave occurs, the angle of incidence equals the angle of reflection.
- Refraction occurs when a wave changes direction as it moves from one medium to another.
- Diffraction occurs when a wave changes direction by bending around an obstacle.
- Interference occurs when two or more waves overlap and form a new wave.

- refraction
- diffraction
- interference
- standing wave
- node
- resonance

THREE-DIMENSIONAL THINKING
Module Wrap-Up

REVISIT THE PHENOMENON

How do buildings move during an earthquake?

CER Claim, Evidence, Reasoning

Explain Your Reasoning Revisit the claim you made when you encountered the phenomenon. Summarize the evidence you gathered from your investigations and research and finalize your Summary Table. Does your evidence support your claim? If not, revise your claim. Explain why your evidence supports your claim.

STEM UNIT PROJECT

Now that you've completed the module, revisit your STEM unit project. You will summarize your evidence and apply it to the project.

GO FURTHER

SEP Data Analysis Lab

Superposition Principle

What happens when waves traveling on the water meet? The amplitudes of the waves add together. As the waves overlap, the amplitude of each point in the overlap region is the sum of the amplitudes of the two waves. In other words, a wave with a 2-m amplitude crossing a wave with a 3-m amplitude makes a wave with a 5-m amplitude at one instant. This property is called the superposition principle. The diagram below shows the superposition of two waves at point P.

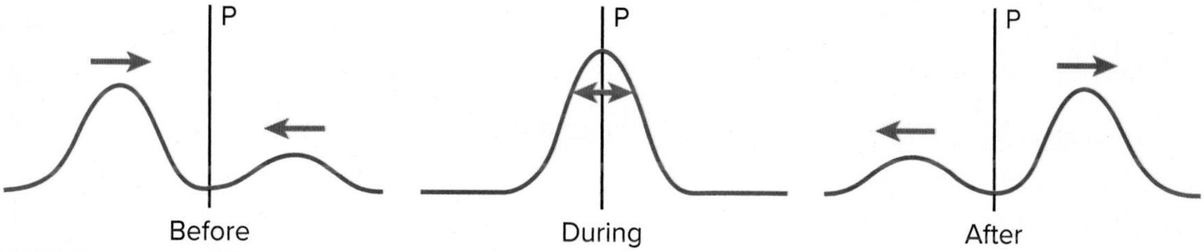

CER Analyze and Interpret Data

1. **Claim** If two waves with amplitudes of +4 cm and +2 cm pass through point P, what is the maximum possible displacement of point P? Draw three scale diagrams showing the waves before they meet, when they meet, and after they meet.
2. **Claim** The amplitudes of two waves are +5 cm and −3 cm. What is the amplitude of the new wave formed after the two waves meet? Make a drawing showing the waves before, during, and after their interaction.
3. **Claim, Reasoning** Two water waves, one with an amplitude of +3 m and another with an amplitude of −3 m, meet each other. Describe what happens as the waves meet.

Trevor Mogg/Alamy Stock Photo

Anton Gvozdikov/Shutterstock

SOUND

ENCOUNTER THE PHENOMENON

Why do two instruments playing the same note sound different?

GO ONLINE to play a video about how an instrument sounds different when it is travelling toward you verses when it is travelling away you.

SEP Ask Questions

Do you have other questions about the phenomenon? If so, add them to the driving question board.

CER Claim, Evidence, Reasoning

Make Your Claim Use your CER chart to make a claim about why two instruments playing the same note sound different. Explain your reasoning.

Collect Evidence Use the lessons in this module to collect evidence to support your claim. Record your evidence as you move through the module.

Explain Your Reasoning You will revisit your claim and explain your reasoning at the end of the module.

GO ONLINE to access your CER chart and explore resources that can help you collect evidence.

LESSON 2: Explore & Explain: Pitch

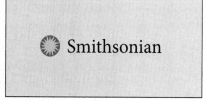

LESSON 4: Explore & Explain: Sonar

Smithsonian

Additional Resources

LESSON 1

THE NATURE OF SOUND

FOCUS QUESTION

How does sound change when traveling through different mediums?

Vibrations and Sound

An amusement park can be a noisy place. The sounds from the rides and games can make it hard to hear what your friends say. All of these sounds have something in common: a vibrating object produces each one. For example, your friends' vibrating vocal cords produce their voices. Vibrating speakers produce the music from a carousel.

Sound Waves

Recall that air is composed of matter. When an object such as a radio speaker vibrates, it collides with some of the particles that make up the nearby air, transferring some energy to those particles. These particles then collide with other particles, passing the energy on further. The energy originally transferred by the vibrating object continues to travel through the air in this way. This process of energy transfer forms a sound wave.

Sound waves are longitudinal waves. Remember that a longitudinal wave is composed of two types of regions–compressions and rarefactions. If you look at **Figure 1,** you will see that when a radio speaker vibrates outward, the particles near the speaker are pushed together, forming a compression. When the speaker moves inward, the particles near the speaker are spread apart, and a rarefaction forms. As long as the speaker continues to vibrate back and forth, sound waves are produced.

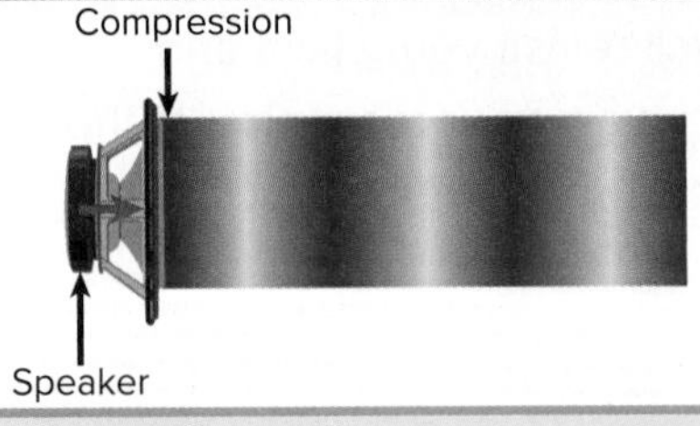

When the speaker vibrates outward, particles in the air next to it are pushed together to form a compression.

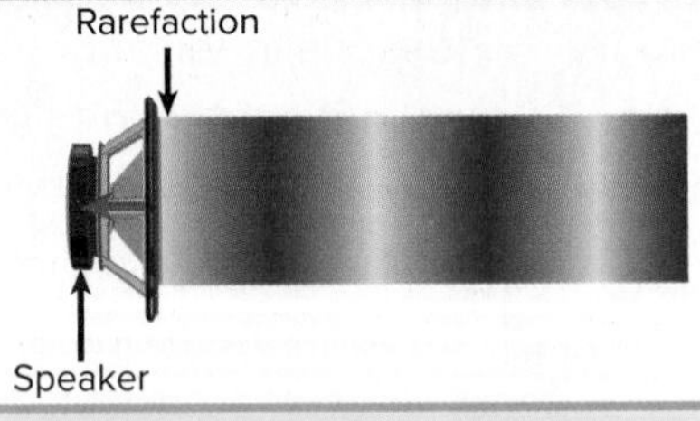

When the speaker vibrates inward, the particles spread apart to form a rarefaction.

Figure 1 A vibrating speaker cone produces longitudinal waves.

3D THINKING — DCI Disciplinary Core Ideas — CCC Crosscutting Concepts — SEP Science & Engineering Practices

COLLECT EVIDENCE

Use your Science Journal to record the evidence you collect as you complete the readings and activities in this lesson.

INVESTIGATE

GO ONLINE to find these activities and more resources.

Quick Investigation: Compare Sounds
Use a model to visualize the effect of different media on sound waves.

Review the News
Obtain information from a current news story about how sound is used to explore Earth's crust. Evaluate your source and communicate your findings to your class.

Figure 2 A vacuum is a region where no matter is present. Outer space is not a perfect vacuum, but there is too little matter for sound waves to travel through outer space. Therefore, these astronauts must use radios to communicate with each other.

Explain *why the astronauts need radios in order to talk to each other.*

Sound in liquids and solids

If you have been underwater and heard garbled voices, you know that sound travels through water. If you have put your ear to your desk and heard sounds, you know that sound travels through solids. Sound waves travel through any type of matter—solid, liquid, or gas. The matter through which a wave travels is called the medium. A sound wave produces compressions and rarefactions in a medium as it travels through that medium.

What happens when there is no matter through which sound can travel? Could sound be transmitted without matter to compress, expand, and collide? In order for astronauts, such as the ones in **Figure 2,** to talk to each other, they must use electronic communication equipment because there is no atmosphere to transmit sound waves. Sound waves cannot travel through the vacuum of outer space.

The speed of sound

The speed of a sound wave through a medium depends on that medium's composition and whether that medium is solid, liquid, or gas. **Table 1** shows the speed of sound through some common mediums. The temperature, density, and elasticity of the medium also affect the speed of sound.

Table 1 Speed of Sound in Different Mediums

Medium	Speed of Sound (m/s)
Air (0°C)	330
Air (20°C)	340
Cork	500
Water (0°C)	1400
Water (20°C)	1500
Copper	3600
Bone	4000
Steel	5800

Temperature and sound speed The speed of sound through a fluid depends on the temperature of that fluid. This relationship is particularly pronounced for gases but also exists for liquids, such as water. As the temperature of a fluid increases, the particles that make up that fluid move faster. This makes them more likely to collide with each other. If the particles that make up a medium collide more often, more energy can be transferred in a shorter amount of time. As a result, the sound waves travel faster.

When houses are close together, you can walk quickly from house to house. When particles are close together, such as in a solid or a liquid, energy can be transferred quickly from particle to particle.

When houses are far apart, it takes longer to walk from house to house. When particles are far apart, such as in a gas, it takes longer to transfer energy from particle to particle.

Figure 3 Walking between houses is a model for particles transferring the energy of a sound wave.
Explain *why sound would travel more slowly in cork than in water.*

Density and sound speed Sound usually travels fastest in solids and slowest in gases. One reason for this is that the individual particles that make up liquids and solids are usually closer together than the particles that make up gases. **Figure 3** helps to illustrate why sound travels quickly in solids and liquids. Imagine that you are going door to door to raise money for a charity. In a city, it doesn't take very long to walk from house to house. In a rural area, it takes longer to walk from house to house. Similarly, when the particles that make up a medium are farther apart, sound travels more slowly through that medium.

Elasticity and sound speed Elasticity is the tendency of an object to rebound to its original state when it is deformed. A rubber ball is elastic. It tends to spring back to its original shape after someone squeezes it. A ball of clay is much less elastic.

Elastic objects also rebound more quickly when sound waves travel through them. Therefore, sound waves travel more quickly through elastic objects. Usually, solids are more elastic than liquids, and liquids are more elastic than gases. This is another reason why sound waves typically travel fastest through solids, slower through liquids, and slowest through gases.

Get It?

Identify two reasons why sounds usually travel faster through solids than through gases.

ACADEMIC VOCABULARY

section
a part set off as distinct
The outer ear is a section of the ear.

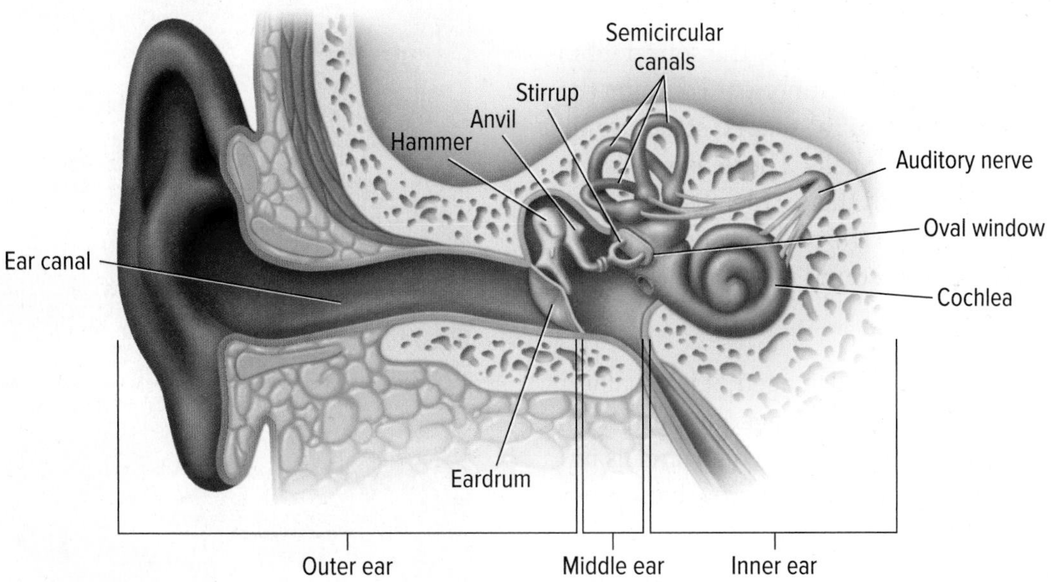

Figure 4 The three sections of the ear gather sound, amplify sound, and convert sound into an electric signal.

Identify *the three sections of the human ear.*

The Ear

Your ears and brain work together to interpret sound waves. When you think of your ear, you probably picture just the fleshy, visible, outer part. But, as shown in **Figure 4,** the human ear has three sections–the outer ear, the middle ear, and the inner ear. Each section of the ear has a different function.

The outer ear

The visible part of your ear, the ear canal, and the eardrum make up the outer ear. The outer ear gathers sound waves. The gathering process starts with the outer part of your ear, which is shaped to help capture and direct sound waves into the ear canal. The ear canal is a passageway that is 2 cm to 3 cm long and is a little narrower than your index finger. The sound waves travel along this passageway, which leads to the eardrum. The **eardrum** is a tough membrane that is about 0.1 mm thick and transmits sound from the outer ear to the middle ear.

Get It?

Identify what makes the eardrum vibrate.

The middle ear

When the eardrum vibrates, it passes the sound waves into the middle ear, where three tiny bones start to vibrate. These bones are the hammer, the anvil, and the stirrup. They make a lever system that increases the force and pressure exerted by the sound waves. The bones amplify the sound wave. The stirrup is connected to a membrane on a structure called the oval window, which vibrates as the stirrup vibrates.

The inner ear

When the membrane in the oval window vibrates, the sound vibrations are transmitted into the inner ear. The inner ear contains the cochlea (KOH klee uh). The **cochlea** is a spiral-shaped structure that is filled with liquid and contains tiny hair cells. These hair cells are shown in **Figure 5.** When the hair cells in the cochlea begin to vibrate, nerve impulses are sent through the auditory nerve and to the brain. It is the cochlea that converts sound waves to nerve impulses.

Hearing loss When a person's hearing is damaged, it is usually because the tiny hair cells in the cochlea are damaged or destroyed, often by loud sounds. This damage is permanent. The hair cells in the cochlea of humans and other mammals do not grow back when damaged or destroyed.

However, current research suggests that doctors may be able to repair damaged or even destroyed hair cells in the future. Much of this research centers on birds. Unlike in mammals, the hair cells in birds do grow back when damaged or destroyed.

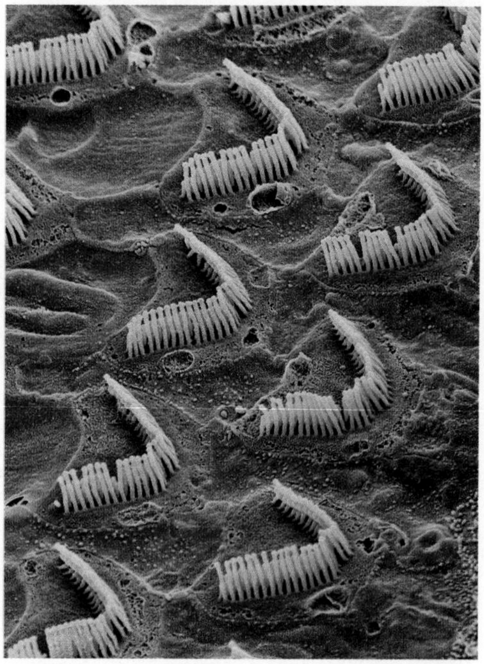

Figure 5 Hair cells in the human ear send nerve impulses to the brain when sound waves cause them to vibrate. In this photo, the hair cells are magnified 5500 times.

Check Your Progress

Summary

- Sound waves are longitudinal waves produced by vibrating objects.
- The compressions and rarefactions of sound waves transfer energy.
- Sound cannot travel through a vacuum.
- The ear detects sound waves and converts them to electrical impulses.

Demonstrate Understanding

1. **Explain** how sound travels from your vocal cords to your friend's ears when you talk.
2. **Summarize** the physical reasons that sound waves travel at different speeds through different mediums.
3. **Explain** why sound speeds up when temperature increases.
4. **Describe** each section of the human ear and its role in hearing.

Explain Your Thinking

5. **Hypothesize** Form a hypothesis to explain why some people hear a ringing in their ears (tinnitus) in the absence of sound.
6. **MATH Connection** Using **Table 1,** calculate how long it takes a sound wave to travel 1.0 km through air when the temperature is 0.0°C.
7. **MATH Connection** Using **Table 1,** calculate how long the same wave takes to travel 1.0 km in air when the temperature is 20.0°C.

LEARNSMART® Go online to follow your personalized learning path to review, practice, and reinforce your understanding.

SPL/Science Source

LESSON 2

PROPERTIES OF SOUND

FOCUS QUESTION

Why can you feel the bass of the music at a concert?

Intensity and Loudness

Recall that the degree of disturbance from a wave corresponds to its amplitude. For a longitudinal wave, amplitude is related to how close together the particles of the medium are in the compressions.

Figure 6 compares longitudinal waves of low amplitude and high amplitude. Increasing the amplitude of a longitudinal wave pushes the particles in that wave's compressions closer together.

To produce a sound wave with greater amplitude, more energy must be transferred from the vibrating object to the medium. This greater energy is then transferred through the medium as the sound wave is transmitted.

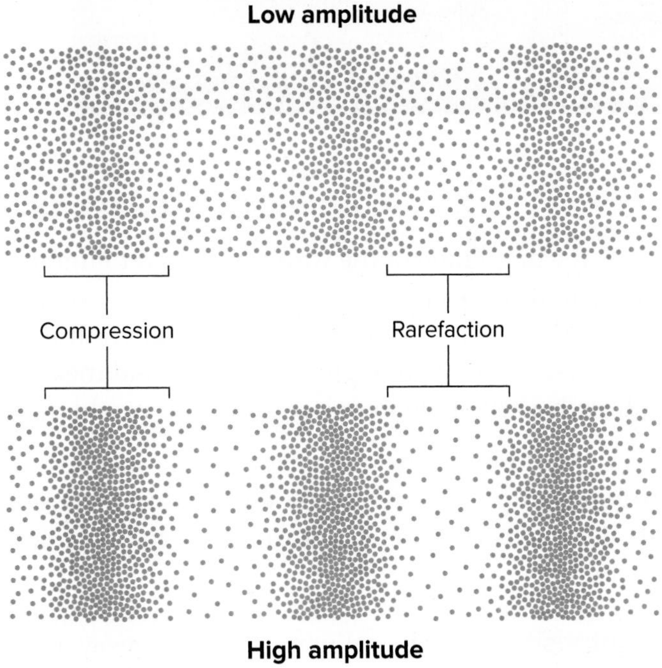

Figure 6 The amplitude of a sound wave depends on the density of the medium in the compressions and rarefactions.
Identify *the areas of highest and lowest density for each wave.*

What happens to the sound waves from a speaker when you turn up the volume? The notes sound the same, but the amplitude of the sound waves increases. This, in turn, causes your eardrums to vibrate with a higher amplitude when the sound waves reach your ears.

Get It?

Describe what happens to sound waves from a speaker when you turn up the volume.

3D THINKING DCI Disciplinary Core Ideas CCC Crosscutting Concepts SEP Science & Engineering Practices

COLLECT EVIDENCE

Use your Science Journal to record the evidence you collect as you complete the readings and activities in this lesson.

INVESTIGATE

GO ONLINE to find these activities and more resources.

LabA: Volume Settings and Loudness
Plan and carry out an investigation to visualize the effect of a changing medium on sound passing through it.

Laboratory: Sound Waves and Pitch
Develop and use a model to visualize the effect of a changing me dium on sound waves.

Figure 7 The intensity of the sound waves from each computer speaker is related to the rate at which energy passes through an imaginary rectangle and how far the listener is from the speaker.

Describe *how the intensity would be different for a person standing 10 m away from the computer speakers.*

Intensity

For sound waves, the amplitude of the waves is related to intensity. **Intensity** is the amount of energy that passes through a certain area in a specific amount of time. Picture sound waves traveling from a computer speaker, through an imaginary rectangle, and to your ears, as shown in **Figure 7.** If you could measure how much energy passed through such a rectangle in one second, you would measure intensity.

When you turn up the volume on your computer, the speaker cone moves in and out through a greater distance. In turn, greater energy is transferred through the medium, resulting in greater intensity. When you turn down the volume, you reduce the energy carried by the sound waves, so you also reduce the intensity.

Distance and intensity Intensity influences how far away a sound can be heard. If you and a friend whisper a conversation, the sound waves you produce have low intensity and are not heard at a far distance. However, when you shout, the sound waves have high intensity and can be heard farther away.

Sound intensity decreases with distance for two reasons. First, the energy that a sound wave carries spreads out as the sound wave spreads out. Second, some of a sound wave's energy converts to other forms of energy, usually thermal energy, as the sound travels through matter. As the sound wave travels farther, more of its energy converts into other forms. Some materials, such as soft, thick curtains, are very effective at converting sound energy to other forms of energy.

Loudness

Some sounds are so loud that they can be painful to hear. **Loudness** is the human perception of sound volume and primarily depends on sound intensity. When sound waves of high intensity reach your ear, they cause your eardrum to move back and forth a greater distance than when sound waves of low intensity reach your ear. As a result, you hear a loud sound.

Get It?

Relate intensity and loudness.

The decibel scale It is hard to say how loud is too loud. Two people are unlikely to agree on what is too loud because people vary in their perceptions of loudness. A sound that seems fine to you may seem earsplitting to your teacher. Even so, the intensity of sound can be described using a measurement scale.

A **decibel** (DE suh bel), abbreviated dB, is a unit of sound intensity. The loudest sounds that you hear are probably more than 10 billion times more intense than the softest sounds that you can hear. In order to handle this wide range of intensities, the decibel measurement scale is set up in a special way.

Every increase in 10 dB on the decibel scale represents a tenfold increase in intensity. This means that a 50-dB sound is 10 times more intense than a 40-dB sound. You might think that this means a 60-dB sound is 20 times more intense than a 40-dB sound. However, a 60-dB sound is 10×10, or 100, times more intense than a 40-dB sound. A 100-dB sound is 10^7, or 10 million, times more intense than a 30-dB sound. A sound can also have an intensity of less than 0 dB. However, people generally cannot hear these sounds.

Sustained sounds above about 90 dB can cause permanent hearing loss. Even short, sudden sounds with intensity levels above 120 dB may cause pain and permanent hearing loss. During some rock concerts, sounds reach this damaging intensity level. Wearing ear protection, such as earplugs, around loud sounds can help protect against hearing loss. **Figure 8** shows some sounds and their intensity levels in decibels.

Loudness in Decibels

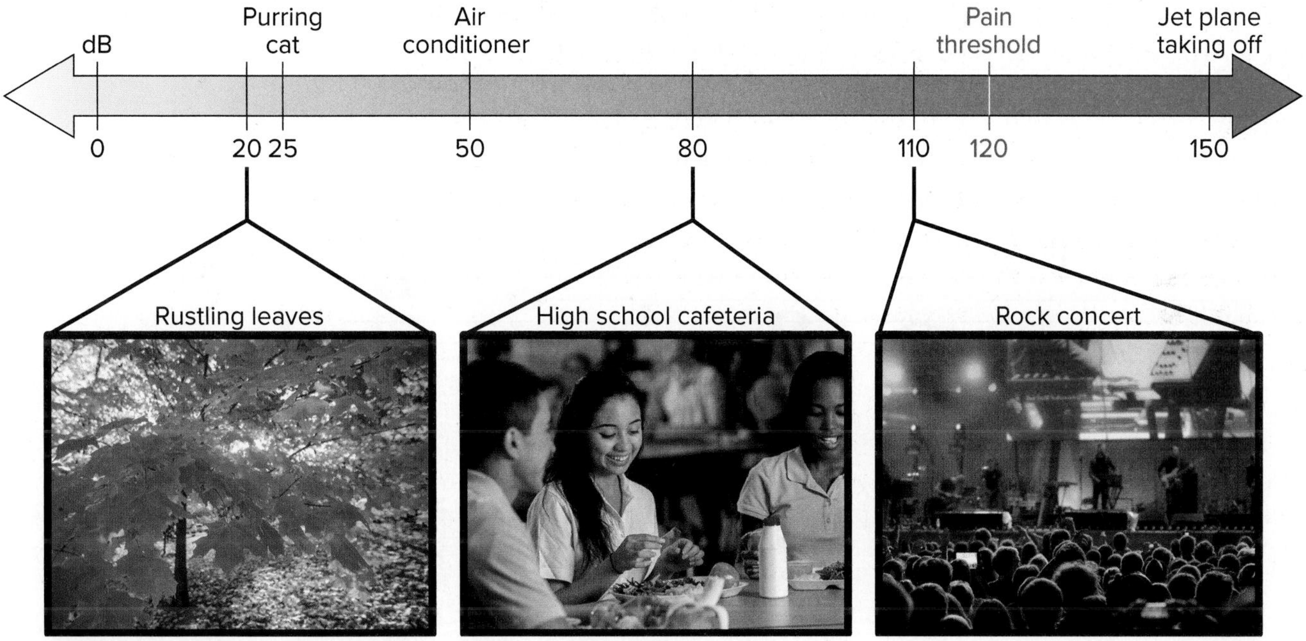

Figure 8 The volumes of different sounds are often measured in decibels.
Identify *where a normal speaking voice would fall on the decibel scale.*

STEM CAREER Connection

Sound Engineering Technician

Do you love both music and technology? You might enjoy working as a sound engineering technician. Sound engineering technicians set up and operate equipment for mixing, amplifying, and recording sound. These are the people you would see operating a sound board at a concert or in the control room at a recording studio.

Pitch

If you have ever studied music, you are probably familiar with the musical scale *do, re, mi, fa, so, la, ti, do.* If you were to sing this scale, your voice would start low and become higher with each note. As you sang, you would hear a change in pitch. **Pitch** is how high or low a sound seems to be. The notes on the right side of a piano have high pitches. The pitch of a sound is primarily related to the frequency of the sound waves.

Frequency and pitch

Frequency is a measure of how many wavelengths pass a particular point each second. For a longitudinal wave, such as sound, the frequency is the number of compressions or the number of rarefactions that pass by each second. Frequency is measured in hertz (Hz). One Hz means that one wavelength passes by in one second.

When sound waves with high frequency reach your ears, many compressions reach your eardrums each second. The waves cause your eardrums and all the other parts of your ears to vibrate more quickly than a sound wave with a low frequency. Your brain interprets these fast vibrations caused by high-frequency waves as a sound with a high pitch.

As the frequency of sound waves decreases, the pitch becomes lower. **Figure 9** shows different notes and their frequencies. Notice that when you sing *do, re, mi, fa, so, la, ti, do,* the second *do* has twice the frequency of the first *do.*

Animals differ in their sensitivities to different ranges of sound frequencies. A human teenager with typical hearing can hear sounds with frequencies from about 20 Hz to 20,000 Hz. However, the highest frequency that a typical human can hear decreases with age. The human ear is most sensitive to sounds in the range of 440 Hz to about 7000 Hz. This roughly corresponds to the notes on the upper half of a piano. Dogs can hear sounds with frequencies up to about 35,000 Hz, and bats can detect frequencies higher than 100,000 Hz.

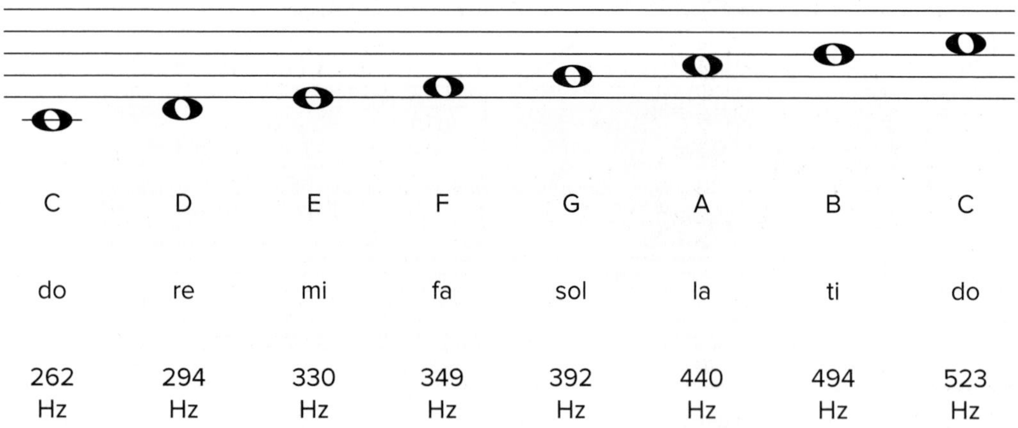

Figure 9 Every musical note has a distinct frequency, which gives that note a distinct pitch.

Describe *how pitch changes when frequency increases.*

SCIENCE USAGE v. COMMON USAGE

pitch

Science usage: how high or low a sound seems to be

The highest note on a violin has a higher pitch than the highest note on a cello.

Common usage: a throw, usually toward a particular person or point

The batter struck out when he swung at a fast pitch.

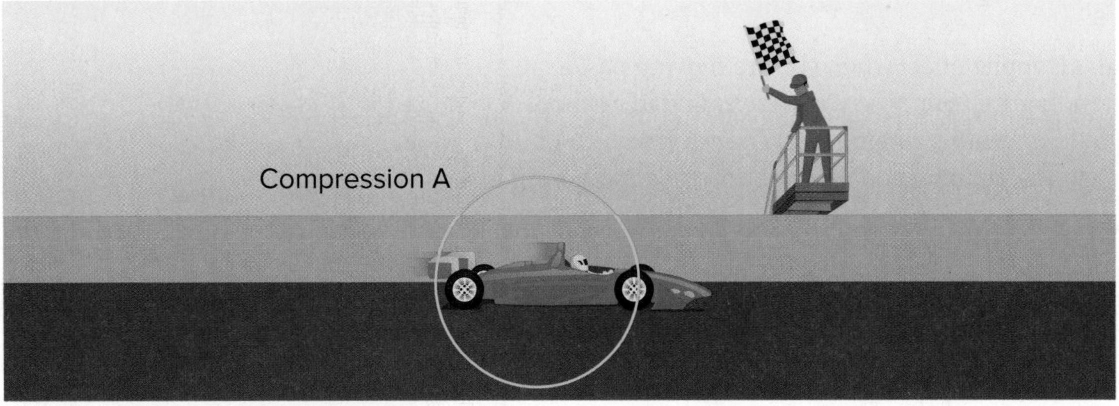

The race car sends out a sound wave as it moves, producing compression A. Compression A continues to move outward, and the car continues to move forward.

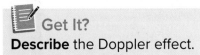

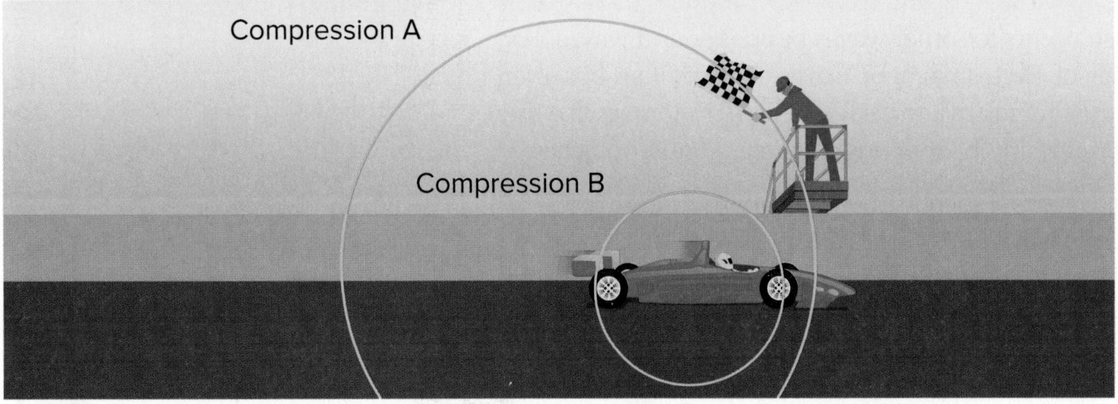

The car is closer to the flagger when it creates compression B. Compressions A and B are closer together in front of the car, so the flagger hears a higher-pitched sound.

Figure 10 The Doppler effect occurs when the source of a sound wave is moving relative to a listener.
Explain *why the flagger will hear a lower-pitched sound once the car passes him.*

The Doppler Effect

Imagine that you are standing at the side of a racetrack with race cars zooming past. When the cars are moving toward you, the pitches of their engines are higher. When the cars are moving away from you, the pitches are lower. The **Doppler effect** is the change in wave frequency due to a wave source moving relative to an observer or an observer moving relative to a wave source. **Figure 10** shows how the Doppler effect occurs.

Get It?
Describe the Doppler effect.

Moving sound sources

As a race car moves, it sends out sound waves in the form of compressions and rarefactions. In the top panel of **Figure 10,** the race car produces a compression labeled A as that race car speeds toward the flagger. Compression A travels through the air toward the flagger.

By the time compression B leaves the race car in the bottom panel of **Figure 10,** the car has moved forward. Because the car has moved since the time it created compression A, compressions A and B are closer together in front of the car. Because the compressions are closer together, the frequency is higher and the flagger hears a higher pitch. The compressions behind the moving car are farther apart, resulting in the flagger observing a lower pitch after the car passes.

Moving observers

You can also observe the Doppler effect when you are moving past a sound source that is standing still. Suppose you were riding in a school bus and you passed a building with a ringing bell. The pitch would sound higher as you approached the building and lower as you rode away from it.

The Doppler effect happens any time a sound source is moving relative to an observer. It occurs whether the sound source or the observer is moving. The faster the change in position, the greater the change in frequency and pitch.

Electromagnetic waves and the Doppler effect

The Doppler effect also occurs for other waves besides sound waves. For example, the frequency of electromagnetic waves changes if an observer and wave source are moving relative to each other. Astronomers use the Doppler effect to help measure the motions of stars and other objects.

In addition, police radar guns, such as the one shown in **Figure 11,** use the Doppler effect to measure the speeds of cars. The radar gun sends radar waves toward a moving car. The waves are reflected from the car and their frequency is shifted, depending on the speed and direction of the car. From the Doppler shift of the reflected waves, the radar gun determines the car's speed.

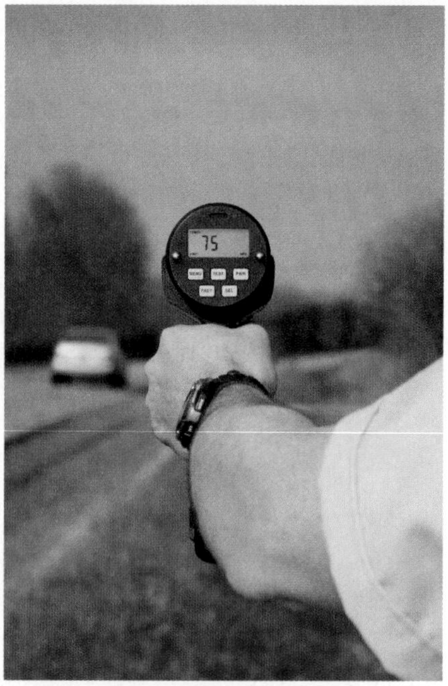

Figure 11 Police use radar guns to measure the speeds of motorists on highways. Radar guns function based on the Doppler effect.

Check Your Progress

Summary

- Tight, dense compressions in a sound wave mean greater intensity, loudness, and energy.
- Sound intensity is measured in decibels.
- Pitch is most strongly related to frequency.
- The Doppler effect is the change in wave frequency due to a wave source moving relative to an observer or an observer moving relative to a wave source.

Demonstrate Understanding

8. **Determine** which will change if you turn up a radio's volume: *wave velocity, intensity, pitch, frequency, wavelength, loudness*. Explain.
9. **Identify** the range of human hearing in decibels and the level at which sound can damage human ears.
10. **Compare and contrast** frequency and pitch.
11. **Draw and label** a diagram that explains the Doppler effect.

Explain Your Thinking

12. **Explain** why a passing car would exhibit a greater sound frequency change when it moves at 30 m/s than when it moves at 12 m/s.
13. **MATH Connection** Using the musical scale in **Figure 9,** make a table showing how many wavelengths will pass you in one minute for each note. What is the relationship between frequency and the number of wavelengths per minute?

LEARNSMART® Go online to follow your personalized learning path to review, practice, and reinforce your understanding.

LESSON 3
MUSIC

FOCUS QUESTION
What are beats, and why do they occur?

Making Music

To someone else, your favorite music might sound like a jumble of noise. A sound wave from noise and a sound wave from music are both shown in **Figure 12.** Noise has random patterns and pitches. **Music** is any collection of sounds that are deliberately used in a regular pattern.

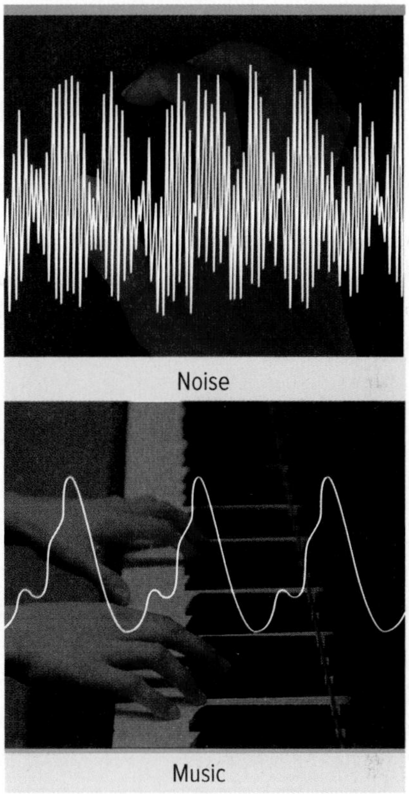

Figure 12 The sound waves that make up noise have a random pattern. The sound waves that make up music have deliberate patterns.

Natural frequencies

Every material or object has a particular set of frequencies at which it vibrates. These frequencies are called natural frequencies. Musical instruments employ the natural frequencies of various objects, such as strings, to control pitch. When you pluck a guitar string, the pitch you hear depends on the string's natural frequencies. Each string on a guitar has a different set of natural frequencies.

Resonance

The sound produced by musical instruments is amplified by resonance. Recall that resonance occurs when a material or an object is made to vibrate at its natural frequencies by absorbing energy from something that is also vibrating at those frequencies. The vibrations of a person's lips in a brass instrument or the reed in a wind instrument cause the air inside the instrument to absorb energy and vibrate at its natural frequencies. The vibrating air makes the instrument sound louder.

Get It?
Explain the difference between music and noise.

3D THINKING | DCI Disciplinary Core Ideas | CCC Crosscutting Concepts | SEP Science & Engineering Practices

COLLECT EVIDENCE
Use your Science Journal to record the evidence you collect as you complete the readings and activities in this lesson.

INVESTIGATE
GO ONLINE to find these activities and more resources.

Lab: A Simple Musical Instrument
Use mathematical thinking to determine the relationship between of the pitch and frequency in a medium.

Laboratory: Musical Instruments
Develop and use a model to visualize the structure and function of musical instruments and how they manipulate sound.

Sound Quality

Suppose your friend played a note on a flute and then a note of the same pitch and loudness on a piano. Even if you closed your eyes, you could tell the difference between the two instruments. Their sounds would not be the same. Each of these instruments has a unique sound quality. **Sound quality** describes the differences between sounds of the same pitch and loudness. Sound quality results from overtones.

Get It?

Compare sound quality and pitch.

Overtones

You might think that a musical instrument vibrates at only one frequency when you play only one note on that instrument. In fact, a musical instrument vibrates at many different frequencies when you play it, even when you play just one note.

The lowest of these frequencies is called the fundamental frequency. On a guitar, for example, the fundamental frequency is produced by the entire string vibrating back and forth, as shown in **Figure 13.** The fundamental frequency primarily determines the pitch of the note that you hear.

The other frequencies are called overtones. An **overtone** is a vibration whose frequency is a multiple of the fundamental frequency. Overtones determine the sound quality of the note that you hear. The first two guitar-string overtones are also shown in **Figure 13.**

Consider again the example of your friend playing the same note on a flute and on a piano. Your friend produces the same fundamental frequency on both instruments. However, the loudness of each overtone differs between the two instruments.

Figure 13 The vibration patterns on a plucked guitar string for the fundamental frequency, first overtone, and second overtone are shown. A plucked guitar string usually vibrates at multiple frequencies simultaneously.

Infer *how the string would vibrate to produce the third overtone.*

Musical Instruments

A musical instrument is any device used to produce a musical sound. You might be familiar with violins, cellos, horns, flutes, and kettledrums from your school orchestra; clarinets, trumpets, and saxophones from a jazz band; or guitars and keyboards from a rock band. These familiar examples form just a small sample of the diverse assortment of instruments that people play throughout the world. For example, Australian Aborigines accompany their songs with a woodwind instrument called a didgeridoo (DIH juh ree dew). Caribbean musicians use rubber-tipped mallets to play steel drums, and a flutelike instrument called a nay is played throughout the Arab world.

Strings

Compared to the overall history of musical instruments, string instruments are quite modern. The first manufactured string instruments probably did not appear until about 6000 years ago. In string instruments, sound is produced by plucking, striking, or drawing a bow across tightly stretched strings.

Because the sound of a vibrating string is soft, almost all string instruments include a way to amplify sound. Many string instruments, such as the violin in **Figure 14,** have a resonator. A **resonator** (RE zuh nay tur) is a hollow chamber that amplifies sound when the air inside of it vibrates. Other string instruments, such as electric guitars, have pick-ups that convert sound into electric signals. These signals are then amplified before being converted back into sound.

Figure 14 Violins, acoustic guitars, and grand pianos use resonators to amplify sound. A violin's resonator causes air to vibrate much more than that air would from the violin string alone.

Identify *what causes the air to vibrate. What amplifies the vibrations?*

ACADEMIC VOCABULARY

diverse

differing from one another

Between chess club, soccer practice, and baking class, her diverse interests have been filling up her schedule.

Figure 15 A pipe organ is a wind instrument. A pipe organ typically has hundreds of pipes controlled from a central console with a keyboard. Pressing a key activates a mechanism that directs air into one of the pipes, producing a sound of the desired pitch and quality. Longer pipes correspond to lower pitches.

Wind and brass instruments

The vibrations of air inside wind and brass instruments determine the frequencies that those instruments produce. These instruments have been around for much longer than string instruments. Humans created the first wind instruments at least 30,000 years ago. Some scientists think that the first wind instruments may have been created more than 45,000 years ago.

Some of the largest instruments in use today are wind instruments. Pipe organs, like the one shown in **Figure 15,** can have hundreds or thousands of pipes. The Wanamaker Organ in Philadelphia, Pennsylvania, has more than 28,000 pipes. Other modern brass and wind instruments include tubas, horns, oboes, and flutes.

In all brass and wind instruments, sound is produced by air vibrating inside the instrument. For example, a flute player blows a stream of air over the edge of the flute's mouth hole. This causes the air inside the flute to vibrate. For an instrument such as a saxophone, the player blows across a reed, causing it to vibrate. The vibrating reed then causes the air in the instrument to vibrate. For a brass instrument, such as a trumpet, the musician vibrates his or her lips to make the air inside the instrument vibrate.

In brass and wind instruments, the length of the vibrating tube of air determines the pitch of the sound produced. In flutes and trumpets, the musician changes the length of the resonator by opening and closing finger holes or valves. In a trombone, the tubing slides in and out to become shorter or longer.

Figure 16 The air inside the resonators of these drums amplifies the sounds produced.

Describe *how the natural frequencies of the air in the drums affect the sounds those drums produce.*

Percussion

Does the sound of a bass drum make your heart pound? Percussion instruments are probably the oldest of all instruments other than the human voice. Since ancient times, people have used drums and other percussion instruments to send signals, accompany important rituals, and entertain one another.

Percussion instruments are struck, shaken, rubbed, or brushed to produce sound. Some, such as the marching band drums shown in **Figure 16,** have a membrane stretched over a resonator. When the drummer strikes the membrane, the membrane vibrates and causes the air inside the resonator to vibrate. The resonator amplifies the sound made when the membrane is struck.

Some drums have a fixed pitch, but others have a pitch that can be changed by tightening or loosening the membrane. Caribbean steel drums and marimbas are both examples of percussion instruments for which you can control the pitch of the instrument. Caribbean steel drums were developed in the 1940s in Trinidad. As many as 32 different striking surfaces hammered from the ends of 55-gallon oil barrels create different pitches of sound. The side of a drum acts as the resonator.

Figure 17 Marimbas are made with many wooden bars, each of which has its own resonator tube. The size of each wooden bar helps to determine the pitch that the musicians produce when they strike that wooden bar.

Explain *why the bars on a marimba are different sizes.*

A marimba, shown in **Figure 17,** is also a percussion instrument. It has a series of wooden bars, each with its own tube-shaped resonator. The musician strikes the bars with mallets that affect the sound quality. Hard mallets make crisp sounds, while softer rubber mallets produce sounds that are more like the drops of water from a leaky faucet. Other types of percussion instruments include cymbals, rattles, and even old-fashioned washboards.

Whether or not they have a fixed pitch, the musicality of percussion instruments comes in large part from rhythm, which is the pattern of sounds in time. When percussion instruments accompany other instruments, they are often used to establish and maintain rhythm for the other instruments to follow.

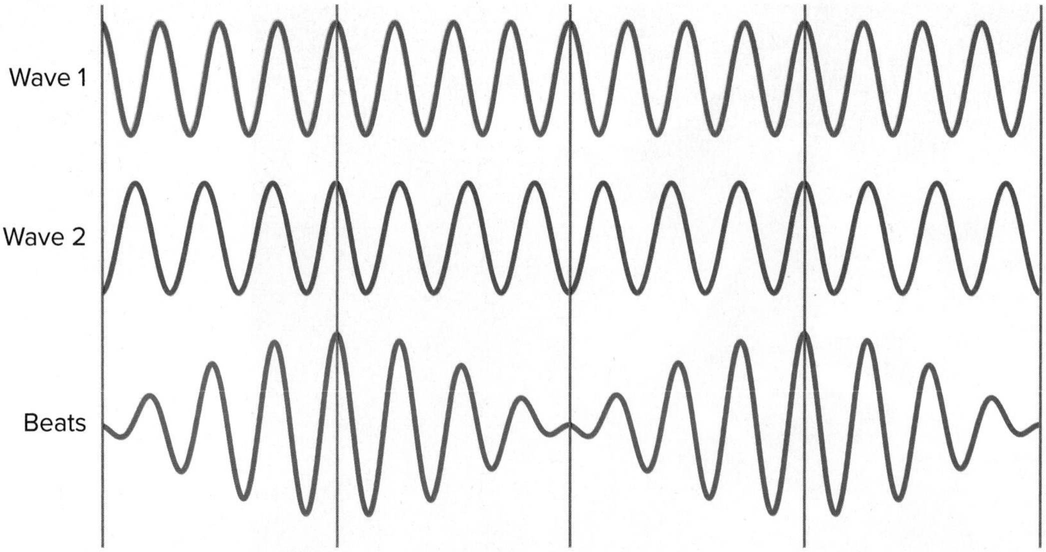

Figure 18 Beats occur when sound waves with slightly different frequencies combine. These sound waves interfere with each other, producing a low-frequency, pulsing variation in loudness called beats. The top two panels illustrate two slightly different sound waves. The bottom panel illustrates the beats that occur when those two sound waves interfere.

Beats

Have you ever heard two flutes play the same note when they were not properly tuned? You might have heard a pulsing variation in loudness. This variation in loudness is called beats and can be unpleasant to the listener. As shown in **Figure 18,** when compressions and rarefactions overlap each other, loudness decreases. When compressions overlap compressions, loudness increases.

If two waves of different frequencies interfere, they produce a new wave that has a different frequency. The frequency of this wave is the difference between the frequencies of the two component waves. The frequency of the beats that you hear decreases as the two waves become closer in frequency. If two flutes play a note at the same frequency, no beats are heard.

Check Your Progress

Summary

- Sound quality refers to the differences between sounds of the same pitch and loudness.
- Sound quality results from specific combinations of overtone frequencies produced in various musical instruments.
- The interference of two waves with different frequencies produces beats.

Demonstrate Understanding

14. **Compare and contrast** music and noise.
15. **Explain** how two instruments could be used to produce a pulsing sound, and identify the name for this pulsing sound.
16. **Describe** how a flute, violin, and drum each produce sound.

Explain Your Thinking

17. **Explain** how two musical notes that have the same pitch and volume could sound very different from each other.
18. **MATH Connection** A guitar string vibrates with a frequency of 440 Hz. When a second string is played at the same time, two beats per second are heard. What are the possible frequencies of vibration of the second string?

LEARNSMART® Go online to follow your personalized learning path to review, practice, and reinforce your understanding.

LESSON 4

USING SOUND

FOCUS QUESTION

What does echolocation and ultrasonic imaging have in common?

Acoustics

When an orchestra stops playing, does it seem as if the sound of the music lingers for a couple of seconds? The sounds and their reflections reach your ears at different times, so you hear echoes. This echoing effect produced by many reflections of sound is called reverberation (re vur bu RAY shun).

During an orchestra performance, reverberation can ruin the sound of the music. To prevent this problem, the people who design concert halls must understand how the size, shape, and furnishings of the room affect the reflection of sound waves.

These scientists and engineers specialize in acoustics (uh KEWS tihks). **Acoustics** is the study of sound. People who study acoustics know that soft, porous materials, such as curtains, can reduce excess reverberation. **Figure 19** shows a concert hall that has been designed to produce a good listening environment.

Figure 19 Cloth drapes, cushioned seats, and carpeted floors help reduce reverberations in this concert hall. The amount of reverberation can be controlled slightly by opening and closing the drapes.

Explain *whether the drapes are more likely to absorb or to reflect sound energy.*

Echolocation

At night, bats swoop around in darkness without bumping into anything. They even manage to find insects and other prey in the dark. Most species of bats depend on echolocation. **Echolocation** is the process of locating objects by emitting sounds and then interpreting the sound waves that are reflected from those objects. **Figure 20** explains more about echolocation.

3D THINKING DCI Disciplinary Core Ideas CCC Crosscutting Concepts SEP Science & Engineering Practices

COLLECT EVIDENCE

Use your Science Journal to record the evidence you collect as you complete the readings and activities in this lesson.

INVESTIGATE

GO ONLINE to find these activities and more resources.

Review the News

Obtain information from a current news story about acoustics. **Evaluate** your source and **communicate** your findings to your class.

Revisit the Encounter the Phenomenon Question

What information from this lesson can help you asnwer the Unit and Module questions?

Figure 20 Visualizing Bat Echolocation

Bats and dolphins navigate and locate prey by using echolocation. The diagrams below show how a bat uses echolocation in hunting for prey. Some species of moths are able to jam a bat's echolocation by emitting their own sounds.

▲ A. The bat emits sound waves.

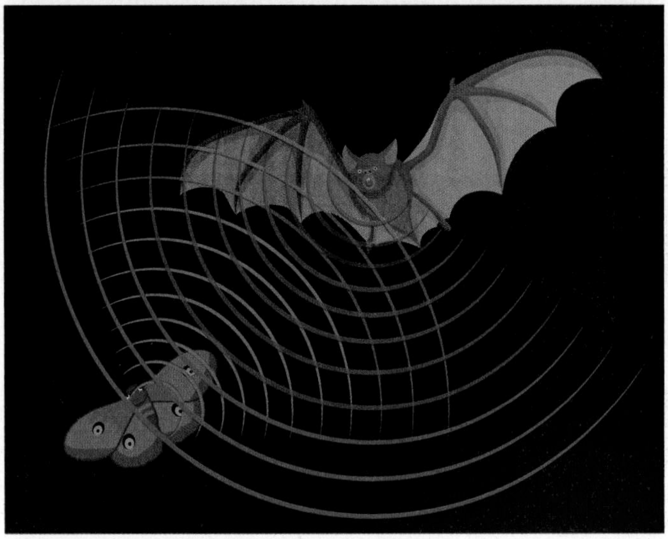

▲ B. Some of the sound waves reflect off a moth.

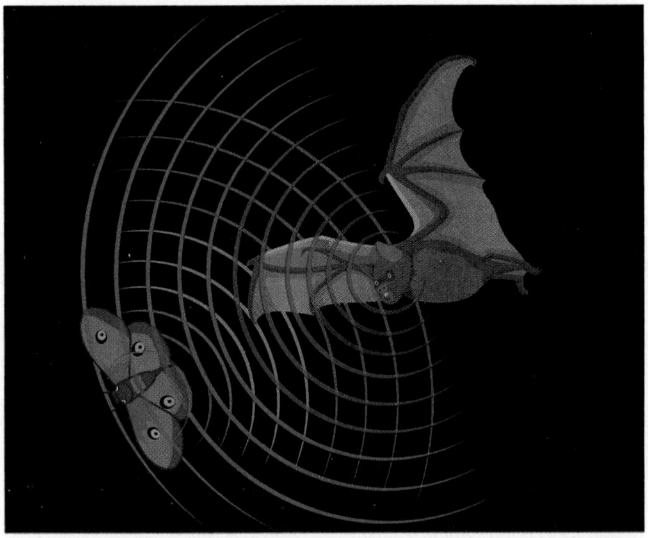

▲ C. The bat determines the moth's location and velocity from the reflected waves.

▲ D. The bat continues to emit sound waves and locate the moth until the moth is captured.

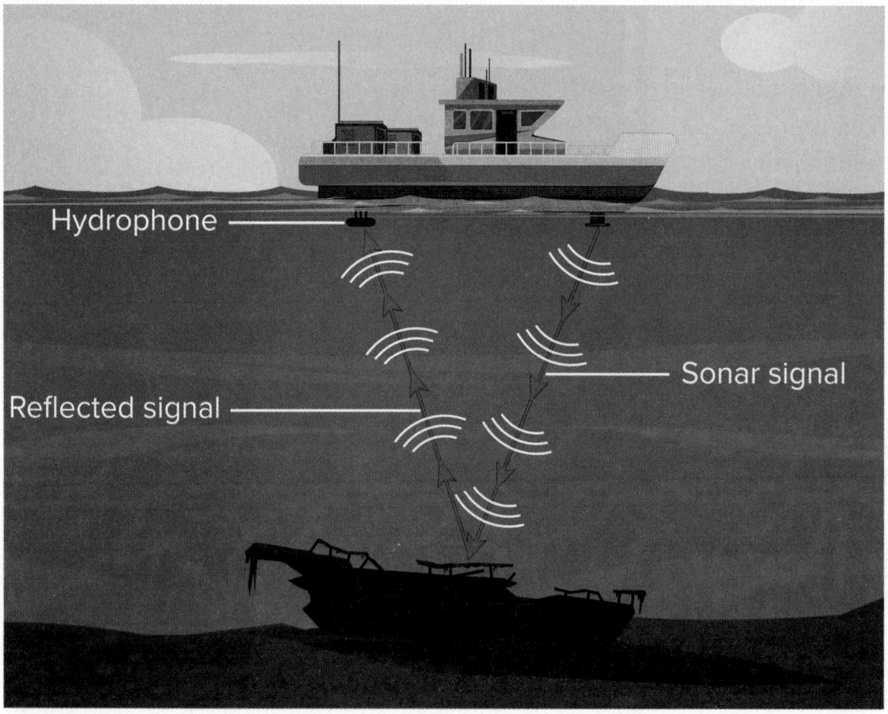

Figure 21 People use sonar to find objects that are underwater. Here, scientists use sonar to locate a sunken ship.

Sonar

More than 140 years ago, a ship named the *SS Central America* disappeared in a hurricane off the coast of South Carolina. In its hold lay ten tons of newly minted gold coins and bars. When the shipwreck occurred, there was no way to search for the ship in the deep water where it sank. The *SS Central America* and its treasures lay at the bottom of the ocean until 1988, when crews used sonar to locate the wreck under 2400 m of water. Over $100 million in gold was eventually recovered.

Sonar, which stands for **SO**und **N**avigation **A**nd **R**anging, is a system that uses the reflection of underwater sound waves to detect objects. First, a sound pulse is emitted toward the bottom of the ocean. The sound travels through the water and is reflected when it hits something solid, as shown in **Figure 21.** A sensitive underwater microphone, called a hydrophone, picks up the reflected signal. Because the speed of sound in water is known, the distance to the object can be calculated by measuring how much time passes between emitting the sound pulse and receiving the reflected signal.

The idea of using sonar to detect underwater objects was first suggested as a way to avoid icebergs, but many other uses have been developed for it. Navy ships use sonar for detecting, identifying, and locating submarines. Fishing crews also use sonar to find schools of fish, and scientists use it to map the ocean floor.

More detail can be revealed by using sound waves of high frequency. As a result, most sonar systems are ultrasonic. **Ultrasound** is sound with frequencies above the range of human hearing, or more specifically, sound with frequencies above 20,000 Hz.

Get It?

Describe how sonar detects underwater objects.

Ultrasound in Medicine

Ultrasonic waves are commonly used in medicine. Medical professionals use ultrasound to examine many parts of the body, including the heart, liver, gallbladder, pancreas, spleen, kidneys, breasts, and eyes. Medical professionals can also use ultrasonic imaging, which is much safer than X-ray imaging, to monitor a human fetus.

When ultrasound is used for medical imaging, an ultrasound technician directs the ultrasound waves toward a target area of a patient's body. The sound waves reflect off the targeted area, and the reflected waves are used to produce electronic signals. A computer program converts these signals into video images called sonograms. A sonogram of a human fetus is shown in **Figure 22.**

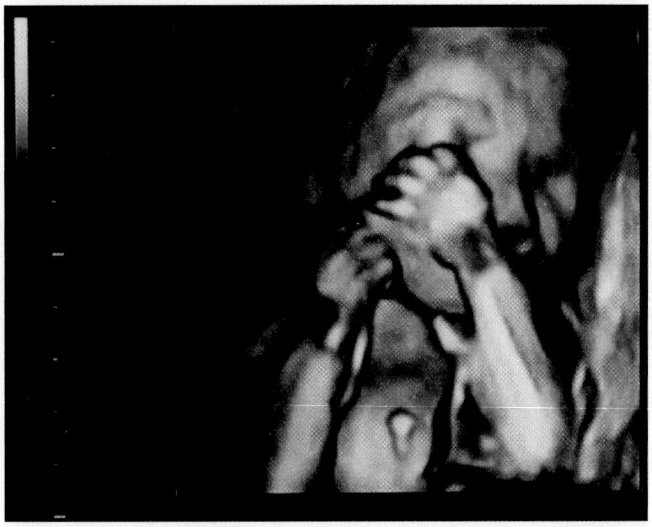

Figure 22 Ultrasonic waves are directed into a pregnant woman's uterus to form images of her fetus. This allows doctors to safely monitor the fetus's growth.

Kidney stones and ultrasound

Medical professionals can also use ultrasound to treat certain medical problems. For example, ultrasound can be used to break up kidney stones. Bursts of ultrasound create vibrations that cause the stones to break into small pieces. These fragments then pass out of the body with the urine. Without ultrasound, surgery would be necessary to remove the kidney stones.

Check Your Progress

Summary

- Acoustics is the study of sound.
- Bats locate objects by emitting sounds and then interpreting the reflected sound waves.
- Humans using sonar can interpret reflected sound to locate objects underwater.
- High-frequency sound waves are useful for detecting and monitoring certain medical conditions.

Demonstrate Understanding

19. **Describe** at least three different ways that people use sound.
20. **Describe** some differences between a gym and a concert hall that might affect the amount of reverberation in each.
21. **Compare and contrast** echolocation and sonar.
22. **Explain** how ultrasonic imaging works.

Explain Your Thinking

23. **Apply** How might sonar technology be useful in locating deposits of oil and minerals?
24. **MATH Connection** Sound travels at about 1500 m/s in seawater. How far will a sonar pulse travel in 46 s?
25. **MATH Connection** How long will it take for an undersea sonar pulse to travel 3 km?

LEARNSMART® Go online to follow your personalized learning path to review, practice, and reinforce your understanding.

"The Menace of Mechanical Music"

In 1906, John Philip Sousa, a musical composer known for his marches, wrote an essay entitled "The Menace of Mechanical Music" to warn Americans of the danger posed by a new invention—the phonograph. Thomas Edison had first demonstrated the phonograph in 1877. It was the first device to allow people to play back a wide variety of recorded sounds.

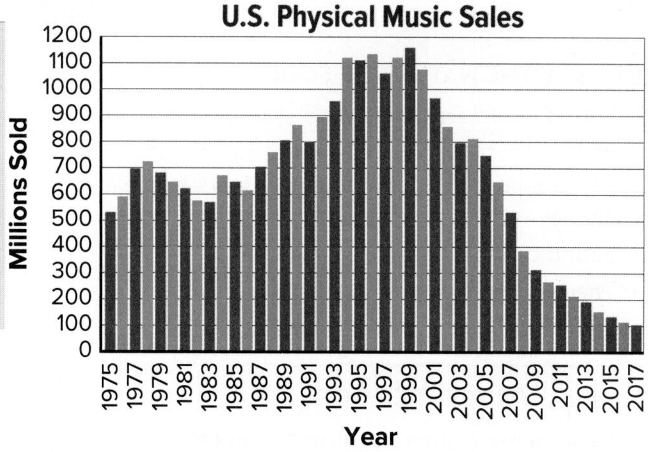

Album sales have fallen as online music has become the music format of choice.

Early concerns The first phonograph parlor opened in 1889. Phonograph parlors were places where people could pay to listen to recorded music. By 1906, Sousa was writing that the phonograph would lead to the death of music as an art form. He was probably also concerned with how he would make a living as a composer in an age of readily available recorded music. However, the fear he expressed would reappear often as music technology changed over time.

Record players Over time, Edison's phonograph evolved into the turntable-style record player. Mass-produced records made recorded music available in people's homes for the first time.

The limitations of early records changed the nature of music. For example, early records could play for only three minutes, placing a time limit on the lengths of recorded songs. Later, long-playing records could play up to 30 minutes per side, making way for the record album format. Record players also allowed only the playback of music. Another invention would be necessary before the average person would be able to record, edit, and remix music.

Cassette tapes During World War II, the invention of magnetic tape recording once again changed how music is produced and enjoyed. Magnetic tape was inexpensive and easy to use. It allowed the average person to record, play back, and edit sounds without bulky equipment. By the 1980s, consumers could record music from records, from the radio, or live onto magnetic cassette tapes, mixing artists, genres, and time periods.

Digital music Digital storage for music first became popular in the 1990s. At first, music was stored digitally on compact discs (CDs) and played back on CD players. When downloadable music files were developed, people began to listen to, record, and edit music without the need for CDs. The graph above shows how downloadable and streaming music has affected music sales in the United States. Some music artists express worries about how downloading or streaming music might affect their art, while others embrace it.

OBTAIN, EVALUATE, AND COMMUNICATE INFORMATION

Research the current laws governing the sale and distribution of music in the United States. Then hold a class debate on the effectiveness and validity of these laws.

MODULE 10

STUDY GUIDE

GO ONLINE to study with your Science Notebook.

Lesson 1 THE NATURE OF SOUND

- Sound waves are longitudinal waves produced by vibrating objects.
- The compressions and rarefactions of sound waves transfer energy.
- Sound cannot travel through a vacuum.
- The ear detects sound waves and converts them to electrical impulses.

- eardrum
- cochlea

Lesson 2 PROPERTIES OF SOUND

- Tight, dense compressions in a sound wave mean greater intensity, loudness, and energy.
- Sound intensity is measured in decibels.
- Pitch is most strongly related to frequency.
- The Doppler effect is the change in wave frequency due to a wave source moving relative to an observer or an observer moving relative to a wave source.

- intensity
- loudness
- decibel
- pitch
- Doppler effect

Lesson 3 MUSIC

- Sound quality refers to the differences between sounds of the same pitch and loudness.
- Sound quality results from specific combinations of overtone frequencies produced in various musical instruments.
- The interference of two waves with different frequencies produces beats.

- music
- sound quality
- overtone
- resonator

Lesson 4 USING SOUND

- Acoustics is the study of sound.
- Bats locate objects by emitting sounds and then interpreting the reflected sound waves.
- Humans using sonar can interpret reflected sound to locate objects underwater.
- High-frequency sound waves are useful for detecting and monitoring certain medical conditions.

- acoustics
- echolocation
- sonar
- ultrasound

THREE-DIMENSIONAL THINKING

Module Wrap-Up

REVISIT THE PHENOMENON

Why do two instruments playing the same note sound different?

CER Claim, Evidence, Reasoning

Explain Your Reasoning Revisit the claim you made when you encountered the phenomenon. Summarize the evidence you gathered from your investigations and research and finalize your Summary Table. Does your evidence support your claim? If not, revise your claim. Explain why your evidence supports your claim.

STEM UNIT PROJECT

Now that you've completed the module, revisit your STEM unit project. You will summarize your evidence and apply it to the project.

GO FURTHER

SEP Data Analysis Lab

What is music?

You have learned how different musical instruments make sounds based on their shape and material. Musical instruments are designed to produce vibrations. These vibrations become the compression waves we hear as sound.

In this activity, you will design a musical instrument and draw a diagram of it. Your instrument can be any shape you choose. However, your instrument must have at least three of the following components: a vibrating string or strings, vibrating air or a vibration caused by blowing air, finger holes, production of a percussion sound, and production of an overtone. Be sure to label these components on your diagram. Your instrument can sit on the ground or be held.

CER Analyze and Interpret Data

1. **Claim, Evidence, Reasoning** How does your instrument produce a string vibration?
2. **Claim, Evidence, Reasoning** How does it produce a wind vibration?
3. **Claim, Evidence, Reasoning** Are there any percussion sounds? How are they produced?

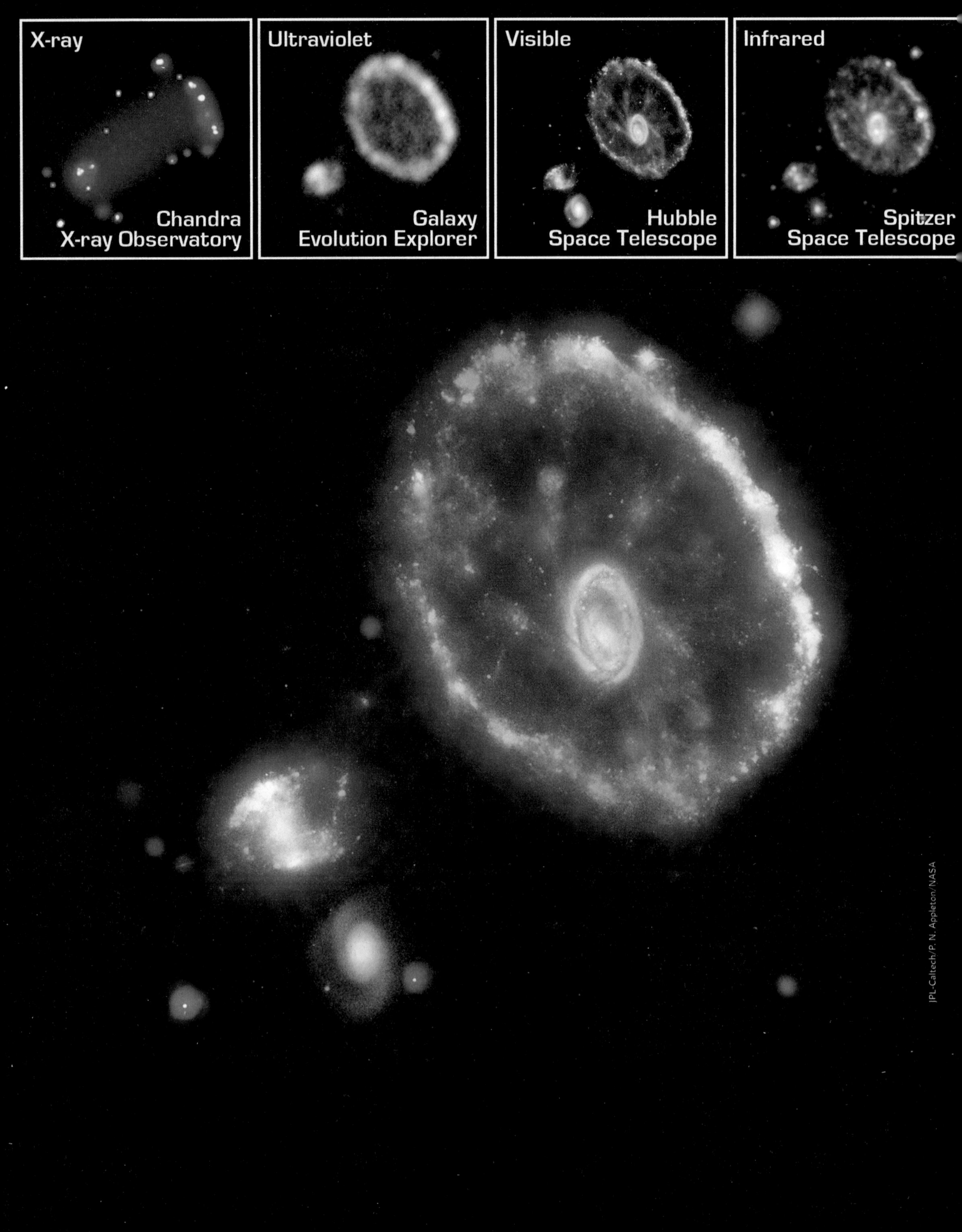

JPL-Caltech/P. N. Appleton/NASA

MODULE 11

ELECTROMAGNETIC WAVES

ENCOUNTER THE PHENOMENON

How do we see X-rays?

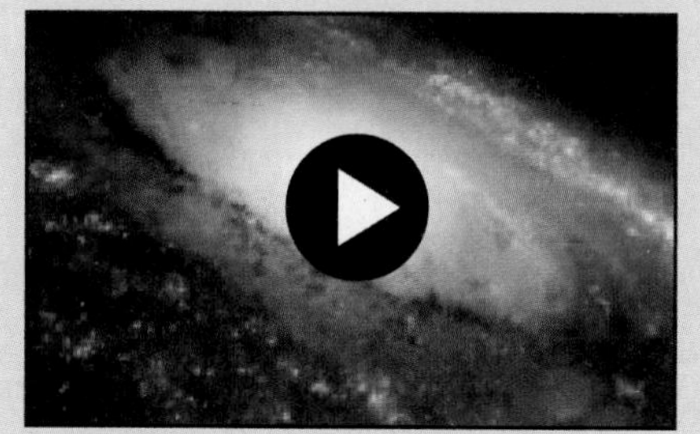

GO ONLINE to play a video about how we use light to figure out what stars are made of.

SEP Ask Questions

Do you have other questions about the phenomenon? If so, add them to the driving question board.

CER Claim, Evidence, Reasoning

Make Your Claim Use your CER chart to make a claim about how we can see in x-ray. Explain your reasoning.

Collect Evidence Use the lessons in this module to collect evidence to support your claim. Record your evidence as you move through the module.

Explain Your Reasoning You will revisit your claim and explain your reasoning at the end of the module.

GO ONLINE to access your CER chart and explore resources that can help you collect evidence.

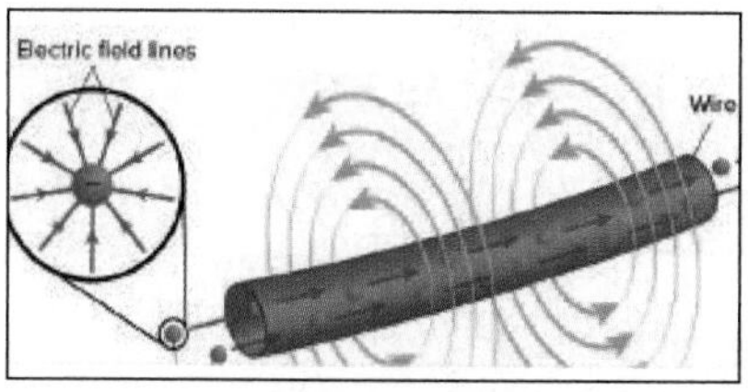

LESSON 1: Explore & Explain: Electromagnetic Waves

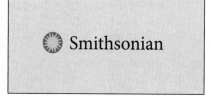

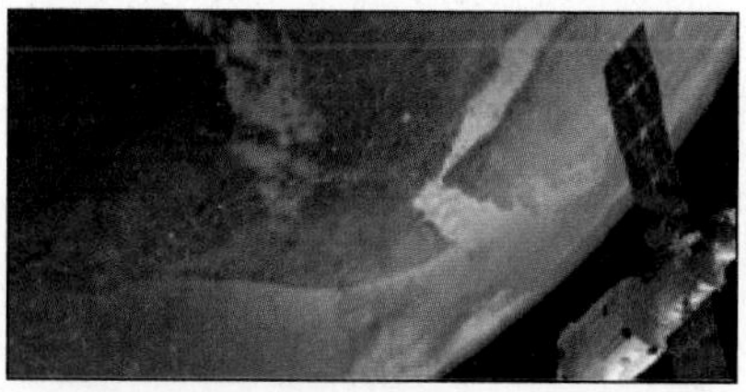

LESSON 3: Explore & Explain: Communication Satellites

Smithsonian

Additional Resources

(t)Video Supplied by BBC Worldwide Learning, (b)koto_feja/Getty Images

LESSON 1
WHAT ARE ELECTROMAGNETIC WAVES?

FOCUS QUESTION
How do electromagnetic waves interact with the things around us?

Waves in Matter

Waves are produced by something that vibrates. By moving an inflatable float up and down in a pool, you can create a wave in the water. When you speak, the vibrations of your vocal cords create a sound wave.

Waves carry energy from one place to another. Look at the water wave and the sound wave in **Figure 1.** Both waves are moving through matter. The water wave is moving through water, and the sound wave is moving through air.

These waves travel because energy is transferred from particle to particle. Without matter to transfer the energy, these waves cannot move. However, there is another type of wave that does not require matter to transfer energy.

Get It?
Identify What produces waves, and what do waves carry?

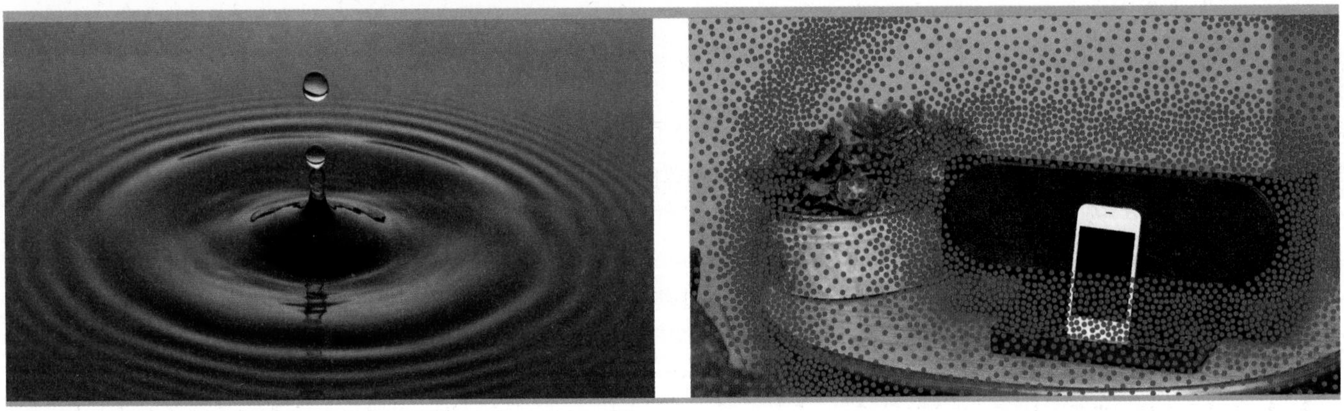

Figure 1 Some waves, such as water waves and sound waves, require matter to move. As the wave travels through the matter, energy is transferred from one particle to the next.

3D THINKING
DCI Disciplinary Core Ideas | CCC Crosscutting Concepts | SEP Science & Engineering Practices

COLLECT EVIDENCE

Use your Science Journal to record the evidence you collect as you complete the readings and activities in this lesson.

INVESTIGATE

GO ONLINE to find these activities and more resources.

Applying Practices: Is light a wave or a particle?
HS-PS4-3. Evaluate the claims, evidence, and reasoning behind the idea that electromagnetic radiation can be described either by a wave model or a particle model, and that for some situations one model is more useful than the other.

Lab: The Speed of Light
Use mathematics to determine the relationships between the wave properties of light.

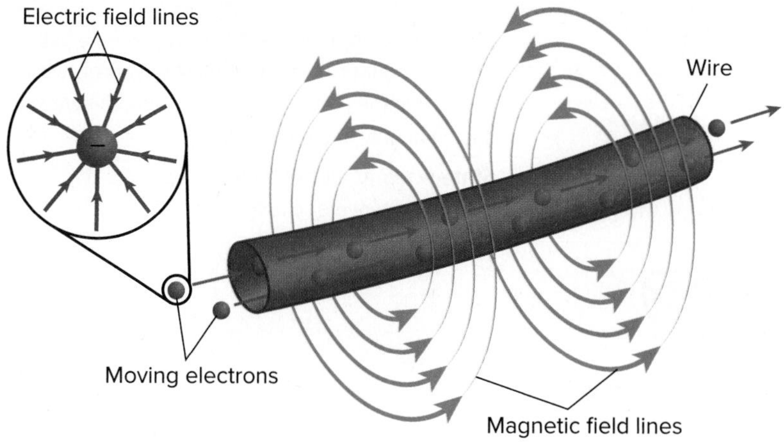

Figure 2 All moving electric charges, such as the electrons in this wire, are surrounded by an electric field and a magnetic field.

Electromagnetic Waves

Electromagnetic waves are made by vibrating electric charges. Electromagnetic waves can be modeled as a wave of changing electric fields and magnetic fields. Instead of transferring energy from particle to particle, electromagnetic waves travel by transferring energy between the electric and magnetic fields. Electromagnetic waves do not require matter to travel because electric and magnetic fields can exist where there is no matter.

Electric and magnetic fields

Recall that electric charges are surrounded by electric fields and that magnets are surrounded by magnetic fields. These fields exert a force even when the charge or magnet is not in contact with an object. Fields exist around an electric charge or a magnet even in a vacuum. A vacuum is a volume of space that contains little or no matter. You might also recall that a moving electric charge, such as the current in the wire shown in **Figure 2,** is surrounded by a magnetic field. Similarly, a moving magnet is surrounded by an electric field. A changing electric field creates a magnetic field, and a changing magnetic field creates an electric field.

Making electromagnetic waves

When an electric charge vibrates, the electric field around it changes. Because the electric charge is in motion, it has a magnetic field around it. This magnetic field also changes as the charge vibrates. As a result, the vibrating electric charge is surrounded by a changing electric field and a changing magnetic field. How do the vibrating fields around the charge become a wave that travels through space? The changing electric field around the charge creates a changing magnetic field. This changing magnetic field then creates a changing electric field. This process continues, with the magnetic field and electric field continually creating each other.

SCIENCE USAGE v. COMMON USAGE

vacuum

Science usage: a volume where there is no or very little matter
Light from distant stars travels to Earth through the vacuum of space.

Common usage: a household appliance that uses suction to clean carpets and upholstery
The carpet has become very dirty since our vacuum cleaner broke.

CCC CROSSCUTTING CONCEPTS

Systems and System Models Research how electromagnetic waves are produced and travel through a vacuum. Write a set of storyboards for a short animation to model the system and explain the process to your classmates. Include a script to accompany the animation. If software is available, create the animation based on your plan.

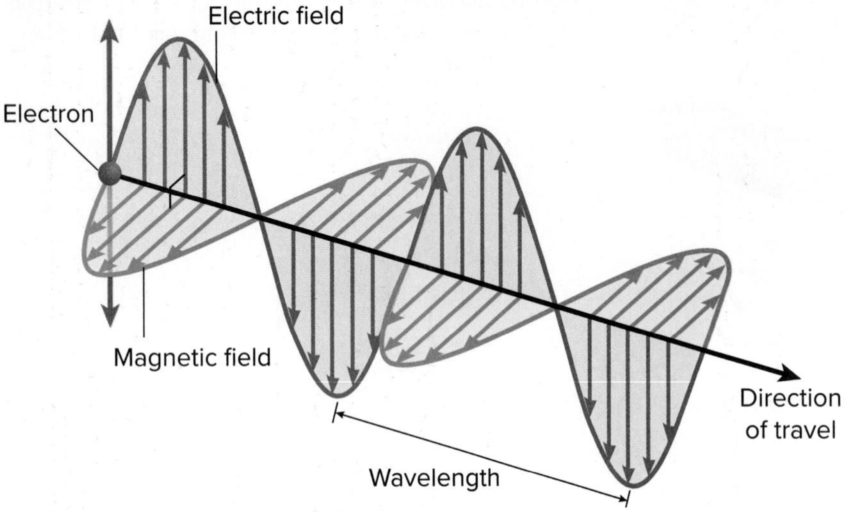

Figure 3 A vibrating electric charge creates an electromagnetic wave. The wave travels outward in all directions from the charge. Here, the wave is shown in only one direction.

State *whether an electromagnetic wave is a transverse wave or a longitudinal wave.*

Properties of Electromagnetic Waves

The vibrating electric and magnetic fields of an electromagnetic wave are perpendicular to each other. That is, they are at right angles (90°) to each other. They travel outward from the moving charge, as shown in **Figure 3.** Because the electric and magnetic fields vibrate at right angles to the direction the wave travels, an electromagnetic wave is a transverse wave.

Speed

In a vacuum, all electromagnetic waves travel at 300,000 km/s. Because light is a type of electromagnetic wave, the speed of electromagnetic waves in a vacuum is usually called the "speed of light." The speed of light is nature's speed limit—nothing travels faster than the speed of light. The speed of an electromagnetic wave in matter depends on the material through which the wave travels. However, it is always slower than the speed of light in a vacuum. In matter, electromagnetic waves are usually the slowest in solids and faster in gases. **Table 1** lists the speed of electromagnetic waves in a vacuum and several common materials. **Figure 4** illustrates that light travels slower and refracts when it enters glass.

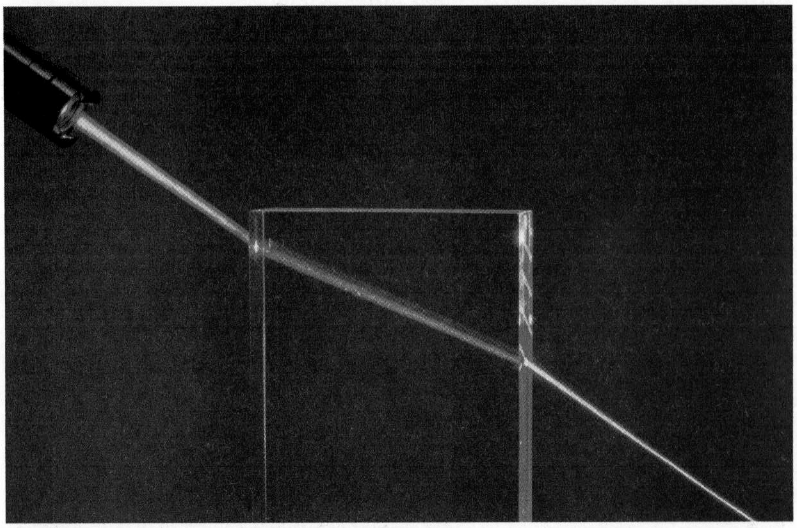

Figure 4 Electromagnetic waves travel slower in glass than in air. This difference in speed leads to refraction when the wave passes from air into glass.

Table 1 Speed of Electromagnetic Waves

Material	Speed (km/s)
None (Vacuum)	300,000
Air	299,000
Water	226,000
Glass	200,000
Diamond	124,000

APPLY SCIENCE

How do scientists work with large numbers?

The speed of light in a vacuum is 300,000,000 m/s. That is a lot of zeros to keep track of when doing a calculation! To make large numbers easier to work with, scientists use a system called scientific notation.

A number written in scientific notation has the form $M \times 10^N$. N is the number of places the decimal point in the number has to be moved so that the number M that results has only one digit to the left of the decimal point.

For example, the speed of light in a vacuum is 300,000,000 m/s = 3.00000000×10^8 m/s. We moved the decimal 8 spaces to the left, so $N = 8$ and $M = 3.00000000$. In this case, the zeros are just place holders, so we can drop them and write the speed of light in a vacuum as 3×10^8 m/s.

Very small numbers can also be written in scientific notation. For example, $0.00000045 = 4.5 \times 10^{-7}$. The minus sign in the exponent indicates that the decimal was moved to the right instead of to the left.

Identify the Problem

In order to make calculations neat and efficient, place the values for the speed of light in different materials into scientific notation.

Speed of Light in Various Materials

Material	Speed (m/s)	Speed (m/s) in Scientific Notation
None (Vacuum)	300,000,000	3×10^8
Air	299,000,000	
Water	226,000,000	
Glass	200,000,000	
Diamond	124,000,000	

Solve the Problem

1. Copy the table, and write each speed in scientific notation.
2. Would it be helpful to write your mass (about 50 kg) in scientific notation? Explain.
3. Write the following measurements in scientific notation:

   a) 0.00005291 s

   b) 0.000000246 m

   c) 0.030042 kg

Frequency and wavelength

Like all waves, electromagnetic waves can be described by their wavelengths and frequencies. The wavelength of an electromagnetic wave is the distance from one crest to another, as shown in **Figure 3.** The frequency is the number of wavelengths that pass a point in one second. The unit for frequency is the hertz (Hz). The frequency of an electromagnetic wave equals the frequency of the vibrating charge that produces the wave. This frequency is the number of vibrations of the charge in one second. Electromagnetic waves follow the wave speed equation, $v = f\lambda$. As the frequency (f) increases, the wavelength (λ) becomes smaller.

Matter and Electromagnetic Waves

All matter contains charged particles that are always in motion. As a result, all objects emit electromagnetic waves. Objects can emit electromagnetic waves at many wavelengths. However, the dominant wavelength emitted becomes shorter as the temperature of the material increases.

Get It?

Describe the relationship between the temperature of a material and the dominant wavelength the material emits.

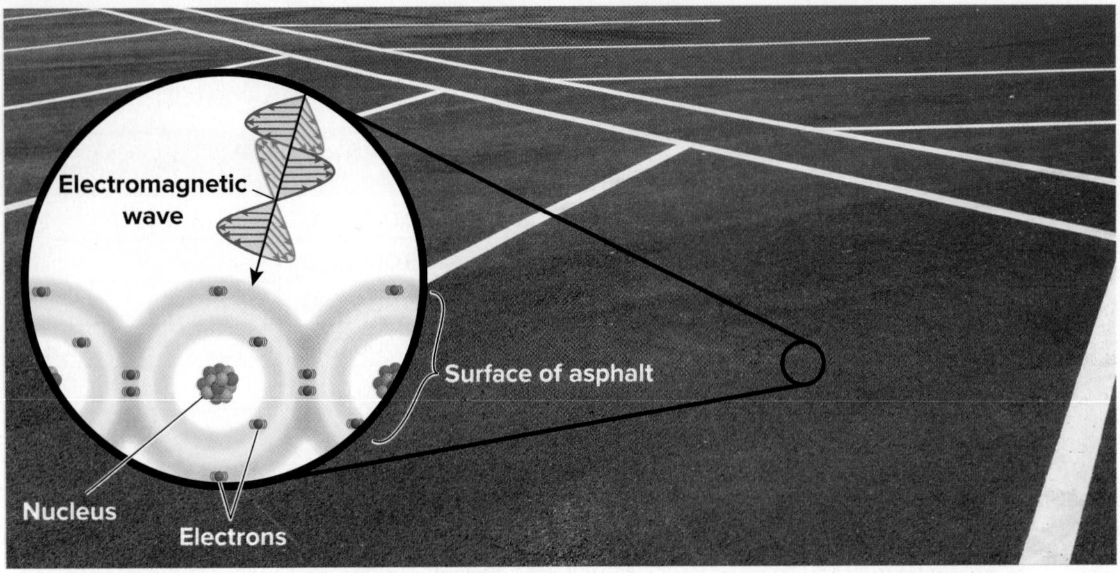

Figure 5 As an electromagnetic wave from the Sun strikes asphalt, charged particles in the asphalt gain energy from the vibrating electric and magnetic fields. This transfer of energy is why asphalt gets hot on sunny summer days.

Electromagnetic waves interact with matter

As an electromagnetic wave moves, it encounters objects. The vibrating electric and magnetic fields of the wave exert forces on the charged particles and magnetic materials that make up the object. This interaction causes the particles in the object to gain energy. For example, electromagnetic waves from the Sun cause electrons in asphalt to vibrate and gain energy, as shown in **Figure 5.** This energy can make the asphalt hot. The energy carried by an electromagnetic wave is called radiant energy. Radiant energy makes fire feel warm and enables you to see.

Waves and Particles

The difference between a wave and a particle might seem obvious—a wave is a disturbance that carries energy, and a particle is a piece of matter. However, in reality, the difference is not so clear.

Waves as particles

In 1887, Heinrich Hertz found that he could create a spark by shining light on a metal. (Today, we know that this spark means that electrons were ejected from the metal.) Hertz found that whether sparks occurred depended on the frequency of the light and not the amplitude. Because the energy carried by a sound wave or water wave depends on its amplitude and not its frequency, this result was mysterious.

Get It?

Identify What determines whether sparks are ejected from a metal when light shines on it?

In 1905, Albert Einstein provided an explanation. An electromagnetic wave can behave and can be modeled as a particle called a photon. A **photon** is a massless bundle of energy that behaves like a particle. The photon's energy depends on the frequency of the wave and increases as the wave's frequency increases. Although many features of electromagnetic waves are explained well by using the wave model, others are better explained using the particle model.

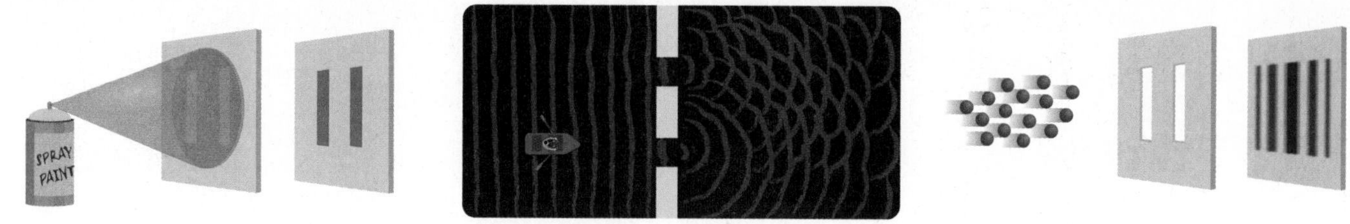

Paint particles sprayed through two slits coat only the area behind the slits.

Water waves produce an interference pattern after passing through two slits.

Electrons fired at two slits form a wavelike interference pattern.

Figure 6 When sent through two narrow slits, electrons behave as a wave, not as particles.

Particles as waves

Because electromagnetic waves could behave as particles, other scientists wondered whether particles, such as electrons, could behave as waves. If a beam of electrons were sprayed at two tiny slits, you might expect that the electrons would strike only the area behind the slits, like the spray paint on the left in **Figure 6.**

But scientists found that the electrons formed an interference pattern typical of waves, as seen on the right in **Figure 6.** When waves pass through narrow slits, they interfere with each other. This experiment showed that electrons can behave like waves. It is now known that all particles can behave like waves. However, this does not mean that particles travel in wavy lines. Rather, this means that they display behavior, such as interference, that was once associated only with waves.

Check Your Progress

Summary

- An electromagnetic wave consists of a vibrating electric field and a vibrating magnetic field.
- Electromagnetic waves carry radiant energy.
- In empty space, electro-magnetic waves travel at 300,000 km/s, the speed of light.
- Electromagnetic waves travel more slowly in matter, with a speed that depends on the material.
- Electromagnetic waves can behave as particles that are called photons.

Demonstrate Understanding

1. **Infer** Would a vibrating proton produce an electromagnetic wave? Would a vibrating neutron? Explain.
2. **Compare** the frequency of an electromagnetic wave with the frequency of the vibrating charge that produces the wave.
3. **Describe** how electromagnetic waves transfer energy to matter.
4. **Explain** how an electromagnetic wave can travel through space that contains no matter.

Explain Your Thinking

5. **Explain** Would a stationary electron produce an electromagnetic wave? Would a stationary magnet? Explain.
6. **MATH Connection** How many minutes does it take an electromagnetic wave to travel from the Sun to Earth (150,000,000 km)?

LEARNSMART® Go online to follow your personalized learning path to review, practice, and reinforce your understanding.

LESSON 2

THE ELECTROMAGNETIC SPECTRUM

FOCUS QUESTION

Why can I only see part of the electromagnetic spectrum?

A Range of Frequencies

Electromagnetic waves have a wide variety of frequencies. They might vibrate once each second or trillions of times each second. The entire range of electromagnetic wave frequencies is called the electromagnetic spectrum. A spectrum is a continuous sequence arranged by a particular property. Each region of the electromagnetic spectrum has a specific name, as shown in **Figure 7.** Each region interacts with matter differently. The human eye detects only a small portion of the electromagnetic spectrum called visible light. Many technologies have been developed based on how waves interact with matter. For example, the antenna of your radio detects radio waves.

Radio Waves

Even though you cannot see them, radio waves are all around you. **Radio waves** are electromagnetic waves with wavelengths longer than 10 cm. Radio waves have long wavelengths and low frequencies, and their photons have low energies. Radio waves have many uses, including communications and medical imaging.

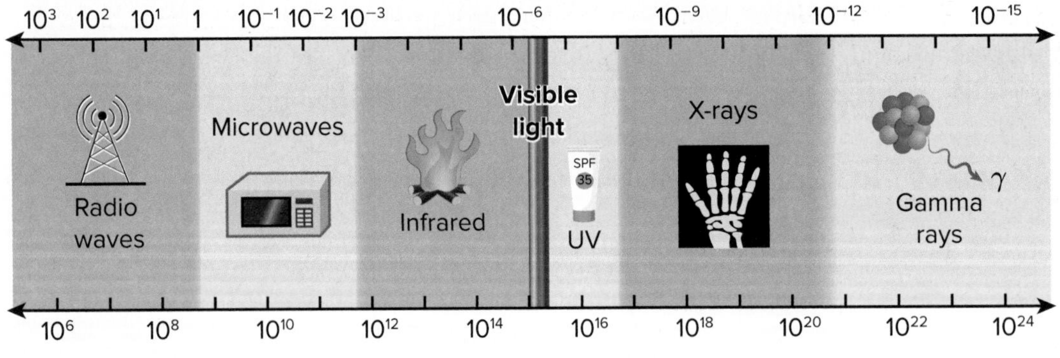

Figure 7 Each region of the electromagnetic spectrum spans a range of frequencies.

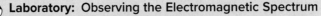

3D THINKING DCI Disciplinary Core Ideas CCC Crosscutting Concepts SEP Science & Engineering Practices

COLLECT EVIDENCE

Use your Science Journal to record the evidence you collect as you complete the readings and activities in this lesson.

INVESTIGATE

GO ONLINE to find these activities and more resources.

Applying Practices: Human Health and Radiation Frequency
HS-PS4-4. Evaluate the validity and reliability of claims in published materials of the effects that different frequencies of electromagnetic radiation have when absorbed by matter.

Laboratory: Observing the Electromagnetic Spectrum
Use mathematics and computational thinking to visualize the range of frequencies of light that interact with our eyes.

Audio transmission

Some radio waves carry an audio signal from a radio station to a radio. However, even though these radio waves carry information that a radio uses to create sound, you cannot hear radio waves. You hear sound when your radio changes the radio wave into a sound wave.

Radar

Another use for radio waves is to find the position and movement of objects by a method called radar. Radar stands for **RA**dio **D**etecting **A**nd **R**anging. With radar, radio waves are transmitted toward an object. By measuring the time required for the waves to bounce off the object and return to a receiving antenna, the location of the object can be found. Radar is used for tracking the movement of aircraft, watercraft, and spacecraft, as shown in **Figure 8.** Law enforcement officers also use radar to measure how fast a vehicle is moving.

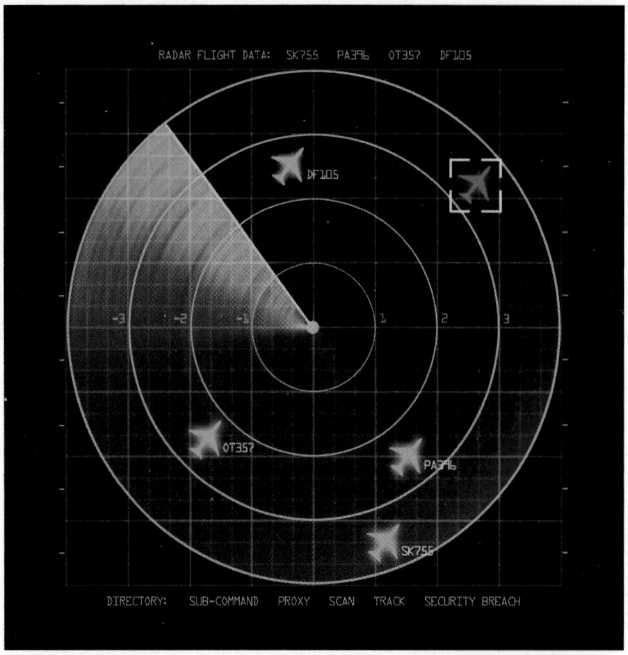

Figure 8 Radar uses radio waves to track airplanes and ships.

Magnetic resonance imaging (MRI)

In the early 1980s, medical researchers developed a technique called magnetic resonance imaging, which uses radio waves to help diagnose illness. The patient lies inside a large cylinder, like the one shown in **Figure 9.** The cylinder contains a powerful magnet, a radio wave emitter, and a radio wave detector.

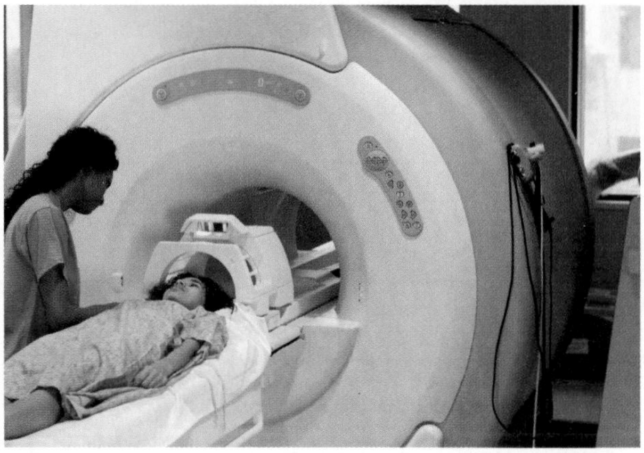

Figure 9 Magnetic resonance imaging is used to produce images of soft tissues.

Protons in hydrogen atoms in bones and soft tissue behave like magnets and align with the strong magnetic field created by the machine's magnet. Some of the protons absorb energy from the radio waves and flip their alignments. The amount of energy a proton absorbs and then re-emits depends on the type of tissue it is part of. A radio receiver detects this released energy. This information is then used to create a map of the different tissues. A picture of the inside of the patient's body is produced painlessly.

ACADEMIC VOCABULARY

transmit

to send from one place to another

My cell phone transmitted the message to my sister.

STEM CAREER Connection

Magnetic Resonance Imaging Technologist

If you are interested in technology in the medical field, you might explore a career as a magnetic resonance imaging (MRI) technologist. MRI technologists maintain and run MRI scanners to create medical images of patients. A clear image is an important tool for physicians to diagnose problems and help determine treatments.

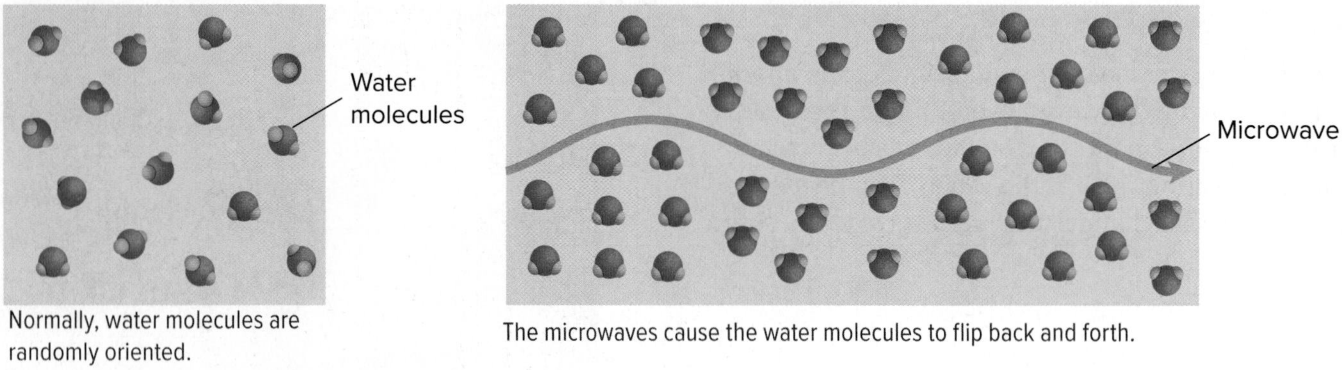

Water molecules
Microwave
Normally, water molecules are randomly oriented.
The microwaves cause the water molecules to flip back and forth.

Figure 10 Microwave ovens use electromagnetic waves to transfer energy to water particles in food.

Microwaves

Electromagnetic waves that have wavelengths between 0.1 mm and 30 cm are called **microwaves.** Microwaves with wavelengths of about 1 cm to 20 cm are widely used for communication, such as for cellular telephones and satellite signals. However, you are probably most familiar with microwaves because of their use in microwave ovens.

Get It?

Describe the differences between microwaves and radio waves.

Microwave ovens

In a microwave oven, microwaves interact with the water molecules in food, as shown in **Figure 10.** Each water molecule has a slight positive charge on one side and a slight negative charge on the other side, so it will align in an electric field. The vibrating electric field inside a microwave oven causes water molecules in food to rotate back and forth billions of times each second. This rotation causes a type of friction between water molecules that generates thermal energy. The thermal energy produced by the water molecules' interactions causes food to cook.

Get It?

Describe the steps by which a microwave oven heats food.

Foods with plenty of water cook well in a microwave oven. Frozen water, however, cannot be warmed using microwaves because the water molecules are bound in a crystallized structure and cannot rotate. On many microwave ovens, there is a special defrost setting. This setting heats the partly melted water on the surface of the food. The inside of the food is then warmed by conduction until all the water is liquid again.

Infrared Waves

Most of the warm air in a fireplace moves up the chimney, yet you feel the warmth of the blazing fire when you stand in front of a fireplace. Why do you feel the warmth? The warmth you feel is thermal energy transmitted to you by infrared waves. **Infrared waves** are electromagnetic waves with wavelengths between about one-thousandth of a meter and about 700-billionths of a meter.

Figure 11 Infrared images of homes and other buildings can provide information about the structure's energy efficiency.

Identify *where energy is escaping from the house.*

Using infrared waves

Every object emits infrared waves. Hotter objects emit more infrared waves than do cooler objects. Infrared detectors can form images of objects from the infrared waves they emit. These images, like the one in **Figure 11,** can help determine how energy-efficient a structure is. Other devices that use infrared waves include television remote controls and CD-ROM drives.

Visible Light

Visible light is the range of electromagnetic waves that you detect with your eyes. Visible light differs from radio waves, microwaves, and infrared waves only by its frequency and wavelength. Visible light has wavelengths around 700-billionths to 400-billionths of a meter.

Color is the brain's interpretation of the wavelengths of the light absorbed by substances in the eye. These colors range from short-wavelength violet to long-wavelength red, as illustrated in **Figure 12.** If all colors of light are present in the same place, you see the light as white.

When light or longer wavelength electromagnetic radiation is absorbed in matter, it is generally converted into thermal energy (heat). As you will see, shorter wavelength electromagnetic radiation (ultraviolet waves, X-rays, gamma rays) can ionize atoms and cause damage to living cells.

Get It?

Compare the effects on matter caused by the absorption of electromagnetic radiation of different wavelengths.

Red (7.00×10^{-7} m) Violet (4.00×10^{-7} m)

Figure 12 This spectrum of visible light shows all colors of light from long-wavelength red to short-wavelength violet.

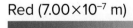

Ultraviolet Waves

Ultraviolet waves are electromagnetic waves with wavelengths from about 400-billionths to 10-billionths of a meter. Ultraviolet waves (UV waves) can enter cells, making ultraviolet waves both useful and harmful.

Useful UVs

You probably know that UV waves cause sunburn. But some exposure to ultraviolet waves is healthy. Ultraviolet waves striking the skin enable your body to make vitamin D, which is needed for healthy bones and teeth.

Ultraviolet waves are also used to disinfect food, water, and medical supplies, as shown in **Figure 13.** When ultraviolet light enters a cell, it damages protein and DNA. For some single-celled organisms, such as bacteria, this damage can mean death.

Ultraviolet waves make some materials fluoresce (floo RES). Materials that fluoresce absorb ultraviolet waves and reemit the energy as visible light. Police detectives sometimes use fluorescent powder to reveal fingerprints.

Harmful UVs

When you spend time in the Sun, you might wear sunscreen to prevent sunburn. Most of the UV waves that reach Earth's surface are longer-wavelength UVA rays. The shorter-wavelength UVB rays are the primary cause of sunburn and skin cancers, but UVA rays contribute to skin cancers and skin damage, such as wrinkling.

The ozone layer About 20 to 50 km above Earth's surface is a region called the ozone layer. Ozone is a molecule composed of three oxygen atoms. The ozone layer is vital to life on Earth because it absorbs most of the Sun's harmful ultraviolet waves and prevents them from reaching Earth's surface, as shown in **Figure 14.**

Figure 13 Ultraviolet light is used to kill bacteria in the water supply to make it safe for drinking. This purifier also gives off visible blue light to let the user know that the device is working.

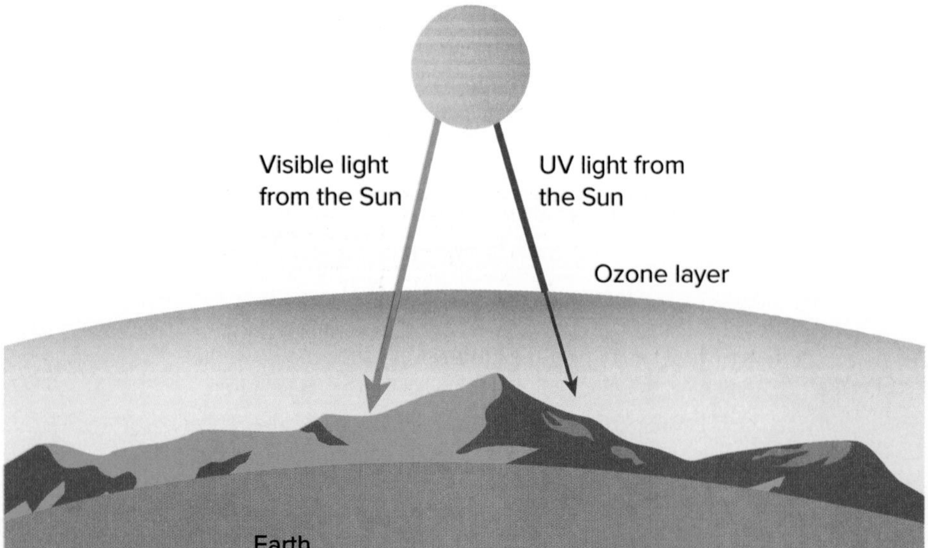

Figure 14 The ozone layer absorbs most of the Sun's UV waves. As a result, few UV waves reach Earth's surface, where they are harmful to organisms.

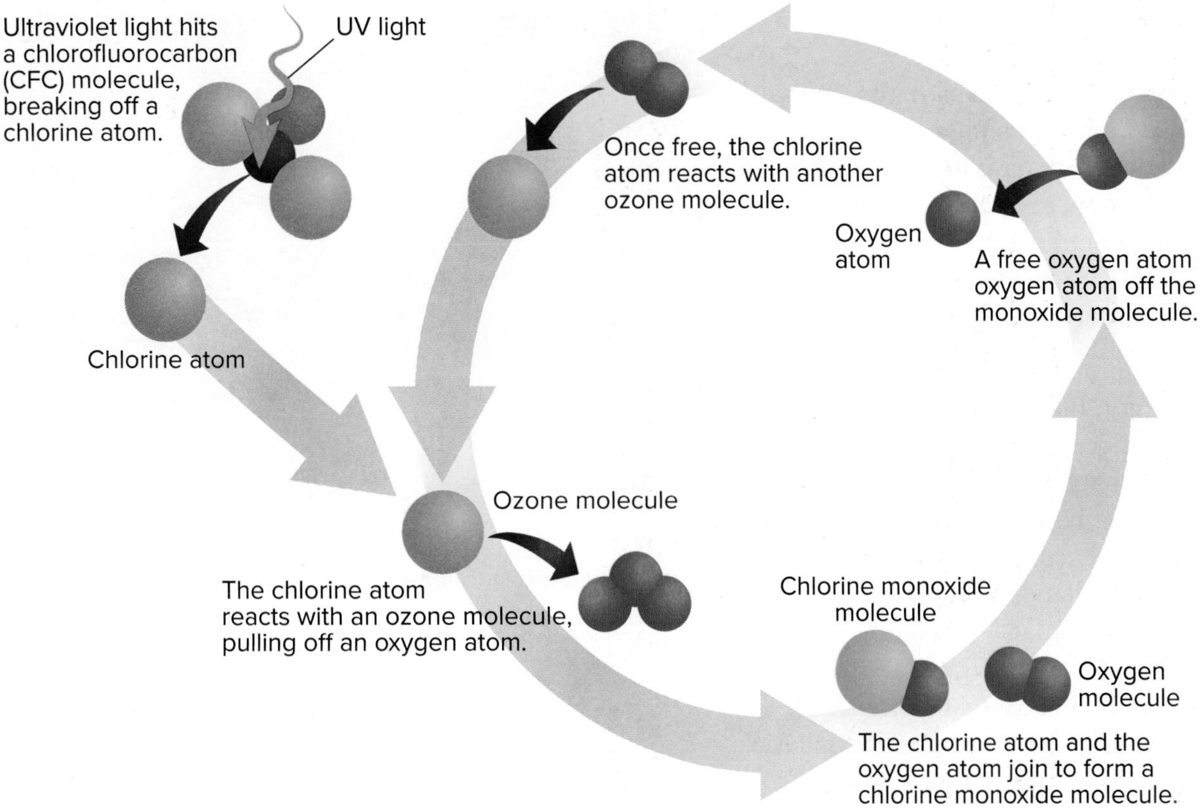

Ultraviolet light hits a chlorofluorocarbon (CFC) molecule, breaking off a chlorine atom.
UV light
Chlorine atom
Once free, the chlorine atom reacts with another ozone molecule.
Oxygen atom
A free oxygen atom pulls the oxygen atom off the chlorine monoxide molecule.
Ozone molecule
The chlorine atom reacts with an ozone molecule, pulling off an oxygen atom.
Chlorine monoxide molecule
Oxygen molecule
The chlorine atom and the oxygen atom join to form a chlorine monoxide molecule.

Figure 15 Ultraviolet radiation breaks apart CFCs, which produces single chlorine atoms. The highly-reactive chlorine atoms destroy ozone molecules.

Damage to the ozone layer The ozone layer changes naturally with the seasons. But in the 1980s and early 1990s, scientists noticed an overall decrease in the amount of ozone in the ozone layer. Averaged globally, the decrease is about 4 percent, but it is greater at higher latitudes. Many scientists think that certain chemicals, such as chlorofluorocarbons (CFCs), caused the reduction of ozone in the ozone layer.

CFCs were widely used in air conditioners, refrigerators, and cleaning fluids, and some were released into the air. CFCs can rise over time into the upper regions of Earth's atmosphere. When CFCs reach the ozone layer, they might react with ozone molecules, as illustrated in **Figure 15.** One chlorine atom from a CFC molecule can break apart thousands of ozone molecules. To prevent the loss of more ozone, many countries worked together to reduce the use of CFCs and other ozone-depleting substances.

X-rays

Electromagnetic waves with wavelengths between about ten-billionths of a meter and ten-trillionths of a meter are called **X-rays.** X-rays have shorter wavelengths than UV waves, and their photons have larger energies. X-rays penetrate skin and soft tissue but not denser materials, such as teeth and bones. Doctors and dentists use low doses of X-rays to form images, like the one in **Figure 16,** of bones and teeth. X-rays are also used in airport screening devices to examine the contents of luggage.

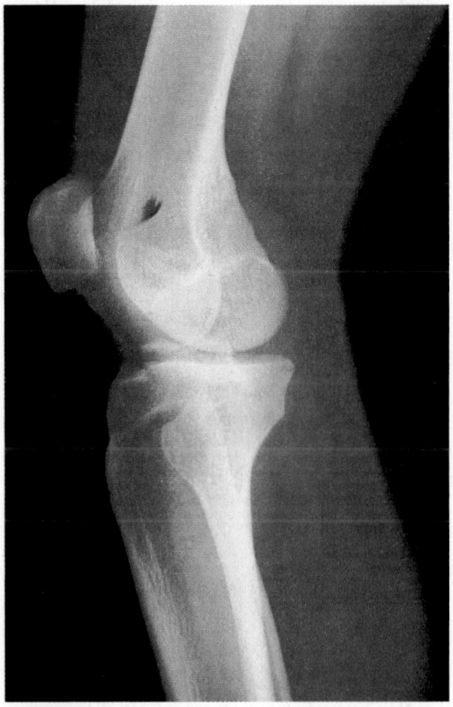

Figure 16 X-rays pass through soft tissue, such as skin and muscle, but they are absorbed by the denser bones. The image of a bone on an X-ray is the shadow cast by the bone.

Gamma Rays

Electromagnetic waves with wavelengths shorter than about 100-trillionths of a meter are called **gamma rays.** Gamma rays have high frequencies and have the highest-energy photons. They have enough energy to penetrate several centimeters of lead. Gamma rays are produced by processes that occur in the nuclei of atoms.

Both X-rays and gamma rays are used in a technique called radiation therapy to kill diseased cells in the human body. A beam of X-rays or gamma rays can damage the biological molecules in living cells, causing both healthy and diseased cells to die. By carefully controlling the amount of X-ray or gamma ray radiation and focusing it on the diseased area, the damage to healthy cells can be reduced during treatment. **Figure 17** shows a patient receiving radiation to treat cancer. The gamma rays are focused on the tumor and kill the cancer cells, while doing little damage to the surrounding healthy cells.

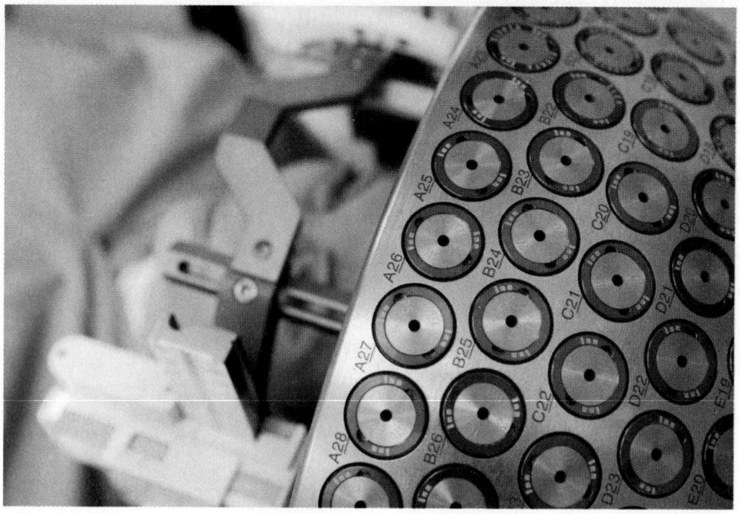

Figure 17 This patient is undergoing radiation therapy. Cancer cells can be killed with carefully controlled beams of X-rays and gamma rays.

Check Your Progress

Summary

- The entire range of frequencies of electromagnetic waves is called the electromagnetic spectrum.
- Radio waves and microwaves have the longest wavelengths.
- All objects emit infrared waves.
- The human eye can detect visible light.
- Ultraviolet waves, X-rays, and gamma rays are both helpful and harmful to humans.

Demonstrate Understanding

7. **Compare and contrast** the properties and uses of radio waves, infrared waves, and ultraviolet waves.
8. **Explain** A mug of tea is heated in a microwave oven. Explain why the tea gets hotter than the mug.
9. **Identify** the beneficial effects and the harmful effects of human exposure to ultraviolet waves.
10. **Name** three household objects that produce electromagnetic waves, and describe how the electromagnetic waves are used.

Explain Your Thinking

11. **Explain** How could infrared imaging be used to find a lost hiker?
12. **MATH Connection** Use scientific notation to express the range of wavelengths corresponding to visible light, ultraviolet waves, and X-rays.
13. **MATH Connection** A nanometer, abbreviated nm, equals one-billionth of a meter, or 10^{-9} m. Express the range of wavelengths corresponding to visible light, ultraviolet waves, and X-rays in nanometers.

LEARNSMART® Go online to follow your personalized learning path to review, practice, and reinforce your understanding.

LESSON 3

RADIO COMMUNICATION

FOCUS QUESTION

Why do I lose cell signal when I run the microwave oven?

Radio Transmissions

Radio is a technology based on how waves interact with matter that is part of everyday experience in the modern world. Each radio station uses an assigned frequency to avoid interfering with other radio broadcasts. Television stations and cell phone companies are also assigned specific frequencies. These frequency ranges are shown in **Figure 18.** The remaining radio frequencies are assigned for other purposes, such as navigation and radio astronomy. Changing the channel on your radio or television allows you to select a particular frequency carrying the information you want to listen to or watch. An electromagnetic wave with the specific frequency that a station is assigned is called a **carrier wave.**

Modulation

The station must do more than simply transmit a carrier wave. It must also send information about the sounds that you are to receive. The sounds produced at the radio station are converted into electric signals. This electric signal is called the signal wave and is used to modify the carrier wave. The process of adding the signal wave to the carrier wave is called **modulation.** There are two ways to modulate carrier waves: amplitude modulation (AM) and frequency modulation (FM).

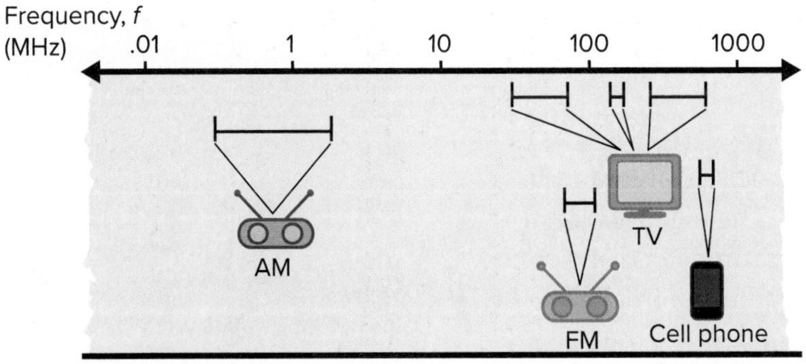

Figure 18 To avoid interference, cell phones, TVs, and radios broadcast at assigned frequencies between 500,000 Hz and 1 billion Hz.

3D THINKING — DCI Disciplinary Core Ideas — CCC Crosscutting Concepts — SEP Science & Engineering Practices

COLLECT EVIDENCE

Use your Science Journal to record the evidence you collect as you complete the readings and activities in this lesson.

INVESTIGATE

GO ONLINE to find these activities and more resources.

Applying Practices: Catching Waves

HS-PS4-5. Communicate technical information about how some technological devices use the principles of wave behavior and wave interactions with matter to transmit and capture information and energy.

Applying Practices: Digital Storage and Transmission of Information

HS-PS4-2. Evaluate questions about the advantages of using a digital transmission and storage of information.

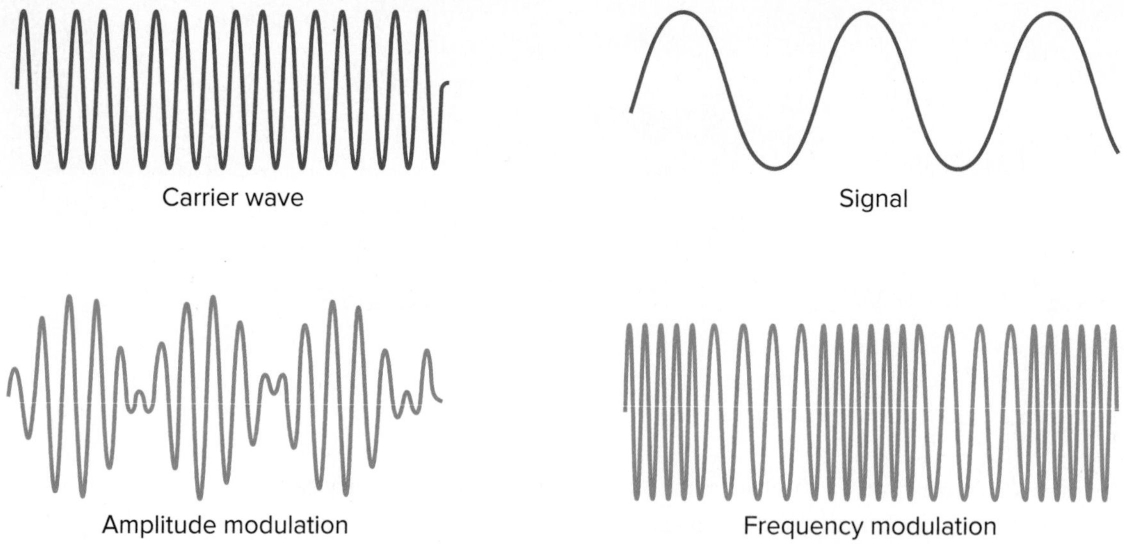

Figure 19 The carrier wave is a wave with the frequency assigned to the station. The signal wave contains information about what you will hear. Modulation is the process of adding the signal wave to the carrier wave. A carrier wave can be modulated with amplitude modulation (AM) or frequency modulation (FM).

AM radio An AM radio station broadcasts information by varying the amplitude of the carrier wave, as shown at the left in **Figure 19.** AM carrier wave frequencies range from 540,000 to 1,600,000 Hz.

FM radio In FM radio signals, the signal wave is used to vary the frequency of the carrier wave, as shown at the right in **Figure 19.** Because the strength of the FM waves is kept fixed, FM signals tend to be clearer than AM signals. FM carrier frequencies range from 88 million to 108 million Hz. These frequencies are much higher than AM frequencies.

Get It?

Compare and contrast AM and FM radio signals.

Broadcasting radio waves

The modified carrier wave is converted from an electric signal to a radio wave by using an antenna, like the one in **Figure 20.** The electric signal causes electrons in the antenna to vibrate. These vibrating electrons create electromagnetic waves that travel outward from the antenna in all directions.

The signal from the radio station is strongest closer to the broadcasting antenna and becomes weaker as you move away. Eventually, the signal will be too weak to be detected by your radio. This is why radios in New York City do not pick up FM radio stations broadcast in Los Angeles. Bad weather, surrounding mountains, and artificial structures can also interfere with radio transmissions.

Get It?

Describe how a radio signal's strength changes as you move away from the tower.

Figure 20 In a broadcast antenna, vibrating electrons produce an electromagnetic wave that travels outward in all directions.

Ingram Publishing/SuperStock

Receiving radio waves

As electromagnetic waves pass by your radio's antenna, the electrons in the metal vibrate, as illustrated in **Figure 21.** These vibrating electrons produce a changing electric current that contains the information about the music and words. This current is used to make the speakers vibrate, creating the sound waves that you hear.

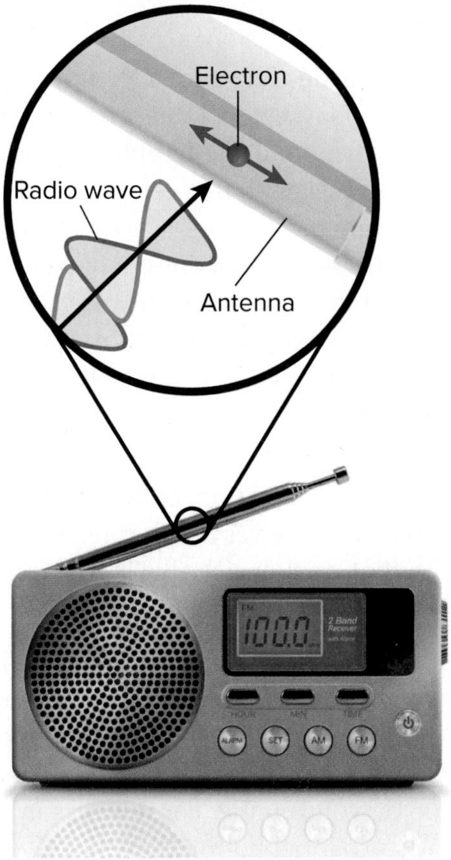

Figure 21 Radio waves exert a force on the electrons in a receiving antenna, causing the electrons to vibrate. The radio then filters out the carrier wave and converts the signal to a sound wave for listeners to hear.

The Digital Revolution

Until the early 21st century, information was sent to TV sets in the same way it was sent to radios. The audio information was sent using FM, and the visual information was sent using AM. The information signals were analog signals. **Analog signals** are electric signals whose values change smoothly over time.

In 2009, full-power television stations in the United States began broadcasting only digital signals. A **digital signal** is an electric signal for which there are only two possible values: ON and OFF. This is similar to a light switch for which the light can be on or off, but it cannot be half-on or half-off.

There are many ways to modulate radio waves using this on-and-off information. The simplest methods, however, resemble traditional AM and FM and are called Amplitude-Shift Keying (ASK) and Frequency-Shift Keying (FSK). These types of digital modulation are shown in **Figure 22.** More complex ways of digital modulation allow more information to be carried by a single wave. In the United States, television stations use multiple amplitude modulations to encode data on the carrier wave.

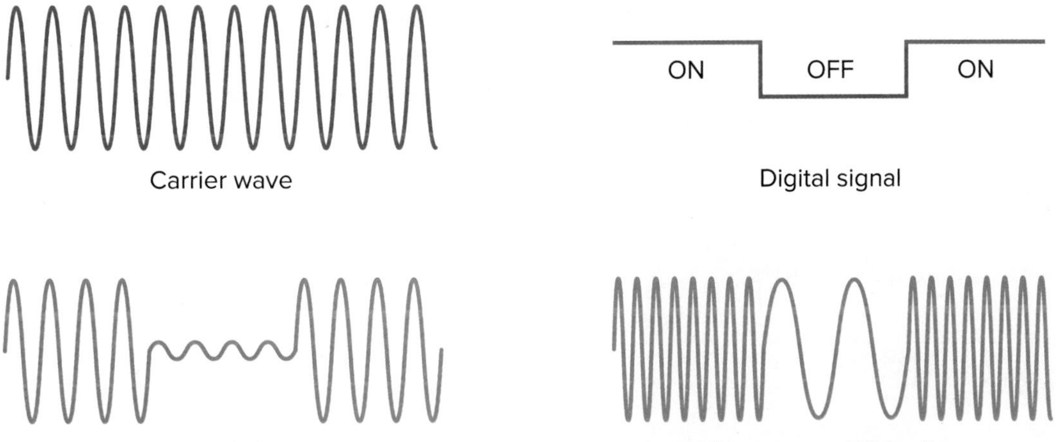

Figure 22 A digital signal can be used to modulate the amplitude or frequency of the carrier wave.
Compare and contrast *amplitude-shift keying with amplitude modulation.*

Telephones

Just a few decades ago, telephones had to be connected with wires. Today, cell phones are seen everywhere. When you speak into a telephone, a microphone converts the sound waves into an electric signal. In a cell phone, this signal is transmitted to and from microwave towers using microwaves or radio waves. The towers, like the ones in **Figure 23,** are several kilometers apart, and each covers an area called a cell. If you move from one cell to another, an automated control station transfers the signal to the new cell and its tower.

Figure 23 The antennae on these microwave towers are transceivers that send signals to and receive signals from nearby cell phones.

Transceivers

A cell phone is a transceiver. A **transceiver** transmits one radio signal and receives another radio signal. Using two signals with different frequencies allows you to talk and listen at the same time without interference.

Cordless telephones are also transceivers. However, you must remain close to the base unit when using a cordless phone. Another drawback is that if someone nearby is using a cordless telephone at the same frequency, you could hear that conversation on your phone. For this reason, many cordless phones have a channel button that allows you to switch to another frequency.

Pagers

Some hospitals ban cell phone use because there are concerns that transceivers might interfere with medical equipment. So, many doctors carry small, portable radio receivers called pagers. To contact the doctor, a caller leaves a callback number or a text message at a central terminal. The message is changed into an electronic signal and transmitted by radio waves along with the identification number of the desired pager. The pager receives all messages transmitted at its assigned frequency, but it only responds to messages with its identification number. Restaurants also use pagers, like the one in **Figure 24,** to notify customers that their tables are ready.

Figure 24 Each pager at a restaurant responds to its own assigned frequency to alert customers that a table is ready.

(t)Russell Illig/Photodisc/Getty Images, (b)Matt Meadows/McGraw-Hill Education

Figure 25 Communications satellites are transceivers. They receive signals at one frequency and send signals at a different frequency. The solar panels on either side of the satellite allow the satellite to obtain energy from the Sun.

Communications Satellites

Since satellites were first developed, thousands have been launched into orbit around Earth. Many of these, like the one in **Figure 25,** are used for communication. The sender broadcasts a microwave signal to the satellite. The satellite receives the signal, amplifies it, and transmits it to a particular region on Earth. Like cell phones, satellites are transceivers. To avoid interference, the satellite receives signals at one frequency and broadcasts signals at a different frequency.

Satellite telephone systems

Some mobile telephones can be used when sailing across the ocean, even though there are no nearby cell phone towers. The telephone transmits the signal directly to a satellite. The satellite relays the signal to a ground station, and the call is passed on to the telephone network. Satellite links work well for one-way transmissions, but two-way communications can have a delay caused by the large distance the signals travel to and from the satellite.

Television satellites

The satellite-reception dishes that you sometimes see in yards or attached to houses are receivers for television satellite signals. Satellite television is used as an alternative to ground-based transmission. Communications satellites use microwaves rather than the radio waves used for normal television broadcasts. Microwaves have shorter wavelengths and travel more easily through the atmosphere. The ground receivers are dish-shaped to help focus the microwaves onto an antenna.

WORD ORIGIN

satellite

comes from the Latin word *satelles*, which means "attendant"

The Moon is a satellite that travels around Earth.

The Global Positioning System

Getting lost while hiking is not uncommon, but if you are carrying a Global Positioning System receiver, it is much less likely to happen. The **Global Positioning System (GPS)** is a system of satellites, ground monitoring stations, and receivers that determine your exact location on or above Earth's surface. The 24 satellites necessary for 24-hour, around-the-world coverage became fully operational in 1995. **Figure 26** illustrates how these satellites are arranged in orbit. Signals from four satellites are used to determine the location of an object using a GPS receiver.

GPS satellites are owned and operated by the United States Department of Defense, but the microwave signals they send out can be used by anyone. Several other countries are working to develop similar systems. Airplanes, ships, cars, and cell phones can use GPS for navigation. Some pet collars also contain GPS receivers. If the pet runs away or is lost, the GPS receiver in the collar can be used to locate the animal.

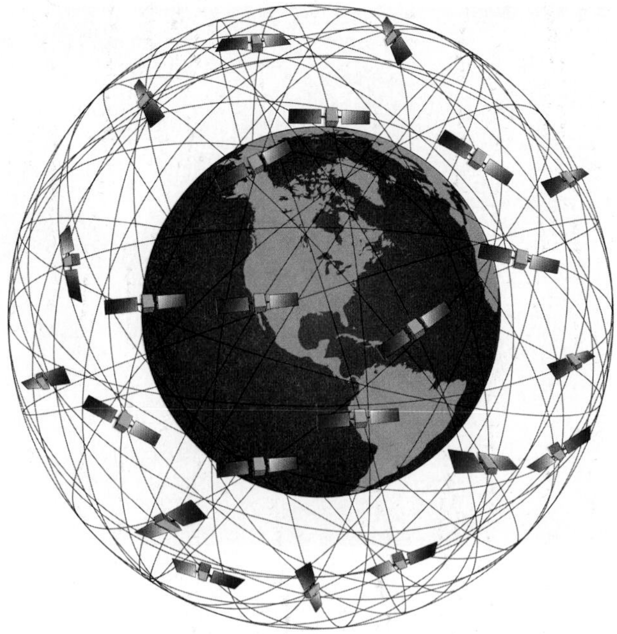

Figure 26 A GPS receiver uses signals from four of the 24 orbiting satellites to determine the receiver's location.

Check Your Progress

Summary

- Radio stations transmit radio waves that receivers convert to sound waves.
- Modulation is the process of adding an information signal to a carrier wave.
- Telephones contain transceivers and convert sound waves into electric signals and electric signals into sound waves.
- Microwave towers and satellites are used to transmit telephone signals.
- The Global Positioning System uses a system of satellites to determine one's exact position.

Demonstrate Understanding

14. **Identify and describe** the steps that a radio station uses to broadcast sounds to your radio receiver.
15. **Explain** the difference between AM and FM radio. Make a sketch of how a carrier wave is modulated in AM and FM radio signals.
16. **Describe** what happens to your signal when you are talking on a cell phone and you travel from one cell to another cell.
17. **Describe** some of the uses of the Global Positioning System. Why might emergency vehicles be equipped with GPS receivers?

Explain Your Thinking

18. **Explain** Why do cordless telephones stop working when you move too far from the base unit?
19. **MATH Connection** A TV screen is composed of many points of light called pixels. A standard TV has 460 pixels horizontally and 360 pixels vertically. A high-definition TV has 1920 horizontal and 1080 vertical pixels. What is the ratio of the number of pixels in a high-definition TV to the number in a standard TV?

LEARNSMART® Go online to follow your personalized learning path to review, practice, and reinforce your understanding.

A New Kind of Ray

Have you ever broken a bone? Visited the dentist? If so, you were exposed to a powerful, invisible form of electromagnetic radiation. Well-understood and widely applied today, this phenomenon was once so mysterious that it was simply called "X"—a mathematical symbol representing something unknown. The discovery of X-rays shocked and inspired scientists around the world.

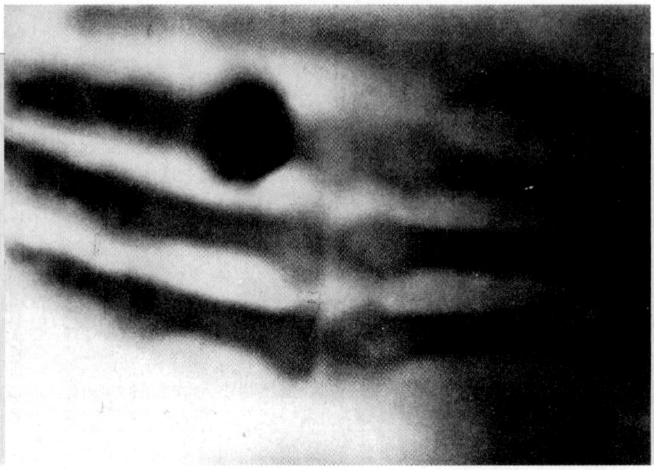

The image of Bertha Roentgen's hand clearly shows the shadows cast by her bones and rings as the X-rays traveled through her flesh.

A ghostly, green glow

On a November afternoon in 1895, German physicist William Roentgen was studying the effects of passing electric current through gas in a device called a cathode ray tube. To eliminate light that would interfere with his experiment, Roentgen covered the tube with thick black paper. Roentgen observed something he could not explain: a screen, which was coated with chemicals, was shimmering with a green glow.

Why did the chemical-coated screen glow? Roentgen knew it was impossible for light produced by the cathode ray tube to penetrate the heavy paper and travel the distance to the screen. He concluded that some unknown, invisible ray—quickly dubbed "X"—had penetrated the glass walls of the gas tube and the heavy black paper, crossed the air, and contacted the screen, causing it to glow.

Famous fingers

Roentgen conducted weeks of experiments and was amazed to find that the X-rays could pass through wood, skin, most metal objects, and even the thick walls of his laboratory. In late December, Roentgen recorded an X-ray image, shown above, of his wife Bertha's hand. In December 1895, Roentgen published his findings in a paper entitled "On a New Kind of Rays." Scientists worldwide quickly and enthusiastically duplicated, expanded upon, and applied Roentgen's findings.

A few months later, Eddie McCarthy fell and injured his wrist. Physicians recognized the potential of X-rays to study internal injuries and used X-rays to photograph the fracture. The medical application of X-ray technology was born.

COMMUNICATE SCIENTIFIC INFORMATION

Research ways that X-ray technology is applied in the medical field. Create a spreadsheet showing several specific uses of the technology, the branch or branches of medicine to which each applies, and a description of the benefits of using X-ray technology in each setting.

MODULE 11

STUDY GUIDE

GO ONLINE to study with your Science Notebook.

Lesson 1 WHAT ARE ELECTROMAGNETIC WAVES?

- An electromagnetic wave consists of a vibrating electric field and a vibrating magnetic field.
- Electromagnetic waves carry radiant energy.
- In empty space, electromagnetic waves travel at 300,000 km/s, the speed of light.
- Electromagnetic waves travel more slowly in matter, with a speed that depends on the material.
- Electromagnetic waves can behave as particles called photons.

- electromagnetic wave
- photon

Lesson 2 THE ELECTROMAGNETIC SPECTRUM

- The entire range of frequencies of electromagnetic waves is called the electromagnetic spectrum.
- Radio waves and microwaves have the longest wavelengths.
- All objects emit infrared waves.
- The human eye can detect visible light.
- Ultraviolet waves, X-rays, and gamma rays are both helpful and harmful to humans.

- radio wave
- microwave
- infrared wave
- visible light
- ultraviolet wave
- X-ray
- gamma ray

Lesson 3 RADIO COMMUNICATION

- Radio stations transmit radio waves that receivers convert to sound waves.
- Modulation is the process of adding an information signal to a carrier wave.
- Telephones contain transceivers and convert sound waves into electric signals and electric signals into sound waves.
- Microwave towers and satellites are used to transmit telephone signals.
- The Global Positioning System uses a system of satellites to determine one's exact position.

- carrier wave
- modulation
- analog signal
- digital signal
- transceiver
- Global Positioning System (GPS)

THREE-DIMENSIONAL THINKING
Module Wrap-Up

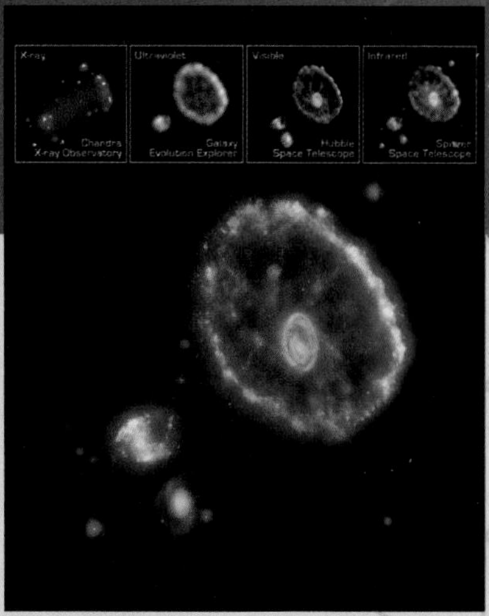

REVISIT THE PHENOMENON

How do we see x-rays?

CER Claim, Evidence, Reasoning

Explain Your Reasoning Revisit the claim you made when you encountered the phenomenon. Summarize the evidence you gathered from your investigations and research and finalize your Summary Table. Does your evidence support your claim? If not, revise your claim. Explain why your evidence supports your claim.

STEM UNIT PROJECT

Now that you've completed the module, revisit your STEM unit project. You will summarize your evidence and apply it to the project.

GO FURTHER

SEP Data Analysis Lab

Radioactive Glass?

Did you know that some of the world's most beautiful glass is radioactive? Around 1880, a very popular glass was made with uranium. Uranium is a radioactive element that gives off gamma rays in high amounts. You might wonder why people would ever use uranium to make glass, but keep in mind that people did not know about the dangers of radioactivity back in the 1800s.

Uranium, in its natural state, is an attractive green-yellow color and was added to glass to give it color. The presence of uranium causes the glass to glow under an ultraviolet lamp, or "black light" as it is called today. The glow is a distinctive shade of green. Antiques dealers test for this glass by seeing if it glows under ultraviolet light.

There are stories that the factories making this glass were shut down in the mid-1900s because workers became ill and died from exposure to uranium. However, this was not entirely true. The outbreak of World War II stopped the glass production, as uranium was used to make atomic bombs. Uranium became very valuable, and governments were afraid it would fall into the wrong hands. It became illegal to make uranium glass.

CER Analyze and Interpret Data

1. **Claim, Evidence, Reasoning** What radioactive chemical is used to give a special color to the glass described?
2. **Claim, Evidence, Reasoning** What property of the chemical makes it dangerous?
3. **Claim, Evidence, Reasoning** If you wanted to test whether a piece of glass was really uranium glass, how would you do it? Explain your reasoning.
4. Why was it illegal to make uranium glass for a while?

JPL-Caltech/P. N. Appleton/NASA

Asharkyu/Shutterstock

LIGHT

ENCOUNTER THE PHENOMENON

How does light transmit information?

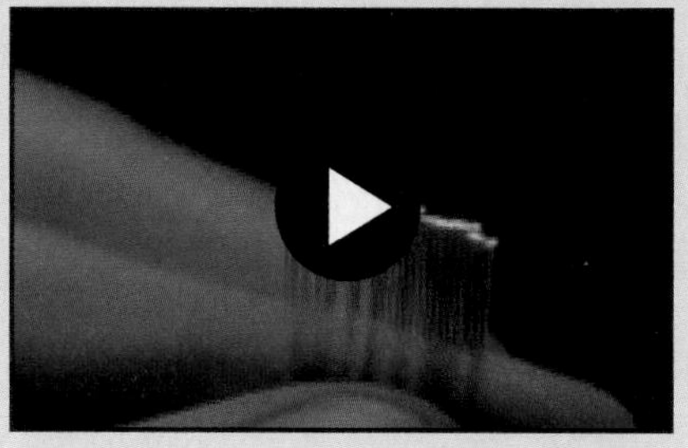

GO ONLINE to play a video about how fiber optic cables work.

SEP Ask Questions

Do you have other questions about the phenomenon? If so, add them to the driving question board.

CER Claim, Evidence, Reasoning

Make Your Claim Use your CER chart to make a claim about how light can transmit information. Explain your reasoning.

Collect Evidence Use the lessons in this module to collect evidence to support your claim. Record your evidence as you move through the module.

Explain Your Reasoning You will revisit your claim and explain your reasoning at the end of the module.

GO ONLINE to access your CER chart and explore resources that can help you collect evidence.

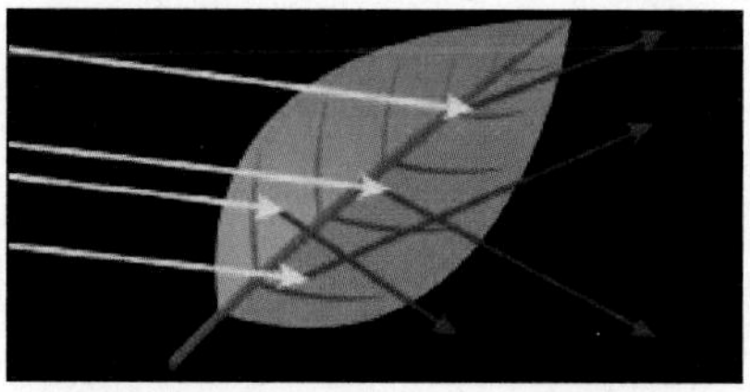

LESSON 2: Explore & Explain: Light and Color

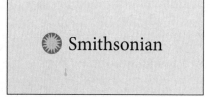

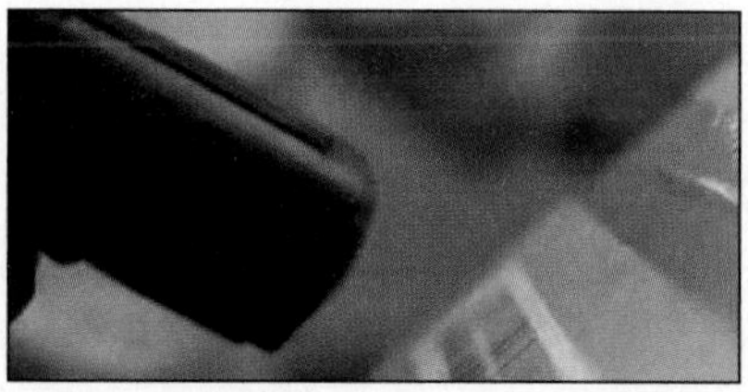

LESSON 4: Explore & Explain: Using Light

Smithsonian

Additional Resources

LESSON 1

THE BEHAVIOR OF LIGHT

FOCUS QUESTION

How are rainbows and mirages formed?

Light and Matter

After your eyes adjust to a dark room, you will see that brightly colored objects look gray or black in the dim light. With the lights on, however, you will see all the objects in the room, including their colors. What you see depends on the amount of light in the room and the color of the objects. To see an object, it must reflect some light into to your eyes.

Opaque, translucent, and transparent

Objects can absorb, reflect, and transmit light. Objects that transmit light allow light to pass through them. An object's material determines the amount of light it absorbs, reflects, and transmits. The material in the candleholder on the right in **Figure 1** is opaque (oh PAYK). **Opaque** materials only absorb and reflect light; no light passes through them. As a result, you cannot see the candle inside.

Figure 1 Different materials interact differently with light. Materials can absorb, reflect, scatter, and transmit light.

3D THINKING DCI Disciplinary Core Ideas CCC Crosscutting Concepts SEP Science & Engineering Practices

COLLECT EVIDENCE

Use your Science Journal to record the evidence you collect as you complete the readings and activities in this lesson.

INVESTIGATE

GO ONLINE to find these activities and more resources.

Quick Investigation: Observe Refraction in Water

Carry out an investigation to determine how light changes direction as it travels from air to water.

Laboratory: Producing a Spectrum

Carry out an investigation to determine the effects of sending light through a prism.

Some materials, such as the candleholder in the middle in **Figure 1,** are translucent (trans LEW sunt). **Translucent** materials transmit light but also scatter it. You cannot see clearly through translucent materials, and objects appear blurry.

The candleholder on the left in **Figure 1** is transparent. **Transparent** materials transmit light without scattering it, so you can see objects clearly through them.

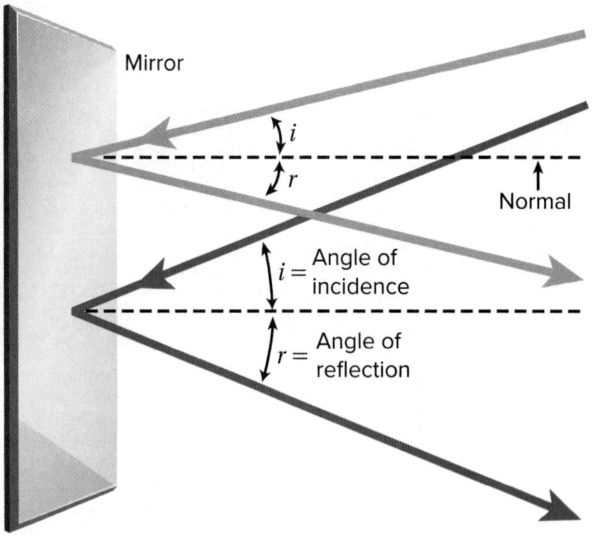

Figure 2 The law of reflection says that when light is reflected, the angle of incidence equals the angle of reflection.

Reflection of Light

Before you left for school this morning, did you look in a mirror to check your appearance? For you to see your reflection in the mirror, light had to reflect off you, hit the mirror, and reflect off the mirror and into your eye. Reflection occurs when a light wave strikes an object and bounces off.

The law of reflection

Like all waves, light obeys the law of reflection. According to the law of reflection, the angle at which a light wave strikes a surface is the same as the angle at which it is reflected. This law is illustrated in **Figure 2.** Light reflected from any surface—a mirror or a sheet of paper—behaves this way.

Regular and diffuse reflection

If light always obeys the law of reflection, why can you see your reflection in a store window but not in a brick wall? The answer involves the smoothness of the surfaces. A smooth, even surface like a pane of glass produces a sharp image by reflecting parallel light waves in only one direction. Reflection of light waves from a smooth surface is regular reflection. A brick wall has an uneven surface that causes incoming parallel light waves to be reflected in many directions, as shown in **Figure 3.** The reflection of light from a rough surface is diffuse reflection. Diffuse reflection does not produce an image.

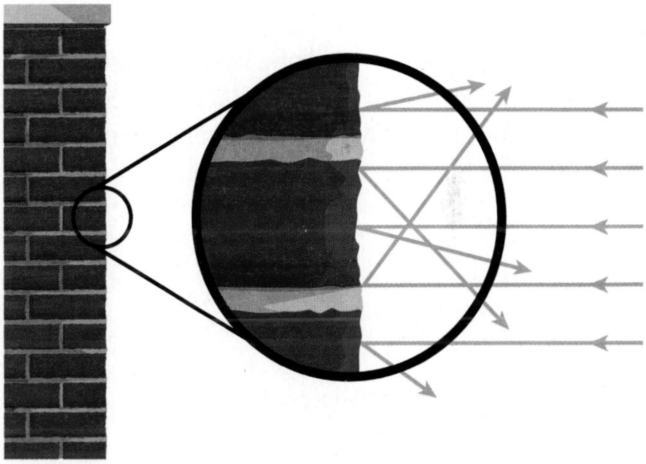

Figure 3 The uneven surface of this brick wall produces a diffuse reflection.

Explain *Use the law of reflection to explain why a rough surface causes parallel light waves to be reflected in many directions.*

Get It?
Identify some objects that produce regular reflections and some objects that produce diffuse reflections.

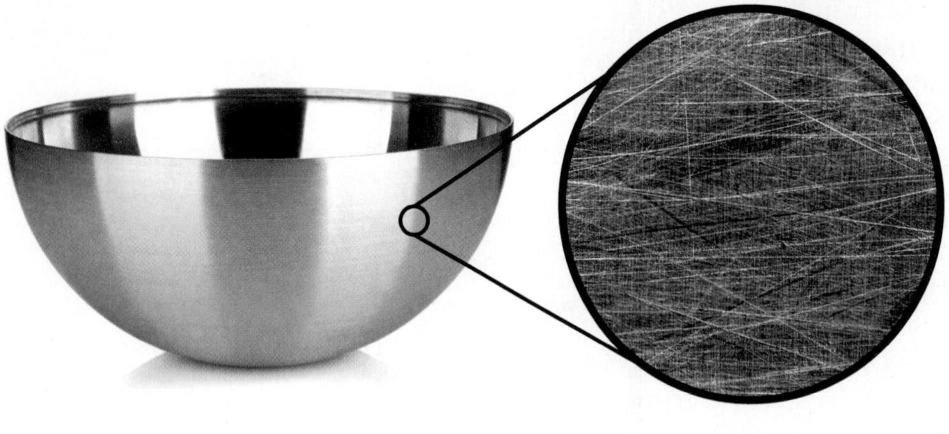

Figure 4 Surfaces that appear smooth to the human eye are seen to be rough at high magnification. These surfaces produce diffuse reflections.

Microscopic roughness

Even a surface that appears to be smooth can be rough enough to cause diffuse reflection. For example, a metal bowl might seem to be smooth, but the surface shows rough spots at high magnification, as shown in **Figure 4.** To cause a regular reflection, the sizes of surface irregularities must be less than the wavelengths of the light that the surface reflects.

Refraction of Light

What occurs when a light wave passes from one material to another—from air to water, for example? Light rays refract, or bend. Refraction is caused by a change in the speed of a wave when it passes from one material to another. If the light wave is traveling at an angle other than 90° to the boundary between the materials, and the speed that light travels is different in the two materials, then the wave will bend. If light hits a boundary at 90°, it will change speed but not bend.

Get It?
Identify when refraction occurs.

The index of refraction

The amount of bending depends on the speed of light in each material. The greater the difference in speeds, the more the light is bent as it crosses the boundary at an angle. Every material has an index of refraction. The **index of refraction** is a property of a material that indicates how much the speed of light in the material is reduced compared to the speed of light in a vacuum. In most materials, the index of refraction also depends on the light's wavelength. Longer wavelengths have smaller indices of refraction.

The larger the index of refraction, the slower the speed of light will be in the material. For example, because glass has a larger index of refraction than air, light moves more slowly in glass than in air. Many useful devices, such as eyeglasses, binoculars, cameras, and microscopes, form images by using glass lenses to refract light.

Figure 5 When white light enters a prism, each color of light is bent a different amount. As a result, the white light is split into the different colors of light.

Prisms

A sparkling glass prism hangs in a sunny window, refracting the sunlight and projecting a colorful pattern onto the walls of the room. How does the bending of light create these colors? It occurs because the amount of bending usually depends on the wavelength of the light. Wavelengths of visible light range from the longer red waves to the shorter violet waves. White light, such as sunlight, is composed of this whole range of wavelengths.

Figure 5 shows what occurs when white light passes through a prism. The triangular prism refracts the light twice—once when it enters the prism and again when it leaves the prism and reenters the air. Because a longer wavelength of light has a smaller index of refraction, it is refracted less than a shorter wavelength. As a result of these different amounts of bending, the different colors are separated when they emerge from the prism. Red light is bent the least.

Get It?

Predict which color of light you would expect to bend the most.

Rainbows

Does the light leaving the prism in **Figure 5** remind you of a rainbow? Like prisms, rain droplets also refract light. The refraction of the different wavelengths can cause white light from the Sun to separate into the individual colors of visible light, as shown in **Figure 6.** In a rainbow, the human eye can usually distinguish only about seven colors clearly. In order of decreasing wavelength, these colors are red, orange, yellow, green, blue, indigo, and violet.

Figure 6 Water droplets can act as prisms. As white light passes through the water droplet, different wavelengths are refracted by different amounts. This produces the separate colors seen in a rainbow.

Identify *Which color of light is refracted the most as it leaves the water droplet? Which color is refracted the least?*

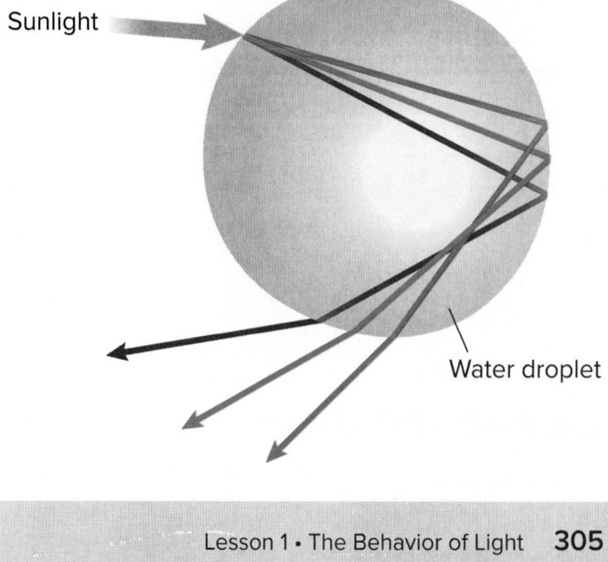

Figure 7 Light waves reflected from an object are refracted when air near the ground is much warmer or cooler than the air above. These refracted waves create one or more additional images.

Mirages

You might have seen what looks like the reflection of an oncoming car in a pool of water on the road ahead. As you get closer, the water seems to disappear. You saw a **mirage,** an image of a distant object produced by the refraction of light through air layers of different densities.

Mirages, such as those shown in **Figure 7,** occur when the air at ground level is much warmer or cooler than the air above it. The density of air increases as air cools. Light waves travel slower as the density of air increases, so light travels slower in cooler air. As a result, light waves refract as they pass through air layers with different temperatures. These refracted light waves form additional images of objects.

Check Your Progress

Summary

- Objects can absorb, reflect, or transmit light.
- Light waves always follow the law of reflection.
- Reflection can be regular or diffuse.
- Refraction occurs when light changes speed in moving from one material to another at an angle to the normal.
- Different wavelengths of light are refracted by different amounts.

Demonstrate Understanding

1. **Describe** two ways that you could direct a light wave around a corner.
2. **Predict** how rubbing a mirror with sandpaper will affect how the mirror reflects light.
3. **Identify** what an object's index of refraction indicates.
4. **Explain** what happens to white light when it passes through a prism.

Explain Your Thinking

5. **Classify** Decide whether the lens of your eye, your fingernails, your skin, and your teeth are opaque, translucent, or transparent. Explain.
6. **MATH Connection** A light ray strikes a mirror at an angle of 42° from the surface of the mirror. What angle does the reflected ray make with the normal?
7. **MATH Connection** A ray of light hits a mirror at 27° from the normal. What is the angle between the reflected ray and the normal?

LEARNSMART® Go online to follow your personalized learning path to review, practice, and reinforce your understanding.

LESSON 2

LIGHT AND COLOR

FOCUS QUESTION

What are the similarities and differences between light and pigments?

Colors

Why do some apples appear to be red, and others look green or yellow? An object's color depends on the wavelengths of light that it reflects and that our eyes detect. You know that white light is a blend of all colors of visible light. When a red apple is struck by white light, it reflects more red light than green or blue light. **Figure 8** shows white light striking a green leaf. The leaf reflects more green light than other colors and appears green.

Although some objects appear to be black, black is not a color that is present in visible light. Objects that are black absorb all colors of light and reflect little or no light back to your eye. White objects are white because they reflect all colors of visible light.

Get It?
Explain why a white object is white.

Seeing Color

As you approach a busy intersection, the color of the traffic light changes from green to yellow to red. On the cross street, the color changes from red to green. How do your eyes detect the differences between red, yellow, and green light?

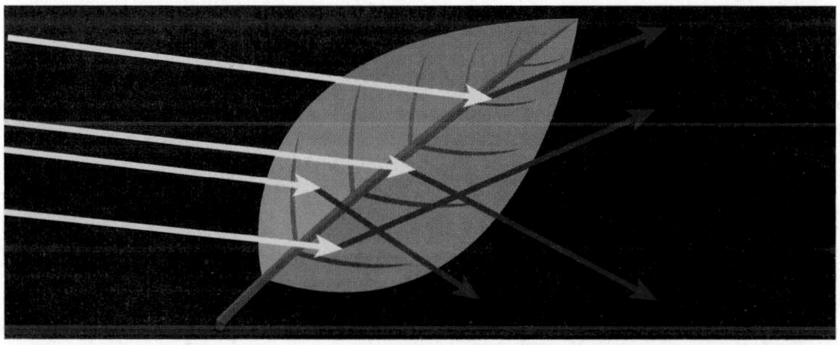

Figure 8 When white light hits this green leaf, more green light is reflected than red or blue light. The leaf absorbs more red and blue light than green light, so the leaf appears green.

3D THINKING | DCI Disciplinary Core Ideas | CCC Crosscutting Concepts | SEP Science & Engineering Practices

COLLECT EVIDENCE

Use your Science Journal to record the evidence you collect as you complete the readings and activities in this lesson.

INVESTIGATE

GO ONLINE to find these activities and more resources.

Laboratory: Light Intensity
Use a model to visualize the effects of distance and direction on light intensity.

Revisit the Encounter the Phenomenon Question
What information from this lesson can help you answer the Unit and Module questions?

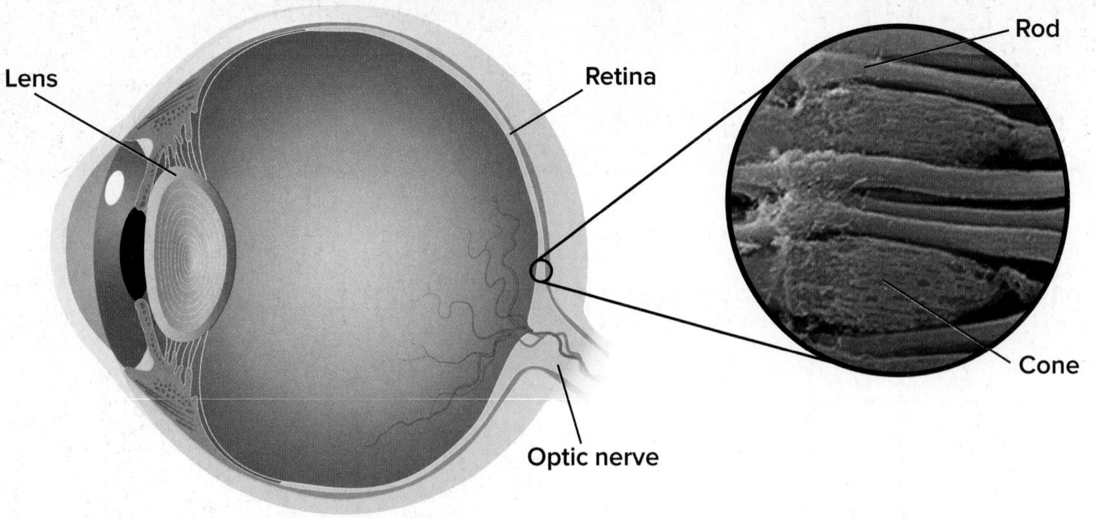

Figure 9 Light enters the eye and hits the retina. The retina is made up of two types of light-detecting cells. Rod cells are sensitive to dim light. Cone cells detect different wavelengths of light. The cone cells send nerve impulses to the brain, which interprets the different combinations of wavelengths as different colors.

Light and the eye

As shown in **Figure 9,** light enters a healthy eye through the lens and is focused on the retina, an area on the inside of your eyeball. The retina contains two types of cells that absorb light. When these cells absorb light, chemical reactions convert light's radiant energy into nerve impulses that are transmitted to the brain. One type of cell in the retina, called a cone, allows you to distinguish colors and detailed shapes of objects. Cones are most effective in daytime vision. The second type of cell, called a rod, is sensitive to dim light and is useful for night vision.

Your eyes have three types of cones, each of which responds to a different range of wavelengths. Red cones respond mostly to red and yellow, green cones respond mostly to yellow and green, and blue cones respond mostly to blue and violet.

Get It?

Identify the colors of light detected by each type of cone cell.

Interpreting color

Why does a banana appear to be yellow? The yellow light reflected by the banana causes the cone cells that are sensitive to red and green light to send signals to your brain. Your brain would get the same signal if a mixture of red light and green light reached your eye. Again, your red and green cones would respond, and you would see yellow light because your brain cannot perceive the difference between incoming yellow light and yellow light produced by combining red and green light. The next time that you are at a play or a concert, look at the lighting above the stage. Observe how the colored lights combine to produce effects onstage.

Color blindness

If one or more of your sets of cones did not function properly, you would not be able to distinguish between certain colors. About 8 percent of men and 0.5 percent of women have a form of color blindness. Most people who are said to be color-blind are not truly blind to color, but they have difficulty distinguishing between a few colors, most commonly red and green. **Figure 10** shows an example of a red-green color blindness test. Because these two colors are used in traffic signals, severely color-blind drivers and pedestrians must use the position of the light, instead of color, to know when to stop and go.

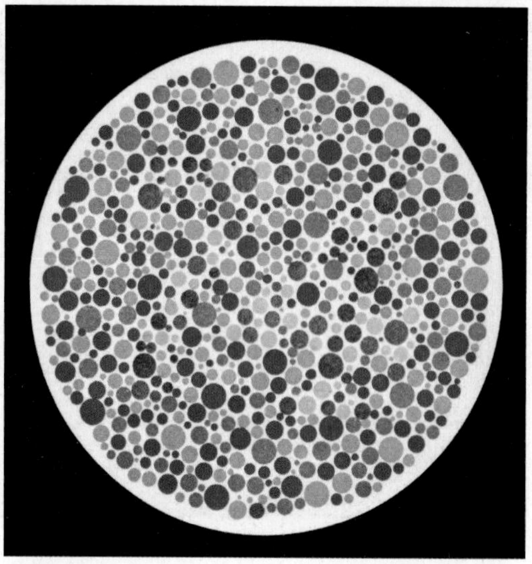

Figure 10 This circle of red and green dots is a common, simple test for red-green color blindness.

Describe *what you see in the dots. What might this tell you about your ability to distinguish between red and green?*

Filtering Colors

Wearing tinted glasses changes the color of almost everything that you see. If the lenses are yellow, the world takes on a golden glow. If they are rose-colored, everything looks pinkish. Something similar would occur if you placed a colored, transparent plastic sheet over this white page. The paper would appear to be the same color as the plastic. The plastic sheet and the tinted lenses are filters. A **filter** is a transparent material that selectively transmits light. For example, color filters transmit one or more colors of light but absorb all others. The color of a filter is the color of the light that it transmits.

Figure 11 shows what happens when you look at a colored object through various colored filters. On the left in **Figure 11,** a blue bowl looks blue because it primarily reflects blue light and absorbs more of the other colors of light. If you look at the bowl through a blue filter, as in the center of **Figure 11,** the bowl still looks blue because the filter transmits the reflected blue light. The right image in **Figure 11** shows how the bowl looks when you examine it through a red filter.

Figure 11 The color of the bowl seems to change when it is viewed through different colored filters.

Explain *Why does the blue bowl appear to be black when viewed through a red filter?*

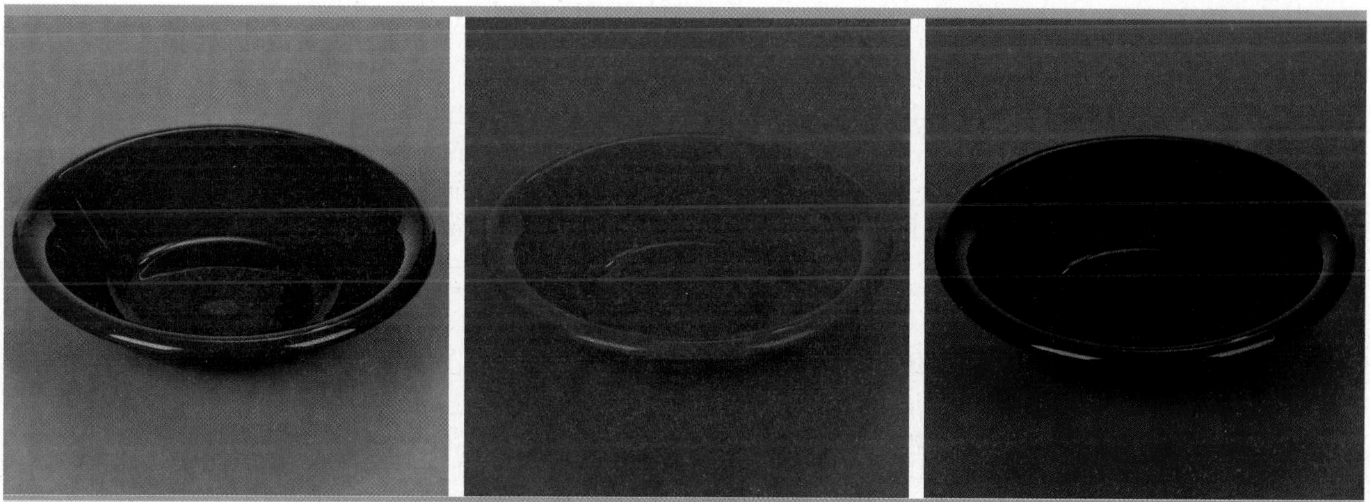

The bowl appears to be blue in white light.

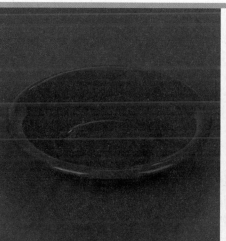

The bowl appears to be blue when viewed through a blue filter.

The bowl appears to be black when viewed through a red filter.

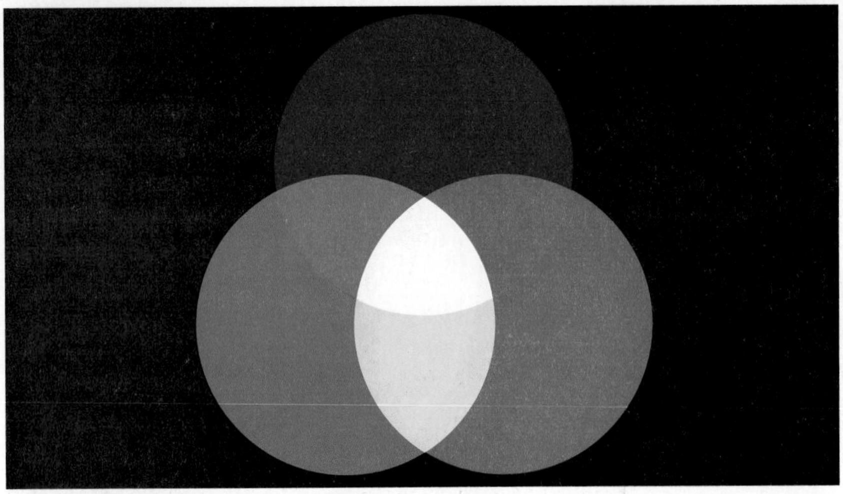

Figure 12 The three primary colors of light are red, green, and blue. White light is produced when the three primary colors of light are mixed in equal amounts.

Mixing Colors

If you have ever browsed through a paint store, you have probably seen displays where customers can select paint samples of almost every imaginable color. To get these colors, the paint store workers mix different colors together. For example, they might mix blue and yellow paint to produce green paint. What would happen if you mixed blue and yellow light? Would you get green light?

Mixing colored lights

From the glowing orange of a sunset to the deep blue of a mountain lake, all the colors that you see can be made by mixing three colors of light. These three colors–red, green, and blue–are the primary colors of light. They correspond to the three different types of cones in the retina of your eye. When they are mixed together in equal amounts, they produce white light, as **Figure 12** shows. Mixing the primary colors in different proportions produces all the colors that you see.

Paint pigments

If you mixed equal amounts of red, green, and blue paint, would you get white paint? If mixing colors of paint were like mixing colors of light, you would. But mixing paint is different. The variety of colors of paint is a result of mixtures of pigments. A **pigment** is a colored material that is used to change the color of other substances.

The color of a pigment results from the different wavelengths of light that the pigment reflects. Paint pigments are usually made of chemical compounds. For example, titanium oxide is a bright white pigment that reflects all colors of light. Another example is lead chromate, which is used to make paint for yellow lines on highways. Other colors can be obtained by mixing various pigments together.

WORD ORIGINS

pigment
from the Latin *pigmentum*, meaning "paint"

The painter made green paint by adding blue and yellow pigments to white paint.

CCC CROSSCUTTING CONCEPTS

Cause and Effect Figure 12 shows that mixing equal amounts of the primary colors of light produces white light. With a partner, use a color addition (RGB) simulator to see what colors you can make using different combinations of red, blue, and green light. Create a chart showing what colors are produced by different combinations. As an extension, find a color subtraction (CMY) simulator and compare how different colors can be made by mixing magenta, cyan, and yellow pigments, as shown in **Figure 13**.

Mixing pigments You can make any pigment color by mixing different amounts of the three primary pigments—magenta (bluish red), cyan (greenish blue), and yellow. In fact, color printers use those pigments to make full-color prints like the pages in this book. However, color printers also use black ink to produce a true black color. A primary pigment's color depends on the color of light that it reflects.

Pigments absorb and reflect a range of colors in sending a single color message to your eye. For example, a yellow pigment viewed in white light appears to be yellow because it reflects red and green light but absorbs blue light. The color of a mixture of two primary pigments is determined by the primary colors of light that both pigments reflect.

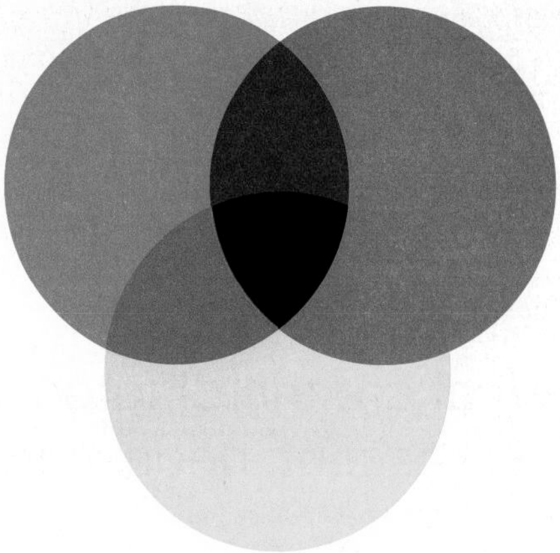

Figure 13 The three primary pigment colors are magenta, yellow, and cyan. When the three primary pigment colors are mixed in equal amounts, they appear to be black.

In **Figure 13,** the area in the center where the colors overlap appears to be black because the three blended primary pigments absorb all the primary colors of light. Recall that the primary colors of light combine to produce white light. They are called additive colors. However, the primary pigment colors combine to produce black. So, the primary pigments are called subtractive colors.

Check Your Progress

Summary

- The color of an object depends on the wavelengths of light it reflects.
- Rod and cone cells are light-sensitive cells found in the human eye.
- The color of a filter is the color of the light that the filter transmits.
- All light colors can be created by mixing the primary light colors—red, green, and blue.
- All pigment colors can be formed by mixing the primary pigment colors—magenta, cyan, and yellow.

Demonstrate Understanding

8. **Explain** why a white fence appears to be white. In your answer, include the colors of light that your eye detects, and tell how your brain interprets those colors.
9. **Identify** what color would be seen if equal amounts of red light and green light were mixed.
10. **Compare and contrast** the primary colors of light and the primary pigment colors.
11. **Describe** how your eyes detect color.

Explain Your Thinking

12. **Predict** Light reflected from an object passes through a green filter, then a red filter, and finally a blue filter. What color will the object appear to be?
13. **MATH Connection** In the human eye, there are about 120,000,000 rods. If 90,000,000 rods trigger at once, what percent of the total number of rods are triggered?
14. **MATH Connection** The wavelengths of a color are measured in nanometers (nm), which is 0.000000001 meters (one-billionth of a meter). Find the wavelength in meters of a light wave that has a wavelength of 690 nm.

LEARNSMART® Go online to follow your personalized learning path to review, practice, and reinforce your understanding.

LESSON 3
PRODUCING LIGHT

FOCUS QUESTION
What are the differences between incandescent, fluorescent, neon, and laser lights?

Incandescent Lights

Incandescent light is generated by heating a piece of metal until it glows. Incandescent lightbulbs have small wire coils called filaments that are usually made of tungsten. When there is an electric current in a filament, its electric resistance makes it become hot enough to emit light. However, about 90 percent of the energy given off by incandescent bulbs is thermal energy. Incandescent lightbulbs are being phased out in many places.

Fluorescent Lights

Fluorescent light is light generated by using phosphors to convert ultraviolet radiation to visible light. A phosphor is a substance that absorbs ultraviolet radiation and then emits visible light. Fluorescent bulbs are filled with a gas at low pressure and coated on the inside with phosphors, as shown in **Figure 14.** An electrode is at each end of the tube. Electrons are given off when the electrodes are connected in a circuit. These electrons collide with the gas atoms, which then emit ultraviolet radiation. The wavelength of light emitted depends on the gas used. The phosphors absorb this radiation and convert it to visible light.

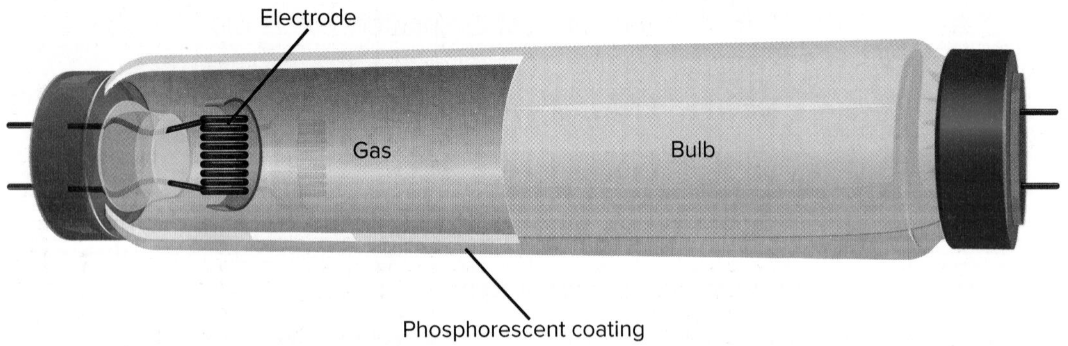

Figure 14 Fluorescent lightbulbs use phosphors to convert ultraviolet light to visible light. They are commonly used in houses, schools, and offices.

3D THINKING | DCI Disciplinary Core Ideas | CCC Crosscutting Concepts | SEP Science & Engineering Practices

COLLECT EVIDENCE
Use your Science Journal to record the evidence you collect as you complete the readings and activities in this lesson.

INVESTIGATE
GO ONLINE to find these activities and more resources.

Quick Investigation: Discover Energy Waste in Lightbulbs
Carry out an investigation to discover the differences between two types of lightbulbs.

Review the News
Obtain information from a current news story about lighting or lightbulbs. Evaluate your source and communicate your findings to your class.

Efficient lighting

Fluorescent lights use as little as one-fifth of the electrical energy to produce the same amount of light as incandescent bulbs. Fluorescent bulbs also last much longer than incandescent bulbs. This higher efficiency can mean lower energy costs over the life of the bulb. Reduced energy usage could reduce the amount of fossil fuels that are burned to generate electricity. This would also decrease the amount of carbon dioxide and pollutants released into Earth's atmosphere.

Hospitals, schools, office buildings, and factories have been using long, tube-shaped fluorescent bulbs for many years. Many homes are now using compact fluorescent bulbs, which can be screwed into traditional lightbulb sockets. This gives consumers one way to save energy without having to replace the entire light fixture.

Neon Lights

The vivid, glowing colors of neon lights, such as the one shown in **Figure 15,** make them a popular choice for signs and eye-catching decorations on buildings. These lighting devices are glass tubes filled with gas—often neon—and work similarly to fluorescent lights. When there is an electric current through the tube, electrons collide with the gas molecules. In this case, however, the gas molecules emit visible light. If the tube contains only neon, the light is bright red-orange. Other gases and phosphor coatings are used to make other colors.

Get It?

Identify what causes the color in a neon light.

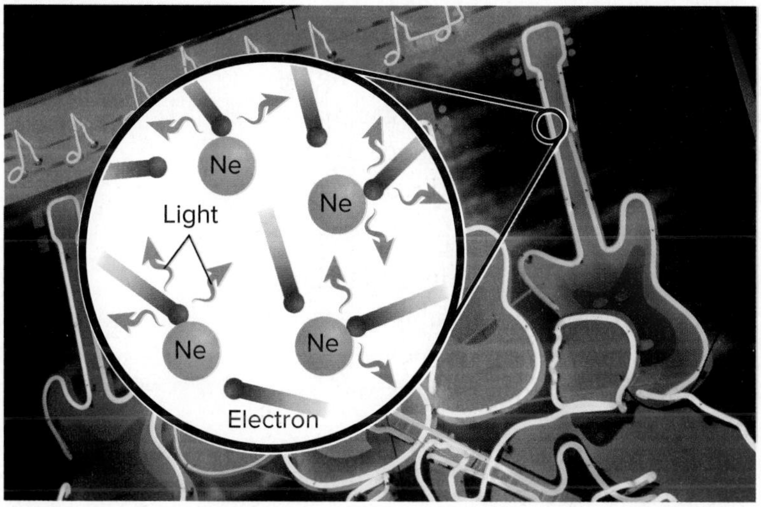

Figure 15 The bright color of this neon light is produced when a current is passed through the neon tube. Electrons collide with neon atoms, causing them to emit a red-orange light.

Tetra Images/Getty Images

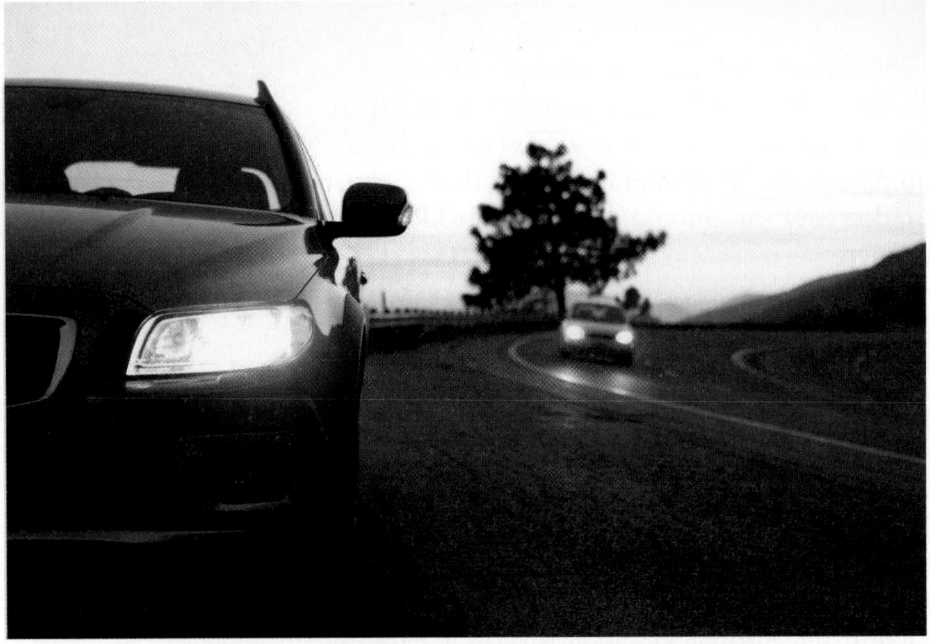

Figure 16 Tungsten-halogen lightbulbs are often used in car headlights because they are brighter than incandescent lightbulbs.

Identify *some advantages and disadvantages of tungsten-halogen headlights.*

Sodium-Vapor Lights

Sodium-vapor lights are often used for streetlights and other outdoor lighting. Inside a sodium-vapor lamp is a tube that contains a mixture of neon gas, a small amount of argon gas, and a small amount of sodium metal. When the lamp is turned on, the gas mixture becomes hot. The hot gases cause the sodium metal to turn to vapor. The hot sodium vapor emits a yellow-orange glow.

Tungsten-Halogen Lights

Tungsten-halogen lights are sometimes used to create intensely bright light. These lights have a tungsten filament inside a quartz bulb or tube. The tube is filled with a gas that contains one of the halogen elements, such as fluorine or chlorine. The presence of this gas enables the filament to become much hotter than the filament in an ordinary incandescent bulb. As a result, the light is much brighter.

Another advantage of tungsten-halogen bulbs is that they last longer than incandescent bulbs. Their long lifetime is due to the chemical interactions between the halogen gas and the tungsten filament. Tungsten-halogen lights are sometimes used on movie sets and in underwater photography. They are also used in many headlights for cars, as shown in **Figure 16.**

Lasers

From laser surgery to laser light shows, lasers have become a large part of the world in which you live. Lasers can be made with many different materials, including gases, liquids, and solids. The wavelength of the laser depends on the materials used. One of the most common is the helium-neon laser, which produces a beam of red light with a wavelength of 632 billionths of a meter.

Amplifying light

In a helium-neon laser, a mixture of helium and neon gases is sealed in a tube with mirrors at both ends. When a voltage is applied across the helium-neon gas, the smaller helium atoms gain enough energy to collide with the larger neon atoms. The neon atoms then release the energy gained from the collisions by emitting light waves of a particular wavelength. The light waves are emitted in all directions. Most of the waves escape through the sides of the tube, but those traveling along the tube are reflected between the two mirrors. As they travel, these light waves stimulate other atoms to emit more light waves. The result is an increase in the amount of light traveling back and forth along the tube.

To allow a small amount of the light out of the tube, one of the mirrors is coated only partially with reflective material. A narrow, intense laser beam is emitted from the partially reflective end of the tube. **Figure 17** illustrates how a laser creates a beam of light.

Figure 17 Lasers produce narrow, intense beams of light.

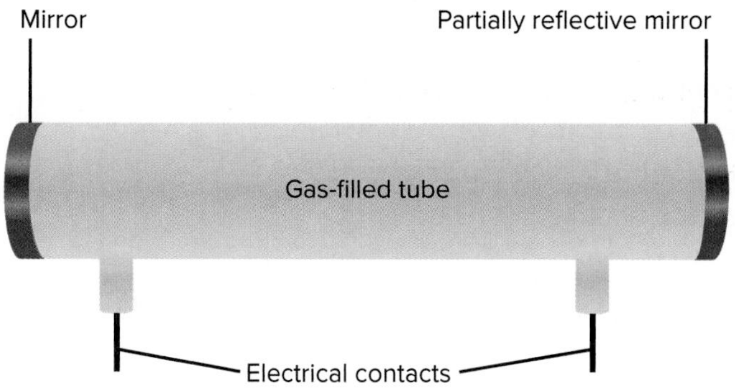

A helium-neon laser consists of a glass tube filled with 90% helium and 10% neon gas. Mirrors are located on each end of the tube. The mirror on the right transmits a small amount of light.

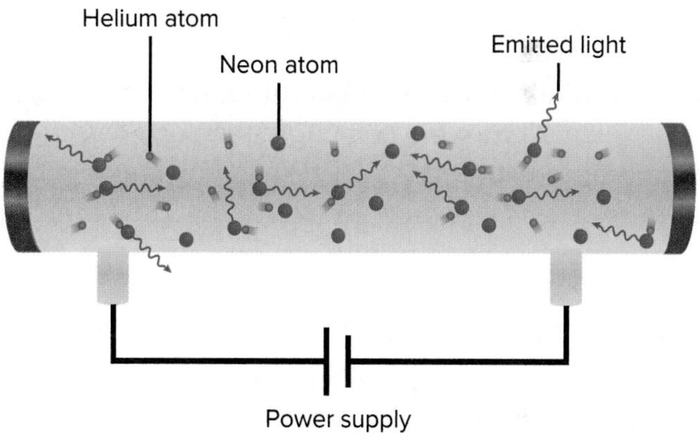

When a voltage is applied to the electrical contacts, an electric field is created throughout the helium-neon gas. The smaller helium atoms gain enough energy to collide with the larger neon atoms. The neon atoms then emit the energy as a single wavelength of light.

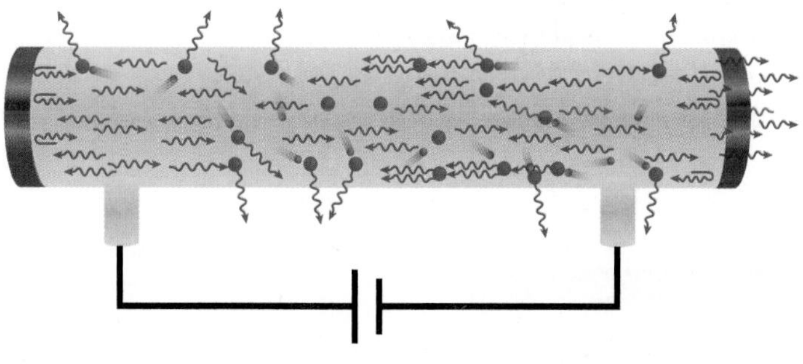

Most of the emitted light waves escape through the sides of the tube and are wasted. However, some waves are reflected back and forth between the mirrors. About 1 percent of light waves that hit the mirror on the right are transmitted.

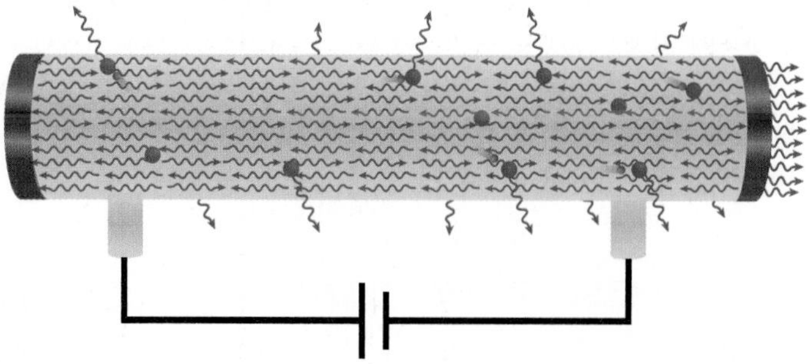

The reflected light can cause other atoms to emit light waves. This results in an increase in the amount of light traveling back and forth in the tube. The laser now emits a narrow, intense beam of light in a single direction.

These waves are coherent because they have the same wavelength and they travel in one direction with a constant distance between their corresponding crests. They combine to form a single wave.

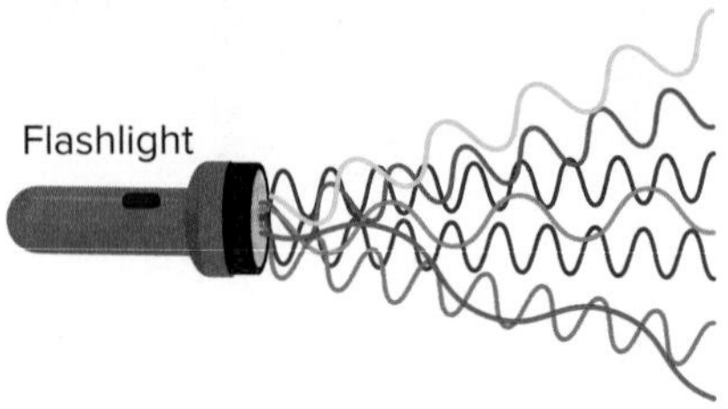

Incoherent waves can have more than one wavelength and do not travel in one direction with their crests at constant distances.

Figure 18 Light waves can be coherent or incoherent.

Coherent and incoherent light

The beams from a laser light do not spread out because laser light is coherent. **Coherent light** is light of only one wavelength that travels in one direction with a constant distance between the corresponding crests of the waves. This is illustrated in **Figure 18.** Notice that the coherent waves combine to form a wave with constant wavelength and frequency.

The light from an ordinary lightbulb is incoherent. **Incoherent light** can have more than one wavelength, can travel in more than one direction, and does not travel with a constant distance between the corresponding crests of the waves. This is also illustrated in **Figure 18.** The waves do not travel in the same direction, so the beam spreads out, and the energy carried by the light waves is spread over a large area.

Using lasers

A laser beam does not spread out as it travels over long distances. Therefore, it can apply large amounts of energy to small areas. DVD burners, surgical tools, and many other useful devices take advantage of this property. In industry, powerful lasers are used for cutting and welding materials. Surveyors and builders use lasers for measuring distances and for leveling. In communications, information can be coded in pulses of light from lasers.

Lasers in medicine Surgeons can use lasers in place of scalpels to cut through tissues in the eye and in other parts of the body. Lasers are routinely used to remove cataracts, reshape the cornea, and repair the retina. The energy from the laser seals blood vessels in the incision and reduces bleeding. Most lasers do not penetrate deeply through the skin, so they can be used to remove small tumors or birthmarks on the surface without damaging deeper tissues.

SCIENCE USAGE v. COMMON USAGE

coherent

Science usage: relating to light waves with a single wavelength that travel in one direction with a constant distance between their corresponding crests

Lasers produce coherent light.

Common usage: orderly, logical, staying together

Please give a coherent explanation of what occurred this morning.

STEM CAREER Connection

Medical Equipment Repairer

Do you have excellent technical skills and an interest in healthcare? Medical equipment repairers, or biomedical equipment technicians, install, maintain, and repair a wide range of medical equipment so that physicians can confidently run accurate tests. Medical equipment repairers generally have an associate's degree in biomedical technology or engineering, but some might also have a bachelor's degree.

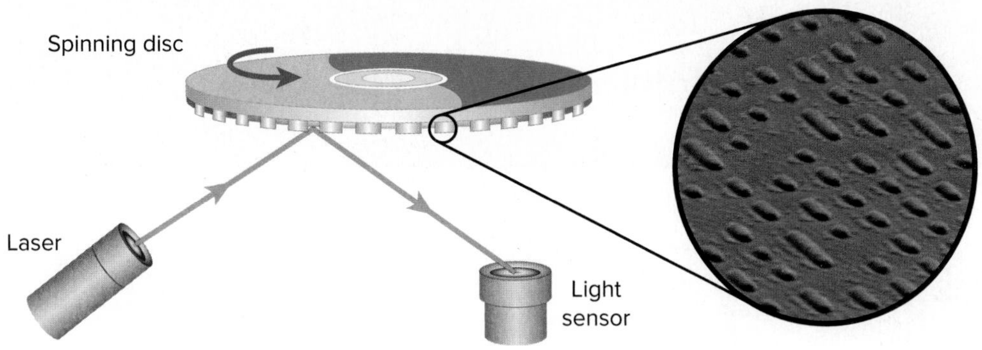

Figure 19 The pits on the bottom of an optical disc store information. An optical disc player uses a laser to detect the pits on the disc. The information is then converted to an electric signal.

Optical discs Optical discs, such as Blu-rays, DVDs, and CDs, are plastic discs with reflective surfaces that are used to store video, sound, and other types of data in digital form. When an optical disc is produced, the information is burned into the surface of the disc with a laser. The laser creates millions of tiny pits in a spiral pattern that starts at the center of the disc and moves out to the edge. A device that plays optical discs also uses a laser to read the disc. As illustrated in **Figure 19,** as the laser beam strikes a pit or flat spot, different amounts of light are reflected to a light sensor. The reflected light is converted to an electric signal that can be converted into sound or images.

Check Your Progress

Summary

- Incandescent and fluorescent lightbulbs are often used in homes, schools, and offices.
- When electrons collide with neon gas, red light is emitted.
- A tungsten-halogen bulb is brighter and hotter than an ordinary incandescent bulb.
- Lasers emit narrow beams of coherent light.

Demonstrate Understanding

15. **Compare and contrast** the two main types of bulbs found in your home. Explain how they produce light.
16. **Discuss** the advantages of using a fluorescent bulb instead of an incandescent bulb.
17. **Describe** the difference between coherent and incoherent light.
18. **Describe** the processes used to produce light in a laser.
19. **Identify** several uses of lasers.

Explain Your Thinking

20. **Apply** Which type of lighting device would you use for each of the following needs: an economical light source in a manufacturing plant, an eye-catching sign that will be visible at night, and a baseball stadium? Explain.
21. **MATH Connection** A 25-W fluorescent light emits 5.0 J of thermal energy each second. What is the efficiency of the fluorescent light?
22. **MATH Connection** If 90 percent of the energy emitted by incandescent bulbs is thermal energy, how much thermal energy is emitted by a 60-W bulb each second?

LEARNSMART® Go online to follow your personalized learning path to review, practice, and reinforce your understanding.

LESSON 4

USING LIGHT

FOCUS QUESTION

How are holograms made?

Polarized Light

You can make transverse waves in a rope vibrate in any direction—horizontal, vertical, or anywhere in between. Light also is a transverse wave and can vibrate in any direction. The direction of vibration for a light wave refers to the direction that its electric or magnetic field vibrates. **Linearly polarized light** is light with a magnetic field that vibrates in only one direction. As always, the electric field is perpendicular to the magnetic field.

Polarizing filters

Light can be polarized by using a special filter with lines of crystals that act like a group of parallel slits. Only light waves with magnetic fields that vibrate in the same direction as the lines of crystals can pass through. This is similar to the rope wave traveling through the fence in **Figure 20.**

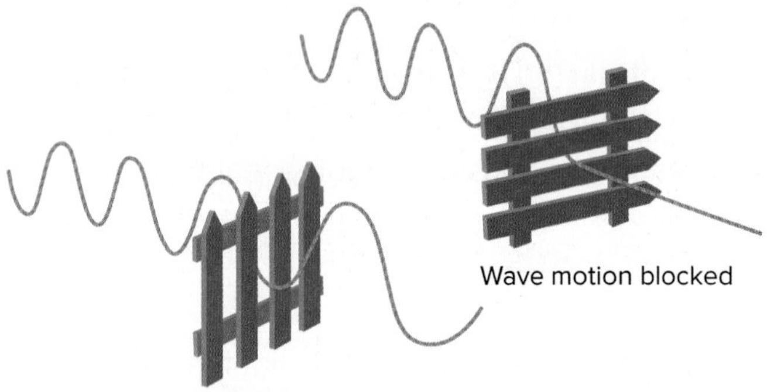

The wave in the rope can travel only through slits that are positioned in the same direction as its vibrations.

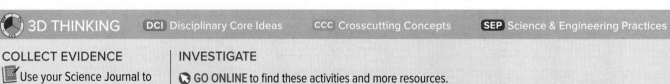

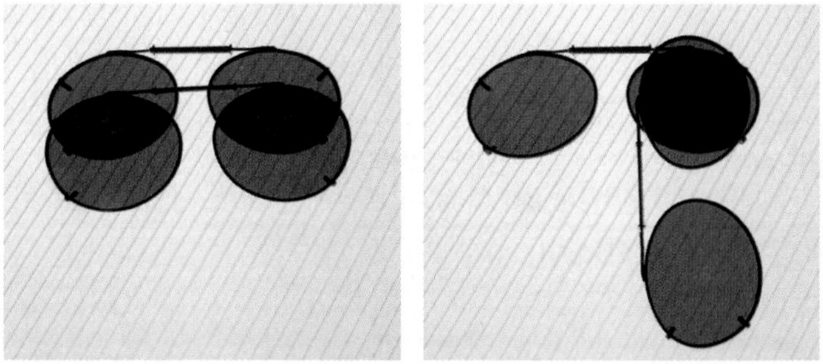

When polarizing filters are stacked with their lines of crystals in the same direction, light can pass through them. When the lines are positioned at right angles, the light is blocked.

Figure 20 A polarizing filter transmits only light waves with magnetic fields vibrating parallel to the lines of crystals in the filter. When multiple filters are used, the amount of light transmitted depends on how the two filters are oriented.

3D THINKING — DCI Disciplinary Core Ideas — CCC Crosscutting Concepts — SEP Science & Engineering Practices

COLLECT EVIDENCE

Use your Science Journal to record the evidence you collect as you complete the readings and activities in this lesson.

INVESTIGATE

GO ONLINE to find these activities and more resources.

Lab: Polarizing Filters
Carry out an investigation into the effects of passing light through polarizing filters.

Revisit the Encounter the Phenomenon Question
What information from this lesson can help you answer the Module question?

Polarized lenses

When light is reflected from a horizontal surface, such as a lake, some of the light is polarized with its magnetic field vibrating vertically. The lenses of polarizing sunglasses are polarizing filters that block the polarized reflected light, thus reducing glare.

Holography

Science museums often have exhibits in which three-dimensional images seem to float in space, like the one shown in **Figure 21.** You can see the image from different angles, just as you would if you viewed the real object. These images are produced by holography. **Holography** is a technique that produces a hologram, a complete three-dimensional photographic image of an object. The three-dimensional images on some credit cards also use holography.

Making holograms

Lasers are needed to produce holograms. The laser beam is split into two parts. One part illuminates the object and reflects onto photographic film. At the same time, the second part of the beam is also directed at the film. The light from the two beams creates an interference pattern on the film. The pattern looks nothing like the original object, but when laser light shines on the pattern on the film, a holographic image is produced.

Get It?
Describe how holographic images are produced.

Information in light

Ordinary photographic images capture only the brightness or intensity of light reflected from an object's surface, while a hologram records both the intensity and the direction. Therefore, holograms convey more information to your eye than does a two-dimensional photograph. They are also more difficult to copy. In addition to credit cards, holographic images are used on identification cards and on some product labels to help prevent counterfeiting. Using X-ray lasers, scientists can produce holographic images of microscopic objects. It may be possible to use a similar technique to create three-dimensional views of biological cells.

Figure 21 Holograms can be displayed as three-dimensional art. An embossed copy of a hologram can be placed over a reflective background to create the security features on important documents.

Philippe Plailly/Science Source

Optical Fibers

When laser light must travel long distances or be sent into hard-to-reach places, optical fibers often are used. These transparent glass fibers transmit light from one place to another by a process called total internal reflection.

Partial reflection

Remember that refraction can happen when light changes speed as it travels from one medium to another. When light travels from water to air, it speeds up, and the direction of the light ray is bent away from the normal, as shown in **Figure 22.** However, not all the light passes through the surface. Some light is reflected back into the water.

As the underwater light ray makes larger angles with the normal, the refracted light rays in the air bend closer to the surface of the water. Additionally, more light is reflected back into the water. At a certain angle, called the critical angle, the refracted ray has been bent so that it is traveling along the surface of the water, as shown in **Figure 22.** For a light ray traveling from water into air, the critical angle is about 49°.

Total internal reflection

Figure 22 also shows what happens if the underwater light ray strikes the boundary between the air and water at an angle greater than the critical angle. There is no longer any refraction, and the light ray does not travel in the air. Instead, the light ray is reflected at the boundary, just as if a mirror were there. This complete reflection of light at the boundary between two different materials is called **total internal reflection.** The light wave follows the law of reflection. For total internal reflection to occur, light must travel slower in the first medium and must strike the boundary at an angle that is greater than the critical angle.

Get It?
Identify when total internal reflection occurs.

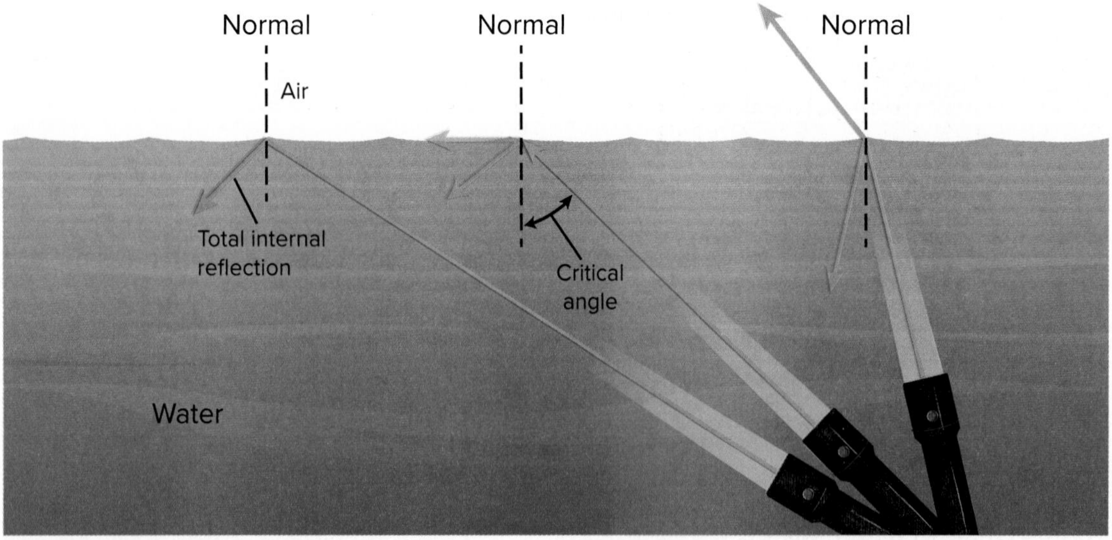

Figure 22 At the surface, where water and air meet, some light is refracted and some is reflected. The greater the angle of incidence is, the greater both the angle of refraction and the amount of reflected light will be. At the critical angle, the refracted wave travels along the water's surface. At angles greater than the critical angle, total internal reflection occurs.

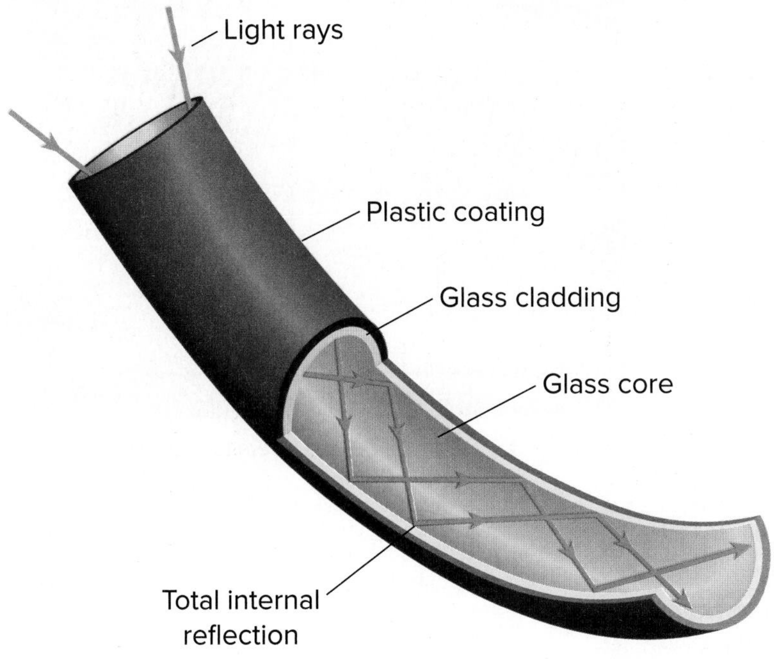

Figure 23 An optical fiber uses total internal reflection. Light travels through the fiber without leaving it, except at the ends.

Using total internal reflection

Total internal reflection makes light transmission in optical fibers possible. As shown in **Figure 23,** light entering one end of the fiber is reflected continuously from the sides of the fiber until it emerges from the other end. The light moves like water through a pipe—almost no light is lost or absorbed in optical fibers. The core of the fiber and the surrounding cladding are made from glasses with two different indexes of refraction. Light moves more slowly in the core than in the surrounding layer. Therefore, total internal reflection can occur at the surface between the two layers.

Using optical fibers

Optical fibers are most often used in communications. Telephone conversations, television programs, and computer data can be coded in light beams. Signals cannot leak from one fiber to another and interfere with other messages, so the signal is transmitted clearly. To send telephone conversations through an optical fiber, sound is converted into digital signals consisting of pulses of light from a light-emitting diode (LED) or from a laser. Some systems use multiple lasers, each with its own wavelength, to fit multiple signals into the same fiber. You could send a million copies of your favorite book in one second on a single fiber. **Figure 24** shows the size of typical optical fibers.

Optical fibers also have medical uses. Doctors use them to explore the inside of the human body. One bundle of fibers transmits light while another carries the reflected light back to the doctor. Physicians can also treat blocked arteries by sending laser light into the body through an optical fiber.

Figure 24 Despite its small size, just one of these optical fibers can carry thousands of phone conversations at the same time.

Optical Scanners

Figure 25 Optical scanners use lasers to read bar codes.

In supermarkets and many other kinds of stores, a cashier passes your purchases over a glass window in the checkout counter. Or, the cashier might hold a handheld device up to each item. In an instant, the optical scanner beeps, and the price of the item appears on a screen. An **optical scanner** is a device that reads intensities of reflected light and converts the information to digital signals. A supermarket scanner detects a bar code, a pattern of thick and thin stripes like the one shown in **Figure 25.** The resulting electrical signal is sent to a computer that searches its database for the item's price and sends the information to the cash register.

You may have used another type of optical scanner to convert pictures or text into forms you can use in computer programs. With a flatbed scanner, for example, you lay a document or picture face down on a sheet of glass and close the cover. An optical scanner passes underneath the glass and reads the pattern of colors. The scanner converts the pattern to an electronic file that can be stored on a computer. Many new photocopiers use scanners like these. The photocopier scans a document and then prints a copy.

Check Your Progress

Summary

- The magnetic fields in linearly polarized light vibrate in only one direction.
- Polarizing filters transmit light polarized in one direction.
- Lasers are used to produce holograms.
- The complete reflection of light at the boundary between two materials is called total internal reflection.

Demonstrate Understanding

23. **Discuss** how optical fibers are used to transmit telephone conversations.
24. **Contrast** polarized and unpolarized light.
25. **Describe** how a hologram is made.
26. **Identify** all the conditions that are necessary for total internal reflection to occur.

Explain Your Thinking

27. **Investigate** On a sunny day, you are looking at the surface of a lake through polarized sunglasses. How could you use your sunglasses to tell if the light reflected from the lake is polarized?
28. **MATH Connection** An optical fiber has a diameter of 0.3 mm. How many fibers would be needed to form a cable with a square cross section if the cross section was 1.5 cm on a side?

LEARNSMART® Go online to follow your personalized learning path to review, practice, and reinforce your understanding.

STEM AT WORK

Shining New Light on an Old Text

A team of scientists and literature scholars from around the world are racing to create a digital version of the oldest surviving copy of Homer's *Iliad*. The thousand-year-old manuscript, known as *Venetus A,* is fading away as light and humidity slowly decompose the fragile pages.

A researcher photographs the pages with a digital camera.

Visible light photographs

To prevent harm to *Venetus A,* the librarians responsible for the manuscript photographed every page of the warped and faded book more than 100 years ago. However, because they are not color photographs, researchers could not read all the details found in the original text.

Beyond the visible spectrum

As photographic technology improved, scientists continued to look for ways to see details that were invisible to the unaided human eye. Illuminating the manuscript with ultraviolet light and using a camera that responds to ultraviolet light can reveal lost text, as shown below left. The faded ink reflects ultraviolet light, but the parchment absorbs ultraviolet wavelengths. The contrast between the ink and the parchment allows faded text to be read.

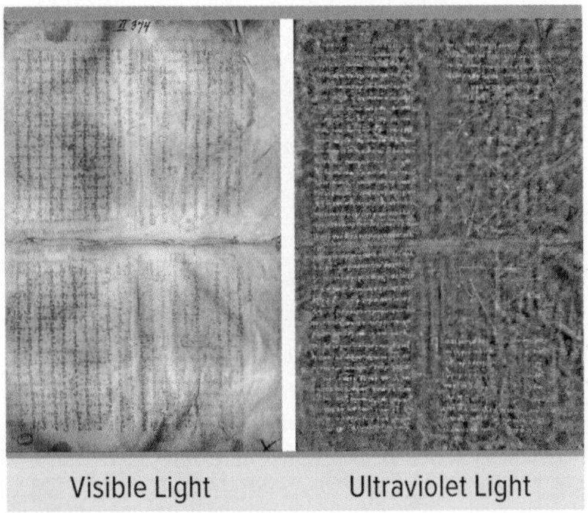

Ultraviolet light can expose text not visible under normal light conditions.

3-D imaging with lasers

Over time, the pages of *Venetus A* have become warped and faded. The warping deformed the original text, making individual letters and entire words illegible. An international team of researchers is correcting this deformation by making a 3-D map of the page and then digitally flattening the page back to its original appearance. First, as shown above, a digital camera takes a high-resolution image of each page. Next, a laser scanner measures the exact deformation of the page. Researchers use the measurements to correct the image captured by the digital camera. This process creates a flat image of the page as it appeared more than 1000 years ago.

SEP DCI CCC

COMMUNICATE SCIENTIFIC INFORMATION

This feature describes an application of lasers. Create a poster or computer presentation describing another application of lasers. Your poster or presentation should clearly explain how the technology is used.

MODULE 12

STUDY GUIDE

GO ONLINE to study with your Science Notebook.

Lesson 1 THE BEHAVIOR OF LIGHT

- Objects can absorb, reflect, or transmit light.
- Light waves always follow the law of reflection.
- Reflection can be regular or diffuse.
- Refraction occurs when light changes speed in moving from one material to another at an angle to the normal.
- Different wavelengths of light are refracted by different amounts.

- opaque
- translucent
- transparent
- index of refraction
- mirage

Lesson 2 LIGHT AND COLOR

- The color of an object depends on the wavelengths of light it reflects.
- Rod and cone cells are light-sensitive cells found in the human eye.
- The color of a filter is the color of the light that the filter transmits.
- All light colors can be created by mixing the primary light colors—red, green, and blue.
- All pigment colors can be formed by mixing the primary pigment colors—magenta, cyan, and yellow.

- filter
- pigment

Lesson 3 PRODUCING LIGHT

- Incandescent and fluorescent lightbulbs are often used in homes, schools, and offices.
- When electrons collide with neon gas, red light is emitted.
- A tungsten-halogen bulb is brighter and hotter than an ordinary incandescent bulb.
- Lasers emit narrow beams of coherent light.

- incandescent light
- fluorescent light
- coherent light
- incoherent light

Lesson 4 USING LIGHT

- The magnetic fields in linearly polarized light vibrate in only one direction.
- Polarizing filters transmit light polarized in one direction.
- Lasers are used to produce holograms.
- The complete reflection of light at the boundary between two materials is called total internal reflection.

- linearly polarized light
- holography
- total internal reflection
- optical scanner

THREE-DIMENSIONAL THINKING

Module Wrap-Up

REVISIT THE PHENOMENON

How does light transmit information?

CER Claim, Evidence, Reasoning

Explain Your Reasoning Revisit the claim you made when you encountered the phenomenon. Summarize the evidence you gathered from your investigations and research and finalize your Summary Table. Does your evidence support your claim? If not, revise your claim. Explain why your evidence supports your claim.

STEM UNIT PROJECT

Now that you've completed the module, revisit your STEM unit project. You will summarize your evidence and apply it to the project.

GO FURTHER

SEP Data Analysis Lab

Light Theory and Everyday Phenomena

Use a library to find information to explain the following phenomena.

CER Analyze and Interpret Data

1. **Claim** How do we perceive the colors in a rainbow?
2. **Claim** Why are sunrises and sunsets sometimes red?
3. **Claim** What causes eyes to appear blue or brown?
4. **Claim and Reasoning** Look at the side of a compact disc that is read by the laser. What do you see? Explain your observation.

Jess Alford/Photodisc/Getty Images

MODULE 13
MIRRORS AND LENSES

ENCOUNTER THE PHENOMENON

What is the farthest you can see with a telescope like this?

GO ONLINE to play a video about how Hubble photographed the Andromeda Galaxy in 1923.

SEP Ask Questions

Do you have other questions about the phenomenon? If so, add them to the driving question board.

CER Claim, Evidence, Reasoning

Make Your Claim Use your CER chart to make a claim about how far you can see with a large telescope. Explain your reasoning.

Collect Evidence Use the lessons in this module to collect evidence to support your claim. Record your evidence as you move through the module.

Explain Your Reasoning You will revisit your claim and explain your reasoning at the end of the module.

GO ONLINE to access your CER chart and explore resources that can help you collect evidence.

LESSON 1: Explore & Explain: Plane Mirrors

LESSON 2: Explore & Explain: Convex Lenses

Smithsonian

Additional Resources

(t)Video Supplied by BBC Worldwide Learning, (bl)Allison Jones, (br)Don Farrall/Photographer's Choice/Getty Images

LESSON 1

MIRRORS

FOCUS QUESTION

How do different types of mirrors form images?

Light and Vision

Have you ever tried to find an address on a house or an apartment at night on a poorly lit street? It is harder to do those activities in the dark than it is when there is plenty of light. Your eyes see by detecting light, so when you can see something, it is because light came from that object to your eyes. Light is emitted from a light source, such as the Sun or a lightbulb, and then reflects off an object, such as the page of a book, as shown in **Figure 1.**

Figure 1 Light from the lamp reflects off the book and into the student's eyes. People see objects when their eyes detect light emitted or reflected from objects.

When light travels from an object to your eye, you see the object. Light can reflect more than once. For example, light can reflect off an object, into a mirror, and then into your eyes. When no light is available to reflect off objects and into your eyes, you cannot see anything. This is why it is hard to see an address in the dark.

Light rays

Light sources send out light waves that travel in all directions. These waves spread out from the light source, just as ripples on the surface of water spread out from the point of impact of a pebble.

You could also think of the light coming from the source as traveling in narrow beams. Each narrow beam travels in a straight line and is called a light ray. Even though light rays can change direction when they are reflected or refracted, your brain interprets images as if light rays travel in a straight line.

Get It?

Identify two ways to imagine light coming from a source.

3D THINKING DCI Disciplinary Core Ideas CCC Crosscutting Concepts SEP Science & Engineering Practices

COLLECT EVIDENCE

Use your Science Journal to record the evidence you collect as you complete the readings and activities in this lesson.

INVESTIGATE

GO ONLINE to find these activities and more resources.

Lab: Reflection of Reflections

Carry out an investigation to discover the effect of using two mirrors at the same time.

Laboratory: Reflection of Light

Carry out an investigation to visualize the effect of reflecting light.

Plane Mirrors

Greek mythology tells the story of a handsome young man named Narcissus who noticed his image in a pond and fell in love with himself. Like pools of water, mirrors are smooth surfaces that reflect light to form images. Just as Narcissus did, you can see yourself as you glance into a quiet pool of water or walk past a shop window. Most of the time, however, you probably look for your image in a flat, smooth mirror. A flat, smooth mirror is a **plane mirror.**

Get It?
Define What is a plane mirror?

Reflections from plane mirrors

What do you see when you look into a plane mirror? Your reflection is upright. If you were one meter in front of the mirror, your image would appear to be one meter behind the mirror, or two meters from you. You might notice that the reflection of any writing in a plane mirror appears backward.

Figure 2 shows how you see yourself in a plane mirror. First, light rays from a light source strike you. Every point that is struck by the light rays reflects these rays so they travel outward in all directions. If your friend were looking at you, these reflected light rays coming from you would enter her eyes so she could see you. However, if a mirror is placed between you and your friend, the light rays are reflected from the mirror into your eyes.

Get It?
Describe the steps that allow you to see your face in a plane mirror.

Figure 2 Light reflects off of the girl's forehead and then off of the mirror before entering the girl's eyes.

SCIENCE USAGE v. COMMON USAGE

plane

Science usage: having no elevations or depressions

A plane mirror is a flat mirror.

Common usage: short for *airplane*; a powered, heavier-than-air aircraft with fixed wings

We took a plane from Houston to Orlando.

Virtual images

You can understand your brain's interpretation of your reflection in a mirror by looking at **Figure 3.** The light waves that are reflected off you travel in all directions. Light rays reflected from your chin strike the mirror at different places. Then, they reflect off the mirror in different directions. A few of these light rays reflect off the mirror in just the right way to enter your eyes.

Recall that your brain always interprets light rays as if they have traveled in a straight line. It does not realize that the light rays have been reflected and that they changed direction. Your reflected image appears to be behind the mirror.

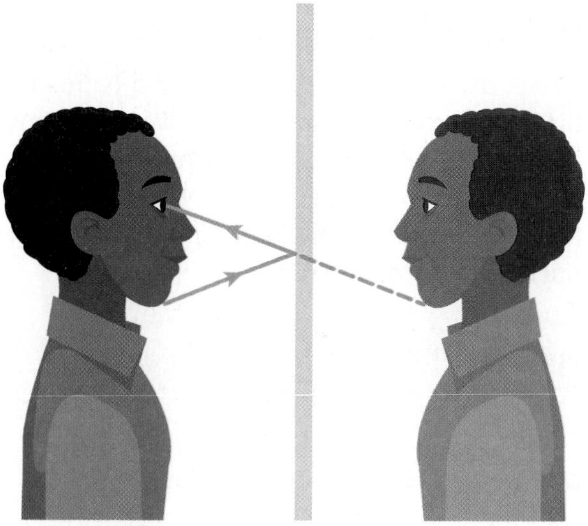

Figure 3 The light rays that reflect off of a plane mirror appear to originate behind that mirror. This gives the illusion that objects exist behind the plane mirror.

Get It?

Explain Why does your reflected image in a plane mirror appear to be behind the mirror?

An image that your brain perceives even though no light rays pass through the location of that image is a **virtual image.** The imaginary light rays that appear to come from virtual images are called virtual rays. The dashed line in **Figure 3** is a virtual ray. Plane mirrors always form upright, virtual images.

Concave Mirrors

Not all mirrors are flat like plane mirrors. A **concave mirror** is a mirror whose surface curves inward. Concave mirrors, like plane mirrors, reflect light waves to form images. However, a concave mirror's curved surface produces different images from a plane mirror's flat surface.

Features of concave mirrors

A concave mirror has an optical axis. The **optical axis** is an imaginary straight line drawn perpendicular to the surface of the mirror at the mirror's center. Concave mirrors are made so that every light ray traveling toward the mirror parallel to the optical axis is reflected through a point on the optical axis called the focal point.

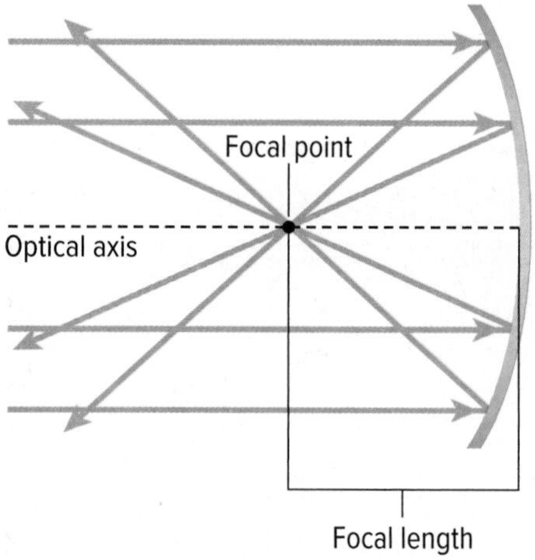

Figure 4 A concave mirror has an optical axis and a focal point. When light rays travel toward the mirror parallel to the optical axis, they reflect through the focal point. Light rays that travel through the focal point before hitting the mirror are reflected parallel to the optical axis.

The **focal point** for a concave mirror is the point on the optical axis on which light rays that are initially parallel to the optical axis converge after they reflect off the mirror. The distance from the center of the mirror to the focal point is the **focal length.** Using the focal point and the optical axis, you can diagram how some of the light rays that travel to a concave mirror are reflected, as shown in **Figure 4.**

Get It?

Describe the relationship between the focal point and the focal length for a concave mirror.

Ray tracing for concave mirrors

You can diagram how concave mirrors form images by tracing some of the light rays involved. Suppose that the distance between an object, such as the candle in **Figure 5,** and the mirror is greater than the focal length. Light rays bounce off the candle in all directions. One light ray, labeled Ray A, starts from a point on the flame of the candle and passes through the focal point on its way to the mirror. Ray A is then reflected parallel to the optical axis.

Another ray, Ray B, starts from the same point on the candle's flame, but it travels parallel to the optical axis as it moves toward the mirror. The mirror then reflects Ray B through the focal point. The place where Ray A and Ray B meet after they are reflected is a point on the reflected image of the flame.

More points on the reflected image can be located in this way. From each point on the candle, one ray can be drawn that passes through the focal point and is reflected parallel to the optical axis. Another ray can be drawn that travels parallel to the optical axis and then reflects through the focal point. The point where the two rays meet is on the reflected image.

Real images

The image that is diagrammed in **Figure 5** is not virtual. Rays of light pass through the location of the image. A **real image** is an image that is formed when light rays converge to form the image. You could hold a sheet of paper at the location of a real image and see the image projected on the paper.

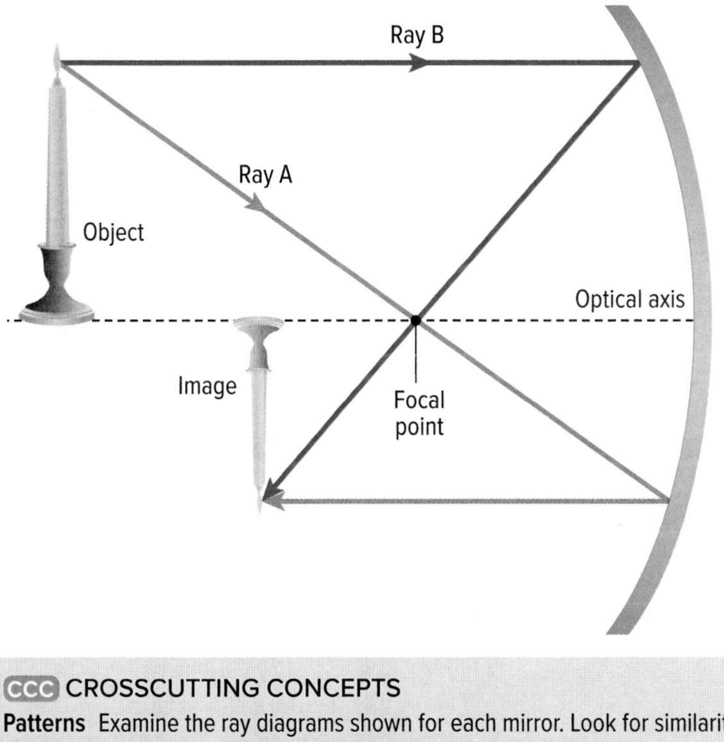

Figure 5 Ray A first passes through the focal point and then reflects parallel to the optical axis. Ray B is first parallel to the optical axis and then reflects through the focal point. An image of the candle forms where the two rays converge.

Diagram *how other points on the image of the candle are formed.*

CCC CROSSCUTTING CONCEPTS

Patterns Examine the ray diagrams shown for each mirror. Look for similarities and differences in the diagrams, and identify any patterns you observe.

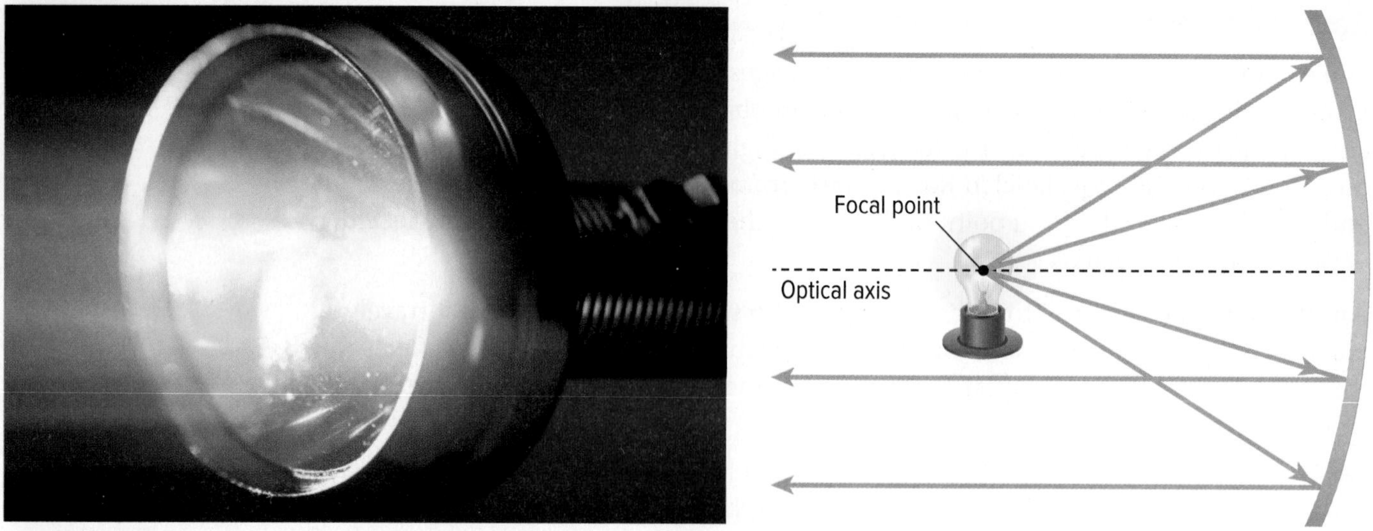

Figure 6 A light beam forms whenever someone places a light source at a concave mirror's focal point.
Explain *why the reflected rays of light in the diagram are parallel to each other.*

Spotlights

What happens when you place an object exactly at the focal point of a concave mirror? **Figure 6** shows that when the object is at the focal point, the mirror reflects all light rays parallel to the optical axis. The rays never meet, and no image forms. Even the virtual rays that extend behind the mirror do not meet. Therefore, a light placed at the focal point is reflected in a beam. Car headlights, flashlights, spotlights, and other devices use concave mirrors in this way to produce light beams with nearly parallel rays.

Virtual images

A concave mirror forms a virtual image when you place an object between the concave mirror and that mirror's focal point. **Figure 7** shows that the reflected rays diverge, and a magnified virtual image forms.

Just as it does with a plane mirror, your brain interprets the diverging rays as if they came from one point behind the mirror. You can find this point by imagining virtual rays that extend behind the mirror. The resulting image is magnified. Shaving mirrors and makeup mirrors are concave mirrors that are used for magnification. They form enlarged, upright images of a person's face so that it is easier to see small details.

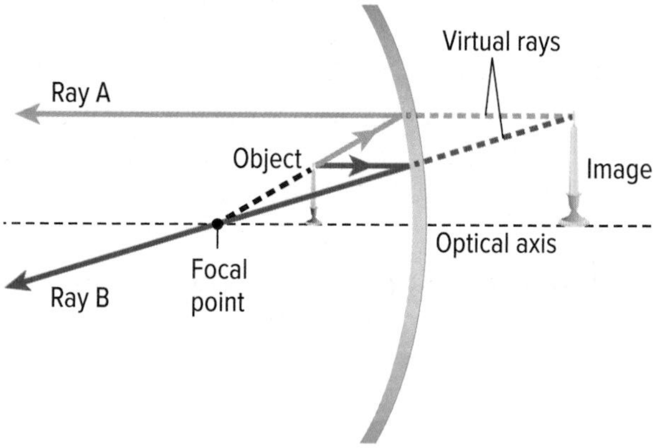

Figure 7 An enlarged and virtual image forms where the virtual rays converge when an object is placed between a concave mirror and that mirror's focal point.
Infer *why this image could not be projected on a screen.*

Nicholas Eveleigh/Stockbyte/Getty Images

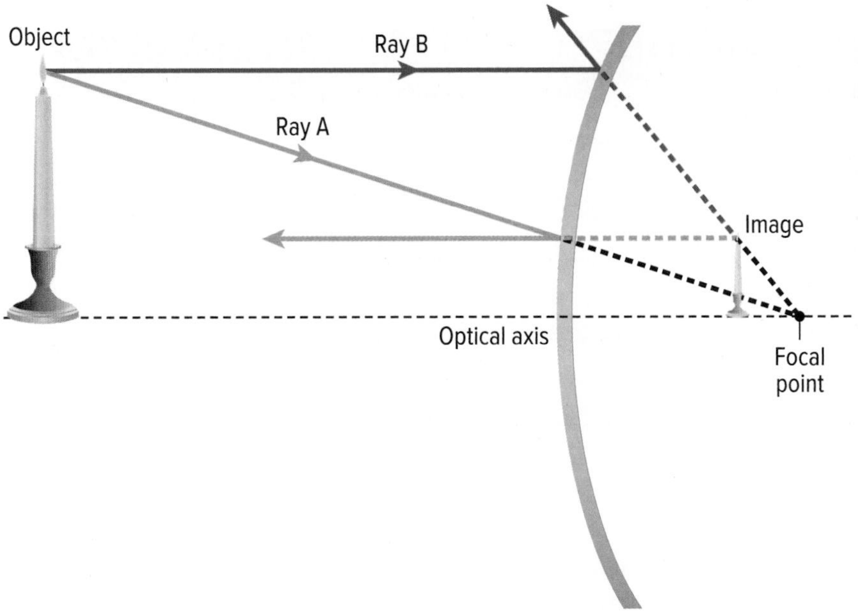

Figure 8 Convex mirrors always form reduced, upright, virtual images.

Convex Mirrors

Why do you think the security mirrors in banks and stores are shaped the way that they are? The next time that you are in a store, look at one of the back corners or at the end of an aisle to see if a large, rounded mirror is mounted there. You can see a large area of the store in the mirror. A **convex mirror** is a mirror that curves outward, like the back of a spoon.

Light rays that hit a convex mirror spread apart after they are reflected. Look at **Figure 8** to see how the rays from an object are reflected off a convex mirror to form an image. The reflected rays diverge and never meet, so the image formed by a convex mirror is a virtual image. The image is also always upright and is smaller than the actual object.

Get It?
Describe the image formed by a convex mirror.

Uses of convex mirrors

Because convex mirrors cause light rays to diverge, they allow large areas to be viewed. As a result, a convex mirror is said to have a wide field of view. In addition to increasing the field of view in places like grocery stores and factories, convex mirrors can widen the view of traffic that can be seen in rear-view or side-view mirrors of automobiles.

However, because the image that a convex mirror forms is smaller than the object, your perception of distance can be distorted. Objects look farther away than they truly are in a convex mirror. Distances and sizes seen in a convex mirror are not realistic, so most convex side mirrors on cars carry a printed warning that states "Objects in mirror are closer than they appear."

Get It?
Evaluate What are the benefits and drawbacks of using convex mirrors instead of plane mirrors on automobiles?

Table 1 Images Formed by Mirrors

Mirror Shape	Distance of Object from Mirror	Virtual/Real	Image Created Upright/ Upside Down	Size
Plane	any distance	virtual	upright	same as object
Concave	object more than two focal lengths from mirror	real	upside down	smaller than object
	object between one and two focal lengths	real	upside down	larger than object
	object at focal point	none	none	none
	object within focal length	virtual	upright	larger than object
Convex	any distance	virtual	upright	smaller than object

Mirror images

The different shapes of plane, concave, and convex mirrors cause them to reflect light in distinct ways. For example, concave mirrors are the only mirrors that magnify images. Convex mirrors always make objects appear to be smaller and farther away than they actually are. Each type of mirror has different uses. Most wall mirrors are plane mirrors. Most makeup and shaving mirrors are concave mirrors. Most store security mirrors are convex mirrors. **Table 1** summarizes the characteristics of plane mirrors, concave mirrors, and convex mirrors.

Get It?

Identify which mirrors form only virtual images.

Check Your Progress

Summary

- You see an object because your eyes detect the light reflected from that object.
- Plane mirrors are smooth and flat.
- No light rays pass through the location of a virtual image.
- A concave mirror curves inward.
- A convex mirror curves outward.

Demonstrate Understanding

1. **Diagram** how both concave mirrors and convex mirrors form images.
2. **Identify** at least one example of a plane mirror, one example of a concave mirror, and one example of a convex mirror.
3. **Describe** the image of an object that is 38 cm from a concave mirror that has a focal length of 10 cm.
4. **Infer** whether a virtual image can be photographed.

Explain Your Thinking

5. **Describe** An object is less than one focal length from a concave mirror. How does the size of the image change as the object gets closer to the mirror?
6. **MATH Connection** If you stand 2 m away from a plane mirror, how far away does your reflection appear to be from you?

LEARNSMART® Go online to follow your personalized learning path to review, practice, and reinforce your understanding.

LESSON 2
LENSES

FOCUS QUESTION
How do different types of lenses form images?

What is a lens?

What do your eyes have in common with cameras and eyeglasses? Each of these things contains at least one lens. A lens is a transparent material with at least one curved surface that causes light rays to bend, or refract, as those rays pass through the lens. The image that a lens forms depends on the shape of the lens. Like curved mirrors, a lens can be convex or concave.

Convex Lenses

A **convex lens** is a lens that is thicker in the middle than at the edges. Its optical axis is an imaginary straight line that is perpendicular to the surface of the lens at its thickest point. When light rays approach a convex lens traveling parallel to its optical axis, the rays are refracted toward the center of the lens, as shown in **Figure 9.**

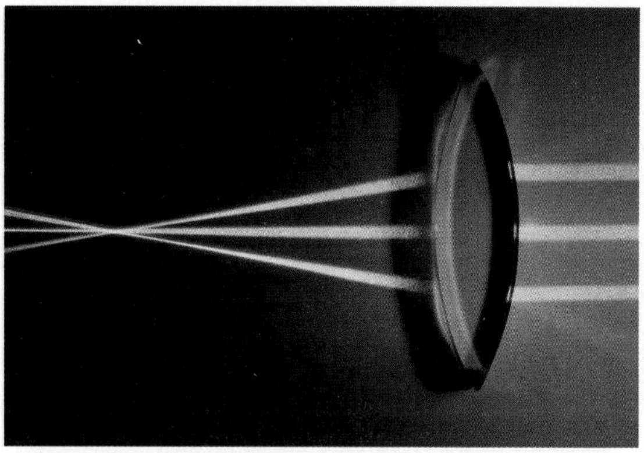

Figure 9 A convex lens bends light inward, toward the optical axis. A light ray that passes straight through the center of the lens does not refract.

Identify *the focal point of the lens.*

All light rays traveling parallel to the optical axis in **Figure 9** are refracted so they pass through a single point, which is the focal point of the lens. The focal length of the lens depends on the shape of the lens. If the sides of a convex lens are less curved, light rays are bent less. As a result, lenses with flatter sides have longer focal lengths. **Figure 9** also shows that light rays traveling along the optical axis are not bent at all.

3D THINKING DCI Disciplinary Core Ideas CCC Crosscutting Concepts SEP Science & Engineering Practices

COLLECT EVIDENCE

Use your Science Journal to record the evidence you collect as you complete the readings and activities in this lesson.

INVESTIGATE

GO ONLINE to find these activities and more resources.

Laboratory: Magnifying Power
Carry out an investigation to determine the relationship between the magnifying power and focal length of a lens.

Virtual Investigation: Mirrors and Lenses
Use a computer model to investigate how the structure of a lens relates to its function in correcting vision.

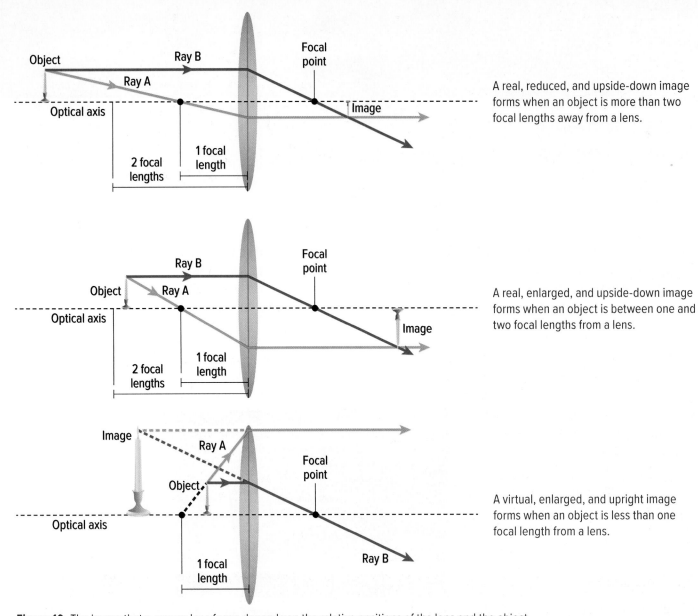

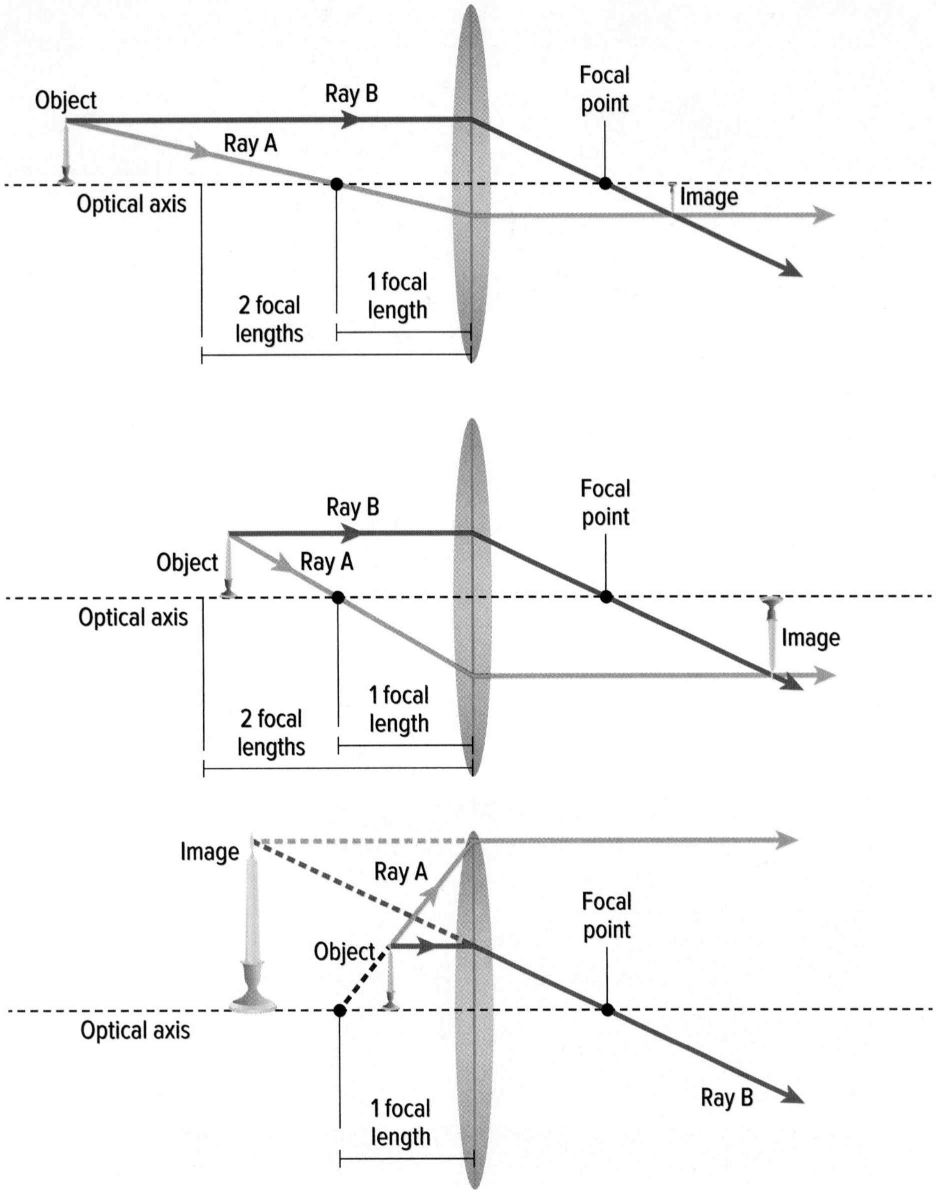

A real, reduced, and upside-down image forms when an object is more than two focal lengths away from a lens.

A real, enlarged, and upside-down image forms when an object is between one and two focal lengths from a lens.

A virtual, enlarged, and upright image forms when an object is less than one focal length from a lens.

Figure 10 The image that a convex lens forms depends on the relative positions of the lens and the object.
Identify *the type of mirror that produces images that are similar to the images produced by a convex lens.*

Forming images with convex lenses

The type of image that a convex lens forms depends on where the object is relative to the focal point of the lens. If an object is more than two focal lengths from the lens, as in the top panel of **Figure 10,** the image is real, reduced, upside-down (inverted), and on the opposite side of the lens from the object.

As the object moves closer to the lens, the image gets larger. The middle panel of **Figure 10** shows the image formed when the object is between one and two focal lengths from the lens. Now the image is larger than the object but is still upside-down.

When an object is less than one focal length from the lens, as shown in the bottom panel of **Figure 10,** the image becomes an enlarged, virtual image. The image is virtual because the light rays from the object are not converging after they have passed through the lens. When you use a magnifying glass, you move a convex lens so that it is less than one focal length from an object. This causes the image of the object to be magnified.

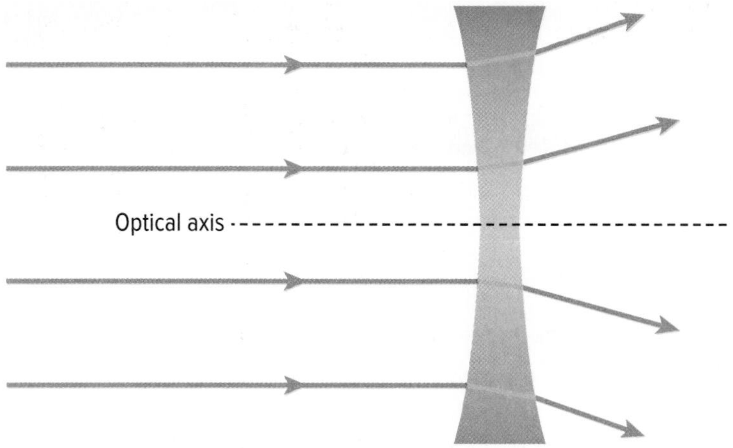

Figure 11 A concave lens causes light rays to diverge.

Classify *Does a concave lens behave more like a concave mirror or a convex mirror?*

Concave Lenses

A **concave lens** is a lens that is thinner in the middle and thicker at the edges. As shown in **Figure 11,** light rays that pass through a concave lens bend outward, away from the optical axis. The rays spread out and never meet at a focal point, so they never form a real image. However, a concave lens can form virtual images. These virtual images are always upright and smaller than the actual object. Notice that concave lenses and convex mirrors both produce the same types of images.

Concave lenses are used in some types of eyeglasses and in some microscopes. Concave lenses are usually placed in combination with other lenses. A summary of the images formed by concave and convex lenses is shown in **Table 2.**

Table 2 Images Formed by Lenses

Lens Shape	Location of Object	Virtual/Real	Image Created Upright/Upside Down	Size
Convex 	object beyond 2 focal lengths from lens	real	upside down	smaller than object
	object between 1 and 2 focal lengths	real	upside down	larger than object
	object within 1 focal length	virtual	upright	larger than object
Concave 	object at any position	virtual	upright	smaller than object

APPLY SCIENCE

How do object distance and image distance compare?

The size and orientation of an image formed by a lens depends on the location of the object and on the nature of the lens. Convex lenses form both real images and virtual images. Concave lenses can form only virtual images. What happens to the location of the image formed by a lens as the object moves closer to or farther from the lens? The distance from the lens to the object is the object distance, and the distance from the lens to the image is the image distance. How are the focal length, object distance, and image distance related to each other?

Identify the Problem

A 5-cm-tall object is placed at different distances from a convex lens with a focal length of 15 cm. The table at the right shows the different object and image distances. How are these two measurements related?

Object and Image Distances

Focal Length (cm)	Object Distance (cm)	Image Distance (cm)
15.0	45.0	22.5
15.0	30.0	30.0
15.0	20.0	60.0

Solve the Problem

1. Describe the relationship between the object distance and the image distance.
2. The lens equation describes the relationship between the focal length and the image and object distances.

$$\frac{1}{\textit{focal length}} = \frac{1}{\textit{object distance}} + \frac{1}{\textit{image distance}}$$

Using this equation, calculate the image distance when the object is placed at a distance of 60.0 cm from the lens.

Eyesight and Lenses

What determines how well you can see the words on this page? If you do not need eyeglasses, the structure of your eye gives you the ability to focus on these words and on other objects around you. Look at **Figure 12.** Light enters your eye through a transparent covering on your eyeball called the **cornea** (KOR nee uh). The cornea causes light rays to bend so that they converge.

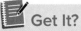

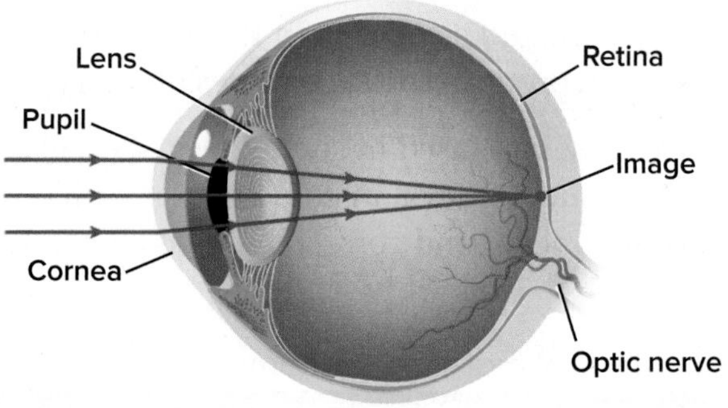

Figure 12 The cornea is a convex lens that causes light rays from distant objects to converge on the retina. The eye lens helps bring closer objects into focus.

Get It?

Describe the function of the cornea.

After passing through the cornea, the light then passes through an opening called the pupil. Behind the pupil is a flexible, convex lens called the eye lens. The eye lens helps focus light rays so that a sharp image is formed on your retina. Your **retina** is the inner lining of your eye, which has cells that convert the light image into electrical signals. These electrical signals are then carried along the optic nerve to your brain, where they can be interpreted.

Get It?

Describe the function of the retina.

Focusing on near and far

How can your eyes focus both on close objects, such as a watch on your wrist, and distant objects, such as a clock across the room? For you to see an object clearly, its image must be focused sharply on your retina. However, the retina is always a fixed distance from the lens. Remember that the location of an image formed by a convex lens depends on the focal length of the lens and the location of the object.

For an image to be formed on the retina, the focal length of the lens needs to be able to change as the distance to the object changes. The lens in your eye is flexible, and muscles attached to it change its shape and its focal length. This is why you can see both near and far away objects.

Look at **Figure 13.** When you focus on an object far from your eye, the muscles around the lens relax. This pulls the lens into a less convex shape. When you focus on a nearby object, these muscles make the lens more curved, causing the focal length to decrease.

Get It?

Describe how the shape of the lens in your eye changes when you focus on a nearby object.

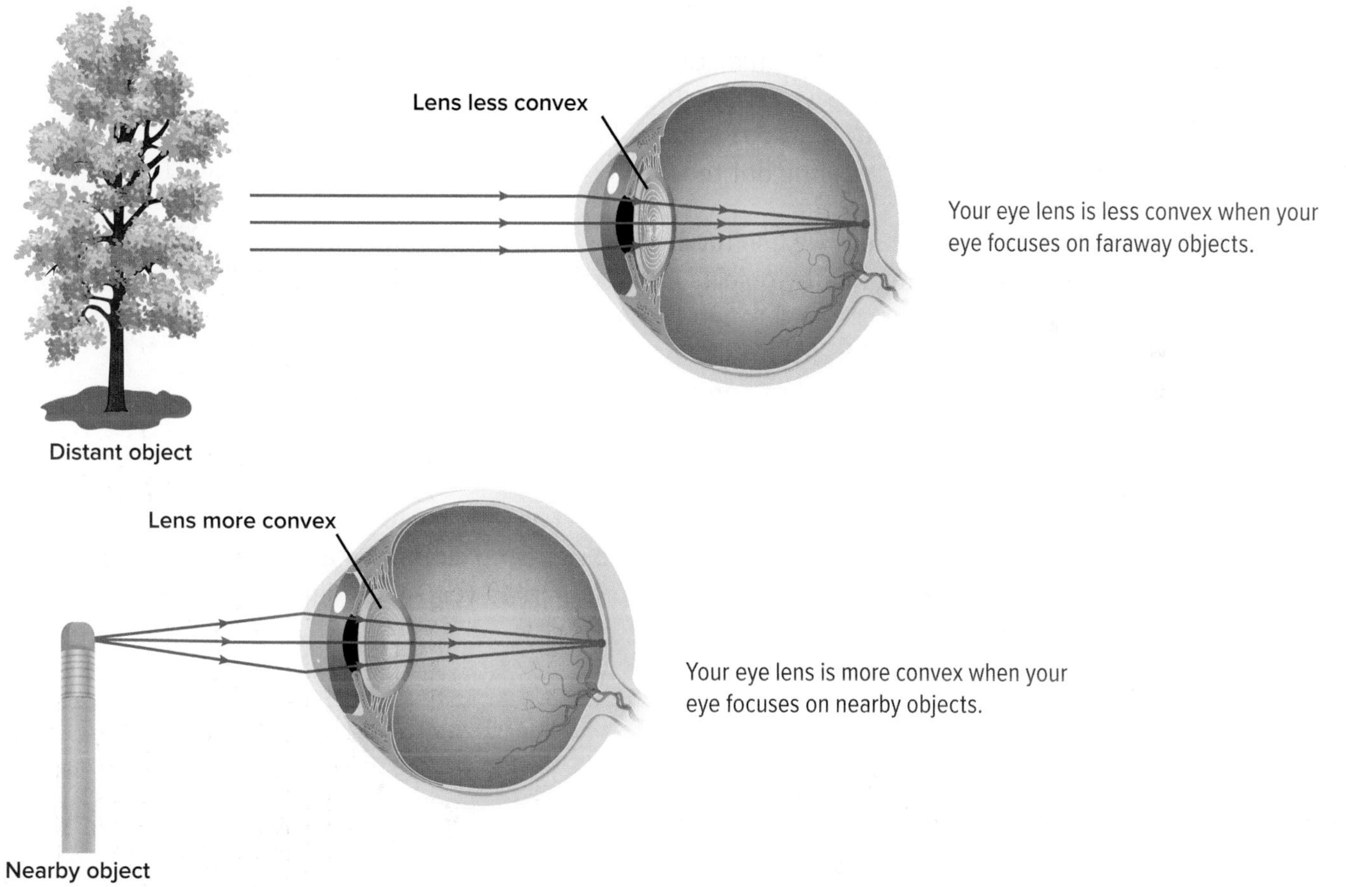

Lens less convex
Your eye lens is less convex when your eye focuses on faraway objects.
Distant object
Lens more convex
Your eye lens is more convex when your eye focuses on nearby objects.
Nearby object

Figure 13 The eye lens changes shape so that you can focus on objects at different distances.

Infer *why you are more likely to get eyestrain by looking at a nearby object than by looking at a faraway object.*

Farsighted

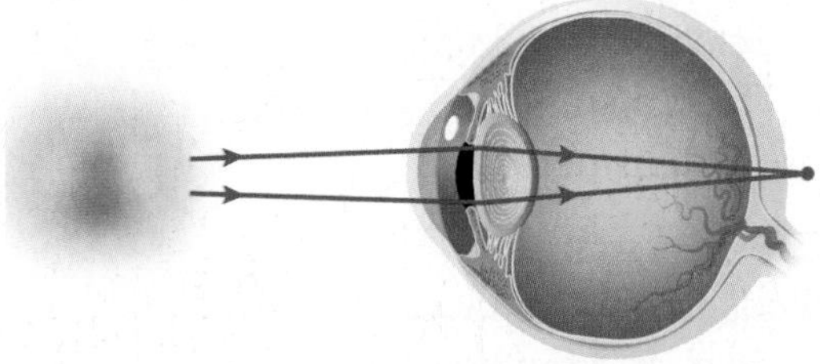

In a farsighted eye, light rays from nearby objects do not converge enough to form a sharp image on the retina.

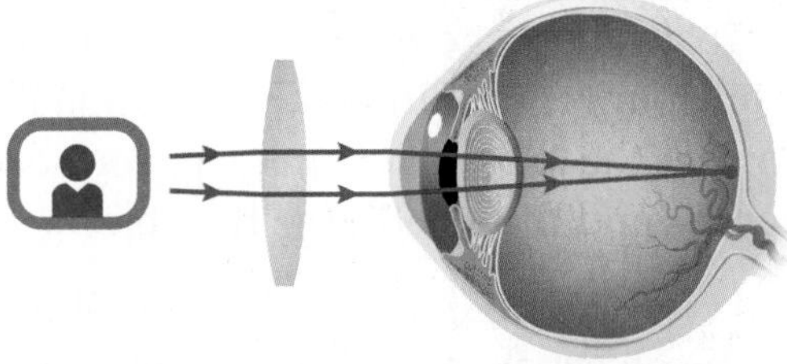

A convex lens causes light rays to converge slightly before entering the eye. As a result, sharp images can be formed on the retina.

Figure 14 A farsighted person can see faraway objects clearly, but he or she has trouble focusing on nearby objects. For example, a farsighted person could watch a football game from the stands without glasses. However, a farsighted person would have more difficulty reading without proper glasses. Farsightedness can be corrected by convex lenses.

Vision problems

People with good vision can clearly see objects that are about 25 cm or farther away from their eyes. However, people with the most common vision problems see objects clearly only at some distances, or they see all objects as being blurry.

Astigmatism One vision problem, called astigmatism, occurs when the surface of the cornea is unevenly curved. When people have astigmatism, their corneas are more oval than round in shape. Astigmatism causes blurry vision at all distances. Corrective lenses must also have an uneven curvature, canceling out the effect of an uneven cornea.

Farsightedness Another vision problem is farsightedness. A farsighted person can see distant objects clearly but cannot bring nearby objects into focus. Light rays from nearby objects do not converge enough after passing through the cornea and the lens to form a sharp image on the retina, as shown in **Figure 14.** The problem can be corrected with a convex lens that bends light rays so they are less spread out before they enter the eye, also shown in **Figure 14.**

Farsightedness is often related to age. As many people age, the lenses in their eyes become less flexible. The muscles around the lenses still contract as they try to change the shape of the lens. However, the lenses have become more rigid and cannot be made curved enough to focus on close objects. People who are more than 40 years old might not be able to focus on objects closer than 1 m from their eyes.

STEM CAREER Connection

Optometrist

Do you have a desire to help improve the quality of life for others? One way to do this is by improving their vision. An optometrist provides vision care, including vision testing and writing prescriptions for corrective lenses.

Nearsighted

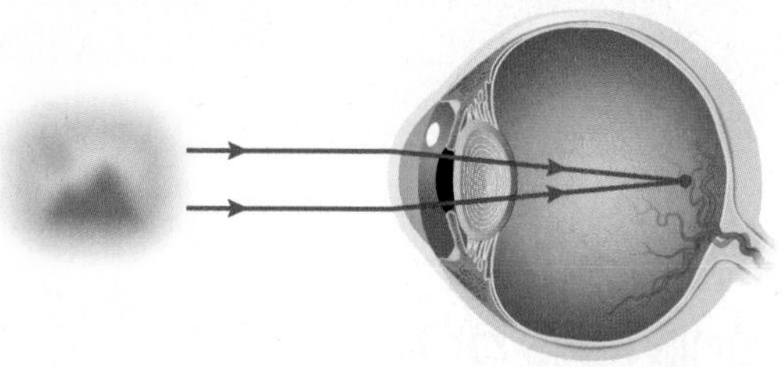

In a nearsighted eye, light rays from distant objects converge too much and form a sharp image in front of the retina.

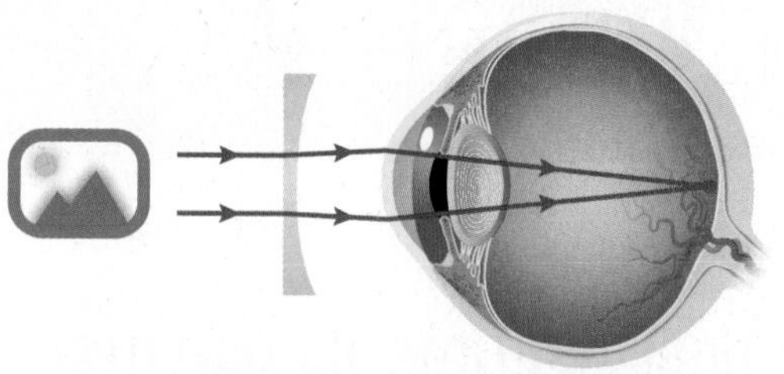

A concave lens makes the light rays spread out before entering the eye, enabling a sharp image to be formed on the retina.

Figure 15 People use concave lenses to correct nearsightedness.

Nearsightedness A person who is nearsighted can see objects clearly only when those objects are nearby. Objects that are far away appear blurred. In a nearsighted eye, the cornea and the lens form a sharp image of a distant object before the light reaches the retina, as shown in **Figure 15.**

To correct this problem, a nearsighted person can wear concave lenses. **Figure 15** shows how a concave lens causes incoming light rays to diverge before they enter the eye. Then the light rays from distant objects can be focused by the eye to form a sharp image on the retina and not in front of it.

Check Your Progress

Summary

- A convex lens is thicker in the middle than at the edges. Light rays are refracted toward the optical axis.
- The image formed by a convex lens depends on the distance of the object from the lens.
- A concave lens is thinner in the middle and thicker at the edges. Light rays are refracted away from the optical axis.
- The cornea and the lens focus light onto the retina.

Demonstrate Understanding

7. **Sketch** light rays as they pass through a convex lens and then through a concave lens.
8. **Compare** the image of an object less than one focal length from a convex lens with the image of an object more than two focal lengths from the lens.
9. **Describe** the image formed by a concave lens.
10. **Explain** how lenses are used to correct vision problems.

Explain Your Thinking

11. **Describe** If image formation by a convex lens is similar to that by a concave mirror, describe the image formed by a light source placed at the focal point of a convex lens.
12. **MATH Connection** If you looked through a convex lens with a focal length of 15 cm and saw a real, upside-down, enlarged image, what is the maximum distance between the lens and the object?

LEARNSMART® Go online to follow your personalized learning path to review, practice, and reinforce your understanding.

LESSON 3

OPTICAL INSTRUMENTS

FOCUS QUESTION

Why do some optical instruments use more than one lens or mirror?

Telescopes

You know from your experience that it is difficult to see faraway objects clearly. When you look at an object, only some of the light reflected from its surface enters your eyes. As you move farther away from the object, the amount of light entering your eyes decreases, as shown in **Figure 16.** As a result, the object appears dimmer and less detailed.

A telescope uses a lens or a concave mirror that is much larger than your eye to gather more of the light from distant objects. The largest telescopes can gather more than a million times more light than the human eye. As a result, objects such as distant galaxies appear much brighter. Because the image formed by a telescope is so much brighter, more detail can be seen when the image is magnified. Telescopes and many other technologies based on the understanding of waves and their interactions with matter are part of everyday experiences in the modern world and in scientific research.

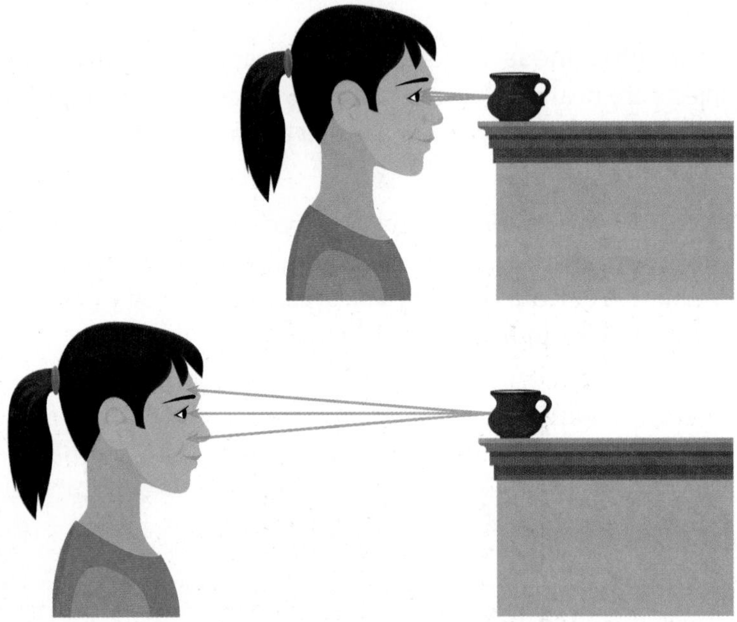

Figure 16 All three of the light rays reflected off the mug enter the girl's eye in the top panel, but only one of those light rays enters the girl's eye in the bottom panel. As the girl gets farther away, fewer light rays reflected from the mug enter the girl's eyes.

Refracting telescopes

One common type of telescope is the refracting telescope. A telescope that uses lenses to gather light from distant objects is called a **refracting telescope.** A simple refracting telescope uses two convex lenses to gather and focus light from distant objects.

3D THINKING DCI Disciplinary Core Ideas CCC Crosscutting Concepts SEP Science & Engineering Practices

COLLECT EVIDENCE

Use your Science Journal to record the evidence you collect as you complete the readings and activities in this lesson.

INVESTIGATE

GO ONLINE to find these activities and more resources.

LabA: Telescopes Today
Obtain and communicate information about the function of large telescopes.

Review the News
Obtain information from a current news story about optical instruments. Evaluate your source and communicate your findings to your class.

Look at the telescope shown in **Figure 17.** Incoming light from distant objects passes through the first lens, called the objective lens. Light rays from distant objects are nearly parallel to the optical axis of the lens. As a result, the objective lens forms a real image at the focal point of the lens, within the body of the telescope.

The second convex lens, called the eyepiece lens, magnifies this real image. When you look through the eyepiece lens, you see an enlarged, upside-down, virtual image of the real image formed by the objective lens.

In order to form detailed images of distant objects, the objective lens of a refracting telescope must be as large as possible. A telescope lens can be supported only around its edge. A large lens can sag or flex due to its own weight, distorting the image that it forms. Another class of telescopes, called reflecting telescopes, do not have this problem.

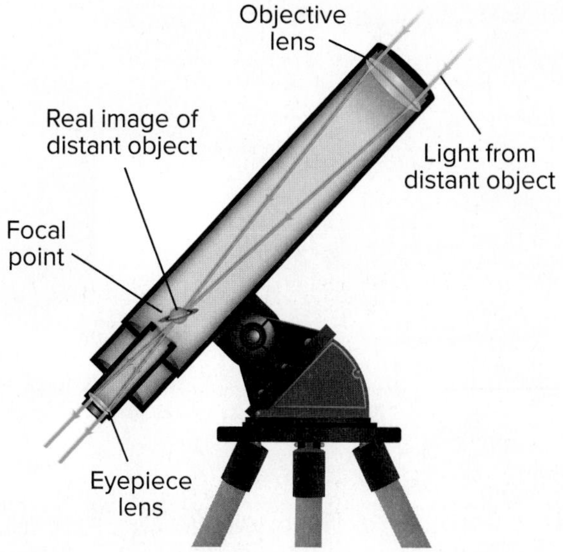

Figure 17 A refracting telescope uses an objective lens and an eyepiece lens to gather light from distant objects.

Reflecting telescopes

A telescope that uses mirrors and lenses to collect and focus light from distant objects is a **reflecting telescope.** Mirrors, unlike lenses, can be supported from behind. This additional support for mirrors prevents mirrors from sagging inside reflecting telescopes. As a result, reflecting telescopes can be much larger than refracting telescopes. **Figure 18** shows a reflecting telescope.

For this reflecting telescope, light from a distant object enters one end of the telescope and strikes a concave mirror at the opposite end. The light reflects off this mirror and converges. Before it converges at a focal point, the light hits a plane mirror inside the telescope tube. The light is then reflected from the plane mirror and toward the telescope's eyepiece. The light rays converge at the focal point, creating a real image of the distant object. Just like a refracting telescope, a convex lens in the eyepiece then magnifies this image.

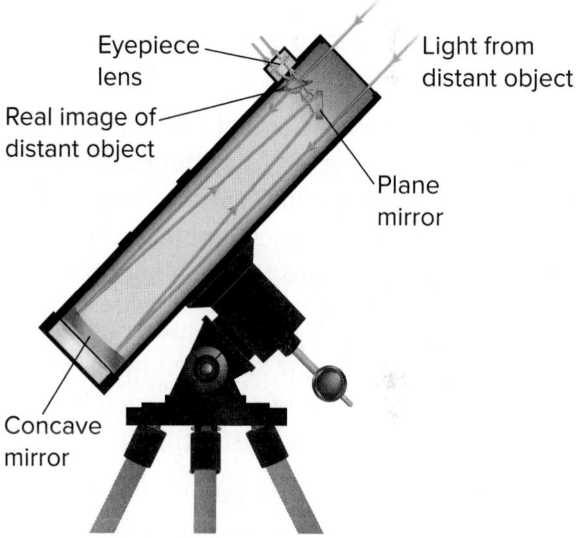

Figure 18 A reflecting telescope uses two mirrors and an eyepiece lens to gather light from distant objects.

Get It?
Identify the interactions of waves and matter that allow reflecting telescopes to function.

Space telescopes

Imagine being at the bottom of a swimming pool and trying to read a sign by the pool's edge. The motion of the water in the pool would distort your view of any object beyond the water's surface. In a similar way, Earth's atmosphere distorts the view of objects in space.

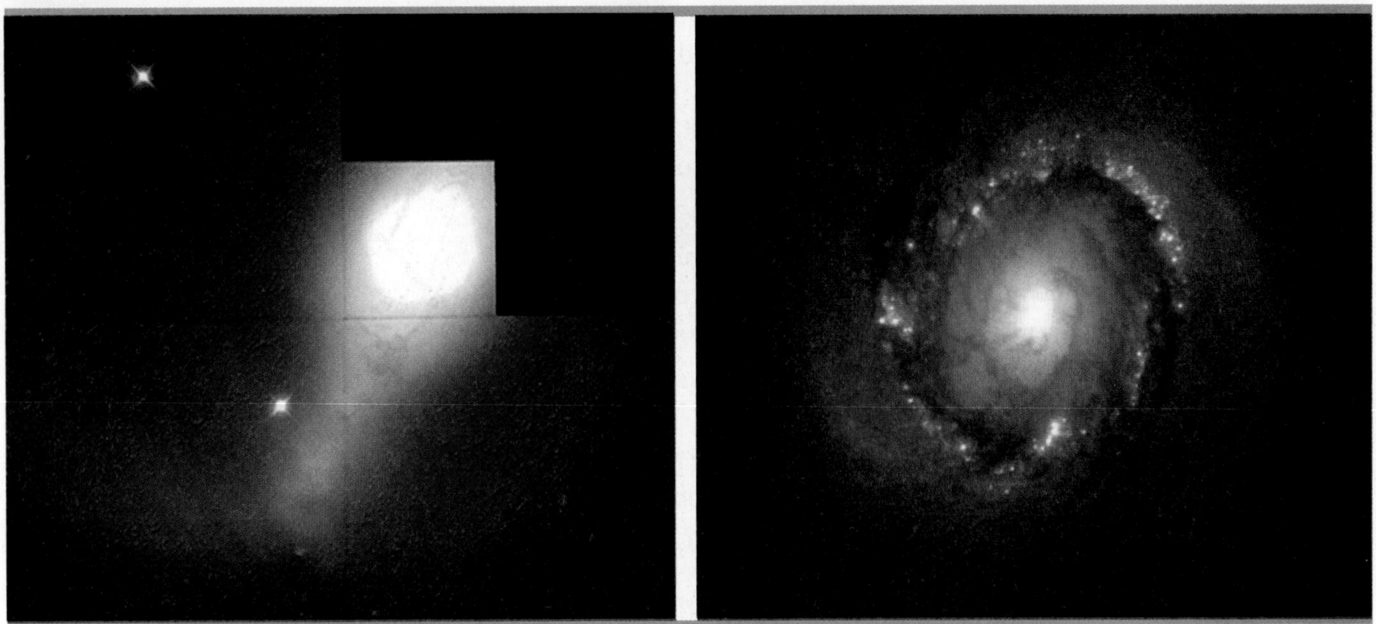

The distorting effects of Earth's atmosphere can cause telescopes on Earth to form blurry images.

The *Hubble Space Telescope* is above Earth's atmosphere and forms clearer images of objects in space.

Figure 19 The view from telescopes on Earth is different from the view from telescopes in space. Both of these images show the core of the same galaxy, and both were taken through telescopes of similar size. However, the photograph on the left was taken at the McDonald Observatory on Earth, and the photograph on the right was taken by the *Hubble Space Telescope*.

To overcome the blurriness of humans' view into space, the National Aeronautics and Space Administration (NASA) built a telescope called the *Hubble Space Telescope* and launched it into space, high above Earth's atmosphere. Because *Hubble* is above Earth's atmosphere, it has produced incredibly sharp and detailed images. **Figure 19** shows the difference in images produced by telescopes on Earth and the *Hubble* telescope.

With the *Hubble Space Telescope,* scientists can detect planets, stars, and galaxies whose light would otherwise be scattered by Earth's atmosphere. *Hubble* is not the only space telescope. Other space telescopes, such as the *Chandra X-Ray Observatory* and the *Spitzer Space Telescope,* help scientists study the universe through X-ray and infrared radiation.

The *Hubble* telescope is a type of reflecting telescope that uses two mirrors to collect and focus light to form an image. The primary mirror in the telescope is 2.4 m across. A next-generation space telescope, called the *James Webb Space Telescope* (*JWST*), is due to be launched in 2021. The primary mirror on *JWST* will be 6.5 m across.

Get It?

Explain why a space telescope is able to produce clearer images than telescopes on Earth.

ACADEMIC VOCABULARY

distort

to twist out of the true meaning or proportion

People who lie distort the truth.

Microscopes

A telescope would be useless if you were trying to study the cells in a butterfly wing, a sample of pond scum, or the differences between a human hair and a horse hair. You would need a microscope to look at such small objects.

A **microscope** is a device that uses two convex lenses with relatively short focal lengths to magnify small, close objects. A microscope, like a telescope, has an objective lens and an eyepiece lens. However, it is designed differently because the objects viewed are close to the objective lens.

Figure 20 shows a simple microscope. The object to be viewed is placed on a transparent slide and is illuminated from below. The light passes by or through the object on the slide and then travels through the objective lens. The objective lens is a convex lens.

The objective lens forms a real, enlarged image of the object because the distance from the object to the lens is between one and two focal lengths. The real image is then magnified again by the eyepiece lens (another convex lens) to create a virtual, enlarged image. This final image can be hundreds of times larger than the actual object, depending on the focal lengths of the two lenses. The total magnification is the magnification of the objective lens times the magnification of the eyepiece lens.

Get It?

Compare the image formed by the objective lens of a microscope with the image formed by the eyepiece lens.

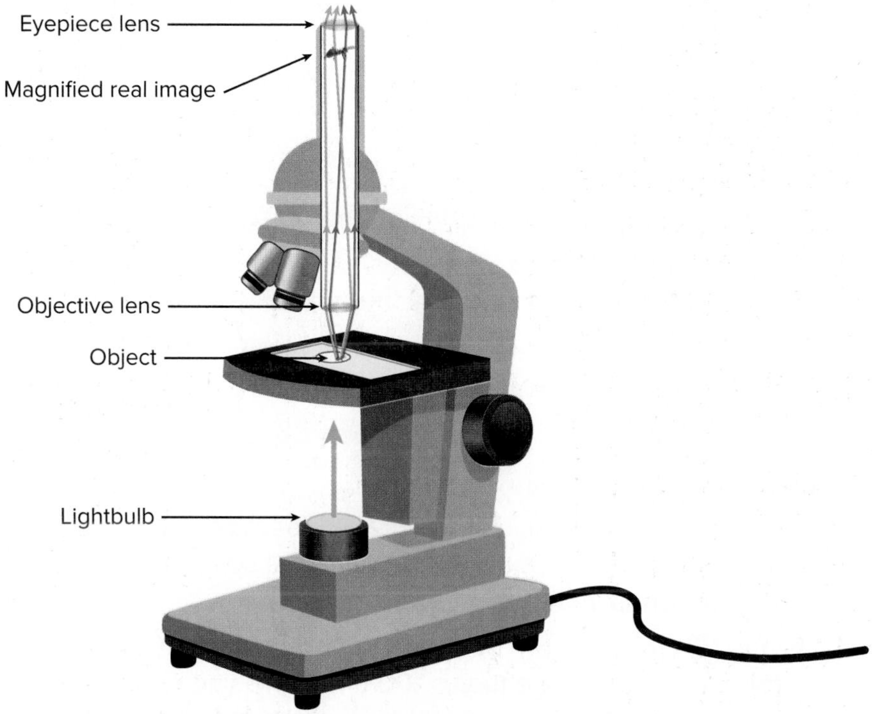

Figure 20 A microscope uses two convex lenses to magnify small objects. The lens closer to the object that is being studied is called the objective lens. In a microscope, unlike in a refracting telescope, more than one lens magnifies the object.

Explain *why this microscope's light source is placed below the object being studied instead of above the object.*

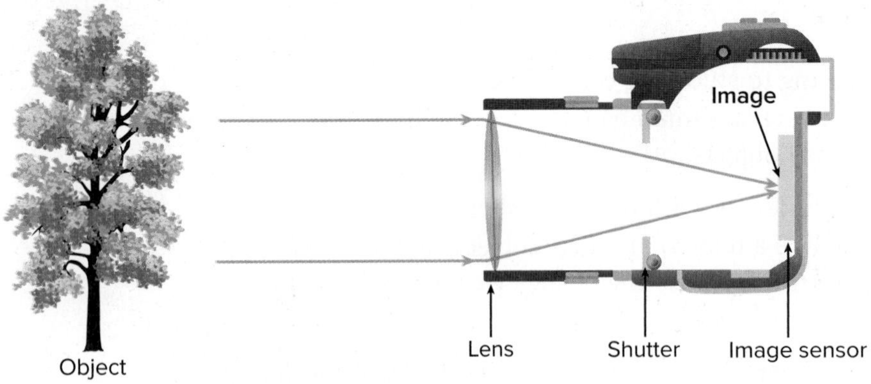

Figure 21 A camera's lens focuses an image onto the image sensor. An image sensor converts the light from an image into a set of electrical signals.

Compare *a digital camera with the human eye.*

Cameras

With the click of a button, you can capture a beautiful scene in a photo. How does a digital camera make a reduced image of a life-sized scene? **Figure 21** shows the path that light follows as it enters a camera from a distant object. The light rays from distant objects are almost parallel to each other. When you take a picture with a camera, a shutter opens to allow light to enter the camera for a specific length of time.

The light reflected off the object enters the camera through an opening called the aperture. The camera lens focuses the image onto an image sensor, which converts light into electric signals. A computer then processes these signals into an image that can be displayed on a screen or printed.

Check Your Progress

Summary

- Refracting telescopes use two convex lenses to gather and focus light.
- Reflecting telescopes use a concave mirror, a plane mirror, and a convex lens to collect, reflect, and focus light.
- A telescope in orbit avoids the distorting effects of Earth's atmosphere.
- A microscope uses two convex lenses with short focal lengths to magnify small, close objects.
- A camera lens focuses light onto an image sensor.

Demonstrate Understanding

13. **Identify** the advantage to making the objective lens larger in a refracting telescope.
14. **Describe** the image formed by the objective lens in a microscope.
15. **Explain** why the largest telescopes are reflecting telescopes instead of refracting telescopes.

Explain Your Thinking

16. **Explain** which optical instrument—a telescope, a microscope, or a camera—forms images in a way most similar to the way your eye forms images.
17. **MATH Connection** Suppose the objective lens in a microscope forms an image that is 100 times the size of an object. The eyepiece lens magnifies this image 10 times. What is the total magnification?

LEARNSMART® Go online to follow your personalized learning path to review, practice, and reinforce your understanding.

The Next Telescopes

Light from distant stars and galaxies reaches Earth day and night. Telescopes capture this light, helping astronomers study the universe. The next generation of telescopes might be able to detect Earth-like planets and uncover secrets of the ancient universe. Three teams are racing to build the world's next giant telescope.

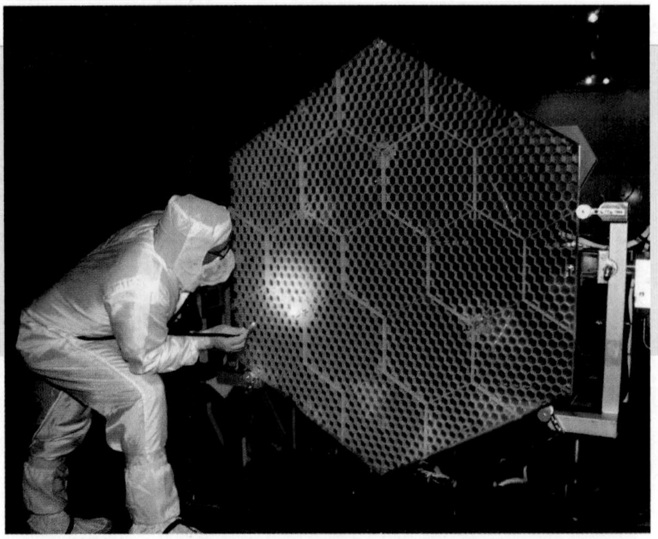

The next generation of giant telescopes will use carefully fitted, segmented mirrors rather than a single, continuous mirror to reflect ancient light.

Light buckets

Telescopes are light buckets, and telescope builders want to catch as much light in their buckets as possible. The larger the telescope's mirror, the more light that the telescope catches. The more light captured, the fainter the objects that the telescope can detect.

Instead of collecting light from a single, continuous mirror, the next generation of telescopes will have mirrors made from many segments, as shown at right. A segmented mirror is more stable than a continuous mirror. As a result, a telescope with a segmented mirror can be much larger than a telescope with a continuous mirror.

Potential New Telescopes

Telescope	Area of Primary Mirror (m^2)
Gran Telescopio Canaria	85 (This is about the area of six parking lot spaces.)
Giant Magellan Telescope	470 (This is about the area of a basketball court.)
Thirty Meter Telescope	700 (This is about the area of a baseball infield.)
European Extremely Large Telescope	1400 (This is about the area of five tennis courts.)

Possible new telescopes

The largest optical telescope in operation today is the Gran Telescopio Canaria. The primary mirror for this telescope has an area of 85 m^2. The smallest candidate for the world's next great telescope is the Giant Magellan Telescope (GMT), which would have a segmented mirror with a total area of 470 m^2. The table details the area of the GMT as well as other possible future telescopes.

Race to first light

The first of the new telescopes to peer into the sky will open a new universe of discovery. The race is not so much to be the biggest but to be first. Even the smallest of these new telescopes would be large enough to spot planets circling nearby stars as well as detect the oldest and most distant objects ever seen.

CONSTRUCT AN ARGUMENT FROM EVIDENCE

Draft a persuasive letter to a prospective donor, describing why your giant telescope deserves the several hundred million dollars in funding needed for construction and operation.

MODULE 13
STUDY GUIDE

GO ONLINE to study with your Science Notebook.

Lesson 1 MIRRORS

- You see an object because your eyes detect the light reflected from that object.
- Plane mirrors are smooth and flat.
- No light rays pass through the location of a virtual image.
- A concave mirror curves inward.
- A convex mirror curves outward.

- plane mirror
- virtual image
- concave mirror
- optical axis
- focal point
- focal length
- real image
- convex mirror

Lesson 2 LENSES

- A convex lens is thicker in the middle than at the edges. Light rays are refracted toward the optical axis.
- The image formed by a convex lens depends on the distance of the object from the lens.
- A concave lens is thinner in the middle and thicker at the edges. Light rays are refracted away from the optical axis.
- The cornea and the lens focus light onto the retina.

- convex lens
- concave lens
- cornea
- retina

Lesson 3 OPTICAL INSTRUMENTS

- Refracting telescopes use two convex lenses to gather and focus light.
- Reflecting telescopes use a concave mirror, a plane mirror, and a convex lens to collect, reflect, and focus light.
- A telescope in orbit avoids the distorting effects of Earth's atmosphere.
- A microscope uses two convex lenses with short focal lengths to magnify small, close objects.
- A camera lens focuses light onto an image sensor.

- refracting telescope
- reflecting telescope
- microscope

THREE-DIMENSIONAL THINKING

Module Wrap-Up

REVISIT THE PHENOMENON

What is the farthest you can see with a telescope like this?

CER Claim, Evidence, Reasoning

Explain Your Reasoning Revisit the claim you made when you encountered the phenomenon. Summarize the evidence you gathered from your investigations and research and finalize your Summary Table. Does your evidence support your claim? If not, revise your claim. Explain why your evidence supports your claim.

STEM UNIT PROJECT

Now that you've completed the module, revisit your STEM unit project. You will apply your evidence from this module and complete your project.

GO FURTHER

SEP Data Analysis Lab

Lenses and Ray Diagrams

Lenses refract light rays twice. Light is refracted first as it enters the lens and again as it leaves the lens. When drawing these diagrams, use the simplified model called the thin lens model. With this model, the lens is replaced by a straight line drawn through the center of the lens. All refraction occurs at this line.

CER Analyze and Interpret Data

1. **Claim** Draw a ray diagram to show why a convex lens is a converging lens.
2. **Claim** Draw a ray diagram to show why a concave lens is a diverging lens.
3. **Claim** Draw a ray diagram to show light passing first through a concave lens, then immediately through a convex lens.

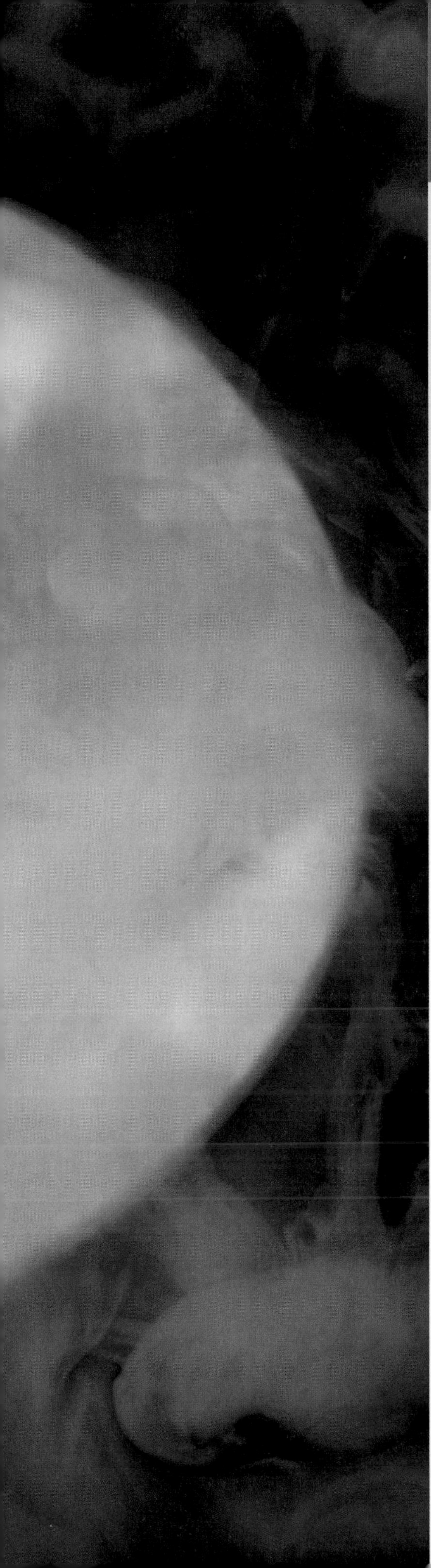

UNIT 4

MATTER

ENCOUNTER THE PHENOMENON

Why can dry ice go directly from a solid to a gas?

SEP Ask Questions

What questions do you have about the phenomenon? Write your questions on sticky notes and add them to the driving question board for this unit.

What happens during a phase change?

Look for Evidence

As you go through this unit, use the information and your experiences to help you answer the phenomenon question as well as your own questions. For each activity, record your observations in a Summary Table, add an explanation, and identify how it connects to the unit and module phenomenon questions.

Solve a Problem

STEM UNIT PROJECT

Model Recycling Rare Earth Elements While rare earth elements are not uniquely rare in Earth's crust, it can be difficult finding deposits of them that are worth mining. Additionally, there is a high environmental cost of extracting them. Investigate where the majority of rare earth elements are found and the implications of this. Research how these elements can be recycled and the benefits of doing so. Use the results of these investigations and the evidence you collected during the unit to model the recycling of rare earth elements.

GO ONLINE In addition to reading the information in your Student Edition, you can find the STEM Unit Project and other useful resources online.

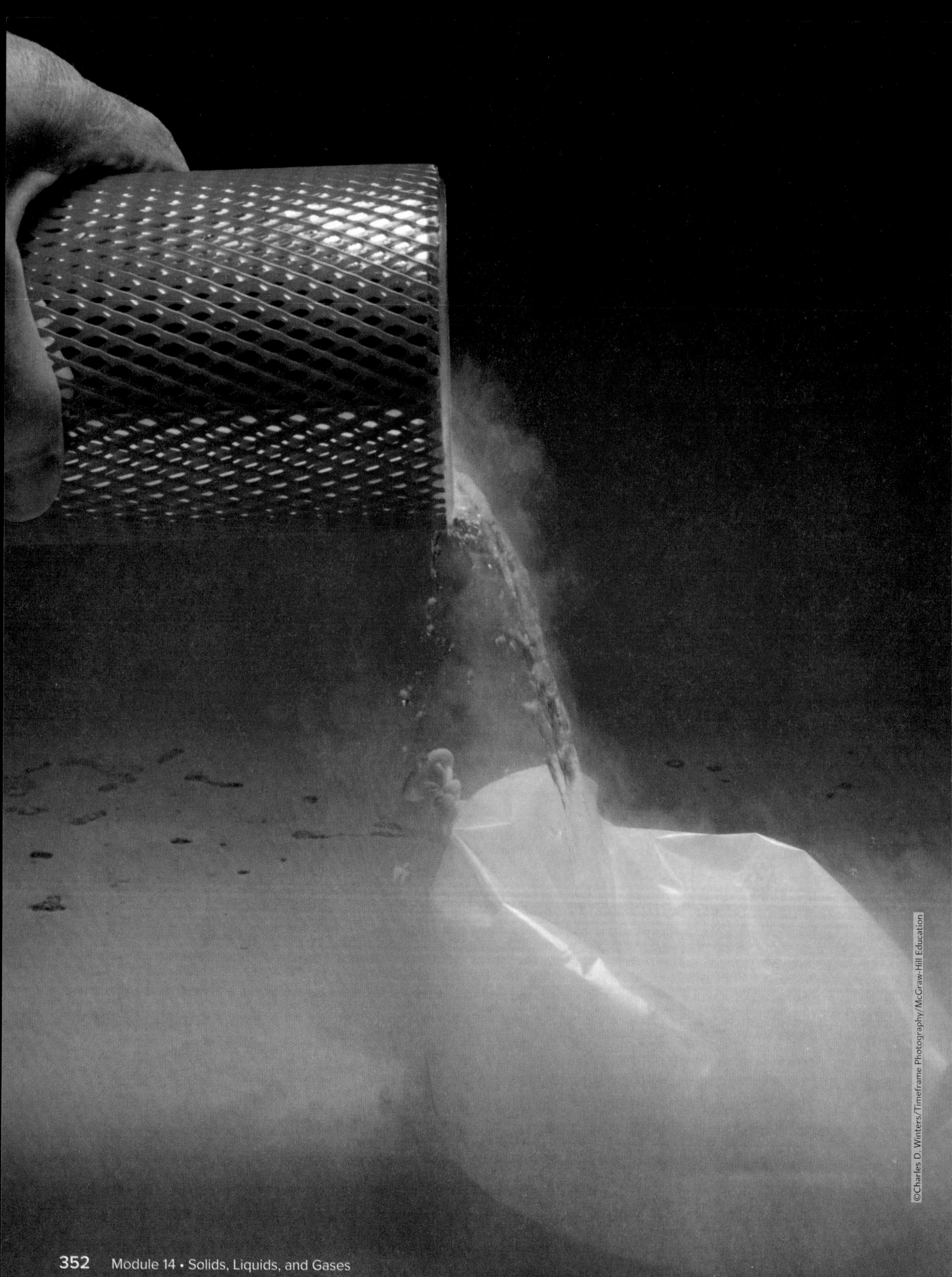

MODULE 14

SOLIDS, LIQUIDS, AND GASES

ENCOUNTER THE PHENOMENON

Why does a balloon crumple and shrink when liquid nitrogen is poured on it?

GO ONLINE to play a video about changing the oxygen in air from a gas to a liquid.

SEP Ask Questions

Do you have other questions about the phenomenon? If so, add them to the driving question board.

CER Claim, Evidence, Reasoning

Make Your Claim Use your CER chart to make a claim about what happens at the particle level as matter changes from a solid to a liquid to a gas.

Collect Evidence Use the lessons in this module to collect evidence to support your claim. Record your evidence as you move through the module.

Explain Your Reasoning You will revisit your claim and explain your reasoning at the end of the module.

GO ONLINE to access your CER chart and explore resources that can help you collect evidence.

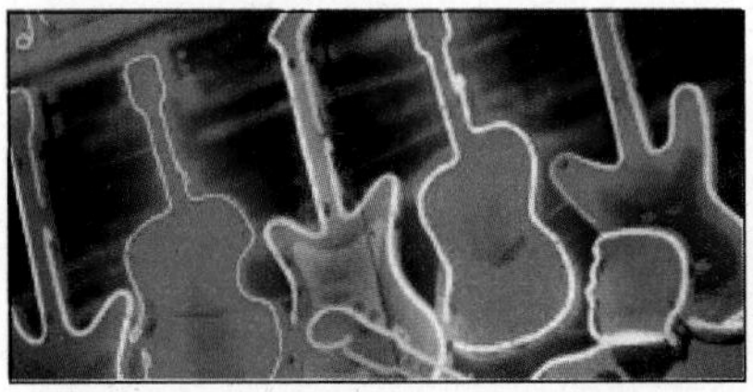

LESSON 1: Explore & Explain: Plasma State

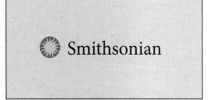

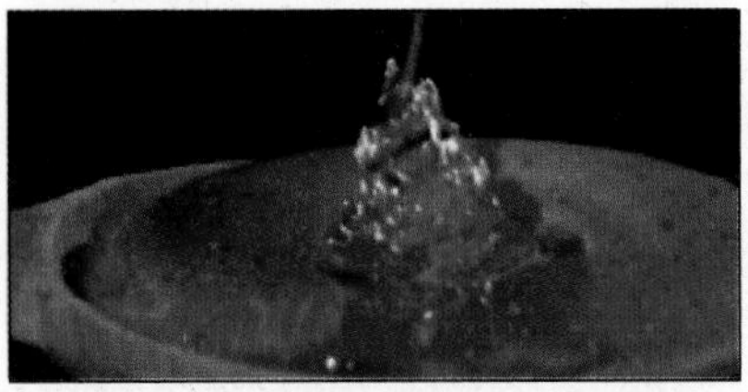

LESSON 2: Explore & Explain: Viscosity

Smithsonian

Additional Resources

LESSON 1

MATTER AND THERMAL ENERGY

FOCUS QUESTION

How do changes in thermal energy affect the particles that make up matter?

Kinetic Theory

You encounter solids, liquids, and gases every day. Look at **Figure 1.** Can you identify the states of matter present? The tea is in the liquid state. The ice cubes dropped into the tea to cool it are in the solid state. Surrounding the glass, as part of the air, is water in the gas state. How do these states compare?

Gas state

To understand the states of matter, we must think about the particles that make up matter. Consider the air around you: it is composed of nitrogen, oxygen, and water, along with other gases. These atoms and molecules—the particles that make up the air—are constantly moving. The **kinetic theory,** also known as kinetic molecular theory, is an explanation of how the particles in gases behave. To explain the behavior of particles, it is necessary to make the following assumptions:

1. All matter is composed of tiny particles (atoms, molecules, and ions).
2. These particles are in constant, random motion.
3. The particles collide with each other and with the walls of any container in which they are held.
4. The amount of energy that the particles lose from these collisions is negligible.

Figure 1 Water is a substance that can exist in all three common states of matter at the same time.
Identify *the solid and liquid states of water in this photo.*

Figure 2, on the next page, illustrates the kinetic theory. Because the particles of a substance in the gas state are in constant motion, colliding with each other and with the walls of their container, gases do not have a fixed volume or shape. The particles that make up a gas spread out so that they fill whatever container they are in.

3D THINKING — DCI Disciplinary Core Ideas — CCC Crosscutting Concepts — SEP Science & Engineering Practices

COLLECT EVIDENCE

Use your Science Journal to record the evidence you collect as you complete the readings and activities in this lesson.

INVESTIGATE

GO ONLINE to find these activities and more resources.

Lab: Phase Changes
Carry out an investigation into the energy absorbed or released from a system when matter changes state.

Revisit the Encounter the Phenomenon Question
What information from this lesson can help you answer the Module question?

Liquid state

Although the kinetic theory explains the behaviors of gas particles, some of its assumptions also apply to liquids and solids. The particles of a substance in the liquid state, shown in **Figure 2,** are also constantly moving, although they are not moving as quickly as they would be if the substance were in the gas state. Particles that make up a substance have less kinetic energy when in its liquid state than when in its gas state.

Because they have less energy, the particles are less able to overcome their attractions to each other. They can slide past each other, allowing a liquid to flow and take the shape of its container. However, because the particles that make up a liquid have not completely overcome the attractive forces between them, the particles cling together, giving the liquid a definite volume. The structure of matter at the bulk scale is determined by forces within and between particles.

Solid state

A solid has a definite shape and volume. The particles that make up a solid are closely packed together, as shown in **Figure 2.** They are still in motion, but they have so little kinetic energy that the particles are unable to overcome their attractions to each other.

Many solids are crystalline, which means their particles have specific geometric arrangements. **Figure 3** shows the geometric arrangement of ice. The hydrogen and oxygen atoms alternate in the arrangement.

Thermal energy

Think about the ice in **Figure 3.** How can frozen, solid ice have motion? The particles that make up solids are held tightly in place by the attractions between the particles. Those attractions give solids a definite shape and volume. However, the particles that make up a solid are still in constant motion. The particles' thermal energy causes them to vibrate.

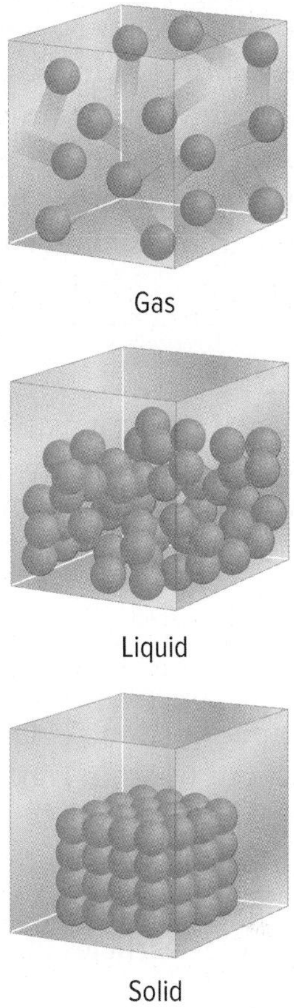

Figure 2 Solids, liquids, and gases differ in how their particles move. These differences account for their physical properties.

Compare and contrast *the shape and volume of each state of matter.*

Figure 3 Ice is a crystalline solid—its particles have a specific geometric arrangement. Although ice does not look like it is moving, its molecules are vibrating in place.

Design Pics Inc./Alamy Stock Photo

Thermal energy is the total energy of a material's particles. Thermal energy is one of the ways energy manifests at the macroscopic scale. It includes both the kinetic energy of the particles as well as their potential energy. Energy from the motions of individual particles and energy from forces that act within or between particles are both forms of thermal energy. Energy from the motion of an object as a whole and energy from its interactions with its surroundings are not thermal energy.

Temperature

Temperature is the term used to explain how hot or cold an object is. Temperature represents the average kinetic energy of the particles that make up a substance. On average, molecules of water at 0°C have less kinetic energy than molecules of water at 100°C.

Changes of State

What happens to a solid when thermal energy is added to it? Think about the iced water in **Figure 4.** The particles that make up the water are moving fast and colliding with the particles that make up the ice cubes. Those collisions transfer energy from the water to the ice. The particles at the surface of an ice cube vibrate faster, transferring energy to other particles in the ice cube.

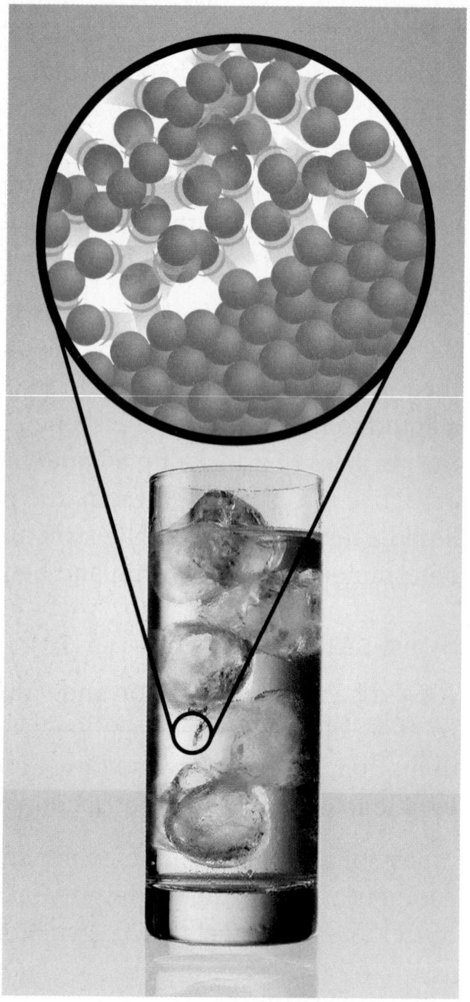

Figure 4 When ice is placed in water, energy from the particles of the liquid water is transferred to the particles of solid ice, melting the ice and cooling the water.

Melting and freezing

Soon, the particles that make up the ice have enough kinetic energy to overcome the attractive forces holding them in their crystalline structure. The ice melts. The **melting point** is the temperature at which a solid becomes a liquid. Energy is required for the particles to slip out of the ordered arrangement of a solid. The **heat of fusion** is the energy required to change a substance from solid to liquid at its melting point.

The transfer of energy between particles of liquid and particles of solid causes the ice to melt, but what happens to the particles of liquid after they collide with the solid? They slow down because they have less kinetic energy. As more of these collisions occur, the average kinetic energy of the particles of the liquid decreases, and the liquid cools.

Freezing is the reverse of melting. When a liquid's temperature is lowered, the average kinetic energy of the molecules decreases. When enough energy has been removed, the molecules become fixed into position. The freezing point is the temperature at which a liquid turns into a solid.

ACADEMIC VOCABULARY

definite

having distinct or certain limits

The teacher set definite standards for the students to meet.

CCC CROSSCUTTING CONCEPTS

Energy and Matter Energy cannot be created or destroyed. It moves between one place and another place, between objects and/or fields, or between systems. Create a graphic organizer that shows the movement of energy during different changes of state.

Vaporization and condensation

How does a liquid become a gas? Remember that the particles that make up a liquid are constantly moving. When particles move fast enough to escape the attractive forces of other particles, they enter the gas state. This process is called vaporization. Vaporization can occur in two ways: evaporation and boiling. The process in which a gas becomes a liquid is called condensation. Condensation is the reverse of vaporization.

Figure 5 As temperature increases, the particles that make up a substance in its liquid state move faster. When their energy creates sufficient pressure to surpass the air pressure above the liquid, the liquid boils.

Infer *What is inside the bubbles of the boiling liquid?*

Evaporation Evaporation occurs at the surface of a liquid and can happen at nearly any temperature. To evaporate, particles must be at the liquid's surface and have enough kinetic energy to escape the attractive forces of the liquid.

Boiling Shown in **Figure 5,** boiling is the second way that a liquid can vaporize. Unlike evaporation, boiling occurs throughout a liquid at a specific temperature, depending on the pressure on the surface of the liquid.

The **boiling point** of a liquid is the temperature at which the pressure of the vapor in the liquid is equal to the external pressure acting on the surface of the liquid. This external pressure pushes down on the liquid, keeping particles from escaping. Particles require energy to overcome this pressure. The **heat of vaporization** is the amount of energy required for the liquid at its boiling point to become a gas.

Sublimation At certain pressures, some substances can change directly from solids into gases without going through the liquid phase. **Sublimation** is the process of a solid changing directly to a gas without forming a liquid. **Figure 6** shows frozen carbon dioxide, also known as dry ice, which is a common substance that undergoes sublimation.

Figure 6 Carbon dioxide (CO_2) turns from a solid directly to a gas. Because this gas is very cold, it causes water in the air to condense, forming a white fog.

Explain, *at the molecular level, the behavior of CO_2 as it undergoes this phase transition.*

(t)Richard Hutchings/Digital Light/McGraw-Hill Education, (b)IrinaK/Shutterstock

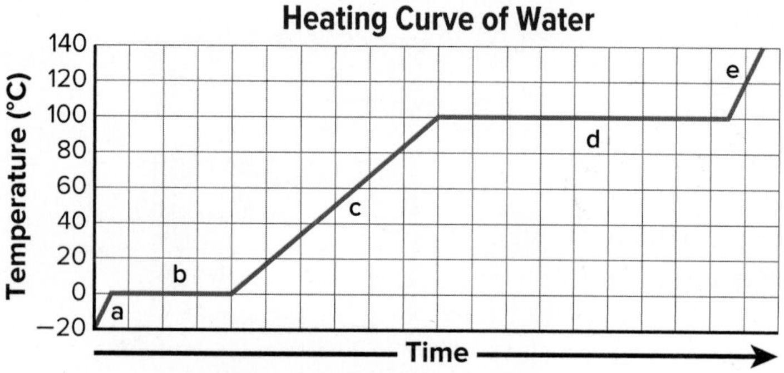

Figure 7 Although thermal energy is added at a constant rate, the temperature of the water increases only at *a*, *c*, and *e*. At *b* and *d*, the added energy is used to overcome the attractions between the particles.

Infer *how this graph would be different if 2.0 kg of water were being heated instead of 1.0 kg of water. How would this graph be different if 0.5 kg of water were being heated?*

Heating curves

A graph of temperature v. time for heating of 1.0 kg of water is shown in **Figure 7.** This type of graph is called a heating curve. It shows how temperature changes over time as thermal energy is continuously added. Notice the two areas on the graph where the temperature does not change. At 0°C, ice is melting. All of the energy put into the ice at this temperature is used to overcome the attractive forces between the particles. The flat line on the graph indicates that temperature remains constant during melting.

After the attractive forces between the particles of the solid are overcome, the particles move more freely, and their temperature increases. At 100°C, water is boiling, the temperature remains constant again, and the graph is flat. All of the energy that is put into the water goes to overcoming the remaining attractive forces between the particles. When all of the attractive forces between the particles of the liquid are overcome, the energy goes into increasing the temperature of the gas.

Get It?

Explain why the graph in Figure 7 is flat at b.

Plasma State

So far, you have learned about the three familiar states of matter—solids, liquids, and gases. However, there is a state of matter beyond the gas state. **Plasma** is matter that has enough energy to overcome not just the attractive forces between its particles but also the attractive forces within its atoms. The atoms that make up a plasma collide with such force that the electrons are stripped off the atoms. As a result, plasmas are made up of electrons and other charged particles.

You may be surprised to learn that most of the ordinary matter in the universe is in the plasma state. Every star that you can see in the sky, including the Sun, is composed of matter in the plasma state. Most of the matter between the stars and galaxies is also in the plasma state. The familiar states of matter—solid, liquid, and gas—are extremely rare in the universe.

Thermal Expansion

Have you ever wondered why a concrete sidewalk has seams? When thermal energy is transferred to a concrete sidewalk, the concrete expands. Without the seams, a concrete sidewalk would crack in hot weather. The kinetic theory can help to explain this behavior.

Recall that particles move faster and farther apart as the temperature rises. This separation of particles results in an expansion of the entire object, known as thermal expansion. **Thermal expansion** is an increase in the size of a substance when the temperature is increased. Substances also contract when they cool.

Thermometers

A common example of liquids undergoing thermal expansion occurs in thermometers, like the one shown in **Figure 8.** The addition of energy causes the particles that make up the liquid in the thermometer to move faster. As their motion increases, the particles that make up the liquid in the narrow thermometer tube start to move farther apart. This causes the liquid in a thermometer to expand and rise as the temperature increases.

Figure 8 As the air temperature goes up, the liquid in the thermometer is heated and expands. As a result, the level of the liquid rises. The liquid in the thermometer contracts with a decrease in temperature.

Hot-air balloons

An application of gases undergoing thermal expansion is shown in **Figure 9.** Hot-air balloons are able to rise due to the thermal expansion of air. The air in the balloon is heated, causing the distance between the particles that make up the air to increase. As the hot-air balloon expands, the number of particles per cubic centimeter decreases. This expansion results in a decreased density of hot air. Because the density of the air in the hot-air balloon is lower than the density of the cooler air outside, the balloon will rise.

Figure 9 When the air inside a hot-air balloon is heated, its particles move farther apart. The balloon rises because the air inside it is less dense than the surrounding air.

Water's strange behavior

Ordinarily, substances contract as their temperatures decrease. However, an exception to this rule is water. Over a small range of temperatures, water expands as the temperature decreases. At first, it behaves like other substances. As the temperature begins to drop, the particles that make up water move closer together. This continues until the water reaches 4°C.

Water molecules are unusual in that they have highly positive and highly negative areas. These charged regions affect the behavior of water. As the temperature of water continues to drop under 4°C, the molecules line up so that only positive and negative areas are near each other, as shown in **Figure 10.** As a result, empty spaces occur in the structure. Water expands as it cools from about 4°C to 0°C and becomes less dense than liquid water. That is why ice floats in liquid water.

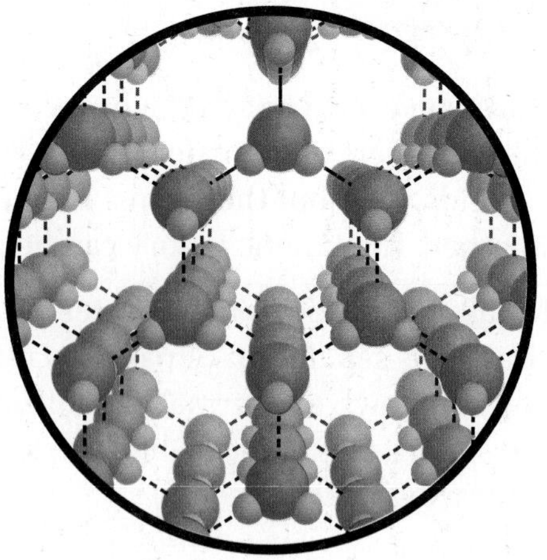

Figure 10 When water freezes, the positively and negatively charged ends of the water molecules interact, creating empty spaces in the crystal lattice.

Explain *why ice floats in water.*

Get It?

Describe how forces between and within water molecules cause ice to float.

Solid or Liquid?

Other substances also show unusual behaviors when changing states. Amorphous solids and liquid crystals are two classes of materials that do not react as you would expect when they are changing states.

Amorphous solids

Ice melts at 0°C, and lead melts at 327°C. But not all solids have a specific temperature at which they melt. Consider a stick of butter. Instead of having a specific melting point, butter softens and melts over a range of temperatures.

Other solids are similar to butter in this way. Instead of having a specific melting point, they soften and gradually turn into a liquid over a temperature range. These solids lack a crystalline structure and are called amorphous solids. One common amorphous solid is glass, shown in **Figure 11.**

Figure 11 Glass lacks the repeating crystalline structure of solids like ice. Rather than melting at a specific temperature, glass becomes increasingly soft and malleable as temperature increases.

WORD ORIGINS

amorphous

comes from the Greek word *amorphos*, which means *without form, shapeless, deformed Clay, which can be shaped and formed easily, is amorphous.*

STEM CAREER Connection

Glass Artist

Glass artists take advantage of the amorphous properties of glass to create artistic pieces. They use different techniques to blow, shape, and join glass under high temperatures. These artists refine their skills through practice. Many study art at the postsecondary level. Others learn through workshops or privately, alongside an experienced artist.

Liquid crystals

Liquid crystals form another group of materials that do not change states in the usual manner. Normally, the ordered geometric arrangement of a solid is lost when the substance goes from the solid state to the liquid state. Liquid crystals start to flow during the melting phase, similar to a liquid. But they do not lose their ordered arrangement completely, as most substances do. Liquid crystals retain their geometric order in specific directions.

Liquid crystals are placed in classes, depending upon the type of order they maintain when they liquify. They are highly responsive to temperature changes and electric fields. Scientists use the unique properties of liquid crystals to make liquid crystal displays (LCDs) for cell phones, calculators, and netbooks, shown in **Figure 12.** LCD screens are composed of individual crystal picture elements, or "pixels" for short. Varying the amount of electricity that passes through the pixel determines how the crystals are aligned and whether light is able to pass through them.

Figure 12 Many small computing devices and electronics, such as cell phones, TVs, and tablets, utilize liquid crystal displays (LCDs).

Check Your Progress

Summary

- There are four main states of matter: solid, liquid, gas, and plasma.
- The kinetic theory is an explanation of how the particles that make up gases behave.
- Thermal energy is the total energy of the particles that make up a material, including kinetic and potential energy.
- Temperature is the average kinetic energy of a substance.

Demonstrate Understanding

1. **Describe** the movement of particles in solids, liquids, and gases.
2. **State** the basic assumptions of the kinetic theory.
3. **Describe,** in terms of kinetic theory, how the particles of a substance behave at its melting point.
4. **Describe,** in terms of kinetic theory, how the particles of a substance behave at its boiling point.

Explain Your Thinking

5. **Infer** How would the heating curve for glass be different from the heating curve for water?
6. **MATH Connection** Using the graph in **Figure 7,** describe the energy changes that occur when water goes from –15°C to 120°C.
7. **MATH Connection** The melting point of acetic acid is 17°C, and the boiling point is 118°C. Draw a graph similar to the graph in **Figure 7** showing the phase changes for acetic acid. Clearly mark the three phases, the boiling point, and the melting point.

LEARNSMART® Go online to follow your personalized learning path to review, practice, and reinforce your understanding.

LESSON 2
PROPERTIES OF FLUIDS

FOCUS QUESTION

What principles describe the behavior of fluids?

Archimedes' Principle and Buoyancy

Some ships are like floating cities. For example, aircraft carriers are large enough to allow airplanes to take off and land on their decks. Despite their weights, these ships float. There is a force pushing up on the ship that opposes the gravitational force pulling the ship down.

What is the force pushing up on the ship? It is called the buoyant force. If the buoyant force is equal to an object's weight, the object will float. If the buoyant force is less than an object's weight, the object will sink. **Buoyancy** is the ability of a fluid—a liquid or a gas—to exert an upward force on an object immersed in it.

Archimedes' principle

In the third century B.C., a Greek mathematician named Archimedes made a discovery about buoyancy. Archimedes found that the buoyant force on an object is equal to the weight of the fluid displaced by the object. For example, if you place a block of wood in water, it will push water out of the way as it begins to sink—but only until the weight of the water displaced equals the block's weight.

When the weight of water displaced—the buoyant force—becomes equal to the weight of the block, the block floats. If the weight of the water displaced is less than that of the block, the block sinks. **Figure 13** shows the forces that affect objects in fluids.

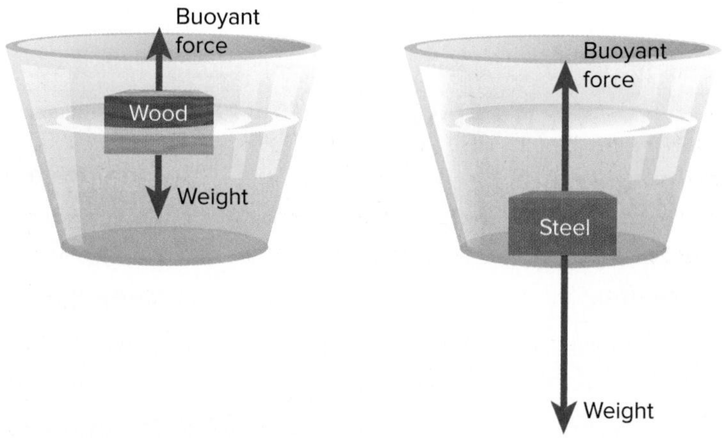

Figure 13 The steel block sinks because the buoyant force from the fluid is less than the gravitational force on the object. When the buoyant force is the same as or greater than the gravitational force—as with the wooden block—the object floats.

Compare *the volumes of the wood block and the steel block. How do the masses of the wood block and the steel block compare?*

Get It?

Infer why rocks sink and rubber balls float in water.

3D THINKING — DCI Disciplinary Core Ideas — CCC Crosscutting Concepts — SEP Science & Engineering Practices

COLLECT EVIDENCE

Use your Science Journal to record the evidence you collect as you complete the readings and activities in this lesson.

INVESTIGATE

GO ONLINE to find these activities and more resources.

Laboratory: Density of a Liquid
Analyze and interpret data to calculate and compare the densities of several liquids.

CCC **Identify Crosscutting Concepts**
Create a table of the crosscutting concepts and fill in examples you find as you read.

Comparing buoyancy and weight

Look again at the wood and steel blocks in **Figure 13.** They both displace the same volume and weight of water when submerged. Therefore, the buoyant forces on the blocks are equal. Yet, the steel block sinks and the wood block floats. What is different?

The steel block weighs much more than the wood block. The gravitational force on the steel block is enough to make the steel block sink. The gravitational force on the wood block is not enough to make the wood block sink.

Density and buoyancy

One way to know whether an object will float or sink is to compare its density to the density of the fluid in which it is placed. An object floats if its density is less than that of the fluid. Remember that density is mass per unit volume. The density of the steel block is greater than the density of water. The wood block's density is less than that of water.

Suppose you formed the steel block into the shape of the hull of a ship filled with air, as shown in **Figure 14.** Now the same mass takes up a larger volume. The overall density of the steel boat and air is less than the density of water. The boat will now float.

Get It?
Explain why a steel block sinks but a steel ship floats.

Pascal's Principle and Pressure

Blaise Pascal (1623–1662), a French scientist, discovered that pressure applied to a fluid is transmitted throughout the fluid. This is why when you squeeze one end of a toothpaste tube, toothpaste emerges from the other end. The pressure has been transmitted through the toothpaste. In order to understand Pascal's principle, you must first understand pressure.

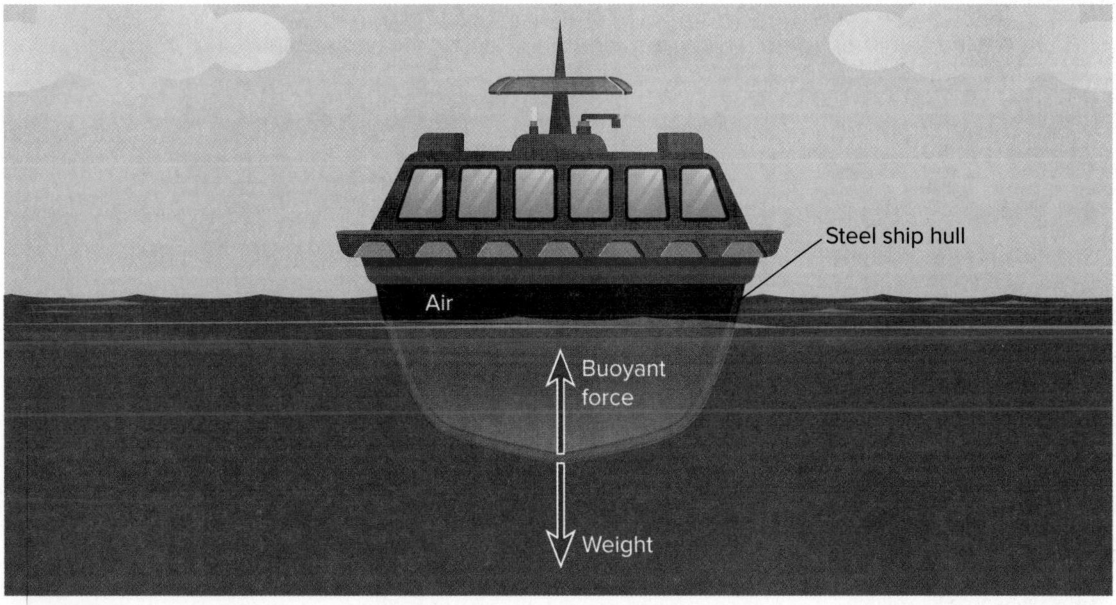

Figure 14 The overall density of a large ship is lower than the density of water because its empty hull contains mostly air.

Infer *why a boat cannot be made of solid steel.*

Pressure

Right now, pressure from the air is pushing on you from all sides, like the pressure you feel underwater in a swimming pool. **Pressure** is force exerted per unit area.

Pressure Equation

$$\text{Pressure (Pa)} = \frac{\text{force (N)}}{\text{area (m}^2\text{)}}$$

$$P = \frac{F}{A}$$

The SI unit of pressure is the pascal (Pa). Because pressure is the amount of force divided by area, one pascal is one newton per square meter (N/m^2). Most pressures are given in kilopascals (kPa) because 1 Pa is a very small amount of pressure.

EXAMPLE Problem 1

CALCULATE FORCE Atmospheric pressure at sea level is about 101 kPa. With how much total force does Earth's atmosphere push on an average human being at sea level? Assume that the surface area of an average human is 1.80 m^2.

List the Unknown: force: F

List the Knowns: pressure: P = 101 kPa = 101,000 Pa

area: A = 1.80 m^2

Set Up the Problem: $P = \frac{F}{A}$

Solve the Problem: $101{,}000 \text{ Pa} = \frac{F}{1.80 \text{ m}^2}$

$F = 101{,}000 \text{ Pa} \times 1.80 \text{ m}^2$

$= 182{,}000 \text{ Pa} \cdot \text{m}^2 = 182{,}000 \frac{\text{N}}{\cancel{\text{m}^2}} \cdot \cancel{\text{m}^2} = 182{,}000 \text{ N}$

Check the Answer: You've set up your equation correctly if the units are the same on both sides: units of pressure = Pa = N/m^2, and (units of force)/(units of area) = N/m^2. The units on both sides of the equation match. Now, just double-check the calculation.

PRACTICE Problems ADDITIONAL PRACTICE

8. A diver who is 10.0 m underwater experiences a pressure of 202 kPa. If the diver's surface area is 1.50 m^2, with how much total force does the water push on the diver?

9. A car weighs 15,000 N, and its tires are inflated to a pressure of 190 kPa. How large is the area of the car's tires that are in contact with the road?

10. CHALLENGE The atmospheric pressure at the surface of Venus is 91 times the pressure at sea level on Earth. With approximately how much total force would Venus's atmosphere push on an average human being at sea level? Assume that the surface area of an average human being is 1.8 m^2.

SCIENCE USAGE v. COMMON USAGE

pressure

Science usage: force per unit area

Increasing the pressure on a gas decreases its volume.

Common usage: the burden of physical or mental stress

Teachers often feel a lot of pressure to help their students do well in school.

Pascal's principle

The idea that pressure is transferred through a fluid can be written as an equation: pressure in = pressure out. Since pressure is force over area, Pascal's principle can be written another way.

Pascal's Principle

$$\frac{\text{input force (N)}}{\text{input area (m}^2)} = \frac{\text{output force (N)}}{\text{output area (m}^2)}$$

$$\frac{F_{in}}{A_{in}} = \frac{F_{out}}{A_{out}}$$

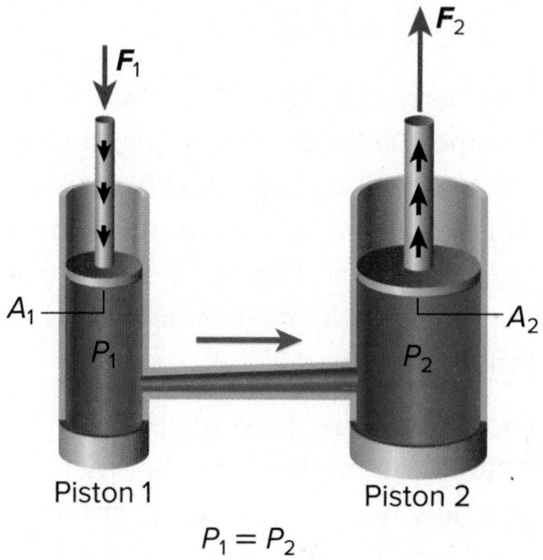

Figure 15 The pressure of the fluid on one side of a hydraulic lift is equal to the pressure on the other side.

Hydraulic lifts Auto repair shops often make use of hydraulic lifts, which move heavy loads in accordance with Pascal's principle. A pipe that is filled with fluid connects small and large cylinders, as shown in **Figure 15.** Pressure applied to the small cylinder is transferred through the fluid and to the large cylinder. With a hydraulic lift, you could use your weight to lift something much heavier than you.

EXAMPLE Problem 2

CALCULATE FORCES A hydraulic lift is used to lift a heavy machine that is pushing down on a 2.8-m² platform with a force of 3700 N. What force must be exerted on a 0.072-m² piston to lift the heavy machine?

List the Unknowns: force on piston: F_{in}

List the Knowns: force on platform: F_{out} = **3700 N**
area of platform: A_{out} = **2.8 m²**
area of piston: A_{in} = **0.072 m²**

Set Up the Problem: $\frac{F_{in}}{A_{in}} = \frac{F_{out}}{A_{out}}$

Solve the Problem: $F_{in} = \left(\frac{F_{out}}{A_{out}}\right)A_{in} = \left(\frac{3700\ \text{N}}{2.8\ \cancel{\text{m}^2}}\right)0.072\ \cancel{\text{m}^2} = 95\ \text{N}$

Check the Answer: The ratio of the forces should be the same as the ratio of the areas. The area of the platform is about 40 times the area of the piston. Therefore, the force on the platform should be about 40 times the force on the piston. 3700 N is about 40 times greater than 95 N, so the answer is reasonable.

PRACTICE Problems

ADDITIONAL PRACTICE

11. A car weighing 15,000 N is on a hydraulic lift platform measuring 10 m². What is the area of the smaller piston if a force of 1100 N is used to lift the car?
12. **CHALLENGE** A heavy crate applies a force of 1500 N on a 25-m² piston. The smaller piston is 1/30 the size of the larger one. What force is needed to lift the crate?

Bernoulli's Principle

Daniel Bernoulli (1700–1782) was a Swiss scientist who studied the properties of moving fluids, such as water and air. Bernoulli found that fluid velocity increases when the flow of the fluid is restricted. Placing your thumb over a running garden hose, as shown in **Figure 16,** demonstrates this effect. When the opening of the hose is decreased in size, the water flows out more quickly.

Figure 16 Bernoulli's principle explains why covering the end of a hose causes the water to flow faster. When the flow of a fluid is restricted, its velocity increases.

Bernoulli examined the relationship between fluid flow and pressure. You might think that increasing the velocity of a fluid's flow would increase its pressure, but Bernoulli found the opposite to be true. According to Bernoulli's principle, as the velocity of a fluid increases, the pressure exerted by that fluid decreases. He published this discovery in 1738.

One application of Bernoulli's principle is the hose-end sprayer. This sprayer is used to apply fertilizers and pesticides to yards and gardens. To use this sprayer, a concentrated solution of the chemical that is to be applied is placed in the sprayer. The sprayer is attached to a garden hose, as shown in **Figure 17.** A strawlike tube is attached to the lid of the unit. The end of the tube is submerged into the concentrated chemical. The water to the garden hose is turned to a high flow rate.

When you are ready to apply the chemicals to the lawn or plant area, you must push a trigger on the sprayer attachment. This allows the water in the hose to flow at a high rate of speed, creating a low pressure area above the strawlike tube. The concentrated chemical solution is sucked up through the straw and into the stream of water. The concentrated solution mixes with water, reducing the concentration to the appropriate level and creating a spray that is easy to apply.

Get It?

Describe how pressure changes as the velocity of a fluid increases.

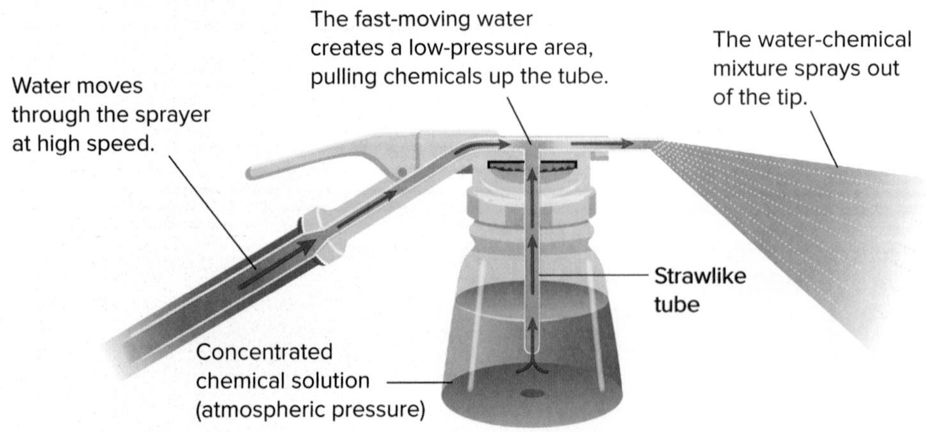

The fast-moving water creates a low-pressure area, pulling chemicals up the tube.
Water moves through the sprayer at high speed.
The water-chemical mixture sprays out of the tip.
Strawlike tube
Concentrated chemical solution (atmospheric pressure)

Figure 17 A hose-end sprayer makes use of Bernoulli's principle.

Val Loh/Photolibrary/Getty Images

Viscosity

Another property exhibited by a fluid is its tendency to flow. While all fluids flow, they vary in the rates at which they flow. **Viscosity** is the resistance of a fluid to flowing. For example, when you take syrup out of the refrigerator and pour it, as shown in **Figure 18,** the flow of syrup is slow. But if this syrup were heated, it would flow much faster. Water flows easily because it has low viscosity. Cold syrup flows slowly because it has high viscosity.

Figure 18 This maple syrup flows slowly because it has high viscosity.

Identify *other examples of liquids with high viscosities.*

What causes viscosity? When a container of liquid is tilted to allow flow to begin, the flowing portion of the liquid transfers energy to the portion of the liquid that is stationary. In effect, the flowing portion of the liquid is pulling the stationary portion of the liquid, causing it to flow, too.

If the flowing portion of the liquid does not effectively pull the other portions of the liquid into motion, then the liquid has a high viscosity, which is a high resistance to flow. If the flowing portion of the liquid pulls the other portions of the liquid into motion easily, then the liquid has low viscosity, or a low resistance to flow.

Check Your Progress

Summary

- If the buoyant force on an object is equal to or greater than the gravitational force on that object, the object will float. If the buoyant force on an object is less than the gravitational force on that object, the object will sink.
- Pascal's principle states that pressure applied to a fluid is transmitted throughout the fluid.
- Bernoulli's principle states that as the velocity of a fluid increases, the pressure exerted by the fluid decreases.
- The resistance to flow by a fluid is called viscosity.

Demonstrate Understanding

13. **Describe** how fluids exert forces on objects.
14. **Explain** why a steel boat floats on water, but a steel block does not.
15. **Explain** why squeezing a plastic mustard bottle forces mustard out the top.
16. **Describe,** using Bernoulli's principle, how tornadoes lift roofs off of buildings.

Explain Your Thinking

17. **Infer** If you blow up a balloon, tie it off, and release it, it will fall to the floor. Why does it fall instead of float? Explain what would happen if the balloon contained helium instead of air.
18. **MATH Connection** The density of water is 1.0 g/cm^3. How many kilograms of water does a submerged 120-cm^3 block displace? Recall that 1.0 kg weighs 9.8 N on Earth. What is the buoyant force on the block?
19. **MATH Connection** To lift an object weighing 21,000 N, how much force is needed on a piston with an area of 0.060 m^2 if the platform being lifted has an area of 3.0 m^2?

LEARNSMART® Go online to follow your personalized learning path to review, practice, and reinforce your understanding.

LESSON 3
BEHAVIOR OF GASES

FOCUS QUESTION
How do gases respond to changes in pressure and temperature?

Boyle's Law–Volume and Pressure

Have you ever seen a weather balloon, like the one shown in **Figure 19?** They carry sensing instruments to very high altitudes to detect weather information. A weather balloon is inflated near Earth's surface with a low-density gas.

Recall that a gas completely fills its container. The balloon remains inflated because of collisions between the gas particles inside the balloon and the balloon itself. In other words, these collisions between gas particles and the container wall cause the gas to exert pressure on the container. As the balloon rises, the atmospheric pressure outside the balloon decreases. This decrease in pressure allows the balloon to expand, eventually reaching a volume between 30 and 200 times its original size. Boyle's law describes the relationship between gas pressure and volume that explains the behavior of weather balloons.

Get It?
Describe what happens to weather balloons as they rise.

Figure 19 A weather balloon expands as it rises due to decreased external pressure. Eventually, the balloon ruptures, and the instruments fall back to the ground.

3D THINKING DCI Disciplinary Core Ideas CCC Crosscutting Concepts SEP Science & Engineering Practices

COLLECT EVIDENCE
Use your Science Journal to record the evidence you collect as you complete the readings and activities in this lesson.

INVESTIGATE
GO ONLINE to find these activities and more resources.

Laboratory: The Behavior of Gases
Carry out an investigation into the effects of changes in pressure and temperature on a sample of a gas.

Revisit the Encounter the Phenomenon Question
What information from this lesson can help you answer the Unit and Module questions?

Volume and pressure

Because a balloon is flexible, its volume can change. In the case of the weather balloon, the volume increases as the external pressure decreases. The volume of the gas inside the weather balloon continues to increase until the balloon can no longer contain it. At this point, the balloon ruptures, and the sensing instruments it was carrying fall to the ground.

From the weather balloon, we know what happens to volume when you decrease pressure. What happens to the pressure from a gas if you decrease its volume—for example, by decreasing the size of the container in which the gas is held? Think about the kinetic theory of matter. The pressure from a gas depends on how often its particles strike the walls of the container. If you squeeze gas into a smaller space, its particles will strike the walls more often, causing increased pressure. The opposite is also true. If you give the particles that make up the gas more space, increasing the volume, they will hit the walls less often, and the pressure from the gas will be reduced.

Get It?

Explain the relationship between pressure and volume.

Robert Boyle (1627–1691), a British scientist, described this property of gases. According to **Boyle's law,** if you decrease the volume of a container of gas and hold the temperature constant, the pressure from the gas will increase. An increase in the volume of the container causes the pressure to drop if the temperature remains constant. **Figure 20** shows this relationship as the volume of a gas is decreased from 10 L to 5 L to 2.5 L. Note the points on the graph that correspond to each of these volumes.

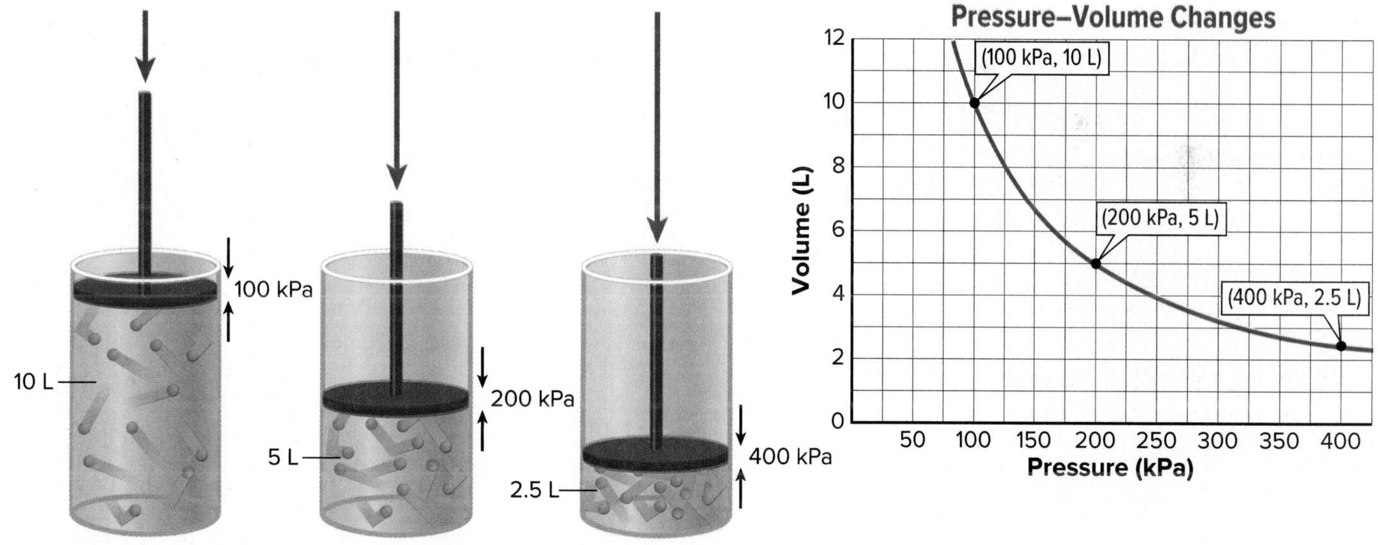

Figure 20 As volume is decreased, a gas exerts increased pressure on the walls of its container. The three canisters are depicted on the graph.

Interpret *What happens to the volume of a gas if the pressure on that gas is doubled?*

An equation for Boyle's law

Boyle's law can be expressed with a mathematical equation. When the temperature of a gas is constant, then the product of the pressure and volume of that gas does not change.

Boyle's Law Equation

initial pressure × initial volume = **final pressure × final volume**

$$P_iV_i = P_fV_f$$

The product of the initial pressure and volume—designated with the subscript *i*—is equal to the product of the final pressure and volume—designated with the subscript *f*. You can use this equation to find one unknown value when you have the other three. The equation will work with any units for either volume or pressure as long as you use the same pressure units for P_i and P_f and the same volume units for V_i and V_f.

Get It?

Show how to write the Boyle's law equation when it is solved for the final pressure of a gas.

EXAMPLE Problem 3

BOYLE'S LAW A weather balloon has a volume of 100.0 L when it is released from sea level, where the pressure is 101 kPa. What will be the balloon's volume when it reaches an altitude where the pressure is 43.0 kPa?

Identify the Unknown: final volume: V_f

List the Knowns:
initial pressure: P_i = **101 kPa**
initial volume: V_i = **100.0 L**
final pressure: P_f = **43.0 kPa**

Set Up the Problem:

$$P_iV_i = P_fV_f$$

$$V_f = V_i\left(\frac{P_i}{P_f}\right)$$

Solve the Problem:

$$V_f = 100.0\text{ L}\left(\frac{101\text{ kPa}}{43.0\text{ kPa}}\right)$$

$$= 235\text{ L}$$

Check the Answer: You can do a quick estimate to check your answer. The pressure was slightly more than halved. Therefore, the volume should slightly more than double. The final volume of 235 L is slightly more than twice the initial volume of 100.0 L. Therefore, the answer seems reasonable.

PRACTICE Problems ADDITIONAL PRACTICE

20. A volume of helium occupies 11.0 L at a pressure of 98.0 kPa. What is the new volume if the pressure drops to 86.2 kPa?
21. **CHALLENGE** A weather balloon has a volume of 90.0 L when it is released from sea level. What is the atmospheric pressure on the balloon when it has grown to a size of 175.0 L?

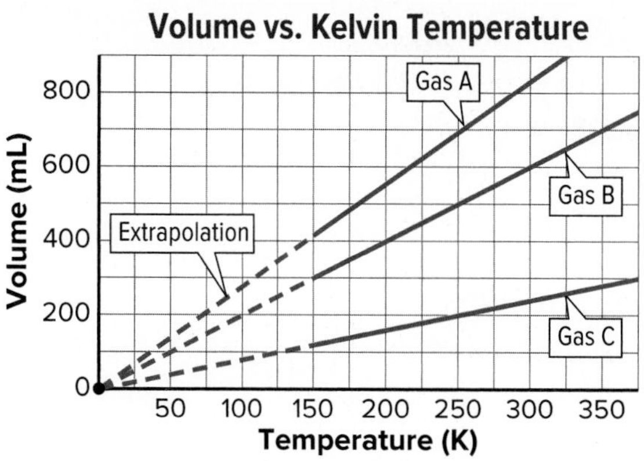

Figure 21 As the temperature of a sample of gas at constant pressure increases, the volume also increases. The dotted lines represent extrapolations of experimental data. Notice that all the extrapolated lines converge at 0 K.

Identify *which gas had the greatest volume change.*

Charles's Law—Temperature and Volume

If you've watched a hot-air balloon being inflated, you know that gases expand when they are heated. Jacques Charles (1746–1823), a French scientist, also noticed this. According to **Charles's law,** the volume of a gas increases with increasing temperature as long as the pressure on the gas does not change. As with Boyle's law, the reverse is also true. The volume of a gas shrinks with decreasing temperature, as shown in **Figure 21.**

Get It?

Explain the relationship between the temperature and volume of a gas.

The kinetic theory and Charles's law

Charles's law can be explained using the kinetic theory of matter. As a gas is heated, the particles that make up that gas move faster and faster. Because the particles that make up the gas move faster, they strike the walls of their container more often and with greater force. In the hot-air balloon, the walls have room to expand. So instead of pressure increasing, the volume increases.

Get It?

Describe How does the kinetic theory of matter explain Charles's law?

An equation for Charles's law

Like Boyle's law, Charles's law can be expressed mathematically. When the pressure on a gas is constant, then the ratio of the volume to the initial temperature does not change. The absolute temperature is the temperature measured in kelvins.

Charles's Law Equation

$$\frac{\text{initial volume}}{\text{initial temperature (K)}} = \frac{\text{final volume}}{\text{final temperature (K)}}$$

$$\frac{V_i}{T_i} = \frac{V_f}{T_f}$$

This shows that the ratio of the initial volume to the initial temperature is equal to the ratio of the final volume to the final temperature. Remember that temperature must be in kelvins.

EXAMPLE Problem 4

USE CHARLES'S LAW A 2.0-L balloon at room temperature (20.0°C) is placed in a refrigerator at 3.0°C. What is the volume of the balloon after it cools in the refrigerator?

Identify the Unknown: final volume: V_f

List the Knowns:

initial volume: $V_i = \mathbf{2.0\ L}$

initial temperature: $T_i = \mathbf{20°C} = 20.0°C + 273 = \mathbf{293\ K}$

final temperature: $T_f = \mathbf{3.0°C} = 3.0°C + 273 = \mathbf{276\ K}$

Set Up the Problem:

$$\frac{V_i}{T_i} = \frac{V_f}{T_f}$$

$$V_f = V_i\left(\frac{T_f}{T_i}\right)$$

Solve the Problem:

$$V_f = 2.0\ \text{L}\left(\frac{276\ \cancel{\text{K}}}{293\ \cancel{\text{K}}}\right)$$

$$= 1.9\ \text{L}$$

Check the Answer: A good way to check your answer here is through experiment! If you place a balloon in a refrigerator, you will notice that the balloon shrinks, but not very much. This is consistent with our answer above.

PRACTICE Problems

ADDITIONAL PRACTICE

22. What would be the final size of the balloon in the example problem above if it were placed in a –18°C freezer?

23. CHALLENGE A gas is heated so that it expands from a volume of 1.0 L to a volume of 1.5 L. If the initial temperature of the gas was 5.0°C, then what is the final temperature of the gas?

Check Your Progress

Summary

- Boyle's law states that if the temperature is constant as the volume of a gas decreases, the pressure increases.
- Charles's law states that at constant pressure, the volume of a gas increases with increasing temperature.
- Both Boyle's law and Charles's law can be expressed as mathematical equations.

Demonstrate Understanding

24. **Describe** what would happen to the volume of a gas if the pressure on it were decreased and then the gas's temperature were increased.

25. **Predict,** using Boyle's law, what will happen to a balloon that an ocean diver takes to a pressure of 202 kPa.

Explain Your Thinking

26. **Predict** what would happen to the volume of a gas if the pressure on that gas were doubled and then the absolute temperature of the gas were doubled.

27. **MATH Connection** A helium balloon has a volume of 2.00 L at 101 kPa. As the balloon rises, the pressure drops to 97.0 kPa. What is the new volume?

28. **MATH Connection** If a 5.0-L balloon at 25°C were gently heated to 30°C, what would be the new volume?

LEARNSMART® Go online to follow your personalized learning path to review, practice, and reinforce your understanding.

NATURE OF SCIENCE

Detecting Dark Matter

The Andromeda Galaxy, shown in the figure below, is a group of approximately one trillion stars held together by the force of gravity. Scientists estimate that the matter we can see, which makes up the stars, planets, and gas clouds in the galaxy, is about 15 percent of the galaxy's mass. The other 85 percent is in a form we do not currently understand.

These detectors attempt to measure rare interactions by searching for antimatter and dark matter and by measuring cosmic rays.

"Missing" mass

Fritz Zwicky, working in the 1930s on gravity's effects on the motion of galaxies, proposed that much of the matter in the universe does not emit or absorb light. While Zwicky's data hinted at mysterious "dark matter," his idea was largely ignored because his measurements were not accurate enough to be conclusive.

In the 1970s, astronomer Vera Rubin used new technology to study light coming from the Andromeda Galaxy. Her accurate data supported the existence of dark matter and renewed interest in Zwicky's proposal. Today, scientists are investigating dark matter to study the 85 percent of the universe that is not explained by current scientific knowledge.

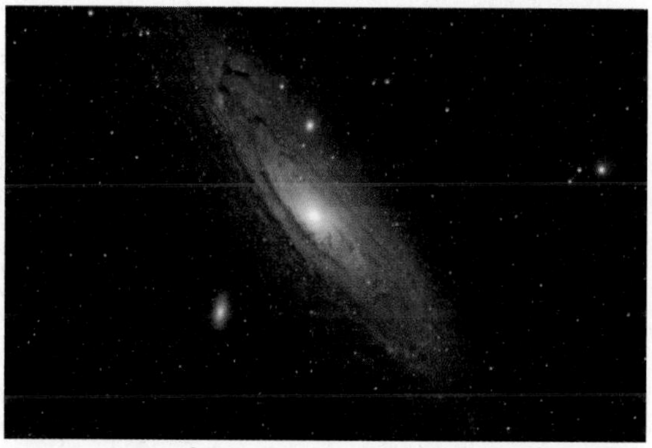

Much of the visible matter is concentrated at the center of the Andromeda Galaxy. Dark matter is spread throughout.

Undiscovered particles

Scientists are trying to detect massive particles that do not usually interact with other particles. This requires sensitive equipment, like that shown in the figure above. Attempts at detecting dark matter particles have been unsuccessful so far.

A new theory of gravity

Some scientists are trying to develop a gravitational theory that can explain the motion of distant galaxies without relying on mysterious undiscovered particles. To date, none of these proposals are able to explain the observed phenomena. Until a new particle is discovered or a new theory is written, dark matter will be a scientific mystery.

OBTAIN, EVALUATE, AND COMMUNICATE INFORMATION

The undiscovered particles of dark matter are commonly known by the descriptive acronym WIMP. Research what WIMP stands for and explain why it accurately describes the particles under consideration.

MODULE 14

STUDY GUIDE

GO ONLINE to study with your Science Notebook.

Lesson 1 MATTER AND THERMAL ENERGY

- There are four main states of matter: solid, liquid, gas, and plasma.
- The kinetic theory is an explanation of how the particles that make up gases behave.
- Thermal energy is the total energy of the particles that make up a material, including kinetic and potential energy.
- Temperature is the average kinetic energy of a substance.

- kinetic theory
- melting point
- heat of fusion
- boiling point
- heat of vaporization
- sublimation
- plasma
- thermal expansion

Lesson 2 PROPERTIES OF FLUIDS

- If the buoyant force on an object is equal to or greater than the gravitational force on that object, the object will float. If the buoyant force on an object is less than the gravitational force on that object, the object will sink.
- Pascal's principle states that pressure applied to a fluid is transmitted throughout the fluid.
- Bernoulli's principle states that as the velocity of a fluid increases, the pressure exerted by the fluid decreases.
- The resistance to flow by a fluid is called viscosity.

- buoyancy
- pressure
- viscosity

Lesson 3 BEHAVIOR OF GASES

- Boyle's law states that if the temperature is constant as the volume of a gas decreases, the pressure increases.
- Charles's law states that at constant pressure, the volume of a gas increases with increasing temperature.
- Both Boyle's law and Charles's law can be expressed as mathematical equations.

- Boyle's law
- Charles's law

THREE-DIMENSIONAL THINKING

Module Wrap-Up

REVISIT THE PHENOMENON

Why does a balloon crumple and shrink when liquid nitrogen is poured on it?

CER Claim, Evidence, Reasoning

Explain Your Reasoning Revisit the claim you made when you encountered the phenomenon. Summarize the evidence you gathered from your investigations and research and finalize your Summary Table. Does your evidence support your claim? If not, revise your claim. Explain why your evidence supports your claim.

STEM UNIT PROJECT

Now that you've completed the module, revisit your STEM unit project. You will summarize your evidence and apply it to the project.

GO FURTHER

SEP Data Analysis Lab

Carbonated Beverages

When you open a container of your favorite soft drink, what do you hear and see? Have you ever wondered why it went "pop" or why it fizzed? Carbon dioxide is added to the drink to give it a lively, tingling sensation. Inside that bottle or can is a lot of chemistry. Research to obtain the information needed to answer the following questions.

CER Analyze and Interpret Data

1. **Identify** What ingredients are in a soft drink?
2. **Identify** How is carbon dioxide added to the water?
3. **Claim** What happens in the space above the soft drink in a closed bottle or can as the soft drink warms to room temperature?
4. **Claim, Evidence, Reasoning** Why are bubbles released when the bottle or can is opened?
5. **Claim, Evidence, Reasoning** When a bottle or can of soft drink is opened, why does it "pop"?

NPS Photo/Alamy Stock Photo

MODULE 15

CLASSIFICATION OF MATTER

ENCOUNTER THE PHENOMENON

Could a person really sink and disappear into a pit of quicksand?

GO ONLINE to play a video about getting stuck in quicksand.

SEP Ask Questions

Do you have other questions about the phenomenon? If so, add them to the driving question board.

CER Claim, Evidence, Reasoning

Make Your Claim Use your CER chart to make a claim about how matter is classified.

Collect Evidence Use the lessons in this module to collect evidence to support your claim. Record your evidence as you move through the module.

Explain Your Reasoning You will revisit your claim and explain your reasoning at the end of the module.

GO ONLINE to access your CER chart and explore resources that can help you collect evidence.

LESSON 1: Explore & Explain: Mixtures

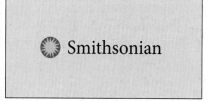

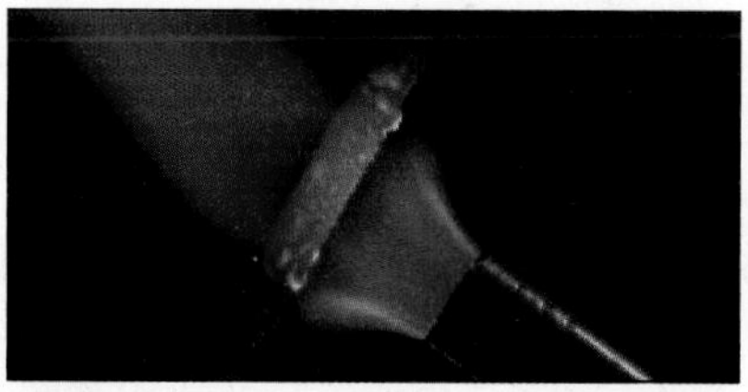

LESSON 2: Explore & Explain: Physical Change

Smithsonian

Additional Resources

LESSON 1

COMPOSITION OF MATTER

FOCUS QUESTION

What are the differences between substances and mixtures?

Substances

Recall that matter is anything that takes up space and has mass. Matter is either a pure substance or a mixture of substances. A pure substance, or simply a **substance,** is a type of matter with a fixed composition. A substance can be an element or a compound.

Elements

All substances are built from atoms. An **element** is a substance made up of atoms that are all alike. The graphite in your pencil point and the copper used in electrical wiring are examples of elements. In graphite, all of the atoms are carbon atoms. In a copper sample, all of the atoms are copper atoms. The familiar elements copper and nitrogen are shown in **Figure 1.** Several other elements and their uses are shown in **Figure 2.** About 90 elements are found naturally on Earth. More than 25 others have been made in laboratories, but most of these are unstable and exist only for short periods of time.

Wire made of copper

Nitrogen gas inside tank

Figure 1 All the atoms of the same element are alike.

3D THINKING | DCI Disciplinary Core Ideas | CCC Crosscutting Concepts | SEP Science & Engineering Practices

COLLECT EVIDENCE

Use your Science Journal to record the evidence you collect as you complete the readings and activities in this lesson.

INVESTIGATE

GO ONLINE to find these activities and more resources.

Probeware Lab: Solutions, Colloids, and Suspensions
Analyze and interpret data to **classify** mixtures as solutions, colloids, or suspensions.

Review the News
Obtain information from a current news story relating to the **composition of matter.** **Evaluate** your source and **communicate** your findings to your class.

Figure 2 Visualizing Elements

When you think of elements, you probably think about those that you see every day, such as silver used to make jewelry or aluminum used to make lightweight bicycles and baseball bats. Many other elements are not as commonly known, but you might see them in everyday objects.

Silicon Present in sand as silicon dioxide, silicon is used to make window glass as well as the silicon chips that run computers.

Americium The synthetic, radioactive element americium is used in smoke detectors.

Titanium Titanium is strong and lightweight and is used for bone or joint replacements and aircraft construction. In rare instances, titanium panels are used in building construction.

Magnesium Chlorophyll—the substance that makes plants green—contains magnesium. Used in metal mixtures, magnesium is lightweight, strong, and resistant to corrosion. Because of these physical properties, it is used in jet engines.

Other applications of elements Lead is an important element due to its high density. Lead blocks harmful radiation because its high density makes it difficult for radiation to pass through. In order to protect the rest of the body from X-ray exposure, lead aprons are used while dental X-rays are being taken, as shown in **Figure 3.**

Aluminum is bendable and resistant to corrosion. It is becoming one of the world's most widely used elements. Because it is strong and lightweight, aluminum is used in automobile parts, airplanes, bicycles, and pots and pans.

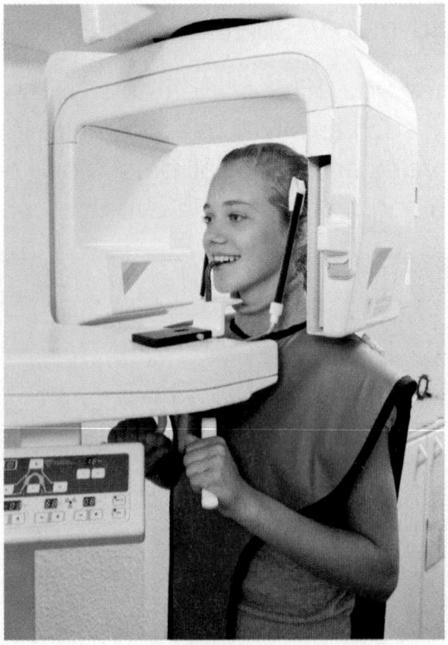

Figure 3 Lead is used as a barrier to protect dental patients from X-ray exposure.

Compounds

Two or more elements can combine to form substances called compounds. A **compound** is a substance in which the atoms of two or more elements are chemically combined in a fixed proportion. For example, water is a compound in which two atoms of the element hydrogen combine with one atom of the element oxygen. Chalk contains one atom each of calcium and carbon and three atoms of oxygen.

Have you put something made from a silvery metal and a greenish-yellow, poisonous gas on your food? Table salt is a chemical compound that fits this description. Even though it is composed of white crystals and adds flavor to food, its components in their elemental forms–sodium and chlorine–are neither crystalline nor salty, as illustrated in **Figure 4.** Like salt, the properties of compounds differ from the properties of the elements that combine to make the compounds. The elements that make up a compound cannot be separated by physical means.

Get It?
Compare How are elements and compounds related?

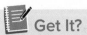

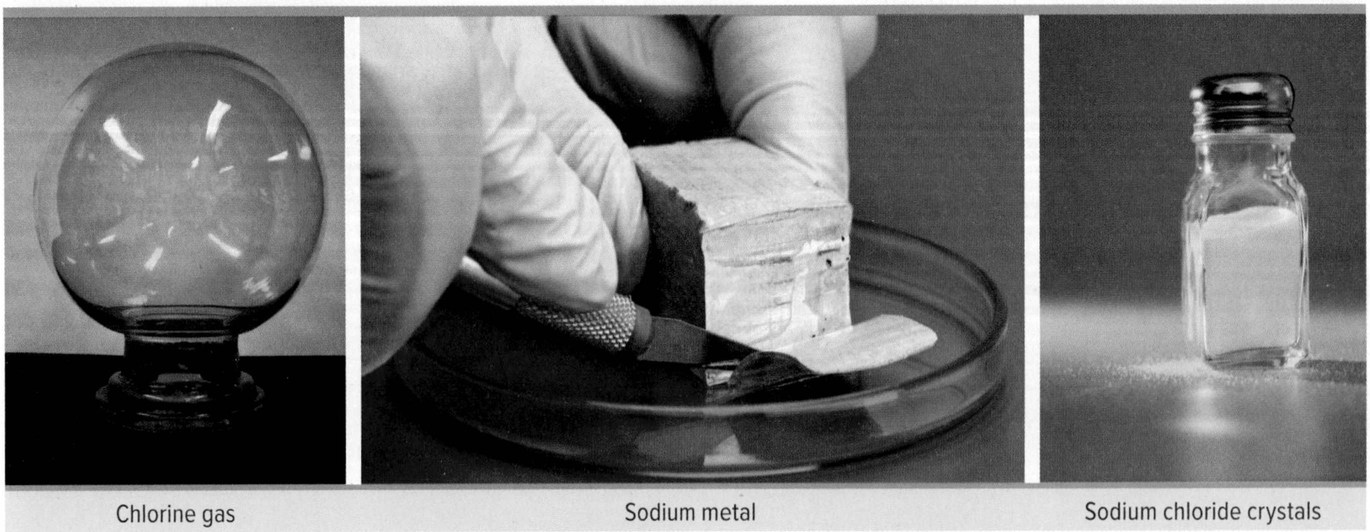

Figure 4 Chlorine, a poisonous gas, and sodium metal combine in a one-to-one ratio to form sodium chloride, or table salt.

Figure 5 Many salad dressings are mixtures of oil, vinegar, and seasoning. Notice the visible herbs and spices floating in the salad dressings.

Mixtures

Salad dressings, such as the examples shown in **Figure 5,** are mixtures. A mixture is matter composed of two or more substances that can be separated by physical means.

Heterogeneous mixtures

In salad dressing made with oil, vinegar, and spices, all of the items in the dressing are in contact, but they do not react with one another. If the dressing is allowed to sit undisturbed long enough, the oil and vinegar will separate. Because the different components remain distinct, this salad dressing is considered an example of a heterogeneous mixture. A mixture in which different materials remain distinct is called a **heterogeneous** (he tuh ruh JEE nee us) **mixture.** Like the dressing, salad is a heterogeneous mixture. The vegetables in a salad are distinct. You can remove specific vegetables if you do not care to eat them.

Some components of heterogeneous mixtures are easy to see, like the components of the salad, but others are not. For example, the shirt shown in **Figure 6** is also a heterogeneous mixture, but you cannot see its individual components. However, with the help of a microscope, you can see the distinct cotton and polyester threads.

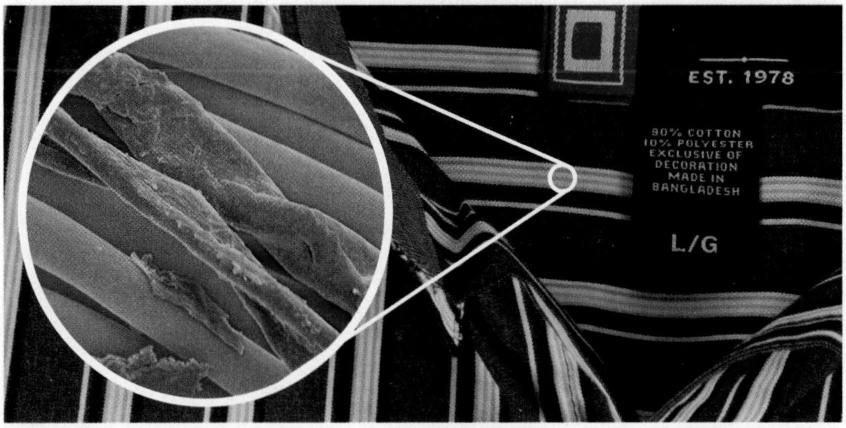

Figure 6 Even though the naked eye cannot detect the individual components that make up this shirt, the different fiber types are clearly visible under a microscope. The shirt is a heterogeneous mixture.

Figure 7 River water is a suspension that carries soil and sediment. If river water slows or sits undisturbed, the suspended particles settle out.

Explain *How can you tell that river water is a suspension?*

Suspensions A **suspension** is a heterogeneous mixture made of a liquid and solid particles that settle. Recall the oil and vinegar salad dressing in **Figure 5.** The dressing also has seasoning particles. The seasoning particles in the liquid will settle to the bottom of the container if allowed to sit undisturbed. River deltas are large-scale examples of how particles in a suspension settle. Rivers flow swiftly through narrow channels, picking up soil and sediment along the way. As the river widens, it flows more slowly, and the suspended particles settle, as shown in **Figure 7.**

Colloids Milk is an example of another kind of heterogeneous mixture called a colloid. It contains water, fats, and proteins in varying proportions. Unlike a suspension, however, its components will not settle if left standing. A **colloid** (KAH loyd) is a heterogeneous mixture with particles that never settle. Paint is a liquid colloid with suspended particles. Gases and solids can contain colloidal particles, too. Fog, shown in **Figure 8,** consists of particles of liquid water suspended in air. Smoke contains solids suspended in air.

Fog—a colloid containing suspended water droplets—scatters the light produced by the vehicle's headlights.

Water droplets suspended in air allow you to see the sunlight as it streams through the forest fog.

Figure 8 Fog is a colloid composed of water droplets suspended in air.

SCIENCE USAGE v. COMMON USAGE

suspend

Science usage: to keep from settling

The seasonings were suspended in the oil and vinegar mixture.

Common usage: to cause to stop temporarily

The game was suspended due to bad weather.

Identifying colloids One way to identify a colloid is by its appearance. Fog appears white because its particles are large enough to scatter light, as shown in **Figure 8.** Sometimes, it is not so obvious that a liquid is a colloid. Colloids can look very much like solutions, which are also mixtures in which the particles cannot be seen.

You can identify whether a liquid is a colloid by passing a beam of light through it. A light beam is invisible as it passes through a solution but can be seen as it passes through a colloid, as shown in **Figure 9.** This occurs because the particles in the colloid are large enough to scatter light, but those in the solution are not. The scattering of a light beam as it passes through a colloid is called the **Tyndall effect.**

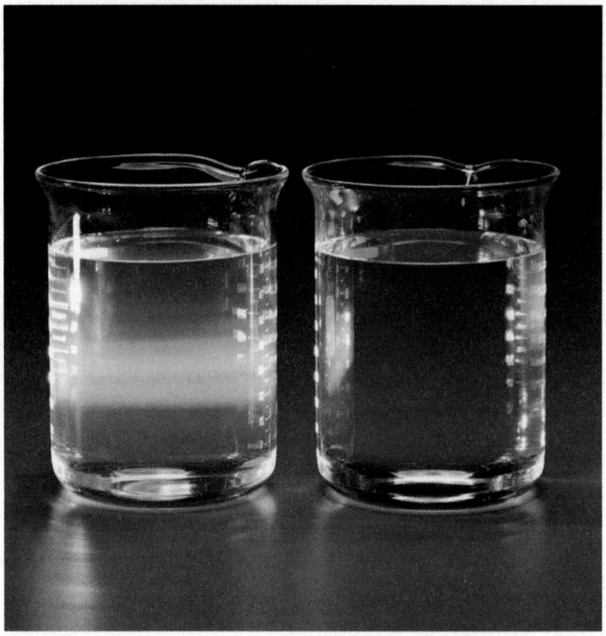

Figure 9 Because of the Tyndall effect, a light beam is scattered by the colloid on the left but passes invisibly through the solution on the right.

Homogeneous mixtures

Soft drinks contain water, sugar, flavoring, coloring, and carbon dioxide gas. **Figure 10** will help you visualize some of these particles in a liquid soft drink. A soft drink in a sealed bottle is an example of a homogeneous mixture. A **homogeneous** (hoh muh JEE nee us) **mixture** is a mixture that remains constantly and uniformly mixed and has particles that are so small that they cannot be seen with a microscope. Due to the interactions between particles, particles in a homogeneous mixture will never settle to the bottom of their container.

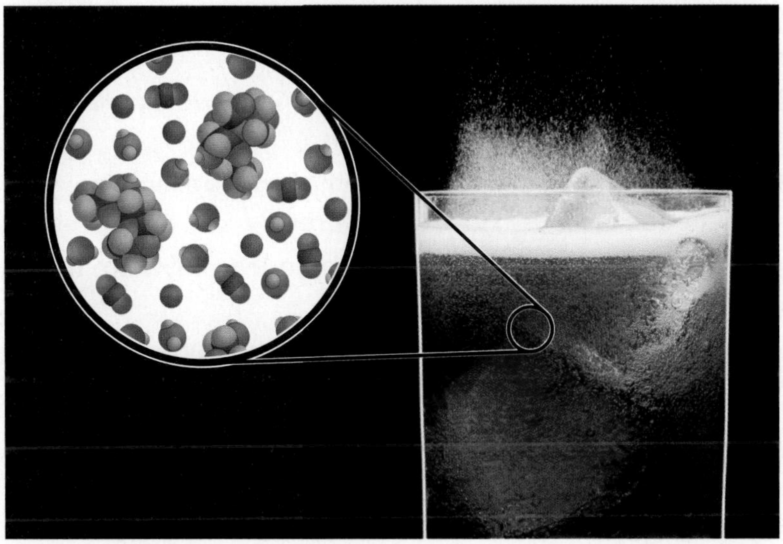

Figure 10 A soft drink can be either heterogeneous or homogeneous. As carbon dioxide fizzes out, the soft drink is a heterogeneous mixture. The resulting flat soft drink is a homogeneous mixture of water, sugar, flavor, color, and some remaining carbon dioxide.

WORD ORIGIN

homogeneous

comes from the Greek *homo*, which means *alike*, and *genea*, which means *source* *Homogeneous mixtures are uniform throughout.*

STEM CAREER Connection

Food Scientist

Salad dressing, mayonnaise, milk, and soft drinks are examples of mixtures that food scientists must understand as they study the processing, preserving, and storage of foods.

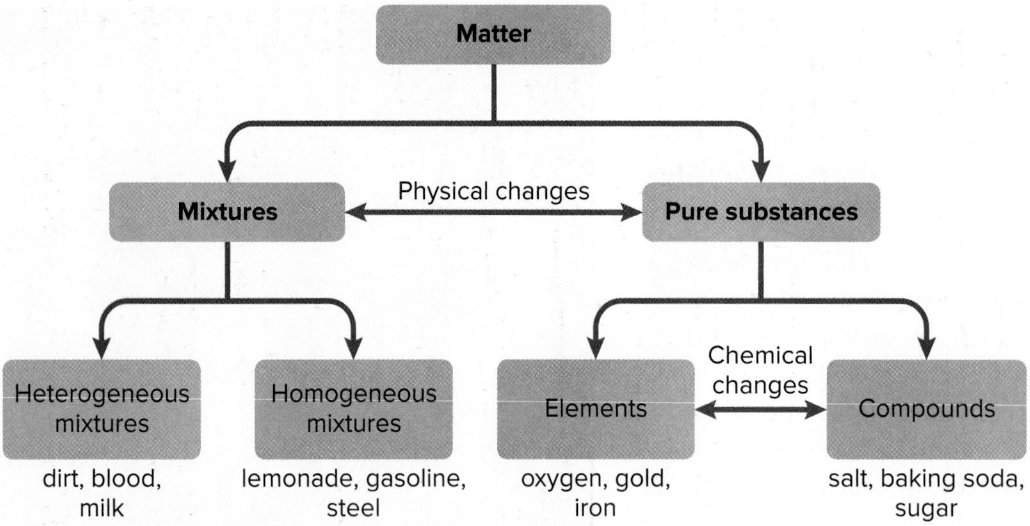

Figure 11 The concept map shows that mixtures can be either heterogeneous or homogeneous. Pure substances can be elements or compounds.

Examine *Where on this chart would you classify pizza?*

Solutions A **solution** is the same thing as a homogeneous mixture. The most familiar solutions might be solids dissolved in liquids, but solutions can also be mixtures of a solid and a gas, a solid and a solid, a gas and a liquid, and so on. Tea, vinegar, steel alloys, and the compressed gas used by divers are all examples of solutions.

Comparing mixtures and substances

Mixtures, unlike compounds, do not always contain the same proportions of the substances of which they are made. Additionally, unlike pure substances, mixtures can be physically separated. A substance has a fixed composition, whereas mixtures can have widely different compositions. These differences are summarized in **Figure 11.**

Check Your Progress

Summary

- An element is a substance with the same kind of atoms.
- There are approximately 90 naturally occurring elements found on Earth and over 25 that have been created in laboratories.
- A compound is a substance that has two or more elements chemically combined in a fixed proportion.
- Mixtures can be heterogeneous or homogeneous and can be separated by physical means.

Demonstrate Understanding

1. **Distinguish** a substance from a mixture. Give two examples of each.
2. **Compare and Contrast** How is a compound similar to a homogeneous mixture? How is it different?
3. **Identify** three elements and three compounds. How are they similar? How are they different?
4. **Summarize** Make a table that compares the properties of suspensions, colloids, and solutions.

Explain Your Thinking

5. **Infer** Why do the words "Shake well before using" indicate that the fruit juice in a carton is a suspension? Why are these words not used on a milk container?
6. **MATH Connection** The weather report this morning stated there is a thick fog in your town. Visibility is less than 500 feet. How many kilometers in front of your vehicle can you see?

LEARNSMART® Go online to follow your personalized learning path to review, practice, and reinforce your understanding.

LESSON 2
PROPERTIES OF MATTER

FOCUS QUESTION
How are the properties and changes of matter classified?

Physical Properties

You can stretch a rubber band, but you cannot stretch a piece of string much, if at all. You can bend a piece of wire, but you cannot easily bend a wooden matchstick. The abilities to stretch and to bend substances are physical properties. The identity of the different substances—rubber, string, wire, wood—does not change. Any characteristic of a material that you can observe without changing the identity of the substance is a **physical property.** Some examples of physical properties are color, shape, size, density, melting point, and boiling point.

Appearance

The appearance of substances, such as the ones shown in **Figure 12,** is a physical property. How would you describe a tennis ball? You could begin by describing its shape, color, and state of matter. You might describe the tennis ball as a brightly colored, fuzzy, hollow sphere. You can measure some physical properties—for example, the diameter of the ball. What physical property of the ball is measured with a balance? How could you measure its volume?

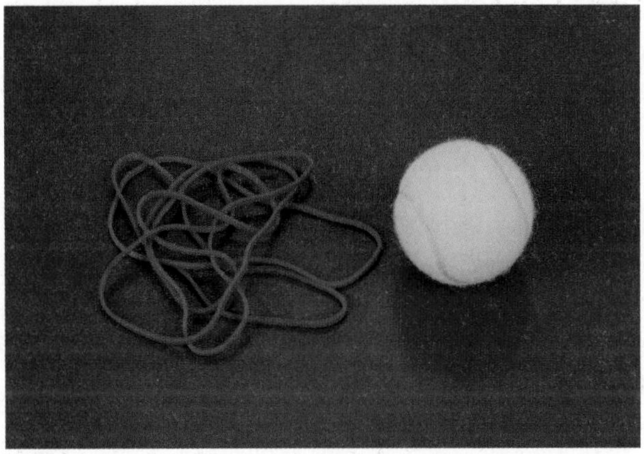

Figure 12 Appearance is a physical property. Appearance includes color, shape, size, texture, and volume.

Compare *the physical properties of the rubber bands and the tennis ball.*

To describe a soft drink in a cup, you could start by calling it a liquid with a brown color. You could measure its volume and temperature. You could describe the fizzy bubbles that appear and burst on its surface, indicating that the soft drink is a mixture that contains a gas. Each of these characteristics is a physical property of that soft drink.

3D THINKING DCI Disciplinary Core Ideas CCC Crosscutting Concepts SEP Science & Engineering Practices

COLLECT EVIDENCE

Use your Science Journal to record the evidence you collect as you complete the readings and activities in this lesson.

INVESTIGATE

GO ONLINE to find these activities and more resources.

Applying Practices: Conservation of Mass
HS-PS2-1. Use mathematics and computational thinking to investigate the conservation of matter in chemical reactions.

Lab: Pure Substances and Mixtures
Carry out an investigation to separate a mixture and identify the pure substances.

Behavior

Some physical properties describe the behavior of a material or a substance. As you might know, objects that contain iron, such as a safety pin, are attracted by a magnet. Attraction to a magnet is a physical property of iron.

Every substance has a specific combination of physical properties that make it useful for certain tasks. Some metals, such as copper, can be drawn out into wires. Others, such as gold, can be pounded into sheets as thin as 0.1 micrometers (μm), about four-millionths of an inch. This property of gold makes it useful for decorating picture frames and other objects. Gold that has been beaten or flattened in this way is called gold leaf.

Think again about a soft drink in a cup. If you knock over the cup, the drink will spread over the table or floor. If you knock over a jar of molasses, however, it does not flow as easily. Viscosity, the resistance to flow, is a physical property of liquids.

Size is the property used to separate sesame seeds from sunflower seeds.

Magnetism easily separates iron from sand.

Figure 13 The best way to separate mixtures depends on their physical properties.

Using physical properties to separate mixtures

Removing the seeds from a watermelon can be done easily based on the physical properties of the seeds compared to the rest of the fruit. **Figure 13** shows a mixture of sesame seeds and sunflower seeds. You can identify the two kinds of seeds by differences in color, shape, and size. By sifting the mixture, you can quickly separate the sesame seeds from the sunflower seeds because their sizes differ.

Now look at the mixture of iron filings and sand shown in **Figure 13.** You probably will not be able to sift out the iron filings because they are similar in size to the sand particles. What you can do is pass a magnet through the mixture. The magnet attracts only the iron filings and pulls them from the sand. This is an example of how a physical property, such as magnetic attraction, can be used to separate substances in a mixture. A similar method is used to separate iron from aluminum and other refuse for recycling. Strong magnets are used in scrapyards and landfills to remove iron for recycling and reuse in an effort to conserve natural resources.

Get It?

Describe how you could use physical properties to separate sand from sugar.

ACADEMIC VOCABULARY

specific

characterized by precise formulation or accurate restriction

Some diseases have specific symptoms.

Physical Change

Physical properties can change while composition remains fixed. If you tear a piece of chewing gum, you change some of its physical properties—its size and shape. However, you have not changed the identity of the materials that make up the gum.

The identity remains the same

When a substance, such as water, freezes, boils, evaporates, or condenses, it undergoes a physical change. A change in size, shape, or state of matter in which the identity of the substance remains the same is called a **physical change.** These changes might involve energy changes, but the kind of substance—the identity of the element or compound—does not change. Because all substances have distinct properties, such as density, specific heat, and melting and boiling points, these properties can often be used to help identify a substance when a particular mixture contains more than one unknown material.

Get It?
Explain why the density of an unknown substance in a mixture can be used to identify the substance.

A substance can change states if it absorbs or releases enough energy. Iron, for example, will melt at high temperatures. Yet, whether in solid or liquid state, iron has physical properties that identify it as iron. Color changes are physical changes, too. For example, when iron is first heated, it glows red. Then, if it is heated to a higher temperature, it turns white, as shown in **Figure 14.**

Get It?
Infer Does a change in state mean that a new substance has formed? Explain.

Using physical changes

A cool drink of water is something most people take for granted; however, in some parts of the world, drinkable water is scarce. Not enough drinkable water can be obtained from wells. Many such areas that lie close to the sea obtain drinking water by using physical properties to separate salt from the water. One method, which uses the property of boiling point, is a type of distillation.

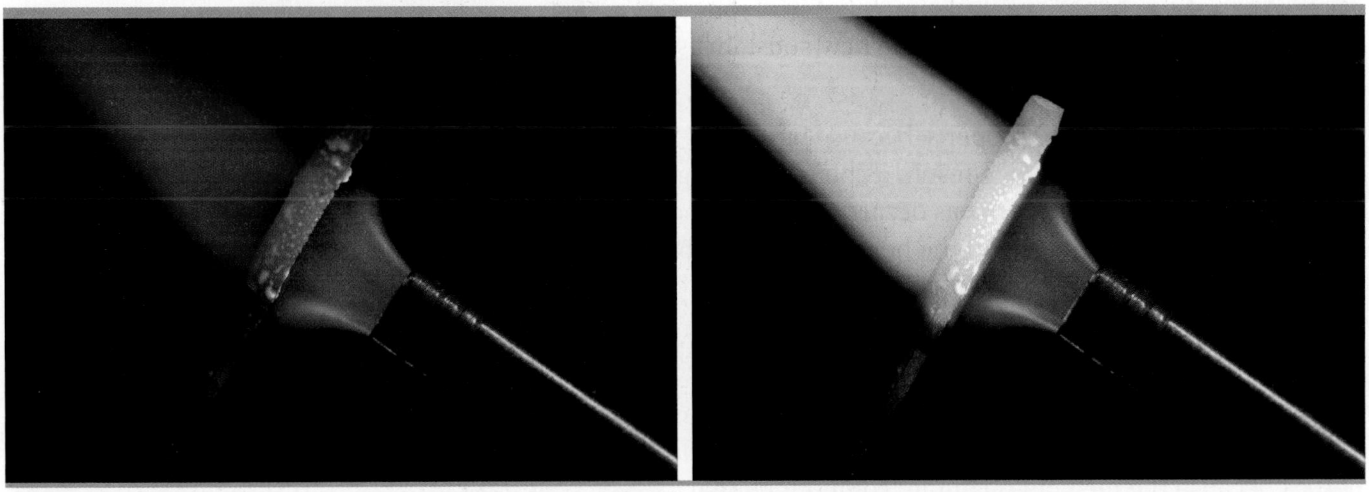

Ted Kinsman/Science Source

Figure 14 Heating iron raises its temperature and changes its color. These changes are physical changes because it is still iron.

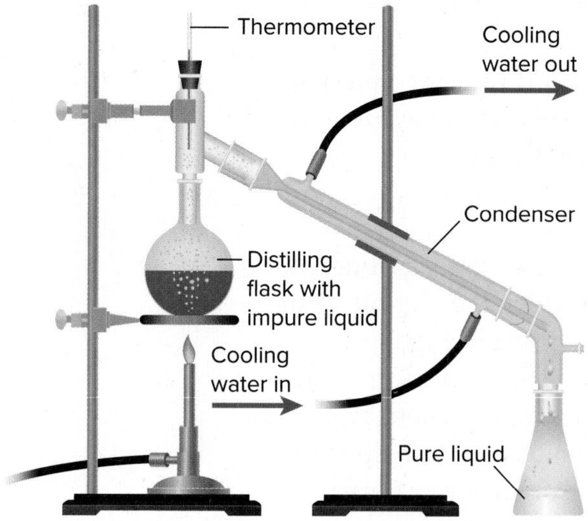

Figure 15 Distillation can separate liquids from solids dissolved in them. The liquid is heated until it evaporates and moves up the column. Then, as it touches the water-cooled surface of the condenser, it becomes liquid again.

Identify *where the solids would be found after distillation is complete.*

Distillation The process of separating substances (for example, salt and water) in a mixture by evaporating a liquid and recondensing its vapor is called **distillation.** Distillation is done in the laboratory using an apparatus similar to the one shown in **Figure 15.**

Two liquids with different boiling points can be separated in this way. The mixture is heated slowly until it begins to boil. The liquid with the lower boiling point vaporizes first and is condensed and collected. Then, as temperature increases the second liquid boils, vaporizes, condenses, and is collected. Distillation is often used in industry. For instance, crude oil obtained from drilling is distilled to separate many different compounds in order to make various products, such as the gasoline used to fuel automobiles.

Chemical Properties and Chemical Changes

You have probably seen warnings on cans of paint thinner and charcoal lighter fluid stating that these liquids are flammable (FLA muh buhl). The tendency of a substance to burn, called its flammability, is an example of a chemical property. Any characteristic of a material that you can observe and that produces one or more new substances is a **chemical property.**

Flammability is a chemical property because burning produces new substances. As a result, a chemical change, also called a chemical reaction, has occurred. Many other substances that are used around the home are flammable. Knowing which ones are flammable helps you to use those items safely.

A less dramatic chemical change can affect some medicines. Look at **Figure 16.** You have probably seen bottles like this in a pharmacy. Many medicines are stored in dark bottles because the medicines contain compounds that can chemically change if they are exposed to light.

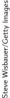

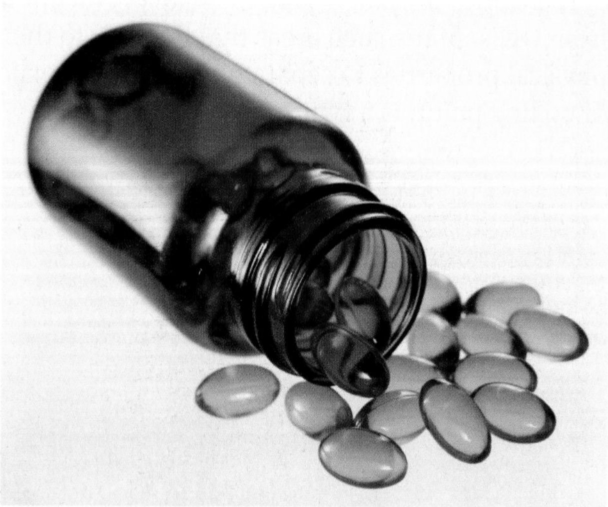

Figure 16 Light can cause chemical changes that ruin the healthy properties of some vitamins. The brown color of this bottle protects the vitamins from light. Reaction to light is a chemical property.

Detecting Chemical Change

If you leave a pan of chili cooking unattended on the stove for too long, your nose soon tells you that something is wrong. Instead of a spicy aroma, you detect an unpleasant smell that alerts you that something is burning. This burnt odor is a clue that a new substance has formed.

The identity changes

The smell of rotten eggs and the formation of rust on bikes and car fenders are also signs that a chemical change has taken place. A change of one substance to another is a **chemical change.** Bubble formation produced by the foaming of an antacid tablet in a glass of water is a sign of new substances being produced. In some chemical changes, a rapid release of energy—detected as heat, light, and sound—is a clue that changes are occurring. A display of fireworks in the night sky is an example. **Figure 17** illustrates another visual clue—the formation of a solid precipitate. What is another example of a chemical change that produces a solid?

Get It?

Define What is a chemical change?

Heating, cooling, and the formation of bubbles or solids in a liquid are all indicators that a reaction is taking place. However, the only sure proof is that a new substance is produced. Consider the following examples. The heat, light, and sound produced when hydrogen gas combines with oxygen in a rocket engine are clear evidence that a chemical reaction has taken place. However, no clues announce the onset of the reaction in which iron and oxygen combine to form rust. The only clue that iron has changed into a new substance is the visible presence of rust. Burning and rusting are chemical changes because new substances form.

Sciencephotos/Alamy Stock Photo

Figure 17 When clear solutions of lead(II) nitrate and potassium iodide mix, a reaction takes place and a yellow solid, lead(II) iodide, appears. The yellow solid that is produced in the chemical reaction is called a precipitate.

Using chemical changes

One case where you might separate substances using a chemical change is in cleaning tarnished silver, such as jewelry. Tarnishing, a chemical reaction between silver metal and sulfur compounds in the air, results in silver sulfide. A chemical reaction in a warm water bath with baking soda and aluminum can change silver sulfide back into silver.

Separating substances using chemical changes is rarely done in the home, but it is commonly done in industrial and laboratory settings. For example, many metals are separated from their ores and then purified using chemical changes.

Flowing water shaped and smoothed these rocks in a physical process.

Both chemical and physical changes shaped the famous White Cliffs of Dover, which line the English Channel.

Figure 18 Weathering can involve physical change and chemical change.

Weathering

The forces of nature continuously shape Earth's surface. Rocks split, deep canyons are carved, sand dunes shift, and limestone formations decorate caves. Do you think these changes, referred to as weathering, are physical or chemical? The answer is both. Geologists, who use the same criteria that you have learned in this chapter, say that some weathering changes are physical and some are chemical.

Get It?

Determine Is weathering a physical change or a chemical change?

Physical weathering

Large rocks can split when water seeps into small cracks, freezes, and expands. Streams can smooth and sculpt hard rock, as shown in **Figure 18.** These are physical changes because the rock does not change into another substance.

Chemical weathering

In other cases, the change is chemical. For example, solid calcium carbonate, a compound found in limestone, reacts with water if it is slightly acidic, such as when it contains some dissolved carbon dioxide. The calcium carbonate reacts to form calcium bicarbonate. This change in limestone is a chemical change because the identity of the substances changes. This chemical change contributes to the weathering of the White Cliffs of Dover, shown in **Figure 18,** and also produces the icicle-shaped rock formations that are found in caves.

CCC CROSSCUTTING CONCEPTS

Energy and Matter Your friend says that chemical weathering makes limestone "disappear". Read about the law of conservation of mass on the next page. Then write a brief message to explain to your friend what really happens to limestone during chemical weathering.

The Conservation of Mass

Wood burns, which means it undergoes combustion. Combustion is a chemical change. Suppose you burn a large log in a fireplace until nothing is left but a small pile of ashes. Smoke, heat, and light are given off, and the changes in the composition of the log confirm that a chemical change took place.

At first, you might think that matter was lost as the log burned because the pile of ashes looks much smaller than the log looked. In fact, the mass of the ashes is less than that of the log. However, suppose that you could collect all of the oxygen in the air that was combined with the log during the burning and all of the smoke and gases that escaped from the burning log and measure their masses too. You would find that no mass was lost after all.

Mass is not gained or lost during any chemical change. In fact, matter is neither created nor destroyed during a chemical change. According to the **law of conservation of mass,** the mass of all substances that are present before a chemical change, the reactants, equals the mass of all of the substances that remain after the change, which are called the products. The number and type of atoms do not change, they are just rearranged. The fact that atoms are conserved, together with knowledge of the chemical properties of the elements involved, can be used to describe and predict chemical reactions.

The Law of Conservation of Mass

total mass of the reactants = total mass of the products

Figure 19 illustrates the law of conservation of mass. Solid sodium bicarbonate in the balloon reacts with liquid hydrochloric acid in the flask. A gas, carbon dioxide, is released and expands the balloon. Without the balloon in place, the gas would escape, and you might think that mass was not conserved. With the balloon to collect the gas, the mass on the scale remains the same. The mass of the reactants is the same as the mass of the products.

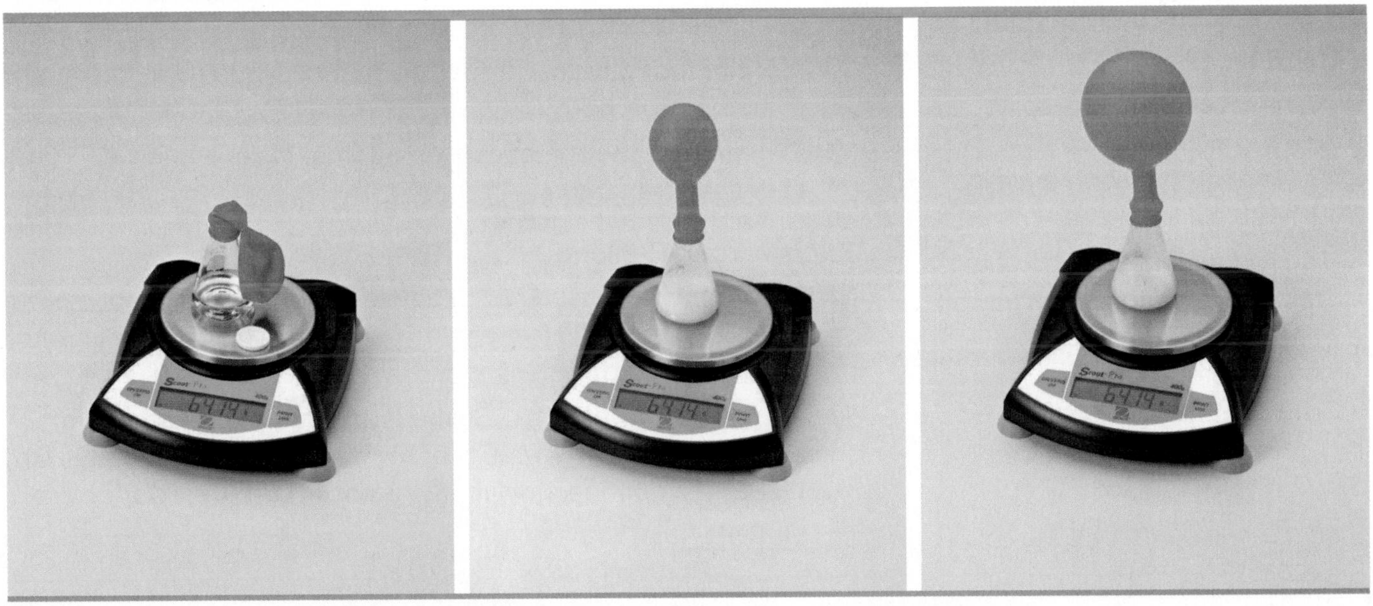

Matt Meadows/McGraw-Hill Education

Figure 19 The reaction between sodium bicarbonate and hydrochloric acid produces carbon dioxide gas, which is collected in the balloon.

Describe *How can you tell that matter was not created or destroyed in this reaction?*

EXAMPLE Problem 1

CALCULATE TOTAL MASS OF PRODUCT When hydrogen reacts with chlorine, the only product is hydrochloric acid. If 18 g of hydrogen react completely with 633 g of chlorine, how many grams of hydrochloric acid are formed?

Identify the Unknown: mass of hydrochloric acid

List the Knowns: mass of hydrogen = 18 g

mass of chlorine = 633 g

Set Up the Problem: total mass of the product = total mass of the reactants

mass of hydrochloric acid = mass of hydrogen + mass of chlorine

Solve the Problem: mass of hydrochloric acid = 18 g + 633 g

The mass of hydrochloric acid is 651 g.

Check the Answer: The mass of reactants and products are equal because the equation was set up according to the law of conservation of mass.

PRACTICE Problems

ADDITIONAL PRACTICE

7. When methane reacts with oxygen, the products are carbon dioxide and water. How many grams of water are formed if 24 g of methane react completely with 96 g of oxygen to form 66 g of carbon dioxide?

8. CHALLENGE Sulfur dioxide reacts with bromine and water to produce hydrogen bromide and sulfuric acid. If 64.1 g of sulfur dioxide react completely with 159.9 g of bromine and an unknown amount of water to form 161.9 g of hydrogen bromide and 98.1 g of sulfuric acid, then how many grams of water react?

Check Your Progress

Summary

- Physical properties can be used to distinguish and separate substances.
- A chemical change is sometimes indicated by cooling, heating, or formation of solids or bubbles.
- The law of conservation of mass states that matter is neither created nor destroyed in a chemical reaction.

Demonstrate Understanding

9. **Explain** why evaporation of water is a physical change and not a chemical change.
10. **Identify** four physical properties that describe a liquid. Identify a chemical property.
11. **Explain** how the law of conservation of mass applies to chemical changes.

Explain Your Thinking

12. **Determine** Does the law of conservation of mass apply to physical changes? How could you test this for melting ice? For the distillation of water?
13. **MATH Connection** Bismuth and fluorine react to form bismuth fluoride. If 417.96 g of bismuth reacts completely with 113.99 g of fluorine, how many grams of bismuth fluoride are formed?

LEARNSMART® Go online to follow your personalized learning path to review, practice, and reinforce your understanding.

Room Temperature Superconductors

Maglev trains have been used as people-movers since 1984. *Maglev* stands for *magnetic levitation*. The first commercial use of high-speed maglev trains began in 2003 in Shanghai, China. Floating on a cushion of air, high-speed maglev trains can travel twice as fast as traditional railroad commuter trains.

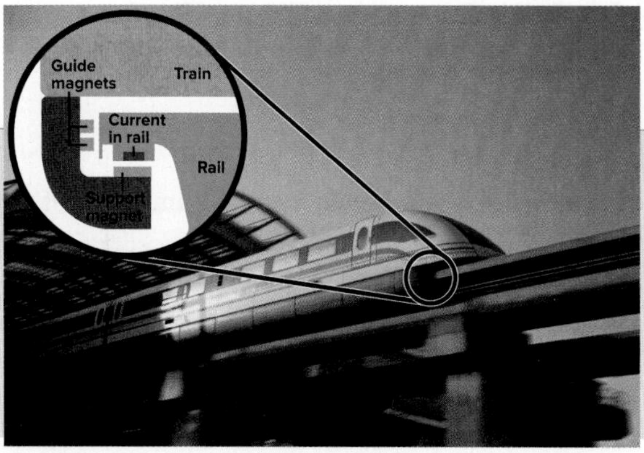

Like two magnets whose north poles push apart, magnetic fields hold the maglev train above the rails and propel the train forward.

Maglev trains do not need the type of engine typical trains use. They are propelled by repulsion between magnetic fields produced by powerful electromagnets, as shown at right. These trains hover over the rails rather than riding directly on them. The next generation of maglev trains rely on electromagnets made with superconductors.

Superconductor electromagnets

A superconductor is a metallic element, compound, or mixture with zero resistance to the flow of electricity. This produces large current in smaller-sized electromagnets and intense magnetic fields. The material must be cooled to near absolute zero, −273°C (0 K), to reach superconductivity. Although energy must be used to cool them, operating costs are lower.

The temperature at which a material becomes a superconductor is called the critical temperature. In recent years, scientists have discovered materials with higher and higher critical temperatures. The recognized world record is −135°C (138 K), although there have been reports of critical temperatures over −73°C (200 K). But scientists are aiming even higher in their pursuit of room-temperature superconductors, which would not need to be cooled.

Benefits

There are many obstacles to overcome in the application of superconducting materials, but the benefits may outweigh the difficulties. Maglev trains are expensive to construct because they are not compatible with traditional rail lines. But maglev trains are very efficient, do not create pollution, operate quietly, and are safer than traditional modes of transportation.

Another application of this technology involves the wires that currently carry electricity to homes and businesses. The wires lose energy while conducting electricity due to resistance. Replacing them with superconducting wires eliminates electrical resistance, but involves many challenges, including the expense of cooling the wires and the brittleness of the superconducting materials. However, room-temperature superconductor wires would make the process much more efficient.

OBTAIN, EVALUATE, AND COMMUNICATE INFORMATION

Research the potential application of room-temperature superconductors used in electrical energy storage devices. These devices are called Superconducting Magnetic Energy Storage (SMES). Report your findings to your class.

MODULE 15
STUDY GUIDE

GO ONLINE to study with your Science Notebook.

Lesson 1 COMPOSITION OF MATTER

- An element is a substance with the same kind of atoms.
- There are approximately 90 naturally occurring elements found on Earth and over 25 that have been created in laboratories.
- A compound is a substance that has two or more elements combined in a fixed proportion.
- Mixtures can be heterogeneous or homogeneous and can be separated by physical means.

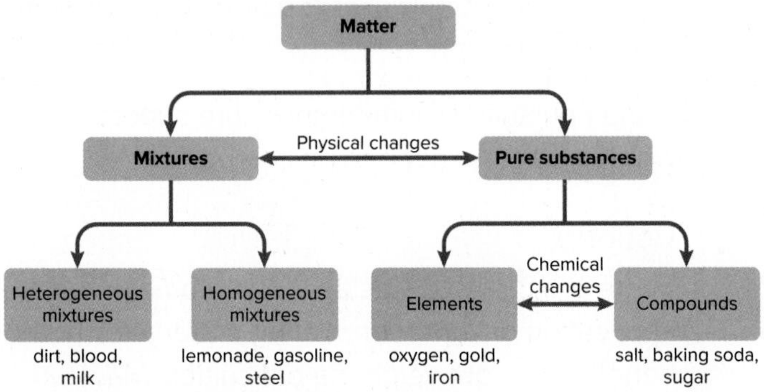

- substance
- element
- compound
- heterogeneous mixture
- suspension
- colloid
- Tyndall effect
- homogeneous mixture
- solution

Lesson 2 PROPERTIES OF MATTER

- Physical properties can be used to distinguish and separate substances.
- A chemical change is sometimes indicated by cooling, heating, or formation of solids or bubbles.
- The law of conservation of mass states that matter is neither created nor destroyed in a chemical reaction.

- physical property
- physical change
- distillation
- chemical property
- chemical change
- law of conservation of mass

THREE-DIMENSIONAL THINKING

Module Wrap-Up

REVISIT THE PHENOMENON

Could a person really sink and disappear into a pit of quicksand?

CER Claim, Evidence, Reasoning

Explain Your Reasoning Revisit the claim you made when you encountered the phenomenon. Summarize the evidence you gathered from your investigations and research and finalize your Summary Table. Does your evidence support your claim? If not, revise your claim. Explain why your evidence supports your claim.

STEM UNIT PROJECT

Now that you've completed the module, revisit your STEM unit project. You will summarize your evidence and apply it to the project.

GO FURTHER

SEP Data Analysis Lab

Classification of Matter

When classifying matter, it is helpful to organize your information in a useful way. One useful way to organize information is to use a dichotomous key. A dichotomous key is a key for identifying items based on a series of choices between alternative characteristics. A dichotomous key showing the terms used to classify matter is drawn to the right. Use the dichotomous key shown here to classify the following types of matter.

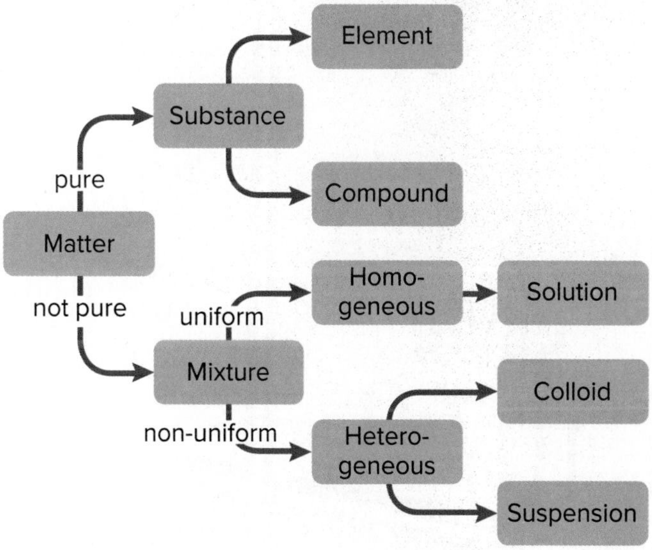

CER Analyze and Interpret Data

1. **Claim** Classify copper.
2. **Claim** Classify sodium chloride.
3. **Claim** Classify muddy water.
4. **Claim** Classify hot coffee.
5. **Claim** Classify milk.

NPS Photo/Alamy Stock Photo

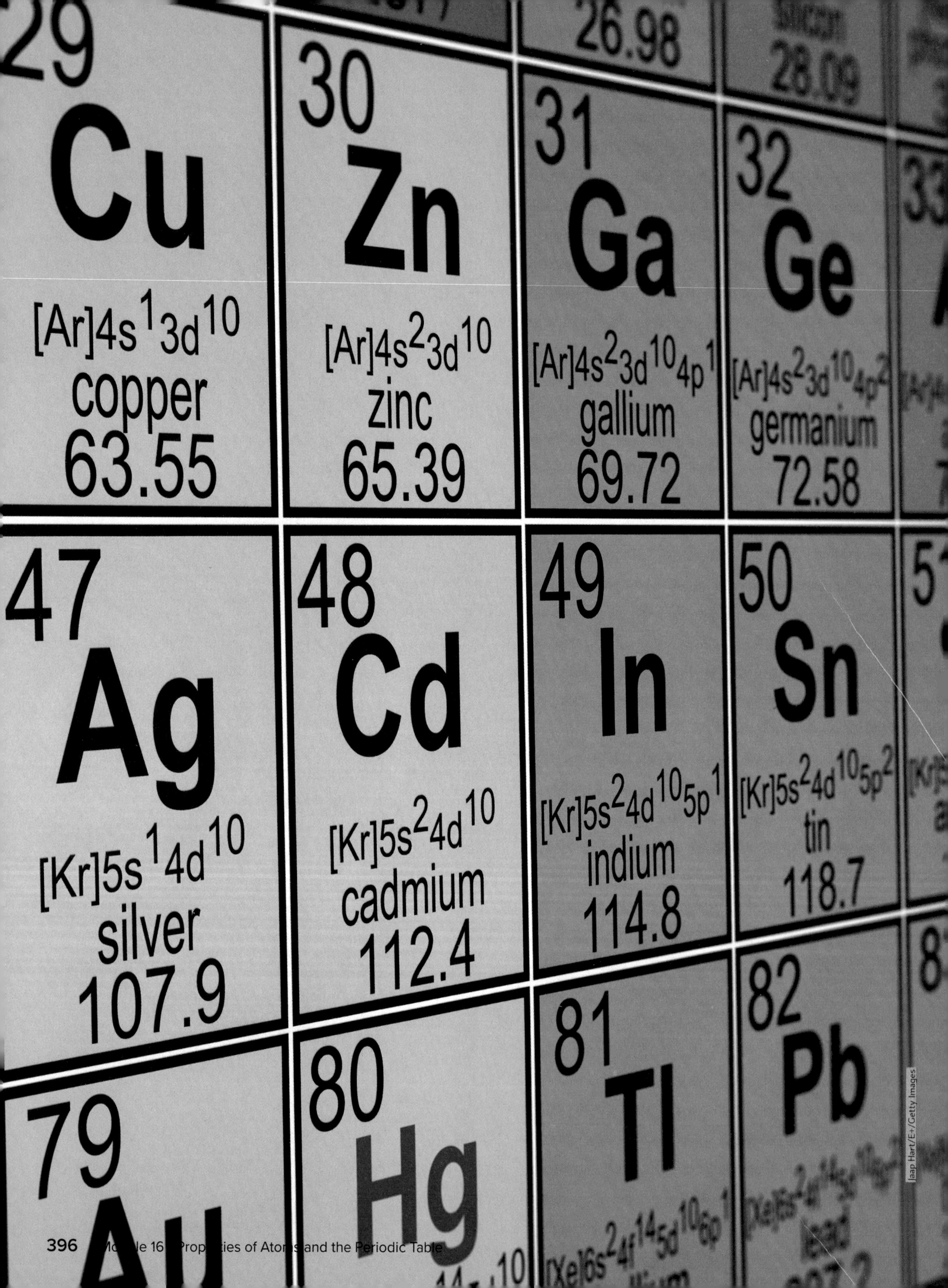

Jaap Hart/E+/Getty Images

MODULE 16

PROPERTIES OF ATOMS AND THE PERIODIC TABLE

ENCOUNTER THE PHENOMENON

Why are there different colors on this periodic table?

GO ONLINE to play a video about how elements were originally grouped in the periodic table.

SEP Ask Questions

Do you have other questions about the phenomenon? If so, add them to the driving question board.

Claim, Evidence, Reasoning

Make Your Claim Use your CER chart to make a claim about why there are different colors on the periodic table. Explain your reasoning.

Collect Evidence Use the lessons in this module to collect evidence to support your claim. Record your evidence as you move through the module.

Explain Your Reasoning You will revisit your claim and explain your reasoning at the end of the module.

GO ONLINE to access your CER chart and explore resources that can help you collect evidence.

LESSON 1: Explore & Explain: Models—Tools for Scientists

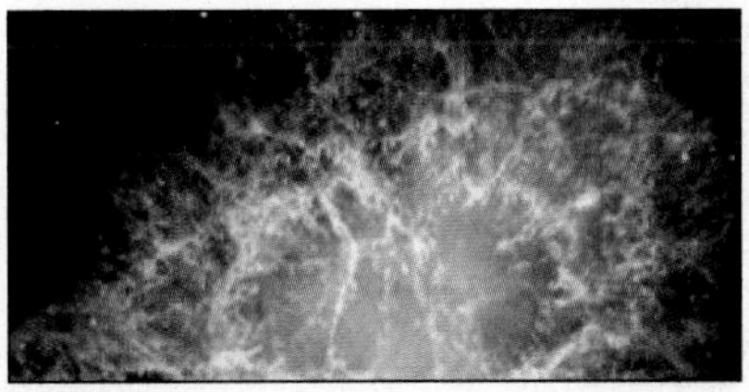

LESSON 2: Explore & Explain: Elements in the Universe

Smithsonian

Additional Resources

(t)Video Supplied by BBC Worldwide Learning, (bl)Science Photo/Shutterstock, (br)NASA, ESA, J. Hester and A. Loll (Arizona State University)

LESSON 1

STRUCTURE OF THE ATOM

FOCUS QUESTION

Can an atom be broken into smaller parts?

Scientific Shorthand

Do you use abbreviations for long words, street addresses, or the names of states? Scientists also use abbreviations. In fact, scientists have developed their own shorthand for naming the elements.

Do the letters C, Al, Ne, and Au mean anything to you? Each letter or pair of letters is a chemical symbol, which is a short or abbreviated name of an element. Chemical symbols, such as those in **Table 1**, consist of one capital letter or a capital letter plus one or two lowercase letters. For some elements, the symbol is the first letter of the element's name. For other elements, the symbol is the first letter of the name plus another letter from its name. Some symbols are derived from Latin. For instance, *argentum* is Latin for silver. The chemical symbol for silver is Ag.

Elements are named in a variety of ways. Some elements are named for their properties, for places, or to honor scientists. For example, the element curium was named to honor Pierre and Marie Curie, scientists who researched radioactivity. Other elements, like germanium, were named after a country. Regardless of the origin of the name, scientists derived the international system of symbols for convenience. It is much easier to write H for hydrogen, O for oxygen, and H_2O for dihydrogen monoxide (water). Because scientists worldwide use this system, everyone recognizes what these symbols represent.

Table 1 Symbols of Common Elements

Element	Symbol	Element	Symbol
Aluminum	Al	Iron	Fe
Calcium	Ca	Mercury	Hg
Carbon	C	Nitrogen	N
Chlorine	Cl	Oxygen	O
Gold	Au	Potassium	K
Hydrogen	H	Sodium	Na

3D THINKING DCI Disciplinary Core Ideas CCC Crosscutting Concepts SEP Science & Engineering Practices

COLLECT EVIDENCE

Use your Science Journal to record the evidence you collect as you complete the readings and activities in this lesson.

INVESTIGATE

GO ONLINE to find these activities and more resources.

Review the News
Obtain information from a current news story about subatomic particles. **Evaluate** your source and **communicate** your findings to your class.

CCC **Identify Crosscutting Concepts**
Create a table of the crosscutting concepts and fill in examples you find as you read.

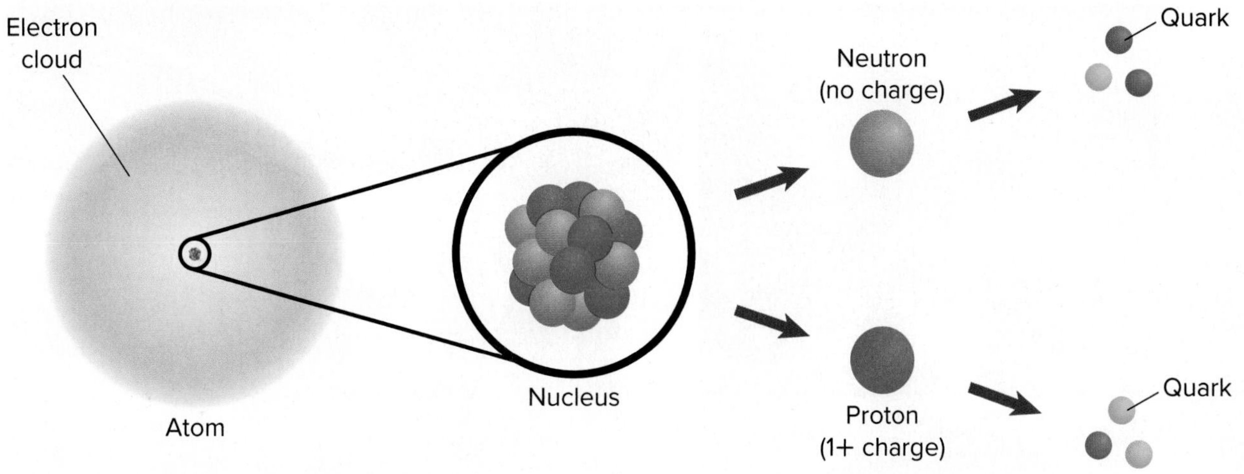

Figure 1 The nucleus of the atom contains protons and neutrons. The proton has a positive charge, and the neutron has no charge. Protons and neutrons are themselves composed of quarks. Electrons occupy a space called the electron cloud, which surrounds the nucleus

Compare and contrast *protons, neutrons, and electrons.*

Subatomic Particles

An element is matter that is composed of only one type of atom. An **atom** is the smallest particle of an element that retains the element's properties. For example, the element iron is composed of only iron atoms, and the element hydrogen is composed of only hydrogen atoms. Atoms are composed of even smaller particles—subatomic particles—called protons, neutrons, and electrons, as shown in **Figure 1.** The small, positively charged center of the atom is called the **nucleus.** The nucleus contains protons and neutrons. **Protons** are particles with an electric charge of 1+. The number of protons in the nucleus is unique for each element. **Neutrons** are electrically neutral particles in the nucleus; they do not have a charge. **Electrons** are particles with an electric charge of 1−. They occupy the space surrounding the nucleus of an atom.

Get It?
Identify the three types of subatomic particles.

Quarks—even smaller particles

Are the protons, electrons, and neutrons that make up atoms the smallest particles that exist? Scientists have inferred that protons and neutrons are composed of smaller particles called **quarks.** Electrons, however, are not made of smaller particles. So far, scientists have confirmed the existence of six unique quarks. A particular arrangement of three of these quarks produces a proton. Another arrangement of three quarks produces a neutron. Identifying the composition of protons and neutrons is a continuing effort.

WORD ORIGINS

atom

comes from the Greek word *atomos*, meaning "indivisible" or "uncuttable"

The basic building block of all matter is the atom.

Figure 2 At 6.4 km in circumference, the Tevatron in Batavia, Illinois, was the world's most powerful particle accelerator when it was completed in 1985. When the Tevatron stopped colliding particles in 2011, the Large Hadron Collider (LHC) had become the largest accelerator. The LHC, which is 27 km in circumference, is designed to use 9300 superconducting magnets.

The search for quarks

To study quarks, scientists accelerate charged particles to tremendous speeds and then force them to collide with—or smash into—protons. These collisions cause the protons to break apart. **Figure 2** shows two particle accelerators. These giant machines use electric and magnetic fields to accelerate, focus, and collide fast-moving particles. The Fermi National Accelerator Laboratory in Batavia, Illinois, housed a machine called the Tevatron that could generate the forces that are required to break apart protons. The Large Hadron Collider (LHC), also shown in **Figure 2,** is a particle accelerator in Geneva, Switzerland. The LHC is capable of even greater forces and, hopefully, will lead to a deeper understanding of the structure of the universe. The ultimate goal is to discover new particles.

Scientists use a variety of collection devices to obtain detailed information about the particles created in a collision. Just as police investigators can reconstruct traffic accidents from tire marks and other evidence at the scene, scientists are able to examine data collectors for evidence of the tiniest of particles. Scientists use inference to identify subatomic particles as well as information about each particle's structure. For example, the wire chamber in **Figure 3** help scientists examine the varying tracks made by different types of particles formed in high-speed collisions.

Figure 3 Scientists use wire chambers to study the tracks left by subatomic particles. The chamber is a complex structure. This image is a view of the inside of a wire chamber.

The sixth quark Finding evidence for the existence of quarks was not an easy task. Scientists discovered five quarks and hypothesized that a sixth quark existed. However, it took several years for a team of nearly 450 scientists from around the world to find the sixth quark. The tracks of the sixth quark were hard to detect because only about one-billionth of a percent of the proton collisions showed the presence of a sixth quark, typically referred to as the top quark.

Models—Tools for Scientists

Scientists and engineers use models to represent objects or ideas that are difficult to visualize, or to picture in your mind. You might have seen models or blueprints of buildings, planetary models of the solar system, or even a model airplane. These are scaled-down models. Scaled-down models allow you to see something that is too large to visualize all at once or something that has not yet been built. Scaled-up models are often used to visualize things that are too small to see. Models of atoms are examples of scaled-up models.

Get It?

Explain the difference between a scaled-up model and a scaled-down model.

To give you an idea of how small an atom is, it would take about 50,000 aluminum atoms stacked one on top of the other to equal the thickness of a sheet of aluminum foil. Scientists have developed scaled-up models that they can use to visualize and study an atom. For a model of the atom to be useful, it must support the accepted ideas about atomic structure and behavior. As new discoveries about atoms are made, scientists must include these new details in the model.

The atomic model

You now know that all matter is composed of atoms, but this was not always accepted as truth. Around 400 B.C., the Greek philosopher Democritus proposed the idea that atoms are tiny particles that make up all matter. Another philosopher, Aristotle, disputed Democritus's idea and proposed that matter was uniform throughout and was not composed of such small particles. Aristotle's incorrect idea was accepted for about 2000 years. In the 1800s, the English scientist John Dalton was able to present evidence to suggest that atoms exist.

Dalton's atomic theory, highlighted in **Table 2,** led to his model of the atom. This model has changed somewhat over time with further investigations by other scientists, as shown on the next page in **Figure 4.** Dalton's modernization of Democritus's idea of the atom provided a physical explanation for chemical reactions. Due to this discovery, scientists could finally express these reactions in quantitative terms, using chemical symbols and equations.

Table 2 Dalton's Atomic Theory

Dalton (1766–1844)

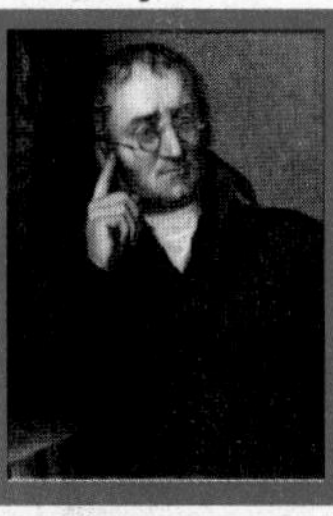

• Matter is composed of extremely small particles called atoms.
• Atoms are indivisible and indestructible.
• Atoms of a given element are identical in size, mass, and chemical properties.
• Atoms of a specific element are different from those of another element.
• Different atoms combine in simple whole-number ratios to form compounds.
• In a chemical reaction, atoms are separated, combined, or rearranged.

CCC CROSSCUTTING CONCEPTS

Scale, Proportion, and Quantity Using the properties of the subatomic particles that make up an atom, develop descriptions or models that compare the properties in terms of scale and quantity.

GeorgiosArt/Getty Images

Figure 4 Visualizing the Early Atomic Models

The currently accepted model of the atom evolved from the ideas and the work of many scientists.

400 B.C. Democritus Model

1904 Thomson Model

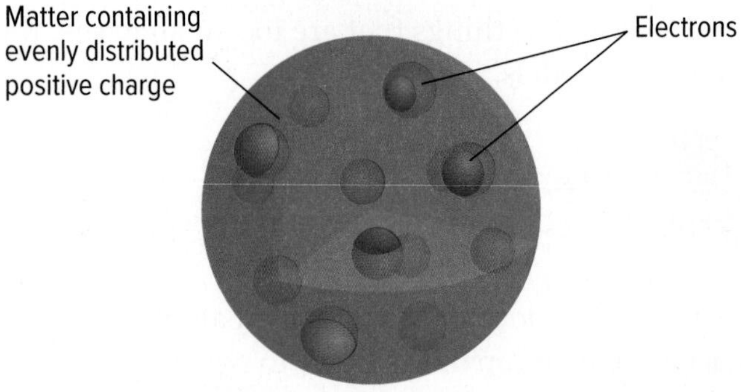

400 B.C. Democritus Model Democritus first proposed that elements consist of tiny, solid particles that cannot be subdivided. He called these particles *atomos*, meaning "uncuttable." Democritus's ideas were criticized by Aristotle, who believed that empty space could not exist. Because Aristotle was one of the most influential philosophers of his time, Democritus's atomic theory was rejected.

1904 Thomson Model English physicist Joseph John Thomson proposed a model that consisted of a spherical atom containing small, negatively charged particles. He thought these "electrons" (in blue) were evenly embedded throughout a positively charged sphere, much like chocolate chips in a ball of cookie dough.

1911 Rutherford Model

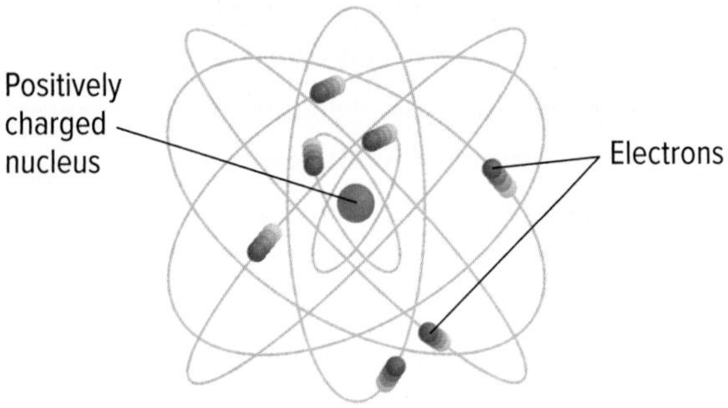

1913 Bohr Model

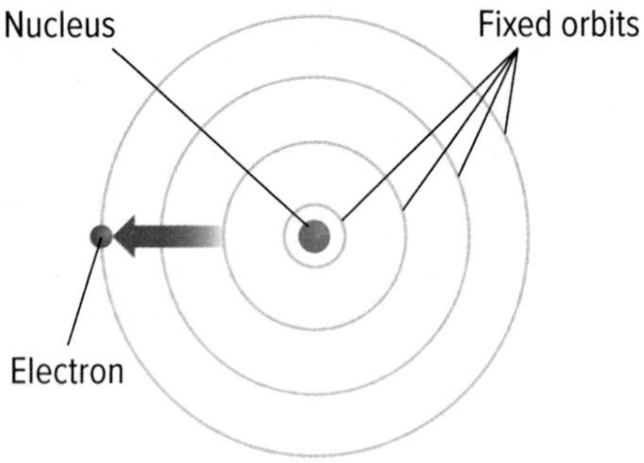

1911 Rutherford Model English physicist Ernest Rutherford proposed the idea that all the positive charge of an atom is concentrated in a central atomic nucleus that is surrounded by electrons.

1913 Bohr Model Danish physicist Niels Bohr hypothesized that electrons travel in fixed orbits. He suggested that electrons could jump between orbits as they absorb or release specific amounts of energy. The Bohr model worked very well for hydrogen, but did not work as well for atoms with many electrons.

The electron cloud model

By 1926, scientists developed the electron cloud model of the atom, which is the model that is accepted today. An **electron cloud** is the area around the nucleus of an atom, where electrons are most likely to be found. The electron cloud is 100,000 times larger in diameter than the nucleus of an atom. In contrast, each electron in the cloud is significantly smaller in mass than a single proton or single neutron.

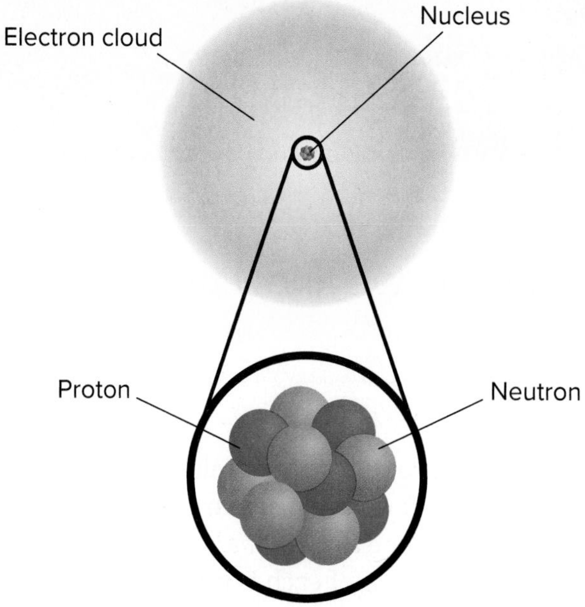

Figure 5 Most of an atom is empty space. The electron cloud represents the area in which the electrons are moving.

Get It?

Explain the difference between the Bohr model and the electron cloud model.

Because an electron's mass is negligible compared to the nucleus and the electron is moving so quickly around the nucleus, it is impossible to describe its exact location in an atom at any moment.

Picture the spokes on a moving bicycle wheel. The spokes are moving so quickly that you cannot pinpoint any single spoke in the wheel. All that you see is a blur that contains all the spokes somewhere within it. In a similar way, an electron cloud is a blur of activity containing all of an atom's electrons somewhere within it. **Figure 5** illustrates the location of the nucleus and the electron cloud in the electron cloud model of the atom.

Check Your Progress

Summary

- Scientists use chemical symbols to abbreviate element names.
- Atoms are composed of protons, neutrons, and electrons.
- Scientists have confirmed the existence of six different quarks.
- The electron cloud model is the current atomic model.

Demonstrate Understanding

1. **Identify** the names, charges, and locations of three types of subatomic particles that make up an atom.
2. **Identify** the chemical symbols for the elements carbon, aluminum, hydrogen, oxygen, and sodium.
3. **Describe** how quarks were discovered.

Explain Your Thinking

4. **Describe** how a rotating electric fan could function as a model of the atom. Explain how the rotating fan is unlike an atom.
5. **MATH Connection** A proton's mass is estimated to be 1.6726×10^{-24} g, and the mass of an electron is estimated to be 9.1093×10^{-28} g. How many times greater is the mass of a proton compared to the mass of an electron?

LEARNSMART® Go online to follow your personalized learning path to review, practice, and reinforce your understanding.

LESSON 2

MASSES OF ATOMS

FOCUS QUESTION

How can we know the mass of a single atom?

Atomic Mass

The nucleus contains almost all of an atom's mass because protons and neutrons are far more massive than electrons. The mass of a proton is roughly the same as that of a neutron—about 1.67×10^{-24} g, as shown in **Table 3.** The mass of each is more than 1800 times greater than the mass of an electron. An electron's mass is so small that it can be ignored when evaluating the mass of an atom.

If you were asked to estimate the height of your school building, you would probably not give an answer in kilometers. Considering the scale of the building, you would more likely give the height in meters. When thinking about the mass of an atom, scientists discovered that even grams were not a small enough unit of measure. Scientists needed a more manageable unit.

The unit of measurement used to quantify an atom's mass is the atomic mass unit (u). The mass of a proton or a neutron is almost equal to 1 amu. This is not coincidence—the unit was defined that way. The atomic mass unit is defined as one-twelfth of the mass of a carbon atom containing six protons and six neutrons, as shown in **Figure 6.** Remember that the mass of an atom is contained almost entirely in the mass of the protons and neutrons in the nucleus. Therefore, each of the 12 particles in the carbon nucleus must have a mass nearly equal to 1 u.

Table 3 Subatomic Particle Masses

Particle	Mass (g)
Proton	1.6726×10^{-24}
Neutron	1.6749×10^{-24}
Electron	9.1093×10^{-28}

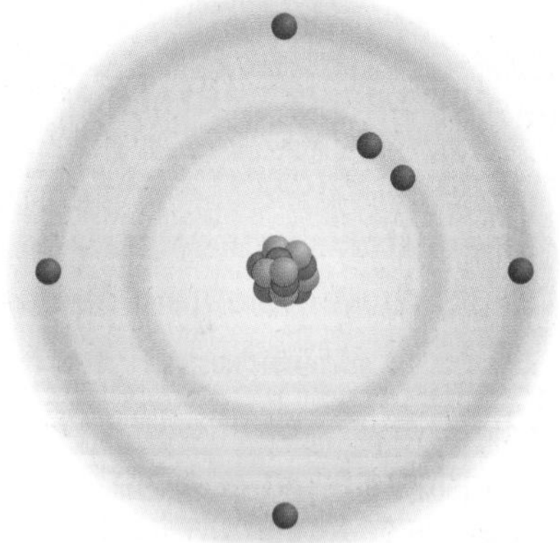

Figure 6 The masses in **Table 3** show how similar the mass of a proton and a neutron are. An electron's mass is negligible compared to an atom's mass. The atomic mass unit is equal to one-twelfth of the mass of the carbon atom shown.

Identify *Where is most of the mass of an atom located?*

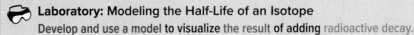

3D THINKING DCI Disciplinary Core Ideas CCC Crosscutting Concepts SEP Science & Engineering Practices

COLLECT EVIDENCE

Use your Science Journal to record the evidence you collect as you complete the readings and activities in this lesson.

INVESTIGATE

GO ONLINE to find these activities and more resources.

Laboratory: Modeling the Half-Life of an Isotope
Develop and use a model to visualize the result of adding radioactive decay.

Revisit the Encounter the Phenomenon Question
What information from this lesson can help you answer the Unit and Module questions?

Table 4 Mass Numbers of Atoms

Element	Symbol	Atomic Number	Protons	Neutrons	Mass Number
Boron	B	5	5	6	11
Carbon	C	6	6	6	12
Oxygen	O	8	8	8	16
Sodium	Na	11	11	12	23
Copper	Cu	29	29	34	63

Atomic number

Recall that an element is made of one type of atom. What determines the type of atom? In fact, the number of protons identifies the type of atom. For example, every carbon atom has six protons. Also, any atom with six protons is a carbon atom. Atoms of different elements have different numbers of protons. For example, atoms with eight protons are oxygen atoms.

Get It?

Describe what determines whether or not an atom is boron.

The number of protons in an atom's nucleus is equal to its **atomic number.** The atomic number of carbon is six. Oxygen's atomic number is eight, as shown in **Table 4.** Therefore, if you are given any one of the following–the name of an element, the number of protons for an element, or the atomic number of an element–you can identify the other two.

For example, if your teacher asked you to identify an atom with an atomic number of 11, you would know that the atom has eleven protons, and it is sodium, as indicated in **Table 4.**

Get It?

Interpret Table 4 to identify the name and the atomic number of the element with 29 protons.

Mass Number

The **mass number** of an atom is the sum of the number of protons and the number of neutrons in the nucleus of the atom.

$$\text{Mass number} = \text{number of protons} + \text{number of neutrons}$$

For example, you can calculate the mass number of the copper atom listed in **Table 4**: 29 protons plus 34 neutrons equals a mass number of 63.

Also, if you know the mass number and the atomic number of an atom, you can calculate the number of neutrons in the nucleus. The number of neutrons is equal to the mass number minus the atomic number. In fact, if you know two of the three numbers–mass number, atomic number, and number of neutrons–you can always calculate the third.

Isotopes

Atoms of the same element can have different mass numbers. For example, some carbon atoms have a mass number of 12, while other carbon atoms have a mass number of 14. The number of protons for each element never changes. So, for an atom's mass number to differ, the number of neutrons must change. Atoms of the same element that have different numbers of neutrons are called **isotopes.**

Get It?

Identify What are isotopes?

To identify isotopes, scientists write the name of the element followed by the isotope's mass number. Carbon with a mass number of 12 is written as carbon-12. Carbon-12 has six protons and six neutrons. Carbon-14 has six protons and eight neutrons. Carbon-12 and carbon-14 are isotopes of the element carbon. Some properties of carbon-12 and carbon-14 are unique due to differences in the number of neutrons that each isotope contains. For example, carbon-14 is radioactive, but carbon-12 is not.

Get It?

Compare the following isotopes of chlorine in terms of mass number, number of protons, and number of neutrons: chlorine-35 and chlorine-37.

Suppose you have a sample of the element boron. Naturally occurring isotopes of boron have mass numbers of 10 or 11. How many neutrons does each isotope contain? Locate boron in **Table 4** on the previous page, and determine the number of protons in an atom of boron. You can then calculate that boron-10 has five neutrons and boron-11 has six neutrons.

APPLY SCIENCE

How can radioactive isotopes help tell time?

Some isotopes are radioactive, which means they decay over time. The time that it takes for half of a radioactive isotope to decay into another isotope is called its half-life. Scientists use the half-lives of radioactive isotopes to measure geologic time.

Identify the Problem

The table to the right shows the half-lives of some radioactive isotopes (parent isotopes) and the isotopes into which they decay (daughter isotopes). For example, it would take 5730 years for half of the carbon-14 atoms in a sample to change into atoms of nitrogen-14. After another 5730 years, half of the remaining carbon-14 atoms will change, and so on. Because the number of carbon-14 atoms changes, while the number of carbon-12 atoms does not, the ratio of the number of carbon-14 atoms to carbon-12 atoms can be used to determine the length of time that has passed.

Half-Lives of Radioactive Isotopes

Parent Isotope	Daughter Isotope	Half-Life
Uranium-238	Lead-206	4.47 billion years
Potassium-40	Argon-40, Calcium-40	1.26 billion years
Rubidium-87	Strontium-87	48.8 billion years
Carbon-14	Nitrogen-14	5730 years

Solve the Problem

1. How many years would it take for half of the rubidium-87 atoms in a piece of rock to change into strontium-87? How many years would it take for three-fourths of the atoms to change?
2. After a long period, only one-fourth of the parent uranium-238 atoms in a sample of rock remain. How many years old would you predict the rock to be?

Average atomic mass

How do scientists account for and represent the different atomic masses of isotopes? As an example, models of two naturally occurring isotopes of boron are shown in **Figure 7.** Because most elements, including boron, naturally occur as more than one isotope, each element can be described by an average atomic mass of the isotopes. The **average atomic mass** of an element is the weighted average mass of all naturally occurring isotopes of an element, measured in atomic mass units (amu), according to their natural abundances.

For example, 80 percent (four out of five) of boron atoms are boron-11, and 20 percent (one out of five) are boron-10. The following calculation gives the weighted average of these two masses.

$$\frac{4}{5}(11 \text{ amu}) + \frac{1}{5}(10 \text{ amu}) = 10.8 \text{ amu}$$

The average atomic mass of the element boron is 10.8 amu. Note that the average atomic mass of boron is closer to the mass of its more abundant isotope, boron-11.

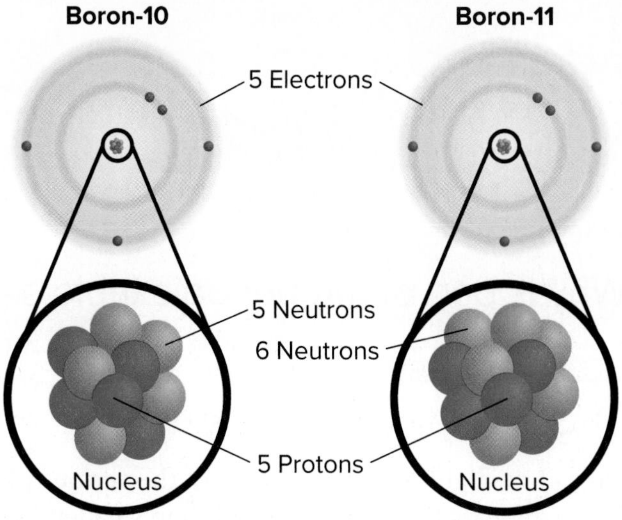

Figure 7 Boron-10 and boron-11 are two isotopes of boron. These two isotopes differ by one neutron. Most naturally occurring elements have more than one naturally occurring isotope.

Explain *why these atoms are isotopes.*

Get It?

Define average atomic mass, and explain how it is calculated.

Check Your Progress

Summary

- Protons and neutrons make up most of an atom's mass.
- Each element has a unique number of protons.
- Atoms of the same element with different numbers of neutrons are called isotopes.
- The average atomic mass of an element is the weighted average mass of all naturally occurring isotopes of that element.

Demonstrate Understanding

6. **Determine** the mass number and the atomic number of a chlorine atom that has 17 protons and 18 neutrons.
7. **Explain** how the isotopes of an element are alike and how they are different.
8. **Explain** why the atomic mass of an element is a weighted-average mass.
9. **Calculate** the number of neutrons in potassium-40.

Explain Your Thinking

10. **Explain** Chlorine has an average atomic mass of 35.45 amu. The two naturally occurring isotopes of chlorine are chlorine-35 and chlorine-37. Do most chlorine atoms contain 18 neutrons or 20 neutrons? Why?
11. **MATH Connection** Use the information in **Table 3** on page 404 to determine the mass in kilograms of each subatomic particle.

LEARNSMART® Go online to follow your personalized learning path to review, practice, and reinforce your understanding.

THE PERIODIC TABLE

FOCUS QUESTION

Why is part of the periodic table separate and below the rest of it?

Organizing the Elements

On a clear night, you can see one of the various phases of the Moon. Each month, the Moon appears to grow larger and then smaller in a predictable pattern. This type of change is periodic. *Periodic* means "repeated in a pattern." For example, a calendar is a periodic table of the days and months of the year. The days of the week are also periodic because they repeat every seven days.

In the late 1800s, a Russian chemist named Dmitri Mendeleev presented a way to organize all the known elements. While studying the physical and chemical properties of the elements, Mendeleev found that these properties repeated in predictable patterns based on an element's atomic mass. Because the pattern repeated, it was considered to be periodic.

Figure 8 shows one of Mendeleev's early periodic charts. Mendeleev arranged elements in rows based on increasing atomic mass and in columns based on elements that shared similar physical and chemical properties. Today, this arrangement is called the periodic table of the elements. In the modern **periodic table,** the elements are arranged by increasing atomic number—not atomic mass—and by periodic changes in physical and chemical properties.

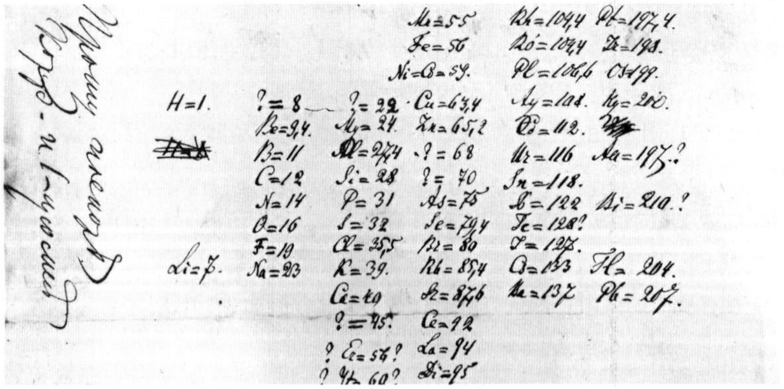

Figure 8 When Mendeleev organized the known elements in order of increasing atomic mass, he discovered that the elements had a periodic pattern in their chemical properties.

Infer *What do the question marks in Mendeleev's chart represent?*

3D THINKING DCI Disciplinary Core Ideas · CCC Crosscutting Concepts · SEP Science & Engineering Practices

COLLECT EVIDENCE

Use your Science Journal to record the evidence you collect as you complete the readings and activities in this lesson.

INVESTIGATE

GO ONLINE to find these activities and more resources.

Applying Practices: Electron Patterns in Atoms

HS-PS1-1. Use the periodic table as a model to predict the relative properties of elements based on the patterns of electrons in the outermost energy level of atoms.

Table 5 Mendeleev's Predictions

Predicted Properties of Ekasilicon (Es)	Actual Properties of Germanium (Ge)
Existence Predicted: 1871	*Actual Discovery: 1886*
Atomic mass = 72	Atomic mass = 72.61
High melting point	Melting point = 938°C
Density = 5.5 g/cm^3	Density = 5.323 g/cm^3
Dark-gray metal	Gray metal
Density of EsO_2 = 4.7 g/cm^3	Density of GeO_2 = 4.23 g/cm^3

Mendeleev's predictions

Mendeleev had to leave blank spaces in his periodic table. He studied the physical and chemical properties and the atomic masses of the elements surrounding all the blank spaces. From this information, he was able to predict the probable properties and the atomic masses for the missing elements that had not yet been discovered.

Get It?

Explain how Mendeleev was able to predict the properties of elements that had not yet been discovered.

Table 5 shows a few of Mendeleev's predicted physical and chemical properties for germanium, which he called ekasilicon. His predictions proved to be accurate when compared to the actual properties of germanium. Scientists later confirmed the identities of the missing elements and found that their properties were similar to what Mendeleev had suggested.

Changes in the periodic table

Although Mendeleev's arrangement of the elements was a success, it required some changes. Atomic mass gradually increased from left to right on Mendeleev's table. If you look at the modern periodic table, in **Figure 9** on the next page, you can locate instances where atomic mass decreases from left to right, such as with nickel and cobalt.

Get It?

Explain how Mendeleev organized his periodic table.

You might also observe that the atomic number always increases from left to right. In 1913, a young English scientist named Henry G. J. Moseley arranged all the known elements based on increasing atomic number (the number of protons) instead of atomic mass. This new arrangement seemed to solve the problem of fluctuating mass. The modern periodic table uses Moseley's arrangement of the elements. The periodic table orders elements horizontally by the number of protons in the atom's nucleus and places those elements with similar chemical properties in columns.

Get It?

Describe how Moseley altered the arrangement of elements on the periodic table and how the change improved the table.

Figure 9 The Periodic Table of the Elements

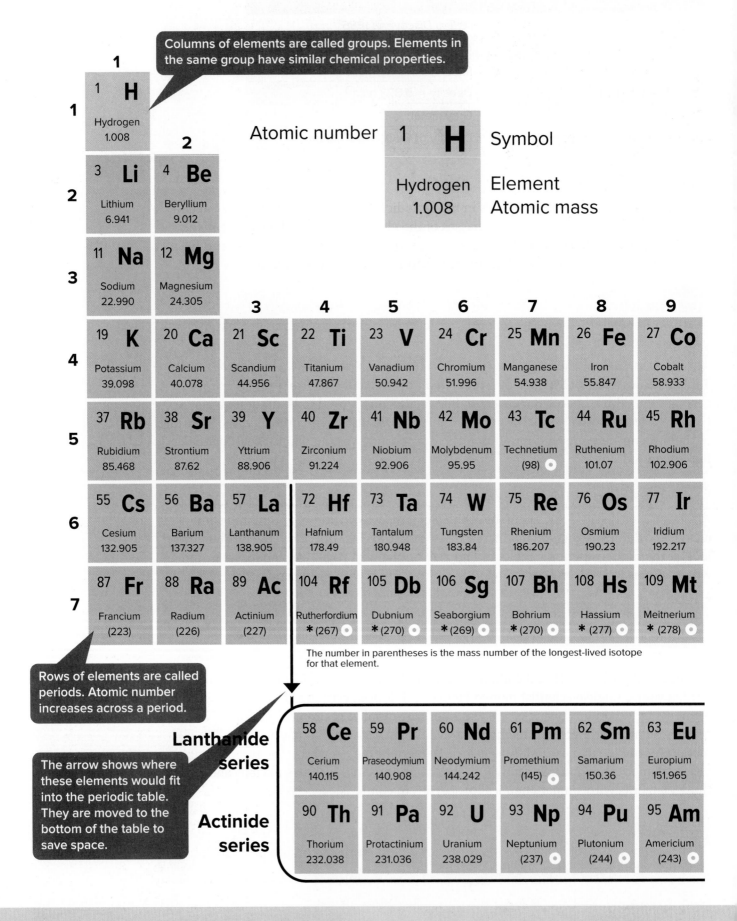

Columns of elements are called groups. Elements in the same group have similar chemical properties.

Atomic number	Symbol	Element	Atomic mass
1	H	Hydrogen	1.008

Period	1	2	3	4	5	6	7	8	9
1	1 H Hydrogen 1.008								
2	3 Li Lithium 6.941	4 Be Beryllium 9.012							
3	11 Na Sodium 22.990	12 Mg Magnesium 24.305							
4	19 K Potassium 39.098	20 Ca Calcium 40.078	21 Sc Scandium 44.956	22 Ti Titanium 47.867	23 V Vanadium 50.942	24 Cr Chromium 51.996	25 Mn Manganese 54.938	26 Fe Iron 55.847	27 Co Cobalt 58.933
5	37 Rb Rubidium 85.468	38 Sr Strontium 87.62	39 Y Yttrium 88.906	40 Zr Zirconium 91.224	41 Nb Niobium 92.906	42 Mo Molybdenum 95.95	43 Tc Technetium (98)	44 Ru Ruthenium 101.07	45 Rh Rhodium 102.906
6	55 Cs Cesium 132.905	56 Ba Barium 137.327	57 La Lanthanum 138.905	72 Hf Hafnium 178.49	73 Ta Tantalum 180.948	74 W Tungsten 183.84	75 Re Rhenium 186.207	76 Os Osmium 190.23	77 Ir Iridium 192.217
7	87 Fr Francium (223)	88 Ra Radium (226)	89 Ac Actinium (227)	104 Rf Rutherfordium * (267)	105 Db Dubnium * (270)	106 Sg Seaborgium * (269)	107 Bh Bohrium * (270)	108 Hs Hassium * (277)	109 Mt Meitnerium * (278)

The number in parentheses is the mass number of the longest-lived isotope for that element.

Rows of elements are called periods. Atomic number increases across a period.

Series						
Lanthanide series	58 Ce Cerium 140.115	59 Pr Praseodymium 140.908	60 Nd Neodymium 144.242	61 Pm Promethium (145)	62 Sm Samarium 150.36	63 Eu Europium 151.965
Actinide series	90 Th Thorium 232.038	91 Pa Protactinium 231.036	92 U Uranium 238.029	93 Np Neptunium (237)	94 Pu Plutonium (244)	95 Am Americium (243)

The arrow shows where these elements would fit into the periodic table. They are moved to the bottom of the table to save space.

The color of an element's block tells you if the element is a metal, nonmetal, or metalloid. The bottom symbol identifies elements that are not present in significant amounts on Earth.

- Metal
- Metalloid
- Nonmetal
- ⊙ Synthetic

10	11	12	13	14	15	16	17	18
								2 He Helium 4.003
			5 B Boron 10.811	6 C Carbon 12.011	7 N Nitrogen 14.007	8 O Oxygen 15.999	9 F Fluorine 18.998	10 Ne Neon 20.180
			13 Al Aluminum 26.982	14 Si Silicon 28.086	15 P Phosphorus 30.974	16 S Sulfur 32.066	17 Cl Chlorine 35.453	18 Ar Argon 39.948
28 Ni Nickel 58.693	29 Cu Copper 63.546	30 Zn Zinc 65.39	31 Ga Gallium 69.723	32 Ge Germanium 72.61	33 As Arsenic 74.922	34 Se Selenium 78.971	35 Br Bromine 79.904	36 Kr Krypton 83.80
46 Pd Palladium 106.42	47 Ag Silver 107.868	48 Cd Cadmium 112.411	49 In Indium 114.82	50 Sn Tin 118.710	51 Sb Antimony 121.757	52 Te Tellurium 127.60	53 I Iodine 126.904	54 Xe Xenon 131.290
78 Pt Platinum 195.08	79 Au Gold 196.967	80 Hg Mercury 200.59	81 Tl Thallium 204.383	82 Pb Lead 207.2	83 Bi Bismuth 208.980	84 Po Polonium 208.982	85 At Astatine 209.987	86 Rn Radon 222.018
110 Ds Darmstadtium * (281) ⊙	111 Rg Roentgenium * (281) ⊙	112 Cn Copernicium * (285) ⊙	113 Nh Nihonium * (286) ⊙	114 Fl Flerovium * (289) ⊙	115 Mc Moscovium * (289) ⊙	116 Lv Livermorium * (293) ⊙	117 Ts Tennessine * (294) ⊙	118 Og Oganesson * (294) ⊙

* Properties are largely predicted.

64 Gd Gadolinium 157.25	65 Tb Terbium 158.925	66 Dy Dysprosium 162.50	67 Ho Holmium 164.930	68 Er Erbium 167.259	69 Tm Thulium 168.934	70 Yb Ytterbium 173.04	71 Lu Lutetium 174.967
96 Cm Curium (247) ⊙	97 Bk Berkelium (247) ⊙	98 Cf Californium (251) ⊙	99 Es Einsteinium * (252) ⊙	100 Fm Fermium * (257) ⊙	101 Md Mendelevium * (258) ⊙	102 No Nobelium * (259) ⊙	103 Lr Lawrencium * (262) ⊙

The Atom and the Periodic Table

The modern periodic table consists of boxes, each containing information such as element name, symbol, atomic number, and atomic mass. A typical box is shown in **Figure 10.** As you have learned, elements on the periodic table are organized based on similarities in their physical and chemical properties. The horizontal rows of elements in the periodic table are called **periods** and are numbered 1 through 7. The vertical columns in the periodic table are called **groups** (also called families), and they are numbered 1 through 18. Elements in each group share similar properties. For example, the elements in group 11—including copper, silver, and gold—are all similar. Each element is a shiny metal and a good conductor of heat and electricity. Why are these elements so similar?

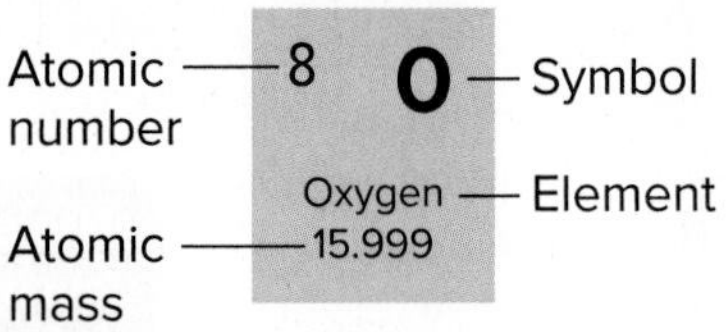

Figure 10 Each box on the periodic table contains the element's name, its atomic number, its chemical symbol, its atomic mass, and its state of matter.

Evaluate *What is the atomic mass of oxygen?*

Electron cloud structure

You have learned that each atom has a charged substructure consisting of a nucleus that is made of protons and neutrons. But where are the electrons? How many are there? Because an atom does not have an overall charge, the number of electrons is equal to the number of protons. Therefore, a carbon atom has six protons and six electrons. An oxygen atom has eight protons and eight electrons. Electrons surround the nucleus and are located in an area called the electron cloud.

Energy Levels Scientists have discovered that electrons within an electron cloud have different amounts of energy. Scientists model the energy differences between electrons by placing electrons in energy levels, as shown in **Figure 11.** Electrons located in energy levels close to the nucleus have less energy than electrons in energy levels farther away. Electrons occupy energy levels in a predictable pattern from the inner to the outer levels.

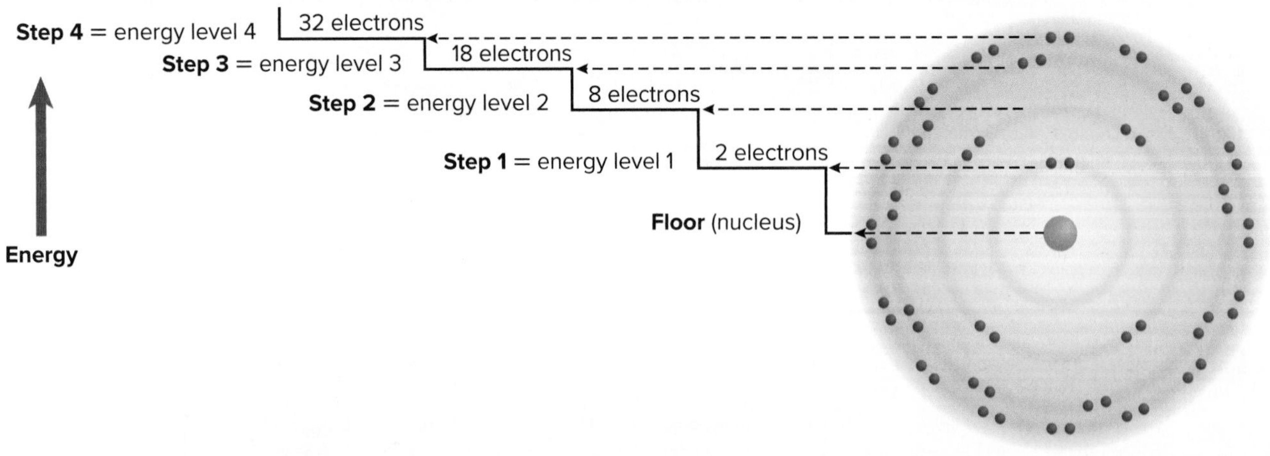

Figure 11 Energy levels in atoms can be represented by stairsteps. Each step away from the nucleus represents an increase in the amount of energy within the electrons. The higher energy levels can hold more electrons.

Elements in the same group have the same number of electrons in their outermost energy levels. These electrons are called valence electrons. It is the number of valence electrons that determines the chemical properties of each individual element. In other words, the repeating patterns of the periodic table reflect patterns of outer electron states. It is important to understand the relationship between the location of an element in the periodic table, the element's chemical properties, and the element's atomic structure.

Rows on the periodic table

Energy levels coincide with the number of rows on the periodic table. These energy levels are named using numbers 1 to 7. The maximum number of electrons that can be placed in each of the first four levels is shown in **Figure 11.** For example, energy level 1 can hold a maximum of two electrons. Energy level 2 can hold a maximum of eight electrons. For energy levels 2 and higher, the outer energy level is stable when it holds eight electrons. Notice, however, that energy levels 3 and 4 can contain more than eight electrons. The way in which energy levels split into sublevels allows for energy levels 3 and higher to contain more than eight electrons. These additional electrons are added to inner sublevels; the outer energy level is still stable when it contains eight electrons.

Filling the first row Remember that the atomic number found on the periodic table is equal to the number of electrons in a neutral atom. Look at the elements in **Figure 12.** The first row has hydrogen, with one electron, and helium, with two electrons, both in energy level 1. Because energy level 1 is the outermost energy level containing an electron, hydrogen has one outer electron. Helium has two outer electrons. Recall from **Figure 11** that energy level 1 can hold a maximum of two electrons. Therefore, helium has a full outer energy level and is chemically stable.

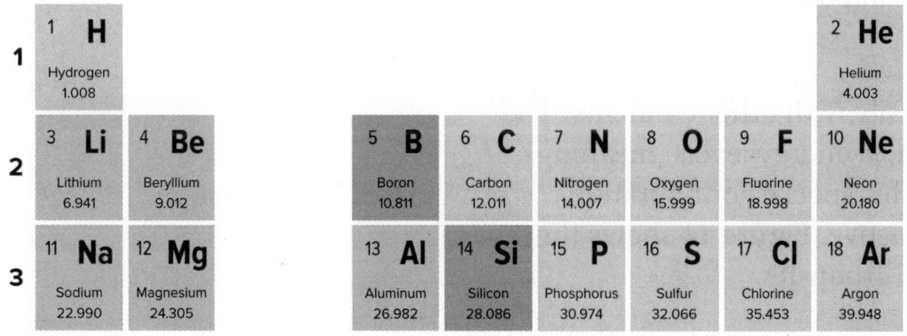

Figure 12 One proton and one electron are added to each element as you travel across a row on the periodic table. When a certain level is complete—as with helium and neon—the next electron added starts a new row.

Explain *how the elements in each group are similar.*

STEM CAREER Connection

Research Physicist

Are you fascinated by the inner workings of atoms? The study of nuclear interactions is shared by chemists and physicists. Research physicists use their knowledge of the physical laws of nature to explain the behavior and composition of atoms. For example, physicists can apply their knowledge of atoms, such as knowing how the amount of energy an electron contains determines its behavior, to the fields of nuclear energy, electronics, optics, and communications.

Filling higher rows The second row starts with lithium, which has three electrons—two in the first energy level and one in the second energy level. Lithium is followed by beryllium, with two outer electrons, boron, with three, and so on. Neon has a complete outermost energy level, with eight outer electrons. Electrons begin filling energy level 3 for elements in the third row. The row ends with argon, which has eight outer electrons.

Electron dot diagrams

Elements in the same group have the same number of electrons in their outermost energy levels. In fact, the repeating patterns in the table reflect these patterns of outer electron states. These electrons determine the chemical properties of an element. They are so significant that American chemist G. N. Lewis created a diagram to represent an element's outermost electrons. An **electron dot diagram** consists of the chemical symbol of an element surrounded by dots to represent the number of electrons in the outermost energy level. **Figure 13** shows the electron dot diagrams for the group 1 elements.

Figure 13 The elements in group 1 have one electron in their outermost energy levels.

Same group—similar properties

The electron dot diagrams for the elements in group 1 show that all members of a group have the same number of outermost electrons. Remember that the number of outermost electrons determines the chemical properties for each element.

A common chemical property of group 1 metals is the tendency to react with nonmetals in group 17. The nonmetals in group 17 have electron dot diagrams similar to chlorine, as shown in **Figure 14.** For example, the group 1 element sodium reacts easily with the group 17 element chlorine. The result is the formation of the compound sodium chloride (NaCl)—ordinary table salt.

Group 18 Not all elements will combine easily with other elements. The elements in group 18 have complete outermost energy levels, meaning that they cannot hold any more electrons. This special configuration makes many of the group 18 elements unreactive. **Figure 14** shows the electron dot diagram for neon, a member of group 18.

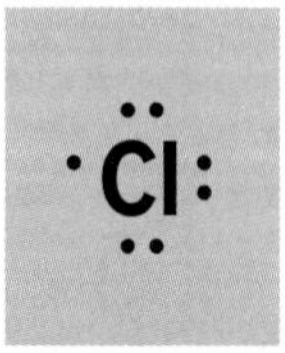

The electron dot diagram for group 17 consists of three sets of paired dots and one single dot.

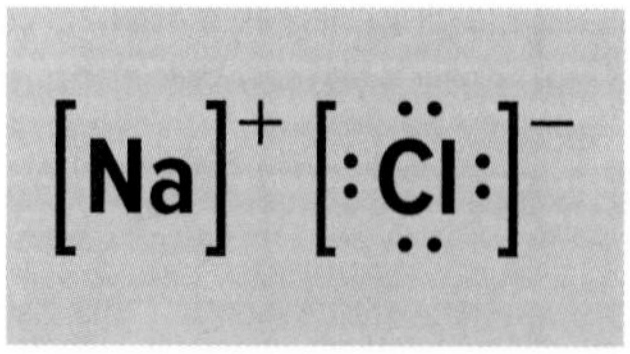

Sodium combines with chlorine to give each element a complete outer energy level in the resulting compound.

Neon, a member of group 18, has a full outer energy level. Neon has eight electrons in its outer energy level, making it unreactive.

Figure 14 Electron dot diagrams show the electrons in an element's outermost energy level.
Relate *the properties of Na, Cl, and Ne atoms and their positions in the periodic table to the arrangement of their electrons.*

Regions of the Periodic Table

The periodic table has areas with specific names. Recall that the horizontal rows of elements are called periods. The elements increase by one proton and one electron as you move from left to right across a period.

All the elements in the blue squares in **Figure 15** are metals. Iron, zinc, and copper are examples of a few common metals. Most metals occur as solids at room temperature. They are shiny, can be drawn into wires, can be pounded into sheets, and are good conductors of heat and electricity.

The elements on the right side of the periodic table, which appear in the yellow squares, are classified as nonmetals. Oxygen, bromine, and carbon are examples of nonmetals. Most nonmetals are gases or brittle solids at room temperature. They are poor conductors of heat and electricity. The elements in the green squares are metalloids. They exhibit properties of metals and nonmetals. Boron and silicon are examples of metalloids.

New elements

Scientists around the world continue their research into the synthesis of elements. In 1994, scientists at the GSI Helmholtz Center for Heavy Ion Research in Darmstadt, Germany, discovered element 111. The International Union of Pure and Applied Chemistry (IUPAC) confirmed the discovery in 2003. The name Roentgenium (Rg) was officially approved in 2004. Element number 112 was discovered at the same laboratory. Synthesis of the element was reported in 1996. IUPAC confirmed the discovery in 2009, and the element was officially named Copernicium (Cn) in 2010. These elements are produced in the laboratory by joining smaller atoms into a single, larger atom. Scientists have synthesized elements 113 through 118. The search for elements with higher atomic numbers continues.

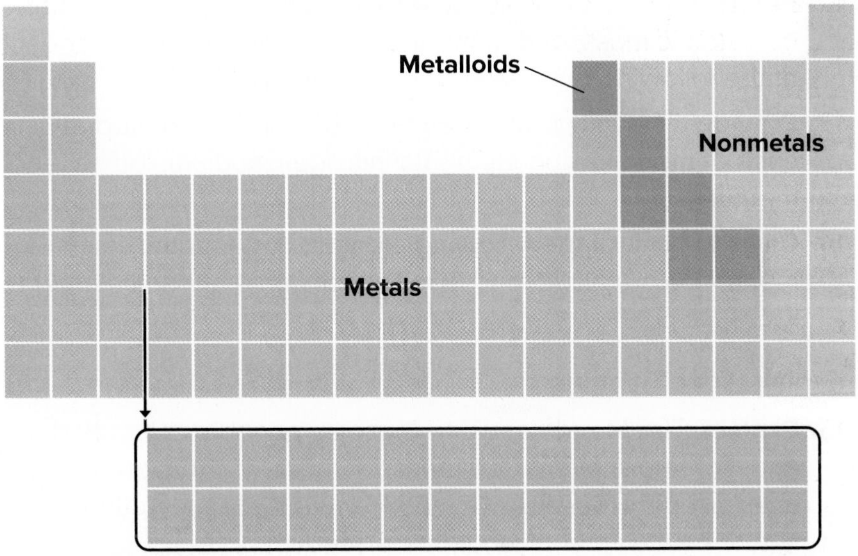

Figure 15 Metalloids are located along the green stair step line in the periodic table. Metals are located to the left of the metalloids and are shown in blue. Except for hydrogen, nonmetals are located to the right of the metalloids and are shown in yellow.

ACADEMIC VOCABULARY

occur

to be found; to come into existence

Many tornadoes occur in the central plains of the United States.

Elements in the Universe

With the development of new technologies, scientists have been able to study the chemistry of the universe. They have learned that many of the same elements are found throughout the universe. These include lightweight elements, such as hydrogen and helium, and heavier elements, such as silicon, oxygen, and iron.

Many scientists think that hydrogen and helium are the building blocks of all other elements. Atoms fuse within stars to produce heavier elements with atomic numbers greater than hydrogen and helium. Exploding stars called supernovas, like the one shown in **Figure 16,** provide evidence to support this theory.

Figure 16 The Crab Nebula is a remnant of a supernova that occurred in A.D. 1054. This is now an area of expanding gas and elements that will be incorporated into newly forming stars.

When stars explode, a mixture of elements, including heavy elements like iron, are expelled into the galaxy. Many scientists think that supernovas have scattered heavy, naturally occurring elements throughout the universe. Promethium, technetium, and elements with atomic numbers greater than 92 are rare or are not found on Earth. Some of these elements, such as neptunium and plutonium, are found only in trace amounts in Earth's crust as a result of uranium decay. Others have been found only in stars.

Check Your Progress

Summary

- Mendeleev organized the elements based on atomic mass and chemical and physical properties.
- Moseley built upon Mendeleev's periodic table by further organizing elements by increasing atomic number.
- Elements in the same vertical column on the periodic table are known as a group. They share similar physical and chemical properties.
- Elements in the same horizontal row on the periodic table are known as a period. They have the same number of energy levels.
- Elements on the periodic table are classified as metals, nonmetals, or metalloids.

Demonstrate Understanding

12. **Relate** Use the periodic table to find the name, atomic number, atomic mass, and number of outermost electrons for each of the following elements: N, P, As, and Sb.
13. **Provide** the symbol, the group number, and the period of each of the following elements: nitrogen, sodium, iodine, and mercury.
14. **Classify** each of the following elements as a metal, a nonmetal, or a metalloid, and give the full name of each element: K, Si, and S.

Explain Your Thinking

15. **Explain** The Mendeleev and Moseley periodic charts had gaps for undiscovered elements. Why do you think Moseley's chart used by Moseley was more accurate at predicting where new elements would be placed?
16. **MATH Connection** Construct a circle graph showing the percentage of elements classified as metals, metalloids, and nonmetals. Use markers or colored pencils to distinguish clearly between each section on the graph.

LEARNSMART® Go online to follow your personalized learning path to review, practice, and reinforce your understanding.

Cassini-Huygens Mission

Since the Cassini-Huygens spacecraft reached Saturn in 2004, the data it has gathered has stunned the scientific community. Two of Saturn's moons, Titan and Enceladus, contain water, and scientists consider them prime places to search for life in our solar system. The spacecraft's first mission extension, the Equinox Mission, revealed new ring structures around Saturn and discovered a tiny moon, Aegaeon, in one of the planet's rings.

The fading bluish hue is a change observed by comparing Cassini's images from 2004, 2009, and 2016.

Solstice Mission

In 2010, the spacecraft embarked on its second mission extension, the Solstice Mission. Using the data gathered by Cassini-Huygens, scientists were able to observe, for the first time, the passing of a season on Saturn and its moons. Cassini observed the Saturnian summer solstice in its northern hemisphere, which occurs only once every 15 Earth years.

As the Saturnian winter turned to summer, Cassini-Huygens recorded images of a huge thunderstorm that raged for more than 200 days and encircled the planet. The spacecraft watched as the color of Saturn's polar region changed as temperatures rose. The changing angle of sunlight on the planet's rings helped scientists learn about the composition, structure, and arrangement of particles in the rings.

Grand Finale

NASA decided to plunge the spacecraft into Saturn to avoid an accidental crash into Titan or Enceladus, which NASA wants to keep pristine for future research. Before its final descent, Cassini embarked on its "Grand Finale." It spent five months making 22 loops through the gap between the planet and its innermost ring. It gathered data about the amount of material in the rings, helping scientists determine how the rings formed. The spacecraft also made gravity and magnetic field maps of Saturn. On September 15, 2017, NASA set Cassini-Huygens on its collision course. As it plunged, the spacecraft's spectrometer sent data about the atmosphere's composition. Data from the 20-year mission will inform space exploration for decades to come.

OBTAIN, EVALUATE, AND COMMUNICATE INFORMATION

Many studies were part of the Cassini-Huygens Mission. Work with a small group to research and create a presentation about one aspect of the mission. Present your findings to the class.

MODULE 16

STUDY GUIDE

GO ONLINE to study with your Science Notebook.

Lesson 1 STRUCTURE OF THE ATOM

- Scientists use chemical symbols to abbreviate element names.
- Atoms are composed of protons, neutrons, and electrons.
- Scientists have confirmed the existence of six different quarks.
- The electron cloud model is the current atomic model.

- atom
- nucleus
- proton
- neutron
- electron
- quark
- electron cloud

Lesson 2 MASSES OF ATOMS

- Protons and neutrons make up most of an atom's mass.
- Each element has a unique number of protons.
- Atoms of the same element with different numbers of neutrons are called isotopes.
- The average atomic mass of an element is the weighted average mass of all naturally occurring isotopes of that element.

- atomic number
- mass number
- isotope
- average atomic mass

Lesson 3 THE PERIODIC TABLE

- Mendeleev organized the elements based on atomic mass and chemical and physical properties.
- Moseley built upon Mendeleev's periodic table by further organizing elements by increasing atomic number.
- Elements in the same vertical column on the periodic table are known as a group. They share similar physical and chemical properties.
- Elements in the same horizontal row on the periodic table are known as a period. They have the same number of energy levels.
- Elements on the periodic table are classified as metals, nonmetals, or metalloids.

- periodic table
- period
- group
- electron dot diagram

THREE-DIMENSIONAL THINKING

Module Wrap-Up

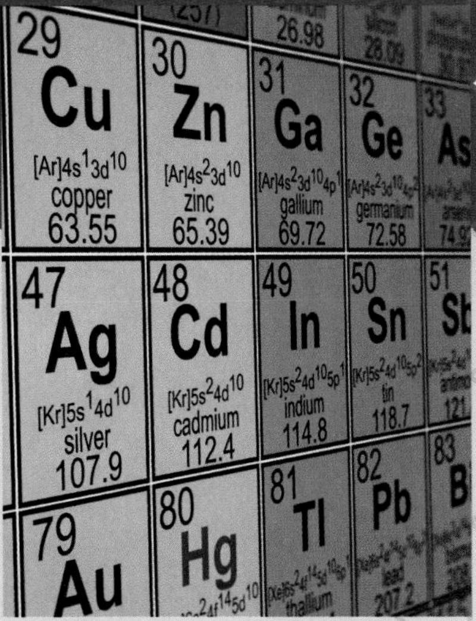

REVISIT THE PHENOMENON

Why are there different colors on this periodic table?

CER Claim, Evidence, Reasoning

Explain Your Reasoning Revisit the claim you made when you encountered the phenomenon. Summarize the evidence you gathered from your investigations and research and finalize your Summary Table. Does your evidence support your claim? If not, revise your claim. Explain why your evidence supports your claim.

STEM UNIT PROJECT

Now that you've completed the module, revisit your STEM unit project. You will summarize your evidence and apply it to the project.

GO FURTHER

SEP Data Analysis Lab

The Building Blocks of Matter

From experiments involving the high-speed collision of particles, scientists have determined that heavy nuclear particles, such as protons and neutrons, are composed of quarks. Quarks are classified into six flavors—up, down, strange, charm, bottom, and top. The flavors are not related to any physical characteristics of the quarks themselves, but are used as classifications. Unlike protons and electrons, which carry whole charges of 1+ or 1–, quarks carry fractional charges. The following table lists some information about quarks.

Quarks

Name	Symbol	Charge	Mass	Name	Symbol	Charge
up	*u*	2/3+	↑	down	*d*	1/3–
strange	*s*	1/3–		charm	*c*	2/3+
bottom	*b*	1/3–		top	*t*	2/3+

CER Analyze and Interpret Data

1. **Identify** Protons and neutrons are composed of triplets of up and down quarks. Use the information in the table to determine which of the following quark triplets represents a proton and which represents a neutron.
   a. udd **b.** uud
2. **Identify** A heavier nuclear particle is composed of a triplet of strange quarks. What is the charge of this particle?

Jaap Hart/E+/Getty Images

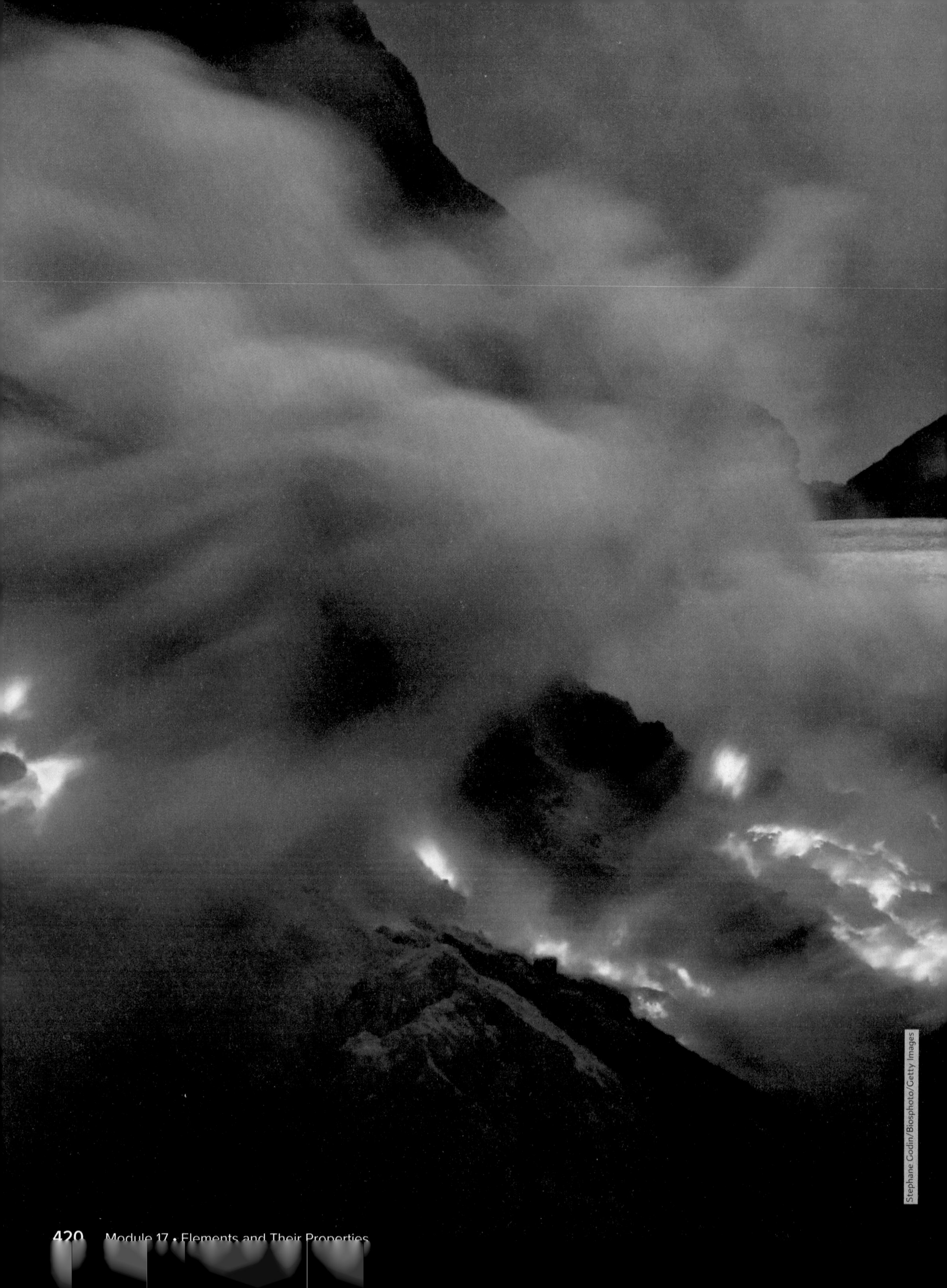

Stephane Codin/Biosphoto/Getty Images

MODULE 17

ELEMENTS AND THEIR PROPERTIES

ENCOUNTER THE PHENOMENON

Why does this lava burn blue?

GO ONLINE to play a video about why this lava is blue.

SEP Ask Questions

Do you have other questions about the phenomenon? If so, add them to the driving question board.

CER Claim, Evidence, Reasoning

Make Your Claim Use your CER chart to make a claim about why this lava is blue. Explain your reasoning.

Collect Evidence Use the lessons in this module to collect evidence to support your claim. Record your evidence as you move through the module.

Explain Your Reasoning You will revisit your claim and explain your reasoning at the end of the module.

GO ONLINE to access your CER chart and explore resources that can help you collect evidence.

LESSON 1: Explore & Explain: The Alkaline Earth Metals

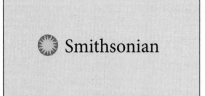

LESSON 3: Explore & Explain: Properties of Metalloids

Smithsonian

Additional Resources

LESSON 1
METALS

FOCUS QUESTION
What are the properties of a typical metal?

Properties of Metals

Coins, paper clips, and some baseball bats are made of metals. **Metals** are elements that are shiny, malleable, ductile, and good conductors of heat and electricity. Except for mercury, metals are solids at room temperature. The shiny property of metals is called metallic luster. Metals are **malleable** (MA lee uh bul), which means they can be hammered or rolled into sheets. Metals are also **ductile,** which means they can be drawn into wires. **Figure 1** shows the malleability and ductility of metals.

These properties make metals suitable for use in a wide range of objects, including eyeglass frames, computers, and buildings. The specific properties of a metal depend on its unique configuration of electrons, protons, and neutrons. However, elements in the same group of the periodic table have similar properties.

Metals and the periodic table

In the periodic table, metals are found to the left of the stairstep line. In the periodic tables in this book, the metal element blocks are colored blue. Notice that most elements are metals. Except for hydrogen, all the elements in groups 1 through 12 are metals, as are the elements under the stairstep line in groups 13 through 15.

Metals are malleable: they can be hammered into thin sheets.

Metals are ductile: they can be drawn into wires.

Figure 1 The various properties of metals make them useful.

Describe *some uses for metal sheets and wires.*

3D THINKING DCI Disciplinary Core Ideas CCC Crosscutting Concepts SEP Science & Engineering Practices

COLLECT EVIDENCE

Use your Science Journal to record the evidence you collect as you complete the readings and activities in this lesson.

INVESTIGATE

GO ONLINE to find these activities and more resources.

Quick Investigation: Discover What's in Cereal
Carry out an investigation to discover the result of adding a magnet to cereal.

Review the News
Obtain information from a current news story about one of the nonmetals. Evaluate your source and communicate your findings to your class.

Bonding in metals

The atoms of metals generally have one to three electrons in their outer energy levels. In chemical reactions, metals tend to give up electrons easily because they are not strongly held by the protons in the nucleus.

Bonding with nonmetals

When metals combine with nonmetals, the atoms of the metals tend to lose electrons to the atoms of nonmetals. The metal atoms become positive ions, and the nonmetal atoms become negative ions. Ions are charged particles with more or fewer electrons than the neutral atom. Both metals and nonmetals become more chemically stable when they form ions.

A positively charged metal ion and a negatively charged nonmetal ion are attracted because of the electric force between them. They form a bond called an ionic bond. For example, a sodium atom can lose an electron to a chlorine atom. As shown in **Figure 2,** the sodium ion and chloride ion bond to form the compound sodium chloride (NaCl), also known as table salt.

Sodium Chloride (NaCl)

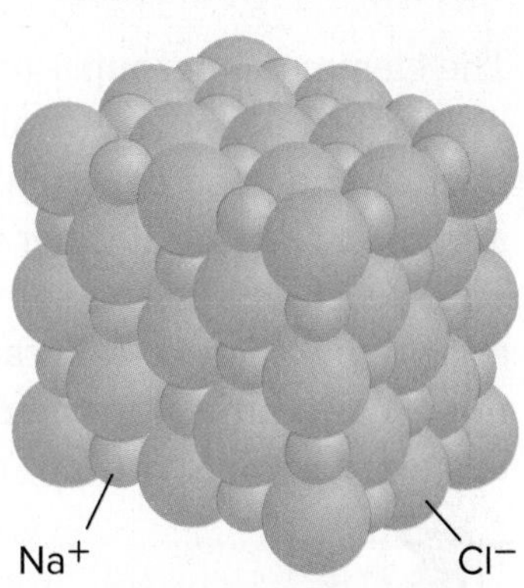

Figure 2 Metals can form ionic bonds with nonmetals. In a crystal of table salt (NaCl), the positive ions come from the metal sodium, and the negative ions come from the nonmetal chlorine.

Metallic bonding

A different type of bonding occurs between the atoms of metals. In **metallic bonding,** positively charged metallic ions are surrounded by a sea of electrons. Outer-level electrons are not held tightly to the nucleus of an atom, but move freely among many positively charged ions, as shown in **Figure 3.**

Metallic bonding explains many of the properties of metals. For example, when a metal is hammered into a sheet or drawn into a wire, it does not break because the ions are in layers that slide past one another without losing their attraction to the electron sea. Metals are also good conductors of heat and electricity because the outer-level electrons are weakly held and travel relatively freely.

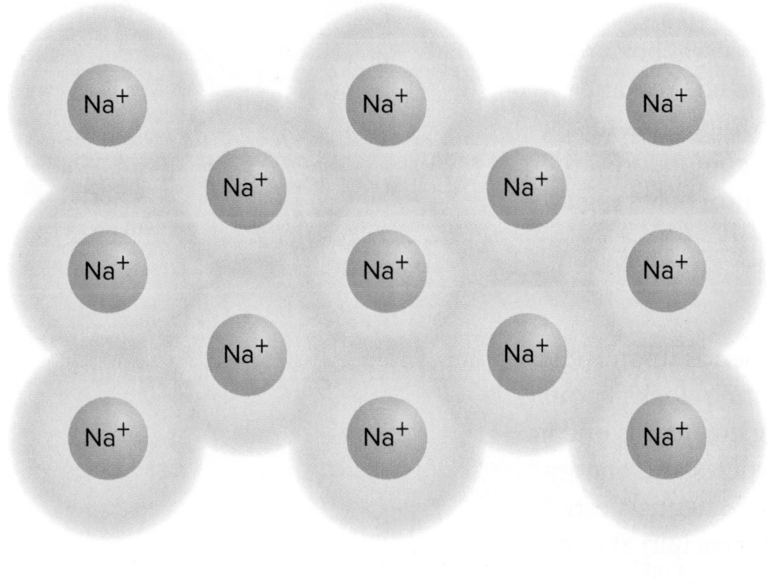

Figure 3 In metallic bonding, the electrons in the shared electron sea are not attached to any one metal ion. This allows the electrons to move.

Explain *Why do metals conduct electricity?*

The Alkali Metals

The elements in group 1 of the periodic table are the alkali (AL kuh li) metals. Like other metals, alkali metals are shiny, malleable, ductile, and good conductors of heat and electricity. However, they are softer than most other metals and are the most reactive metals. They react rapidly and sometimes violently with oxygen and water, as shown in **Figure 4.** Because they are so reactive, alkali metals do not occur naturally in their elemental forms, and pure samples must be stored in oil to prevent reaction with oxygen and water in the air.

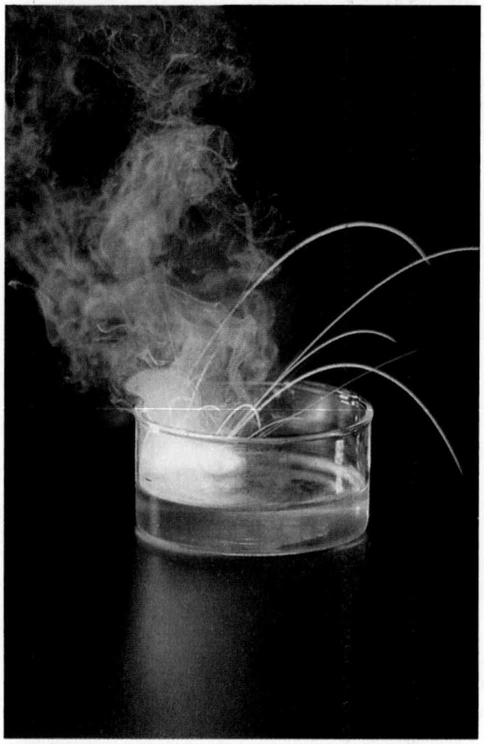

Figure 4 Alkali metals are very reactive. For example, the vigorous reaction between potassium and water releases enough thermal energy to ignite the hydrogen gas that forms.

Get It?

Explain how knowledge of the properties of alkali metals is used to predict how they will react and determine how they should be stored.

Atomic structure explains the reactive nature of alkali metals. Each atom of an alkali metal has one electron in its outer energy level. This electron is easily given up when an alkali metal combines with a nonmetal. As a result, the alkali metal atom becomes a positively charged ion in a compound such as sodium chloride (NaCl) or potassium bromide (KBr).

Get It?

Explain how interactions of electric charges at the atomic scale account for the reactivity of alkali metals.

Get It?

Explain how the location of alkali metals on the periodic table is related to the number of electrons in the outer energy level.

The Alkali Metals

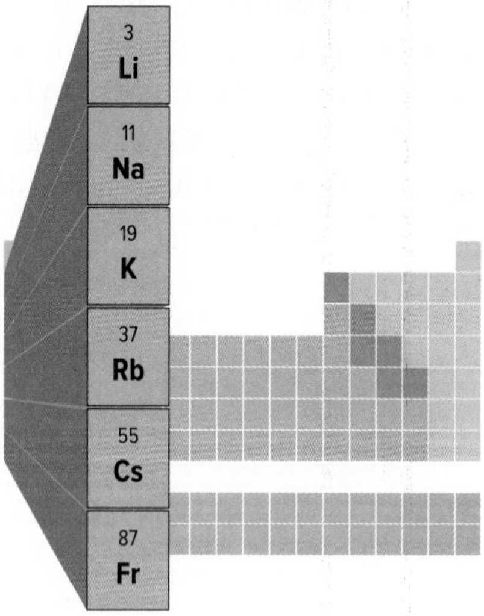

Lithium, sodium, and potassium

Look carefully at the nutritional information on a cereal box. You will notice that sodium and potassium are often listed. You and other living things need potassium and sodium compounds to stay healthy. Lithium can also benefit health. Lithium compounds are sometimes used to treat bipolar disorder. The lithium helps regulate chemical levels that are important to mental health.

Rubidium, cesium, and francium

The operation of some light-detecting sensors depends upon rubidium or cesium compounds. Cesium is used in atomic clocks because some of its isotopes are radioactive. A **radioactive element** is one in which the nucleus breaks down and gives off particles and energy. Francium is also radioactive and is extremely rare. Scientists estimate that Earth's crust contains less than 30 g of francium at one time.

The Alkaline Earth Metals

The alkaline earth metals make up group 2 of the periodic table. Like most metals, these metals are shiny, malleable, and ductile. Like the alkali metals, they combine readily with other elements and are not found as free elements in nature.

Each atom of an alkaline earth metal has two electrons in its outer energy level. These electrons are given up when an alkaline earth metal combines with a nonmetal. The alkaline earth metal atom becomes the positively charged ion in a compound, such as calcium fluoride (CaF_2). Some compounds of the alkaline earth metals are used to color fireworks, like those in **Figure 5.**

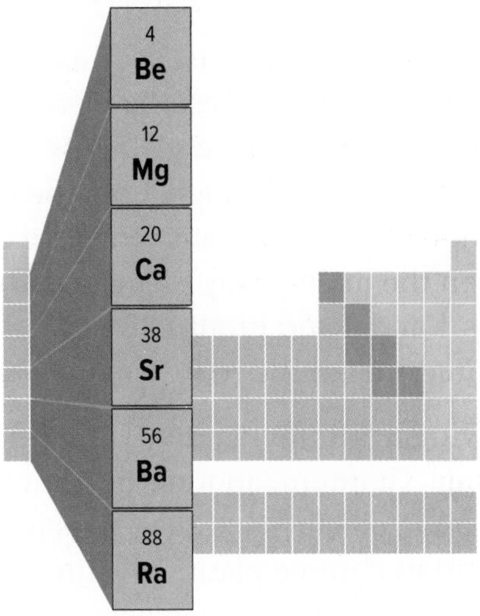

Get It?

Compare and contrast the alkali metals and the alkaline earth metals.

Magnesium

Magnesium's lightness and strength make it a good material for use in cars, planes, spacecraft, household ladders, and baseball and softball bats. Most life on Earth depends upon chlorophyll, a magnesium-containing compound that enables plants to absorb light and make food.

Calcium

Calcium is seldom used as a free metal, but its compounds are useful and essential for life. Marble statues and some countertops are made of calcium carbonate ($CaCO_3$). Calcium carbonate is the major component of the mineral limestone, which is found in many caves. You might take a vitamin with calcium. Calcium phosphate ($CaPO_4$) helps make your bones strong.

Figure 5 Alkaline earth metals make spectacular fireworks. The metals used determine the colors of the displays.

Other alkaline earth metals

The compound barium sulfate ($BaSO_4$) is used to diagnose some digestive disorders because it absorbs X-ray radiation well. First, the patient swallows a barium compound. Next, an X-ray is taken while the barium compound is going through the digestive tract. A doctor can then see where the barium is in the body and can use this information to diagnose internal abnormalities.

Radium, the last element in group 2, is radioactive and is found associated with uranium. It was once used to treat cancers but is being replaced with more readily available radioactive elements.

John Lund/Blend Images LLC

The Transition Elements

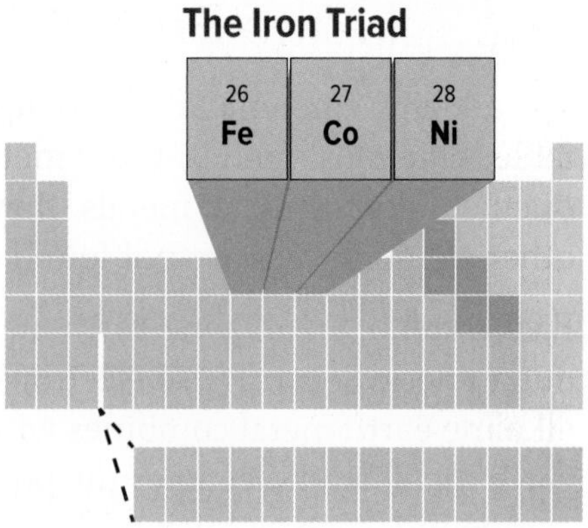

Many transition elements, such as iron and gold, are familiar because they are less reactive than the metals in groups 1 and 2 and often occur in nature as uncombined elements. **Transition elements** are the elements in groups 3 through 12 in the periodic table. They are called transition elements because they are considered to be in transition between the main-group elements. The main-group elements are groups 1 and 2 and groups 13 through 18. Main-group elements are sometimes called the representative elements.

A glowing lightbulb filament is made from the transition element tungsten. Titanium, another transition element, is used in bike frames, ships, and golf clubs because of its lightness and strength. Chromium is used in making steel and in chrome plating. Platinum is used to make jewelry because it is rare, resistant to corrosion, and more valuable than gold.

Transition elements often form colored compounds, as shown in **Figure 6.** The gems' colors come from chromium compounds. The blue glass gets its color from cobalt. Cadmium yellow and cobalt blue paints are made from compounds of transition elements. However, their use is limited because they are so toxic.

Iron, cobalt, and nickel

Iron, cobalt, and nickel form a unique cluster of transition elements sometimes called the iron triad. They are the most common magnetic elements and are used in steel and other metal mixtures.

Iron is the second-most abundant metal in Earth's crust and the most widely used of all metals. It is the main component of steel. Other metals, such as nickel and cobalt, are added to steel to give it various characteristics. Nickel is also used to give a shiny, protective coating to other metals.

Figure 6 Chromium compounds give rubies their red color and emeralds their green color. Cobalt is used to make blue glass.

Figure 7 Copper is often mixed with zinc or nickel to make modern coins. Gold, silver, and bronze—an alloy of copper and tin—are used to make athletic medals.

Copper, silver, and gold

The Coinage Metals

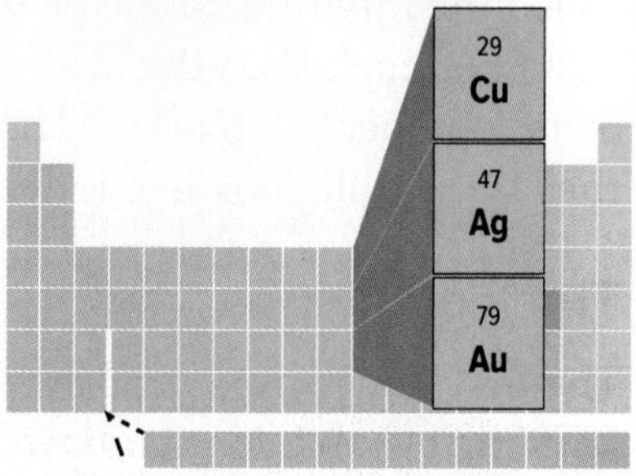

You are probably familiar with copper, silver, and gold–three of the elements in group 11. Because they are so stable and malleable and can be found as free elements in nature, these metals were once used widely to make coins. For this reason, they are known as the coinage metals. Because they are so expensive, silver and gold are rarely used in coins anymore. The United States stopped making everyday coins with gold in 1933 and coins with silver in 1964. Most coins now are mixtures of nickel, zinc, and copper, as shown in **Figure 7.**

The coinage metals have a variety of other uses, such as in the athletic medals in **Figure 7.** Copper is often used in electrical wiring because of its superior ability to conduct electricity and its relatively low cost. The compounds silver iodide (AgI) and silver bromide (AgBr) are used to make photographic film and paper because they break down when they are exposed to light. Silver and gold are used in jewelry because of their attractive colors, relative softness, resistance to corrosion, and rarity.

Get It?

Explain why gold's relative softness makes it a good material for use in jewelry.

Zinc, cadmium, and mercury

Zinc, Cadmium, and Mercury

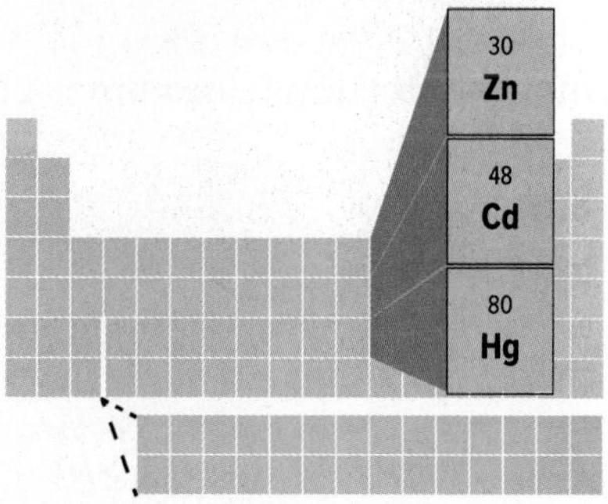

Zinc, cadmium, and mercury are found in group 12 of the periodic table. Zinc combines with oxygen in the air to form a thin, protective coating of zinc oxide on its surface. Zinc and cadmium, which also forms a protective coating, are often used to coat metals such as iron. Cadmium also is used in rechargeable batteries.

Mercury is the only metal that is a liquid at room temperature. It is used in thermostats, switches, and batteries. Mercury is toxic and can accumulate in the body, so it is rarely used in modern thermometers. People have died from mercury poisoning due to repeatedly eating fish that lived in mercury-contaminated water.

Metal

Metalloid

Nonmetal

H																															He
Li	Be																									B	C	N	O	F	Ne
Na	Mg																									Al	Si	P	S	Cl	Ar
K	Ca															Sc	Ti	V	Cr	Mn	Fe	Co	Ni	Cu	Zn	Ga	Ge	As	Se	Br	Kr
Rb	Sr															Y	Zr	Nb	Mo	Tc	Ru	Rh	Pd	Ag	Cd	In	Sn	Sb	Te	I	Xe
Cs	Ba	La	Ce	Pr	Nd	Pm	Sm	Eu	Gd	Tb	Dy	Ho	Er	Tm	Yb	Lu	Hf	Ta	W	Re	Os	Ir	Pt	Au	Hg	Tl	Pb	Bi	Po	At	Rn
Fr	Ra	Ac	Th	Pa	U	Np	Pu	Am	Cm	Bk	Cf	Es	Fm	Md	No	Lr	Rf	Db	Sg	Bh	Hs	Mt	Ds	Rg	Cn	Nh	Fl	Mc	Lv	Ts	Og

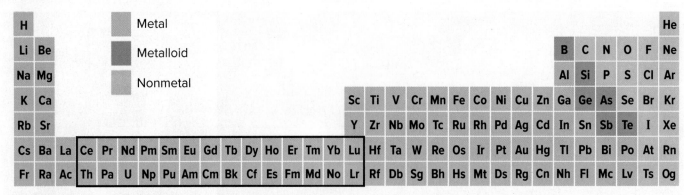

Figure 8 This is what the periodic table would look like if the inner transition elements were positioned where they should be. To save space, they are usually placed below the periodic table.

Identify *an advantage and a disadvantage to the above arrangement of the periodic table.*

The Inner Transition Elements

The two rows of elements that seem disconnected from the rest of the periodic table are called the inner transition elements. They are called this because they fit within the transition metals on the table. The inner transition elements are located between groups 3 and 4 in periods 6 and 7. They are usually listed below the table. **Figure 8** shows what the table would look like if these elements were not positioned below the table.

The lanthanides

The first row of inner transition elements is called the lanthanide series because it follows the element lanthanum. Lanthanum, cerium, praseodymium, and samarium are used with carbon to make a compound that is used for movie lighting. Compounds of europium, gadolinium, and terbium are used as colored phosphors. Phosphors change ultraviolet light into visible light.

Get It?

Explain why the first row of inner transition elements is called the lanthanide series.

The actinides

The second row of inner transition elements includes elements with atomic numbers 90 to 103. These elements are called the actinide series because they follow the element actinium. All of the actinides are radioactive and unstable. They are rare or nonexistent in nature. Their instability also makes them difficult to research.

Thorium and uranium are the only actinides found in Earth's crust in usable quantities. Thorium is used in making the glass for camera lenses because it bends light without much distortion. Uranium is best known for its use in nuclear reactors and in weapons.

ACADEMIC VOCABULARY

transition

a movement or evolution from one stage to another

The transition from middle school to high school is difficult for many students.

CCC CROSSCUTTING CONCEPTS

Patterns Different patterns can be observed in the periodic table. The patterns can predict the properties of elements. Compare and contrast alkali metals, alkaline Earth metals, transition elements, and inner transition elements. Create a graphic organizer that will help you remember the patterns.

Metals in Earth's Crust

Earth's hardened outer layer, called the crust, contains many compounds and a few uncombined metals, such as gold. Metals that are found in Earth's crust are minerals. Minerals are often found in ores, as shown in **Figure 9.** Ores are mixtures of minerals, clay, and rock, and they occur naturally in Earth's crust.

Most metals must be mined and separated from their ores. After an ore is mined, the mineral is separated from the rock and clay. Then the mineral is often converted to another form. This step usually involves heat and is called roasting. The metal is then refined into a pure form. Later, it can be alloyed with other metals. Removing the waste rock can be expensive. If the cost becomes greater than the value of the desired material, the mineral mixture is no longer classified as an ore.

The western United States has several copper mines. Most of the world's platinum is found in South Africa. The United States imports most of its chromium from South Africa, the Philippines, and Turkey. Chromium is used to harden steel, to manufacture stainless steel, and to form other alloys.

Figure 9 Ores containing metallic copper and copper compounds are mined throughout the world. This sample contains turquoise blue grandviewite, which is a compound of copper, aluminum, hydrogen, sulfur, and oxygen.

Check Your Progress

Summary

- Metals tend to form ionic and metallic bonds.
- Group 1 elements are called alkali metals and are the most reactive metals.
- Group 2 elements are called alkaline earth metals and are very reactive.
- Transition elements are elements in groups 3–12 in the periodic table. They have a wide variety of properties and uses.
- Inner transition elements fit in the periodic table between groups 3 and 4 in periods 6 and 7.

Demonstrate Understanding

1. **Describe** how to test a sample of an element to see if it is a metal.
2. **Compare and contrast** the uses of the iron triad and of the coinage metals.
3. **Classify** the following as alkali metals, alkaline earth metals, transition elements, or inner transition elements: calcium, gold, iron, magnesium, plutonium, potassium, sodium, and uranium.
4. **Discuss** how metallic bonding accounts for the common properties of metals.

Explain Your Thinking

5. **Predict** Suppose you discovered a new element with 120 protons and 2 electrons in its outer level. In what group does this new element belong? What properties would you expect it to have?
6. **MATH Connection** Pennies used to be made of copper and zinc and had a mass of 3.1 g. Today, pennies are made of copper-plated zinc and have a mass of 2.5 g. A new penny's mass is what percent of an old penny's mass?

LEARNSMART® Go online to follow your personalized learning path to review, practice, and reinforce your understanding.

LESSON 2

NONMETALS

FOCUS QUESTION

What are the properties of a typical nonmetal?

Properties of Nonmetals

Most of your body's mass is made of oxygen, carbon, hydrogen, and nitrogen, as shown in **Figure 10.** Calcium, phosphorus, sulfur, and chlorine are among the other elements found in your body. Except for the metal calcium, these elements are nonmetals. **Nonmetals** are elements that are usually gases or solids at room temperature. Solid nonmetals are not malleable or ductile, but are brittle or powdery. Nonmetals are poor conductors of heat and electricity because the electrons in nonmetals are not free to move as they do in metals.

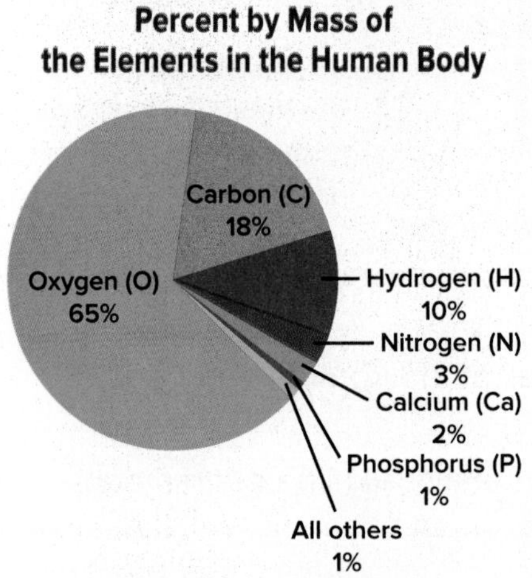

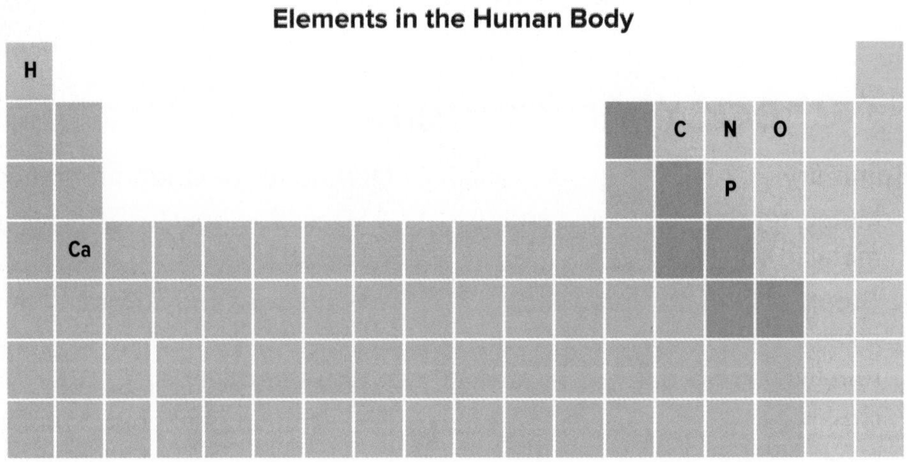

Figure 10 Humans are composed of mostly nonmetals. The pie chart breaks down the ratio of elements in the human body by mass (left). These elements mostly fall in the nonmetal portion of the periodic table (right).

3D THINKING DCI Disciplinary Core Ideas CCC Crosscutting Concepts SEP Science & Engineering Practices

COLLECT EVIDENCE

Use your Science Journal to record the evidence you collect as you complete the readings and activities in this lesson.

INVESTIGATE

GO ONLINE to find these activities and more resources.

Quick Investigation: Identify Chlorine Compounds in Your Water
Carry out an investigation to determine the effect of adding silver nitrate to water.

Revisit the Encounter the Phenomenon Question
What information from this lesson can help you answer the Unit and Module questions?

Calcium and fluorine bond ionically to form calcium fluoride (CaF_2).

Carbon and oxygen bond covalently to form carbon dioxide (CO_2).

Figure 11 Nonmetals form ionic bonds with metals and form covalent bonds with other nonmetals.

Nonmetals and the periodic table

In the periodic table, all nonmetals except hydrogen are found to the right of the stair-step line. On the table in the back of your book, the nonmetal element blocks are colored yellow. The noble gases, group 18, make up the only group of elements that are all nonmetals. Group 17 elements are called the halogens. The halogens, except astatine, are also nonmetals. The other nonmetals are found in groups 13 through 16.

Bonding in nonmetals

Nonmetals become negative ions when they gain electrons from metals. Calcium fluoride (CaF_2), which is shown in **Figure 11,** is an ionic compound. It forms from the nonmetal fluoride and the metal calcium.

When bonded with other nonmetals, atoms of nonmetals usually share electrons to form covalent bonds. Compounds made of atoms that are covalently bonded are called covalent compounds. The covalent compound carbon dioxide (CO_2) is shown in **Figure 11.** Carbon dioxide is a gas that you exhale and that plants need to survive.

Hydrogen

If you could count all the atoms in the universe, you would find that about 90 percent of them are hydrogen atoms. Most hydrogen on Earth is found in the compound water. The word *hydrogen* comes from the Greek word *hydro,* which means "water." When water is broken down into its elements, hydrogen becomes a gas composed of diatomic molecules. A **diatomic molecule** consists of two atoms of the same element in a covalent bond.

Hydrogen is highly reactive. A hydrogen atom has a single electron, which the atom shares when it combines with other nonmetals. For example, hydrogen bonds with oxygen to form water (H_2O), in which hydrogen shares electrons with oxygen. Hydrogen can gain an electron when it combines with alkali and alkaline earth metals. The compounds formed are hydrides, such as sodium hydride (NaH).

Hydrogen

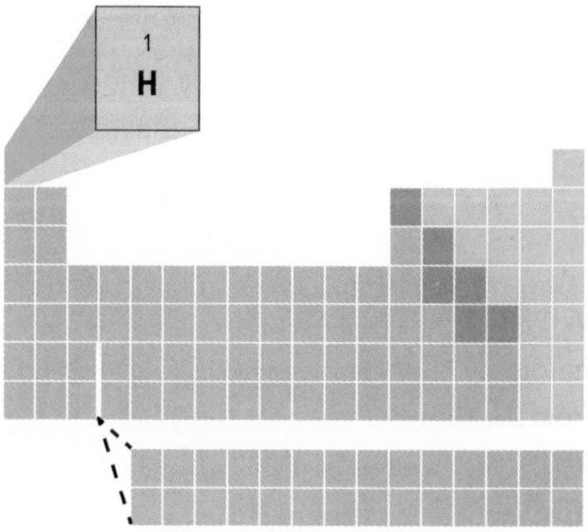

(l)Jiri Vaclavek/Shutterstock, (r)IrinaK/Shutterstock

The Halogens

Fluorine, chlorine, bromine, iodine, and astatine are called halogens and make up group 17. They are very reactive in their elemental forms, and their compounds have many uses. For example, halogen lightbulbs contain small amounts of bromine or iodine vapor.

Because an atom of a halogen has seven electrons in its outer energy level, only one electron is needed to complete this energy level. If a halogen gains an electron from a metal, an ionic compound called a salt is formed. An example of this is sodium chloride (NaCl). You know this compound as table salt. In the gaseous state, the halogens form reactive diatomic molecules and can be identified by their distinctive colors. Chlorine is greenish-yellow, bromine is reddish-orange, and iodine is violet.

The Halogens

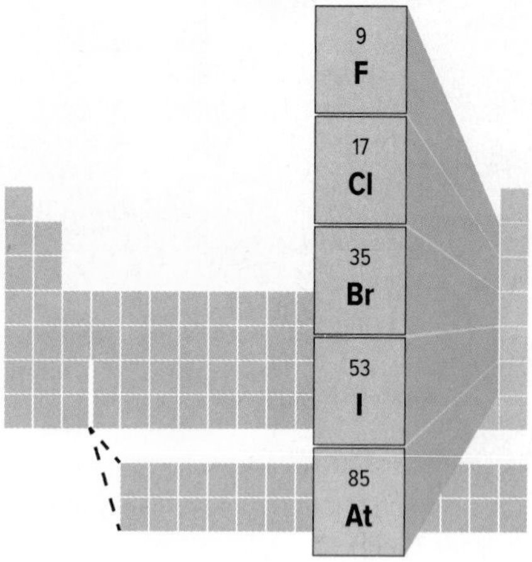

Fluorine

Fluorine is the most chemically active of the nonmetal elements. As you can see in **Figure 12,** fluorine compounds have varied uses. Fluorine compounds, called fluorides, are added to toothpastes and to city water systems to help prevent tooth decay. Hydrofluoric acid, a mixture of hydrogen fluoride and water, is used to etch glass and to frost the inner surfaces of lightbulbs. It is also used in the fabrication of semiconductors.

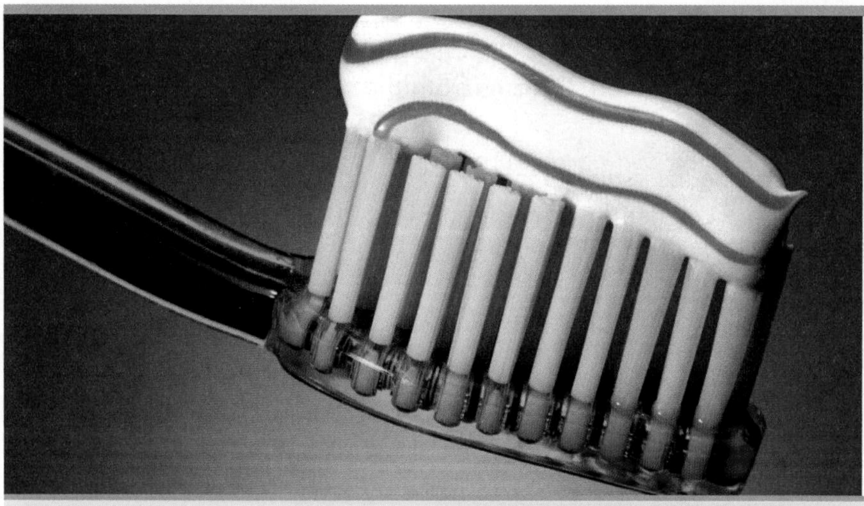

Fluorides are added to toothpaste to prevent tooth decay.

Hydrofluoric acid is used to etch circuit boards.

Figure 12 Fluorine's compounds have many uses.

WORD ORIGINS

halogen

from the Greek words *hals*, meaning "salt," and *-gen*, meaning "making"

Chlorine is a halogen found in table salt.

CCC CROSSCUTTING CONCEPTS

Patterns Patterns in the periodic table can be used to predict the properties of nonmetals. Create a poster that illustrates the relationship between patterns in the periodic table and properties of nonmetals.

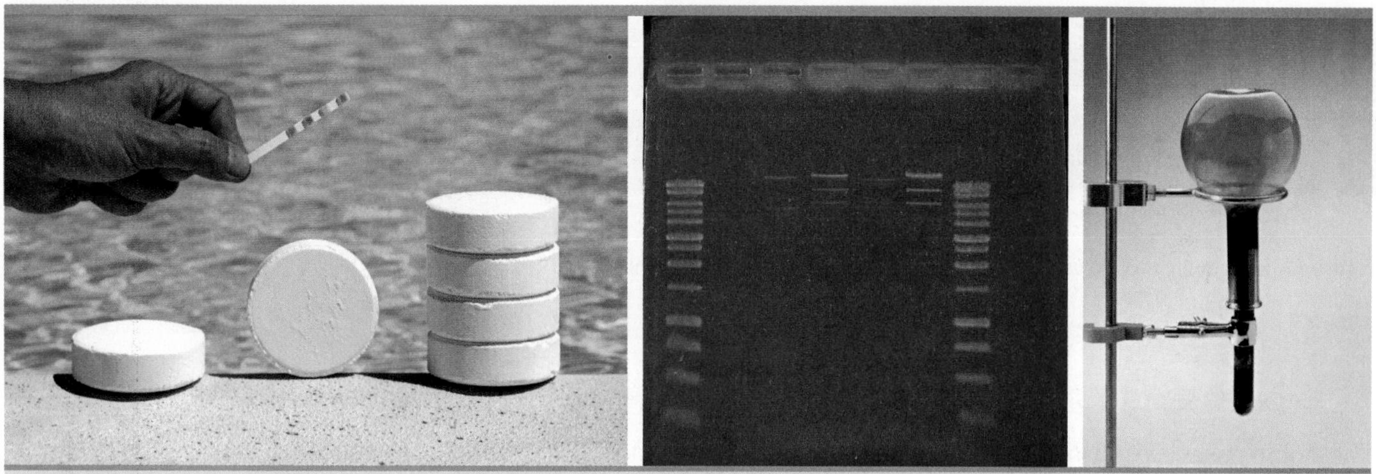

Chlorine compounds are used to disinfect water in swimming pools.

Scientists use a bromine compound to stain DNA samples.

Iodine will sublime at room temperature.

Figure 13 Halogens have a wide variety of uses and properties.

Chlorine and bromine

The odor that you sometimes smell near a swimming pool is chlorine. Chlorine compounds, like those found in the tablet shown in **Figure 13,** disinfect water. Chlorine, the most abundant halogen, is obtained from seawater at ocean-salt recovery sites. Household and industrial bleaches that are used to whiten flour, clothing, and paper also contain chlorine compounds.

Bromine, the only nonmetal that is a liquid at room temperature, also is extracted from compounds in seawater. Some hot tubs use bromine compounds instead of chlorine compounds to disinfect water. Bromine compounds were once used in cosmetics and as flame retardants. But, due to health concerns, these compounds are being used less frequently.

Another bromine compound is used to study genetic material, such as DNA. The bromine compound binds to the DNA in certain areas and acts as a sort of tag. Under fluorescent light, the bromine compound absorbs the fluorescent light and emits a reddish visible light, as shown in **Figure 13.**

Get It?

Name some uses of chlorine and bromine compounds.

Iodine and astatine

Iodine, a shiny purple-gray solid at room temperature, is obtained from seawater. When heated, iodine sublimes to a purple vapor, as seen in **Figure 13.** Iodine is essential in your diet for production of the hormone thyroxin and to prevent goiter, an enlarging of the thyroid gland in the neck. Iodine compounds are also used as disinfectants.

Astatine is the last member of group 17. It is radioactive and rare but has many properties similar to those of the other halogens. Because it is so rare in nature, scientists usually make astatine for research purposes. Medical researchers are investigating the possibility of using astatine's radioactive properties to treat cancer.

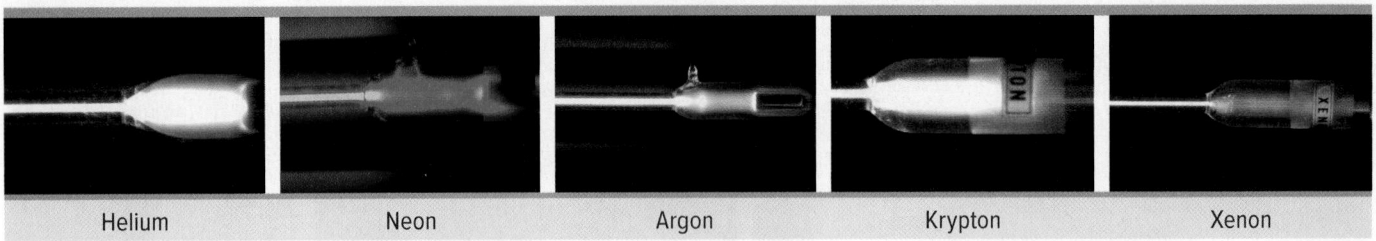

Figure 14 Each noble gas glows a different color when a current is passed through it.

The Noble Gases

The noble gases exist as isolated atoms. They are stable because their outermost energy levels are full. No naturally occurring noble gas compounds are known, but several compounds of argon, krypton, and xenon, primarily with fluorine, have been created in a laboratory.

The stability of noble gases makes them useful. For example, helium is less dense than air but does not burn in oxygen. This makes it safer than hydrogen to use in blimps and balloons. An electric current will cause the noble gases to glow, as shown in **Figure 14.** For this reason, some noble gases, such as neon and argon, are used for brightly colored signs. The noble gases are also used in many lasers. One common type of laser is a helium-neon laser. Helium-neon lasers produce beams of intense, red light.

The Noble Gases

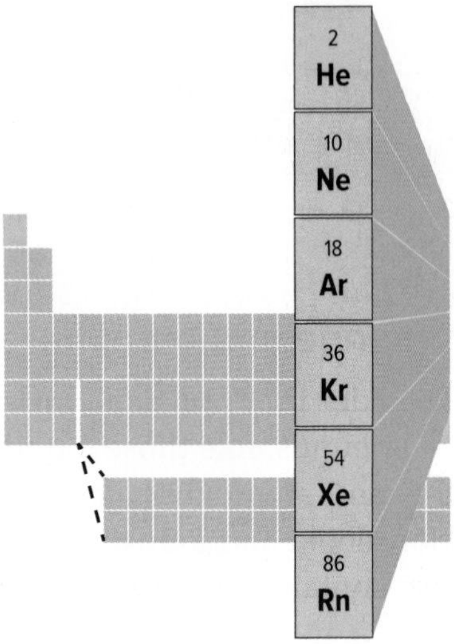

Check Your Progress

Summary

- Nonmetals are usually gases or brittle solids that are not shiny and do not conduct heat or electricity.
- Hydrogen is the most common element in the universe and is highly reactive.
- Halogens are in group 17 and are highly reactive.
- Noble gases exist as isolated atoms in nature.

Demonstrate Understanding

7. **Explain** how solid nonmetals are different from solid metals.
8. **Describe** two ways in which nonmetals combine with other elements.
9. **Identify** the nonmetal in these compounds: MgO, NaH, $AlBr_3$, and FeS.
10. **Identify** some common uses for the halogen compounds.
11. **Explain** why the noble gases are unreactive.

Explain Your Thinking

12. How you can tell that a gas is a halogen?
13. **MATH Connection** If a chlorine atom has a mass of 35.5 atomic mass units and a sodium atom has a mass of 23.0 atomic mass units, what is the mass of one $NaCl$ unit?

LEARNSMART® Go online to follow your personalized learning path to review, practice, and reinforce your understanding.

LESSON 3
MIXED GROUPS

FOCUS QUESTION

What are the differences between metals, nonmetals, and metalloids?

Properties of Metalloids

Can an element be both a metal and a nonmetal? In a sense, the metalloid elements are. **Metalloids** are elements that have some properties of metals and some properties of nonmetals. They can form ionic bonds and covalent bonds. Some metalloids can conduct electricity better than most nonmetals but not as well as many metals. The metalloids are located along the stairstep line on the periodic table. In this book, the metalloid element blocks are green. Groups 13 through 17 are mixed groups and contain metals, nonmetals, and metalloids.

Figure 15 Boron is added to glass to make heat-resistant lab equipment.

The Boron Group

Boron, a metalloid, is the first element in group 13. You might find the boron compounds borax and boric acid in your home. Borax is a water softener used in some laundry products. Boric acid is a mild antiseptic. Boron also is used to make heat-resistant glass, such as the lab equipment in **Figure 15.**

Aluminum, a metal in group 13, is the most abundant metal in Earth's crust. Because it is strong and light, aluminum is used in soft-drink cans, foil, pans, bicycle frames, and airplanes. Gallium is a metal used in electronic components. The last two group 13 elements are the rare metals indium and thallium.

The Boron Group

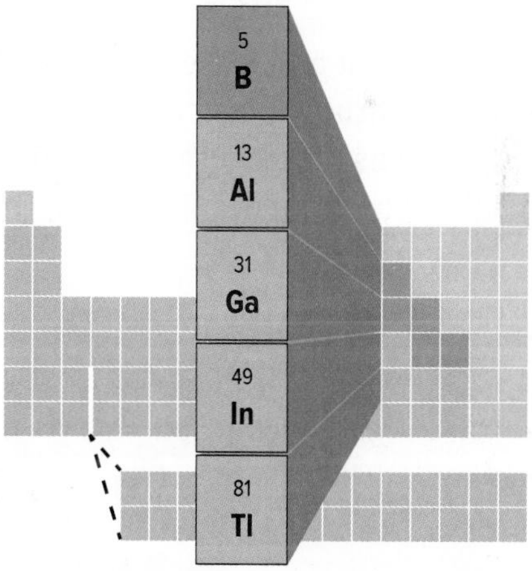

3D THINKING DCI Disciplinary Core Ideas CCC Crosscutting Concepts SEP Science & Engineering Practices

COLLECT EVIDENCE

Use your Science Journal to record the evidence you collect as you complete the readings and activities in this lesson.

INVESTIGATE

GO ONLINE to find these activities and more resources.

Applying Practices: Touching the Future

HS-PS2-6. Communicate scientific and technical information about why the molecular-level structure is important in the functioning of designed materials.

The Carbon Group

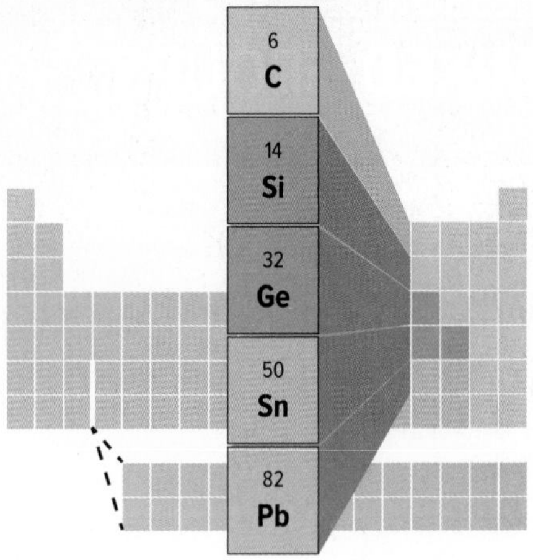

Each element in group 14, the carbon family, has four electrons in its outer energy level, but this is where much of the similarity ends. Carbon is a nonmetal, silicon and germanium are metalloids, and tin and lead are metals.

Carbon

Carbon, a nonmetal, occurs as an element in coal and as a compound in oil, natural gas, and foods. Carbon in these materials can combine with oxygen to produce carbon dioxide (CO_2). In the presence of sunlight, plants utilize CO_2 to make food. Carbon compounds, many of which are essential to life, can be found in you and all around you. All organic compounds contain carbon, but not all carbon compounds are organic.

Allotropes of carbon What do the diamond in a diamond ring and the graphite in your pencil have in common? Both are carbon. Diamond and graphite, as shown in **Figure 16,** are allotropes of carbon. **Allotropes** are different molecular structures of the same element.

Get It?
Define the term *allotrope.*

Graphite and diamond Graphite is a black powder that consists of layers of hexagonal structures of carbon atoms. In the hexagons, each carbon atom is bonded to three other carbon atoms. The fourth electron of each atom is bonded weakly to the layer next to it. This structure allows the layers to slide easily past one another, making graphite an excellent lubricant.

A diamond is transparent and extremely hard. In a diamond, each carbon atom is bonded to four other carbon atoms at the vertices, or corner points, of a tetrahedron. In turn, many tetrahedrons join together to form a giant molecule in which the atoms are held tightly in a strong crystalline structure. This structure accounts for the hardness of diamond.

Figure 16 Different arrangements of carbon atoms give graphite and diamond different properties.
Identify *the geometric shapes that make up each allotrope.*

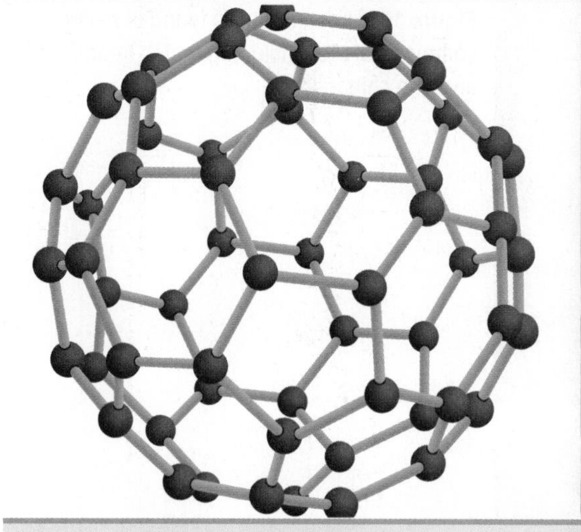

This soccer-ball-shaped allotrope of carbon is informally called a buckyball.

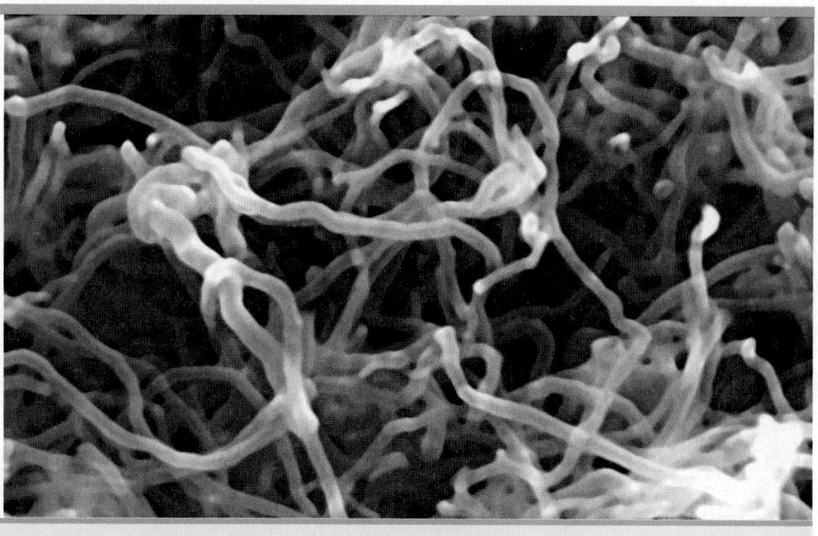

Each nanotube is about one-billionth of a meter in diameter.

Figure 17 The carbon allotrope buckminsterfullerene is used to make nanotubes.
Explain *How do you know that buckyballs are an allotrope of carbon?*

Buckyballs In the mid-1980s, a new allotrope of carbon called buckminsterfullerene was discovered. This soccer-ball-shaped molecule is shown in **Figure 17** and is informally called a buckyball. It was named after R. Buckminster Fuller, an architect-engineer who designed buildings with similar shapes.

In 1991, scientists were able to use buckyballs to synthesize extremely thin, graphite-like tubes, like those in **Figure 17.** These tubes, called nanotubes, are about one-billionth of a meter (1 nanometer) in diameter. You could stack tens of thousands of nanotubes to get the thickness of one piece of paper. Nanotubes might be used someday to make stronger building materials and to make computers that are smaller and faster.

Silicon and germanium

Silicon, a metalloid, is second only to oxygen in abundance in Earth's crust. Silicon is found in sand (SiO_2) and in almost all rocks and soil. The crystal structure of silicon dioxide is similar to the structure of diamond. Silicon occurs as two allotropes. One allotrope of silicon is a hard, gray substance, and the other is a brown powder.

Silicon is the main component in semiconductors. **Semiconductors** are elements that conduct an electric current under certain conditions. Many of the electronics that you use every day, such as computers, need semiconductors to run. The diode in **Figure 18** contains silicon. Germanium, the other metalloid in group 14, is also used to make semiconductors.

Figure 18 Silicon is a semiconductor that is often used to make diodes and other electronic components for computers.

Tin and lead

Tin, a metal, is used to coat other metals in order to prevent corrosion. Steel cans that are used for storing food are coated with tin. Tin also is combined with other metals to make bronze and pewter. The metal lead was once widely used in paint; however, it is no longer used due to its toxicity. Lead is also used in car batteries.

Figure 19 Phosphorous is found in many common objects, including match heads and fine china.

The Nitrogen Group

The nitrogen family makes up group 15. Each element has five electrons in its outer energy level. These elements tend to share electrons and form covalent compounds with other nonmetallic elements.

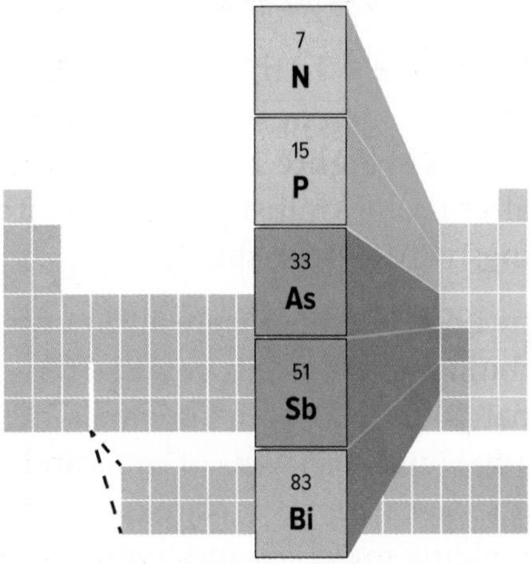

Nitrogen and phosphorus

Nitrogen is a nonmetal that is used to make ammonia (NH_3) and nitrates, which are compounds that contain the nitrate ion (NO_3^-). Nitrates and ammonia are used in fertilizers. Nitrogen is the fourth-most abundant element in your body. Each breath that you take is about 80 percent gaseous nitrogen in the form of diatomic molecules (N_2). But animals and plants cannot use nitrogen in its diatomic form. The nitrogen must be combined into compounds, such as amino acids.

Phosphorus is a nonmetal that has three allotropes. Phosphorus compounds can be used for many things, from water softeners to fertilizers. Phosphorus compounds are also used in match heads and fine china, as shown in **Figure 19.**

Arsenic, antimony, and bismuth

Arsenic and antimony are metalloids, and bismuth is a metal. Antimony and bismuth are used with other metals to lower their melting points. Because of this property, bismuth is used in automatic fire-sprinkler heads, such as the one in **Figure 20.** The thermal energy from fire melts the metal, which releases the plug and allows water to escape and put out the fire. Arsenic is used in some semiconductors. Arsenic was once used as a pigment in paints, but it is no longer used because many arsenic compounds are toxic.

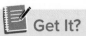

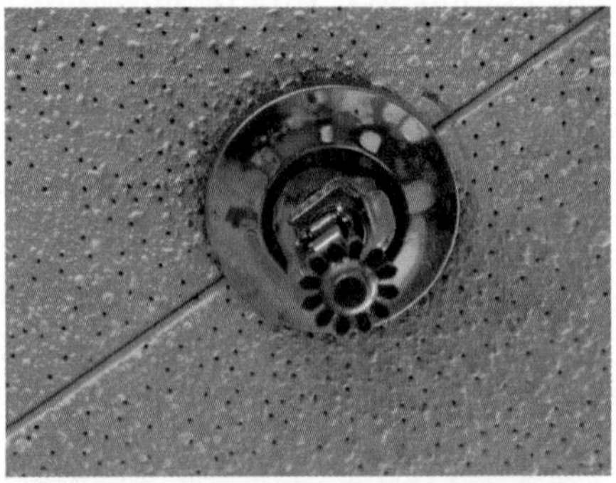

Figure 20 The bismuth in this sprinkler head lowers the melting point of the metal so that water will be released in the event of a fire.

Get It?
Explain why bismuth is used in fire-sprinkler heads.

The Oxygen Group

Group 16 is the oxygen group. These elements have six electrons in their outer energy levels. They will accept two electrons from a metal or bond covalently with other nonmetals.

The Oxygen Group

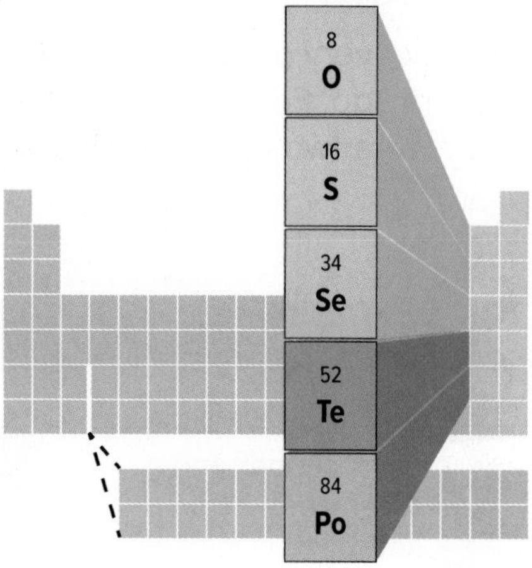

Oxygen

Oxygen, a nonmetal, exists in the air as diatomic molecules (O_2). Nearly all living things on Earth, including people, need O_2 for respiration. During electrical storms, some oxygen molecules change into ozone molecules (O_3). A layer of ozone around Earth protects living things from some of the Sun's radiation.

There are also many useful compounds that contain oxygen. Water (H_2O) is another essential substance for living organisms. Hydrogen peroxide (H_2O_2), which is used as a disinfectant, is another oxygen compound. Plants absorb carbon dioxide (CO_2) and give off the oxygen we breathe.

Other group 16 elements

The second element in the oxygen group is sulfur. Sulfur is a nonmetal that has several allotropes. It exists as different-shaped crystals and as a noncrystalline solid. Sulfur combines with metals to form sulfides of distinctive colors that are used as pigments in paints.

The nonmetal selenium and the metalloids tellurium and polonium are the other group 16 elements. Selenium is the most common of these three. You need trace amounts of selenium in your diet. Selenium is found in many multivitamins. However, selenium is toxic if too much of it gets into your system.

APPLY SCIENCE

What is in Earth's crust?

Earth's crust is composed of oxygen, silicon, aluminum, iron, calcium, sodium, potassium, magnesium, and small amounts of other elements. Scientists have determined the percent by mass of each element in Earth's crust, as seen in the table to the right. What would be the best way to present these data?

Identify the Problem

The data about Earth's crust need to be put into a more convenient format. A circle graph would be a good way to present the data.

Solve the Problem

1. Copy the table. Compute how many degrees of the circle will represent each element.
2. Use the information from your completed table to draw the circle graph.

Elements in Earth's Crust

Element	Percent by Mass	Degrees
Oxygen	46.6	
Silicon	27.7	
Aluminum	8.1	
Iron	5.0	
Calcium	3.6	
Sodium	2.8	
Potassium	2.6	
Magnesium	2.1	
Other elements	1.5	

Figure 21 Visualizing the Discovery of the Elements

The discovery of naturally occurring elements began thousands of years ago. For example, carbon—in the form of coal—and sulfur were two of the first known elements.

6 **C** | 16 **S**

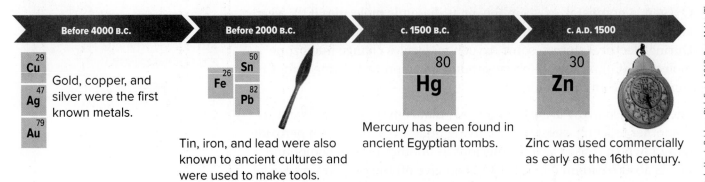

Before 4000 B.C.

29 **Cu**
47 **Ag**
79 **Au**

Gold, copper, and silver were the first known metals.

Before 2000 B.C.

26 **Fe**
50 **Sn**
82 **Pb**

Tin, iron, and lead were also known to ancient cultures and were used to make tools.

c. 1500 B.C.

80 **Hg**

Mercury has been found in ancient Egyptian tombs.

c. A.D. 1500

30 **Zn**

Zinc was used commercially as early as the 16th century.

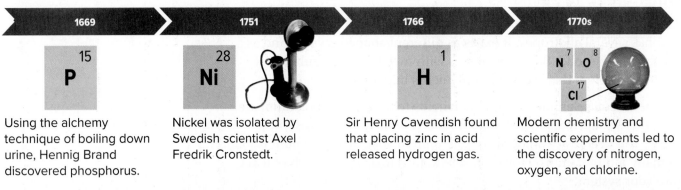

1669

15 **P**

Using the alchemy technique of boiling down urine, Hennig Brand discovered phosphorus.

1751

28 **Ni**

Nickel was isolated by Swedish scientist Axel Fredrik Cronstedt.

1766

1 **H**

Sir Henry Cavendish found that placing zinc in acid released hydrogen gas.

1770s

7 **N**
8 **O**
17 **Cl**

Modern chemistry and scientific experiments led to the discovery of nitrogen, oxygen, and chlorine.

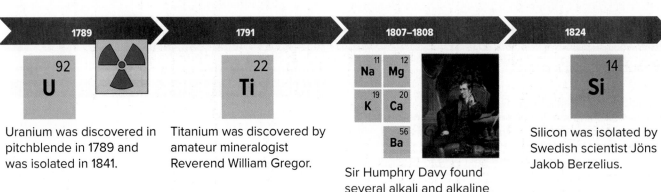

1789

92 **U**

Uranium was discovered in pitchblende in 1789 and was isolated in 1841.

1791

22 **Ti**

Titanium was discovered by amateur mineralogist Reverend William Gregor.

1807–1808

11 **Na**
12 **Mg**
19 **K**
20 **Ca**
56 **Ba**

Sir Humphry Davy found several alkali and alkaline earth metals.

1824

14 **Si**

Silicon was isolated by Swedish scientist Jöns Jakob Berzelius.

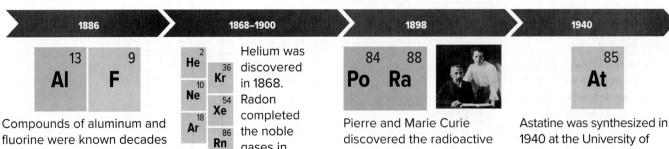

1886

13 **Al**
9 **F**

Compounds of aluminum and fluorine were known decades before scientists isolated the elements in 1886.

1868–1900

2 **He**
10 **Ne**
18 **Ar**
36 **Kr**
54 **Xe**
86 **Rn**

Helium was discovered in 1868. Radon completed the noble gases in 1900.

1898

84 **Po**
88 **Ra**

Pierre and Marie Curie discovered the radioactive elements polonium and radium.

1940

85 **At**

Astatine was synthesized in 1940 at the University of California, Berkeley. In 1943, it was found to occur naturally in very small amounts.

(l to r, t to b)Alasdair James/Getty Images, GoodWin777/Shutterstock, Muhammad Zaman al-Munajjim al-Asturlabi (Iranian active 1643-1689)/The Metropolitan Museum of Art, New York, Harris Brisbane Dick Fund, 1963, Ryan McVay/Photodisc/Getty Images, ©Joe Franek/McGraw-Hill Education, James Lonsdale (British painter 1777-1839);William Henry Worthington (English painter ca. 1790-after 1839)Digital Image/Yale Center for British Art, Everett Historical/Shutterstock

Figure 22 Since the 1940s, nuclear scientists have been using particle accelerators to make new elements that are not found naturally on Earth.

Discovering and Making Elements

The first elements known were those that occur naturally in their elemental forms, such as gold, lead, tin, and carbon. Gold, for example, has been known for at least 6000 years. Most elements, however, were discovered after the birth of modern chemistry in the 1700s. **Figure 21,** on the previous page, shows the history of the discovery of many common elements.

Synthetic elements

By 1935, only four elements with fewer than 92 protons were missing from the periodic table. In 1939, Francium (87) was discovered. Elements with 43, 61, and 85 protons could not be found. Scientists believed that such elements could exist and began trying to create them in the lab. Elements created in the lab are called synthetic elements.

By smashing existing elements with particles accelerated in a heavy ion accelerator, like the one in **Figure 22,** scientists have created elements that are not typically found on Earth. The first synthetic element, technetium (43), was created in 1939. Astatine (85) was created in 1940. Later, it was found that small amounts of astatine occur in nature. Promethium (61), the last missing element, was synthesized in 1947.

Why make elements? Bombarding uranium with neutrons can make neptunium, element 93. Half of the synthesized atoms of neptunium disintegrate in about two days. This might not sound useful, but when neptunium atoms disintegrate, they form plutonium. Plutonium has been produced in control rods of nuclear reactors and has been used in bombs. Plutonium also can be changed to americium, element 95, which is used in home smoke detectors. Synthetic elements are also useful because they are radioactive. For example, technetium's radioactivity makes it ideal for many medical applications.

STEM CAREER Connection

Nuclear Physicist

Do you have a strong affinity for physics and chemistry? Are you interested in a research career that allows you to study the properties of atomic and subatomic particles? If you are, you may be interested in becoming a nuclear physicist.

CCC CROSSCUTTING CONCEPTS

Patterns With a partner, discuss how scientists used what they already knew about patterns in the periodic table to discover new elements. Write a script for dialogue between two scientists. The dialogue should explain how a synthetic element was discovered.

Transuranium elements Since 1940, scientists have been working to find elements with more and more protons. Except for technetium (43) and promethium (61), each synthetic element has more than 92 protons. Elements that have more than 92 protons, the atomic number of uranium, are called **transuranium elements.**

These elements do not belong exclusively to the metal, nonmetal, or metalloid group. These elements are found toward the bottom of the periodic table. Some are in the actinide series, and some are on the bottom row of the main periodic table. All the transuranium elements are synthetic and radioactive, and many of them disintegrate quickly.

95
Am
Americium
(243)

Figure 23 Americium, a transuranium element, was first synthesized in 1944 as part of the Manhattan Project.

Seeking stability If you made something that fell apart less than a second after you created it, you might think you were not successful. However, nuclear scientists do just that when they synthesize atoms. Many synthetic elements last for only a fraction of a second after they are constructed and can be made only in small amounts.

Making new elements is beneficial, however. By studying how the synthesized elements form and disintegrate, scientists can gain an understanding of the forces holding the nucleus together. In the 1960s, scientists theorized that more-stable synthetic elements could exist. Finding a large, stable synthetic element might help scientists understand how the forces inside the atom work.

Check Your Progress

Summary

- Metalloids are elements that can have metallic and nonmetallic properties.
- The elements in group 14 have four electrons in their outer energy levels and have many uses.
- The elements in group 15 tend to share electrons and form covalent bonds.
- The elements in group 16 have six electrons in their outer energy levels and can form covalent and ionic bonds.
- By synthesizing elements, scientists might better understand how the forces inside an atom behave.

Demonstrate Understanding

14. **Explain** why groups 14 and 15 are better representatives of mixed groups than are groups 13 and 16.
15. **Define** the term *semiconductor,* and give an example of a metalloid that is used to make semiconductors.
16. **Compare and contrast** natural elements and synthetic elements.
17. **Describe** the differences and similarities between metals, nonmetals, and metalloids.

Explain Your Thinking

18. **Explain** Graphite and diamond are made of the element carbon. Explain why graphite is a lubricant and diamond is the hardest gem known.
19. **MATH Connection** An isotope of copernicium has 112 protons and 173 neutrons. How many particles are in the nucleus of the atom?

LEARNSMART® Go online to follow your personalized learning path to review, practice, and reinforce your understanding.

The Power of Peer Review

You might not think that scientific research is a highly competitive endeavor, but one area of intense scientific competition is the discovery of new elements. Scientists hope that creating and studying new elements will give them insight into the atom's structure.

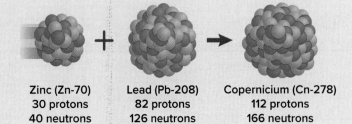

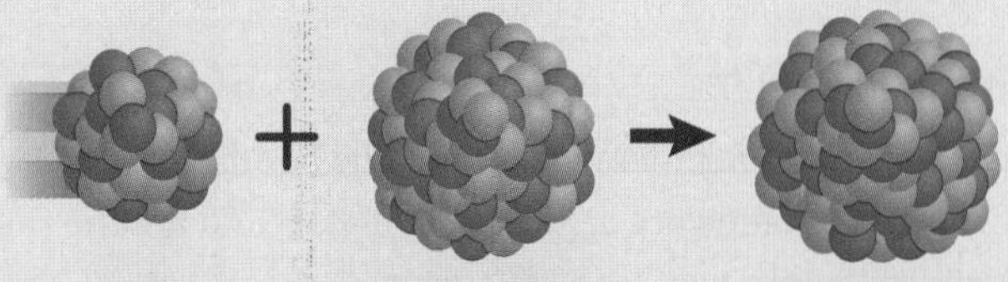

Element 112 is created by fusing lead and zinc nuclei.

A race to be first

Elements with an atomic number greater than 92 do not occur naturally on Earth. Teams of scientists in labs around the world use elaborate machines and cutting-edge computing power to produce and study these extremely rare elements. These elements are generally made by fusing atoms of lighter elements together.

The first team to create a new element wins the right to name the element. More important, however, is the prestige associated with being the first to create the element. This honor can help the team earn more funding for further research.

Peer review

After a team publishes a paper claiming discovery, other researchers work to confirm the results. New elements are added to the periodic table only after the original results are repeated and thoroughly reviewed by other scientists. Peer review helps increase accuracy and reduce bias in scientific publications.

The process can sometimes take years. In 1996, a group at the Center for Heavy Ion Research in Darmstadt, Germany, created element 112, copernicium. Copernicium is made by fusing a zinc nucleus and a lead nucleus, as shown in the figure.

After careful review by labs around the world, element 112 was officially added to the periodic table in 2009 and was officially named in 2010.

Alleged scientific fraud

In 1999, the race to discover new elements was interrupted by a scandal involving alleged data fabrication. A team of researchers from Lawrence Berkeley National Laboratory in California published a paper claiming the production of element 118.

Labs in Germany, France, and Japan, as well as the team in California, tried unsuccessfully to repeat the procedure and its results. Later, an investigating committee concluded that a scientist had falsified the original data. In 2001, Lawrence Berkeley retracted its discovery claim and fired the accused scientist. In this case, the strong peer review process helped reveal fraud.

OBTAIN, EVALUATE, AND COMMUNICATE INFORMATION

The 18 elements from atomic number 95 to 112 are named after places or famous scientists. Research one of these elements, and write a short paragraph detailing its discovery and the origin of its name.

STUDY GUIDE

GO ONLINE to study with your Science Notebook.

Lesson 1 METALS

- Metals tend to form ionic and metallic bonds.
- Group 1 elements are called alkali metals and are the most reactive metals.
- Group 2 elements are called alkaline earth metals and are very reactive.
- Transition elements are elements in groups 3–12 in the periodic table. They have a wide variety of properties and uses.
- Inner transition elements fit in the periodic table between groups 3 and 4, in periods 6 and 7.

- metal
- malleable
- ductile
- metallic bonding
- radioactive element
- transition element

Lesson 2 NONMETALS

- Nonmetals are usually gases or brittle solids that are not shiny and do not conduct heat or electricity.
- Hydrogen is the most common element in the universe and is highly reactive.
- Halogens are in group 17 and are highly reactive.
- Noble gases exist as isolated atoms in nature.

- nonmetal
- diatomic molecule

Lesson 3 MIXED GROUPS

- Metalloids are elements that can have metallic and nonmetallic properties.
- The elements in group 14 have four electrons in their outer energy levels and have many uses.
- The elements in group 15 tend to share electrons and form covalent bonds.
- The elements in group 16 have six electrons in their outer energy levels and can form covalent and ionic bonds.
- By synthesizing elements, scientists might better understand how the forces inside an atom behave.

- metalloid
- allotrope
- semiconductor
- transuranium element

THREE-DIMENSIONAL THINKING

Module Wrap-Up

REVISIT THE PHENOMENON

Why does this lava burn blue?

CER Claim, Evidence, Reasoning

Explain Your Reasoning Revisit the claim you made when you encountered the phenomenon. Summarize the evidence you gathered from your investigations and research and finalize your Summary Table. Does your evidence support your claim? If not, revise your claim. Explain why your evidence supports your claim.

STEM UNIT PROJECT

Now that you've completed the module, revisit your STEM unit project. You will apply your evidence from this module and complete your project.

GO FURTHER

SEP Data Analysis Lab

The Halogens

The halogens are the most reactive nonmetallic group. Fluorine is the most reactive nonmetal. All the halogens in the gas state are highly toxic and hazardous to handle.

It is possible to relate the melting and boiling points of the halogens to the forces acting between the molecules. Make a graph of the melting point versus atomic number and the boiling point versus atomic number from the information in the table.

Element	Atomic number	MP (K)	BP (K)
Fluorine	9	55	85
Chlorine	17	172	239
Bromine	35	266	332
Iodine	53	387	457

CER Analyze and Interpret Data

1. **Reasoning** The temperatures given above are in kelvin (K) units. Why are kelvins easier to graph than are Celsius degrees?
2. **Claim** Which of the halogens exist as a gas at room temperature (295 K)? As a liquid? As a solid?
3. **Evidence** What does the graph tell you about the atomic number of a halogen and the forces that act between molecules?

Emil O/Shutterstock

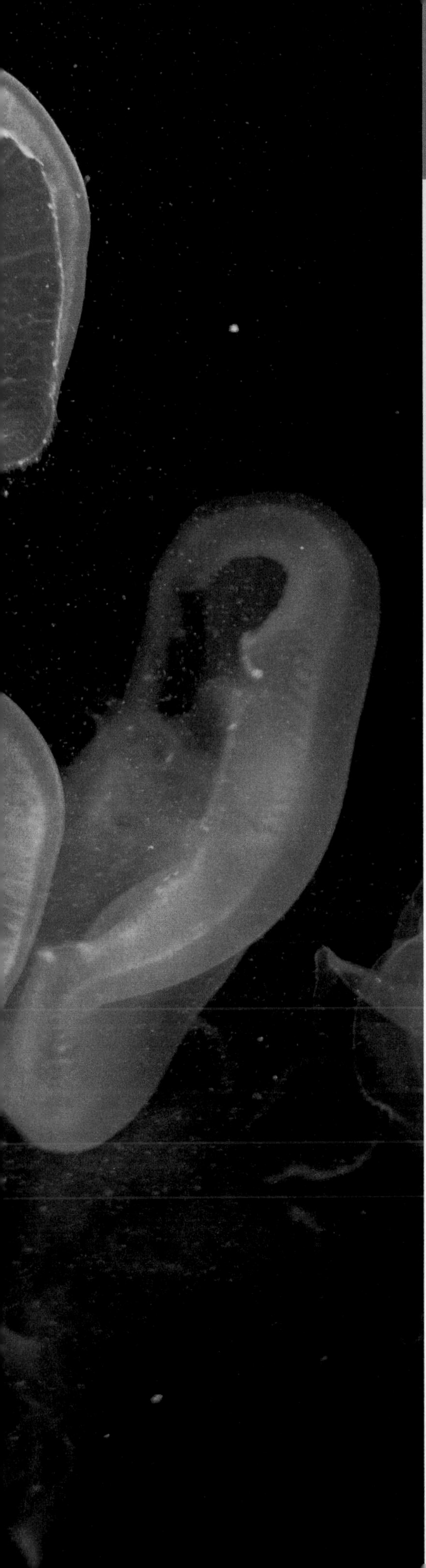

UNIT 5

REACTIONS

ENCOUNTER THE PHENOMENON

Why are the jellyfish glowing?

SEP Ask Questions

What questions do you have about the phenomenon? Write your questions on sticky notes and add them to the driving question board for this unit.

What happens during a chemical reaction?

Look for Evidence

As you go through this unit, use the information and your experiences to help you answer the phenomenon question as well as your own questions. For each activity, record your observations in a Summary Table, add an explanation, and identify how it connects to the unit and module phenomenon questions.

Solve a Problem

PERFORMANCE EXPECTATION

STEM UNIT PROJECT

Model Bioremediation There are many methods for dealing with pollution. One of those methods is bioremediation—using living organisms to detoxify the polluted area. Research the chemistry of bioremediation.

Choose a process to model. Model the process in a real-life situation. How would you use the bioremediation process to improve the situation? Use the results of these investigations and the evidence you collected during the unit to model a bioremediation process in a real-life situation.

GO ONLINE In addition to reading the information in your Student Edition, you can find the STEM Unit Project and other useful resources online.

Florea Marius Catalin/iStockphoto/Getty Images

MODULE 18

CHEMICAL BONDS

ENCOUNTER THE PHENOMENON

What makes these rock crystals form perfect cubes?

GO ONLINE to play a video about cubic halite salt crystals in Merkers Mine.

SEP Ask Questions

Do you have other questions about the phenomenon? If so, add them to the driving question board.

CER Claim, Evidence, Reasoning

Make Your Claim Use your CER chart to make a claim about why these rock crystals form perfect cubes. Explain your reasoning.

Collect Evidence Use the lessons in this module to collect evidence to support your claim. Record your evidence as you move through the module.

Explain Your Reasoning You will revisit your claim and explain your reasoning at the end of the module.

GO ONLINE to access your CER chart and explore resources that can help you collect evidence.

LESSON 2: Explore & Explain: Molecules

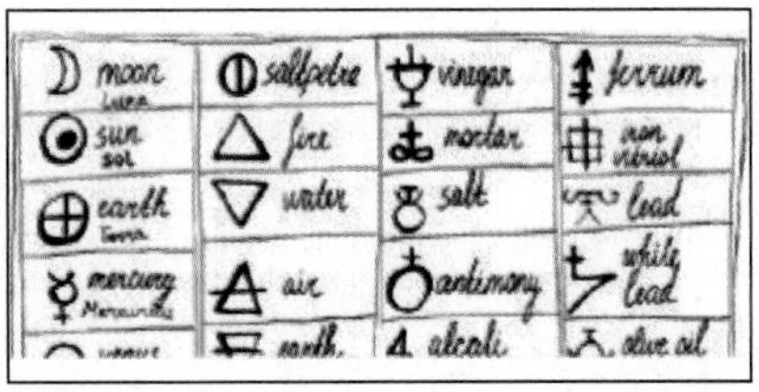

LESSON 3: Explore and Explain: Writing Chemical Formulas

Smithsonian

Additional Resources

(t)Video Supplied by BBC Worldwide Learning, (bl)Martin Rogers/Photographer's Choice/Getty Images, (br)Daria Rosen/Shutterstock

LESSON 1

STABILITY IN BONDING

FOCUS QUESTION

Can you name some common compounds and the individual elements that make them up?

Combined Elements

Why is the Statue of Liberty green? Was it painted that way? No, the Statue of Liberty was not painted. It is made of the metal copper, which is a bright, shiny, reddish-brown element. So, again, we ask: Why is the Statue of Liberty green?

Compounds

The matter around you includes elements such as copper, sulfur, and oxygen. Elements may combine chemically to form compounds. For example, the copper in the Statue of Liberty combines with oxygen and sulfur in the air to form a number of different compounds, including copper sulfate. Copper sulfate is one of the compounds responsible for the blue-green patina seen on the Statue of Liberty, as shown in **Figure 1.** Although copper sulfate includes copper and sulfur, it is very different from both elements.

Copper | Sulfur | Copper sulfate

Figure 1 Copper sulfate isn't shiny and reddish-brown like elemental copper. Nor is it a pale yellow solid like sulfur, or a colorless, odorless gas, like oxygen.

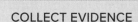

3D THINKING **DCI** Disciplinary Core Ideas **CCC** Crosscutting Concepts **SEP** Science & Engineering Practices

COLLECT EVIDENCE

Use your Science Journal to record the evidence you collect as you complete the readings and activities in this lesson.

INVESTIGATE

GO ONLINE to find these activities and more resources.

Lab: Atomic Trading Cards
Develop and use a model to visualize the properties of different elements from the periodic table.

Review the News
Obtain information from a current news story about a chemical compound. Evaluate your source and communicate your findings to your class.

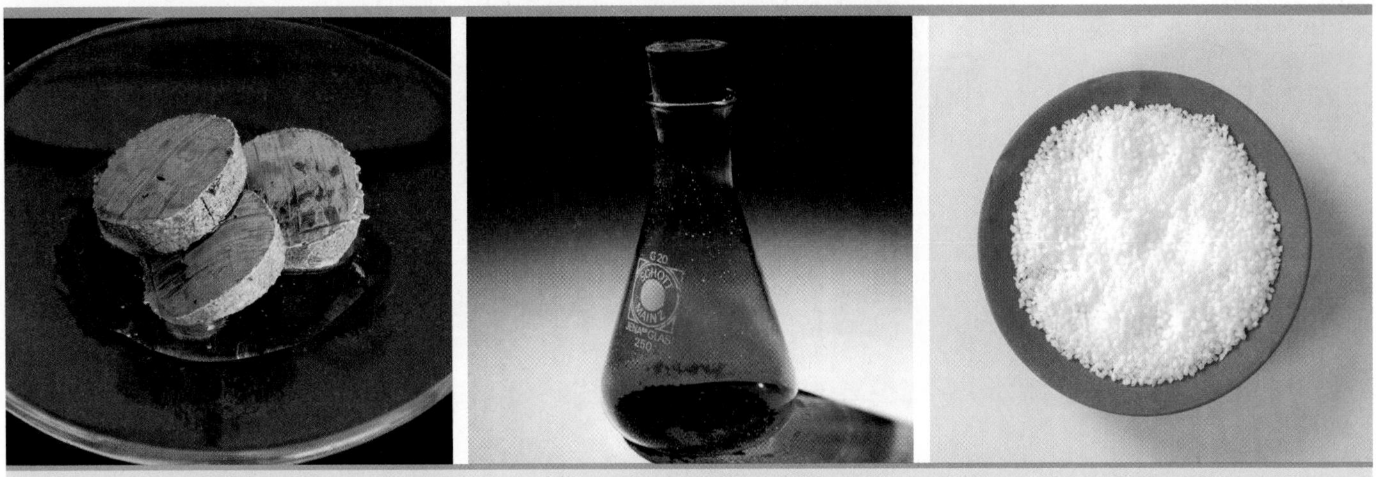

Potassium — Iodine — Potassium iodide

Figure 2 The properties of a compound can be very different from those of its component elements.
Describe *how the properties of potassium iodide are different from those of potassium and of iodine.*

New properties

A compound formed when elements combine often has properties that aren't anything like those of the individual elements. For example, potassium iodide, shown in **Figure 2,** is a compound made from the elements potassium and iodine. Potassium is a shiny, soft, silvery metal that reacts violently with water. Iodine is a blue-black solid that turns to a purple gas at room temperature. Would you have guessed that these elements combine to make a white, crystalline salt?

Formulas

Every element has a chemical symbol. For example, the chemical symbol Na represents the element sodium, and the symbol Cl represents the element chlorine. When written as NaCl, these symbols make up the formula for the compound sodium chloride. A **chemical formula** shows what elements a compound contains and the exact number of the atoms of each element in a unit of that compound.

Another compound with which you are familiar is H_2O, which is more commonly known as water. This formula contains the symbols H for the element hydrogen and O for the element oxygen. Notice the subscript number 2 written after the H for hydrogen. *Subscript* means "written below." A subscript written after a symbol tells how many atoms of that element are in a single unit of the compound. If a symbol has no subscript, the unit contains only one atom of that element. A unit of H_2O contains two hydrogen atoms and one oxygen atom.

Look at the formula for each compound listed in **Table 1.** What elements combine to form each compound? How many atoms of each element are required to form each of the compounds?

Table 1 Some Familiar Compounds

Common Name	Chemical Name	Chemical Formula
Sand	silicon dioxide	SiO_2
Milk of magnesia	magnesium hydroxide	$Mg(OH)_2$
Cane sugar	sucrose	$C_{12}H_{22}O_{11}$
Lime	calcium oxide	CaO
Vinegar	acetic acid	CH_3COOH
Laughing gas	dinitrogen monoxide	N_2O
Grain alcohol	ethanol	C_2H_5OH
Battery acid	sulfuric acid	H_2SO_4
Stomach acid	hydrochloric acid	HCl

Get It?
Interpret What does a chemical formula show you?

(l)©Jill Braaten/McGraw-Hill Education, (c)Ruddy Gold/Age Fotostock, (r)Viktor1/Shutterstock

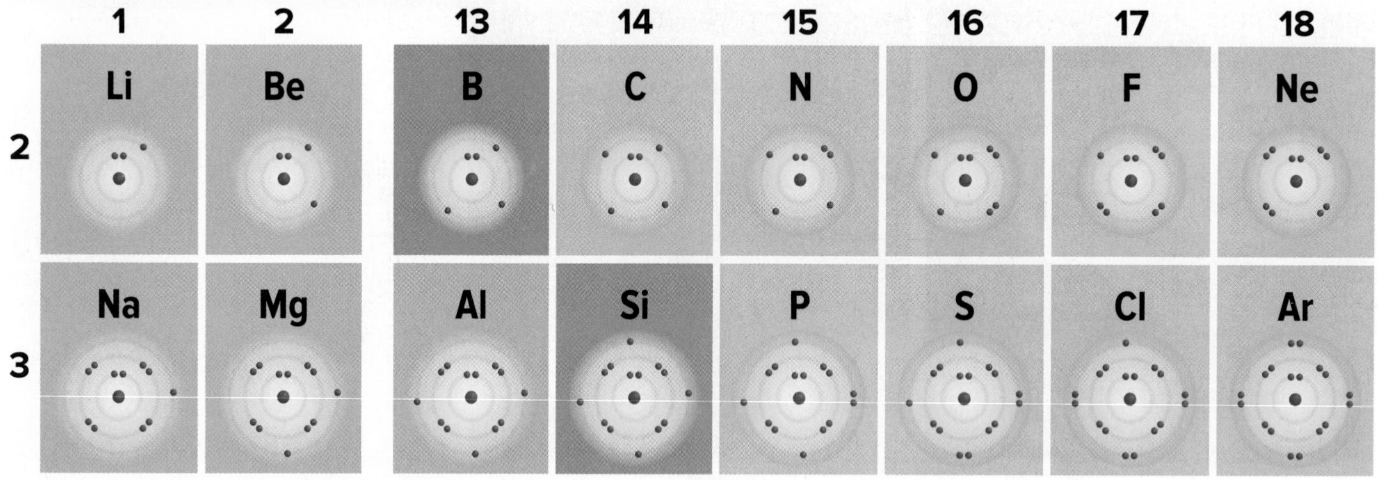

Figure 3 Simplified representations of electron distribution can help you understand why atoms bond. Note that the number of electrons in each group's outer level increases across the table. The noble gases in group 18 have filled outer levels.

Chemical Bond Formation

Why do atoms form compounds? The answer to that question is in the last column on the periodic table. The six noble gases in group 18 seldom form compounds. In fact, the first compound containing a noble gas was not made until 1962. Why? Atoms of noble gases are unusually stable. The reason for this stability lies in the arrangement of the electrons.

Electron dot diagrams

To understand the stability of atoms, it is helpful to show the electrons in the outer energy level of an atom—its valence electrons—in electron dot diagrams. Electron dot diagrams contain the chemical symbol for an element, surrounded by dots representing the element's valence electrons.

How do you know how many dots to make in electron dot diagrams? Recall that the repeating patterns of the periodic table reflect patterns of outer electron states. For groups 1, 2, and 13 through 18, you can use a periodic table or the portion of it shown in **Figure 3.** Look at the ring depicting the outer energy level of each of the elements. Group 1 elements have one outer electron each. The elements in group 2 have two. Group 13 elements have three, group 14 have four, and so on to group 18, the noble gases, which have eight each.

The unique noble gases An atom is chemically stable when its outer energy level is complete. The outer energy levels of helium and hydrogen are stable with two electrons. The outer energy levels of nearly all other elements are stable with eight.

The noble gases are stable because they have a full outer energy level. **Figure 4** shows electron dot diagrams of the noble gases. Notice that eight dots surround Kr, Ne, Xe, Ar, and Rn, but only two dots surround He.

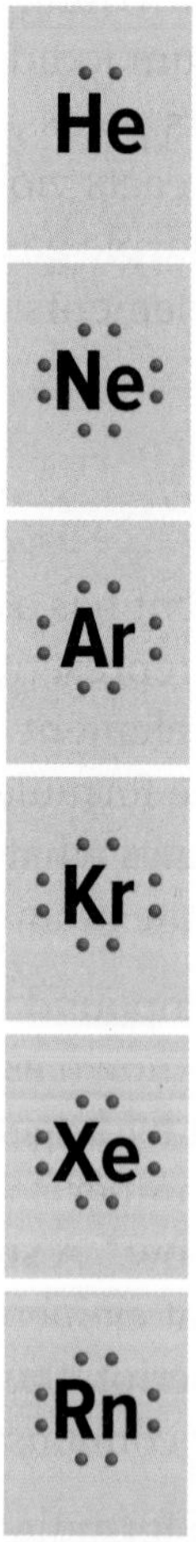

Figure 4 Electron dot diagrams of noble gases show that they all have a stable, filled outer energy level.

Get It?

Explain why the noble gases are unusually stable.

Unfilled and filled energy levels How do electron dot diagrams represent elements, and how does that relate to the elements' abilities to form compounds? Hydrogen and helium, the elements in period 1 of the periodic table, can hold a maximum of two electrons in their outer energy levels. Hydrogen contains one electron in its lone energy level. The dot diagram for hydrogen has a single dot next to its symbol. This means that hydrogen's outer energy level is not full. It is more stable when it is part of a compound.

In contrast, helium's outer energy level contains two electrons, as represented by its electron dot diagram with two dots—a pair of electrons—next to its symbol. Helium already has a full outer energy level by itself and is chemically stable. As a result, helium rarely forms compounds. Compare the electron dot diagrams of helium and hydrogen below.

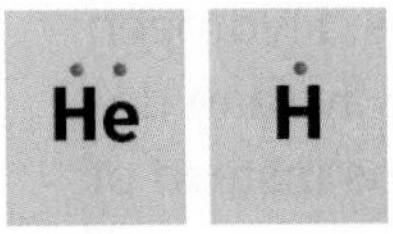

Outer levels—getting their fill How does hydrogen, or any other element, change its number of outer electrons to become stable? Atoms with unstable outer energy levels can lose, gain, or share electrons to obtain a stable outer energy level. They do this by combining with other atoms that also have partially complete outer energy levels. As a result, each achieves stability.

Gaining and losing electrons **Figure 5** shows electron dot diagrams for sodium (Na) and chlorine (Cl). When they combine, sodium loses one electron and chlorine gains one electron. You can see from the electron dot diagram that chlorine now has a stable outer energy level, similar to a noble gas.

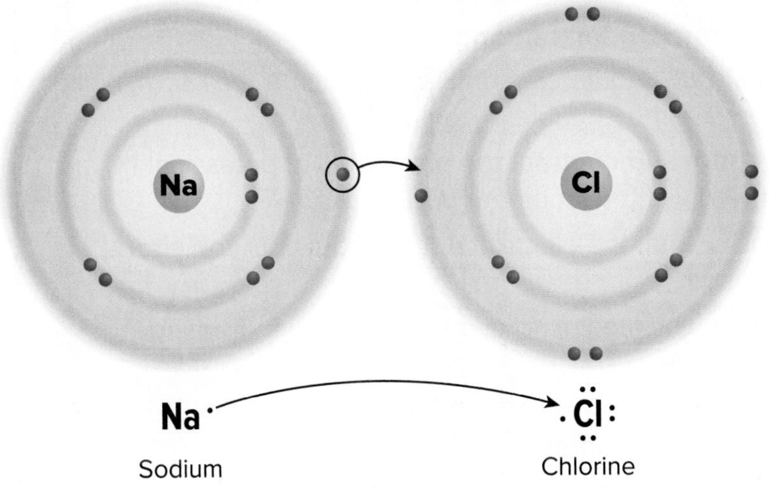

Figure 5 Chlorine has seven electrons in its outer energy level; sodium has only one. When sodium loses an electron and chlorine gains one, they each have a stable outer energy level.

CCC CROSSCUTTING CONCEPTS

Patterns Atoms of group 1 elements and atoms of group 17 elements tend to react with one another in a one-to-one ratio. Describe how this observed pattern at the macroscopic scale provides evidence that supports the explanation of the atomic scale shown in **Figure 5.**

As shown in **Figure 5** on the previous page, sodium has only one electron in its outer energy level, which it loses to combine with chlorine in sodium chloride. Sodium now has an outer energy level that is stable with eight electrons. When the outer electron of sodium is removed, a complete inner energy level is revealed and becomes the outer energy level. Sodium and chlorine are now stable because of the exchange of an electron.

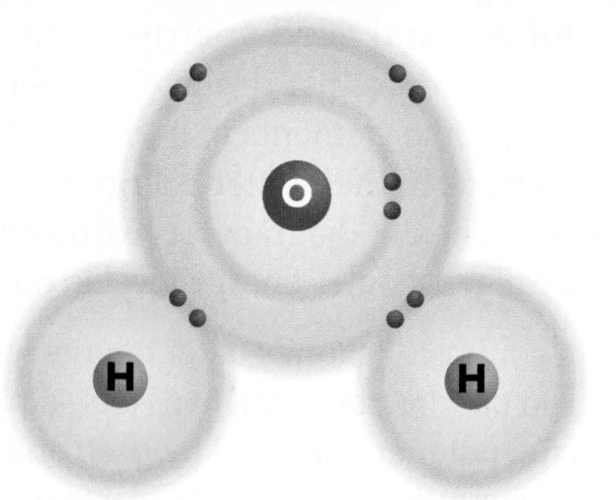

Figure 6 In water, hydrogen contributes one electron and oxygen contributes the other in each hydrogen-oxygen bond. The atoms share the electrons, instead of giving them up, to achieve a complete outer energy level for each atom in the compound.

Sharing electrons A hydrogen atom has one electron in its outer energy level. So, it needs one electron to fill its outer energy level. An oxygen atom has six electrons in its outer energy level. It needs two electrons for its outer level to be stable, with eight electrons. Hydrogen and oxygen become stable and form bonds in a different way than do sodium and chlorine. Hydrogen and oxygen share electrons instead of gaining or losing them. **Figure 6** shows how hydrogen and oxygen share electrons to achieve a more stable arrangement and to form water.

Chemical bond formation When atoms gain, lose, or share electrons, an attraction forms between the atoms, pulling them together to form a compound. This attraction is called a chemical bond. A **chemical bond** is the force that holds atoms together in a compound.

Check Your Progress

Summary

- A chemical formula describes the number and type of atoms in a compound.
- The elements of group 18, the noble gases, rarely combine with other elements.
- Most atoms need eight electrons to complete their outer energy levels.
- Electron dot diagrams show the electrons in the outer energy level of an atom.
- A chemical bond is the force that holds atoms together in a compound.

Demonstrate Understanding

1. **Explain** why some elements are stable on their own, while others are more stable in compounds.
2. **Compare and contrast** the properties of potassium (K) and iodine (I) with the compound KI.
3. **Identify** what the electron dot diagram tells you about bonding.
4. **Explain** why electric forces are essential to forming compounds.
5. **Describe** why chemical bonding occurs. Give two examples of how bonds can form.

Explain Your Thinking

6. **Interpret** The label on a box of cleanser states that it contains CH_3COOH. What elements are in this compound? How many atoms of each element can be found in a unit of CH_3COOH?
7. **MATH Connection** Given that the molecular mass of magnesium hydroxide ($Mg(OH)_2$) is 58.32 amu and the atomic mass of an atom of oxygen is 15.999 amu, what percentage of this compound is oxygen?

LEARNSMART® Go online to follow your personalized learning path to review, practice, and reinforce your understanding.

LESSON 2

TYPES OF BONDS

FOCUS QUESTION

What is the difference between ionic and covalent bonds?

Ions

A neutral atom has the same number of electrons as protons. But not all atoms are neutral. Atoms can lose or gain electrons to become more stable. An atom that has gained or lost an electron is called an ion. An **ion** is a charged particle that has either more or fewer electrons than protons. When an atom loses electrons, it becomes a positively charged ion known as a *cation*. When an atom gains electrons, it becomes a negatively charged ion known as an *anion*. The electric forces between oppositely charged particles can hold ions together.

Some of the most common compounds are made by the loss and gain of just one electron. These compounds contain an element from group 1 and an element from group 17 on the periodic table. Some examples are sodium chloride (NaCl) and potassium iodide (KI), an ingredient in iodized salt. **Figure 7** shows the importance of iodine in nutrition.

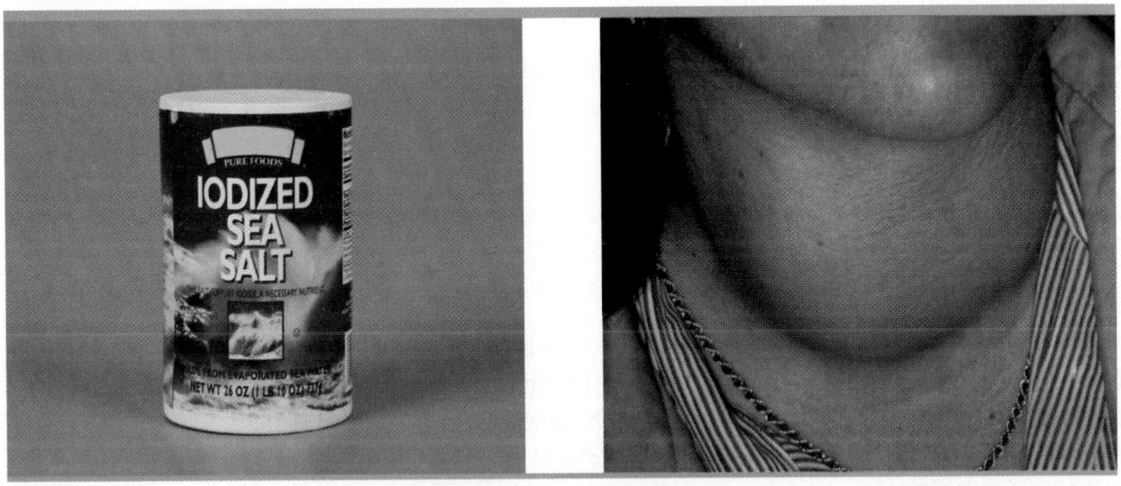

Figure 7 Iodized salt is an important dietary source of iodine. A lack of iodine causes health problems. For example, a goiter, an enlargement of the thyroid gland in the neck, can be caused by iodine deficiency.

3D THINKING DCI Disciplinary Core Ideas CCC Crosscutting Concepts SEP Science & Engineering Practices

COLLECT EVIDENCE

Use your Science Journal to record the evidence you collect as you complete the readings and activities in this lesson.

INVESTIGATE

GO ONLINE to find these activities and more resources.

Applying Practices: Investigate Interparticle Forces
HS-PS1-3. Plan and conduct an investigation to gather evidence to compare the structure of substances at the bulk scale to infer the strength of electrical forces between particles.

Revisit the Encounter the Phenomenon Question
What information from this lesson can help you answer the Unit and Module questions?

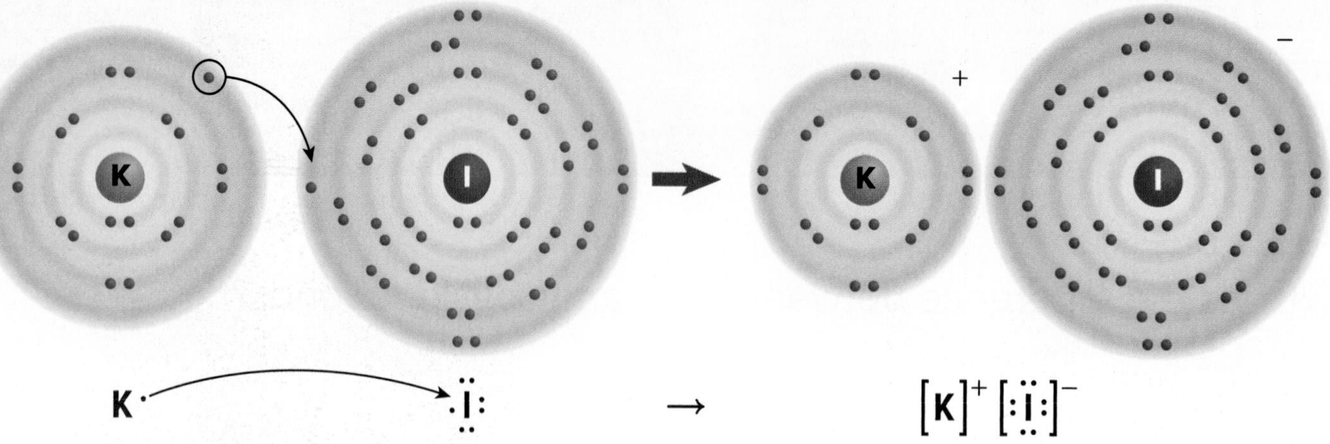

Figure 8 Potassium must transfer one electron to iodine in order for both to achieve stable outer energy levels.
Compare and contrast *the outer energy level of each atom before and after the transfer of the electron.*

Transfer of electrons

What happens when potassium and iodine atoms come together? A neutral atom of potassium has one electron in its outer energy level. This is not a stable outer energy level. Recall that, for most elements, a stable outer energy level contains eight electrons. When it forms a compound with iodine, potassium loses the one electron from its fourth level. With the fourth level gone, the third level is a complete outer energy level.

Although the complete outer energy level means the atom is now stable, because it has lost an electron, it is no longer neutral. The potassium atom has become an ion. When a potassium atom loses an electron, the atom becomes a positively charged ion because there is one electron fewer in the atom than there are protons in the nucleus. The 1+ charge of the potassium cation is shown as a superscript written after the element's symbol, K^+, to indicate its charge. *Superscript* means "written above."

Get It?
Explain What part of an ion's symbol indicates its charge?

An iodine atom has seven electrons in its outer energy level. It needs one more electron in order to have a stable outer energy level. During the reaction with potassium, the iodine atom gains an electron, giving its outer energy level eight electrons. This atom is no longer neutral because it has gained an extra negative particle. It now has a charge of 1− and is called an iodide anion, written as I^-.

The compound formed between potassium and iodine is called potassium iodide. The electron dot diagrams for the process are shown in **Figure 8.** As they lose or gain electrons, the two atoms become stable ions. Notice that the resulting compound has a neutral charge because the 1+ charge of K^+ and the 1− charge of I^- cancel each other.

Get It?
Explain why an atom of iodine tends to react to gain one electron.

The ionic bond

When ions combine in this way, a bond is formed. An **ionic bond** is the force of attraction between the opposite charges of the ions in an ionic compound. The number of positive charges must equal the number of negative charges in order to form a neutral compound.

The formation of magnesium chloride ($MgCl_2$) is another example of ionic bonding. When magnesium reacts with chlorine, a magnesium atom loses the two electrons in its outer energy level and becomes the positively charged ion Mg^{2+}. At the same time, two chlorine atoms gain one electron each to become negatively charged chloride ions, Cl^-. In this case, a magnesium atom has two electrons to donate, but a single chlorine atom needs to accept only one electron. Therefore, it takes two chlorine atoms, as shown in **Figure 9,** to accept the two electrons that the magnesium atom donates.

Get It?

Explain What is the charge of an ionic compound?

Zero net charge As with potassium iodide, the result of these ionic bonds is a neutral compound. The compound as a whole is neutral because the sum of the charges on the ions is zero. The positive charge of the magnesium ion is exactly equal to the negative charge of the two chloride ions. In other words, when atoms form an ionic compound, their electrons are shifted between the individual atoms, but the overall number of protons and electrons of the combined atoms remains equal and unchanged. Therefore, the compound is neutral.

Ionic bonds are usually formed between metals and nonmetals. Ionic compounds are often formed by elements across the periodic table from each other. They are typically crystalline solids with high melting points.

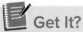

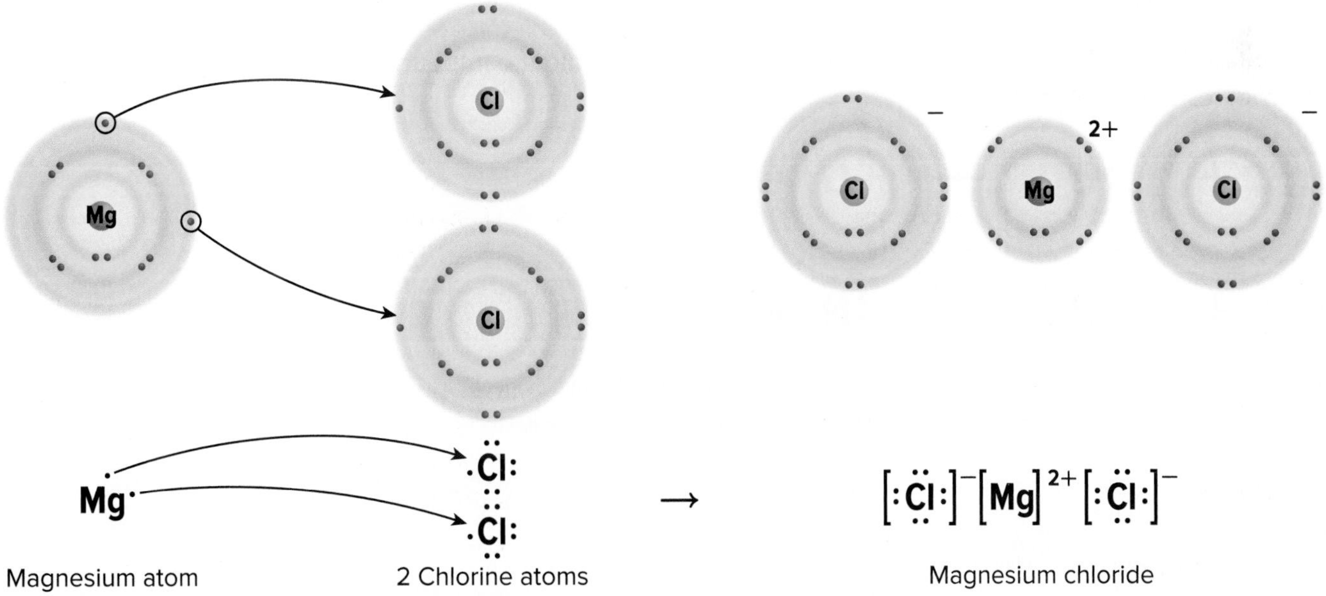

Figure 9 A magnesium atom gives an electron to each of two chlorine atoms to form $MgCl_2$.

Compare and contrast the *formation of KI with the formation of $MgCl_2$.*

Molecules

Some atoms of nonmetals are unlikely to lose or gain electrons. For example, group 14 elements would have to either gain or lose four electrons to achieve a stable outer energy level. The loss of this many electrons takes a great deal of energy. Each time an electron is removed, the nucleus holds the remaining electrons even more tightly. Therefore, these atoms become more chemically stable by sharing electrons rather than by becoming ions.

The attraction that forms between atoms when they share electrons is known as a **covalent bond.** The neutral particle that forms as a result of electron sharing is called a **molecule.**

Figure 10 Hydrogen and oxygen each contribute one electron to the hydrogen-oxygen bonds in a water molecule.

The covalent bond

A single covalent bond is composed of two shared electrons. Usually, one shared electron comes from each atom in the bond. A water molecule contains two single bonds, as shown in **Figure 10.** In each bond, a hydrogen atom contributes one electron to the bond, and the oxygen atom contributes the other. The two electrons are shared, forming a single bond. The result of this type of bonding is a stable outer energy level for each atom in the molecule.

Multiple covalent bonds

A covalent bond can also contain more than one pair of electrons. In the diatomic molecule of oxygen (O_2), each oxygen atom has six electrons in its outer energy level and needs to gain two electrons to become stable. It can do this by sharing two of its electrons with another oxygen atom. When each atom contributes two electrons to the bond, the bond contains four electrons, or two pairs of electrons. Each pair of electrons represents a bond. Therefore, two pairs of electrons represent two bonds, called a double bond. Each oxygen atom is stable, with eight electrons in its outer energy level. Similarly, a bond that contains three shared pairs of electrons is a triple bond. The diatomic molecule N_2 is an example shown in **Figure 11.**

Covalent bonds form between nonmetallic elements. Many covalent compounds are solids or gases at room temperature.

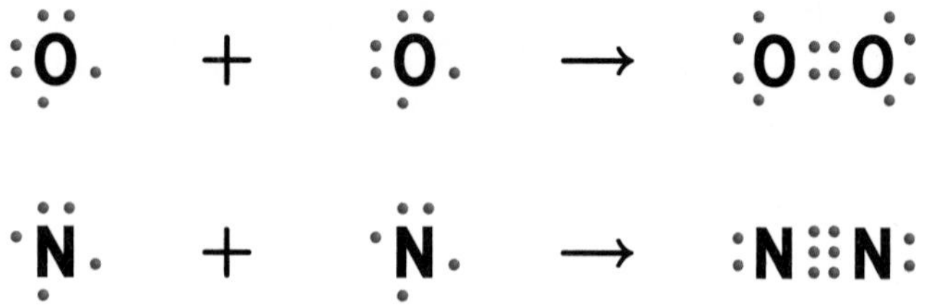

Figure 11 Electron dot diagrams for O_2 and N_2 show that each shares multiple pairs of electrons.
Interpret *What do the electron pairs represent in terms of the structure of the molecules?*

WORD ORIGINS

molecule

comes from the Latin word *moles*, which means "mass"

A water molecule is composed of two hydrogen atoms and one oxygen atom.

Equal sharing

When electrons are shared in covalent bonds formed by similar or identical atoms, such as in the N_2 or O_2 molecules just discussed, the electron charge is shared equally across the bond. A **nonpolar bond** is a covalent bond in which electrons are shared equally by both atoms.

Get It?

Describe the atoms involved in a nonpolar bond.

Unequal sharing

In some molecules, however, electrons are not shared equally, and the electron charge is concentrated more on one end of the molecule than the other. Why aren't electrons always shared equally? Different types of atoms exert different levels of attraction for the electrons in a covalent bond. The strength of the attraction of each atom to its electrons is related to the size of the atom, the charge of the nucleus, and the total number of electrons the atom contains. Recall that a magnet has a stronger pull when it is right next to a piece of metal. In a similar way, a nucleus has a stronger attraction to electrons nearby. In addition, you know that a strong magnet holds metal more firmly than does a weak magnet. Similarly, more positive charge in a nucleus attracts electrons more strongly.

Partial charges One example of unequal sharing in a covalent bond is found in a molecule of hydrogen chloride (HCl), shown in **Figure 12.** Chlorine atoms have a stronger attraction for electrons than hydrogen atoms have. As a result, the shared electrons will spend more time near the chlorine atom. The chlorine atom has a partial negative charge, which is represented by a lowercase Greek letter delta with a negative superscript (δ^-). Because the electrons spend less time near the hydrogen atom, it has a partial positive charge, represented by δ^+.

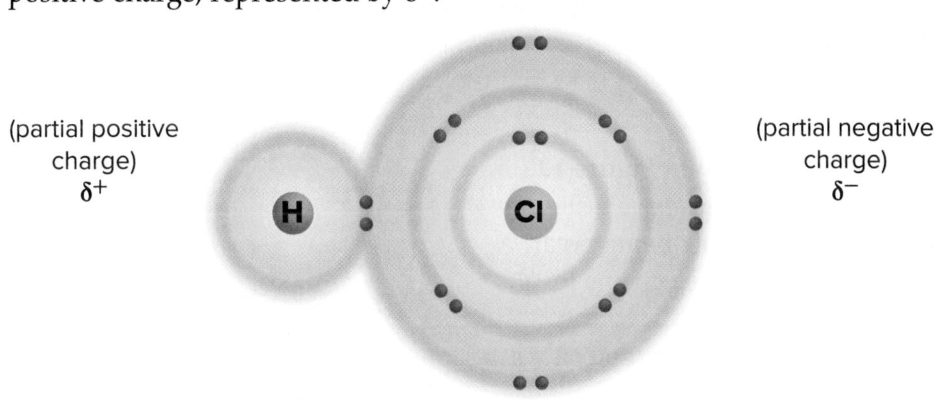

Figure 12 The electrons shared by hydrogen and chlorine spend more time near the chlorine atom, giving the hydrogen chloride molecule partial charges on its ends.

CCC CROSSCUTTING CONCEPTS

Patterns Would you expect elements that form polar bonds to be close together or far apart on the periodic table? Write an explanation for your prediction based on what you know about the repeating patterns of the periodic table.

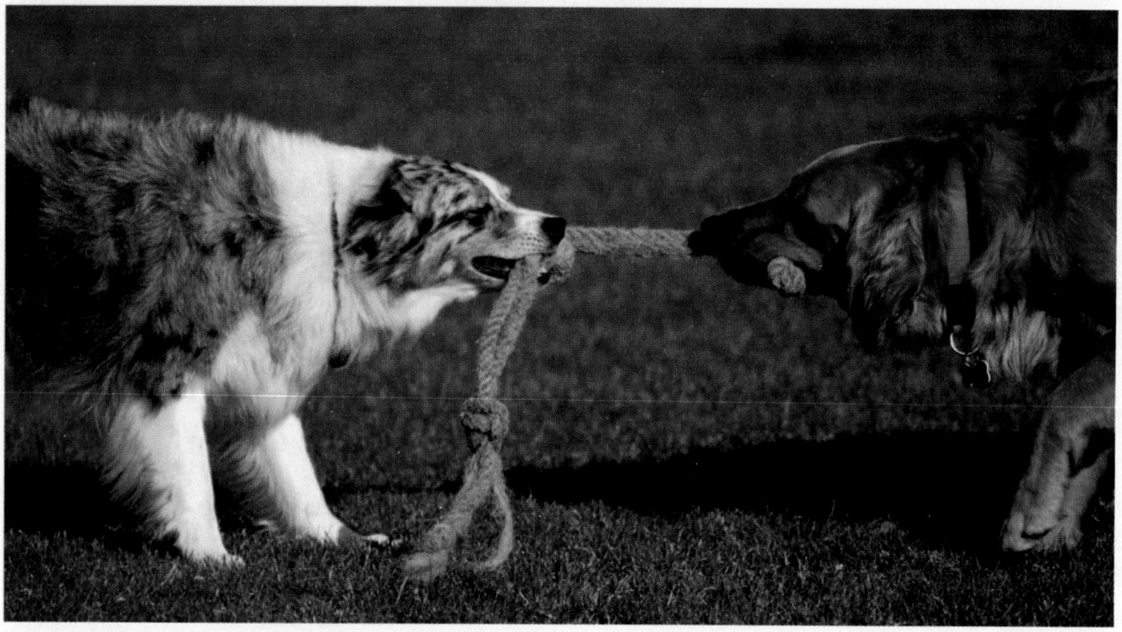

Figure 13 In a tug-of-war, the stronger dog can pull harder and has the advantage.
Compare and Contrast *How is a tug-of-war similar to unequal sharing of electrons?*

Tug-of-war When electrons are shared unequally, the bond is said to be polar. The term *polar* means "having opposite ends." A **polar bond** is a bond in which electrons are shared unequally, resulting in a slightly positive end and a slightly negative end. It might help you to visualize a polar bond as the rope in a tug-of-war, as shown in **Figure 13.** Think of the shared electrons as being in the space in the center. As the two dogs in the tug-of-war pull, the middle of the rope ends up closer to the stronger dog. Similarly, each atom in a bond attracts the electrons that they share, but the electrons will be held more closely to the atom with the stronger pull. These attractions between electric charges at the atomic scale help explain the structure and properties of matter when you look at how they affect the molecules they hold together.

Polar and nonpolar molecules

Look again at the molecule of hydrogen chloride (HCl) in **Figure 12.** The atom holding the electrons more closely will always have a slightly negative charge. This polar bond results in the molecule being polar. A **polar molecule** is one in which the unequal sharing of electrons results in a slightly positive end and a slightly negative end, although the overall molecule is neutral. On the other hand, a **nonpolar molecule** is a molecule that does not have oppositely charged ends.

Polarity and geometry Just because a molecule is nonpolar doesn't mean its electrons are all shared equally. When a molecule contains just two atoms, such as HCl, it is easy to see how the polar bond creates a polar molecule. When a molecule contains three or more atoms, however, like the molecule of H_2O shown in **Figure 14,** you need to consider both the polarity of the bonds and the shape of the molecule to determine whether the molecule is polar or nonpolar. **Figure 15,** on the next page, shows that determining polarity is dependent on multiple factors. Polar bonds can be found in nonpolar molecules.

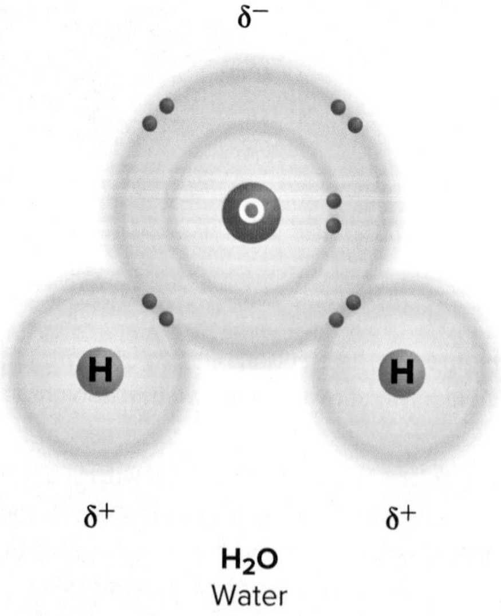

Figure 14 Water is a polar molecule. Water's polarity accounts for many of its unique properties.

Polarity and molecular shape In molecules with a single bond, polarity is determined by the bond. However, in molecules with multiple bonds, polarity is determined by the shape of the molecule as well as the polarity of each of the bonds. As shown in **Figure 15,** it is possible to have a nonpolar molecule with polar bonds.

Figure 15 Trichloromethane and carbon tetrachloride have similar molecules. Due to their differing symmetries, however, one is nonpolar and one is polar.

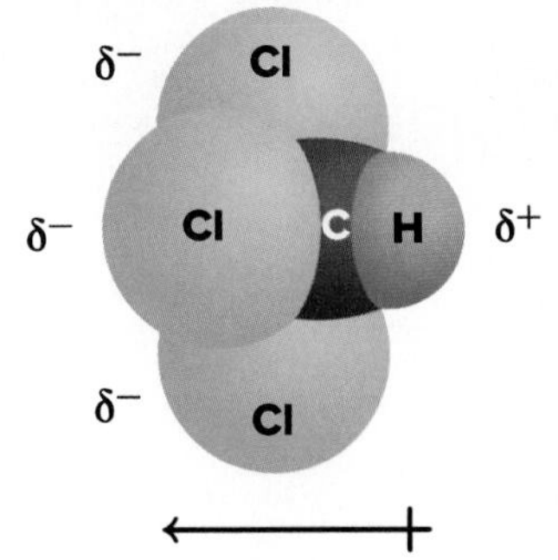

$CHCl_3$
Trichloromethane

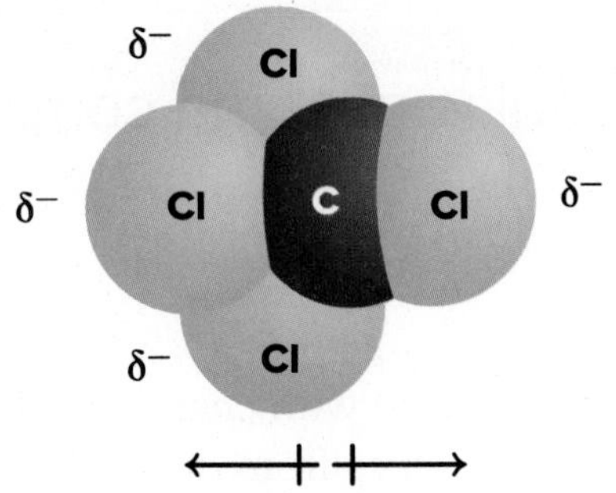

CCl_4
Carbon tetrachloride

In a molecule of trichloromethane ($CHCl_3$), each carbon-chlorine bond is polar, with a partial negative charge on the chlorine atom. Because of the shape of the molecule, there is an overall negative charge on the chlorine end and an overall positive charge on the hydrogen end.

Carbon tetrachloride (CCl_4) is very similar to trichloromethane. As with $CHCl_3$, each of the carbon-chlorine bonds is polar. But because the chlorines are evenly distributed around the carbon, the charges are also evenly distributed. As a result, the molecule as a whole is nonpolar.

Check Your Progress

Summary

- An ion is a charged particle that has either fewer or more electrons than protons, resulting in a net positive or negative charge.
- An ionic bond is the force of attraction between oppositely charged ions.
- Some atoms share electrons instead of losing or gaining them. These atoms form covalent bonds with other atoms.
- When the electrons in a covalent bond are shared unequally, the result is a polar bond.

Demonstrate Understanding

8. **Compare and contrast** ionic and covalent bonds.
9. **Determine** the type of bonding in CaO and in SO_2.
10. **Name** the type of particle formed by covalent bonds.
11. **Identify** the following compounds as polar or nonpolar: HBr, Cl_2, and H_2O.

Explain Your Thinking

12. **Predict** From the given symbols, choose two elements that are likely to form an ionic bond: O, Ne, S, Ca, and K. Next, select two elements that would likely form a covalent bond. Explain your predictions.
13. **MATH Connection** Aluminum oxide (Al_2O_3) can be produced during rocket launches. Show that the sum of the positive and negative charges in a unit of Al_2O_3 equals zero.

LEARNSMART® Go online to follow your personalized learning path to review, practice, and reinforce your understanding.

LESSON 3

WRITING FORMULAS AND NAMING COMPOUNDS

FOCUS QUESTION

Can you name the chemical formulas of some common compounds?

Writing Chemical Formulas

Does the table in **Figure 16** look like it has anything to do with chemistry? It is an early table of the elements and was made for alchemy, a practice from the Middle Ages that laid the foundations for modern chemistry. Alchemists used symbols like these to write the formulas of substances created when individual elements combined. Modern chemists use a different system of symbols to communicate about substances. They use chemical formulas. Before you can write a chemical formula, you must have all the needed information at your fingertips. What will you need to know?

Oxidation numbers

To write a chemical formula, you need to know the elements involved and the number of electrons they gain, lose, or share to become stable. This last piece of information is what chemists refer to as an element's oxidation number. An **oxidation number** is a positive or negative number that indicates how many electrons an atom has gained, lost, or shared to become stable.

For ionic compounds, the oxidation number is the same as the charge on the ion. For example, a sodium ion has a charge of 1+ and an oxidation number of 1+. A chloride ion has a charge of 1− and an oxidation number of 1−. Notice that, as with charges, the sign of the oxidation number is written after, not before, the number itself.

Get It?

Determine the oxidation number for each of the ions in the ionic compound calcium bromide, $CaBr_2$.

Figure 16 Early representations of elements used pictorial symbols to identify known substances.

3D THINKING | DCI Disciplinary Core Ideas | CCC Crosscutting Concepts | SEP Science & Engineering Practices

COLLECT EVIDENCE

Use your Science Journal to record the evidence you collect as you complete the readings and activities in this lesson.

INVESTIGATE

GO ONLINE to find these activities and more resources.

Quick Investigation: Make a Hydrate

Carry out an invesitgation to determine the effects of adding and removing water in a compound.

CCC Identify Crosscutting Concepts

Create a table of the crosscutting concepts and fill in examples you find as you read.

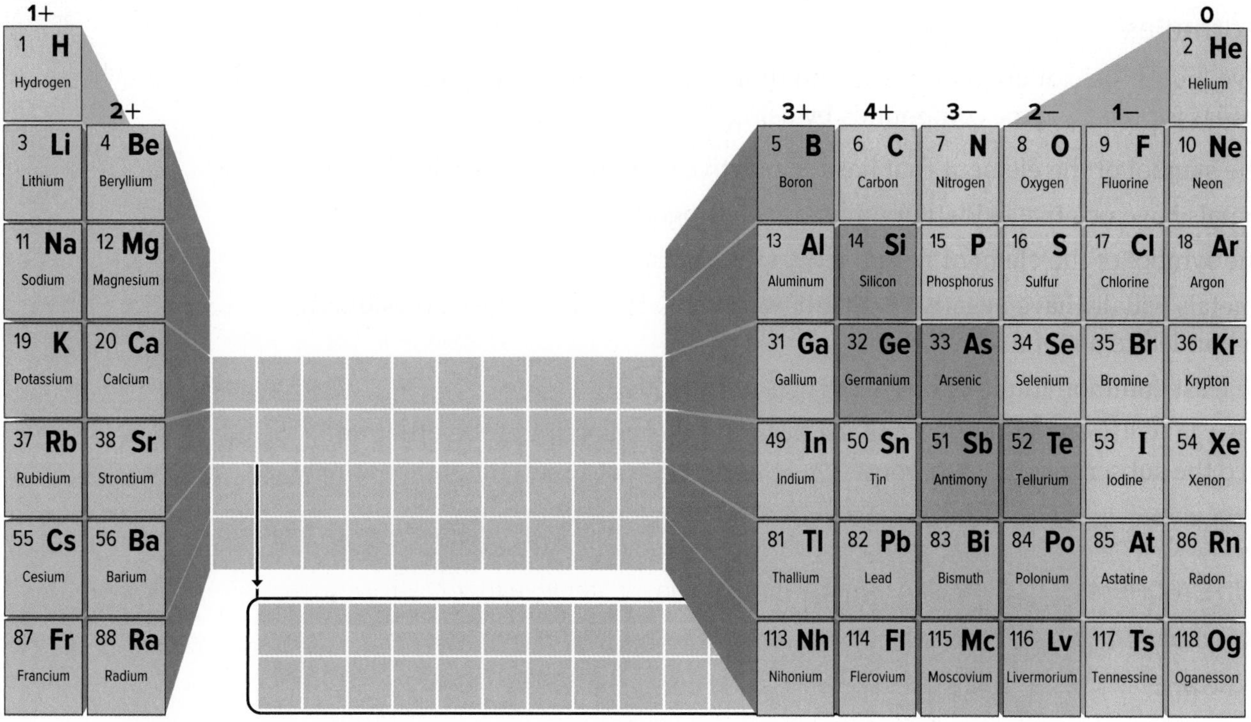

Figure 17 The number at the top of each column is the most common oxidation number of the elements in that group.

Oxidation numbers and the periodic table

Many elements have the same oxidation number, as shown in **Figure 17.** Notice how they fit with the periodic table groupings.

Many of the transition elements can have more than one oxidation number, as shown in **Table 2.** When naming these compounds, the oxidation number is expressed by a roman numeral that is added to the name. For example, the oxidation number of iron in iron(III) oxide is 3+.

Table 2 Special Ions

Ion Name	Oxidation Number
Copper(I)	1+
Copper(II)	2+
Iron(II)	2+
Iron(III)	3+
Chromium(II)	2+
Chromium(III)	3+
Lead(II)	2+
Lead(IV)	4+

Get It?

Determine the oxidation number of lead in the ion lead(IV).

Binary ionic compounds

The easiest compounds to write formulas for are **binary compounds,** which are composed of two elements. When writing formulas, remember that, although the individual ions that make up an ionic compound carry charges, the compound itself is neutral. For example, lithium fluoride is composed of a lithium ion with a 1+ charge and a fluoride ion with a 1− charge. Combining one of each ion makes a neutral compound with the formula LiF.

Some compounds require more figuring. Aluminum oxide contains an aluminum ion with a 3+ charge and an oxide ion with a 2− charge. To determine the overall positive and negative charge, you must find the least common multiple of 3 and 2, which is 6. In order to have a 6+ charge, you need two aluminum ions. In order to have a 6− charge, you need three oxygen ions. This gives the neutral compound Al_2O_3.

Writing formulas

Once you've found the oxidation numbers and their least common multiple, you can write formulas for binary ionic compounds by using the rules below.

1. Write the symbol of the element that has the positive oxidation number or charge.
   - All metals have positive oxidation numbers. Hydrogen often does.
2. Write the symbol of the element with the negative oxidation number.
   - Nonmetals usually have negative oxidation numbers. Hydrogen does occasionally, when bonded to a metal.
3. Find the least common multiple of the charges of the ions.
   - The charge (without the sign) of one ion becomes the subscript of the other ion. Reduce the subscripts to the smallest whole numbers that retain the ratio of the ions.

EXAMPLE Problem 1

DETERMINE A CHEMICAL FORMULA What is the formula for lithium nitride?

List the Unknown: formula for lithium nitride

List the Knowns: symbol and oxidation number of the positive ion:

lithium Li^{1+}

symbol and oxidation number of the negative ion:

nitrogen N^{3-}

Set Up the Problem: Write the symbol of the element with the positive charge, followed by the symbol of the element with the negative charge:

$Li^{1+}N^{3-}$

Solve the Problem: The charge (without the sign) of one ion becomes the subscript of the other:

Li_3N_1 or Li_3N

In this case, the subscripts already reflect the smallest whole numbers that retain the ratio of the ions.

Check the Answer: Check the answer by determining whether your compound is neutral. The charge on three lithium ions is 3+, and the charge on one nitride ion is 3–. The compound is neutral, so the formula is correct.

PRACTICE Problems ADDITIONAL PRACTICE

14. What is the formula for potassium sulfide?

15. What is the formula for calcium chloride?

16. CHALLENGE What is the formula for lead(IV) phosphide?

ACADEMIC VOCABULARY

ratio

the relationship in quantity, amount, or size between two or more things; proportion

In the formula Al_2O_3, the ratio of aluminum atoms to oxygen atoms is 2 to 3.

Naming

Name binary ionic compounds with these rules.

1. Write the name of the positive ion.
2. Using **Table 2** on page 463, check to see whether the positive ion forms more than one oxidation number. If so, determine the oxidation number of the ion from the formula of the compound. Remember, the overall charge of the compound is zero, and the negative ion has only one possible charge. Write the charge of the positive ion using roman numerals in parentheses after the ion's name.
3. Write the root name of the anion. The root is the first part of the element's name. Chlorine is *chlor-*; oxygen is *ox-*.
4. Add the ending *-ide* to the root, as shown in **Table 3.** Chlorine becomes chlor*ide,* and oxygen becomes ox*ide.*

Subscripts are not part of the name for ionic compounds.

Table 3 Elements in Binary Compounds

Element	*–ide* Name
Oxygen	oxide
Phosphorus	phosphide
Nitrogen	nitride
Sulfur	sulfide

EXAMPLE Problem 2

NAME A BINARY IONIC COMPOUND What would a chemist name the compound CuCl?

List the Unknown: compound name for CuCl

List the Knowns: the names of the atoms in the compound: copper and chlorine

Set Up the Problem:

1. Name the positive ion in the compound.
2. Check **Table 2** on page 463 to determine whether the positive ion can have more than one oxidation number. If it can, determine which one to use; name the positive ion; and write the charge, using roman numerals in parentheses.
3. Write the root name of the negative ion.
4. Add the ending *-ide* to the root of the negative ion.

Solve the Problem:

1. The positive ion is copper.
2. Copper has two oxidation numbers. The oxidation number for chlorine is 1–, so copper has to be 1+.
   The positive ion is *copper(I).*
3. The negative atom is chlorine. The root is *chlor-*.
4. Adding *-ide* to the negative root name makes *chloride.*
   The compound name is copper(I) chloride.

Check the Answer: In an ionic compound, the positive ion is a metal, and the negative ion is a nonmetal—the name should be in this order. The name *copper(I) chloride* has the metal first and the nonmetal second.

PRACTICE Problems

ADDITIONAL PRACTICE

17. What is the name of $AlCl_3$?
18. What is the name of Li_2S?
19. CHALLENGE What is the name of Cr_2O_3?

Compounds with Complex Ions

Not all compounds are binary. Many common compounds contain more than two atoms. Baking soda—used in cooking, as a medicine, and for brushing your teeth—has the formula $NaHCO_3$. This is an example of an ionic compound that is not binary. Some compounds, including baking soda, contain polyatomic ions. The prefix *poly-* means "many," so the term *polyatomic* means "having many atoms." A **polyatomic ion** is a positively or negatively charged, covalently bonded group of atoms. Polyatomic ions as a whole contain two or more elements. Even though polyatomic ions contain more than one element, they act like individual ions in forming compounds. The polyatomic ion in baking soda is the hydrogen carbonate ion (HCO_3^-), which is commonly called bicarbonate.

Writing formulas

Table 4 lists several common polyatomic ions. To write formulas for compounds containing these ions, follow the rules for binary compounds, with one addition. When more than one polyatomic ion is needed to balance the charges of the ions, write parentheses around the polyatomic ion before adding the subscript.

How would you write the formula of barium chlorate? First, identify the symbol of the positive ion. Barium has a symbol of Ba and forms a 2+ ion, Ba^{2+}. Next, identify the negative chlorate ion. **Table 4** shows that it is ClO_3^-. Finally, you need to balance the charges of the ions to make the compound neutral. It will take two chlorate ions with a 1− charge to balance the 2+ charge of the barium ion. Because the chlorate ion is polyatomic, you use parentheses before adding the subscript. Therefore, the formula is $Ba(ClO_3)_2$. Another example of naming complex compounds is shown in **Table 5.**

Table 4 Polyatomic Ions

Charge	Name	Formula
1+	ammonium	NH_4^+
1−	acetate chlorate hydroxide nitrate	$C_2H_3O_2^-$ ClO_3^- OH^- NO_3^-
2−	carbonate sulfate	CO_3^{2-} SO_4^{2-}
3−	phosphate	PO_4^{3-}

Table 5 Naming Complex Compounds

To write the chemical formula for ammonium phosphate, answer these questions:
1. What is the positive ion and its charge? The positive ion is NH_4^+, and its charge is 1+.
2. What is the negative ion, and its charge? The negative ion is PO_4^{3-}, and its charge is 3−.
3. How can the charges be balanced in order to make the compound neutral? Three NH_4^+ ions (3+) balance one PO_4^{3-} (3−) ion. Add parentheses for subscripts greater than 1. The chemical formula for ammonium phosphate is $(NH_4)_3PO_4$.

Naming

First, write the name of the positive ion. Then write the name of the negative ion. What is the name of $Sr(OH)_2$? Begin by writing the name of the positive ion, strontium. Then find the name of the polyatomic ion OH^-. **Table 4** lists it as hydroxide. Thus, the name is strontium hydroxide.

Get It?

Determine What is the name of Na_2CO_3?

STEM CAREER CONNECTION

Pharmacist

Pharmacists dispense presciption medicine, but that is not all they do. They can also give medical advice relating to prescription medicines and healthy living, conduct health screenings, and provide immunizations. Pharmacists must have an excellent knowledge of a wide variety of different compounds, their names and formulas, and their effects on the body—both helpful and harmful.

Naming Binary Covalent Compounds

Covalent compounds are formed between elements that are nonmetals. Some pairs of nonmetals can form more than one compound with each other. For example, nitrogen and oxygen can combine to form a number of different compounds, including N_2O, NO, NO_2, and N_2O_5. In the system of naming using oxidation numbers, which you have just learned, each of these compounds would be called nitrogen oxide. From that name, you would not know the composition of the compound. So, a different system of naming must be used for covalent compounds.

Using prefixes

Scientists use the Greek prefixes in **Table 6** to indicate how many atoms of each element are in a binary covalent compound. The nitrogen and oxygen compounds N_2O, NO, NO_2, and N_2O_5 would be named dinitrogen monoxide, nitrogen monoxide, nitrogen dioxide, and dinitrogen pentoxide, respectively. Notice that the last vowel of the prefix is dropped when the second element begins with a vowel, as in *monoxide* and *pentoxide*. The prefix *mono-* is always omitted from the name of the first element of the compound. For example, CO is carbon monoxide as opposed to monocarbon monoxide.

Table 6 Prefixes for Covalent Compounds

Number of Atoms	Prefix	Example
1	mono–	carbon monoxide
2	di–	sulfur dioxide
3	tri–	phosphorus trichloride
4	tetra–	carbon tetrachloride
5	penta–	dinitrogen pentoxide
6	hexa–	uranium hexafluoride
7	hepta–	dichlorine heptoxide
8	octa–	xenon octafluoride

Get It?

Write the name of the compound S_2O_3.

Compounds with Added Water

Some compounds have water molecules as part of their structures. These compounds are called hydrates. A **hydrate** is a compound that has water chemically attached to its atoms and written into its chemical formula.

Common hydrates

The term *hydrate* comes from a word that means "water." When a solution of cobalt chloride evaporates, pink crystals that contain six water molecules for each unit of cobalt chloride are formed. The formula for this compound is $CoCl_2 \cdot 6H_2O$, and it is called cobalt chloride hexahydrate. You can remove water from these crystals by heating them. The resulting blue compound is called anhydrous cobalt chloride. The word *anhydrous* means "without water." When anhydrous (blue) $CoCl_2$ is exposed to water, even from the air, it will revert back to its hydrated state.

SCIENCE USAGE v. COMMON USAGE

hydrate

Science usage: a compound that has water chemically attached to its atoms
Gypsum is a hydrate.

Common usage: to supply water to a person in order to maintain proper fluid balance
It's important to hydrate before and after physical activity.

The plaster of paris shown in **Figure 18** is a building material that relies on a hydration reaction for its function. It is produced by heating calcium sulfate dihydrate, also called gypsum, to a high temperature to remove most of the water. When the plaster of paris is ready to be used, water is added to form a mixture that flows and can be poured into a mold. The plaster of paris reacts quickly with the water to produce gypsum, the hydrate. The mixture quickly sets and hardens to make a strong and light structure, releasing energy in the form of heat.

When writing a formula for a hydrate, the water is shown after a " • ". Following the dot, write the number of water molecules attached to the compound. For example, calcium sulfate dihydrate (gypsum) is written $CaSO_4 \cdot 2H_2O$. Another common hydrate is magnesium sulfate heptahydrate, also called Epsom salt. The formula for this hydrate is $MgSO_4 \cdot 7H_2O$. Notice that when naming hydrates, you use the same prefixes listed in **Table 6,** on the previous page, to indicate the number of water molecules.

Figure 18 Gypsum is used in building materials. The fact that it is a hydrate means that it can be heated for some time before it has released all of its water.

Infer *How does gypsum's ability to hold water make it a useful material for building construction?*

Check Your Progress

Summary

- A binary compound is composed of two elements.
- The oxidation number tells how many electrons an atom has gained, lost, or shared to become stable.
- An ionic compound is made of ions, but the compound as a whole is neutral.
- A polyatomic ion is a positively or negatively charged, covalently bonded group of atoms.
- Greek prefixes are used to indicate how many atoms of each element are in a binary covalent compound.
- A hydrate is a compound that has water chemically attached to its atoms.

Demonstrate Understanding

20. **Predict** formulas for the following compounds: potassium iodide, phosphorus pentachloride, magnesium hydroxide, aluminum sulfate, dichlorine heptoxide, and calcium nitrate trihydrate.
21. **Write** the names of these compounds: Al_2O_3, $Ba(ClO_3)_2$, SO_2, NH_4Cl, PCl_3, and $Mg_3(PO_4)_2 \cdot 4H_2O$.
22. **Determine** the oxidation number of each atom in the following compounds: sodium chloride and iron(II) oxide.

Explain Your Thinking

23. **Explain** whether sodium and potassium will react to form a bond with each other.
24. **MATH Connection** The overall charge on the polyatomic sulfate ion is 2–. Its formula is SO_4^{2-}. If oxygen has a 2– oxidation number, determine the oxidation number of sulfur in this polyatomic ion.

LEARNSMART® Go online to follow your personalized learning path to review, practice, and reinforce your understanding.

ENGINEERING & TECHNOLOGY

Nonstick Surfaces

In your kitchen, there is probably a pan coated with the same nonstick substance used in artificial hearts and in NASA spacecraft. The substance, known as polytetrafluoroethylene (PTFE), has been used in hundreds of applications.

Long PTFE molecules create a smooth, low-friction surface on which food can slide easily.

Losing the ability to bond

PTFE, represented in the model below, is a long string of carbon atoms surrounded by fluorine atoms. The bonds between carbon and fluorine result in a compound that is very stable. PTFE is so stable that it will not bond with most other chemicals. PTFE can be used as a protective barrier between materials that would otherwise bond, such as laboratory equipment that comes in contact with corrosive chemicals.

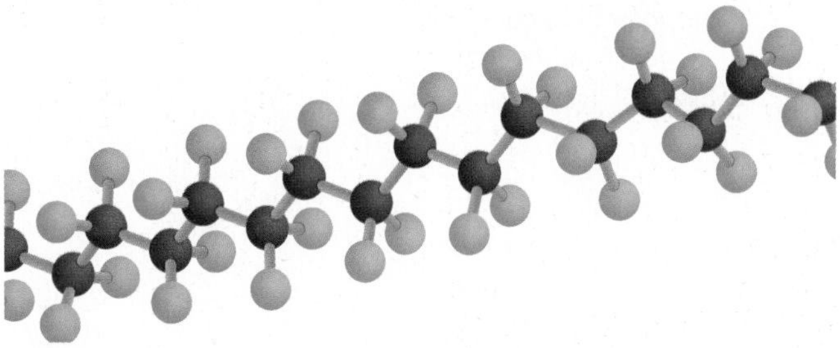

PTFE can contain thousands of carbon atoms, each bonded to two fluorine atoms.

Nonstick surface

The long chain structure of PTFE forms a low-friction surface. Materials coated with PTFE become smooth and slippery because the long molecules fill in the rough crevices of the underlying surface, as shown in the image at the top right.

These two properties of PTFE—resistance to bonding and low friction—are what make the nonstick pan in your kitchen so useful. Not only will food not bond to the surface of the pan, but it also slides easily across the smooth surface. As a result, less oil can be used, and there is less mess to clean up after using a pan coated with PTFE to cook a meal.

Medical uses

PTFE has many medical uses because it will not deteriorate over time. It is used in attaching artificial heart valves to existing tissue and can also be woven into a mesh that can be surgically implanted into arteries. These strong, flexible PTFE implants reinforce and repair damaged blood vessels.

PTFE in space

Reactive ions in space bombard the *Hubble Space Telescope* and can bond to and degrade the telescope's sensitive equipment. A PTFE coating helps prevent the damaging ions from bonding to the telescope. Research continues into the next generation of nonstick coatings that will help protect future space missions.

OBTAIN, EVALUATE, AND COMMUNICATE INFORMATION

Some scientific advances have had significant environmental impacts. Work with a partner to research potential negative impacts caused by PTFE. Create a presentation that weighs the benefits of PTFE against its drawbacks.

MODULE 18
STUDY GUIDE

GO ONLINE to study with your Science Notebook.

Lesson 1 STABILITY IN BONDING

- A chemical formula describes the number and type of atoms in a compound.
- The elements of group 18, the noble gases, rarely combine with other elements.
- Most atoms need eight electrons to complete their outer energy levels.
- Electron dot diagrams show the electrons in the outer energy level of an atom.
- A chemical bond is the force that holds atoms together in a compound.

- chemical formula
- chemical bond

Lesson 2 TYPES OF BONDS

- An ion is a charged particle that has either fewer or more electrons than protons, resulting in a net positive or negative charge.
- An ionic bond is the force of attraction between oppositely charged ions.
- Some atoms share electrons instead of losing or gaining them. These atoms form covalent bonds with other atoms.
- When the electrons in a covalent bond are shared unequally, the result is a polar bond.

- ion
- ionic bond
- covalent bond
- molecule
- nonpolar bond
- polar bond
- polar molecule
- nonpolar molecule

Lesson 3 WRITING FORMULAS AND NAMING COMPOUNDS

- A binary compound is composed of two elements.
- The oxidation number tells how many electrons an atom has gained, lost, or shared to become stable.
- An ionic compound is made of ions, but the compound as a whole is neutral.
- A polyatomic ion is a positively or negatively charged, covalently bonded group of atoms.
- Greek prefixes are used to indicate how many atoms of each element are in a binary covalent compound.
- A hydrate is a compound that has water chemically attached to its atoms.

- oxidation number
- binary compound
- polyatomic ion
- hydrate

THREE-DIMENSIONAL THINKING

Module Wrap-Up

REVISIT THE PHENOMENON

What makes these rock crystals form perfect cubes?

CER Claim, Evidence, Reasoning

Explain Your Reasoning Revisit the claim you made when you encountered the phenomenon. Summarize the evidence you gathered from your investigations and research and finalize your Summary Table. Does your evidence support your claim? If not, revise your claim. Explain why your evidence supports your claim.

STEM UNIT PROJECT

Now that you've completed the module, revisit your STEM unit project. You will summarize your evidence and apply it to the project.

GO FURTHER

SEP Data Analysis Lab

Electron Dot Diagrams

The electrons in an atom's outer energy level are the electrons that are important to consider in chemical bonds and chemical reactions. These electrons can be represented in a diagram called an electron dot diagram. The outermost electrons are drawn as dots around the chemical symbol. In this activity, you will draw electron dot diagrams for several elements.

Procedure

1. Write the symbol for the element. For electron dot diagrams, this symbol represents the nucleus and all of the electrons of the atom except the outermost electrons.
2. Use the periodic table to determine how many outermost electrons the element has. Do this by finding to which group the element belongs.
3. Draw a dot to represent each electron in the outer level of the element. Two electrons can be placed on each side of the symbol. The first two electrons should be paired on the right side of the symbol. The rest of the outer electrons should be distributed counterclockwise one by one around the other sides of the symbol.

CER Analyze and Interpret Data

1. **Claim** Write electron dot diagrams for the following elements: hydrogen, neon, sodium, calcium, aluminum, fluorine, argon, and potassium.
2. **Evidence and Reasoning** Why do sodium and potassium have the same number of dots in their electron dot diagrams? What does this tell you about the chemistry of these two elements?

Florea Marius Catalin/iStockphoto/Getty Images

McGraw-Hill Education

MODULE 19

CHEMICAL REACTIONS

ENCOUNTER THE PHENOMENON

What chemical reactions occur when you bake cupcakes?

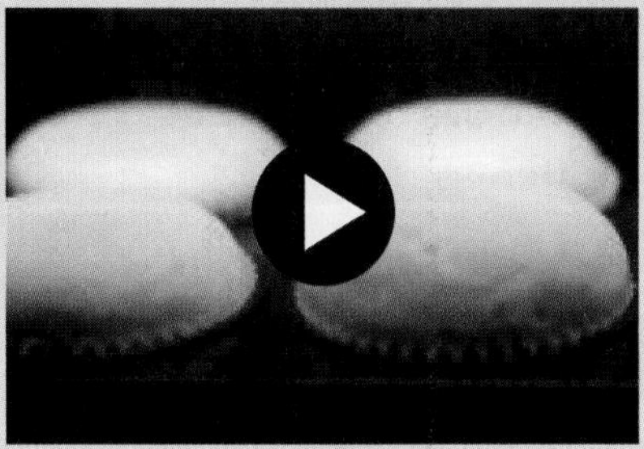

GO ONLINE to play a video about how chemical reactions cause changes in cake batter.

SEP Ask Questions

Do you have other questions about the phenomenon? If so, add them to the driving question board.

CER Claim, Evidence, Reasoning

Make Your Claim Use your CER chart to make a claim about what chemical reactions occur when baking cupcakes. Explain your reasoning.

Collect Evidence Use the lessons in this module to collect evidence to support your claim. Record your evidence as you move through the module.

Explain Your Reasoning You will revisit your claim and explain your reasoning at the end of the module.

GO ONLINE to access your CER chart and explore resources that can help you collect evidence.

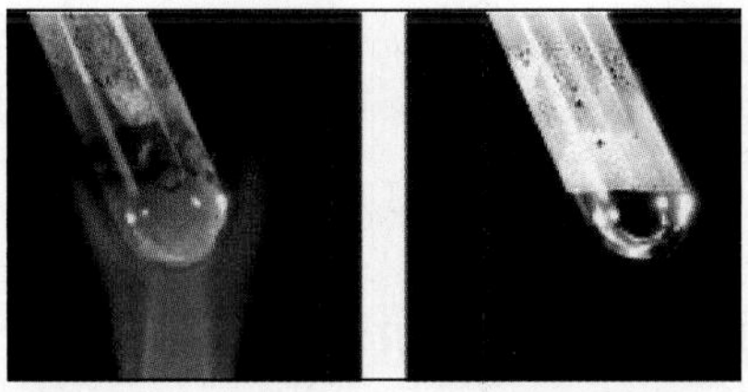

LESSON 1: Explore & Explain: Balancing Equations

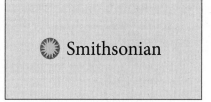

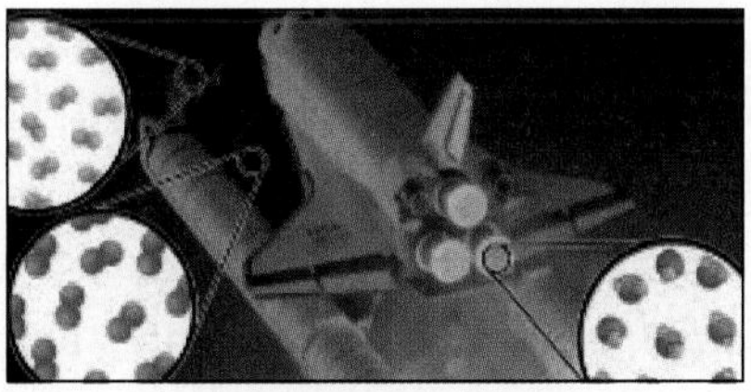

LESSON 4: Explore & Explain: Reaction Rates

Smithsonian

Additional Resources

LESSON 1
CHEMICAL CHANGES

FOCUS QUESTION
How do you write a chemical equation for a chemical reaction?

Lavoisier and the Conservation of Mass

Chemical reactions take place all around you, and even within you. A **chemical reaction** is a change in which one or more substances are converted into new substances. The starting substances that react are called **reactants.** The new substances produced are called **products.** This relationship between reactants and products can be written as follows: reactants → products.

By the 1770s, the pseudoscience of alchemy was starting to be replaced by chemistry. While alchemy imitated science, alchemists did not provide science-based explanations about the natural world. However, scientists, such as the French chemist Antoine Lavoisier, studied chemical reactions using scientific methods. As a result of such study, Lavoisier established that the total mass of the products always equals the total mass of the reactants. This principle is demonstrated in **Figure 1.**

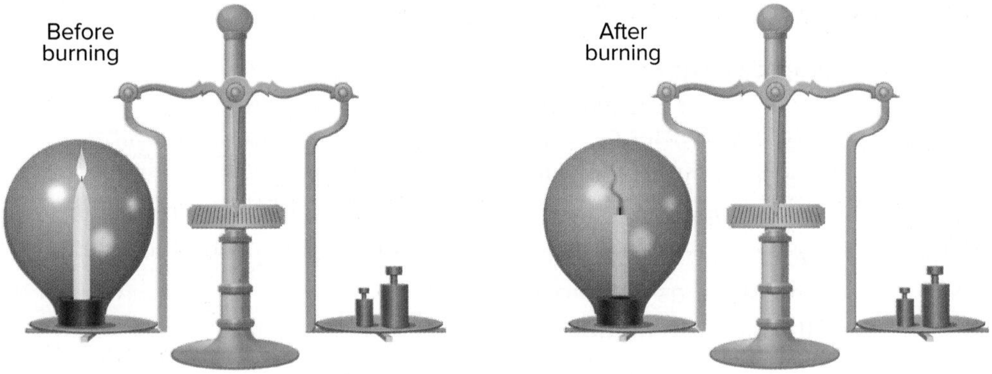

Figure 1 The mass of the candle and oxygen before burning equals the mass of the candle and gaseous products after burning.

3D THINKING DCI Disciplinary Core Ideas CCC Crosscutting Concepts SEP Science & Engineering Practices

COLLECT EVIDENCE

Use your Science Journal to record the evidence you collect as you complete the readings and activities in this lesson.

INVESTIGATE

GO ONLINE to find these activities and more resources.

Applying Practices: Electron States and Simple Chemical Reactions

HS-PS1-2. Construct and revise an explanation for the outcome of a simple chemical reaction based on the outermost electron states of atoms, trends on the periodic table, and knowledge of the patterns of chemical properties.

The mystery of what happens when substances change made Lavoisier curious. In one experiment, Lavoisier placed a carefully measured mass of solid mercury(II) oxide into a sealed container. When he heated this container, he noted a dramatic change. The red powder transformed into a silvery liquid that he recognized as mercury metal, and a gas was produced. When he determined the mass of the liquid mercury and the gas, their combined masses were exactly the same as the mass of the red powder he started with.

Lavoisier's experiment also established that the gas produced by heating mercury(II) oxide was a component of air. He did this by heating mercury metal with air and saw that a portion of the air—what we now know as oxygen gas—combined with the mercury to produce red mercury(II) oxide.

mercury(II) oxide		oxygen		mercury
10.0 g	=	0.7 g	+	9.3 g

Notice that the mass of the mercury(II) oxide reactant is equal to the combined masses of the mercury metal and the oxygen gas products. Hundreds of experiments carried out in Lavoisier's laboratory confirmed that matter is not created or destroyed, but conserved in a chemical reaction. This important principle became known as the law of conservation of mass. This means that in any chemical reaction, the total starting mass of all reactants is equal to the total final mass of all products.

Get It?

Explain the law of conservation of mass.

Writing Equations

If you wanted to describe the chemical reaction shown in **Figure 2,** you might write something like this:

Nickel(II) chloride, dissolved in water, plus sodium hydroxide, dissolved in water, produces solid nickel(II) hydroxide plus sodium chloride, dissolved in water.

This series of words is rather cumbersome, but all of the information is important. The same is true of descriptions of most chemical reactions.

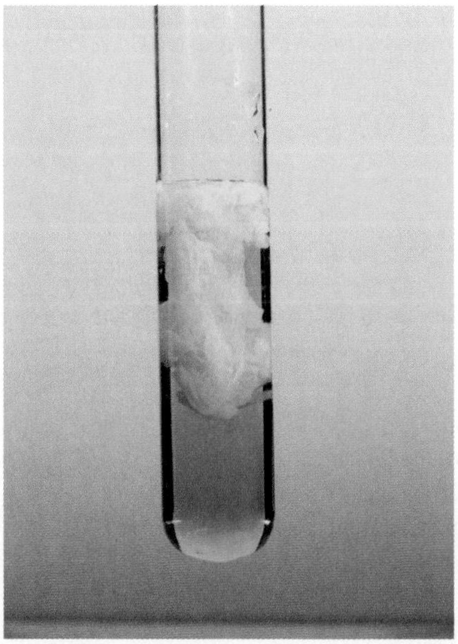

Figure 2 When solutions of nickel(II) chloride and sodium hydroxide mix, a chemical reaction occurs that produces nickel(II) hydroxide as a white solid.

Many words are needed to state all the important information about reactions. Therefore, scientists use a shorthand method to describe chemical reactions. A **chemical equation** is a way to describe a chemical reaction using chemical formulas and other symbols. Some of the symbols used in chemical equations are listed in **Table 1.**

Table 1 Symbols Used in Chemical Equations

Symbol	Meaning
$\rightarrow$	produces or yields
$+$	plus
(s)	solid
(l)	liquid
(g)	gas
(aq)	aqueous—a substance is dissolved in water
$\xrightarrow{\text{heat}}$	The reactants are heated.
$\xrightarrow{\text{light}}$	The reactants are exposed to light.
$\xrightarrow{\text{elec.}}$	An electric current is applied to the reactants.

The chemical equation for the reaction shown in **Figure 2** looks like this:

$$NiCl_2(aq) + 2NaOH(aq) \rightarrow Ni(OH)_2(s) + 2NaCl(aq)$$

It is much easier to quickly and clearly identify what is happening by writing the information in this form. Chemical equations quickly convey information such as the states of matter of the reactants and products. Later, you will learn how chemical equations make it easier to calculate the quantities of reactants that are needed and of products that are formed.

Coefficients

A chemical equation lists more than just the chemical formulas of the reactants and products. Notice the numbers next to the formulas for sodium hydroxide (NaOH) and sodium chloride (NaCl) above. What do they mean? These numbers, called **coefficients,** represent the number of units of each substance taking part in a reaction.

When no coefficient appears before a substance in a balanced equation, a coefficient of 1 is assumed. In the equation above, for example, one unit of nickel(II) chloride ($NiCl_2$) combined with two units of sodium hydroxide (NaOH). The products were one unit of nickel(II) hydroxide ($Ni(OH)_2$) and two units of sodium chloride (NaCl). Remember that according to the law of conservation of mass, atoms are rearranged but never destroyed in a chemical reaction. Notice in the equation above that there are equal numbers of each type of atom on each side of the arrow.

Get It?
Summarize Describe the purpose of coefficients in a chemical equation.

ACADEMIC VOCABULARY

shorthand
a method of writing quickly using abbreviations or special symbols for certain words or phrases
Taking notes in shorthand can save time.

CCC CROSSCUTTING CONCEPTS

Energy and Matter Lavoisier's discovery that the total mass of products always equals the total mass of the reactants is a manifestation of the law of conservation of mass. Write this law, and then compare it to the law of conservation of energy. As an extension, look up and write the law of conservation of mass-energy. Then explain what it means and how it relates to the equation $E = mc^2$.

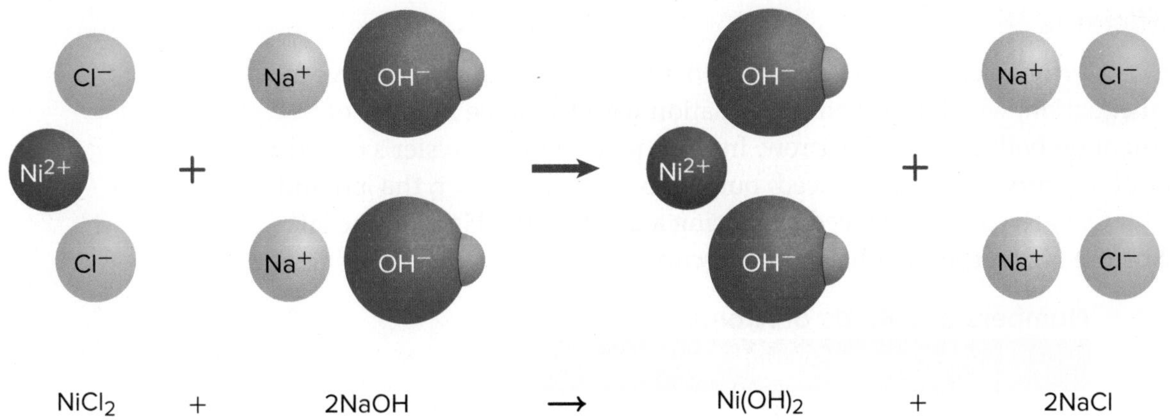

Figure 3 There are equal numbers of each type of atom on both sides of the arrow.

Knowing the number of units of reactants enables chemists to add the correct amounts of reactants to a reaction. Also, these coefficients indicate exactly how much product will form. For example, you would react one unit of $NiCl_2$ with two units of NaOH to produce one unit of $Ni(OH)_2$ and two units of NaCl. You can see this reaction in **Figure 3,** above.

Balancing Equations

Lavoisier's mercury(II) oxide reaction, shown in **Figure 4,** can be written as:

$$HgO(s) \xrightarrow{heat} Hg(l) + O_2(g)$$

Notice that the number of mercury atoms is the same on both sides of the equation but that the number of oxygen atoms is not:

Numbers and Kinds of Atoms

	HgO(s) →	Hg(l) +	$O_2(g)$
Hg	1	1	
O	1		2

In this equation, one oxygen atom appears on the reactant side of the equation and two appear on the product side. But according to the law of conservation of mass, one oxygen atom cannot just become two. Nor can you simply add the subscript 2 and write HgO_2 instead of HgO. The formulas HgO_2 and HgO do not represent the same compound. In fact, HgO_2 does not exist. The formulas in a chemical equation must accurately represent the compounds that react. Fixing this equation requires a process called balancing. Balancing an equation doesn't change what happens in a reaction—it simply changes the way the reaction is represented.

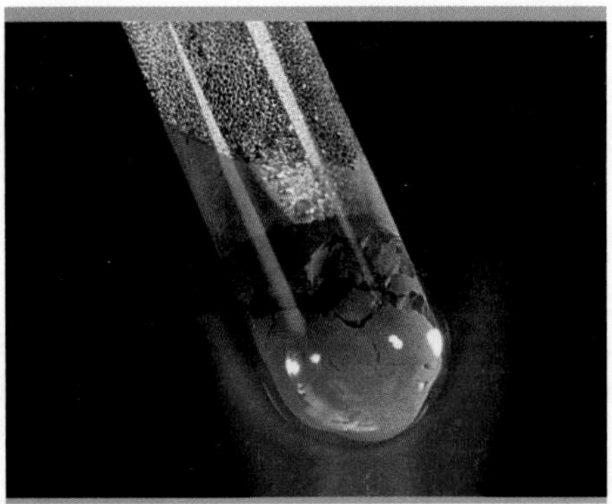

Mercury(II) oxide

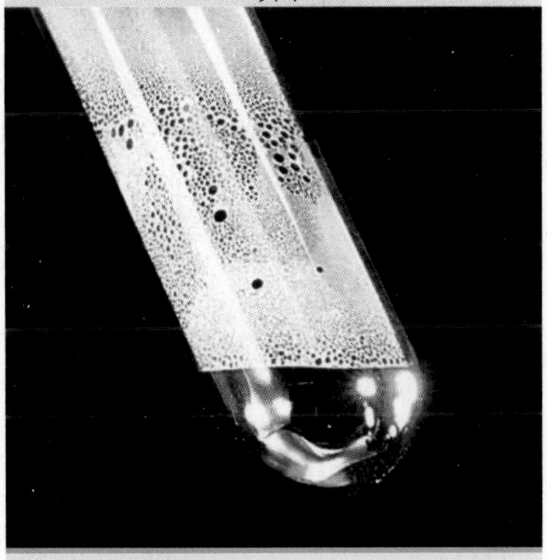

Liquid mercury and oxygen

Figure 4 Mercury metal and oxygen form when mercury oxide is heated.

Choosing coefficients

The balancing process involves changing coefficients in a reaction in order to achieve a **balanced chemical equation,** which is a chemical equation with the same number of atoms of each element on both sides of the arrow. In the equation for Lavoisier's experiment, the number of mercury atoms is balanced, but one oxygen atom is on the left and two are on the right. Oxygen can be balanced by placing a 2 before the HgO on the left. Then, placing a 2 before Hg on the right balances mercury.

Numbers and Kinds of Atoms

	$2HgO(s)$ →	$2Hg(l)$ +	$O_2(g)$
Hg	2	2	
O	2		2

Try your balancing act

Magnesium (Mg) burns with such a brilliant white light that it is often used in fireworks, as shown in **Figure 5.** Burning leaves a white powder called magnesium oxide (MgO). To write a balanced chemical equation, follow these steps.

Figure 5 When magnesium combines with oxygen, it produces a bright white flame, making it ideal for applications such as sparklers, fireworks, and flares.

Step 1 Write a chemical equation for the reaction of magnesium with oxygen. Recall that oxygen is a diatomic molecule.

$$Mg(s) + O_2(g) \rightarrow MgO(s)$$

Step 2 Count the atoms in reactants and products. The magnesium atoms are balanced, but the oxygen atoms are not.

Numbers and Kinds of Atoms

	$Mg(s)$ +	$O_2(g)$ →	$MgO(s)$
Mg	1		1
O		2	1

Step 3 Choose coefficients that balance the equation. Remember, never change subscripts of a correct formula to balance an equation. Try putting a coefficient of 2 before MgO.

$$Mg(s) + O_2(g) \rightarrow 2MgO(s)$$

Step 4 Recheck the numbers of each atom on each side of the equation, and adjust coefficients again if necessary. Now two Mg atoms are on the right side and only one is on the left side. So a coefficient of 2 is needed for Mg to balance the equation.

$$2Mg(s) + O_2(g) \rightarrow 2MgO(s)$$

Get It?

Summarize How can you tell whether a chemical equation is balanced or not?

EXAMPLE Problem 1

BALANCE EQUATIONS A sample of barium sulfate ($BaSO_4$) is placed on a piece of paper, which is then ignited. Barium sulfate reacts with the carbon (C) from the burned paper, producing barium sulfide (BaS) and carbon monoxide (CO). Write a balanced equation for this reaction.

List the Knowns: We know the substances that are involved in the reaction: barium sulfate ($BaSO_4$), carbon (C), barium sulfide (BaS), and carbon monoxide (CO).

Set Up the Problem: From this, we can write a chemical equation showing reactants and products:

$$BaSO_4(s) + C(s) \rightarrow BaS(s) + CO(g)$$

Then, count and list the atoms found on both sides of the equation.

Numbers and Kinds of Atoms

	$BaSO_4(s)$ +	C(s) →	BaS(s) +	CO(g)
Ba	1		1	
S	1		1	
O	4			1
C		1		1

Solve the Problem: The oxygen atoms are unbalanced. Try putting a 4 in front of CO. Now you have 4 oxygen atoms on the right, which balances oxygen, but the carbon atoms become unbalanced. To fix this, add a 4 in front of the C in the reactants. The balanced equation looks like this:

$$BaSO_4(s) + 4C(s) \rightarrow BaS(s) + 4CO(g)$$

Check the Answer: Count the number of atoms on each side of the equation and verify that they are equal.

PRACTICE Problems

ADDITIONAL PRACTICE

1. Balance this equation: $MgCl_2(aq) + AgNO_3(aq) \rightarrow Mg(NO_3)_2(aq) + AgCl(s)$.
2. Balance this equation: $NaOH(aq) + CaBr_2(s) \rightarrow Ca(OH)_2(s) + NaBr(aq)$.
3. HCl is slowly added to aqueous Na_2CO_3, forming NaCl, H_2O, and CO_2. Write a balanced equation for this reaction.
4. **CHALLENGE** Write the balanced equation for the reaction of hydrogen gas and oxygen gas that forms water.

STEM CAREER Connection

Chemical Engineer

Chemistry is not just for classrooms. Chemical engineers use their knowledge of chemistry and other sciences to develop processes and solve problems in a wide range of industries, including the production of food, drugs, fuels, fabrics, plastics, electronics, and many other products and materials.

Understanding Chemical Equations

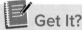

Figure 6 When charcoal burns, carbon (C) reacts with oxygen (O_2). The product of the reaction is carbon dioxide (CO_2).

Have you watched someone cook food outdoors on a charcoal grill? When charcoal burns, as shown in **Figure 6,** heat is liberated by the chemical reaction between carbon in the charcoal and oxygen in the air. Molecules of carbon dioxide gas are produced. Note in the balanced equation for this reaction that one oxygen molecule (O_2) is required for each carbon atom (C) and that one carbon dioxide molecule (CO_2) is produced. Also, we know from the periodic table that the average mass of a carbon atom is 12.01 amu, and the average mass of an oxygen atom is 16.00 amu. Therefore, the average mass of an O_2 molecule is 32.00 amu (2 × 16.00 amu), and the average mass of a CO_2 molecule is 44.01 amu (12.01 amu + (2 × 16.00 amu)).

C(s)	+	O_2(g)	→	CO_2(g)
1 atom		1 molecule		1 molecule
12.01 amu		32.00 amu		44.01 amu

In the laboratory, selecting a single carbon atom (which has a mass of 12.01 amu) and reacting it with a single oxygen molecule (mass 32.00 amu) is virtually impossible. Instead, chemists more commonly consider reactions of large numbers of particles, with masses expressed in grams, rather than reactions of single atoms and molecules with masses in amu. For example, 12.01 grams of carbon reacting with 32.00 grams of oxygen has the same ratio of masses as the balanced equation. That's because the number of carbon atoms in 12.01 grams of carbon is very nearly equal to the number of oxygen molecules in 32.00 grams of oxygen. In fact, 12.01 grams of carbon contain 6.02×10^{23} carbon atoms, and 32.00 grams of oxygen contain 6.02×10^{23} oxygen molecules.

Get It?
Explain why chemists use masses in grams instead of amu.

Moles

Because the number of particles involved in most chemical reactions is so large, chemists use a counting unit called the mole (mol). One **mole** is the amount of a substance that contains 6.02×10^{23} particles of that substance. The reaction between one mole of carbon and one mole of oxygen, yielding one mole of carbon dioxide, is summarized in **Table 2.**

Table 2 Moles, Mass, and Particles

Equation	C(s) +	O_2(g) →	CO_2(g)
Number of moles	1	1	1
Mass	12.01 g	32.00 g	44.01 g
Number of particles	6.02×10^{23} atoms	6.02×10^{23} molecules	6.02×10^{23} molecules

Molar mass

The mass in grams of one mole of a substance is called its **molar mass.** Just as the mass of a dozen eggs is different from the mass of a dozen watermelons, different substances have different molar masses. The atomic mass of titanium (Ti), for example, is 47.87 amu, and the molar mass is 47.87 g/mol. By comparison, the atomic mass of sodium (Na) is 22.99 amu, and its molar mass is 22.99 g/mol.

For a compound, such as nitrogen dioxide (NO_2), the molar mass is the sum of the masses of its component atoms. The nitrogen dioxide (NO_2) molecule contains one nitrogen atom (1 × 14.01 amu) and two oxygen atoms (2 × 16.00 amu = 32.00 amu). So, NO_2 has a molar mass of 46.01 g/mol (14.01 g/mol + 32.00 g/mol = 46.01 g/mol).

Mole-mass conversions

Given the mass of a substance, you can use the molar mass as a conversion factor to calculate the number of moles. The following example uses this method to calculate the number of moles in 50.00 g of NO_2.

$$50.00 \text{ g } NO_2 \times \frac{1 \text{ mol } NO_2}{46.01 \text{ g } NO_2} = 1.087 \text{ mol } NO_2$$

Similarly, given the number of moles of a substance, you can use the molar mass as a conversion factor to calculate the mass. What is the mass of 0.2020 mol of NO_2?

$$0.2020 \text{ mol } NO_2 \times \frac{46.01 \text{ g } NO_2}{1 \text{ mol } NO_2} = 9.294 \text{ g } NO_2$$

Check Your Progress

Summary

- A chemical reaction is a process that involves one or more reactants changing into one or more products.
- A balanced chemical equation indicates relative amounts of reactants and products.
- A mole (mol) is the amount of a substance that contains 6.02×10^{23} particles of that substance.

Demonstrate Understanding

5. **Identify** the reactants and the products in the following chemical equation.

   $$Cd(NO_3)_2(aq) + H_2S(g) \rightarrow CdS(s) + 2HNO_3(aq)$$

6. **Explain** the importance of the law of conservation of mass.
7. **Explain** why oxygen gas must be written as O_2 in a chemical equation.

Explain Your Thinking

8. **Explain** why the sum of the coefficients on the reactant side of a balanced equation does not have to equal the sum of the coefficients on the product side of the equation.
9. **MATH Connection** Balance the following equation:

   $$Fe(s) + Cl_2(g) \rightarrow FeCl_3(s)$$

10. **MATH Connection** Calculate how many moles are in 125 g of water (H_2O).
11. **MATH Connection** Calculate the mass of 3.000 mol of calcium (Ca).

LEARNSMART® Go online to follow your personalized learning path to review, practice, and reinforce your understanding.

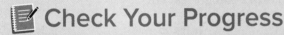

LESSON 2

CLASSIFYING CHEMICAL REACTIONS

FOCUS QUESTION

How do you classify chemical reactions?

Types of Reactions

There are all sorts of chemical reactions. In fact, literally millions of chemical reactions occur every day, and scientists have described many of them and continue to describe more. Chemists have characterized these reactions and organized them into five main categories: combustion, synthesis, decomposition, single displacement (also known as single replacement), and double displacement (also known as double replacement).

Combustion reactions

If you have ever seen something burning, you have seen a combustion reaction. The reaction shown in **Figure 7** creates flames of heat and light as carbon in the wood reacts with oxygen in the air to form carbon dioxide (CO_2). Lavoisier deduced that the process of combustion involves a substance combining with oxygen. Our definition states that a **combustion reaction** occurs when a substance reacts with oxygen to produce energy in the form of heat and light. Many combustion reactions also fit into other categories. For example, the reaction between carbon and oxygen is also a synthesis reaction.

Figure 7 *Burning* is another word for a combustion reaction. The reactants in this reaction are carbon and hydrogen (in the wood) and oxygen (from the air). The products are carbon dioxide (CO_2) and water (H_2O).

3D THINKING DCI Disciplinary Core Ideas CCC Crosscutting Concepts SEP Science & Engineering Practices

COLLECT EVIDENCE

Use your Science Journal to record the evidence you collect as you complete the readings and activities in this lesson.

INVESTIGATE

GO ONLINE to find these activities and more resources.

Laboratory: Chemical Reactions
Carry out an investigation to determine the stability and change of a chemical reaction.

CCC Identify Crosscutting Concepts
Create a table of the crosscutting concepts and fill in examples you find as you read.

Sinelev/Shutterstock

Synthesis reactions

One of the easiest reaction types to recognize is a synthesis reaction. In a **synthesis reaction,** two or more substances combine to form another substance. The generalized formula for this reaction type is A + B → AB. The reaction in which hydrogen gas (H_2) combines with chlorine gas (Cl_2) to form hydrogen chloride (HCl) is an example of a synthesis reaction.

$$H_2(g) + Cl_2(g) \xrightarrow{light} 2HCl(l)$$

This synthesis reaction requires the addition of light. It can be explosive in direct sunlight. Another synthesis reaction with which you may be familiar is the combination of oxygen (O_2) with iron (Fe) in the presence of water to form hydrated iron(III) oxide (Fe_2O_3), which is known as rust.

Figure 8 In the presence of light, hydrogen peroxide (H_2O_2) will decompose, producing oxygen gas (O_2) and water (H_2O).

Infer *Why is the hydrogen peroxide not in a clear bottle?*

Decomposition reactions

A decomposition reaction is just the reverse of a synthesis. Instead of two substances coming together to form a third, a **decomposition reaction** occurs when one substance breaks down, or decomposes, into two or more substances. The general formula for this type of reaction can be expressed as AB → A + B.

Most decomposition reactions require the input of heat, light, or electricity. For example, hydrogen peroxide (H_2O_2), shown in **Figure 8,** will slowly decompose in the presence of light, producing oxygen gas (O_2) and water (H_2O).

$$2H_2O_2(l) \xrightarrow{light} O_2(g) + 2H_2O(l)$$

Single displacement

The chemical reaction in which one element replaces another element in a compound is called a **single-displacement reaction.** Single-displacement reactions—sometimes called *single-replacement reactions*—are described by the general equation A + BC → AC + B. Here you can see that atom A displaces atom B to produce a new molecule, AC. A single displacement reaction is illustrated in **Figure 9,** where a copper wire is put into a solution of silver nitrate. Because copper is a more active metal than silver, it replaces the silver, forming a blue copper(II) nitrate solution. The silver, which is not soluble, forms crystals on the wire.

$$Cu(s) + 2AgNO_3(aq) \rightarrow Cu(NO_3)_2(aq) + 2Ag(s)$$

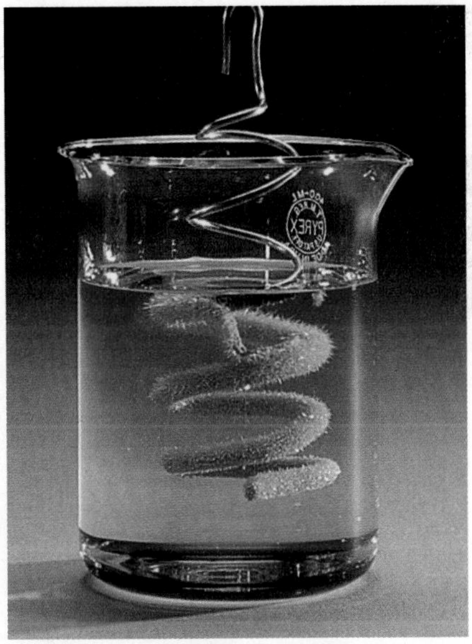

Figure 9 Copper from the wire replaces silver in silver nitrate, forming a blue-tinted solution of copper(II) nitrate.

Get It?

Summarize Describe what happens in a single-displacement reaction.

Sometimes single-displacement reactions can cause problems. For example, if iron-containing vegetables such as spinach are cooked in aluminum pans, aluminum can displace iron from the vegetable. This causes a black deposit of iron to form on the sides of the pan. For this reason, it is better to use stainless steel or enamel cookware when cooking spinach.

We can predict which metal will replace another using an activity series like the one shown in **Figure 10,** which lists metals according to how reactive they are. A metal will replace any less active metal. Notice that silver and gold are two of the least active metals on the list. That is why these elements often occur as deposits of the pure element. More reactive metals often occur as compounds.

Most active ↑

Metals
- Lithium
- Potassium
- Calcium
- Sodium
- Aluminum
- Manganese
- Zinc
- Iron
- Tin
- Lead
- Copper
- Silver
- Platinum
- Gold

Least active

Figure 10 An activity series is a useful tool for determining whether a chemical reaction will occur and for determining the result of a single-replacement reaction.

Double displacement

The positive ion of one compound replaces the positive ion of the other to form two new compounds in a **double-displacement reaction**—sometimes called a *double-replacement reaction.* You know that a double-displacement reaction is taking place if a precipitate, water, or a gas forms when two ionic compounds in solution are combined. A **precipitate** is an insoluble compound that comes out of solution during this type of reaction. The generalized formula for this reaction is AB + CD → AD + CB.

Get It?
Classify What kind of reaction produces a precipitate?

The reaction of copper(II) chloride with sodium hydroxide is an example of this type of reaction. A precipitate—copper(II) hydroxide—forms, as shown in **Figure 11.**

$$CuCl_2(aq) + 2NaOH(aq) \rightarrow Cu(OH)_2(s) + 2NaCl(aq)$$

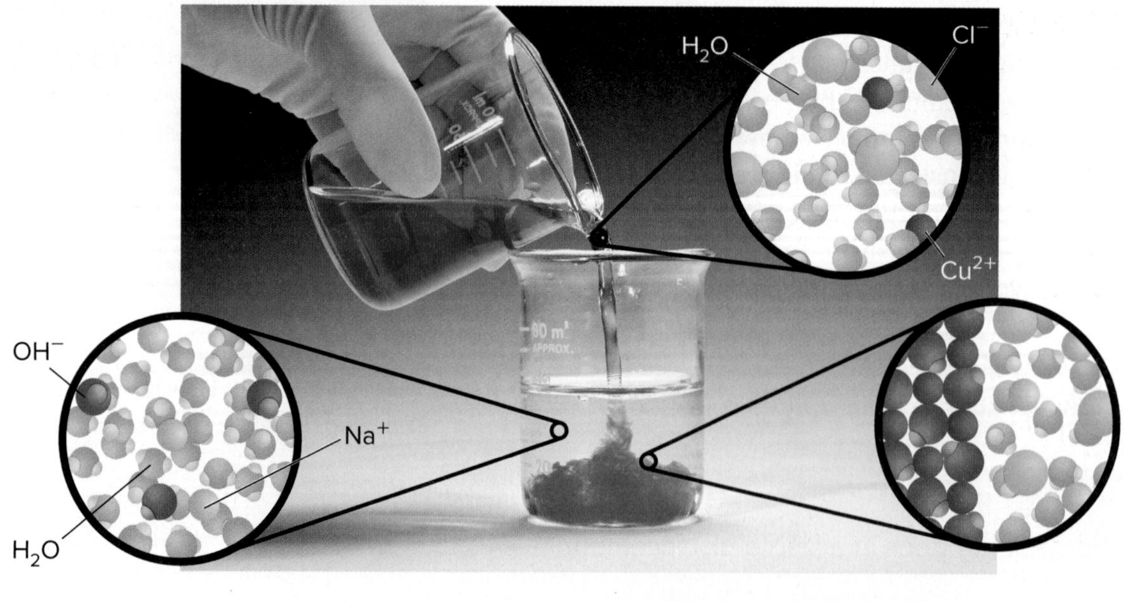

Figure 11 When solutions of copper(II) chloride ($CuCl_2$) and sodium hydroxide (NaOH) are mixed, a solid—copper(II) hydroxide ($Cu(OH)_2$)—is formed. Formation of a precipitate is a sign of a double-displacement reaction.

Oxidation-Reduction Reactions

One characteristic that is common to many chemical reactions is the transfer of electrons. Chemists use the term **oxidation** to describe the loss of electrons and **reduction** to describe the gain of electrons. Chemical reactions involving electron transfer of this sort often involve oxygen, which is very reactive, pulling electrons from metallic elements. Corrosion of metal is a visible result, as shown in **Figure 12.**

Reduction and oxidation always work as a pair. The substance that gains an electron or electrons becomes more negative, and we say it is reduced. The substance that loses an electron or electrons then becomes more positive, and we say it is oxidized. Oxidation-reduction reactions are also commonly called redox reactions.

Figure 12 Rust is a common form of corrosion. Oxygen (O_2) from the air pulls electrons from iron (Fe) to form iron(III) oxide (Fe_2O_3).

Infer *Which element is oxidized, and which is reduced?*

Check Your Progress

Summary

- Chemical reactions are organized into five basic classes: combustion, synthesis, decomposition, single displacement, and double displacement.
- An activity series predicts which metal will replace another in a single-displacement reaction.
- Some reactions produce a solid called a precipitate when two ionic substances are combined.
- Oxidation is the loss of electrons, and reduction is the corresponding gain of electrons.

Demonstrate Understanding

12. **Characterize** each reaction by determining its reaction type.
    a. $CaO(s) + H_2O(l) \rightarrow Ca(OH)_2(aq)$
    b. $Fe(s) + CuSO_4(aq) \rightarrow FeSO_4(aq) + Cu(s)$
    c. $C_{10}H_8(l) + 12O_2(g) \rightarrow 10CO_2(g) + 4H_2O(g)$
    d. $NaCl(aq) + AgNO_3(aq) \rightarrow NaNO_3(aq) + AgCl(s)$
    e. $NH_4NO_3(s) \rightarrow N_2O(g) + 2H_2O(g)$
13. **Describe** what happens in a combustion reaction.
14. **Compare and contrast** synthesis reactions and decomposition reactions.
15. **Determine,** using **Figure 10,** if zinc will displace gold in a chemical reaction and explain why or why not.

Explain Your Thinking

16. **Describe** one possible economic impact of redox reactions. How might that impact be lessened?
17. **MATH Connection** The following chemical reaction is balanced, but the coefficients used are greater than necessary. Rewrite this balanced equation using the smallest coefficients possible.

$$9Fe(s) + 12H_2O(l) \rightarrow 3Fe_3O_4(s) + 12H_2(g)$$

18. **MATH Connection** Sulfur trioxide (SO_3), a pollutant released by coal-burning plants, can react with water (H_2O) in the atmosphere to produce sulfuric acid (H_2SO_4). Write the balanced equation for this reaction.

LEARNSMART® Go online to follow your personalized learning path to review, practice, and reinforce your understanding.

CHEMICAL REACTIONS AND ENERGY

FOCUS QUESTION

How does the energy of reactants and products compare in different reactions?

Energy Exchanges in Chemical Reactions

Crowds often gather to watch a rocket launch. Hundreds of kilograms of solid and liquid rocket fuel are converted into a gas, pushing the enormous rocket skyward. The combustion of rocket fuel is an example of a rapid chemical reaction.

Most chemical reactions proceed more slowly, but all chemical reactions release or absorb energy. This energy can take many forms, such as thermal energy, motion, light, sound, or electricity. The thermal energy produced by a wood fire and the light emitted by a glow stick are examples of energy released by chemical reactions.

Chemical bonds are the source of this energy. When most chemical reactions take place, some chemical bonds in the reactants are broken, which requires energy called activation energy. In order for products to be produced, new bonds must form. Bond formation releases energy. Reactions such as dynamite combustion, shown in **Figure 13,** require much less energy to break chemical bonds than the energy released when new bonds are formed. The result is a release of energy and an explosion.

Figure 13 When new bonds are formed, energy is released. In a reaction like exploding dynamite, the energy that is released from the products' formation is much greater than the energy required to break bonds in the reactants.

3D THINKING DCI Disciplinary Core Ideas CCC Crosscutting Concepts SEP Science & Engineering Practices

COLLECT EVIDENCE

Use your Science Journal to record the evidence you collect as you complete the readings and activities in this lesson.

INVESTIGATE

GO ONLINE to find these activities and more resources.

Applying Practices: Modeling Energy in Chemical Reactions
HS-PS1-4. Develop a model to illustrate that the release or absorption of energy from a chemical reaction system depends upon the changes in total bond energy.

More Energy Out

Many of the reactions with which you are most familiar involve the release of energy. Chemical reactions that release energy are called **exergonic reactions.** In these reactions, the activation energy required to break the original bonds is less than the energy that is released when new bonds form. As a result, some form of energy, such as light or thermal energy, is given off by the reaction. The abdomen of a firefly glows as a result of an exergonic reaction that produces visible light, as shown in **Figure 14.**

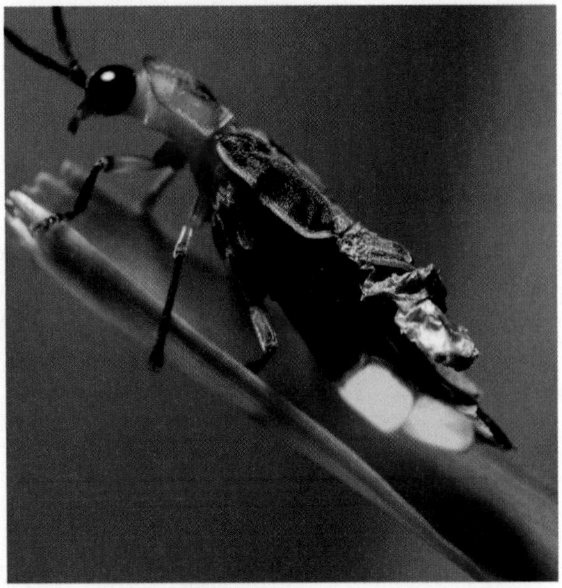

Figure 14 The chemical reactions happening inside the abdomen of a firefly produce light.

Infer *How do you know these are exergonic reactions?*

Thermal energy released

In many reactions, the energy given off is thermal energy. This is the case with some heat packs that are used to treat muscle aches and other problems. When the energy given off is primarily in the form of thermal energy, the reaction is called an **exothermic reaction.** Wood burning and the explosion of dynamite are exothermic reactions. Iron rusting is also exothermic, but, under typical conditions, the reaction proceeds so slowly that it's difficult to detect any temperature change.

Get It?

Infer Why is a log fire considered to be an exothermic reaction?

Exothermic reactions provide most of the power used in homes and industries, as shown in **Figure 15.** Fossil fuels contain carbon. These fuels, such as coal, petroleum, and natural gas, combine with oxygen to yield carbon dioxide gas and energy. Unfortunately, impurities in these fuels, such as sulfur, burn as well, producing pollutants such as sulfur dioxide. Sulfur dioxide combines with water in the atmosphere, producing acid rain.

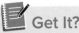

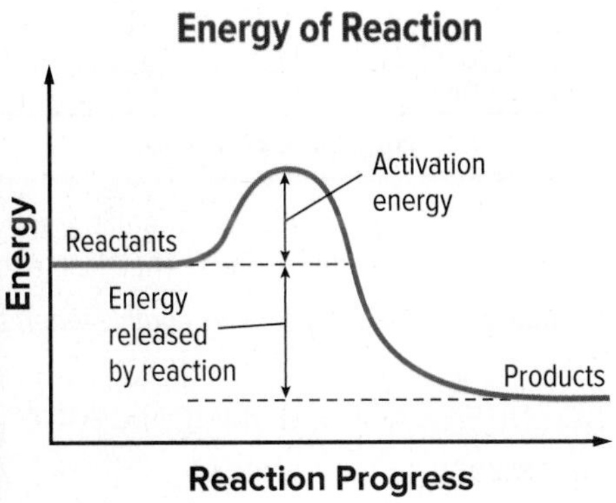

Figure 15 Burning coal is an example of an exothermic reaction. The energy released from burning coal heats water, which becomes steam and rotates an electricity-generating turbine. The graph shows the release of energy as the reaction progresses.

More Energy In

Sometimes a chemical reaction requires more energy to break bonds than is released when new ones are formed. These reactions are called **endergonic reactions.** The energy absorbed can be in the form of light, thermal energy, or electricity.

Electricity is often used to supply energy to endergonic reactions. For example, an electric current passed through water produces hydrogen and oxygen, shown in **Figure 16.** Also, aluminum metal is obtained from its ore using the following endergonic reaction:

$$2Al_2O_3(l) \xrightarrow{elec.} 4Al(l) + 3O_2(g)$$

Figure 16 Water is stable, but the addition of an electrical current will cause it to decompose into hydrogen and oxygen.

Thermal energy absorbed

When the energy needed to keep a reaction going is in the form of thermal energy, the reaction is called an **endothermic reaction.** The terms *exothermic* and *endothermic* are not just related to chemical reactions. They can also describe physical changes. If you ever had to soak a swollen ankle in an Epsom salt solution, you probably noticed that when you mixed the Epsom salt in water, the solution became cold. The dissolving of Epsom salt absorbs thermal energy. Thus, it is a physical change that is endothermic.

Cooking involves the addition of thermal energy to bring about chemical changes in food. When baking cookies, you might add baking soda ($NaHCO_3$) to the dough mixture. Through an endothermic reaction, the baking soda breaks down into sodium carbonate (Na_2CO_3), carbon dioxide gas (CO_2), and water vapor (H_2O). As these gases are released, tiny pockets form in the dough, which causes the cookies to puff up, as shown in **Figure 17.**

Figure 17 Baking involves endothermic reactions such as the decomposition of baking soda. The graph shows how energy is absorbed during these chemical reactions.

Compare *How did the cookies change when they were baked?*

SCIENCE USAGE v. COMMON USAGE

conserve

Science usage: to hold a physical quantity, such as energy or mass, constant during a physical or chemical change
Matter is conserved in chemical reactions.

Common usage: to manage resources wisely; to save resources
It is important to conserve wetlands, a valuable natural resource.

Conservation of Energy in Chemical Reactions

You learned in an earlier module that energy can change from one form to another, but the total amount of energy never changes. This principle is usually stated as the law of conservation of energy. Does the total amount of energy remain constant in chemical reactions, too? Consider the exergonic burning of methane (CH_4), the major component of natural gas, as described by the following equation. Note that energy is included as a product.

$$CH_4(g) + 2O_2(g) \rightarrow CO_2(g) + 2H_2O(g) + \text{energy}$$

During this process, some of the chemical energy of the reactants is released as thermal energy and light. However, the sum of the energy released and the chemical energy of the products is exactly equal to the chemical energy of the reactants in an exergonic chemical reaction.

$$\text{chemical energy of reactants} = \text{chemical energy of products} + \text{energy released}$$

So, the total amount of energy before and after the reaction remains the same. Similarly, the total amount of energy remains the same in endergonic chemical reactions. Summarizing, the law of conservation of energy applies to chemical reactions as well as to other types of energy transformations.

Check Your Progress

Summary

- Breaking chemical bonds absorbs energy.
- Forming chemical bonds releases energy.
- Exergonic chemical reactions release energy. Endergonic chemical reactions absorb energy.
- Exothermic reactions give off thermal energy. Endothermic reactions absorb thermal energy.

Demonstrate Understanding

19. **Classify** the chemical reaction photosynthesis as endergonic or exergonic. Explain.
20. **Explain** why the total amount of energy does not decrease in an exergonic chemical reaction.
21. **Explain** how a reaction could be endergonic but not endothermic.
22. **Classify** the reaction that makes a firefly glow in terms of energy input and output.

Explain Your Thinking

23. **Apply** To develop a product that warms people's hands, would you use an exothermic or endothermic reaction? Why?
24. **MATH Connection** If an endothermic reaction begins at 26°C and decreases by 2°C per minute, how long will it take to reach 0°C?
25. **MATH Connection** Create a graph of the data in question 24. After 5 minutes, what is the temperature of the reaction?

LEARNSMART® Go online to follow your personalized learning path to review, practice, and reinforce your understanding.

REACTION RATES AND EQUILIBRIUM

FOCUS QUESTION

What determines the rate of a reaction?

Reaction Rates

Some chemical reactions, such as the combustion of rocket fuel, take place rapidly and release tremendous amounts of energy in a matter of seconds. Other reactions, such as the rusting of steel, proceed so slowly that you hardly notice any change from one week to the next. The **reaction rate** is the rate at which reactants change into products.

Consider the synthesis reaction shown in **Figure 18,** which has the following equation.

$$2H_2(l) + O_2(l) \rightarrow 2H_2O(g) + \text{energy}$$

A chemist might choose one of several ways to state the rate of this reaction: the rate at which one of the two reactants is used up, the rate at which water is produced, or the rate at which energy is released. Because many factors influence the rates of chemical reactions, the chemist would also state the conditions under which the reaction occurred.

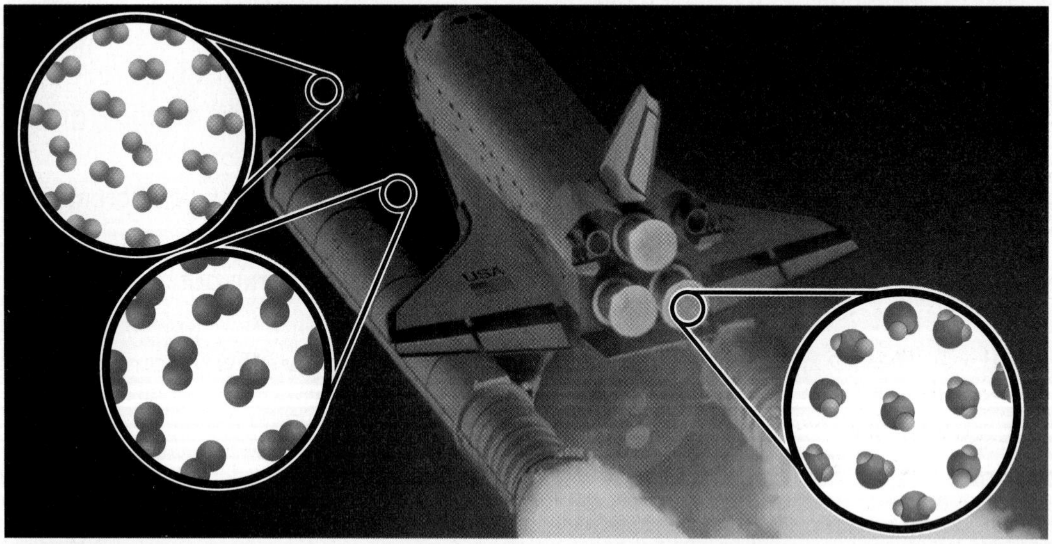

Figure 18 The main engine for the space shuttle combines liquid hydrogen and liquid oxygen to produce water vapor. The shuttle is able to overcome Earth's gravity because this reaction occurs very rapidly.

3D THINKING DCI Disciplinary Core Ideas CCC Crosscutting Concepts SEP Science & Engineering Practices

COLLECT EVIDENCE

Use your Science Journal to record the evidence you collect as you complete the readings and activities in this lesson.

INVESTIGATE

GO ONLINE to find these activities and more resources.

Applying Practices: Concentration and Reaction Rates

HS-PS1-5. Apply scientific principles and evidence to provide an explanation about the effects of changing the temperature or concentration of the reacting particles on the rate at which a reaction occurs.

Factors Affecting Reaction Rates

You already know that sugar dissolves faster in hot water than it does in cold water. Sugar dissolving in water is not a chemical reaction. However, the rates of most chemical reactions, too, vary with temperature. Chemists use a commonsense idea to explain why reaction rates depend upon temperature and other factors, such as concentration and surface area. This idea is called the collision model. The **collision model** states that atoms, ions, and molecules must collide in order to react. The collision model helps explain why changing the conditions of a chemical reaction can have an effect on the reaction rate.

Temperature

You normally store perishable foods such as milk, eggs, and vegetables in a refrigerator. That's because lowering the temperature decreases the rates of the chemical reactions that cause spoilage. Conversely, increasing the temperature of chemical reactions generally increases their reaction rates.

Why does temperature affect reaction rate? Recall that the temperature of a substance is a measure of the average kinetic energy of all of its particles. At higher temperatures, therefore, reacting particles move faster and collide more frequently. A higher collision frequency alone, however, does not completely explain the increase in reaction rate. Because the particles are moving faster at higher temperatures, they collide with greater energy. As a result, a greater percentage of collisions result in a reaction between colliding particles.

Get It?
Use the collision model to explain the effect of increased temperature on reaction rates.

Figure 19 The rate of the reaction between magnesium and hydrochloric acid increases when the concentration of the acid is increased.

Mg in dilute HCl

Mg in concentrated HCl

Concentration

Another way you can change the rate of a chemical reaction is by changing the concentration of one or more of the reactants. Concentration describes the number of particles of a substance per unit volume. Chemists usually express concentration as moles of a substance per liter (mol/L).

Consider the two test tubes shown in **Figure 19.** Each tube has a magnesium (Mg) ribbon immersed in a solution of hydrochloric acid (HCl). The difference between the two tubes is the concentration of the acid solution. As the magnesium and hydrochloric acid react, hydrogen gas (H_2) is released as a product, so you can compare the rates of the two reactions by how rapidly bubbles are formed. Why does magnesium ribbon react faster with the more concentrated hydrochloric acid? The more concentrated acid solution contains more reacting particles per unit volume, resulting in more opportunities for collisions between reacting particles. As a result, the reaction rate is greater.

Get It?
Use the collision model to predict how the reaction rate changes over time as a magnesium ribbon reacts with HCl in a dilute HCl solution.

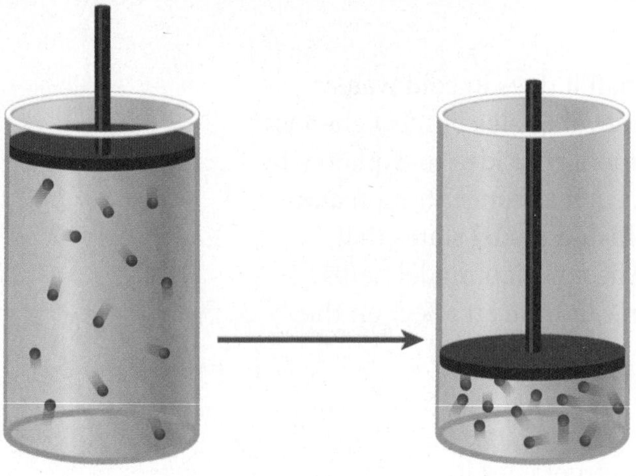

Figure 20 As volume decreases, pressure increases. The decreased volume and increased pressure mean that the particles are closer together and strike each other more often.

Volume and pressure

For chemical reactions involving gases, volume and pressure are important considerations because they relate to the concentrations of the reacting gases. For example, decreasing the volume of a flask containing gases while maintaining a constant temperature increases the concentrations of the gases. Just as with liquid solutions, increasing the concentrations of gases increases the rate at which the particles collide with each other and with the walls of the container. The pressure inside the flask increases. More importantly, the reaction rate increases as well because the reacting gas particles collide with each other more frequently. The effect of increased pressure and decreased volume on gas particles is demonstrated in **Figure 20.**

Get It?

Compare and contrast the effects of increased concentration of liquid reactants and decreased volume of gaseous reactants.

Surface area

Which dissolves more quickly–granulated sugar or a sugar cube? As you probably guessed, the answer is granulated sugar because the individual grains of sugar have much greater total surface area compared to the sugar cube. Dissolving sugar is a physical change, but increased surface area also increases the rate of chemical reactions.

Operators of grain elevators must take measures to ensure that grain dust and oxygen in the air do not combine in a combustion reaction. Even on a scorching-hot day, there is little danger that whole grains of wheat or kernels of corn will react rapidly with oxygen in the air. However, the fine particles that make up grain dust can react explosively on a hot day, as shown in **Figure 21.** The larger total surface area of the grain dust greatly increases the rate at which reacting particles collide. With more collisions per unit time, the rate of the combustion reaction increases dramatically.

Figure 21 Grain dust can be explosive because of its increased surface area.

Catalysts and inhibitors

Some reactions proceed too slowly to be useful. To speed up such a reaction, a catalyst can be added. A **catalyst** is a substance that speeds up a chemical reaction without being permanently changed itself. When you add a catalyst to a reaction, the mass of the product that is formed remains the same, but it will form more rapidly. The catalyst remains unchanged and often is recovered and reused. Catalysts are used to speed many reactions in industry, such as the process of polymerization to make plastics and fibers. In order to break down food, your body utilizes special catalysts called enzymes.

At times, it is worthwhile to prevent certain reactions from occurring. For example, foods often spoil because they react with oxygen from the air. Substances called **inhibitors** are used to slow down the rates of chemical reactions or prevent a reaction from happening at all. Food preservatives are inhibitors that prevent the reactions that lead to the spoilage of certain foods.

One thing to remember when thinking about catalysts and inhibitors is that they do not change the amount of product produced. They only change the rate of production.

Get It?

Compare and contrast catalysts and inhibitors in terms of how they affect reaction rates.

Figure 22 Some reactions move in only one direction—from reactants to products—like a one-way street, such as the Brooklyn Queens Expressway. Other reactions move in both directions—from reactants to products to reactants—like a two-way street.

Equilibrium

Think of a one-way street, on which vehicles may travel in only one direction. Now examine the general formula for a chemical reaction (reactants → products). Do you notice a similarity?

The single reaction arrow may lead you to suppose that every chemical reaction proceeds in only one direction, from reactants to products. Under certain conditions, some reactions do exactly that. And when such a reaction continues until at least one reactant is completely consumed, the reaction is said to "go to completion." The decomposition of potassium chlorate ($KClO_3$) into potassium chloride (KCl) and oxygen (O_2) is just such a reaction.

$$2KClO_3(s) \rightarrow 2KCl(s) + 3O_2(g)$$

Unlike reactions that go to completion, many reactions under certain conditions can occur in both directions. These reactions are said to be "reversible." A **reversible reaction** is one that can occur in both the forward and reverse directions. Think of a reversible reaction as a two-way street, as shown in **Figure 22.** Vehicles may proceed in both directions at the same time.

When a reversible reaction's forward and reverse reactions take place at exactly the same rate, a state of balance, or equilibrium, exists. **Equilibrium** (plural, *equilibria*) is a state in which forward and reverse reactions or processes proceed at equal rates. An equilibrium state is indicated with double reaction arrows, as shown below. Chemists call the left-to-right reaction the forward reaction and the right-to-left reaction the reverse reaction.

$$\text{reactants} \rightleftharpoons \text{products}$$

Get It?

Contrast the forward and reverse reactions in a reversible reaction.

Types of equilibria

Some equilibria involve physical changes, rather than chemical reactions. When opposing physical changes take place at equal rates, a state of physical equilibrium exists. In a sealed bottle of soda, for example, CO_2 molecules are continually escaping from the solution. At the same time—and at an identical rate—CO_2 molecules are reentering the solution. This state of physical equilibrium is shown in **Figure 23.**

$$CO_2(aq) \rightleftharpoons CO_2(g)$$

Likewise, when opposing chemical reactions take place at equal rates, a state of chemical equilibrium exists. An example is the chemical equilibrium in the Haber process, which is used to manufacture ammonia (NH_3) by reacting nitrogen (N_2) with hydrogen (H_2).

$$N_2(g) + 3H_2(g) \rightleftharpoons 2NH_3(g) + \text{energy}$$

When this reaction is at equilibrium, ammonia is constantly being formed. At the same time and at the same rate, nitrogen and hydrogen molecules are being reformed.

Figure 23 In a sealed bottle of soda, dissolved CO_2 molecules are continually coming out of solution and reentering solution.

Infer *What happens to the equilibrium when the bottle is opened?*

WORD ORIGINS

equilibrium

comes from the Latin word *aequus*, meaning *equal*

When reactants and products are created at equal rates, the reaction is said to be in a state of equilibrium.

Factors affecting equilibria

When a state of equilibrium exists, the forward and reverse reactions are taking place at equal rates. The net amounts of both reactants and products remain constant. In the previous example of the sealed bottle of soda, what would happen if the bottle were opened? As you can assume, the system would no longer be at equilibrium and, for a time, physical changes would proceed toward the right in the following equation:

$$CO_2(aq) \rightleftharpoons CO_2(g)$$

Chemical reactions at equilibrium can likewise change. An equilibrium system may be subjected to stresses that speed up or slow down one of the opposing reactions. Instead of remaining constant, the net amounts of reactants and products favor one of the directions of the reaction. The equilibrium becomes temporarily unbalanced. But in time, the forward and reverse reactions again reach a state of balance. A new equilibrium state is established, now with changed amounts of reactants and products.

When a stress is imposed on an equilibrium system, the equilibrium responds to the stress according to a general rule known as Le Châtelier's (luh SHAHT uhl yays) principle. **Le Châtelier's principle** states that if a stress is applied to a system at equilibrium, the equilibrium shifts in the direction that opposes the stress. A stress is any kind of change that disturbs the equilibrium. Common stresses include the following: changing concentration by adding or removing a reactant or product; changing temperature by adding or removing heat, as shown in **Figure 24;** and changing volume and pressure. When a forward or reverse reaction rate increases or decreases in response to a stress, the equilibrium is said to "shift."

Charles D. Winters/Science Source

Figure 24 In the reaction between dinitrogen tetroxide and nitrogen dioxide ($N_2O_4 \rightleftharpoons 2NO_2$), the equilibrium shifts toward the reddish-brown NO_2 when placed in boiling water (left) and toward the colorless N_2O_4 when placed in the ice water bath (right).

How could a chemical engineer apply Le Châtelier's principle to maximize the production of ammonia (NH_3) in the following equilibrium system?

$$N_2(g) + 3H_2(g) \rightleftharpoons 2NH_3(g) + \text{energy}$$

Changing concentration Suppose that the manufacturing process is engineered to remove ammonia (NH_3) as it is formed. The concentration of ammonia decreases, which causes the rate of the reverse reaction to decrease. As a result, the forward reaction is temporarily faster than the reverse reaction—described as a shift to the right—and more ammonia is formed.

Changing temperature Suppose that the engineer lowers the temperature, thus removing energy from the system. The equilibrium responds by reacting to release energy and raise the temperature. A shift to the right occurs, resulting in an increase in ammonia production.

Changing volume and pressure Because the reaction vessel contains gases, decreasing the volume increases the pressure. If possible, the equilibrium responds to reduce the pressure. The pressure can be reduced by decreasing the number of gas molecules. Because the product (NH_3) side of the equation has fewer gas molecules (2) than the reactant side (4), the equilibrium shifts to the right. More ammonia is formed as a result.

Check Your Progress

Summary

- The rates of chemical reactions can be manipulated by changing the conditions under which the reaction takes place.
- A state of equilibrium exists when forward and reverse reactions or processes take place at equal rates.
- Le Châtelier's principle describes how an equilibrium responds to a stress.

Demonstrate Understanding

26. **List** four ways to change the rate of a chemical reaction.
27. **Describe** two ways in which you might state the rate of a chemical reaction.
28. **Explain** what must happen in order for two molecules to react.
29. **Compare and contrast** chemical and physical equilibrium.

Explain Your Thinking

30. **Apply** Describe two ways you could influence the following equilibrium to produce more ethanal (CH_3CHO). Use Le Châtelier's principle to explain why each of your methods would produce the desired result.

$$C_2H_2(g) + H_2O(g) \rightleftharpoons CH_3CHO(g) + \text{energy}$$

31. **MATH Connection** For the reaction described in question 30, the concentration of CH_3CHO is found to increase from 0.0300 mol/L to 0.0500 mol/L in 42.5 seconds. Express the average rate of the reaction in mol CH_3CHO produced/L•s.

LEARNSMART® Go online to follow your personalized learning path to review, practice, and reinforce your understanding.

Lavoisier

Antoine-Laurent Lavoisier, shown at right with his wife, was perhaps the greatest chemist of his generation and one of the most influential scientific thinkers of all time. His accomplishments are even more remarkable considering the chaotic and violent era in which they were achieved.

Lavoisier and his wife spent many long hours working in his laboratory.

Government bureaucrat As a young man, Lavoisier participated in the first comprehensive geologic survey of France. Shortly after this, he led a government committee to reform the gunpowder industry. The resulting surplus of gunpowder helped American revolutionaries win independence from the British. Over the next 30 years, he worked in the French taxation system and became very wealthy.

Scientific research Lavoisier used his fortune to construct one of the greatest laboratories in Europe. State-of-the-art equipment and an unlimited budget allowed him to perform investigations such as the combustion of diamonds depicted at lower left. Lavoisier published *Elements of Chemistry* in 1789. The book marked the birth of modern chemistry, featuring the first table of elements, a system for naming compounds, and extensive descriptions of laboratory techniques.

The lenses of this enormous device focus sunlight, creating enough heat to combust diamonds.

The French Revolution Lavoisier lived much of his adult life in a time of intense social and political upheaval. As the French Revolution became violent, Lavoisier's membership in the Academy of Sciences, and his association with the monarchy and the aristocracy, made him a target of the violence. But it was his role as a tax collector that ultimately cost him his life. By all accounts, Lavoisier was an honest bureaucrat. He did not levy excessive taxes, he did not steal from the government, and he even advocated for a more equitable tax system. But on May 8, 1794, Lavoisier became one of thousands to be beheaded during the Reign of Terror.

SEP DCI CCC

OBTAINING, EVALUATING, AND COMMUNICATING INFORMATION

Biography Lavoisier was a world-renowned scientist. Yet, due to the political atmosphere of the time, his death was marked by little fanfare. Write a biographical article about Lavoisier, highlighting his contributions to chemistry.

MODULE 19
STUDY GUIDE

GO ONLINE to study with your Science Notebook.

Lesson 1 CHEMICAL CHANGES

- A chemical reaction is a process that involves one or more reactants changing into one or more products.
- A balanced chemical equation indicates relative amounts of reactants and products.
- A mole (mol) is the amount of a substance that contains 6.02×10^{23} particles of that substance.

- chemical reaction
- reactants
- products
- chemical equation
- coefficient
- balanced chemical equation
- mole
- molar mass

Lesson 2 CLASSIFYING CHEMICAL REACTIONS

- Chemical reactions are organized into five basic classes: combustion, synthesis, decomposition, single displacement, and double displacement.
- An activity series predicts which metal will replace another in a single-displacement reaction.
- Some reactions produce a solid called a precipitate when two ionic substances are combined.
- Oxidation is the loss of electrons, and reduction is the corresponding gain of electrons.

- combustion reaction
- synthesis reaction
- decomposition reaction
- single-displacement reaction
- double-displacement reaction
- precipitate
- oxidation
- reduction

Lesson 3 CHEMICAL REACTIONS AND ENERGY

- Breaking chemical bonds absorbs energy.
- Forming chemical bonds releases energy.
- Exergonic chemical reactions release energy. Endergonic reactions absorb energy.
- Exothermic reactions give off thermal energy. Endothermic reactions absorb energy.

- exergonic reaction
- exothermic reaction
- endergonic reaction
- endothermic reaction

Lesson 4 REACTION RATES AND EQUILIBRIUM

- The rates of chemical reactions can be manipulated by changing the conditions under which the reaction takes place.
- A state of equilibrium exists when forward and reverse reactions or processes take place at equal rates.
- Le Châtelier's principle describes how an equilibrium responds to a stress.

- reaction rate
- collision model
- catalyst
- inhibitor
- reversible reaction
- equilibrium
- Le Châtelier's principle

THREE-DIMENSIONAL THINKING

Module Wrap-Up

REVISIT THE PHENOMENON

What chemical reactions occur when you bake cupcakes?

CER Claim, Evidence, Reasoning

Explain Your Reasoning Revisit the claim you made when you encountered the phenomenon. Summarize the evidence you gathered from your investigations and research and finalize your Summary Table. Does your evidence support your claim? If not, revise your claim. Explain why your evidence supports your claim.

STEM UNIT PROJECT

Now that you've completed the module, revisit your STEM unit project. You will summarize your evidence and apply it to the project.

GO FURTHER

SEP Data Analysis Lab

Enzyme Catalase

Hydrogen peroxide (H_2O_2) forms naturally in living organisms but can damage the cell if it is not broken down. The enzyme catalase is created by our cells to immediately break down hydrogen peroxide into oxygen and water.

Materials

test tubes, hydrogen peroxide (3%), assorted living material: sliced raw potato, liver, yeast, etc., assorted nonliving material: slice of cooked potato, cooked liver, etc.

Procedure

1. Label each test tube. You should have two test tubes for each material to be tested: one for the living material and one for the nonliving material to serve as a control.
2. Fill each test tube 1/3 full with fresh hydrogen peroxide.
3. Add a small amount of the material to be tested.
4. Note whether or not bubbles are produced.

CER Analyze and Interpret Data

1. **Claim** Which test tubes produced bubbles?
2. **Claim** Write the chemical reaction that occurred when bubbles were produced.
3. **Claim** What product from the chemical reaction was being released in the bubbles that formed?
4. **Evidence, Reasoning** Hydrogen peroxide is often used to treat cuts. Explain why hydrogen peroxide bubbles when it comes into contact with a cut.

H. Mark Weidman Photography/Alamy Stock Photo

RADIOACTIVITY AND NUCLEAR REACTIONS

ENCOUNTER THE PHENOMENON

How does a nuclear reactor work?

GO ONLINE to play a video about the discovery of nuclear fission.

SEP Ask Questions

Do you have other questions about the phenomenon? If so, add them to the driving question board.

CER Claim, Evidence, Reasoning

Make Your Claim Use your CER chart to make a claim about how a nuclear reactor works.

Collect Evidence Use the lessons in this module to collect evidence to support your claim. Record your evidence as you move through the module.

Explain Your Reasoning You will revisit your claim and explain your reasoning at the end of the module.

GO ONLINE to access your CER chart and explore resources that can help you collect evidence.

LESSON 1: Explore and Explain: Describing the Nucleus

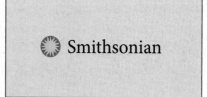

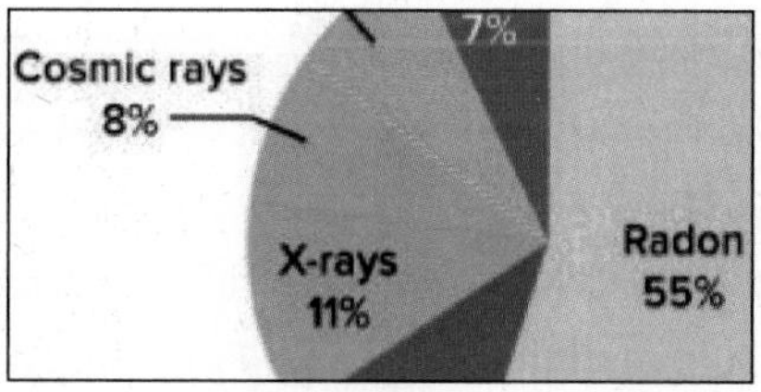

LESSON 3: Explore and Explain: Background Radiation

Smithsonian

Additional Resources

LESSON 1

THE NUCLEUS

FOCUS QUESTION

How do you break an atom?

Describing the Nucleus

Recall that atoms are composed of protons, neutrons, and electrons. The nucleus of an atom is composed of protons and neutrons. Protons have a positive electric charge. Neutrons have no electric charge. So, the number of protons in a nucleus determines that nucleus's total charge. Negatively charged electrons are attracted to the positively charged nucleus and swarm around it. An electron has a charge that is equal but opposite to a proton's charge. Atoms contain the same number of protons as electrons.

Size of the nucleus

The protons and neutrons that make up a nucleus are packed together tightly. The region outside the nucleus where the electrons are located is large compared to the size of the nucleus. As **Figure 1** helps show, the nucleus occupies only a tiny fraction of the space in an atom. But the nucleus has almost all of the atom's mass. Neutrons are slightly more massive than protons. However, each is almost 2000 times as massive as an electron.

Figure 1 The size of a nucleus in an atom can be compared to a marble sitting in the middle of an empty soccer stadium. The diameter of an atom is approximately 10,000 times greater than the diameter of an atomic nucleus.

3D THINKING DCI Disciplinary Core Ideas CCC Crosscutting Concepts SEP Science & Engineering Practices

COLLECT EVIDENCE

Use your Science Journal to record the evidence you collect as you complete the readings and activities in this lesson.

INVESTIGATE

GO ONLINE to find these activities and more resources.

Quick Investigation: Model Forces between Protons
Develop and use a model to identify **the result** of electromagnetic forces.

Review the News
Obtain information from a current news story about nuclear physics. **Evaluate** your source and **communicate** your findings to your class.

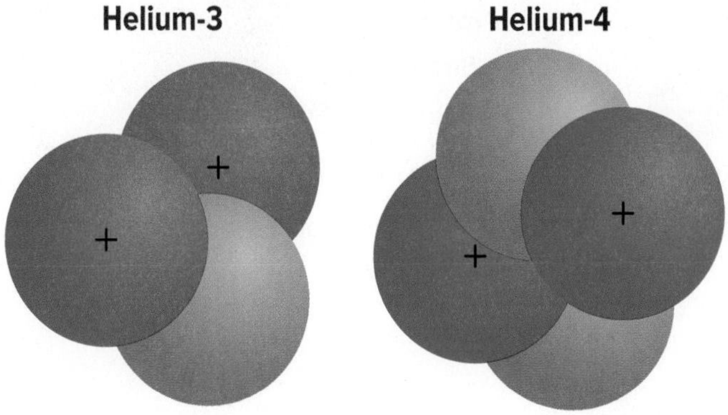

Figure 2 These two isotopes of helium each have the same number of protons, but different numbers of neutrons. Most helium nuclei have two neutrons, but a few have only one neutron.

Calculate *the ratio of neutrons to protons in each of these isotopes of helium.*

Isotopes

The atoms of an element all have the same number of protons in their nuclei. For example, the nuclei of all carbon atoms have six protons. However, naturally occurring carbon nuclei can have six, seven, or eight neutrons.

Isotopes are nuclei that have the same number of protons but different numbers of neutrons. Each element has many different isotopes. For example, the element carbon has three isotopes that occur naturally. The atoms of all isotopes of an element have the same chemical properties. However, each isotope has its own nuclear properties. For example, some carbon isotopes are radioactive, but other carbon isotopes are not radioactive. **Figure 2** shows two isotopes of helium.

Nucleus numbers

You can describe a nucleus with its numbers of protons and neutrons. The atomic number is the number of protons that are a part of a nucleus. The total mass of all the protons and neutrons that make up a nucleus is nearly the same as the mass of the atom. As a result, the number of protons plus neutrons is called the mass number.

Get It?

Define the term *atomic number*.

You can represent a nucleus with its atomic number, mass number, and the symbol of the element to which it belongs. The representation for the nucleus of the most common isotope of carbon is shown below as an example.

$$\begin{matrix}\text{mass number} \rightarrow \\ \text{atomic number} \rightarrow\end{matrix} {}^{12}_{6}\text{C} \leftarrow \text{element symbol}$$

This isotope is called carbon-12. The number of neutrons equals the mass number minus the atomic number. The number of neutrons that are a part of carbon-12 is $12 - 6 = 6$. The nucleus of carbon-12 is composed of six protons and six neutrons. Look at **Figure 2** again. How many protons does helium-4 have? How many neutrons does helium-4 have? What is the total number of protons plus neutrons?

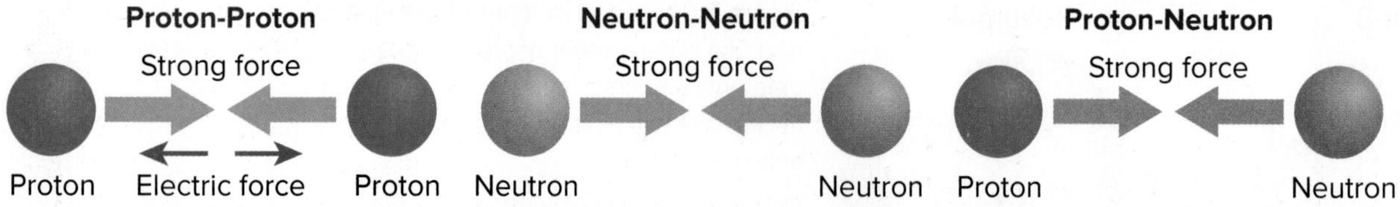

Figure 3 The particles in the nucleus are attracted to each other by the strong force. The strong force is one of four fundamental forces in nature. The other three include the weak force, the electromagnetic force, and gravity.

Forces in the Nucleus

How are the protons and neutrons that make up a nucleus held together so tightly? Positive electric charges repel each other, so why do the protons that are part of a nucleus not push each other away? The **strong force** is the force that causes protons and neutrons to be attracted to each other. **Figure 3** illustrates the strong force between protons and neutrons.

The strong force is one of the four basic forces in nature. It is about 100 times stronger than the electromagnetic force. The attractive forces between all of the protons and neutrons that make up a nucleus keep the nucleus together.

Get It?

Identify the force that produces the attraction between protons and neutrons.

However, it is important to remember that protons and neutrons have to be extremely close to be attracted by the strong force. The strong force is a short-range force that quickly becomes extremely weak as protons and neutrons get farther apart. The electromagnetic force is a long-range force, so protons that are farther apart are still repelled by the electric force, as shown in **Figure 4.**

Get It?

Compare the strong force between two protons that are very close together with the electromagnetic force between the same two protons.

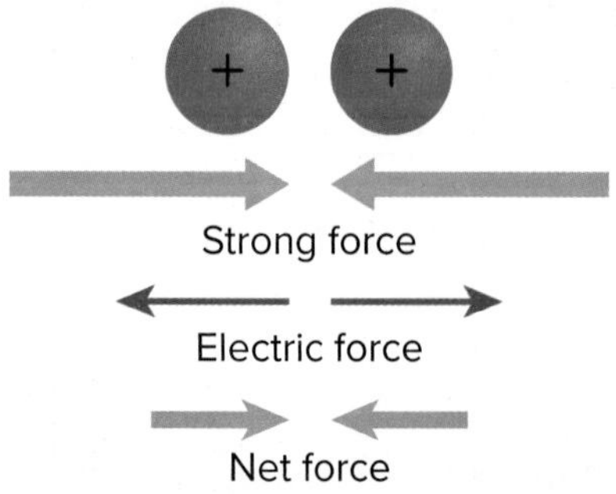

When protons are close together, they are attracted to each other. The attraction due to the short-range strong force is much stronger than the repulsion due to the long-range electric force.

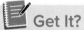

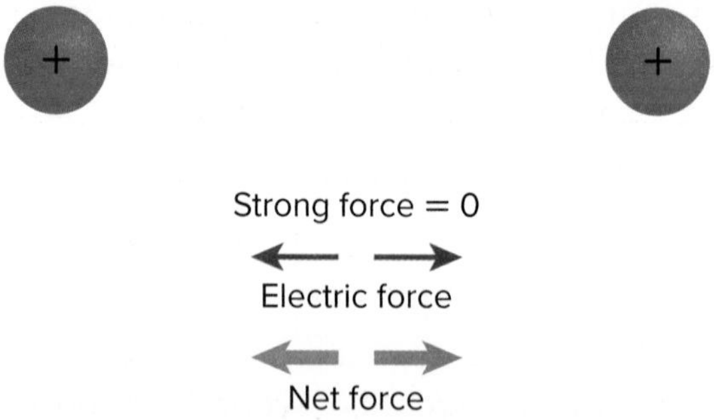

When protons are too far apart to be attracted by the strong force, they still are repelled by the electric force between them. Then the net force between them is repulsive.

Figure 4 The total force between two protons depends on how far apart those protons are. The strong force is very strong but also very short-ranged. The electromagnetic force is much weaker but can act over much greater distances.

Infer *whether the net force between two protons could ever be zero.*

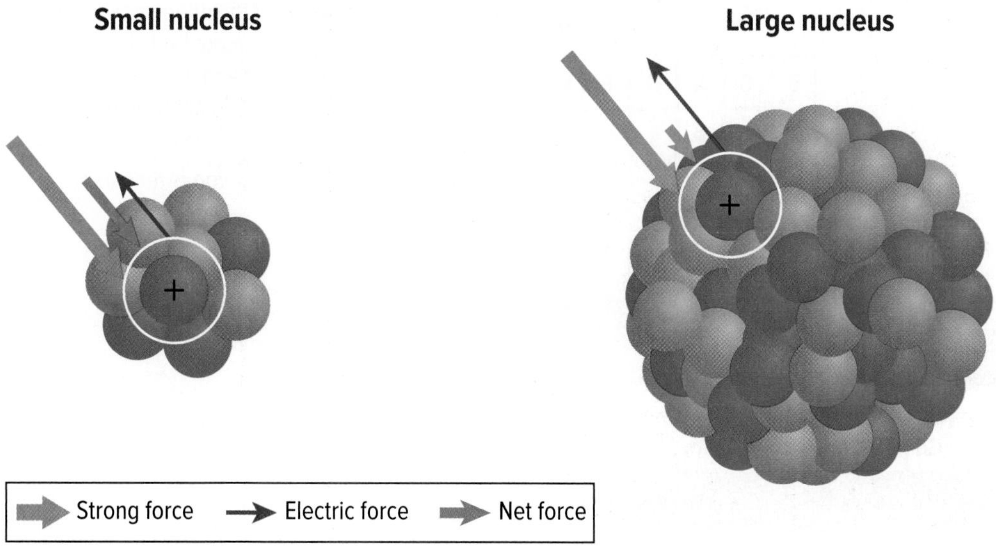

Figure 5 For nuclei with few protons, the repulsive force on a proton due to the other protons is small. For nuclei with many protons, the attractive strong force is exerted only by the nearest neighbors, but all of the protons exert repulsive forces. The total repulsive force is then large.

Compare *the net force holding the smaller nucleus together with the net force holding the larger nucleus together.*

Nuclei with few protons

The left side of **Figure 5** shows the forces in a small nucleus—one with relatively few protons. If a nucleus has only a few protons and neutrons, they are all close enough together to be attracted to each other by the strong force. Because only a few protons are present, the total electric force repelling the protons from each other is small. As a result, the net forces between the protons and the neutrons hold the nucleus together tightly.

Nuclei with many protons

Some nuclei, such as uranium nuclei, are composed of many protons and neutrons. In these cases, each proton or neutron is attracted to only a few neighbors by the strong force, as shown on the right of **Figure 5.** The other protons and neutrons are too far away. Therefore, the strong force holding a proton or neutron in place in a large nucleus is about the same as for a small nucleus.

However, all of the protons in a large nucleus exert a repulsive electric force on each other. Thus, the electric repulsive force on a proton in a nucleus with many protons is large. As a result, a nucleus with many protons is held together less tightly than a nucleus with fewer protons.

Get It?

Explain why, in a large nucleus, the strong force holding a proton in place is about the same as for a small nucleus, but the electric force on the proton is greater.

Neutron to proton ratios

There are no repulsive electric forces between neutrons. Why are there no atomic nuclei composed completely of neutrons? Such an atomic nucleus would not have a stable ratio of neutrons to protons. In less massive elements, an isotope is stable when the ratio is about 1:1.

In more massive elements, an isotope is stable when the ratio of neutrons to protons is about 3:2. Nuclei with too many neutrons compared to the number of protons are unstable. Larger nuclei have a higher neutron to proton ratio because neutrons contribute to the attractive strong force but not to the repulsive electric force within the nucleus.

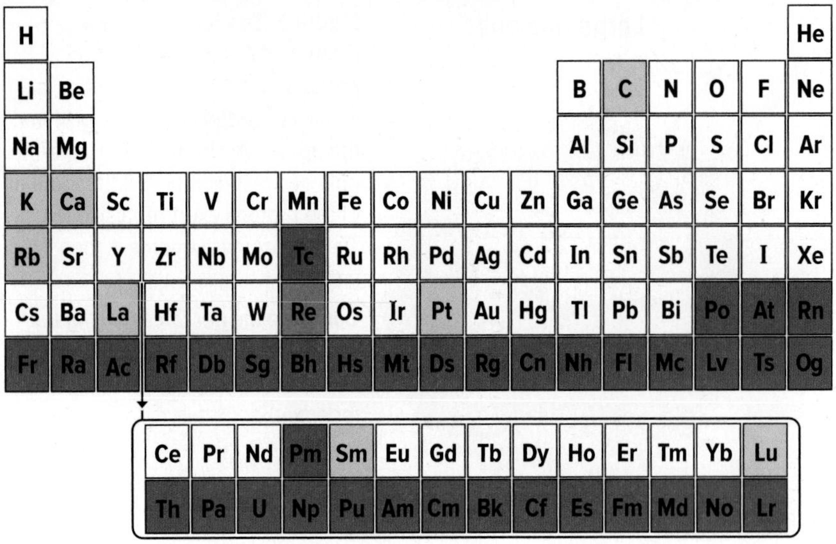

Figure 6 In this diagram of the periodic table, the redder the element box, the more likely it is that an isotope of that element will be radioactive. The white boxes indicate elements that do not have measurable percentages of radioactive isotopes.

Identify *where radioactive isotopes tend to be on the periodic table.*

Radioactivity

In most nuclei, the strong force is able to keep the nucleus permanently together. These nuclei are stable. When the strong force is not large enough to hold a nucleus together tightly, the nucleus can decay. When a nucleus decays, it emits matter and energy. **Radioactivity** is the process of nuclei decaying and emitting matter and energy. Although nuclear processes, such as radioactive decay, involve the release or absorption of energy, they do not change the total number of neutrons plus protons.

All nuclei that contain more than 83 protons are radioactive. However, some nuclei that contain fewer than 83 protons are also radioactive, such as carbon-14. In addition, no nuclei with more than 92 protons is stable enough to occur naturally. Instead, people must synthesize these elements, usually in a lab. **Figure 6** shows which elements have measurable percentages of radioactive isotopes.

Check Your Progress

Summary

- Isotopes of an element have the same number of protons but different numbers of neutrons.
- The atomic number is the number of protons in a nucleus. The mass number is the number of protons and neutrons in a nucleus.
- A nucleus might decay due to its neutron to proton ratio.
- Radioactivity is the process of nuclear decay.

Demonstrate Understanding

1. **Compare** the properties of the strong force to the properties of the electromagnetic force.
2. **Compare** the forces in a small nucleus to the forces in a large nucleus.
3. **Explain** why large nuclei tend to be radioactive.

Explain Your Thinking

4. **Explain** whether you would expect helium-6 to be radioactive or stable.
5. **MATH Connection** What is the approximate ratio of neutrons to protons in a nucleus of radon-222?

LEARNSMART® Go online to follow your personalized learning path to review, practice, and reinforce your understanding.

LESSON 2

NUCLEAR DECAYS AND REACTIONS

FOCUS QUESTION

How can we get electricity from breaking atoms?

Nuclear Decays

When a nucleus decays, the total number of protons and neutrons does not change as particles and energy are emitted from it. We call this emission *nuclear radiation*. Three types of nuclear radiation are alpha, beta, and gamma radiation. Alpha and beta radiation are composed of particles. Gamma radiation is composed of electromagnetic waves.

Alpha particles

When the strong force is not strong enough to hold a nucleus together, that nucleus emits alpha particles. An **alpha particle** is a particle that is composed of two protons and two neutrons. An alpha particle is the same as a helium-4 nucleus.

Get It?

Identify the components of an alpha particle.

Alpha particles are extremely massive when compared with other nuclear radiation. For example, an alpha particle is about 7000 times more massive than a beta particle. Alpha particles also have twice as much charge as beta particles. Because of their high mass and charge, alpha particles interact with other matter frequently. As a result, alpha particles transfer energy to their surroundings very quickly as they travel through solids, liquids, and gases. Alpha particles are the least penetrating form of nuclear radiation. A sheet of paper will stop most alpha particles. **Table 1** summarizes the properties of alpha particles.

Table 1 Alpha Particle

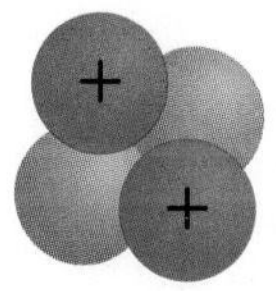	**Description**	high-energy helium-4 nucleus
	Symbol	$^{4}_{2}He$
	Mass	approximately 4 hydrogen atoms
	Charge	+2
	Can be stopped by...	a sheet of paper

3D THINKING DCI Disciplinary Core Ideas CCC Crosscutting Concepts SEP Science & Engineering Practices

COLLECT EVIDENCE

Use your Science Journal to record the evidence you collect as you complete the readings and activities in this lesson.

INVESTIGATE

GO ONLINE to find these activities and more resources.

Revisit the Encounter the Phenomenon Question
What information from this lesson can help you answer the Unit and Module questions?

CCC **Identify Crosscutting Concepts**
Create a table of the crosscutting concepts and fill in examples you find as you read.

Table 2 Beta Particle

	**Description**	high-energy electron
	Symbol	$^{0}_{-1}e$
	Mass	approximately 1/7000th the mass of an alpha particle
	Charge	−1
	Can be stopped by...	a sheet of aluminum that is 3 mm thick

Beta particles and the weak force

So far, you have learned about three fundamental forces in nature. They are the gravitational force, the electromagnetic force, and the strong force. The fourth and final fundamental force is called the weak force. The weak force causes beta decay.

Like the strong force, the weak force is very short-ranged. The weak force is also weaker than all the other fundamental forces except for gravity. However, the weak force can cause neutrons to decay into protons when the neutron-to-proton ratio in a nucleus is too high. When this decay happens, the nucleus emits an electron. A **beta particle** is a high-energy electron that is emitted when a neutron decays into a proton.

Beta particles are much faster and more penetrating than alpha particles. It takes a 3-mm thick sheet of aluminum to absorb most beta radiation. **Table 2** summarizes the properties of beta particles.

Gamma rays

Gamma rays are extremely high-energy electromagnetic waves. Lead bricks or other heavy materials are necessary to stop gamma rays. They are usually emitted from a nucleus along with alpha decay and beta decay. Gamma rays have no mass and no charge and travel at the speed of light. A nucleus releases energy, but no particles, during gamma decay. **Table 3** summarizes the properties of gamma rays.

Table 3 Gamma Ray

	**Description**	high-energy, high-frequency electromagnetic wave
	Symbol	γ
	Mass	0
	Charge	0
	Can be stopped by...	thick blocks of lead

ACADEMIC VOCABULARY

fundamental

of or relating to essential structure, function, or facts

Some scientists study and try to understand the most fundamental laws of nature.

STEM CAREER Connection

Nuclear Technician

As a nuclear technician, you might work with engineers and physicists to conduct nuclear research or in a nuclear power plant. Nuclear technicians must learn to operate specialized equipment to monitor radiation levels and take special precautions to avoid exposure to radiation.

Damage from nuclear decay

Alpha, beta, and gamma radiation can all be dangerous to human tissue. Biological molecules inside your body are large and easily damaged. A single alpha or beta particle or gamma ray can damage many fragile biological molecules. Damage from radiation can cause cells to function improperly, leading to illness and disease.

Transmutation

Another result of nuclear decay is transmutation. Transmutation occurs during alpha and beta decay. **Transmutation** is the process of changing one element into a different element.

During alpha decay, a nucleus loses two protons and two neutrons. The resulting nucleus has two fewer protons and two fewer neutrons. The nucleus transmutes into an entirely different element. The new element has an atomic number two less than that of the original element. The mass number of the new element is four less than that of the original element. The top half of **Figure 7** shows a transmutation caused by alpha decay. Note that in alpha decay, there is a release of energy, but the total number of neutrons plus protons does not change.

Get It?

Identify the changes in the nucleus that result from transmutation caused by alpha decay, and explain how they affect the total number of protons and neutrons.

During beta decay, a neutron emits an electron and becomes a proton. The resulting nucleus has one more proton. It becomes the element with an atomic number one greater than that of the original element. However, the total number of protons and neutrons does not change during beta decay. Therefore, the mass number of the new element is the same as that of the original element. The bottom half of **Figure 7** shows a transmutation caused by beta decay.

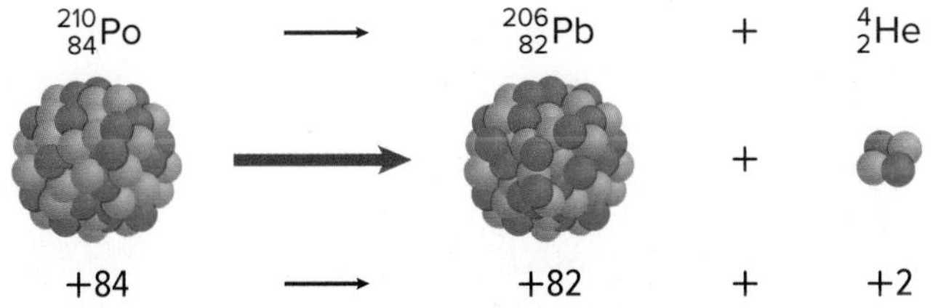

Due to alpha decay, polonium changes into lead.

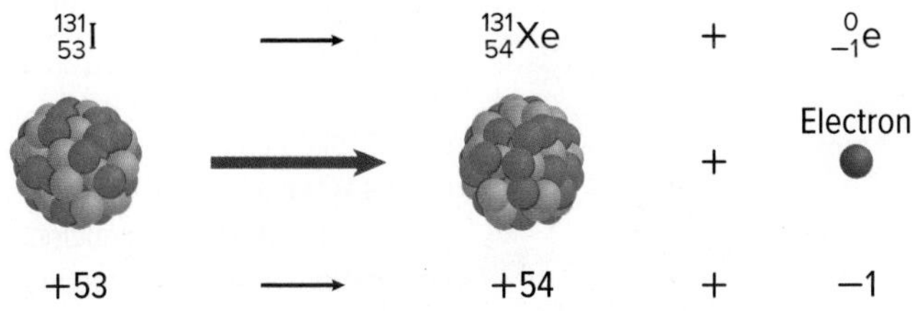

Due to beta decay, iodine changes into xenon.

Figure 7 Elements transmute into new elements whenever they emit alpha or beta particles. Gamma rays do not have mass or charge. Therefore, a nucleus will not transmute into another nucleus if it emits only gamma rays. However, gamma rays are often emitted along with alpha or beta particles.

Compare *the total charge before a radioactive decay with the total charge after a radioactive decay.*

Nuclear Fission

Energy is released whenever a nucleus emits nuclear radiation. However, the amount of energy released during a nuclear decay is very small compared to the amount of energy that can be released during nuclear fission.

Recall that nuclear fission is the process of splitting a nucleus into two or more smaller nuclei. Scientists can do this by bombarding the larger nucleus with neutrons. **Figure 8** shows a nuclear fission reaction for uranium-235. Other isotopes, including plutonium-239 and uranium-233, also undergo nuclear fission. The nuclear bomb that was dropped on Hiroshima during World War II was powered by the fission of uranium-235.

Chain reactions

The products of a fission reaction usually include several neutrons in addition to the smaller nuclei. These neutrons can then strike other nuclei in the sample, causing them to split as well. These reactions then release more neutrons, causing additional nuclei to split. A **chain reaction** is a series of repeated fission reactions caused by the release of additional neutrons in every fission. A chain reaction is shown in **Figure 9.**

If a chain reaction is uncontrolled, an enormous amount of energy is released in an instant. This is what happens when a nuclear fission bomb is detonated. However, a chain reaction can be controlled by adding materials that absorb neutrons. If enough neutrons are absorbed, the reaction will be controlled. This is how a nuclear power plant operates.

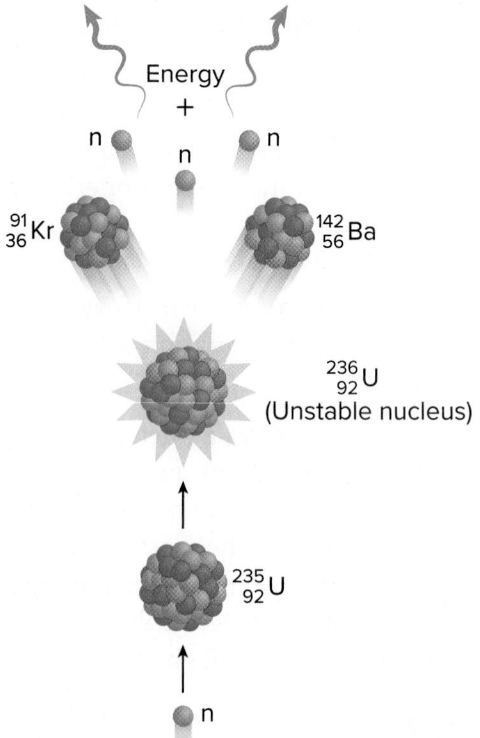

Figure 8 Some large nuclei split apart when they absorb a neutron. Here, a neutron splits a uranium-235 nucleus into two smaller nuclei and three more neutrons.

Predict *whether carbon-12 would be likely to undergo nuclear fission.*

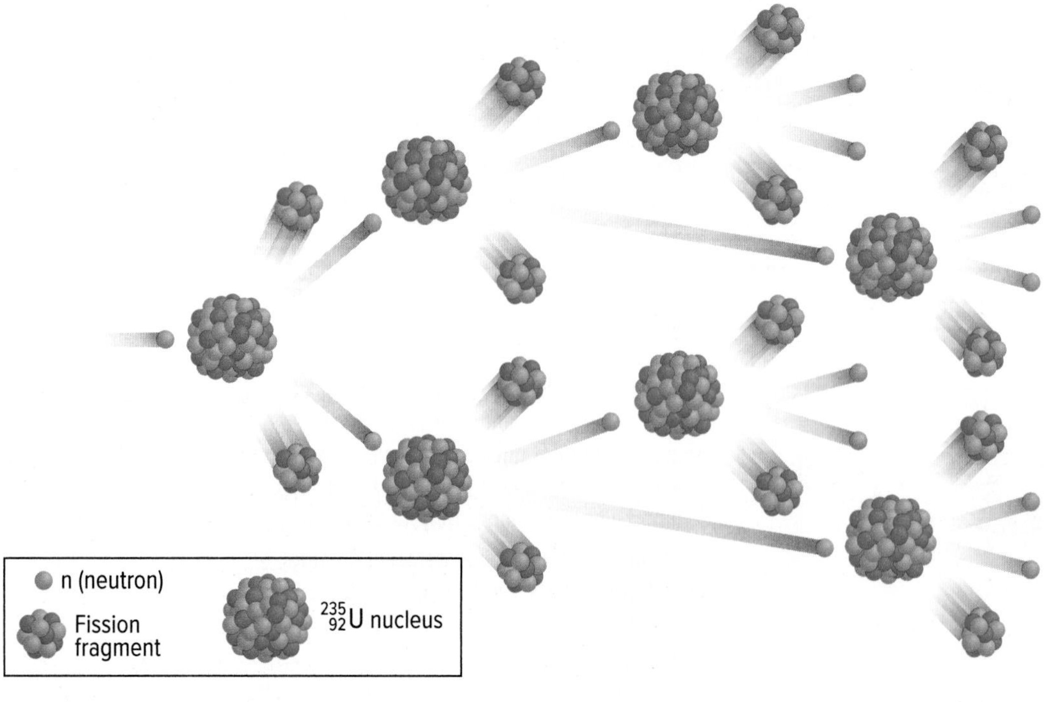

Figure 9 A chain reaction occurs when neutrons emitted from a split nucleus cause other nuclei to split and emit additional neutrons. An uncontrolled chain reaction occurs when a nuclear fission bomb is detonated. In a nuclear power plant, the chain reaction is controlled with rods of material that absorb some of the neutrons.

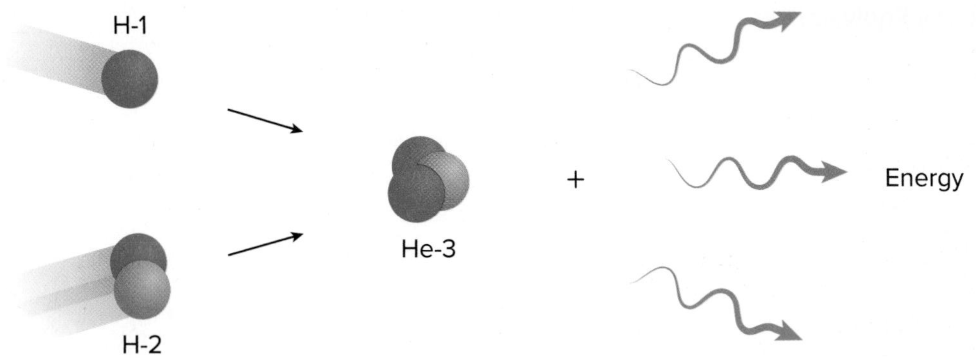

Figure 10 One form of fusion is shown here. Helium-3 is produced when a hydrogen-1 isotope combines with a hydrogen-2 isotope. This reaction is common inside the Sun.

Nuclear Fusion

Nuclear fusion reactions can release even more energy than nuclear fission reactions can. Recall that nuclear fusion is the process of two or more nuclei combining to form a nucleus of greater mass. **Figure 10** shows an example of nuclear fusion that occurs inside the Sun. Fission splits nuclei apart. Fusion fuses nuclei together.

Temperature and fusion

For nuclear fusion to occur, nuclei must get close to each other. However, all nuclei have positive electric charge. Therefore, they repel each other. In order to fuse, the nuclei need to have enough kinetic energy to overcome this repulsion.

Remember that the kinetic energy of atoms increases as their temperature increases. Only at temperatures of millions of degrees Celsius are nuclei moving so fast that they can get close enough for fusion to occur. These extremely high temperatures are found in the centers of stars, including the Sun.

Get It?

Explain why the extremely high temperatures found in the centers of stars are needed for fusion to occur in stars.

Every atom that exists, other than hydrogen-1, was originally constructed through nuclear fusion. This nuclear fusion occurs in the cores of stars and during the explosions of those stars. Nuclear fusion also occurred in the first moments after the Big Bang. Thermal energy and temperature are extremely high in each of these situations.

The Sun and fusion

The Sun emits more than 3.8×10^{26} J of energy every second. This radiant energy is first extracted from hydrogen nuclei in the Sun's core through a series of nuclear fusion reactions. The most important series of fusion reactions that occurs in the Sun is shown in **Figure 11.** This series is also common to other stars that have masses similar to the mass of the Sun. In more massive stars, another series of reactions is more common.

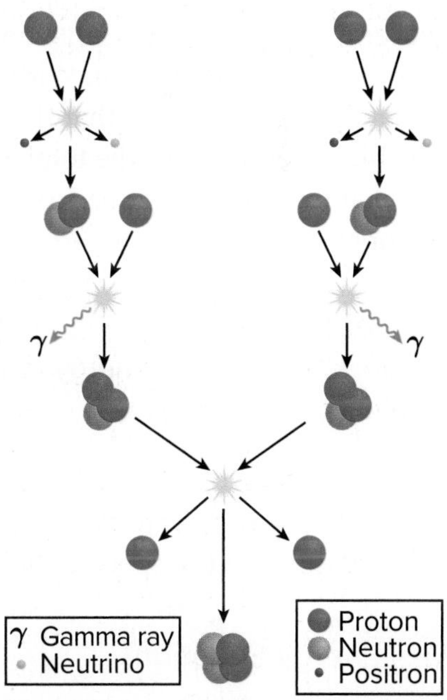

Figure 11 A series of fusion reactions inside the Sun's core provides the Sun's power. A neutrino is a tiny, nearly massless particle. A positron is a particle with the charge of a proton and the mass of an electron.

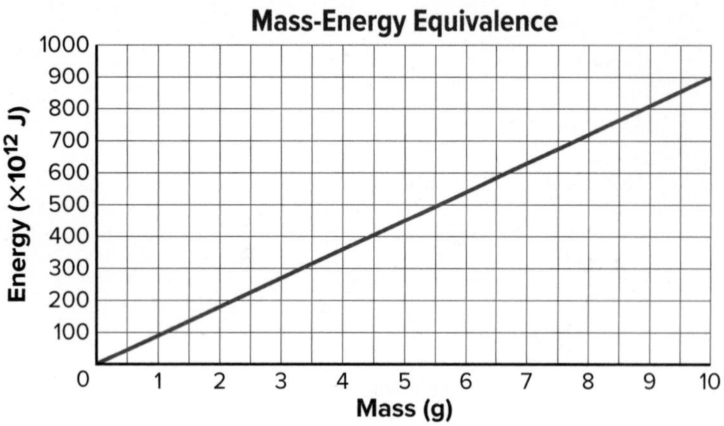

Figure 12 A small amount of mass is the same thing as a very large amount of energy. Albert Einstein first explained mass-energy equivalence in 1905, although others had suggested this connection between mass and energy before then.

Identify *the amount of energy that is equivalent to 2 g of mass.*

Mass and Energy

During the fusion reactions that take place inside the Sun, no matter is ejected from the Sun. However, the total mass of all the particles in the reaction is greater before the reaction than after the reaction. How can there be less mass after a reaction if no matter leaves the Sun? Isn't mass conserved?

Get It?

Compare the masses before and after a nuclear fusion reaction.

According to the theory of special relativity, a tremendous amount of energy is the same thing as a small amount of mass. Mass is energy, and energy is mass. **Figure 12** shows this relationship. One gram of mass is enough energy to launch the Washington Monument into orbit.

However, unless incredibly large energies are involved, the mass-energy relationship is almost impossible to observe. With nuclear fission and fusion reactions, such tremendous energies are involved that the total amount of matter after the reaction can be noticeably different from the total amount of matter before the reaction. This change in the total amount of matter occurs even when the total number of protons plus neutrons does not change.

Get It?

Explain why the mass-energy relationship can be observed in nuclear changes but not in chemical changes.

Converting between mass and energy is just a conversion of units. It is similar to converting from miles per hour to meters per second. To convert from units of mass to units of energy, multiply by the speed of light in a vacuum squared (c^2).

Mass-Energy Equation

units of energy (joules) = [**units of mass** (kg)] × [**speed of light in a vacuum** (m/s)]^2

$$E = mc^2$$

The speed of light in a vacuum is 300,000,000 m/s. The example problem and practice problems on the next page will help you to further explore mass-energy equivalence.

EXAMPLE Problem 1

CONVERT UNITS OF ENERGY TO UNITS OF MASS The Sun emits approximately 3.8×10^{26} J of radiant energy every second. How much mass does the Sun lose every second due to this energy emission?

Identify the Unknown: mass: m

List the Knowns: Energy: $E = 3.8 \times 10^{26}$ J
speed of light in a vacuum: $c = 3.0 \times 10^8$ m/s

Set Up the Problem: $E = mc^2$

Solve the Problem: $3.8 \times 10^{26}\ \text{J} = m(3.0 \times 10^8\ \text{m/s})^2$

$$m = \frac{3.8 \times 10^{26}\ \text{J}}{(3.0 \times 10^8\ \text{m/s})^2}$$

$m = 4.2 \times 10^9$ kg (roughly the mass of 12 Empire State Buildings)

Check the Answer: Put the answer back into the equation $E = mc^2$. Then, $E = 4.2 \times 10^9\ \text{kg} \times (3.0 \times 10^8\ \text{m/s})^2 = 3.8 \times 10^{26}$ J. This is the same as the energy given in the problem, so the math in this problem is correct.

PRACTICE Problems

ADDITIONAL PRACTICE

6. How much energy is equal to 1 kg of mass?
7. **CHALLENGE** The population of the United States releases approximately 17×10^{18} J of energy every year by burning gasoline for trucks and automobiles. Approximately how much mass is this?

Check Your Progress

Summary

- Radioactivity can result in alpha particles, beta particles, or gamma rays.
- Nuclear fission occurs when a neutron strikes a nucleus, causing it to split into smaller nuclei.
- Nuclear fusion occurs when two nuclei combine to form another nucleus.
- A small amount of mass is the same as a tremendous amount of energy.

Demonstrate Understanding

8. **Contrast** the energy that can be released during a nuclear fission reaction with the energy that can be released during a nuclear fusion reaction.
9. **Contrast** alpha particles, beta particles, and gamma rays.
10. **Explain** why mass-energy equivalence is not apparent for chemical reactions.

Explain Your Thinking

11. **Explain** why high temperatures are needed for fusion reactions to occur but not for fission reactions to occur.
12. **MATH Connection** In a chain reaction, two additional fissions occur for each nucleus that is split. If one nucleus is split in the first step of the reaction, how many nuclei will have been split after the fifth step?

LEARNSMART® Go online to follow your personalized learning path to review, practice, and reinforce your understanding.

LESSON 3

RADIATION TECHNOLOGIES AND APPLICATIONS

FOCUS QUESTION

How can radiation both cause and cure cancer?

Detecting Nuclear Radiation

Special equipment is necessary to detect and study nuclear radiation. One such piece of equipment is a Geiger counter. A Geiger counter, as shown in **Figure 13,** has a tube with a positively charged wire running through the center of a negatively charged metal tube. This tube is filled with a low-density gas.

When radiation enters the tube at one end, it knocks electrons from the gas molecules. These electrons then knock more electrons off other gas molecules, producing an electron avalanche. The positive wire attracts these electrons, resulting in a current in that wire. This current is amplified and produces a clicking sound. The frequency of the clicks indicates the intensity of radiation.

Often, scientists want to do more than just detect nuclear radiation. Unlike a Geiger counter, a wire chamber can track the trajectories of subatomic particles as well as detect those particles' presence. A wire chamber works much like a Geiger counter. However, a wire chamber contains an array of positively charged wires instead of just one wire.

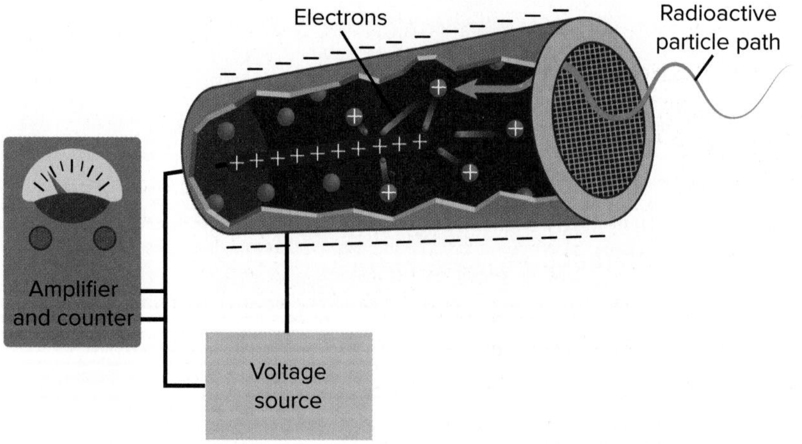

Figure 13 Subatomic particles, including alpha particles and beta particles, can be detected by a Geiger counter. The greater the amount of radiation, the greater the electron avalanche and the greater the number of clicks. Many Geiger counters also include a gauge with a needle that measures the incoming radiation.

3D THINKING DCI Disciplinary Core Ideas CCC Crosscutting Concepts SEP Science & Engineering Practices

COLLECT EVIDENCE

Use your Science Journal to record the evidence you collect as you complete the readings and activities in this lesson.

INVESTIGATE

GO ONLINE to find these activities and more resources.

Applying Practices: Modeling Fission, Fusion, and Radioactive Decay
HS-PS1-8. Develop models to illustrate the changes in composition of the nucleus of the atom and the energy released during the processes of fission, fusion, and radioactive decay.

Background Radiation

You might be surprised to learn that humans have been bathed in radiation for millions of years. This radiation, called background radiation, is not produced by humans. Instead, it is emitted mainly by radioactive isotopes found in Earth's rocks, soils, and atmosphere. This background radiation is low-level but is still detectable.

Building materials, such as bricks, wood, and stones, contain traces of radioactive materials. Traces of naturally occurring radioactive isotopes are also in our food, water, and air. Background radiation is even emitted from inside our own bodies. For example, our bodies contain the isotopes carbon-14 and potassium-40. Both of these isotopes are radioactive.

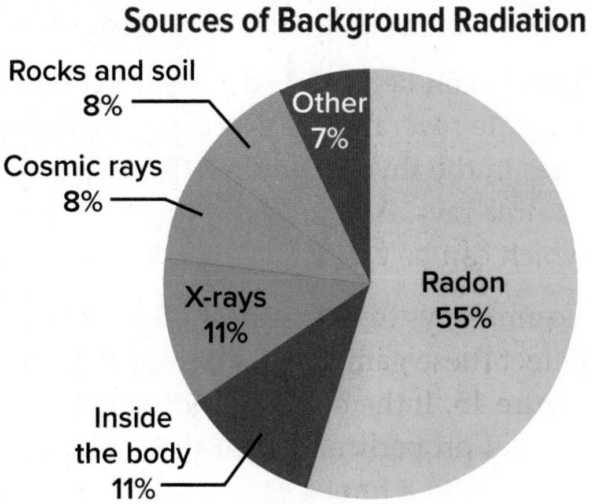

Figure 14 This circle graph shows the sources of background radiation received on average by a person living in the United States. Most of the radiation that originates inside our bodies is from potassium-40, which we consume in our food. Most X-ray exposure is from medical X-rays.

Sources of background radiation

Background radiation comes from several sources, as shown in **Figure 14.** The most common source of background radiation, radon gas, can seep into houses and basements from surrounding soil and rocks. In addition, some background radiation comes from high-speed particles that strike Earth's atmosphere from outer space. These high-speed particles are called cosmic rays.

The amount of background radiation that a person receives can vary greatly. The amount depends on the types of rocks underground, types of materials used to construct a person's home, and the elevation at which a person lives, among other things. However, some amount of background radiation is present for everyone and has been present throughout history and prehistory.

Using Nuclear Radiation in Medicine

It would be easier to find a friend in a crowded area if she told you that she would be wearing a red hat. In a similar way, scientists can find one molecule in a large group of molecules if it is "wearing" something unique. A molecule cannot wear a red hat; however, it can be found easily if it has a radioactive atom in it. A radioactive atom emits radiation that doctors can detect.

A **tracer** is a radioactive isotope that doctors use to locate molecules in an organism. Doctors use tracers to follow where particular molecules go in a human body and to study how organs function. This might seem harmful, but the radiation levels are too low to be harmful or dangerous. Agricultural scientists also use tracers in agriculture to monitor the uptake of nutrients and fertilizers. Common tracers include technetium-99m and iodine-131. These are useful tracers because they emit gamma rays that medical imaging equipment can easily detect.

SCIENCE USAGE v. COMMON USAGE

tracer

Science usage: a substance used to follow the course of a chemical or biological process
With tracers, doctors can track the progress of substances through your digestive system.

Common usage: a person who looks for missing persons or property
She hired a tracer to help find her missing dog.

CCC CROSSCUTTING CONCEPTS

Energy and Matter Research the use of radioactive tracers. Select one and develop a presentation that describes how the energy from the tracer is used to track matter in a system.

Iodine tracers in the thyroid

Tracers can be used to detect problems in your thyroid, which helps regulate several body processes, including growth. Iodine accumulates in the thyroid. Iodine-131, a radioactive isotope of iodine, emits gamma rays. A patient can ingest a capsule that contains iodine-131, which can be easily absorbed by the patient's thyroid.

Gamma rays from the iodine-131 penetrate the skin. Doctors can detect these gamma rays, producing an image like the one shown in **Figure 15.** If the detected radiation is not intense, then the thyroid has not properly absorbed the iodine-131. This could be due to the presence of a tumor.

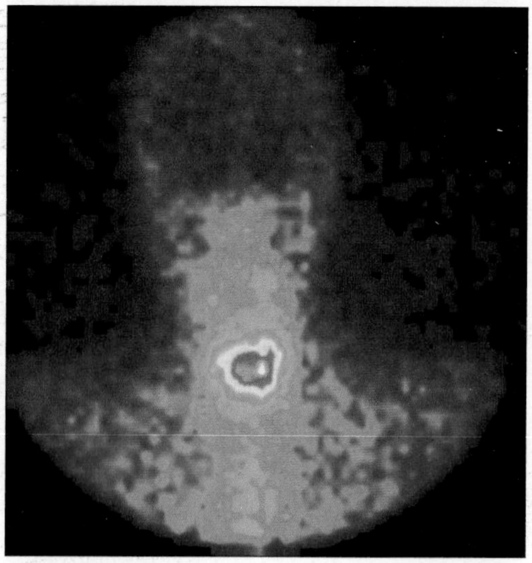

Figure 15 Radioactive iodine-131 builds up in the thyroid gland and emits gamma rays, which can be detected to form an image of this cancer patient's thyroid.

Get It?

Describe a medical use for iodine-131.

Cancer treatments

When a person has cancer, a group of cells in that person's body grows out of control. Cancer is a harmful, and often, fatal disease. The left panel of **Figure 16** shows two cancerous cells. The right panel of **Figure 16** shows a cancer patient undergoing radiation therapy. Doctors can use radiation to stop some types of cancerous cells from growing and dividing.

Remember that radiation can ionize nearby atoms. If a source of radiation is placed near cancer cells, atoms in those cells can be ionized. If the ionized atoms are in a critical molecule, such as DNA or RNA, then the molecule may no longer function properly. The cell then may stop growing or may even die.

Noncancerous cells can also be damaged during radiation therapy. For this reason, doctors must be careful to focus the radiation on the cancer cells as much as possible. However, radiation therapy still often harms healthy cells. Cancer patients often experience severe side effects when they receive radiation therapy.

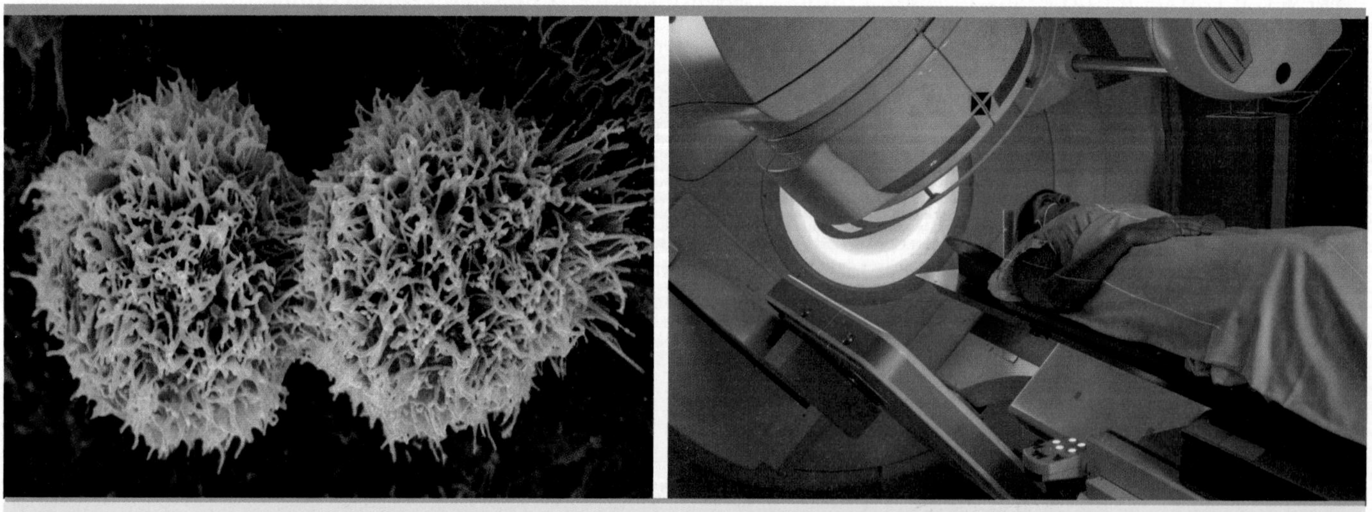

Cancer cells | Radiation therapy

Figure 16 During radiation therapy, a radioactive source is placed close to the tumor. The goal in such a procedure is to damage the tumor as much as possible while minimizing damage to the surrounding tissue.

Half-Life

How can you tell when a radioactive isotope is going to decay? Suppose you shake a box filled with hundreds of pennies. You then remove every penny that comes up tails. You would remove about half the pennies every time you did this.

You could not predict exactly which pennies would come up tails each time. However, you could predict approximately how many pennies would come up tails after each shake. You could also predict how many times you can repeat this process before you have removed all of the pennies from the box.

Radioactive decay works in a similar way. You cannot know when a specific radioactive nucleus will decay. However, you can accurately predict approximately how many radioactive nuclei will decay in a given amount of time.

If a large number of radioactive nuclei are present in a sample, it is possible to consider the half-life of that sample. **Half-life** is the amount of time it takes for half the nuclei in a sample of an isotope to decay. Half-life is similar to the amount of time between shakes for the pennies in the shoe box.

Get It?
Define the term *half-life*.

The radioactive nucleus is called the parent nucleus. The nucleus left after the isotope decays is called the daughter nucleus. **Figure 17** shows the proportion of decaying nuclei left after each half-life. Notice that one-half of the original parent nuclei remain after one half-life. One-quarter of the original parent nuclei remain after two half-lives.

Get It?
Define the term *daughter nucleus*.

Half-lives vary widely. The half-life of polonium-214 is less than a thousandth of a second. Carbon-14 has a half-life of slightly less than 6000 years. Uranium-238 has a half-life of 4.5 billion years. Scientists can use their knowledge of the half-life of an isotope to calculate the ages of rocks, fossils, and artifacts.

Get It?
Identify how many of the original parent nuclei remain after three half-lives.

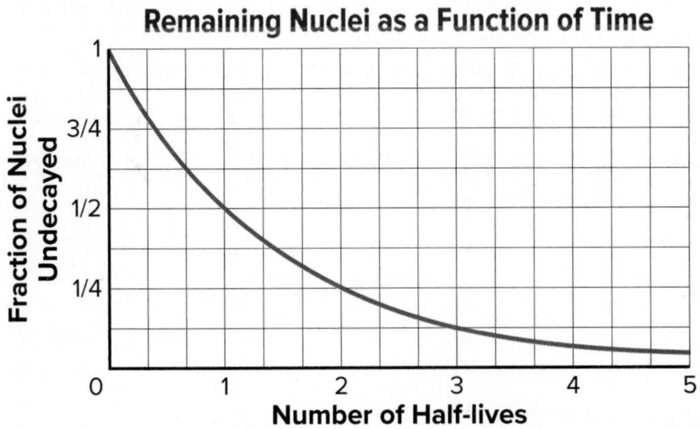

Figure 17 One-half of a radioactive sample decays every half-life.
Identify *how many half-lives would be necessary for three-quarters of a radioactive sample to decay.*

Radiometric dating

The rock shown in **Figure 18** is from the Moon and is more than 4 billion years old. How can scientists learn the age of something that is thousands, millions, or even billions of years old? Scientists use many different methods to determine the ages of samples. One of their most effective methods of dating samples involves an understanding of radioactivity and half-life. Scientists call this method *radiometric dating.*

First, scientists measure the amounts of radioactive isotope and daughter isotope in a sample. Then, they calculate the number of half-lives that need to pass to give the measured amounts. The number of half-lives can then be multiplied by the length of each half-life. This gives the amount of time that has passed since the isotope began to decay. This is usually close to the total amount of time that has passed since the object was formed.

Different isotopes are useful in dating various types of materials. Carbon-14 can be used to date the fossils of once-living organisms that are tens of thousands of years old. However, carbon-14 cannot be used to date materials that were not once a part of a living organism or materials that are more than about 60,000 years old. Uranium-235, which has a much longer half-life, can be used to date rocks and minerals that are billions of years old. Other isotopes that are used in radiometric dating are potassium-40, rubidium-87, and samarium-147.

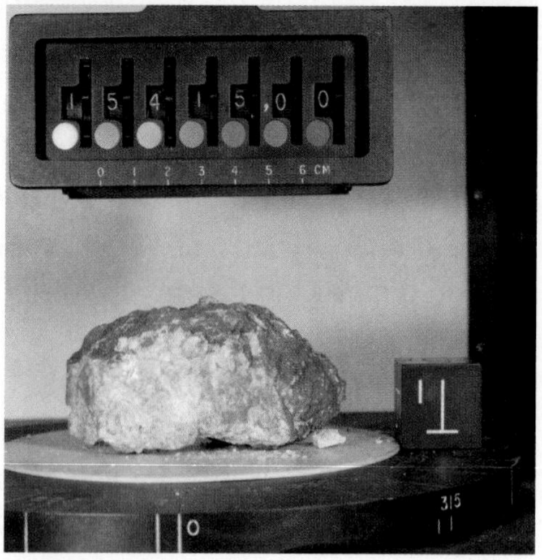

Figure 18 This rock is commonly known as the Genesis Rock. It was returned from the Moon during the Apollo 15 mission. Radiometric dating techniques showed this rock to be more than 4 billion years old.

Check Your Progress

Summary

- Alpha and beta particles can be detected by Geiger counters and in wire chambers.
- Background radiation is low-level radiation emitted mainly by radioactive isotopes in Earth's rocks, soils, and atmosphere.
- Doctors use radioactive isotopes as tracers for medical diagnoses and radiation to kill cancer cells.
- The half-life of an isotope is the length of time that it takes half of a sample of that isotope to decay.

Demonstrate Understanding

13. **Explain** how half-life would help determine which isotopes might be useful for a medical test.
14. **Describe** what happens when beta particles pass through a Geiger counter.
15. **Explain** why background radiation can never be completely eliminated.

Explain Your Thinking

16. **Explain** Why is an archaeologist unable to use carbon-14 to accurately date the age of a skeleton that is millions of years old?
17. **MATH Connection** What is the percentage of radioactive nuclei left after 3 half-lives pass?
18. **MATH Connection** If the half-life of iodine-131 is 8 days, how much of a 5.0-g sample is left after 32 days?

Rodney Grubbs/NASA

LEARNSMART® Go online to follow your personalized learning path to review, practice, and reinforce your understanding.

The Atomic Bomb

Nuclear fission was a well-researched phenomenon by January 1939. Newspapers around the world reported the "splitting of the atom." In late 1939, as Hitler's Germany threatened all of Europe, leading scientists in the U.S. urged the government to begin a top-secret atomic weapons research program. This program became known as the Manhattan Project.

U.S. and U.S.S.R. Total Strategic Warheads

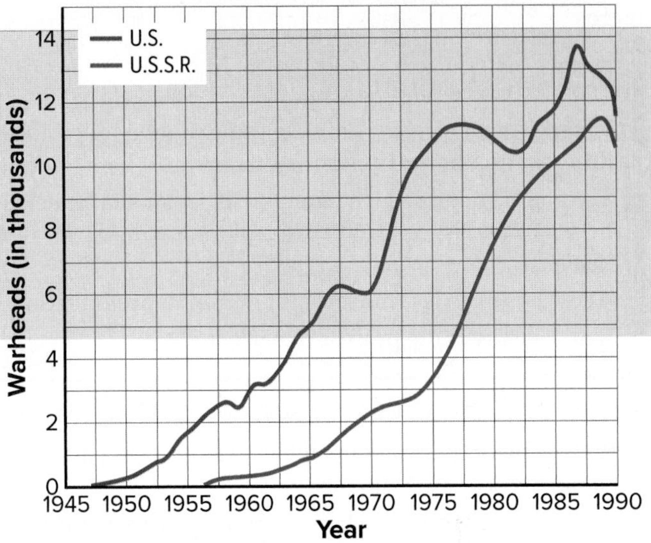

The U.S. and the Soviet Union spent trillions of dollars building stockpiles of nuclear weapons.

War Effort

The Manhattan Project was a massive effort that involved more than 130,000 workers and dozens of facilities across the country at its peak. On July 16, 1945, the work resulted in *Trinity*, the first test explosion of an atomic bomb. By May of 1945, Germany had surrendered, and the U.S. military was concentrating on defeating Japan. The atomic bombs, initially intended for use against Germany, would instead be used against Japan.

Hiroshima, Japan, was the first city destroyed by an atomic bomb.

The End of WW II

The photo at left shows the devastation caused by the Hiroshima bomb. Three days later, a second bomb was dropped on Nagasaki, Japan. One day later, Japan surrendered. These two bombs and the resulting radiation killed about 200,000 people.

Controlling Nuclear Weapons

The U.S. and U.S.S.R. stockpiled nuclear weapons after WW II, as shown above. In the 1970s, arms control agreements limited these weapons and banned nuclear testing, antiballistic missile systems, and weapons in space. Over the years, fewer new nuclear systems were developed, and nuclear stockpiles were reduced. Today, the United Nations and other organizations seek to reduce the number of nuclear weapons and to prevent more countries from building these weapons of mass destruction.

USE A MODEL TO ILLUSTRATE

Construct a model of a modern nuclear power plant to illustrate the components of the systems involved and how they operate to control the release of energy during nuclear fission. Present your model to your classmates.

MODULE 20
STUDY GUIDE

GO ONLINE to study with your Science Notebook.

Lesson 1 THE NUCLEUS

- strong force
- radioactivity

- Isotopes of an element have the same number of protons but different numbers of neutrons.
- The atomic number is the number of protons in a nucleus. The mass number is the number of protons and neutrons in a nucleus.
- A nucleus may decay due to its neutron to proton ratio.
- Radioactivity is the process of nuclear decay.

Lesson 2 NUCLEAR DECAYS AND REACTIONS

- alpha particle
- beta particle
- transmutation
- chain reaction

- Radioactivity can result in alpha particles, beta particles, or gamma rays.
- Nuclear fission occurs when a neutron strikes a nucleus, causing it to split into smaller nuclei.
- Nuclear fusion occurs when two nuclei combine to form another nucleus.
- A small amount of mass is the same as a tremendous amount of energy.

Lesson 3 RADIATION TECHNOLOGIES AND APPLICATIONS

- tracer
- half-life

- Alpha and beta particles can be detected by Geiger counters and in wire chambers.
- Background radiation is low-level radiation emitted mainly by radioactive isotopes in Earth's rocks, soils, and atmosphere.
- Doctors use radioactive isotopes as tracers for medical diagnoses and radiation to kill cancer cells.
- The half-life of an isotope is the length of time that it takes half of a sample of that isotope to decay.

THREE-DIMENSIONAL THINKING

Module Wrap-Up

REVISIT THE PHENOMENON

How does a nuclear reactor work?

CER Claim, Evidence, Reasoning

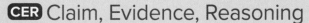

Explain Your Reasoning Revisit the claim you made when you encountered the phenomenon. Summarize the evidence you gathered from your investigations and research and finalize your Summary Table. Does your evidence support your claim? If not, revise your claim. Explain why your evidence supports your claim.

STEM UNIT PROJECT

Now that you've completed the module, revisit your STEM unit project. You will apply your evidence from this module and complete your project.

GO FURTHER

SEP Data Analysis Lab

How do isotopes differ?

Most elements exist in nature as isotopes. Isotopes of an element are almost identical in their chemical properties and reactions. However, the nuclear properties of isotopes are different. Not only do isotopes differ in mass, but some may be radioactive. Using a periodic table, complete the information in the chart below.

Symbol	Atomic number	Number of protons	Mass number	Number of neutrons
1. $^{1}_{1}H$			1	
2. $^{2}_{1}H$	1			1
3. $^{3}_{1}H$			3	
4. $^{4}_{2}He$		2		2
5. $^{12}_{6}C$			12	
6. $^{14}_{6}C$			14	
7. $^{49}_{21}Sc$		21		28
8. $^{63}_{27}Co$			63	
9. $^{222}_{88}Ra$			222	
10. $^{226}_{88}Ra$			226	

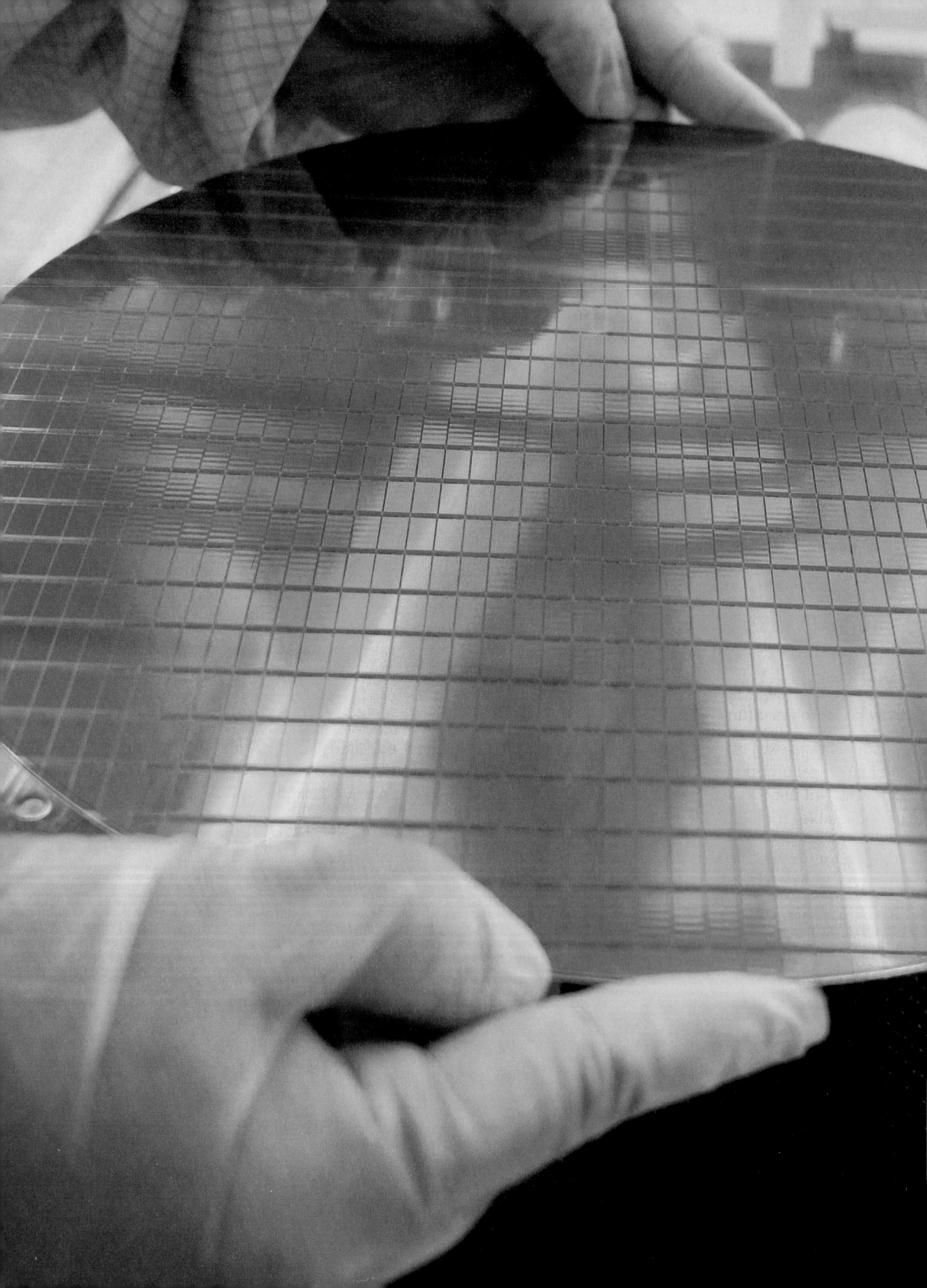

MS Mikel/Shutterstock

UNIT 6

APPLICATIONS OF CHEMISTRY

ENCOUNTER THE PHENOMENON

How are advancements in chemistry related to technology?

SEP Ask Questions

What questions do you have about the phenomenon? Write your questions on sticky notes and add them to the driving question board for this unit.

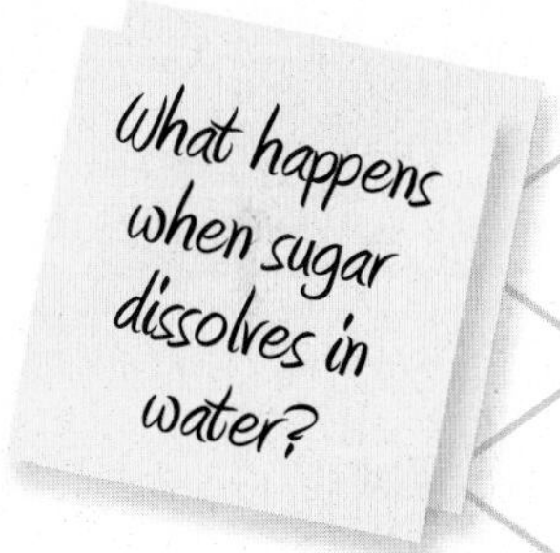

What happens when sugar dissolves in water?

Look for Evidence

As you go through this unit, use the information and your experiences to help you answer the phenomenon question as well as your own questions. For each activity, record your observations in a Summary Table, add an explanation, and identify how it connects to the unit and module phenomenon questions.

Solve a Problem

STEM UNIT PROJECT

Model a Plant-Based Flame Retardant Flame-retardant chemicals prevent fire-related property damage and save lives. However, there are environmental and health concerns about some of these chemicals. Research new flame retardant chemicals that are plant based, environmentally friendly, and biodegradeable. Use the results of these investigations and the evidence you collected during the unit to model how you would make the new environmentally friendly flame retardant.

GO ONLINE In addition to reading the information in your Student Edition, you can find the STEM Unit Project and other useful resources online.

Carsten Peter/Speleoresearch & Films/National Geographic/Getty Images

MODULE 21

SOLUTIONS

ENCOUNTER THE PHENOMENON

How do crystals form?

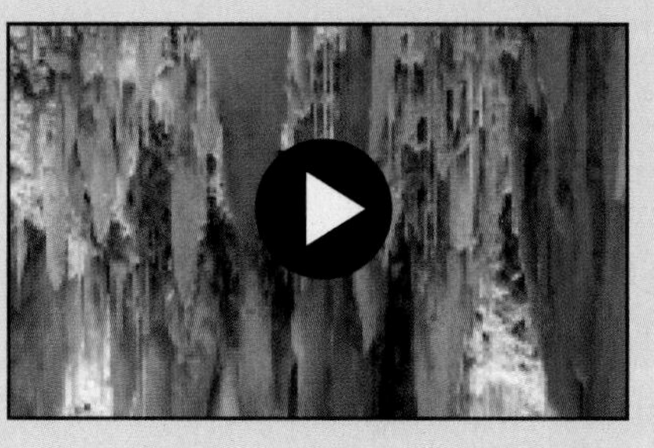

GO ONLINE to play a video about the formation of stalactites, stalagmites, and crystals in caves.

SEP Ask Questions

Do you have other questions about the phenomenon? If so, add them to the driving question board.

CER Claim, Evidence, Reasoning

Make Your Claim Use your CER chart to make a claim about how crystals form.

Collect Evidence Use the lessons in this module to collect evidence to support your claim. Record your evidence as you move through the module.

Explain Your Reasoning You will revisit your claim and explain your reasoning at the end of the module.

GO ONLINE to access your CER chart and explore resources that can help you collect evidence.

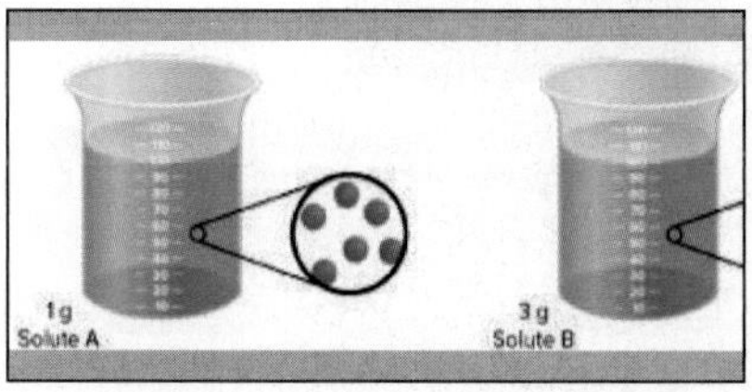

LESSON 2: Explore and Explain: How much can dissolve?

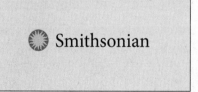

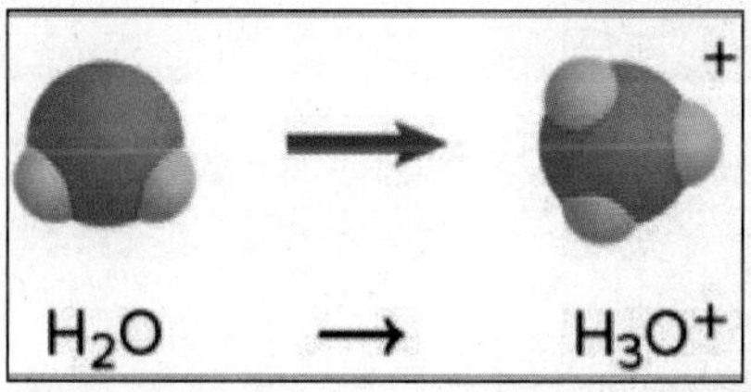

LESSON 3: Explore and Explain: Ion Formation in Solution

Smithsonian

Additional Resources

LESSON 1

HOW SOLUTIONS FORM

FOCUS QUESTION

How do solutions form?

What is a solution?

Hummingbirds are fascinating creatures. They can hover for long periods while they use their long beaks to sip nectar from flowers. To attract hummingbirds, many people use feeder bottles that have brightly colored flowers, as shown in **Figure 1.** The liquid is a solution of sugar and water.

Suppose you are making some hummingbird food. When you add sugar to water and stir, the sugar crystals disappear. Why does this happen?

Hummingbird food is one of many solutions. A solution is a homogeneous mixture, which means it has the same composition throughout the mixture. The reason why you no longer see the sugar crystals is that they have formed a solution. The sugar molecules mixed evenly among the water molecules.

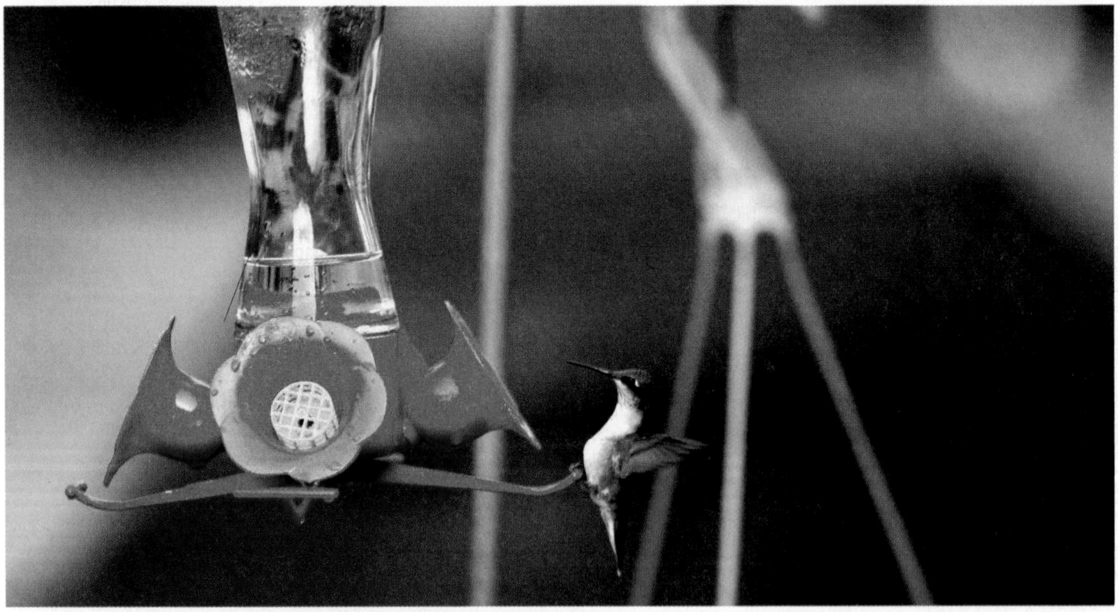

Figure 1 Liquid solutions may contain gases, other liquids, and solids. For example, this hummingbird food contains sugar (solid), and oxygen and nitrogen (gases), all dissolved in water.

Maryellen Valaitis

3D THINKING

DCI Disciplinary Core Ideas   CCC Crosscutting Concepts   SEP Science & Engineering Practices

COLLECT EVIDENCE

Use your Science Journal to record the evidence you collect as you complete the readings and activities in this lesson.

INVESTIGATE

GO ONLINE to find these activities and more resources.

Virtual Investigation: Ratios of Elements
Use a model to determine how the ratios of elements affects an alloys' physical properties.

Quick Investigation: Observe the Effect of Surface Area
Carry out an investigation to determine the effect surface area has on solubilty.

This diver breathes a solution of oxygen, nitrogen, and helium gases.

The bronze used to make this statue is a solid solution of copper and tin.

Figure 2 Solutions can also be mixtures of solids or gases. Gases naturally diffuse within a common container. But solids, such as bronze, must be in a molten state to combine into a solution.

Solutes and Solvents

To describe a solution, you can say that one substance is dissolved in another. The substance being dissolved in a solution is the **solute.** The substance in which a solute is dissolved is the **solvent.** When a solid or gas dissolves in a liquid, the solid or gas is the solute and the liquid is the solvent. Thus, in salt water, salt is the solute and water is the solvent. In carbonated soft drinks, carbon dioxide gas is one of the solutes and water is the solvent. When a liquid dissolves in another liquid, the substance that is present in the larger amount is typically called the solvent.

Get It?

Explain How do you know which substance is the solute in a solution?

Nonliquid solutions

Solutions can also be gaseous or even solid. Examples of a gaseous solution and a solid solution are shown in **Figure 2.** Did you know that the air that you breathe is a solution? In fact, all mixtures of gases are solutions. Air is a solution of 78 percent nitrogen, 21 percent oxygen, and small amounts of other gases, such as argon, carbon dioxide, and water vapor.

The sterling silver and brass used in musical instruments are examples of solid solutions. Sterling silver contains 92.5 percent silver and 7.5 percent copper. Brass is a solution of copper and zinc metals. Solid solutions are known as alloys. An **alloy** is a mixture of elements that has metallic properties. Alloys are made by melting the solid solute and solvent together. Most coins, as shown in **Figure 3** on the next page, are alloys.

SCIENCE USAGE v. COMMON USAGE

solution

Science usage: a homogeneous mixture *Sugar water is a solution of solid sugar particles mixed with water.*

Common usage: an action or process of solving a problem *The solution for getting the car from the mud was a tow truck.*

STEM CAREER Connection

Veterinarian

Veterinarians provide health care to animals and help maintain and improve public health. They frequently work with solutions. Veterinarians need to understand solution chemistry when prescribing and administering medications to sick animals. For example, they might treat a diabetic pet with an intravenous dextrose solution.

Figure 3 Visualizing Metal Alloys

How do vending machines recognize whether the money being placed into them is the correct currency? They recognize coins by size, mass, and, sometimes, electrical conductivity.

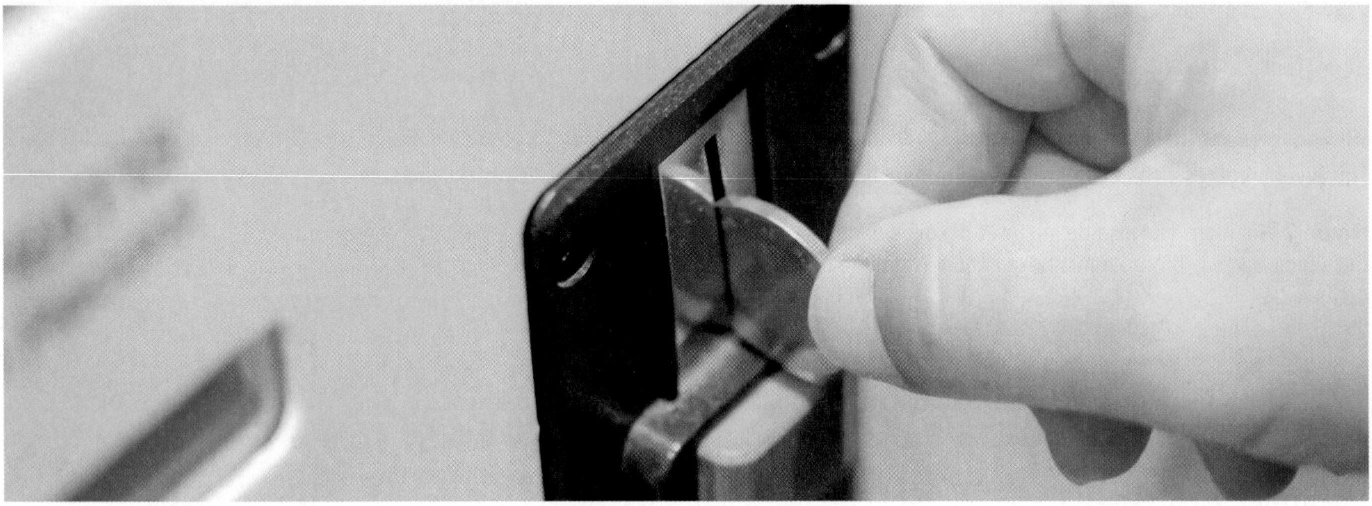

The gold-colored Sacagawea dollar was first issued in 2000 to replace the silver-colored Susan B. Anthony dollar. It is the same size and mass as the Susan B. Anthony dollar, but the electrical conductivities are different.

The newer Sacagawea dollar has a different metal composition than the older Susan B. Anthony dollar.

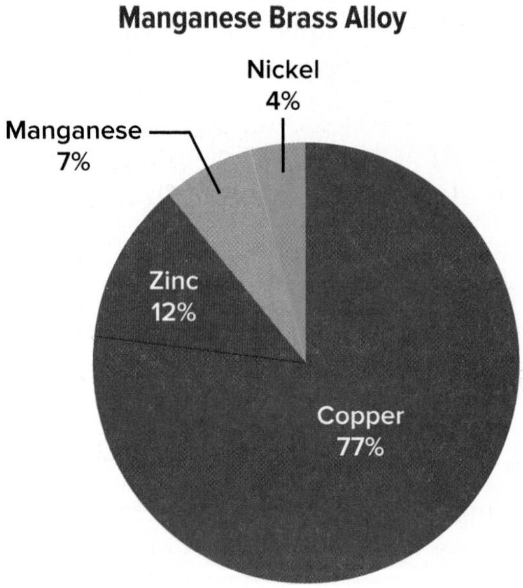

After testing thousands of coins for conductivity, a manganese brass alloy was chosen for the Sacagawea dollar.

An alloy's electrical conductivity is based on the composition of the alloy. Each alloy has its own specific electrical conductivity. If the Sacagawea dollar failed to match the conductivity of the Susan B. Anthony dollar, vending machines would have to be modified or replaced to take the Sacagawea dollar. By adding the right combination of zinc, manganese, and nickel to copper, the correct conductivity was achieved, and the expensive task of replacing vending machines was avoided.

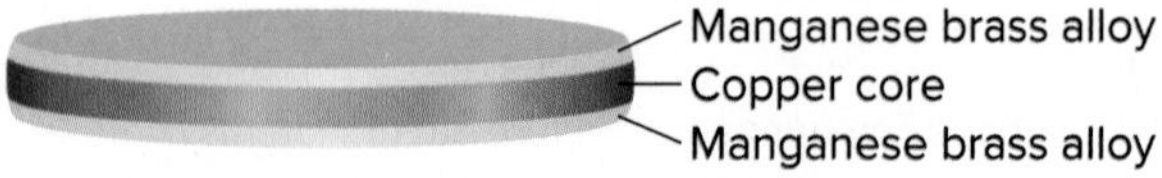

The Sacagawea dollar's copper core is half the coin's thickness. The outer layer is composed of the manganese brass alloy.

How Substances Dissolve

Fruity drinks made from powdered mixes are examples of solutions made by dissolving solids in liquids. Like the hummingbird food shown in **Figure 1,** they contain sugar, as well as other substances, such as food coloring and added flavors.

The dissolving of a solid in a liquid occurs at the surface of the solid. Interactions of matter at the bulk scale are determined by electrical forces within and between atoms. To understand how water solutions form, keep in mind two things that you have learned about water. Like the particles of any substance, water molecules are constantly moving, and water is a polar molecule.

Get It?
Identify where a solid actually dissolves when placed in a liquid.

How it happens

Figure 4 shows the process of sugar dissolving in water. In step 1, the negative ends of water molecules are attracted to the positive ends of sugar crystals as water moves past the surface of the solid sugar. In step 2, the water molecules pull the sugar molecules into solution. Finally, in step 3, the water molecules and the sugar molecules mix evenly.

The process described in the three steps shown in **Figure 4** repeats as layer after layer of sugar molecules moves away from the crystal. The same three steps occur for any solid solute dissolving in a liquid solvent.

Dissolving liquids and gases

A similar but more complex process takes place when liquids and gases dissolve. Liquid and gas particles move much more freely than particles of solids move. When gases dissolve in gases or when liquids and gases dissolve in liquids, particle movement eventually spreads solutes evenly throughout the solvent, resulting in a homogeneous mixture.

Dissolving solids in solids

Solid particles do move a little, but this motion is not enough to spread particles evenly throughout a mixture. Solid metals are first melted and then mixed together while still molten. In the liquid state, the atoms can spread out evenly and will remain mixed after they have cooled.

Figure 4 Dissolving sugar in water is a three-step process.

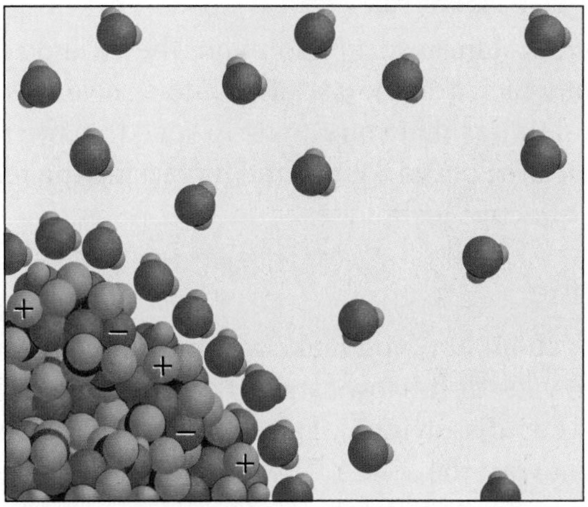

Step 1 At the surface of the sugar crystal, oppositely charged parts of the sugar and water molecules attract each other.

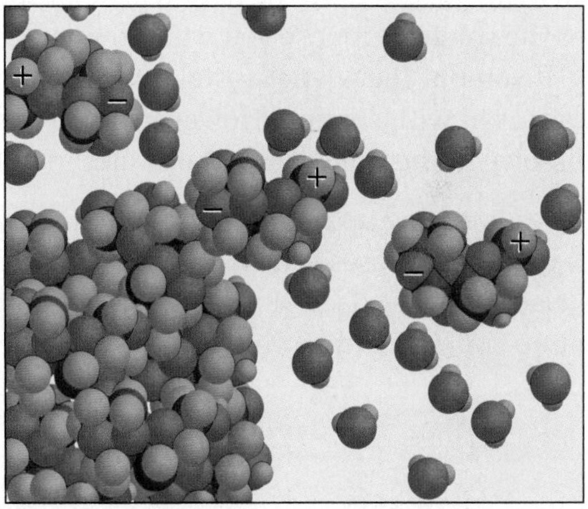

Step 2 Because the water molecules are moving in the liquid, they pull sugar molecules away from the crystal.

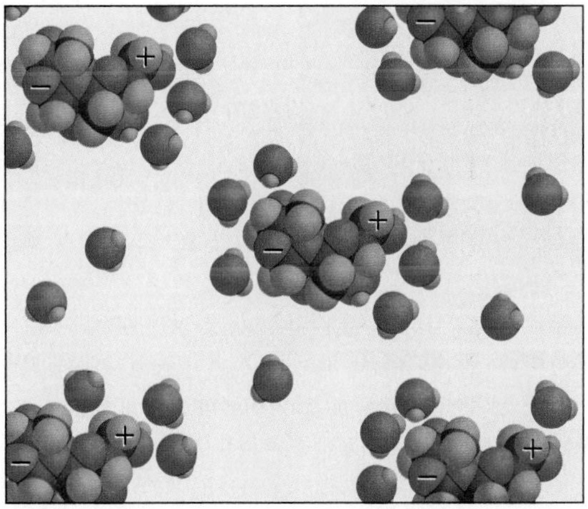

Step 3 Water molecules and sugar molecules continue to spread out until a homogeneous mixture forms.

Rate of Dissolving

If two substances form a solution, they will do so at a measurable rate. Sometimes the rate at which a solute dissolves into a solvent is fast. Other times it is slow. There are several things that you can do to speed up the rate of dissolving. Stirring, increasing the surface area of the solute, and increasing the temperature of the solvent are three of the most effective techniques.

Stirring

Think about how you make a drink from a powdered mix. After you add the mix to water, you stir it. How can stirring speed up the dissolving process? Stirring a solution speeds up dissolving because it moves the solvent around, bringing more solvent into contact with the solute. The solvent attracts the particles of solute, causing the solid solute to dissolve faster.

Surface area

Another way to speed up the dissolving of a solid in a liquid is to increase the surface area of the solute. Suppose you want to sweeten your water with a 5-g crystal of rock candy. If you put the whole crystal into a glass of water, it might take several minutes to dissolve, even with stirring. However, if you first grind the rock candy into a powder, it will dissolve in the same amount of water in only a few seconds.

Why does breaking up a solid cause it to dissolve faster? Breaking the solid into smaller pieces greatly increases its surface area, as you can see in **Figure 5.** Because dissolving takes place at the surface of the solid, increasing the surface area allows more solvent to come into contact with more solid solute. Therefore, the speed of the dissolving process increases.

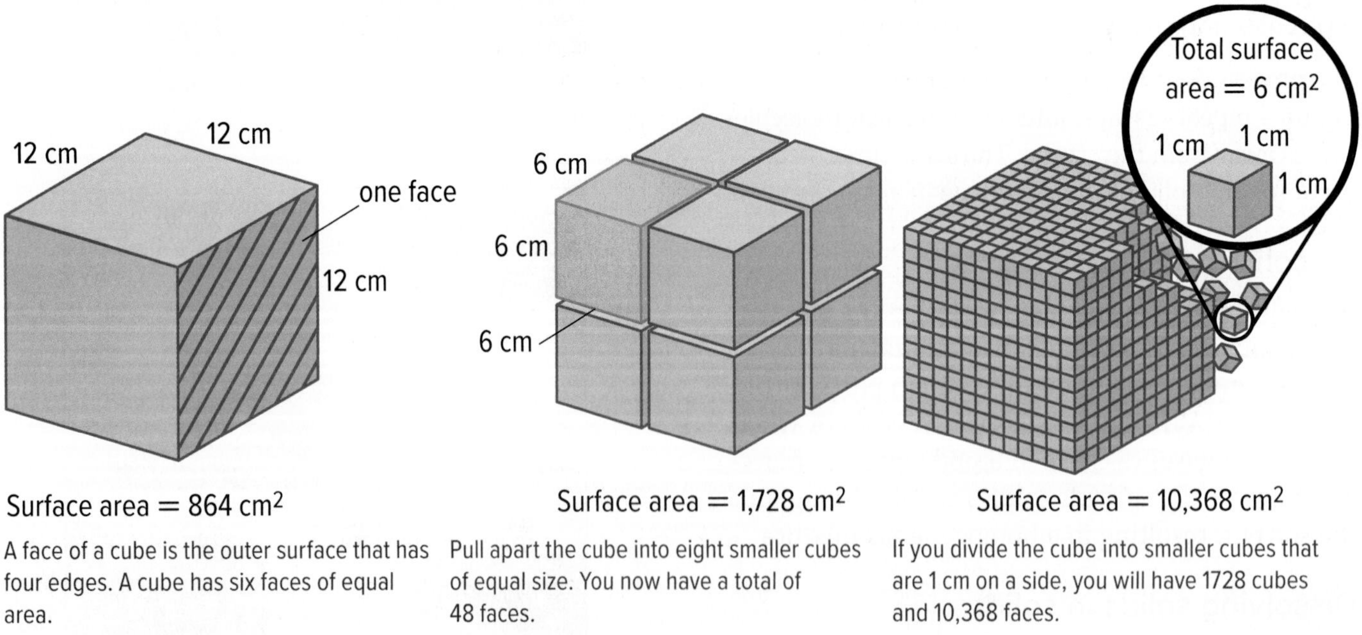
12 cm
12 cm
one face
12 cm
Surface area = 864 cm2
A face of a cube is the outer surface that has four edges. A cube has six faces of equal area.
6 cm
6 cm
6 cm
Surface area = 1,728 cm2
Pull apart the cube into eight smaller cubes of equal size. You now have a total of 48 faces.
Total surface area = 6 cm2
1 cm
1 cm
1 cm
Surface area = 10,368 cm2
If you divide the cube into smaller cubes that are 1 cm on a side, you will have 1728 cubes and 10,368 faces.

Figure 5 Surface area is the area of the exterior surface of an object, measured in square units. Increasing surface area increases the speed at which a solute is dissolved by a solvent.

EXAMPLE Problem 1

CALCULATE SURFACE AREA Suppose the length, height, and width of a cube are each 1 cm. If the cube is cut in half to form two rectangular pieces, what is the total surface area of the new pieces?

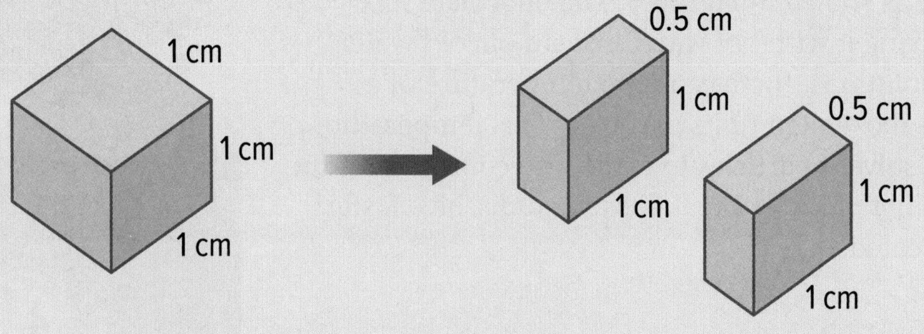

Identify the Unknown: Total surface area of the two new pieces

List Knowns:
length = l = 1 cm
height = h = 1 cm
width = w = 0.5 cm

Set Up the Problem:
The rectangular solids each have six faces.
Surface area front and back = $2(h \times w)$
Surface area left and right = $2(h \times l)$
Surface area top and bottom = $2(w \times l)$
Surface area of one piece = $2(h \times w) + 2(h \times l) + 2(w \times l)$
Total surface area = Number of pieces × Surface area of one piece

Solve the Problem:
Surface area of one piece =
$2(1\text{ cm} \times 0.5\text{ cm}) + 2(1\text{ cm} \times 1\text{ cm}) + 2(0.5\text{ cm} \times 1\text{ cm}) = 4\text{ cm}^2$
The total surface area of the two new pieces = $2(4\text{ cm}^2) = 8\text{ cm}^2$

Check the Answer:
Total surface area of the original cube = $6(w \times h) = 6(1\text{ cm} \times 1\text{ cm}) = 6\text{ cm}^2$
Dividing the cube in two increased the surface area, which is reasonable.

PRACTICE Problems

ADDITIONAL PRACTICE

1. The length, height, and width of a cube are each 3 cm. If the cube is cut in half to form two rectangular pieces, what is the total surface area of the new pieces?
2. If a cube that has a length, height, and width of 4 cm is broken down into 8 cubes of equal size, what is the surface area of the 8 new cubes?
3. **CHALLENGE** A cube of salt with a length, height, and width of 5 cm each is attached along a face to another cube of salt with the same dimensions. How much surface area is lost by combining the cubes to form a rectangular solid?

Temperature

In addition to stirring and increasing the surface area of the solute, a third way to increase the rate at which most solids dissolve is to increase the temperature of the solvent. Think about making hot chocolate from a mix, as shown in **Figure 6.** The chocolate mix dissolves faster by mixing it with hot water instead of cold water. This is true of many solutions. Increasing the temperature of a solvent speeds up the movement of its particles. This temperature increase causes more solvent particles to come into contact with the solute. As a result, solute particles break loose and dissolve faster when the solvent is heated.

Figure 6 Hot water allows powdered chocolate mix to dissolve faster than cold water does and results in fewer solid clumps.

Controlling the process

Think about how the three factors that you just learned affect the rate of dissolving. Can these factors combine to further increase the rate, or perhaps control the rate, of dissolving? Each technique of stirring, increasing the surface area, and heating, is known to increase the rate of dissolving by itself. When two or more techniques are combined, the rate of dissolving increases even more.

Consider a sugar cube placed in cold water. You know that the sugar cube will eventually dissolve. You can predict that heating the water will increase the rate by some amount. You can also predict that a combination of heating and stirring will further increase the rate. Finally, you can predict that increasing the surface area by crushing the cube, combined with heating and stirring, will result in the fastest rate of dissolving.

Check Your Progress

Summary

- A solution is a homogeneous mixture.
- Solutions are composed of solutes and solvents.
- Stirring, surface area, and temperature all affect the rate of dissolving.

Demonstrate Understanding

4. **Summarize** possible ways in which phases of matter could combine to form a solution.
5. **Draw** a diagram that shows how a solid dissolves in a liquid.
6. **Describe** how stirring, surface area, and temperature affect the rate of dissolving.

Explain Your Thinking

7. **Explain** Amalgams are sometimes used in tooth fillings and are made of mercury. Explain why an amalgam is a solution.
8. **MATH Connection** Calculate the surface area of a rectangular solid with dimensions $l = 2$ cm, $w = 1$ cm, and $h = 0.5$ cm.
9. **MATH Connection** If the length of the rectangle in question 8 is increased by 10%, by what percentage will the surface area increase?

LEARNSMART® Go online to follow your personalized learning path to review, practice, and reinforce your understanding.

LESSON 2

CONCENTRATION AND SOLUBILITY

FOCUS QUESTION

What is concentration and solubility?

Concentration

Suppose you add one teaspoon of lemon juice to a glass of water to make lemonade. Your friend adds four teaspoons of lemon juice to the same amount of water in another glass. You could say that your lemonade is diluted and your friend's lemonade is concentrated. Your friend's drink has more lemon than your drink has. A concentrated solution is one in which a large amount of solute is dissolved in the solvent. A dilute solution is one with a small amount of solute in the solvent. These are relative concentrations.

Precise concentrations

How much real fruit juice is in an average fruit drink? *Concentrated* and *diluted* are not precise terms. But solution concentrations can be described precisely. The **concentration** of a solution is the amount of solute dissolved in a given amount of solvent.

One way to state concentration precisely is to give the percentage by volume of the solute. The percentage by volume of the juice in the orange drink shown in **Figure 7** is 25 percent. Adding 25 mL of solute to 75 mL of the solvent makes 100 mL of a 25 percent solution. This means that for a volume of 100 mL of juice drink, there is a volume of 25 mL of juice (the solute) and 75 mL of water (the solvent).

$$\frac{25 \text{ mL juice}}{25 \text{ mL juice} + 75 \text{ mL water}} \times 100 = 25\% \text{ by volume of juice}$$

25% JUICE
NO SUGAR OR COLORS ADDED. NO PRESERVATIVES. VERY LOW SODIUM. GLUTEN FREE.
100% ORANGE JUICE
• A FULL DAY'S SUPPLY OF VITAMIN C
Nutrition Facts
Serving Size 8 fl oz (240 mL)
Servings Per Container 7
Amount Per Serving
Calories 110 Calories from Fat 0
PASTEURIZED
NATURALLY SODIUM FREE
NO WATER OR PRESERVATIVES ADDED
**ORANGES ARE NOT GENETICALLY ENGINEERED.

Figure 7 The concentrations of juice drinks often are given in percent by volume. Concentrations of juice drinks commonly range from 10 percent to 100 percent juice.

Identify *the product that has the higher concentration of orange juice.*

3D THINKING DCI Disciplinary Core Ideas CCC Crosscutting Concepts SEP Science & Engineering Practices

COLLECT EVIDENCE

Use your Science Journal to record the evidence you collect as you complete the readings and activities in this lesson.

INVESTIGATE

GO ONLINE to find these activities and more resources.

Virtual Investigation: Solutions
Use a model to determine the effect temperature has on solubility.

LabA: How soluble are two salts at varying temperatures?
Carry out an investigation to determine the effect varying temperatures have on solubility.

How much can dissolve?

You can stir several teaspoons of sugar into lemonade, and the sugar will dissolve. However, if you continue adding sugar, a point is eventually reached when no more sugar dissolves and the excess sugar sinks to the bottom of the glass. This indicates how soluble sugar is in water. **Solubility** (sol yuh BIH luh tee) is the maximum amount of a solute that can be dissolved in a given amount of solvent at a given temperature. Solubility of substances dissolved in water is often expressed as grams of solute per 100 g of water (g/100 g water).

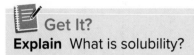

Get It?

Explain What is solubility?

Comparing solubilities

Figure 8 shows two beakers with the same volume of water and two different solutes. In one beaker, one gram of Solute A dissolves completely, but additional solute does not dissolve and falls to the bottom of the beaker. In the other beaker, one gram of Solute B dissolves completely, and two more grams of Solute B also dissolve before additional solute begins to fall to the bottom of the beaker. If you assume that the temperature of the water is the same in both beakers, you can conclude that substance B is more soluble in water than substance A.

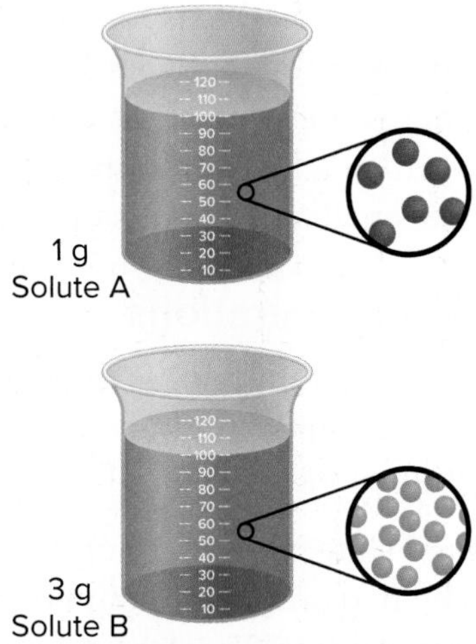

Figure 8 Three grams of solute B dissolve in 100 mL of water. In the same temperature and volume of water, only 1 gram of solute A dissolves. This means that solute B is more soluble in water.

The solubilities of solutes in water vary. **Table 1** shows the solubilities of several substances in water at a temperature of 20°C. For solutes that are gases, such as hydrogen, oxygen, and carbon dioxide, the pressure must be given with the solubility since the solubility can vary at different pressures.

Get It?

Rank the solubilities of salt, washing soda, and table sugar in water at 20°C from most soluble to least soluble using the information in **Table 1.**

Table 1 Solubility in Water at 20°C and Normal Atmospheric Pressure

State of Substance	Substance	Solubility in g/100 g of Water
Solid	salt (sodium chloride)	35.9
	baking soda (sodium bicarbonate)	9.6
	washing soda (sodium carbonate)	21.4
	lye (sodium hydroxide)	109.0
	table sugar (sucrose)	203.9
Gaseous	hydrogen	0.00017
	oxygen	0.005
	carbon dioxide	0.16

Types of Solutions

How much solute can dissolve in a given amount of solvent? That depends on a number of factors, including the solubility of the solute. Here, you will examine three types of solutions that are defined by the amount of a solute dissolved in a solvent.

Saturated solutions

If you add 35 g of copper(II) sulfate ($CuSO_4$) to 100 g of water at 20°C, only 32 g will dissolve. You have a saturated solution because no more copper(II) sulfate can dissolve. A **saturated solution** is a solution that contains all of the solute that it can hold at a given temperature. However, if you heat the mixture to a higher temperature, more copper(II) sulfate dissolves.

Generally, as the temperature of a liquid solvent increases, the amount of solid solute that can dissolve in it also increases. **Table 2** shows the amounts of a few solutes that can dissolve in 100 g of water at different temperatures. Each would form a saturated solution. Some of these are also compounds shown on the graph in **Figure 9.**

Table 2 Solubility of Compounds in g/100 g of Water

Compound	0°C	20°C	100°C
Copper(II) sulfate	23.1	32.0	114
Potassium bromide	53.6	65.3	104
Potassium chloride	28.0	34.0	56.3
Potassium nitrate	13.9	31.6	245
Sodium chlorate	79.6	95.9	204
Sodium chloride	35.7	35.9	39.2
Sucrose (sugar)	179.2	203.9	487.2

Get It?
Explain how the temperature of a liquid solvent affects the solubility of a compound.

Solubility curves Each line on the graph in **Figure 9** is called a solubility curve for a particular substance. You can use a solubility curve to determine how much solute will dissolve at any temperature given on the graph. For example, about 79 g of potassium bromide (KBr) will form a saturated solution in 100 g of water at 50°C. How much sodium chloride (NaCl) will form a saturated solution with 100 g of water at the same temperature? You can also see that as temperature increases, so does the solubility of a substance.

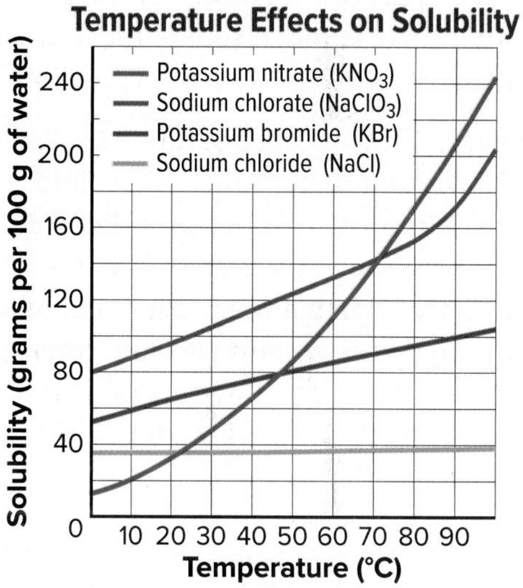

Figure 9 The effect of temperature on the solubility of four different compounds is shown in this solubility curve.

Unsaturated solutions

An **unsaturated solution** is any solution that can dissolve more solute at a particular temperature. Often, when a saturated solution is heated to a higher temperature, it becomes unsaturated and is therefore able to dissolve more solute. The term *unsaturated* is not a precise term. If you look at **Table 2,** you will see that 35.9 g of sodium chloride (NaCl) forms a saturated solution in 100 g of water with a temperature of 20°C. However, an unsaturated solution of sodium chloride could be any amount less than 35.9 g in 100 g of water with a temperature of 20°C.

Get It?
Explain why the term *unsaturated* is not precise.

A seed crystal of sodium acetate is added to a supersaturated solution of sodium acetate.

Excess solute immediately forms a solid.

A solid continues to form until the solution is saturated.

Figure 10 A supersaturated solution is unstable.

Explain *why supersaturated solutions are unstable.*

Supersaturated solutions

If you make a saturated solution of potassium nitrate (KNO_3) at 100°C and then let it cool to 20°C, part of the solute comes out of solution. At the lower temperature, the solvent cannot hold as much solute. Most other saturated solutions behave in a similar way when cooled.

However, if you cool a saturated solution of sodium acetate ($NaC_2H_3O_2$) from 100°C to 20°C without disturbing it, no solute comes out of solution. At this point, the solution is supersaturated. A **supersaturated solution** is one that contains more solute than a saturated solution at the same temperature. Supersaturated solutions are unstable. **Figure 10** shows that when a seed crystal of sodium acetate is dropped into the supersaturated solution, excess sodium acetate comes out of solution.

Solution Energy

The formations of some solutions are exothermic—they give off energy to the surrounding environment. One example is reusable heat packs. The heat packs contain a supersaturated solution of sodium acetate ($NaC_2H_3O_2$). The solution warms when Na^+ and $C_2H_3O_2^-$ ions interact with water molecules.

On the other hand, some substances must draw energy from the surroundings to dissolve. During this endothermic process, the solution becomes colder. Cold packs made of ammonium nitrate (NH_4NO_3) and water operate in this way. When the inner bag is broken, water mixes with ammonium nitrate. The interaction between ammonium nitrate and water draws energy from the surroundings, which causes the pack to cool.

Solubility of Gases

If you shake an opened bottle of a carbonated soft drink, it bubbles up and might squirt out. Shaking, stirring, or pouring a solution of a gas exposes more gas particles to the surface, where they escape from the liquid and come out of solution.

Pressure effects

Carbonated soft drinks are bottled so that the pressure inside of the bottle is greater than the pressure outside of the bottle. This increases the amount of carbon dioxide dissolved in the liquid. When you open the bottle, the pressure inside of the bottle decreases, and the carbon dioxide gas mixed in the soft drink escapes from the soft drink and bubbles out, as shown in **Figure 11.**

Temperature effects

Another way to increase the amount of gas that dissolves in a liquid is to cool the liquid. This is just the opposite of what you do to increase the amounts of most solids dissolved in a liquid. For example, even more carbon dioxide will bubble out of a warm soft drink than out of a cold soft drink.

Figure 11 These carbonated soft drinks are bottled under pressure to keep carbon dioxide in solution. Opening the bottles reduces the pressure on the surface of the gas-liquid solution and carbon dioxide bubbles out of solution.

Check Your Progress

Summary

- Solubility curves help predict how much solute can dissolve at a particular temperature.
- Saturated, unsaturated, and supersaturated solutions are defined by how much solute can dissolve in a solvent.
- Solutions absorb or give off energy as they form.
- Temperature and pressure affect how much gas dissolves in a liquid.

Demonstrate Understanding

10. **Contrast** What is the difference between solubility and concentration?
11. **Compare and contrast** the difference between relative and precise concentrations. Give examples.
12. **Explain** Do all solutes dissolve to the same extent in the same solvent? How do you know?
13. **Identify** the type of solution that you have if solute continues to dissolve as you add more.

Explain Your Thinking

14. **Explain** why keeping a carbonated beverage capped and refrigerated helps keep it from going flat.
15. **MATH Connection** By volume, orange drink is 10 percent each of orange juice and corn syrup. A 1.5-L can of the drink costs $0.95. A 1.5-L can of orange juice is $1.49, and 1.5 L of corn syrup is $1.69. Per serving, does it cost less to make your own orange drink or to buy it?

LEARNSMART® Go online to follow your personalized learning path to review, practice, and reinforce your understanding.

LESSON 3
PARTICLES IN SOLUTION

FOCUS QUESTION

How do ions form in solution?

Ion Formation in Solution

Did you know that there are charged particles in your body that conduct electricity? In fact, you could not live without them. Some help nerve cells transmit messages. Each time that you blink your eyes or wave your hand, nerves control how muscles respond. Recall that these charged particles are called ions. Compounds that produce solutions of ions in water are known as **electrolytes.**

Solutions containing electrolytes conduct electricity. Some substances, such as sodium chloride, are strong electrolytes because they are entirely in the form of ions in solution. Strong electrolytes conduct a strong current. Other substances, such as acetic acid in vinegar, remain mainly in the form of molecules when they dissolve. They produce few ions, conduct a weak current, and are called weak electrolytes. Substances that form no ions in water and do not conduct electricity are called **nonelectrolytes.** Fats and sugars are examples of nonelectrolytes.

Ionization

Solutions of electrolytes form in two ways. One way applies to molecular solutions, such as hydrogen chloride in water. The molecules of hydrogen chloride are composed of neutral atoms. In order to form ions, the molecules must be broken apart so that the atoms take on a charge. The process in which molecular compounds dissolve in water and form charged particles is called **ionization.** The process is shown in **Figure 12,** using hydrogen chloride (HCl) as a model.

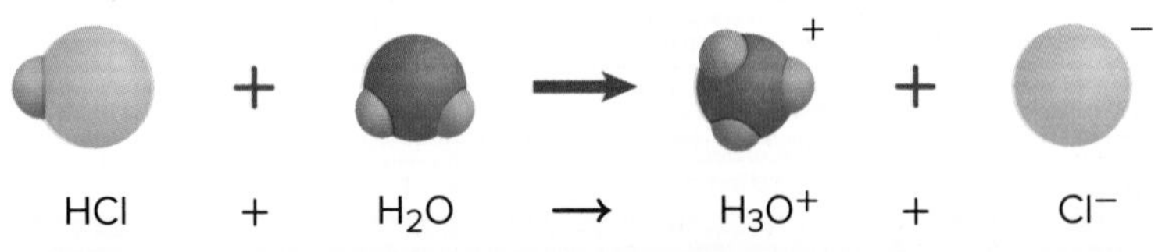

Figure 12 Both hydrogen chloride (HCl) and water (H_2O) are polar molecules. Water surrounds the hydrogen chloride molecules and pulls them apart, forming positive hydrogen ions (H^+) and negative chloride ions (Cl^-). The water molecules attract the hydrogen ions and form hydronium ions (H_3O^+).

3D THINKING DCI Disciplinary Core Ideas | CCC Crosscutting Concepts | SEP Science & Engineering Practices

COLLECT EVIDENCE

Use your Science Journal to record the evidence you collect as you complete the readings and activities in this lesson.

INVESTIGATE

GO ONLINE to find these activities and more resources.

Lab: Boiling Points of Solutions

Carry out an investigation to determine the **change** in boiling point when salt is added to a **solution.**

CCC Identify Crosscutting Concepts

Create a table of the **crosscutting concepts** and fill in examples you find as you read.

Dissociation

The second way that solutions of electrolytes form is the separation of ions in ionic compounds. The ions already exist in the ionic compound. Polar water molecules surround the ionic compound and pull apart the compound into its individual ions. **Dissociation** is the process in which positive and negative ions of an ionic solid mix with the solvent to form a solution.

Get It?

Name the two ways that solutions of electrolytes form.

A model of a sodium chloride (NaCl) crystal is shown in **Figure 13.** In the crystal, each positive sodium ion (Na^+) is attracted to six negative chloride ions (Cl^-). Each of the negative chloride ions is attracted to six positive sodium ions, and these attractions create a continuous pattern that exists throughout the crystalline structure.

When placed in water, the crystalline structure breaks apart. Remember that water molecules are polar, which means that the positive ends of the water molecules—the hydrogen atoms in a water molecule—are attracted to the negative chloride ions. Likewise, the negative end of the water molecule—the oxygen atom—is attracted to the positive sodium ions.

In **Figure 14,** water molecules approach the sodium ions (Na^+) and chloride ions (Cl^-) in the crystal. The water molecules break apart the crystalline structure by pulling the ions away and surrounding them in solution. The sodium and chloride ions have dissociated. The solution now consists of sodium and chloride ions mixed with water. The ions move freely through the solution and are capable of conducting an electric current.

Sodium Chloride (NaCl)

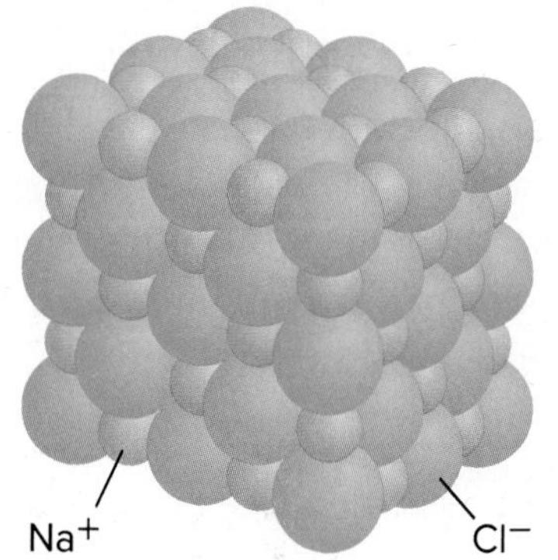

Figure 13 This is a model of a sodium chloride crystal. Each chloride ion is surrounded by six sodium ions, and vice versa.

Get It?

Compare and Contrast What are the differences and similarities between dissociation and ionization?

Solvation Process of NaCl

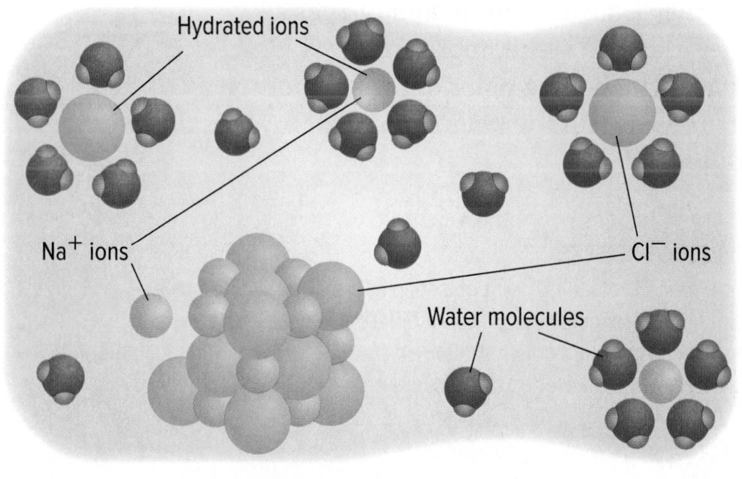

Figure 14 Sodium chloride (NaCl) is an ionic compound, and water is a polar molecule. Sodium chloride dissociates as water molecules attract and pull the sodium and chloride ions from the crystal. Water molecules then surround and separate the Na^+ and Cl^- ions.

Explain *Why will sodium chloride in solution conduct electricity?*

Effects of Solute Particles

All solute particles—polar and nonpolar, electrolyte and nonelectrolyte—affect the physical properties of the solvent. These effects can be useful. For example, adding antifreeze to water in a car radiator lowers the freezing point of the radiator fluid. Salt added to the ice and water mixture in an ice-cream maker lowers the freezing point of the solution, and the ice cream freezes faster. The effect that a solute has on a solvent depends on the number of solute particles in solution and not on the chemical nature of the particles.

Lowering freezing point

Adding a solute, such as antifreeze, to a solvent lowers the freezing point of the solvent. The amount that the freezing point lowers depends upon the concentration of the solute particles that you add.

As a substance freezes and changes state from a liquid to a solid, the particles arrange themselves in an orderly pattern. A solute interferes with the formation of this pattern, making it harder for the solvent to freeze, as shown in **Figure 15.** To overcome this, the temperature of the solvent must decrease to freeze the solution.

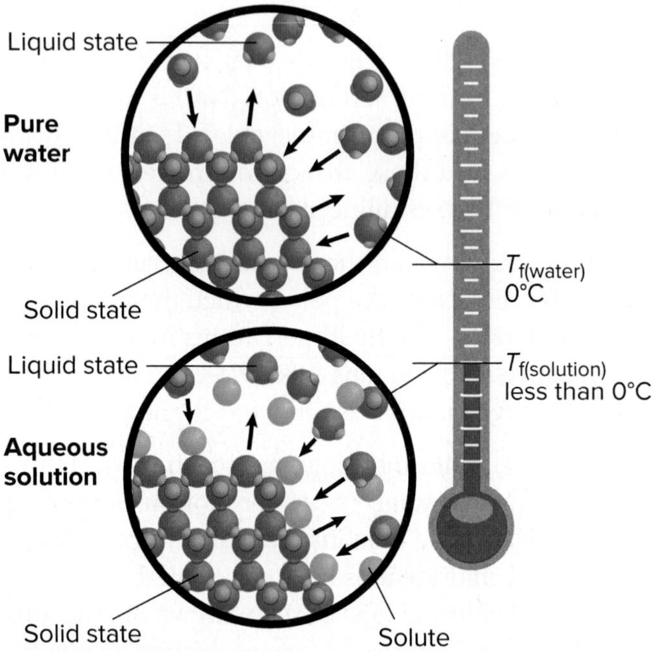

Figure 15 Freezing is the change from the liquid to the solid state. If a solute is added, it can prevent the liquid from forming an orderly solid at its usual freezing point.

Explain *how adding a solute affects the freezing point of a solution.*

Animal antifreeze Animals that live in extremely cold climates have their own internal antifreeze. For example, caribou's lower legs contain substances that prevent freezing in subzero temperatures. The caribou can stand for long periods of time in snow with no harm to their legs.

Fish living in polar waters also have a natural chemical antifreeze called glycoprotein (gli koh PROH teen) in their bodies. Glycoprotein prevents ice crystals from forming in moist tissues in fish.

Raising boiling point

Surprisingly, antifreeze also raises the boiling point of radiator fluid. How can it do this? Solute particles interfere with the solvent particles as they transition from liquid to gas at the surface of the solution. This lowers the vapor pressure. As a result, more energy is needed for the solvent to escape from the liquid surface. The boiling point of the solution will be higher than the boiling point of pure solvent. The amount the boiling point is raised depends upon the concentration of solute present.

ACADEMIC VOCABULARY

affect

to produce a material influence upon or alteration in *The rain affected their plans for a picnic.*

CCC CROSSCUTTING CONCEPTS

Patterns The freezing point and the boiling point of a solvent are both affected by the concentration of the solute particles in a solution, but in different ways. Create a graphic organizer that shows the effects of solute concentration on freezing point and on boiling point. Explain in writing why each shows a different pattern of change

Car radiators The beaker on the left in **Figure 16** represents a car radiator when it contains water only—no antifreeze. Some of the water molecules on the surface will vaporize and change to the gaseous state. The number of water molecules that vaporize depends upon the temperature of the liquid water. As temperature increases, water molecules move more quickly and more molecules vaporize. Finally, when the pressure of the water vapor equals atmospheric pressure, the water begins to boil.

The beaker on the right in **Figure 16** shows the result of adding antifreeze to water. Particles of solute are evenly distributed throughout the solution, including at the surface. Now fewer water molecules can reach the surface and vaporize, making the vapor pressure of the solution lower than that of the pure solvent. This means that it will require a higher temperature to make the water boil. Antifreeze increases the boiling point of the water in a vehicle's radiator and helps prevent overheating.

In a beaker of pure water, water molecules vaporize freely from the surface.

Solute particles block part of the surface, making it more diffcult for solvent to vaporize.

Figure 16 Solute particles raise the boiling point of a solution.

Describe *how antifreeze in a car can help prevent both freezing and overheating.*

Get It?

Describe How does antifreeze affect the vapor pressure of a pure solvent?

Check Your Progress

Summary

- Ionization and dissociation are two ways to form ions in solution.
- Solute particles lower the freezing point of a solution.
- Solute particles raise the boiling point of a solution.

Demonstrate Understanding

16. **Explain** how the concentration of a solute in a solution influences its boiling point and freezing point.
17. **Identify** what kinds of solute particles are present in water solutions of electrolytes and nonelectrolytes.
18. **Determine** whether ionization or dissociation has taken place if calcium phosphate ($Ca_3(PO_4)_2$) breaks into Ca^{2+} and PO_4^{3-}.

Explain Your Thinking

19. **Apply** People often put salt on ice that forms on sidewalks and driveways during the winter. The salt helps melt the ice, forming a saltwater solution. Explain why this solution resists refreezing.
20. **MATH Connection** Use the data points (0, 12), (10, 8), (20, 4), and (30, 0) to graph the effect of a solute on the freezing point of a solvent. Label the x-axis *Solute (g)* and the y-axis *Freezing point (degrees Celsius).* Find the slope of the line that you graph.

LEARNSMART® Go online to follow your personalized learning path to review, practice, and reinforce your understanding.

LESSON 4

DISSOLVING WITHOUT WATER

FOCUS QUESTION

How does polarity affect solubility?

When Water Will Not Work

Water often is referred to as the universal solvent because it can dissolve so many substances. But there are some substances that cannot dissolve in water. Why? Water molecules have positive and negative areas that allow them to attract polar or ionic solutes. Recall that polar molecules have positive areas and negative areas. However, nonpolar molecules have no separated positive and negative areas. Because of this, nonpolar molecules are not attracted to ionic or polar substances, such as water.

Nonpolar solutes

Nonpolar molecules do not dissolve in water or only dissolve a very small amount. **Figure 17** shows how oil, which consists of nonpolar molecules, does not mix with seawater, a polar solution of water, salt, and nutrients.

Oils contain hydrocarbons, which are large molecules made of carbon and hydrogen atoms. The carbon and hydrogen atoms share electrons in a nearly equal manner. The nonpolar oil molecules are not attracted to polar water molecules and will not dissolve.

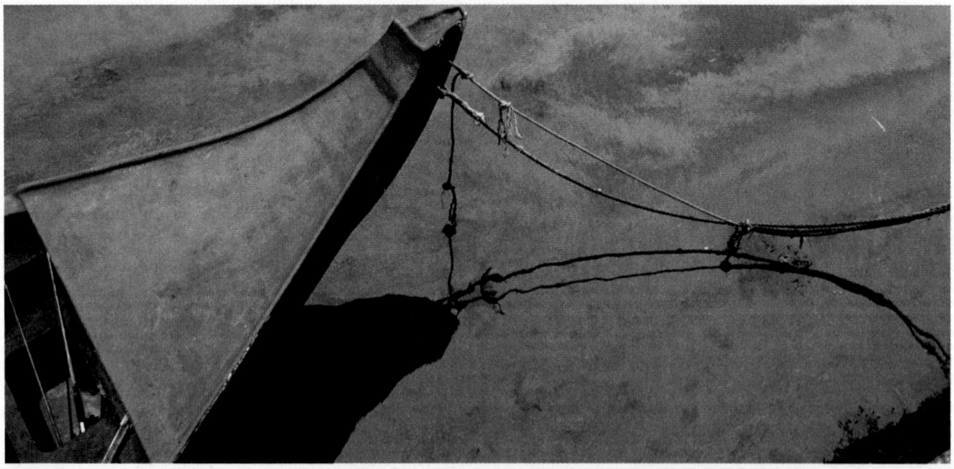

Figure 17 Oil is composed of nonpolar molecules, and water molecules are polar. Oil does not mix with water and remains on the surface of the water.

Keen Press/Photodisc/Getty Images

3D THINKING | DCI Disciplinary Core Ideas | CCC Crosscutting Concepts | SEP Science & Engineering Practices

COLLECT EVIDENCE

Use your Science Journal to record the evidence you collect as you complete the readings and activities in this lesson.

INVESTIGATE

GO ONLINE to find these activities and more resources.

Quick Investigation: Observe Clinging Molecules
Plan and carry out an investigation to determine the effects of polarity on the attraction of molecules.

Review the News
Obtain information from a current news story about nonaqueous solutions. Evaluate your source and communicate your findings to your class.

Nonpolar solvents

Some substances around your house may be useful as nonpolar solvents. For example, mineral oil can dissolve candle wax from glass or metal candleholders. Mineral oil and wax are nonpolar substances. Oily peanut butter, a nonpolar solvent, is sometimes used to remove bubble gum, also a nonpolar substance, from hair.

Many nonpolar solvents are connected with specific jobs. Oil-based paints contain pigments that are dissolved in oil. In order to thin or remove such paints, a nonpolar solvent must be used. People who paint pictures using oil-based paints probably use the solvent turpentine. It comes from the sap of a pine tree. **Figure 18** shows how well turpentine dissolves nonpolar paint.

Dry cleaners also use nonpolar solvents. The word *dry* refers to the fact that no water is used in the process. Because molecules of a nonpolar solute can easily slip in among molecules of a nonpolar solvent, dry cleaning removes oil and grease stains.

Drawbacks of nonpolar solvents Although nonpolar solvents have many uses, they have some drawbacks, too. First, many nonpolar solvents are flammable. Also, some solvents are extremely toxic. These solvents are hazardous if they come into contact with skin or if their vapors are inhaled. For these reasons, you must always be careful when handling these substances and never use them in an enclosed area. Good ventilation is critical because nonpolar solvents tend to evaporate more readily than water. Even small amounts can produce high concentrations of harmful vapor in the air.

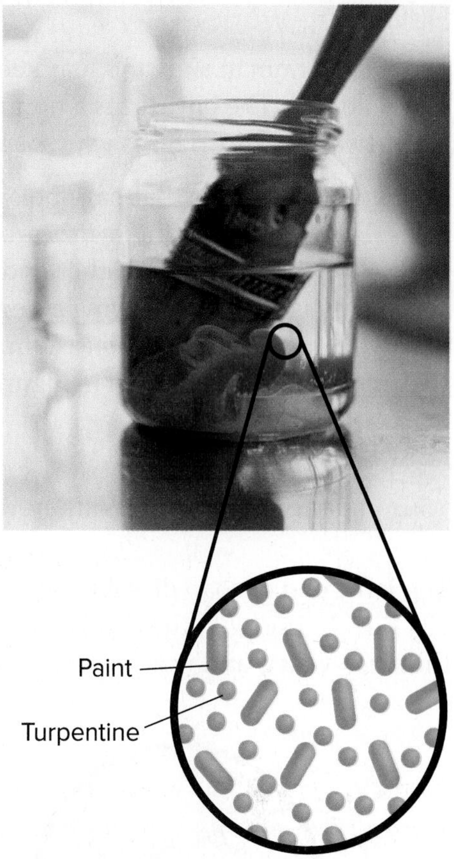

Figure 18 Turpentine, a nonpolar solvent, mixes with oil-based paints as a thinner and can also be used as a brush cleaner.

Versatile Molecules

Some substances are versatile because they have a nonpolar end and a polar end. For example, **Figure 19** shows that the alcohol ethanol has a polar and a nonpolar end. In addition, sodium stearate has a polar and a nonpolar end. As a result, ethanol and sodium stearate can dissolve polar and nonpolar substances. Sodium stearate is an important ingredient in soap.

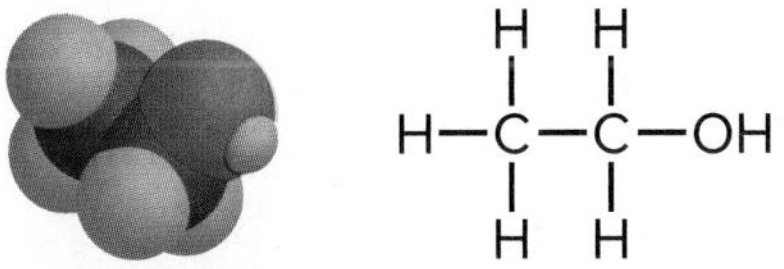

Ethanol (C_2H_5OH) has a polar –OH group and a nonpolar $–C_2H_5$ group.

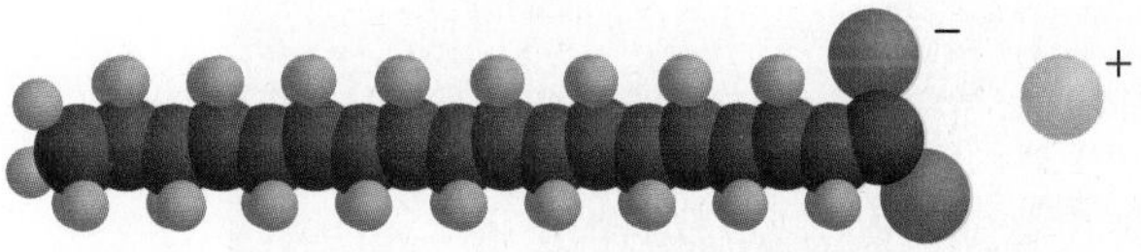

The long end of the hydrocarbon chain for sodium stearate is nonpolar. One end of the molecule is ionic.

Figure 19 Ethanol and sodium stearate are examples of molecules that have both polar and nonpolar ends.

How soap works

The oils on human skin and hair keep them from drying out, but the oils can also attract and hold dirt. The oily dirt is a nonpolar mixture, so washing with water alone will not clean away the dirt. This is why soap is needed.

Soaps are substances that have polar and nonpolar properties. Soaps are salts of fatty acids. Fatty acids are long hydrocarbon molecules with a nonpolar end and a carboxylic acid group –COOH at the other end. When a soap is made, the hydrogen atom of the acid group is removed, leaving a negative charge behind to form an ionic bond with a positive ion of sodium or potassium. For example, when Na^+ bonds with $-COO^-$, the sodium stearate salt, shown in **Figure 19,** is made.

Get It?

Summarize Why is soap required to clean oily dirt?

The ionic end of soap dissolves in water, and the long hydrocarbon portion dissolves in oily dirt. In this way, the dirt is removed from your skin, hair, or a fabric, suspended in the wash water, and washed away, as shown in **Figure 20.**

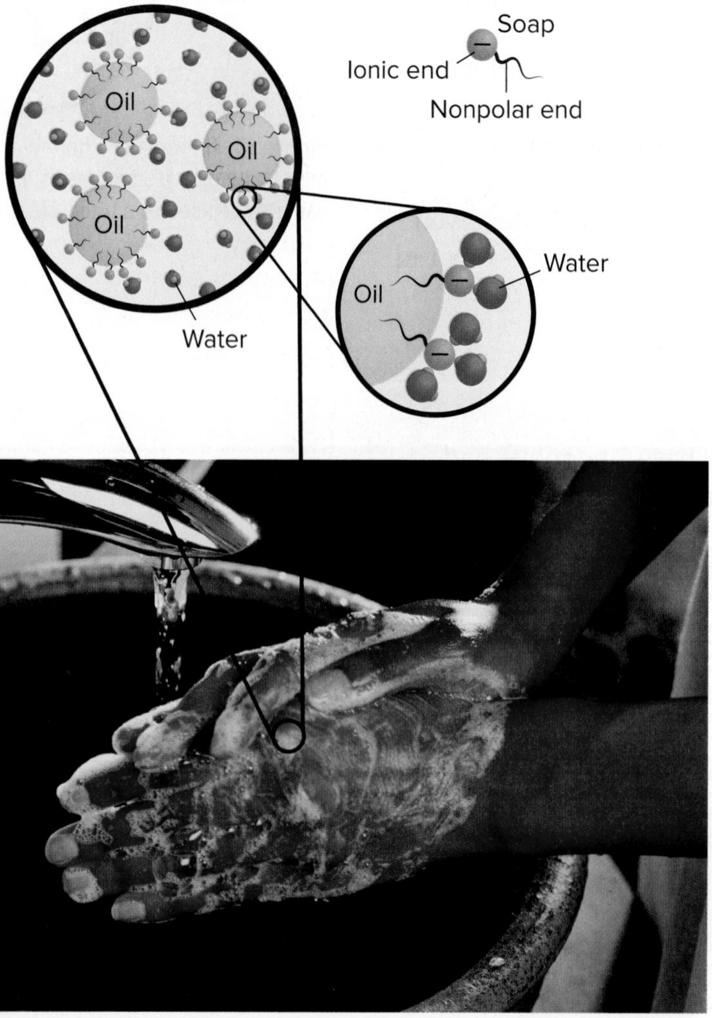

Figure 20 Soap cleans because its nonpolar hydrocarbon part dissolves in oily dirt and its ionic part interacts strongly with water. The soap carries the oily dirt along as you use the water to rinse.

Steve Mason/Photodisc/Getty Images

Figure 21 The structural formula of vitamin A shows a long hydrocarbon chain that makes it nonpolar. Foods such as liver, lettuce, cheese, eggs, carrots, sweet potatoes, and milk are good sources of this fat-soluble vitamin. Vitamins D, E, and K are also fat-soluble vitamins.

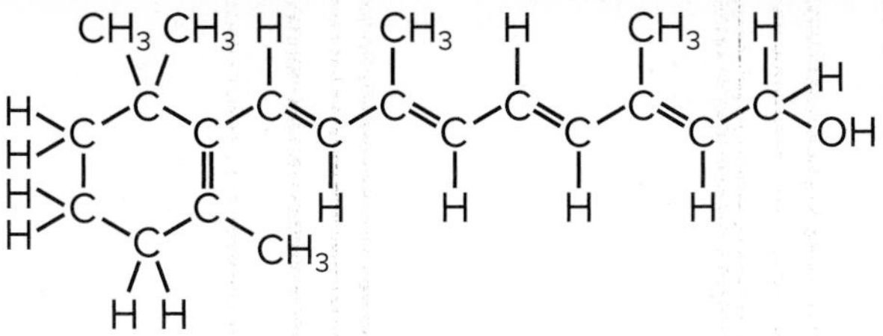

Vitamin A

Polarity and Vitamins

Taking the right types of vitamins in the correct doses is important to your health. Some of the vitamins that you need, such as vitamin A, shown in **Figure 21,** are nonpolar and dissolve in fat, which is another nonpolar substance. Because fat and fat-soluble vitamins do not wash away with the water that is present in the cells throughout your body, the excess vitamins can accumulate in your tissues. Fat-soluble vitamins are toxic in high concentrations, so taking large doses that are not recommended by your physician can be dangerous.

Other vitamins, such as vitamins B and C, are polar molecules, meaning they are water soluble. When you look at the structure of vitamin C, shown in **Figure 22,** you will see that it has several carbon-to-carbon bonds. This might make you think that it is nonpolar. But if you look again, you will see that it also has several oxygen-to-hydrogen bonds that resemble those found in water. This makes vitamin C polar.

Vitamin C

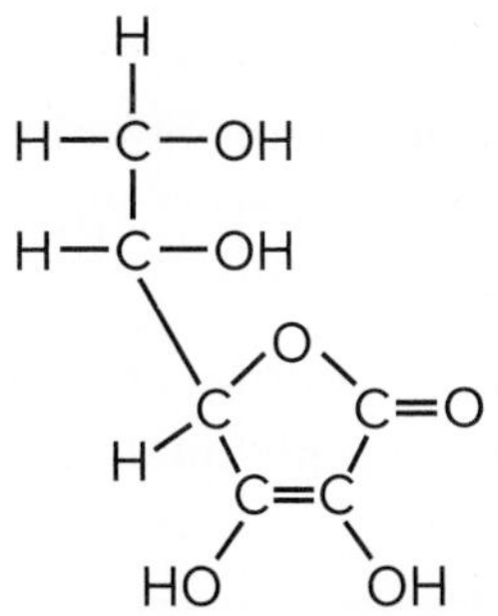

Figure 22 Although vitamin C has carbon-to-carbon bonds, it is water soluble because it also has polar groups. Foods that are good sources of vitamin C help heal wounds and help the body absorb iron.

Explain *Compare the number of oxygen atoms in vitamin C with the number in vitamin A (in Figure 21). What effect does oxygen have in these two molecules?*

(t)Maximilian Stock Ltd./Photolibrary/Getty Images, (b)Pikselstock/Shutterstock

Polar vitamins dissolve readily in the water that is present in your body. These vitamins do not accumulate in tissue because any excess vitamin is washed away. For this reason, you must replace water-soluble vitamins more quickly than fat-soluble vitamins by eating enough of the foods that contain them or by taking vitamin supplements. **Table 3** shows a variety of sources of vitamin C. In general, the best way to stay healthy is to eat a variety of healthy foods. Such a diet will supply the vitamins that you need with no risk of overdoses.

Get It?

Restate Why is it necessary to replace water-soluble vitamins more quickly than fat-soluble vitamins?

Table 3 Sources of Vitamin C

Food	Serving Size	Amount of Vitamin C (mg)
Orange juice, fresh	1 cup	124
Green peppers, raw	$\frac{1}{2}$ cup	96
Broccoli, raw	$\frac{1}{2}$ cup	70
Cantaloupe	$\frac{1}{4}$ melon	70
Strawberries	$\frac{1}{2}$ cup	42

Check Your Progress

Summary

- Polar solvents dissolve polar solutes, and nonpolar solvents dissolve nonpolar solutes.
- Nonpolar solvents have many household and industrial uses.
- Some molecules have both polar and nonpolar parts and are versatile solvents.

Demonstrate Understanding

21. **Explain** how a polar solvent dissolves a polar solute and how a nonpolar solvent dissolves a nonpolar solute.
22. **Describe** polar and nonpolar molecules.
23. **Explain** how one solute can dissolve in both polar and non-polar solvents.
24. **Draw** a diagram to explain how soap cleans your hands.

Explain Your Thinking

25. **Predict** What might happen to your skin if you washed too often?
26. **MATH Connection** If 60 mg of vitamin C in a multivitamin provides only 75 percent of the recommended daily dosage for children, how much is recommended?
27. **MATH Connection** To get the recommended dose of vitamin C, approximately how much fresh orange juice must you drink? (Refer to Table 3.)

LEARNSMART® Go online to follow your personalized learning path to review, practice, and reinforce your understanding.

Hidden Nobel Prize Medals

On April 9, 1940, chemist George de Hevesy faced a dilemma. Outside his laboratory in Copenhagen, Denmark, Nazi forces were occupying the city. Inside, he held in his hand two Nobel Prize medals for physics, belonging to Max von Laue and James Franck, German scientists who had protested Nazi policies. They had left their medals, which resembled the medal shown below, in Copenhagen for safekeeping.

The precipitate lead iodide is formed when lead nitrate reacts with sodium iodide.

Dissolving gold De Hevesy decided that the only way to make sure the Nazis would not seize the medals was to hide them in plain sight. Gold is not a very reactive element. One of the only chemicals that can accomplish the task is aqua regia, a mixture of nitric acid and hydrochloric acid. De Hevesy reacted the Nobel Prize medals with aqua regia, put the resulting solution in a jar, and stored it on a shelf with other liquid solutions.

Nobel Prizes are awarded in several categories including chemistry, medicine, physics, literature, economics, and peace.

Although the Nazis searched the laboratory, they left the solution undisturbed. After the end of World War II, the gold was extracted from the solution and sent to the Royal Swedish Academy of Sciences, the group that awards Nobel Prizes. Franck and von Laue later received newly cast medals to replace the ones that de Hevesy had dissolved.

The process of precipitation In the Nobel Prize medal example, the gold and aqua regia reacted to form substances that dissolve in water. Together, they formed a solution. The gold was extracted from the solution by a process called precipitation, as shown above. Precipitation occurs when a reaction in a solution forms a solid product, called a precipitate, that does not dissolve in the solvent. De Hevesy's gold formed a precipitate that could be filtered from the liquid, cleaned, dried, and molded into the replacement medals.

COMMUNICATE SCIENTIFIC IDEAS

Suppose you are a journalist present the day that Franck and von Laue received their recast Nobel Prize medals. Write a news article describing the scene and your interview with the scientists.

MODULE 21

STUDY GUIDE

GO ONLINE to study with your Science Notebook.

Lesson 1 HOW SOLUTIONS FORM

- A solution is a homogeneous mixture.
- Solutions are composed of solutes and solvents.
- Stirring, surface area, and temperature all affect the rate of dissolving.

- solute
- solvent
- alloy

Lesson 2 CONCENTRATION AND SOLUBILITY

- Solubility curves help predict how much solute can dissolve at a particular temperature.
- Saturated, unsaturated, and supersaturated solutions are defined by how much solute can dissolve in a solvent.
- Solutions absorb or give off energy as they form.
- Temperature and pressure affect how much gas dissolves in a liquid.

- concentration
- solubility
- saturated solution
- unsaturated solution
- supersaturated solution

Lesson 3 PARTICLES IN SOLUTION

- Ionization and dissociation are two ways to form ions in solution.
- Solute particles lower the freezing point of a solution.
- Solute particles raise the boiling point of a solution.

- electrolyte
- nonelectrolyte
- ionization
- dissociation

Lesson 4 DISSOLVING WITHOUT WATER

- Polar solvents dissolve polar solutes, and nonpolar solvents dissolve nonpolar solutes.
- Nonpolar solvents have many household and industrial uses.
- Some molecules have both polar and nonpolar parts and are versatile solvents.

THREE-DIMENSIONAL THINKING

Module Wrap-Up

REVISIT THE PHENOMENON

How do crystals form?

CER Claim, Evidence, Reasoning

Explain Your Reasoning Revisit the claim you made when you encountered the phenomenon. Summarize the evidence you gathered from your investigations and research and finalize your Summary Table. Does your evidence support your claim? If not, revise your claim. Explain why your evidence supports your claim.

STEM UNIT PROJECT

Now that you've completed the module, revisit your STEM unit project. You will summarize your evidence and apply it to the project.

GO FURTHER

SEP Data Analysis Lab

Volatile Organic Compounds

Some nonpolar substances, such as many organic solvents, readily evaporate at or near room temperature. The tendency of a liquid to evaporate at room temperature is called volatility. Nonpolar substances that tend to evaporate at room temperature are called volatile organic compounds (VOCs). The diagram compares the volatility of five organic solvents classified as VOCs with that of water.

Volatility of Organic Compounds Relative to Water

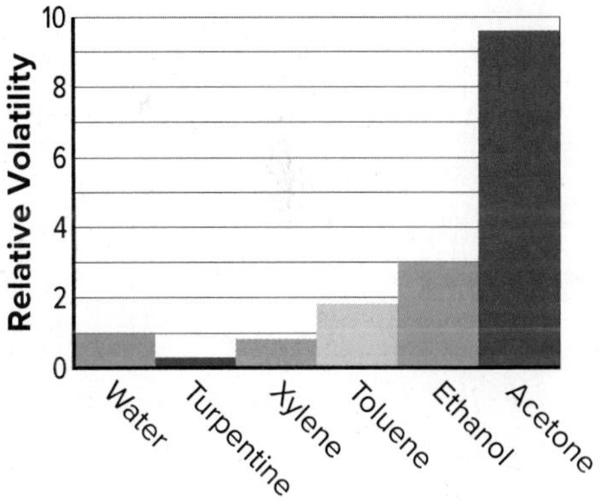

Because volatile organic solvents can be used to dissolve a wide range of nonpolar solutes, they are found in many household products, such as cleaners, stain removers, disinfectants, and insect repellents. VOCs are also used as solvents in materials such as glue, paint, varnishes, and lacquers. Stricter government guidelines on air and water pollution, and growing concerns among consumers, have resulted in an increase in the manufacturing of products with lower VOC levels.

CER Analyze and Interpret Data

1. **Claim** Use a dictionary to find the origin of the term *volatile*.
2. **Claim, Evidence** What properties make volatile organic solvents useful in household products and materials?
3. **Claim** Find out which of the substances in the diagram are compounds. Determine their chemical formulas and molecular mass numbers.
4. **Claim, Evidence, Reasoning** Research volatile organic compounds. Investigate the effects of VOCs on indoor air quality, the ozone layer, drinking water, or smog. Report your findings to the class.

Carsten Peter/Speleoresearch & Films/National Geographic/Getty Images

Evgeny Haritonov/Shutterstock

MODULE 22

ACIDS, BASES, AND SALTS

ENCOUNTER THE PHENOMENON

How do you determine if this stream is acidic or basic?

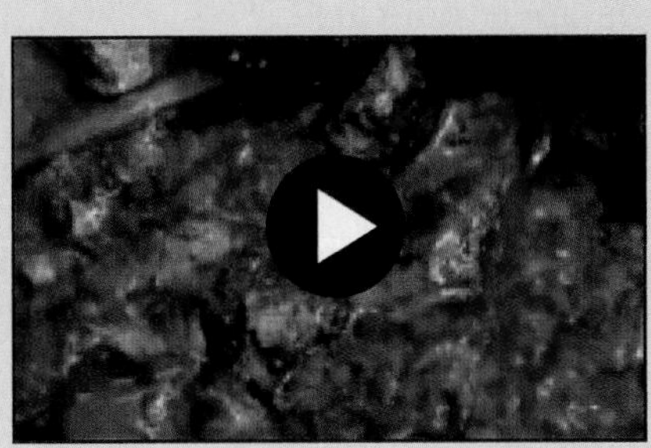

GO ONLINE to play a video about the acidification of water.

SEP Ask Questions

Do you have other questions about the phenomenon? If so, add them to the driving question board.

CER Claim, Evidence, Reasoning

Make Your Claim Use your CER chart to make a claim about how to determine if the stream is acidic or basic.

Collect Evidence Use the lessons in this module to collect evidence to support your claim. Record your evidence as you move through the module.

Explain Your Reasoning You will revisit your claim and explain your reasoning at the end of the module.

GO ONLINE to access your CER chart and explore resources that can help you collect evidence.

LESSON 1: Explore and Explain: Bases

LESSON 3: Explore and Explain: Common Salts

Smithsonian

Additional Resources

LESSON 1

ACIDS AND BASES

FOCUS QUESTION

How can you determine whether a solution is acidic or basic?

Acids

What comes to mind when you hear the word *acid*? Do you think of a substance that can burn your skin or burn a hole through metal? Do you think about sour foods like those shown in **Figure 1?** Although some acids can burn, the acids in foods are safe to eat.

Properties of acids

An **acid** is a substance that produces hydrogen ions (H^+) in a water solution. The ability to produce these ions gives acids their characteristic properties. As acids dissolve in water, H^+ ions interact with water molecules to produce **hydronium ions** (H_3O^+) (hi DROH nee um · I ahnz). Acids have several common properties. Acids cause the sour taste of many foods. However, taste should never be used to test for the presence of acids. Some acids can damage tissue by causing painful burns. Acids are corrosive. Some acids react strongly with certain metals, like zinc, and seem to eat away the metals as metallic compounds and hydrogen gas form. Acids also react with indicators. An **indicator** is an organic compound that changes color in the presence of acids and bases. For example, blue litmus paper is an indicator that turns red in acid.

Figure 1 The acids in these common foods give them their distinctive sour tastes.

3D THINKING DCI Disciplinary Core Ideas CCC Crosscutting Concepts SEP Science & Engineering Practices

COLLECT EVIDENCE

Use your Science Journal to record the evidence you collect as you complete the readings and activities in this lesson.

INVESTIGATE

GO ONLINE to find these activities and more resources.

Quick Investigation: Observe Acid Relief
Carry out an investigation to determine the effect of using an antacid tablet for relief.

Review the News
Obtain information from a current news story about acids and bases. **Evaluate** your source and **communicate** your findings to your class.

Table 1 Common Acids and Their Uses

Name, Formula	Use	Other Information
Acetic acid (CH_3COOH)	food preservation, commercial organic syntheses	vinegar (about 5% acetic acid)
Acetylsalicylic acid ($HOOC–C_6H_4–OOCCH_3$)	pain relief, fever relief; to reduce inflammation	main component of aspirin
Ascorbic acid ($H_2C_6H_6O_6$)	antioxidant, vitamin	Vitamin C occurs naturally in some foods and is added to others.
Carbonic acid (H_2CO_3)	carbonated drinks	involved in cave formation and acid rain
Hydrochloric acid (HCl)	cleans steel in a process called pickling	Gastric juice in the stomach is a solution of HCl and water.
Nitric acid (HNO_3)	to make fertilizers	colorless, yet yellows when it is exposed to light
Phosphoric acid (H_3PO_4)	to make soft drinks, fertilizers, and detergents	slightly sour but pleasant taste; detergents containing phosphates cause water pollution
Sulfuric acid (H_2SO_4)	car batteries; to manufacture fertilizers and other chemicals	dehydrating agent that extracts water from air

Common acids

Many foods contain acids. In addition to citric acid in citrus fruits, another example is lactic acid found in yogurt and buttermilk. It is also produced in your muscles during high levels of exercise when your muscles need oxygen. Any pickled food contains vinegar, also known as acetic acid. Your stomach uses hydrochloric acid to help digest your food.

At least four acids (sulfuric, phosphoric, nitric, and hydrochloric) play vital roles in industrial applications. All four of these acids are used in wastewater and water treatment. Sulfuric, phosphoric, and nitric acids are used in the production of fertilizers. Most of the nitric acid and sulfuric acid and approximately 90 percent of phosphoric acid produced are used for this purpose. Sulfuric acid is used to process pulp and make paper. Phosphoric acid is used in odor control. Nitric acid is involved in the production of nylon. Hydrochloric acid is part of the production of asphalt.

Get It?
Name four acids that are important for industry.

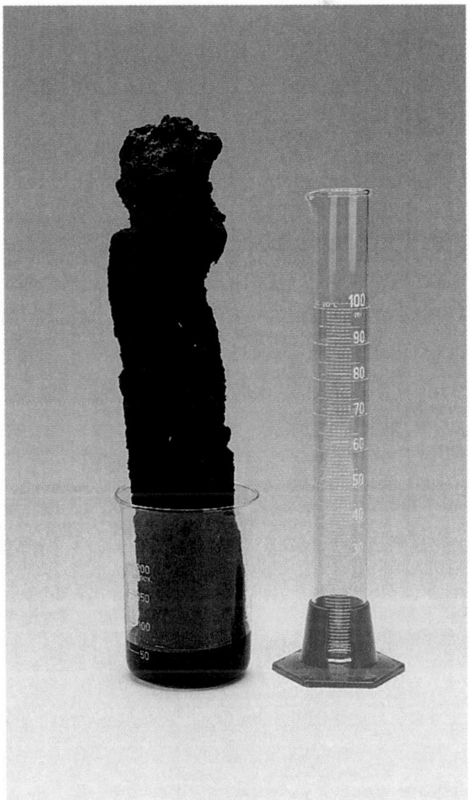

Figure 2 When sulfuric acid from the graduated cylinder is added to sugar in the beaker, a reaction occurs that produces black solid carbon and water.

Table 1 shows the names and formulas of a few acids, their uses, and some of their properties. Many acids can cause a chemical burn. For example, sulfuric acid reacts with skin cells and removes water as easily as it takes water from sugar, as shown in **Figure 2.**

Bases

Most bases contain an OH^-, called a **hydroxide ion,** in their chemical formula. A **base** is a substance that produces hydroxide ions when it is dissolved in water. In addition, a base is any substance that accepts H^+ from acids. The definitions are related because the OH^- ions produced by some bases accept H^+ ions produced by some acids. Acids can be defined using this framework as well. An acid is any substance that donates H^+ to a base. In this definition, acids and bases are complements of one another—the base accepts the ion donated by the acid.

Get It?

Describe how hydrogen ions are associated with both acids and bases.

You might not be as familiar with bases as you are with acids. Although you can eat some foods that contain acids, you do not consume many bases. Some foods, such as egg whites, are slightly basic. Other examples of basic materials are baking powder and amines, organic compounds with a $-NH_2$ group, found in some foods. Some medicines, such as milk of magnesia and antacids, are also basic. Still, you come in contact with many bases every day. For example, each time you wash your hands using soap, you are using a base. Bases remove dirt and grime.

Bases also are important in many types of cleaning products, such as those shown in **Figure 3.** Bases are important in industry. For example, sodium hydroxide is used in the paper industry to separate fibers of cellulose from wood pulp. These fibers are made into paper.

Properties of bases

Although acids and bases share some properties in common, bases have their own characteristic properties. In the pure, undissolved state, many bases are crystalline solids. In solution, bases feel slippery and have a bitter taste. Like strong acids, strong bases are corrosive, and contact with skin can result in severe chemical burns. Therefore, taste and touch should never be used to test for the presence of a base. Finally, like acids, bases react with indicators to produce changes in color. Red litmus paper turns blue in bases.

Figure 3 Bases are found in many cleaning products used in the home. For example, many glass cleaners contain ammonia (NH_3). **Identify** *the property of bases that is evident in soaps.*

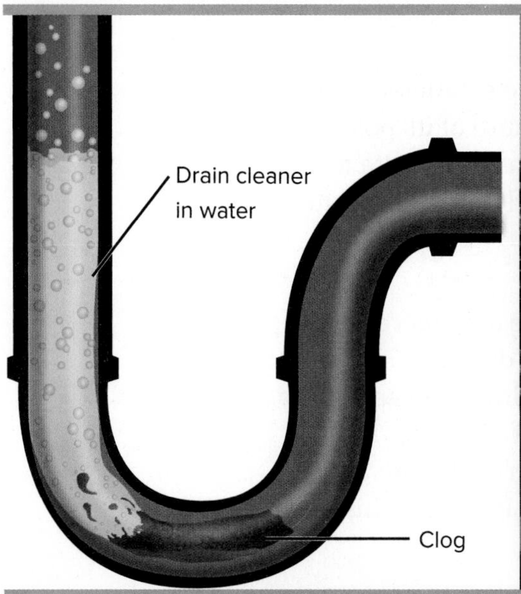

Some drain cleaners contain sodium hydroxide (NaOH), which dissolves grease, and small pieces of aluminum. The aluminum reacts with NaOH, producing hydrogen and dislodging solids, such as hair.

Aluminum hydroxide ($Al(OH)_3$) is a base that is used in water-treatment plants. Its sticky surface collects impurities, making them easier to filter from the water.

Figure 4 Bases are used in a variety of applications, ranging from the home to industrial settings.

Common bases

You are probably familiar with the bases in common cleaning products, but the uses of other bases, such as NaOH and $Al(OH)_3$, as shown in **Figure 4,** might be unfamiliar to you. **Table 2** lists some bases that have household and industrial uses and some additional information about them.

Table 2 Common Bases and Their Uses

Name, Formula	Use	Other Information
Aluminum hydroxide ($Al(OH)_3$)	color-fast fabrics, antacid, water purification	sticky gel that collects suspended clay and dirt particles on its surface
Calcium hydroxide ($Ca(OH)_2$)	to make leather, mortar and plaster; lessen acidity of soil	slaked lime
Magnesium hydroxide ($Mg(OH)_2$)	laxative, antacid	called milk of magnesia when it is mixed with water
Sodium hydroxide (NaOH)	to make soap, oven cleaner, drain cleaner, textiles, and paper	called lye and caustic soda; generates heat when it is combined with water; reacts with metals to form hydrogen (exothermic reaction)
Ammonia (NH_3)	cleaners, fertilizer; to make rayon and nylon	irritating odor that is damaging to nasal passages and lungs

Solutions of Acids and Bases

Most of the products that rely on the chemistry of acids and bases are solutions, such as the cleaning products and food products mentioned previously. Because of its polarity, water is the main solvent in these products. Solutions of acids and bases produce ions that are capable of conducting an electric current. Thus, they are said to be electrolytes.

Ionization of acids

You have learned that substances such as HCl, HNO_3, and H_2SO_4 are acids because of their abilities to produce hydrogen ions (H^+) in water. When an acid dissolves in water, the water molecules surround the neutral molecules of the acid, pulling them apart into ions. The positive hydrogen ions are attracted to the negative ends of the water molecules to form hydronium ions (H_3O^+). Therefore, an acid can be described more accurately as a compound that produces hydronium ions when dissolved in water. This process is shown in **Figure 5.**

Dissociation of bases

Many bases are ionic compounds, formed from a positive metal ion and a negative hydroxide ion (OH^-). If you look at **Table 2,** you will find that most of the substances listed contain the letters OH in their chemical formulas. When bases that contain the letters OH dissolve in water, the negative areas of nearby water molecules attract the positive ion in the base. The positive areas of nearby water molecules attract the OH^- of the base. The base dissociates into a positive metal ion and a negative hydroxide ion (OH^-). This process is also shown in **Figure 5.** Unlike acid ionization, water molecules do not combine with the ions formed from the base.

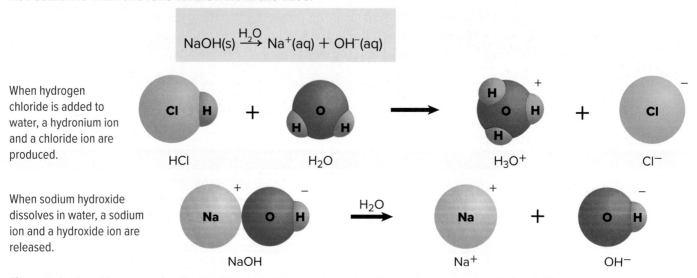

$$NaOH(s) \xrightarrow{H_2O} Na^+(aq) + OH^-(aq)$$

When hydrogen chloride is added to water, a hydronium ion and a chloride ion are produced.

When sodium hydroxide dissolves in water, a sodium ion and a hydroxide ion are released.

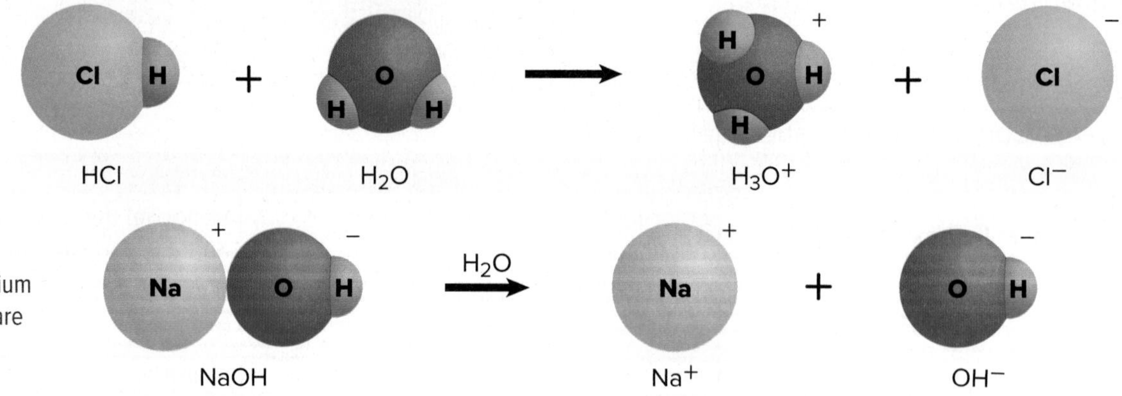

Figure 5 Acids and bases are classified by the ions that they produce when they are in an aqueous solution. Acids produce hydronium ions in water. Bases produce hydroxide ions in water.

WORD ORIGINS

acid

comes from the Latin *acidus* meaning *sour*

A grapefruit tastes sour because it contains ascorbic acid.

CCC CROSSCUTTING CONCEPTS

Structure and Function Research how the structure of water molecules plays an important role in how acids and bases form solutions. Prepare a presentation that explains your findings.

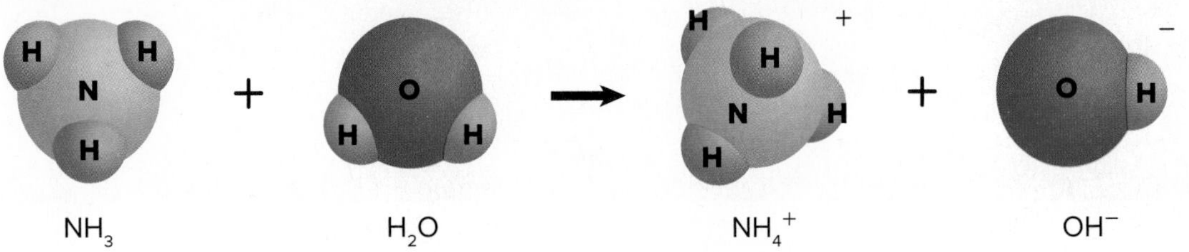

Figure 6 Ammonia reacts with water to produce hydroxide ions in solution; therefore, it is a base.

Ionization of bases

Recall that a base is also any substance that accepts an H^+ from acids. Ammonia (NH_3) is a base, even though it does not contain the letters OH in its chemical formula. Ammonia is a base because it accepts a hydrogen ion (H^+) from water. In a water solution, ionization takes place when the ammonia molecule attracts a hydrogen ion from a water molecule, forming an ammonium ion (NH_4^+). This produces a hydroxide ion (OH^-), as shown in **Figure 6.**

Get It?
Explain how ammonia reacts in a water solution.

Ammonia is a common household cleaner. However, products containing ammonia should never be mixed with other chlorine-based cleaners (sodium hypochlorite), such as some bathroom bowl cleaners and bleach. A reaction between sodium hypochlorite and ammonia produces the toxic gases hydrazine and chloramine. Breathing these gases can severely damage lung tissues and cause death.

Check Your Progress

Summary

- Acids are sour tasting and corrosive, and they make blue litmus turn red.
- Bases exist as crystals in the solid state, are slippery, have a bitter taste, are corrosive, and make red litmus turn blue.
- The polar nature of water allows acids and bases to dissolve in water and form ions.

Demonstrate Understanding

1. **Describe** how an acidic solution forms when HCl is mixed in water and how a basic solution forms when NaOH is mixed in water.
2. **Explain** what an indicator is.
3. **Write** the formulas of three important acids and three important bases, and describe their uses.
4. **Compare and contrast** how NH_3 and $Ca(OH)_2$ form OH^- ions in water.

Explain Your Thinking

5. **Apply** A friend asks you to get something from the kitchen, but he uses chemical formulas to ask for it. He asks for a drink which does not contain H_2CO_3, but does have $H_2C_6H_6O_6$. What might he be asking for?
6. **MATH Connection** Calculate the molecular mass of acetylsalicylic acid ($HOOC\text{-}C_6H_4\text{-}OOCCH_3$).

LEARNSMART® Go online to follow your personalized learning path to review, practice, and reinforce your understanding.

LESSON 2

STRENGTH OF ACIDS AND BASES

FOCUS QUESTION

Can you give an example of an acid that is harmful, and one that is not?

Strong and Weak Acids and Bases

Some acids must be handled with great care. For example, the sulfuric acid found in car batteries can burn your skin. Yet you drink some acids, such as citric acid in orange juice and carbonic acid in soft drinks. Obviously, some acids are stronger than others.

The strength of an acid or a base depends on the degree to which acid or base particles form into ions in water. In a **strong acid,** all the acid ionizes upon dissolving in water. Hydrochloric acid, nitric acid, and sulfuric acid are examples of strong acids. In a **weak acid,** only a small fraction of the molecules ionize upon dissolving in water. Acetic acid and carbonic acid are examples of weak acids.

Ions in solution can conduct an electric current. The more ions that a solution contains, the more current it can conduct. The ability of a solution to conduct a current can be demonstrated by using a lightbulb connected to a battery with leads placed in the solution, as shown in **Figure 7.** The strong acid solution conducts more current, and the lightbulb burns brightly. The weak acid solution does not conduct as much current as a strong acid solution, and the bulb burns less brightly.

Figure 7 Acetic acid, a weak acid, forms few ions, so the bulb is dim. Hydrochloric acid, a strong acid, forms many ions, so the bulb is brighter.

3D THINKING | DCI Disciplinary Core Ideas | CCC Crosscutting Concepts | SEP Science & Engineering Practices

COLLECT EVIDENCE

Use your Science Journal to record the evidence you collect as you complete the readings and activities in this lesson.

INVESTIGATE

GO ONLINE to find these activities and more resources.

Virtual Investigation: Titrations
Use a model to quantify the concentration of acidic solution.

CCC **Identify Crosscutting Concepts**
Create a table of the crosscutting concepts and fill in examples you find as you read.

Acid strength

In strong acids, such as hydrochloric acid (HCl), nearly all the acid ionizes to form hydronium ions (H_3O^+) and chloride ions (Cl^-) and leaves almost no HCl molecules present. This is shown by writing the equation using a single arrow pointing toward the ions that are formed.

$$HCl(g) + H_2O(l) \longrightarrow H_3O^+(aq) + Cl^-(aq)$$

Equations describing the ionization of weak acids, such as acetic acid (CH_3COOH), are written using double arrows pointing in opposite directions. This means that the reaction does not go to completion.

$$CH_3COOH(l) + H_2O(l) \leftrightharpoons H_3O^+(aq) + CH_3COO^-(aq)$$

An acetic acid solution is mostly made of CH_3COOH molecules. Only a relatively few hydronium ions (H_3O^+) and acetate ions (CH_3COO^-) are in solution.

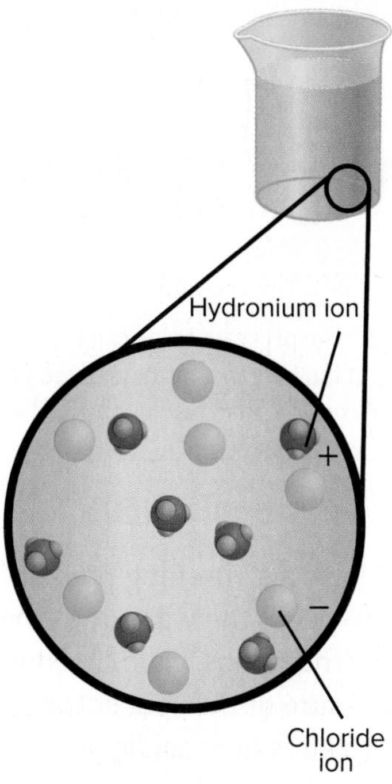

This is a dilute solution of hydrochloric acid, a strong acid.

Base strength

Many bases are ionic compounds that dissociate to produce ions when they dissolve. A **strong base** dissociates completely upon dissolving in water. The dissociation of sodium hydroxide (NaOH) to form sodium ions (Na^+) and hydroxide ions (OH^-) is shown below.

$$NaOH(s) \longrightarrow Na^+(aq) + OH^-(aq)$$

When ammonia (NH_3) is dissolved in water, an ammonium ion (NH_4^+) and a hydroxide ion (OH^-) form. This is shown using double arrows to indicate that not all of the molecules ionize. A **weak base** is one that does not ionize completely. Ammonia produces only a few ions in water, and most of the ammonia remains in the form of NH_3.

$$NH_3(aq) + H_2O(l) \leftrightharpoons NH_4^+(aq) + OH^-(aq)$$

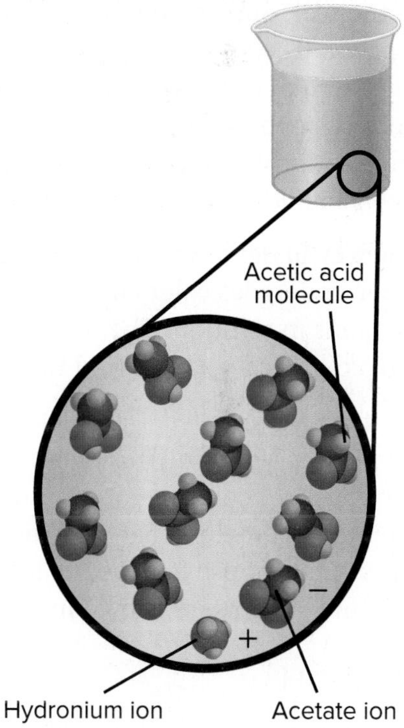

This is a concentrated solution of acetic acid, a weak acid.

Figure 8 You can have a dilute solution of a strong acid and a concentrated solution of a weak acid.

Strength and concentration

The terms *strength* and *concentration* can be confused when describing acids and bases. The terms *strong* and *weak* refer to the degree to which an acid or base ionizes or dissociates in solution. *Strong* acids ionize completely, and *strong* bases dissociate completely. *Weak* acids and bases ionize only partially.

In contrast, the terms *dilute* and *concentrated* indicate the concentration of a solution, which is the amount of acid or base dissolved in the solution. It is possible to have dilute solutions of strong acids and bases and concentrated solutions of weak acids and bases, as shown in **Figure 8.**

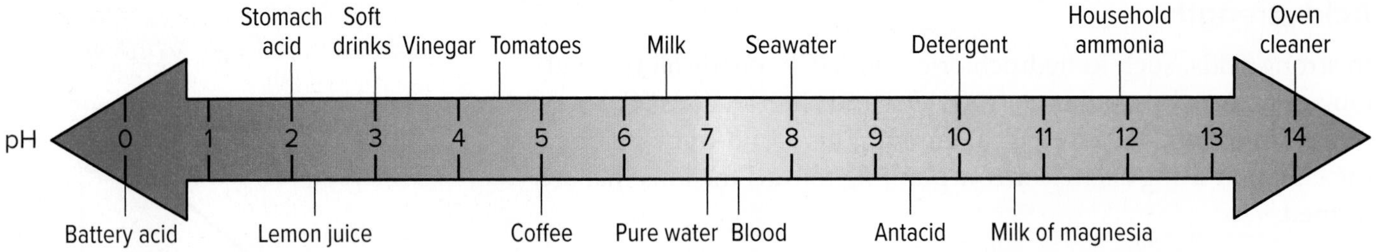

Figure 9 The pH scale helps classify solutions as acidic or basic. Each number on the pH scale represents a change in the H^+ concentration of ten times. For example, a solution of pH = 3 has an H^+ concentration that is ten times greater than a solution of pH = 4.

pH of a Solution

If you have a swimming pool or a tropical fish aquarium, you know that the pH of the water must be controlled. Also, many products, such as shampoos, claim to control pH so it suits your type of hair. **pH** is a measure of the concentration of H^+ (H_3O^+) ions in solution. The pH measures how acidic or basic a solution is. The greater the H_3O^+ concentration is, the lower the pH and the more acidic the solution. A scale to indicate pH has been devised, as shown in **Figure 9.**

As the scale shows, solutions with pHs lower than 7 are described as acidic. The lower the value, the more acidic the solution. Solutions with pHs greater than 7 are basic. The higher the pH is, the more basic the solution. A solution that has a pH of exactly 7 has equal concentrations of H_3O^+ ions and OH^- ions. This solution is considered to be neutral. Pure water at 25°C has a pH of 7.

One way to determine pH is by using universal indicator paper. This paper undergoes color changes in the presence of H_3O^+ ions and OH^- ions in solution. The final color of the pH paper is matched with colors in a chart to find the pH, as shown in **Figure 10.**

An instrument called a pH meter is another tool used to determine the pH of a solution. This meter is operated by immersing the electrodes in the test solution and reading the display. Small, battery-operated pH meters that have digital readouts are precise and convenient for use outside the laboratory when testing the pH of soils and streams, as shown in **Figure 10.**

Blood pH

Your blood circulates throughout your body, carrying oxygen, removing carbon dioxide, and absorbing nutrients from food that you have eaten. For blood to carry out its many functions properly, the pH of blood must remain between 7.0 and 7.8. The main reason for this is that enzymes, which are the protein molecules that act as catalysts for many reactions in the body, cannot work outside this pH range. Yet, you can eat foods that are acidic without changing the pH of your blood. How can this be? The answer is that your blood contains solutions called buffers that resist pH changes when small amounts of acids or bases are added.

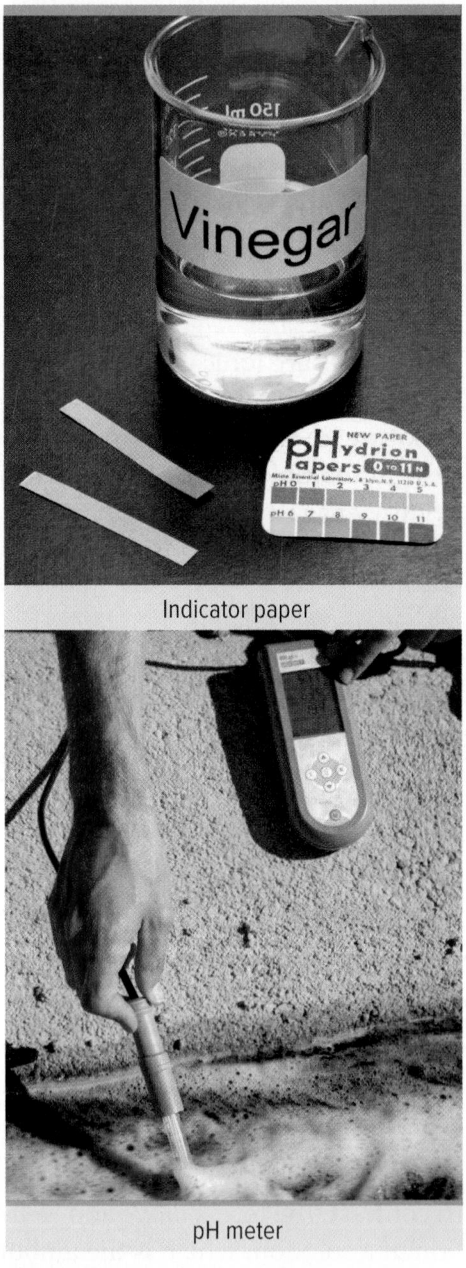

Figure 10 The pH of a sample can be measured in several ways, two of which (indicator paper or a pH meter) are shown here.

Determine *which method provides a more accurate reading.*

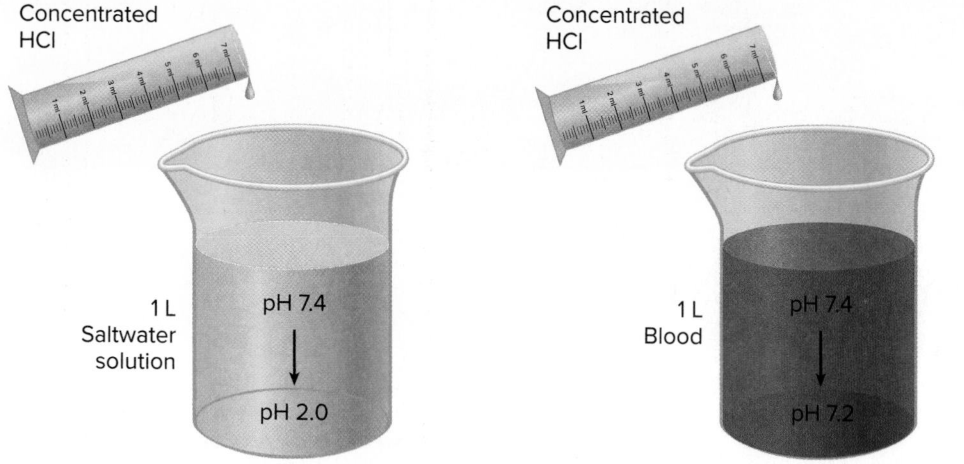

Figure 11 This experiment shows how well buffers in blood work. Adding 1 mL of concentrated HCl to 1 L of salt water changes the pH from 7.4 to 2.0. Adding the same amount of concentrated HCl to 1 L of blood changes the pH from 7.4 to 7.2.

Buffers are solutions containing ions that react with acids or bases to minimize their effects on pH. One buffer system in blood involves a solution of carbonic acid (H_2CO_3) and bicarbonate ions (HCO_3^-). Because of these buffer systems, even small amounts of concentrated acids will not change the pH of blood much, as shown in **Figure 11.** Buffers help keep your blood close to a nearly constant pH of 7.4. If the pH of 7.4 is not maintained, a medical condition called acidosis can occur. This can occur during severe dehydration when excessive levels of acids build up in body fluids.

Get It?

Explain what buffers are and how they are important for health.

Check Your Progress

Summary

- Strength refers to the degree to which an acid or a base forms ions in water. Concentration refers to how much acid or base is present in solution.
- Acids and bases can conduct electricity in solution.
- Acids and bases are classified based on pH.
- Buffers are solutions that minimize the effects of an acid or a base on pH.

Demonstrate Understanding

7. **Compare and contrast** a dilute solution of a strong acid and a concentrated solution of a weak acid.
8. **Describe** two techniques used to measure the pH of a solution.
9. **Explain** how electricity can be conducted by acids and bases.
10. **Relate** pH values of 9.1, 1.2, and 5.7 to hydronium and hydroxide ion concentration and characterize each as basic, acidic, or very acidic.

Explain Your Thinking

11. **Explain** The proper pH range for a swimming pool is between 7.2 and 7.8. Most pools use two substances, Na_2CO_3 and HCl, to maintain this range. How would you adjust the pH if you found it was 8.2? 6.9?
12. **MATH Connection** To determine the difference in pH strength, calculate 10^n, where n is the difference between pHs. How much more acidic is a solution of pH 2.4 than a solution of pH 4.4?

LEARNSMART® Go online to follow your personalized learning path to review, practice, and reinforce your understanding.

LESSON 3
SALTS

FOCUS QUESTION
Can you name different types of salt?

Neutralization

Normally, stomach acid contains a dilute solution of hydrochloric acid and water. Too much acid can cause indigestion. Antacids contain bases or other compounds of sodium, potassium, calcium, magnesium, or aluminum that react with stomach acid to lower acid concentration. **Figure 12** shows antacid tablets reacting in a solution that is similar to stomach acid. The equation for this reaction is:

$$HCl(aq) + NaHCO_3(s) \rightarrow NaCl(aq) + CO_2(g) + H_2O(l)$$

Neutralization is a chemical reaction in which an acid and a base form a salt and water. In the above equation, HCl is neutralized by $NaHCO_3$.

Salt formation

Hydrochloric acid can be neutralized by sodium hydroxide in another common neutralization reaction.

$$HCl(aq) + NaOH(aq) \rightarrow NaCl(aq) + H_2O(l)$$

In the equation above, the hydrogen ion and hydroxide ion form water. The remaining ions form the salt sodium chloride (NaCl). A **salt** is a compound formed when negative ions from an acid combine with positive ions from a base.

Antacids contain bases that neutralize excess stomach acid.

Figure 12 An antacid tablet containing the base sodium bicarbonate ($NaHCO_3$) reacts in your stomach much as it does in this dilute solution of HCl. Usually, people chew antacid tablets before swallowing them.

Explain *how chewing the tablet affects the rate of the reaction.*

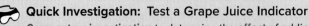

3D THINKING   DCI Disciplinary Core Ideas   CCC Crosscutting Concepts   SEP Science & Engineering Practices

COLLECT EVIDENCE

Use your Science Journal to record the evidence you collect as you complete the readings and activities in this lesson.

INVESTIGATE

GO ONLINE to find these activities and more resources.

Quick Investigation: Test a Grape Juice Indicator
Carry out an investigation to determine the effect of adding an acid or base to a solution.

Revisit the Encounter the Phenomenon Question
What information from this lesson can help you answer the Unit and Module questions?

Table 3 Common Salts and Their Uses

Name, Formula	Common Name	Uses
Sodium chloride (NaCl)	salt	food, manufacture of chemicals
Sodium hydrogen carbonate ($NaHCO_3$)	sodium bicarbonate, baking soda	food, antacids
Calcium carbonate ($CaCO_3$)	calcite, chalk	manufacture of paint and rubber tires
Potassium nitrate (KNO_3)	saltpeter	fertilizers
Potassium carbonate (K_2CO_3)	potash	manufacture of soap and glass
Sodium phosphate (Na_3PO_4)	TSP	detergents
Ammonium chloride (NH_4Cl)	sal ammoniac	dry-cell batteries

Acid-base general equation

The following general equation represents acid-base reactions in water.

$$\text{acid} + \text{base} \rightarrow \text{salt} + \text{water}$$

Another example of a neutralization reaction occurs between HCl, an acid, and $Ca(OH)_2$, a base. This reaction produces the salt $CaCl_2$ and water.

$$2HCl(aq) + Ca(OH)_2(aq) \rightarrow CaCl_2(aq) + 2H_2O(l)$$

Common Salts

There are many salts, a few of which are shown in **Table 3.** Most salts are composed of a positive metal ion and an ion with a negative charge, such as Cl^- or CO_3^{2-}. Ammonium salts contain the ammonium ion (NH_4^+) rather than a metal.

Salts are essential for many animals, large and small. Some domestic animals are supplied salts, as shown in **Figure 13.** Some animals find it at natural deposits. At these natural deposits, animals obtain ions such as sodium ions (Na^+) and calcium ions (Ca^{2+}), that are important for health. Even insects, such as butterflies, need salts and are often found clustered on moist ground. You also need salts, especially because you lose them through perspiration. How humans obtain one salt, sodium chloride, is shown in **Figure 14** on the next page.

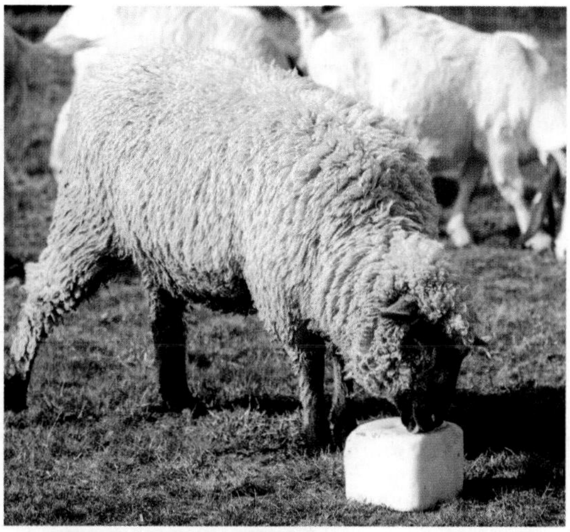

Figure 13 Like many animals, sheep need salts to help maintain body processes.

SCIENCE USAGE v. COMMON USAGE

salt

Science usage: a compound formed when negative ions from an acid combine with positive ions from a base

A salt is a product of the reaction between an acid and a base.

Common usage: an ingredient commonly used in cooking to provide flavor

Carol added salt to the eggs.

Figure 14 Visualizing Salt Production

Table salt (NaCl) is used for a variety of purposes: seasoning and preservation of food, deicing roadways, feeding animals and plants, and softening water. Salt is extracted from land and water. Three different technologies are commonly used.

Sodium Chloride (NaCl)

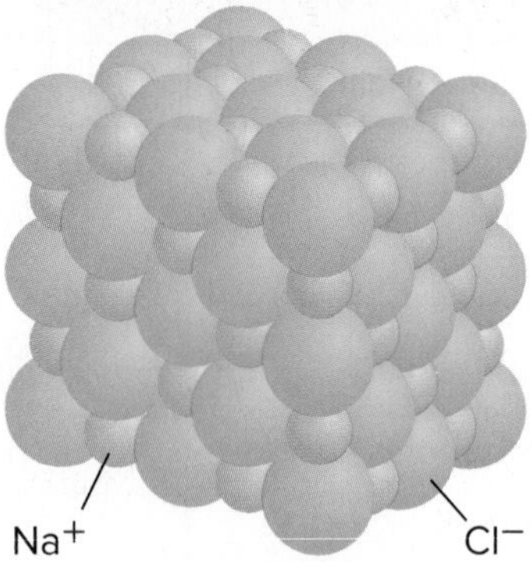

NaCl (Table salt)

Evaporation Salt Shallow ponds are filled with brine (salt water). The brine is moved from pond to pond to remove impurities and to concentrate it through evaporation. When the desired salt concentration has been reached, it is then pumped from evaporation ponds into crystallizing ponds. After the salt crystallizes, the remaining water is drained off. The crystals are washed, crushed, and dried.

Rock Salt Salt deposits exist due to the gradual evaporation of bodies of localized salt water. Salt can be found in deposits that have layers of sedimentary rocks, such as anhydrite, which is a calcium sulfate mineral. Salt deposits are found at relatively shallow depths, averaging 150 m to 600 m below Earth's surface. Salt is mined from these deposits by drilling and blasting.

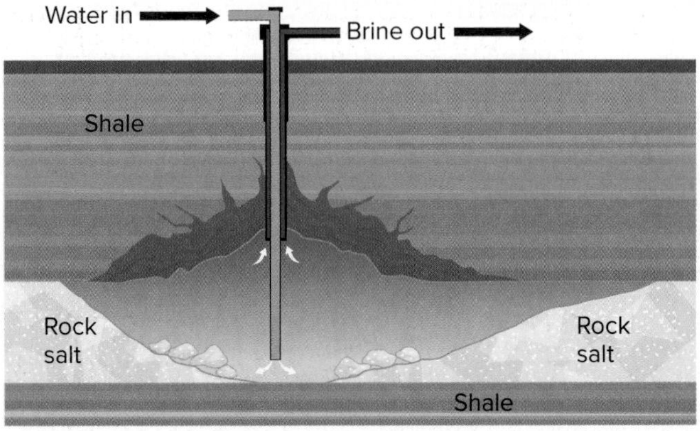

Solution Mining Solution mining is another technology used to obtain the table salt that we sprinkle on our food. A well pumps water into an underground salt deposit to dissolve the salt and to form a saturated solution called a brine. The brine is heated, and the salt crystallizes after the water boils away.

Titration

Sometimes you need to know the amount of acid or base in a solution, for example, to determine the purity of a commercial product. This can be done using titration (ti TRAY shun). **Titration** is the process in which a solution of known concentration is used to determine the concentration of another solution. **Figure 15** shows a titration procedure.

Titration involves a solution of known concentration, called the standard solution. The standard solution is slowly added to a solution of unknown concentration, which contains an acid/base indicator. If the solution of unknown concentration is a base, a standard acid solution is used. If the solution of unknown concentration is an acid, a standard base solution is used.

The endpoint has a color change

The titration shown in **Figure 15** shows how to determine the concentration of an acid solution. First, add a few drops of an indicator, such as phenolphthalein (fee nul THAY leen), to a carefully measured volume of the acidic solution of unknown concentration. Phenolphthalein is colorless in an acid, but it turns bright pink in the presence of a base.

Then, slowly and carefully add a basic solution of known concentration to this acid/indicator mixture. Toward the end of the titration, add the base drop by drop until one last drop of the base turns the solution pink and the color persists. The point at which the color persists is known as the endpoint–the point at which the acid is completely neutralized by the base. When the volume of base used is known, use that value, the known concentration of the base, and the volume of the acid to calculate the concentration of the acid solution.

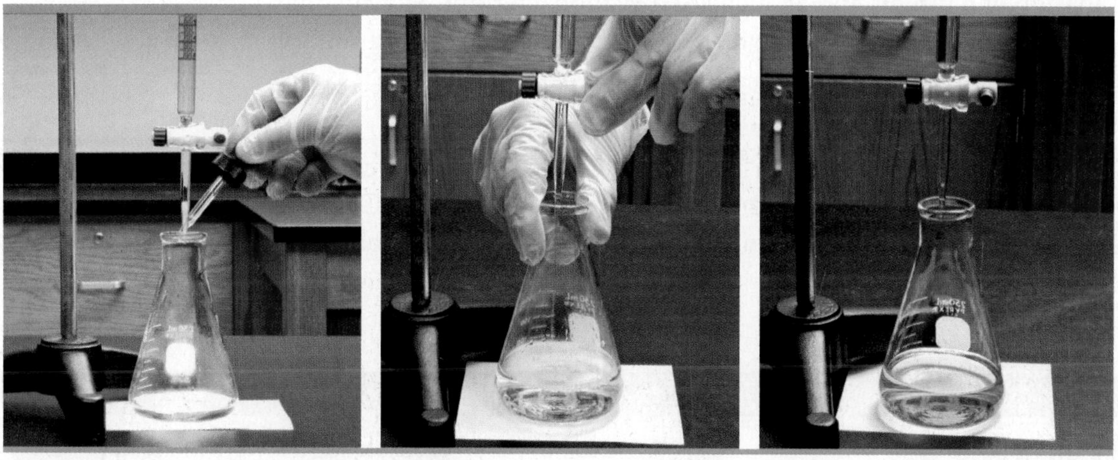

Figure 15 In this titration, a base of known concentration is being added to an acid of unknown concentration. The swirl of pink color shows that the end point is near but disappears until the endpoint has been reached. A permanent light-pink color marks the endpoint of this titration.

Explain *why you must add the base to the acid drop by drop near the end of the titration.*

ACADEMIC VOCABULARY

process
a series of actions or operations leading to an end
To obtain accurate results, the process of titration must be slow.

STEM CAREER Connection

Water and Wastewater Treatment Plant Operator
Providing clean, safe drinking water to homes and businesses is an important task. Those who operate water and wastewater treatment plants must carefully monitor the processes and run frequent tests on the water at the plant to ensure the safety of the water, including its pH and the concentrations of dissolved substances.

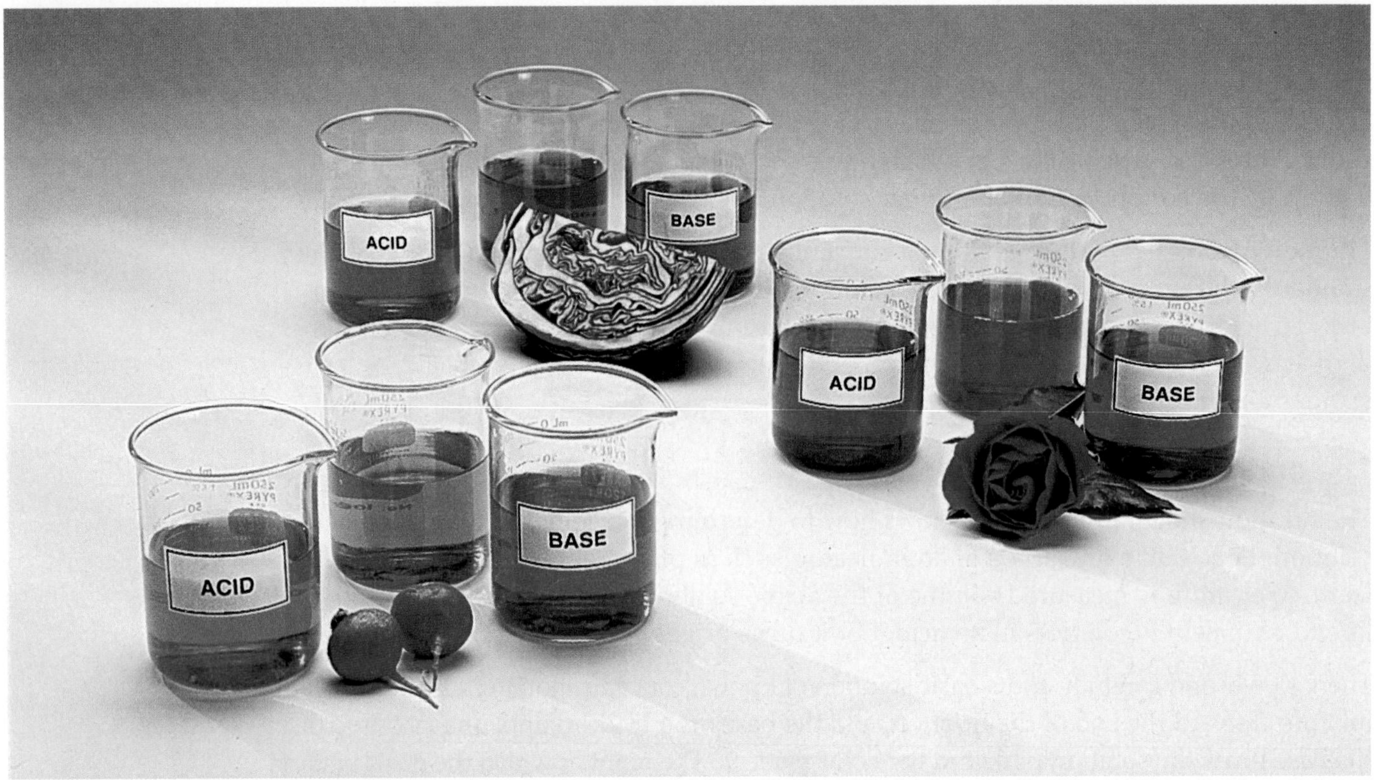

Figure 16 Natural indicators include red cabbage, radishes, and roses.

Many natural substances are acid-base indicators. In fact, the indicator litmus comes from a lichen, a combination of a fungus and an algae or a cyanobacterium. Flowers that are indicators include hydrangeas, which produce blue blossoms when the pH of the soil is acidic and pink blossoms when the soil is basic. The color change of hydrangeas is just the opposite of the color change of litmus.

Other natural indicators possess a range of colors. For example, the color of red cabbage varies from deep red at pH 1 to lavender at pH 7 and yellowish-green at pH 10. Grape juice is also an indicator. **Figure 16** shows common substances that can be used as natural indicators.

APPLY SCIENCE

How can you handle an upsetting situation?

At some time, most of us have experienced an upset stomach. Often, the cause is excess acid within our stomachs. For digestive purposes, our stomachs contain dilute hydrochloric acid with a pH between 1.6 and 3.0. A doctor might recommend an antacid treatment for an upset stomach. What type of compound is "anti acid"?

Identify the Problem

You have learned that neutralization reactions of acids and bases form salts. Antacids typically contain small amounts of $Ca(OH)_2$, $Al(OH)_3$, or $NaHCO_3$, which are bases.

Whereas having an excess of acid lowers the pH of your stomach contents, antacid compounds raise the pH of your stomach contents. How does an antacid change pH to make you feel better?

Solve the Problem

1. Write the chemical equation for the reaction of HCl and $Mg(OH)_2$. Use it to explain why an antacid helps with excess stomach acid.
2. Why is it important to have some acid in your stomach?
3. Design an experiment that compares how well various antacid products neutralize acid.

Matt Meadows/McGraw-Hill Education

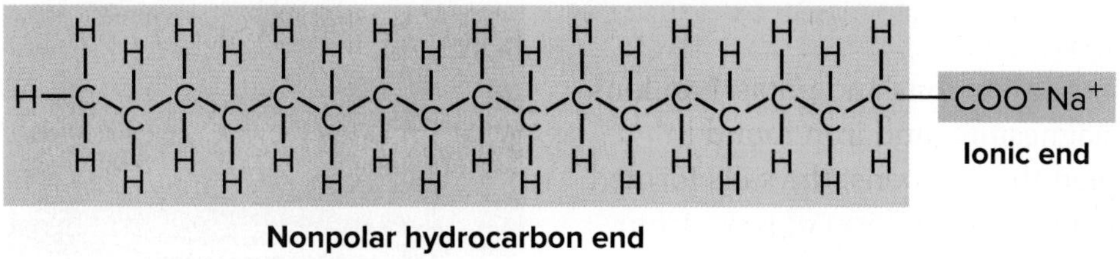

Figure 17 Soaps that contain sodium, like this one made from stearic acid, are solids. Those that contain potassium are liquids.
Write *the chemical formula for this soap.*

Soaps and Detergents

The next time that you are in a supermarket, go to the aisle with soaps and detergents. You will see all kinds of products—solid soaps, liquid soaps, and detergents for washing clothes and dishes. What are all these products? Do they differ from one another? Yes, they differ slightly in how they are made and in the ingredients included for color and aroma. Still, all these products are classified into two types—soaps and detergents.

Soaps

The reason why soaps clean so well is explained by polar and nonpolar molecules. **Soaps** are salts. As shown in **Figure 17,** they have a nonpolar chain of carbon atoms on one end and a sodium or potassium salt of a carboxylic (kar bahk SIH lihk) acid group (−COOH) at the other end. The nonpolar, hydrocarbon end interacts with nonpolar oils and dirt so they can be removed readily, and the polar, ionic end ($-COO^-Na^+$ or $-COO^-K^+$) helps them dissolve in polar water.

To make an effective soap, the acid must contain 12 to 18 carbon atoms. If it contains fewer than 12 atoms, it will not be able to mix well to clean oily dirt. If it has too many carbon atoms, the salt will not be soluble in water. **Figure 18** shows how soap interacts with dirt particles to clean your hands.

Get It?
Explain why soaps must have polar and nonpolar ends.

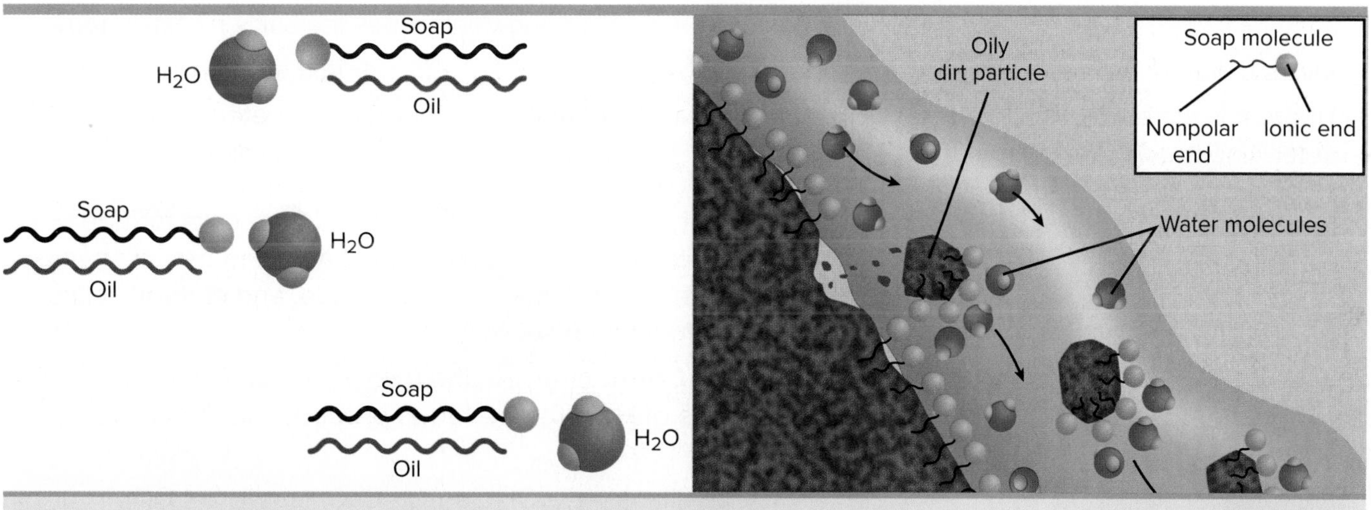

The long, hydrocarbon end of a soap molecule mixes well with oily dirt, and the ionic end attracts water molecules.

Dirt that is linked with the soap rinses away as water flows over it.

Figure 18 Soaps are salts that have a nonpolar end and an ionic end.

Commercial soaps

One problem with all soaps is that the sodium ions and potassium ions can be replaced by ions of calcium, magnesium, and iron found in some water known as hard water. When this happens, the salts formed are insoluble and precipitate out of solution in the form of soap scum. Detergents were developed to avoid this problem.

Get It?

Explain What is soap scum?

Detergents

Similar to soaps, detergents have long hydrocarbon chains. But instead of a salt of a carboxylic acid group at the end, they may contain a salt of a sulfonic acid group. These acids form more soluble salts with the ions in hard water, lessening the problem of soap scum. Detergents can also be used in cold water. Most contain additional ingredients called builders and surfactants to enhance the ability to clean and to produce suds. Despite solving the hard water problem, detergents are not the complete solution to our needs. Some detergents contained phosphates, but are no longer made because they cause water pollution. Certain sulfonic acid detergents cause excess foaming in water treatment plants and streams, as shown in **Figure 19.** These detergents do not break down easily and remain in the environment for long periods of time.

Figure 19 Foam from detergents can build up in waterways. Phosphorus, which caused algal blooms, has been removed from detergents, but such pollution can still be harmful. For example, fish eggs can be destroyed by detergent pollution.

Check Your Progress

Summary

- Neutralization is a chemical reaction between an acid and a base.
- Salt is a dietary essential.
- Titration is a method used to determine the concentration of an acidic or a basic solution.
- Molecules of soaps and detergents have polar and nonpolar ends.

Demonstrate Understanding

13. **Write** the balanced chemical equation for one neutralization reaction. In your equation, which reactant contributes the salt's positive ion? Which one contributes the salt's negative ion?
14. **Identify** the purpose of an indicator in a titration experiment.
15. **Compare and contrast** the composition of detergents and soaps.

Explain Your Thinking

16. **Predict** Give the names and formulas of the salts formed in the following neutralizations: sulfuric acid and calcium hydroxide, nitric acid and potassium hydroxide, and carbonic acid and aluminum hydroxide.
17. **MATH Connection** In the following reaction, how many molecules of HCl are needed to produce four molecules of H_2O?

$$2HCl(aq) + Ca(OH)_2(aq) \rightarrow CaCl_2(aq) + 2H_2O(l)$$

LEARNSMART® Go online to follow your personalized learning path to review, practice, and reinforce your understanding.

NATURE OF SCIENCE

Preventing Acid Precipitation

Fish, shellfish, and insects die. The leaves and needles of trees are damaged. Some trees die. Other animals, such as frogs and birds, grow sick and die. In a nearby city, something is slowly damaging the edges of buildings and statues, as shown below. What has caused this devastation? Acid precipitation is the culprit.

One renewable resource that is widely used today is hydroelectricity.

What is acid precipitation?

Acid precipitation forms when sulfur dioxide (SO_2) and nitrogen oxides (NO_x) in the atmosphere combine with oxygen, other chemicals, and precipitation, such as rain and snow, to form acids. These substances cause the precipitation's pH to fall from its normal pH of approximately 5.6 to between 5.0 and 5.5. In some parts of the United States, the pH of acid precipitation is as low as 4.3.

Damage caused by acid precipitation

Reduction and prevention

To reduce or prevent acid precipitation, the U.S. government requires power plants that burn fossil fuels, especially coal, to keep emissions of these gases below certain levels by installing scrubbers, which remove SO_2 from smoke. Power plants have also started burning coal that contains less sulfur.

For the last 20 years, automobiles made in the U.S. have been required to have catalytic converters, which reduce the emission of NO_x. It also helps to use alternative energy sources, such as solar, wind, and water power, which do not release SO_2 and NO_x. Further reductions occur by turning off lights and appliances when they are not in use and by not overheating or overcooling homes. New appliances are more energy efficient. To reduce pollution from cars, people can carpool, use public transportation, ride bicycles, walk, or drive vehicles that run on alternative fuels, which emit no or low amounts of harmful gases.

CONSTRUCT AN EXPLANATION BASED ON EVIDENCE

Work with a partner to locate information and evidence of the effects of acid precipitation in your community. Explain how scientific knowledge informed decisions made by your local government.

MODULE 22

STUDY GUIDE

GO ONLINE to study with your Science Notebook.

Lesson 1 ACIDS AND BASES

- Acids are sour tasting and corrosive, and they make blue litmus turn red.
- Bases exist as crystals in the solid state, are slippery, have a bitter taste, are corrosive, and make red litmus turn blue.
- The polar nature of water allows acids and bases to dissolve in water and form ions.

- acid
- hydronium ion
- indicator
- hydroxide ion
- base

Lesson 2 STRENGTH OF ACIDS AND BASES

- Strength refers to the degree to which an acid or a base forms ions in water. Concentration refers to how much acid or base is present in solution.
- Acids and bases can conduct electricity in solution.
- Acids and bases are classified based on pH.
- Buffers are solutions that minimize the effects of an acid or a base on pH.

- strong acid
- weak acid
- strong base
- weak base
- pH
- buffer

Lesson 3 SALTS

- Neutralization is a chemical reaction between an acid and a base.
- Salt is a dietary essential.
- Titration is a method used to determine the concentration of an acidic or a basic solution.
- Molecules of soaps and detergents have polar and nonpolar ends.

- neutralization
- salt
- titration
- soap

THREE-DIMENSIONAL THINKING

Module Wrap-Up

REVISIT THE PHENOMENON

How do you determine if this stream is acidic or basic?

CER Claim, Evidence, Reasoning

Explain Your Reasoning Revisit the claim you made when you encountered the phenomenon. Summarize the evidence you gathered from your investigations and research and finalize your Summary Table. Does your evidence support your claim? If not, revise your claim. Explain why your evidence supports your claim.

STEM UNIT PROJECT

Now that you've completed the module, revisit your STEM unit project. You will summarize your evidence and apply it to the project.

GO FURTHER

SEP Data Analysis Lab

White Gloves

Hotels and cleaning services often boast about giving customers the "white glove treatment," an expression long associated with extreme cleanliness. It refers to something being so clean that even a white-gloved hand can't detect any dirt. In reality, the white glove treatment has little to do with dirt and everything to do with acids.

Don't Touch!

Acids are caustic, meaning they can deteriorate or eat away at materials, causing them to wear down over time. Due to the acidic oils on the human hand, museums post signs asking patrons not to touch objects on display. Museum staff sometimes wear white cotton gloves to handle artifacts without damaging them.

Sweat Damage

The human hand is full of sweat glands. Sweat glands keep the surfaces of palms and soles damp so we can grip and feel things. The secretion from a sweat gland is 99 percent water. The remaining 1 percent is potassium, urea, lactic acid, ammonia, and salts. These chemicals can cause irreversible damage to a variety of materials, including paper, metals, fabric, photographs, wood, and paintings.

CER Analyze and Interpret Data

1. **Claim, Evidence, Reasoning** Can paper materials, such as old documents or antique books, escape harm through use of the "white glove treatment"? Explain.
2. **Claim** Name two acidic compounds found in sweat glands.

Rweisswald/iStock/Getty Images

ORGANIC COMPOUNDS

ENCOUNTER THE PHENOMENON

Why is this natural fire always burning?

GO ONLINE to play a video about how the flames of Yanartaş continuously burn.

SEP Ask Questions

Do you have other questions about the phenomenon? If so, add them to the driving question board.

CER Claim, Evidence, Reasoning

Make Your Claim Use your CER chart to make a claim about why this natural fire is always burning.

Collect Evidence Use the lessons in this module to collect evidence to support your claim. Record your evidence as you move through the module.

Explain Your Reasoning You will revisit your claim and explain your reasoning at the end of the module.

GO ONLINE to access your CER chart and explore resources that can help you collect evidence.

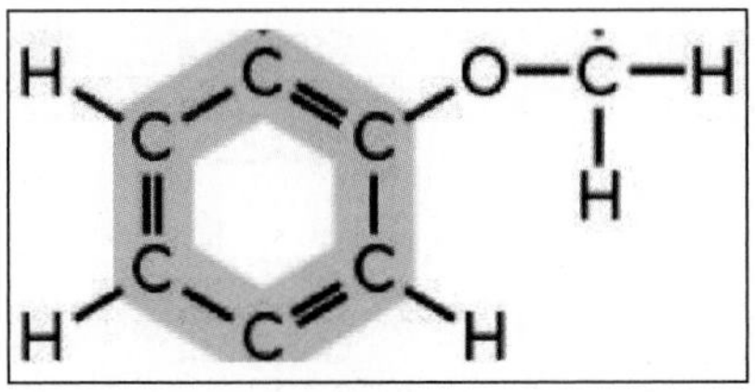

LESSON 1: Explore and Explain: Organic Compounds

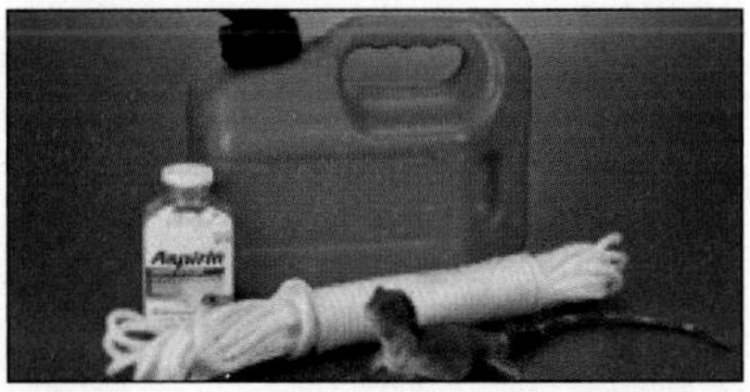

LESSON 3: Explore and Explain: Uses for Petroleum Compounds

Smithsonian

Additional Resources

LESSON 1

SIMPLE ORGANIC COMPOUNDS

FOCUS QUESTION

Why does the quantity and orientation of carbon impact a molecule's boiling point?

Organic Compounds

What do you have in common with gasoline, vanilla flavoring, and natural rubber? These items contain carbon compounds, as shown in **Figure 1.** You also contain carbon compounds. Most compounds containing the element carbon are **organic compounds.**

At one time, scientists thought that only living organisms made organic compounds, which is how they got their name. By 1830, scientists had made organic compounds in laboratories, but they still called them organic compounds. Of the millions of carbon compounds known today, more than 90 percent of them are organic compounds. The others, including carbon dioxide and the carbonates, are inorganic compounds.

Bonding

You may wonder why carbon can form so many different organic compounds. A carbon atom has four electrons in its outer energy level. Therefore, a carbon atom can form four covalent bonds with atoms of carbon or with other elements. Recall that a covalent bond is formed when two atoms share a pair of electrons. The four available bonding sites allow carbon to form single, double, and triple bonds. As a result, carbon can make many types of compounds, including small compounds used as fuel, complex compounds found in medicines, and the long chains used in plastics.

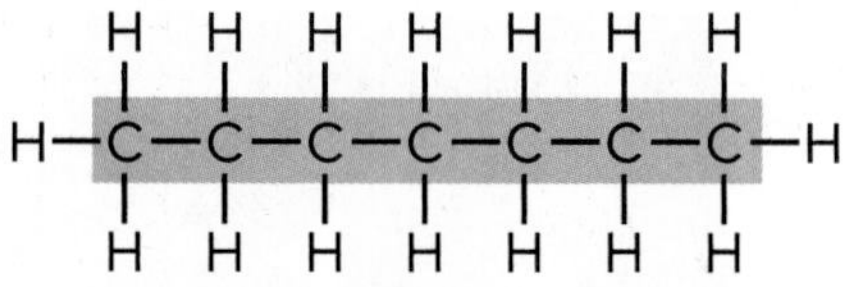

Heptane is a component of gasoline.

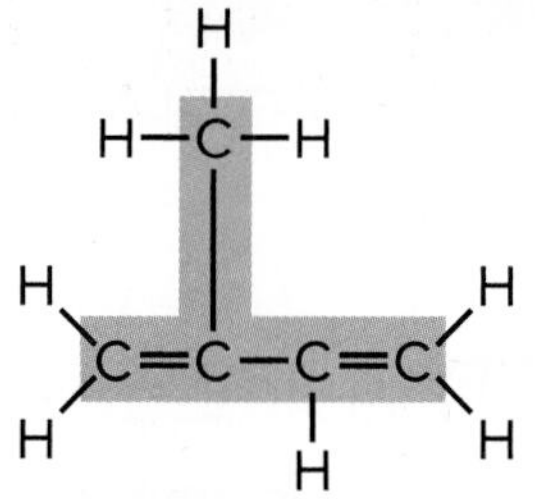

Isoprene exists in natural rubber.

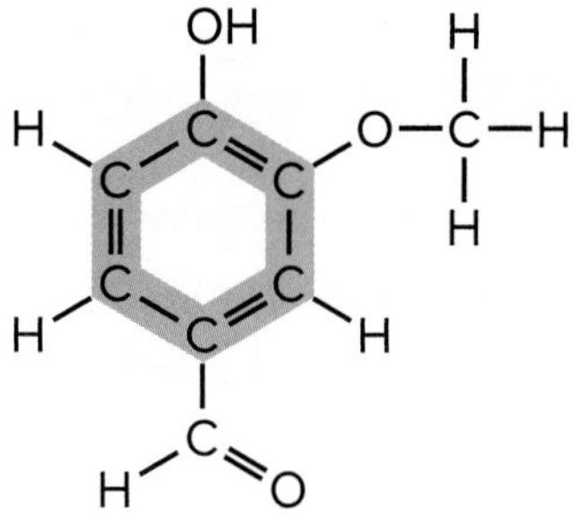

Vanillin is found in vanilla flavoring.

Figure 1 Organic compounds contain carbon. Carbon can form straight chains, branched chains, and rings.

Get It?

Explain why the element carbon can form so many different organic compounds.

3D THINKING DCI Disciplinary Core Ideas CCC Crosscutting Concepts SEP Science & Engineering Practices

COLLECT EVIDENCE

Use your Science Journal to record the evidence you collect as you complete the readings and activities in this lesson.

INVESTIGATE

GO ONLINE to find these activities and more resources.

Virtual Investigation: Carbon Chemistry
Develop and use a model to explore the structure of carbon compounds.

Quick Investigation: Model Hexane Isomers
Develop and use a model to explore the structure of hexane isomers.

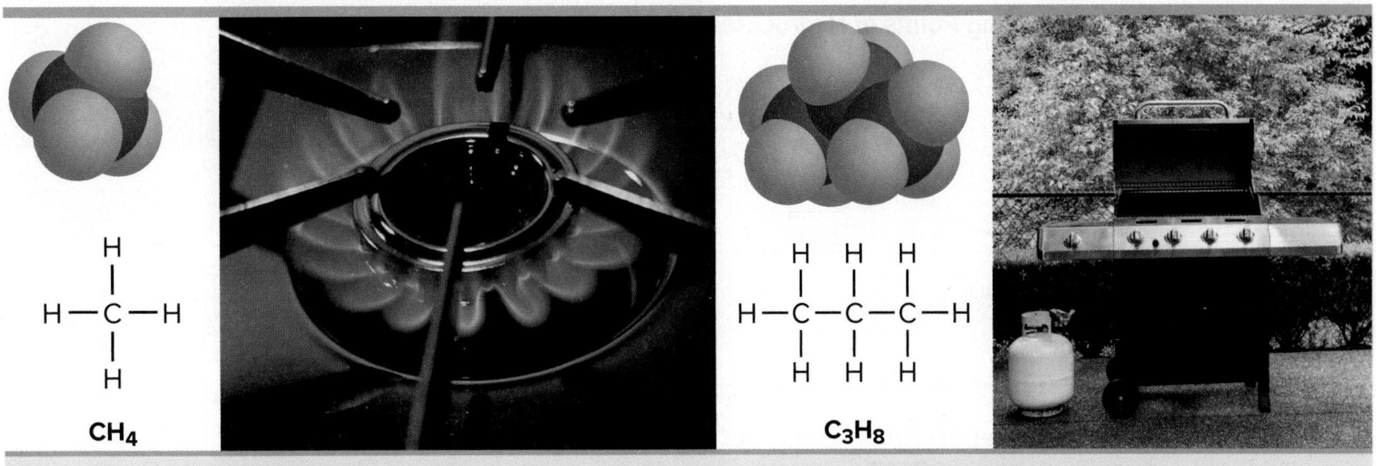

Natural gas is mostly methane (CH_4). Fuel tanks often contain propane (C_3H_8).

Figure 2 Organic molecules can be represented by their chemical formulas, by their structural formulas, or by space-filling models.

Compare and contrast *methane and propane.*

Arrangement

Look back at **Figure 1.** Notice that carbon atoms can form straight chains, branched chains, and even rings. This variety of arrangements is another reason carbon can form so many compounds.

Representations

Organic compounds are commonly represented in three ways, as shown in **Figure 2.** The simplest way is with the chemical formula. For example, a main component of natural gas used in homes is the organic compound methane. Methane's chemical formula is CH_4. The second way to represent the molecule is with the structural formula. The structural formula uses lines to show bonds between atoms. Each line between atoms represents a single covalent bond. The third way, the space-filling model, shows a more realistic picture of the relative size and arrangement of the atoms in the molecule.

Hydrocarbons

Carbon forms an enormous number of compounds with hydrogen alone. A compound composed of only carbon and hydrogen atoms is called a **hydrocarbon.** The natural gas methane is a hydrocarbon. Another hydrocarbon used as fuel is propane. Some stoves, most outdoor grills, and hot-air balloons burn propane. The structural formulas and space-filling models of propane and methane are shown in **Figure 2.** Hydrocarbons produce more than 90 percent of the energy that humans use. Hydrocarbons are also important in medicines, foods, and clothing. To understand how hydrocarbons can play so many roles, you must examine how their structures differ.

SCIENCE USAGE v. COMMON USAGE

organic

Science usage: relating to most carbon-containing compounds
Methane is an organic compound containing carbon and hydrogen.

Common usage: related to a product that was grown or created without using synthetic compounds
The farmer raises organic tomatoes and uses only natural pesticides.

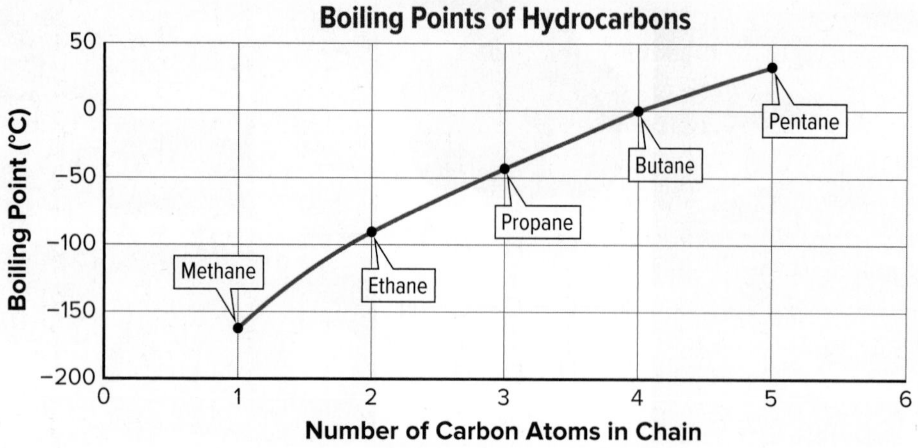

Figure 3 The root of a hydrocarbon's name indicates how many carbons are in the hydrocarbon. Boiling points of hydrocarbons increase as the number of carbon atoms in the chain increases.

Predict *the approximate boiling point of hexane.*

Table 1 Roots for Hydrocarbons

Root	Number of Carbon Atoms
Meth–	1
Eth–	2
Prop–	3
But–	4
Pent–	5
Hex–	6
Hept–	7
Oct–	8

Hydrocarbon lengths

The number of carbon atoms in a straight-chain hydrocarbon is indicated by the root of its name. For example, methane has one carbon atom, ethane has two, and propane has three. **Table 1** summarizes some of these roots. The length of the carbon chain affects the properties of the compound. For example, **Figure 3** shows a graph of the boiling points of some hydrocarbons. Notice the boiling point increases with the addition of carbon atoms.

Bonding in hydrocarbons

Hydrocarbons can contain single, double, and triple bonds between carbon atoms. An easy way to know what type of bond that a hydrocarbon has is to look at the last three letters of the compound's name. The ending *–ane* indicates a single bond only, the ending *–ene* indicates a double bond, and *–yne* indicates a triple bond.

Single bonds In some hydrocarbons, the carbon atoms are joined by single covalent bonds. Hydrocarbons containing only single-bonded carbon atoms are called **saturated hydrocarbons.** Saturated means that a compound holds as many hydrogen atoms as possible—it is saturated with hydrogen atoms. **Table 2** lists four saturated hydrocarbons. Each carbon atom can be considered a link in a chain connected by single covalent bonds.

Table 2 Some Saturated Hydrocarbons

Name	Methane	Ethane	Propane	Butane
Chemical Formula	CH_4	C_2H_6	C_3H_8	C_4H_{10}
Structural Formula	H \| H—C—H \| H	H H \| \| H—C—C—H \| \| H H	H H H \| \| \| H—C—C—C—H \| \| \| H H H	H H H H \| \| \| \| H—C—C—C—C—H \| \| \| \| H H H H

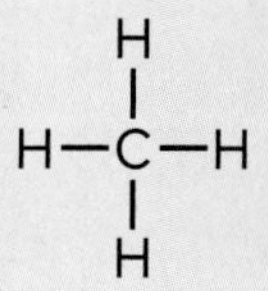

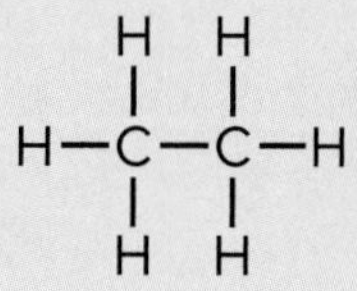

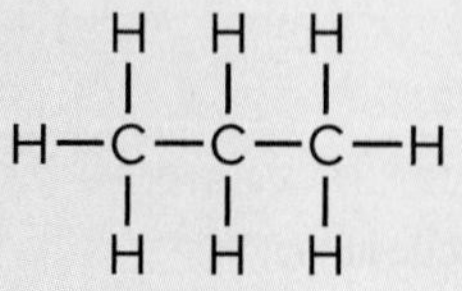

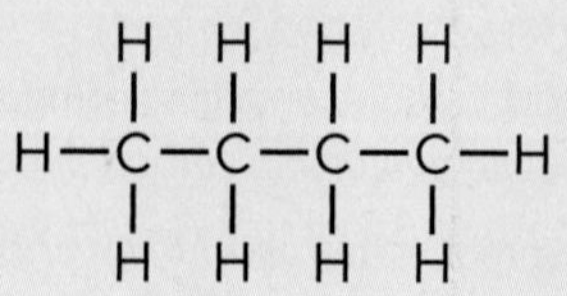

Multiple bonds Hydrocarbons can also have double and triple bonds. Hydrocarbons that contain at least one double or triple bond are called **unsaturated hydrocarbons.** What would happen if the two carbon atoms in ethane (C_2H_6) shared two pairs of electrons and formed a double bond? The resulting compound would be ethene (C_2H_4). Ethene is sometimes called ethylene. Many fruits can form small quantities of ethylene gas, which aids in ripening. The hydrocarbon ethyne (C_2H_2) contains a triple bond in which three pairs of electrons are shared between the two carbon atoms. Ethyne has only two hydrogen atoms and is used in some welding torches. The differences between ethane, ethene, and ethyne are shown in **Figure 4.**

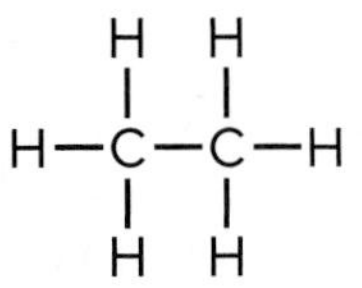

Ethane
C_2H_6

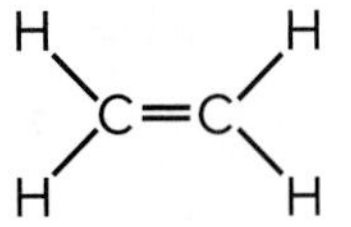

Ethene
C_2H_4

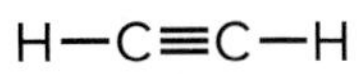

Ethyne
C_2H_2

Figure 4 Saturated hydrocarbons, such as ethane, have only single bonds. Unsaturated hydrocarbons, such as ethene and ethyne, have double or triple bonds between carbon atoms.

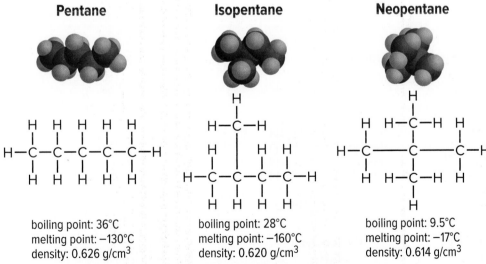

Get It?

State the number of carbon atoms and hydrogen atoms in each molecule in **Figure 4.**

Isomers

Pentane, isopentane, and neopentane have exactly the same chemical formula (C_5H_{12}). However, their carbon atoms are arranged very differently, as shown in **Figure 5.** In a molecule of pentane, the carbon atoms form a continuous chain. Isopentane has one branch, and neopentane has two branches. Pentane, isopentane, and neopentane are isomers. **Isomers** are compounds that have identical chemical formulas but different molecular structures and shapes.

Properties of isomers The arrangement of carbon atoms in each compound changes the shape of the molecule, which affects its physical properties. Generally, melting points and boiling points lower as the amount of branching in an isomer increases. You can see this pattern in **Figure 5.** You may notice that there is an exception to the pattern—the melting point of neopentane is higher than that of pentane or isopentane. In this case, the high melting point results from the symmetry of the molecule and its globular shape.

Pentane | **Isopentane** | **Neopentane**

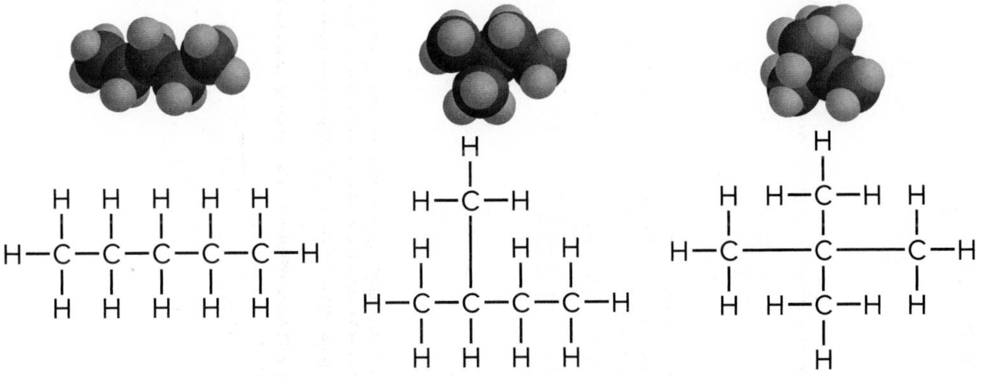

Pentane	Isopentane	Neopentane
boiling point: 36°C	boiling point: 28°C	boiling point: 9.5°C
melting point: −130°C	melting point: −160°C	melting point: −17°C
density: 0.626 g/cm^3	density: 0.620 g/cm^3	density: 0.614 g/cm^3

Figure 5 The amount of branching in the chain affects the compound's physical properties.
Compare *the number of carbon atoms in the isomers of pentane.*

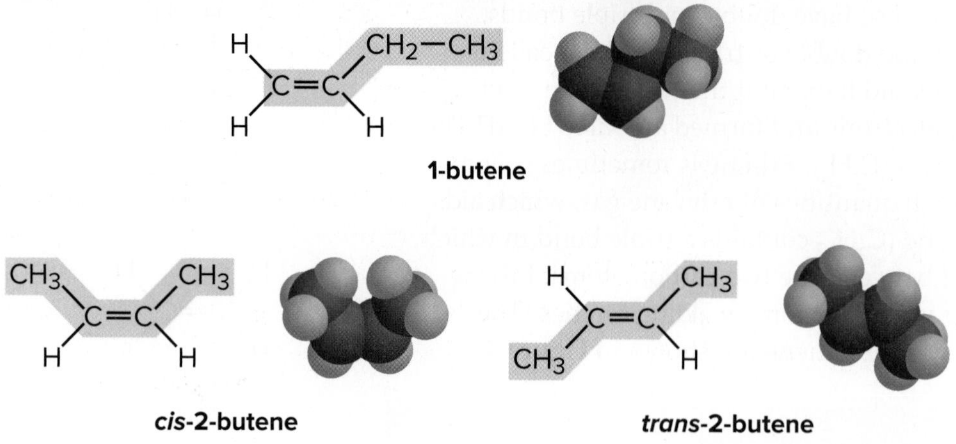

Figure 6 Butene isomers have the chemical formula C_4H_8, but the location of their double bonds and their geometries are different.

Other isomers There are many other kinds of isomers in organic and inorganic chemistry. Examine the three nonbranched isomers of butene shown in **Figure 6.** Notice that the double bond can be located in different places on the chain and that the chain can bend in different ways.

Another type of isomer differs only slightly in how the atoms are arranged in space. Such isomers form what are often called right-handed and left-handed molecules and look like mirror images. Two such isomers may have nearly identical physical and chemical properties.

Carbon Rings

Recall that carbon can also form rings. For example, cyclopropane has three carbon atoms joined into a ring by single bonds, as shown in **Figure 7.** *Cyclo–* means circular. Propane and cyclopropane are not isomers. Propane's chemical formula is C_3H_8, while cyclopropane's chemical formula is C_3H_6. Like straight and branched chains, cyclic carbon chains can have double and triple bonds. For example, cyclopentene (C_5H_8) has a double bond and cyclooctyne (C_8H_{12}) has a triple bond. These molecules are also shown in **Figure 7.**

Get It?
Explain why propane and cyclopropane are not isomers, even though both have three carbon atoms and neither has double or triple bonds.

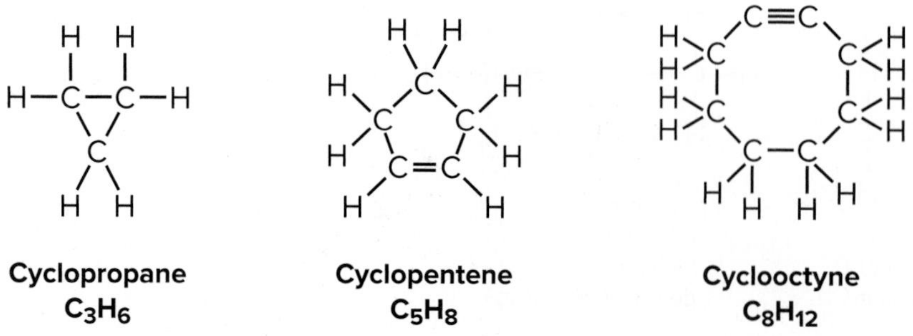

Figure 7 Cyclopropane, cyclopentene, and cyclooctyne are cyclic hydrocarbons.

Benzene

Figure 8 shows a special type of carbon ring called benzene. As you can see, the benzene molecule has six carbon atoms bonded into a ring. **Benzene** (C_6H_6) is a cyclic hydrocarbon with carbon atoms that are joined with alternating single and double bonds. The electrons in the double bonds are shared by all six carbon atoms in the ring. The equal sharing of electrons is represented by the benzene symbol: a circle in a hexagon.

Fused rings The sharing of these six electrons causes the benzene molecule to be very stable. The carbon atoms are bound in a rigid, flat structure. Benzene acts as a framework upon which new molecules can be built. Benzene rings can fuse together, like in the naphthalene (NAF thuh leen) molecule in **Figure 8.** Naphthalene is used in mothballs, which have a distinct odor. Many known compounds contain three or more rings fused together. Tetracycline (teh truh SI kleen) antibiotics are based on a fused-ring system containing four fused rings.

Benzene

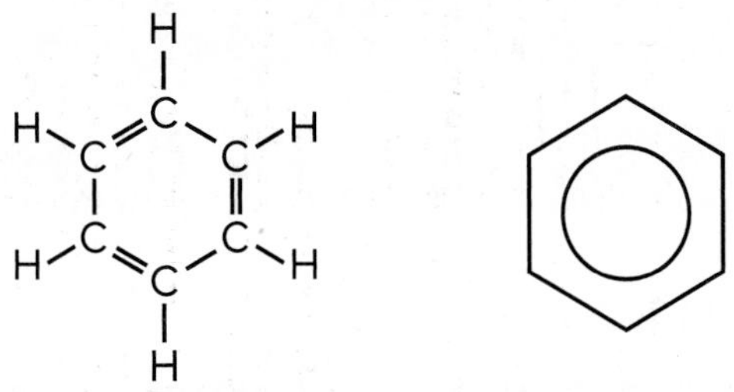

Structural formula **Skeletal formula**

Benzene can be represented in different ways.

Naphthalene

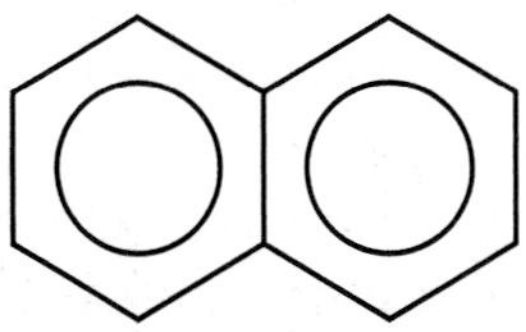

Skeletal formula

Naphthalene, used in mothballs, is a fused-ring system.

Figure 8 Benzene (C_6H_6) is a very stable cyclic hydrocarbon. Benzene rings can fuse to make other compounds.

Check Your Progress

Summary

- Carbon can form many compounds because it has four electrons in its outer energy level.
- Hydrocarbons can be saturated or unsaturated.
- Isomers are compounds that have identical chemical formulas but different molecular structures.
- Benzene contains six carbon atoms bonded into a ring with alternating double and single bonds.

Demonstrate Understanding

1. **Define** the term *organic compounds* and explain how they got this name.
2. **Classify** each of the following compounds as organic or inorganic: C_4H_{10}, H_2O, FeO, CH_3COOH, and CaS.
3. **Compare and contrast** ethane, ethene, and ethyne.
4. **Explain** the term *saturated* in relation to hydrocarbons. What are these compounds saturated with?
5. **Describe** how boiling and melting points generally vary as branching in hydrocarbon isomers increases.

Explain Your Thinking

6. **Analyze** Cyclobutane is a cyclic, saturated hydrocarbon containing four carbon atoms. Draw its structural formula. Are cyclobutane and butane isomers? Explain.
7. **MATH Connection** Adding one double bond to octane (C_8H_{18}) makes the hydrocarbon octene (C_8H_{16}). Write the formulas for adding one, two, and three more double bonds to octane. What is the decrease in the number of hydrogen atoms for each double bond added?

LEARNSMART® Go online to follow your personalized learning path to review, practice, and reinforce your understanding.

LESSON 2
SUBSTITUTED HYDROCARBONS

FOCUS QUESTION

What are the properties and uses of substituted hydrocarbons?

Replacing Hydrogen

Chemists often change hydrocarbons into other compounds having different physical and chemical properties. They may include a double or triple bond or add different atoms or groups of atoms to a hydrocarbon. Some of these changed compounds are substituted hydrocarbons. A **substituted hydrocarbon** has one or more of its hydrogen atoms replaced by atoms, or groups of atoms, of other elements. The groups of atoms used in the substitution are called functional groups. Depending on what properties are needed, chemists decide what functional groups to add. Examples of substituted hydrocarbons are shown in **Figure 9.**

Substituting Oxygen Groups

Oxygen is found in the air, in water, and in many substituted hydrocarbons. Oxygen can form single and double bonds with carbon and single bonds with hydrogen. As a result, there are many compounds containing just carbon, hydrogen, and oxygen. These compounds include alcohols, organic acids, and esters.

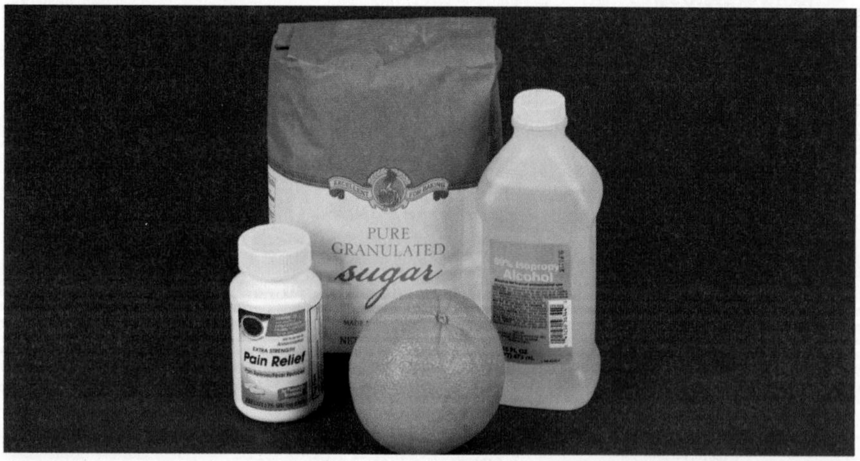

Figure 9 By substituting elements such as oxygen, nitrogen, and chlorine for hydrogen atoms, a wide variety of organic compounds can be made. These diverse compounds have many uses.

3D THINKING DCI Disciplinary Core Ideas CCC Crosscutting Concepts SEP Science & Engineering Practices

COLLECT EVIDENCE

Use your Science Journal to record the evidence you collect as you complete the readings and activities in this lesson.

INVESTIGATE

GO ONLINE to find these activities and more resources.

Review the News
Obtain information from a current news story about substituted hydrocarbons. **Evaluate** your source and **communicate** your findings to your class.

Revisit the Encounter the Phenomenon Question
What information from this lesson can help you answer the Module question?

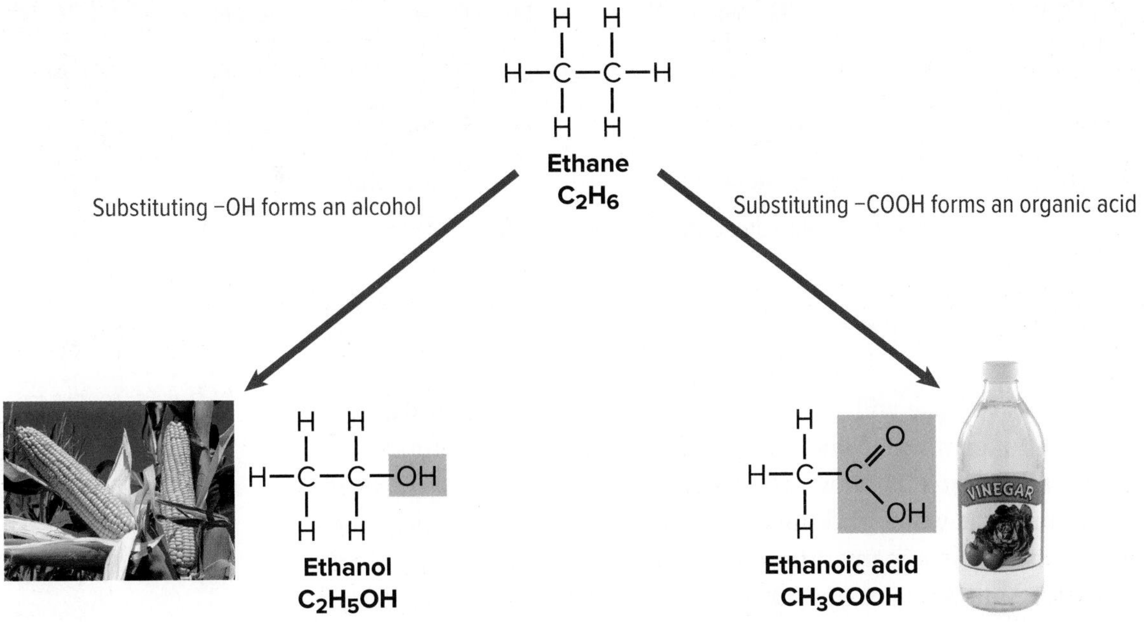

Most ethanol (C_2H_5OH) is obtained from corn.

Ethanoic acid (CH_3COOH) is found in vinegar.

Figure 10 Functional groups containing oxygen are found in alcohols and organic acids.

Describe *how ethane, ethanol, and ethanoic acid are related.*

Alcohols

Rubbing alcohol, used for rubbing aching muscles, is a substituted hydrocarbon. An **alcohol** is formed when –OH groups replace one or more hydrogen atoms in a hydrocarbon. Alcohols are an important group of organic compounds. They often serve as solvents and disinfectants and can be used as pieces to assemble larger molecules. **Figure 10** shows ethanol, an alcohol produced by the fermentation of sugar in grains and in fruit.

Organic acids

Organic acids form when a carboxyl group (–COOH) is substituted for one of the hydrogen atoms in a hydrocarbon. Ethanoic acid, also known as acetic acid, is an organic acid found in vinegar. As shown in **Figure 10,** the structures of ethanol and ethanoic acid are similar. You know some other organic acids, too—citric acid found in citrus fruits, such as oranges and lemons, and lactic acid found in sour milk.

Get It?

Explain why organic acids are considered substituted hydrocarbons.

CCC CROSSCUTTING CONCEPTS

Structure and Function Study **Figure 10.** Create a similar diagram to show the structures of propane, propanol, and propanoic acid. Conduct research to determine uses for the alcohol and the organic acid to include in your diagram.

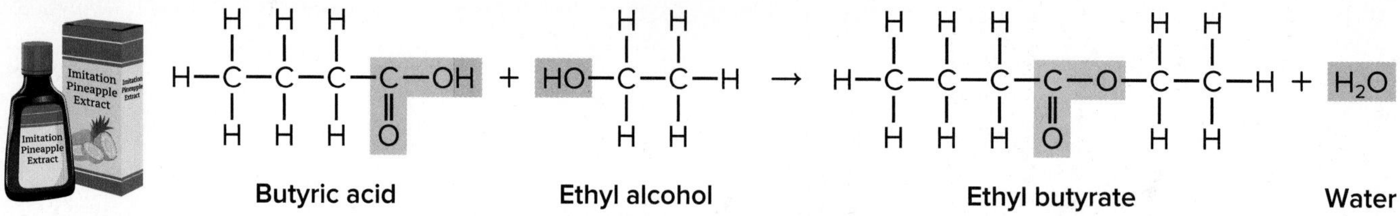

Figure 11 Butyric acid and ethyl alcohol combine to form the ester ethyl butyrate and water. Ethyl butyrate tastes like pineapple.

Esters

Recall that mixing an acid and a base will yield water and a salt. In a similar way, alcohols and organic acids combine to form water and an ester. An **ester** is a substituted hydrocarbon with a −COOC− group. **Figure 11** shows the reaction of butyric (byew TIHR ihk) acid and ethyl alcohol to produce water and the ester ethyl butyrate, which is a component in pineapple flavor. Esters have many different applications. Esters of the alcohol glycerine are used to make commercial soaps. Other esters can be made into fibers for clothing, and still others are used in flavors and perfumes.

Esters for flavor and odors Many fruit-flavored soft drinks and desserts taste like real fruit. In some cases, the flavor is artificial, and not from real fruit. Most likely, this artificial flavor contains esters. The odor of some individual esters immediately makes you think of particular fruits, as shown in **Figure 12.** Many natural and artificial flavors contain a blend of different esters.

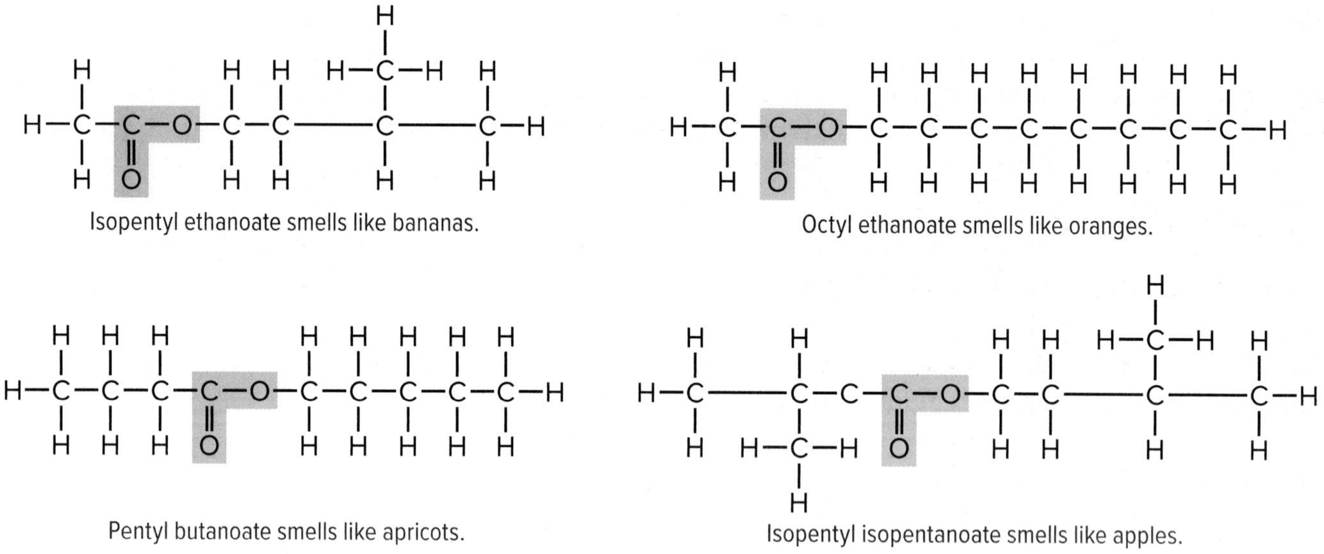

Isopentyl ethanoate smells like bananas.
Octyl ethanoate smells like oranges.
Pentyl butanoate smells like apricots.
Isopentyl isopentanoate smells like apples.

Figure 12 Because of their strong, fruity aromas, esters are used as flavoring and scents.

ACADEMIC VOCABULARY

component

an ingredient or part

The circuit had a large number of components.

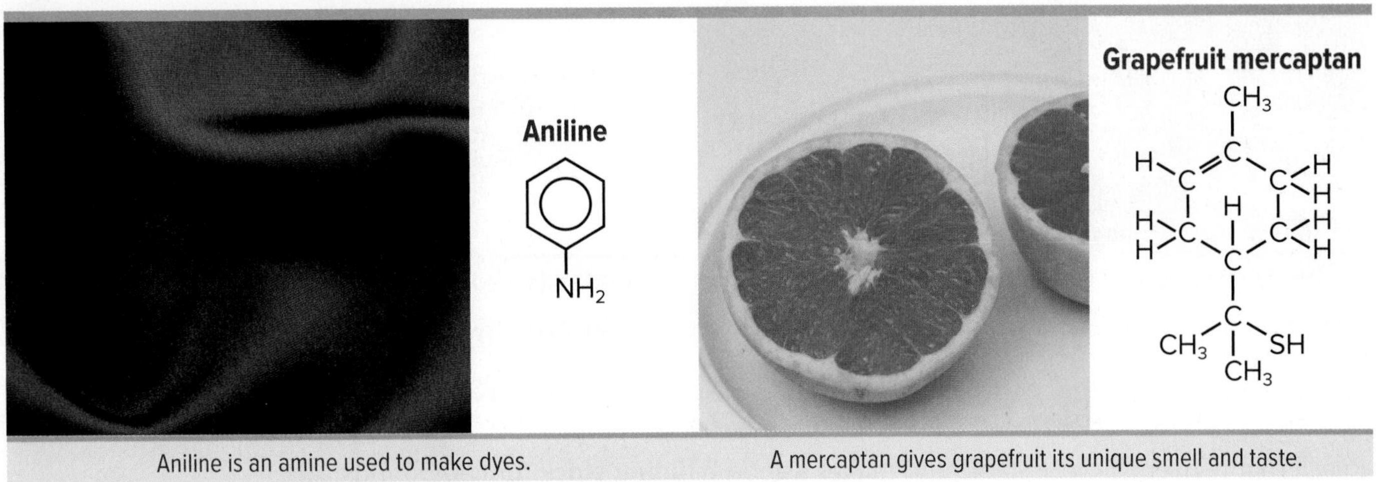
Aniline
NH2
Grapefruit mercaptan
Aniline is an amine used to make dyes.
A mercaptan gives grapefruit its unique smell and taste.

Figure 13 Amines and mercaptans are two types of substituted hydrocarbons.

Substituting Other Elements

Other functional groups can be added to hydrocarbons. Each group has unique properties. Three common groups are amines, mercaptans, and halocarbons.

Amines

An **amine** forms when an amine group ($-NH_2$) replaces a hydrogen atom in a hydrocarbon. For example, aniline is formed when $-NH_2$ replaces a hydrogen atom on a benzene ring. Aniline, shown in **Figure 13,** is used to make dyes. Amines are also essential for life.

Mercaptans

When the group $-SH$ replaces a hydrogen atom in a hydrocarbon, the resulting compound is a thiol. Thiols are commonly called mercaptans. Most mercaptans have unpleasant odors. This can be useful to animals such as skunks. Strangely, small concentrations of foul-smelling mercaptans are often found in pleasant-smelling substances. For example, the odor of grapefruits is due to the mercaptan shown in **Figure 13.**

Mercaptan odors are also powerful. You can smell skunk spray in concentrations as low as 0.5 parts per million. Such a powerful stink can be an asset to people, too. Natural gas has no odor of its own so it is impossible to smell a gas leak. For this reason, gas companies add small amounts of a mercaptan to the gas to make people aware of leaks.

Halocarbons

A substituted hydrocarbon with a halogen, such as chlorine or bromine, in place of a hydrogen atom is called a halocarbon. For example, when four chlorine atoms replace four hydrogen atoms in ethene, the result is tetrachloroethene (teh truh klor uh eh THEEN), a solvent used in dry cleaning. Adding four fluorine atoms to ethene makes a compound that is a starting material for making nonstick coatings on cookware, as shown in **Figure 14.**

Tetrafluoroethene

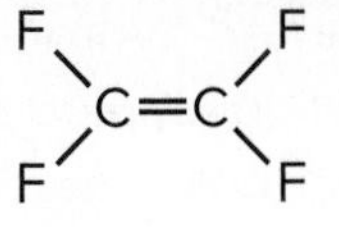

Figure 14 Tetrafluoroethene is a halocarbon used to make nonstick coatings on pans.

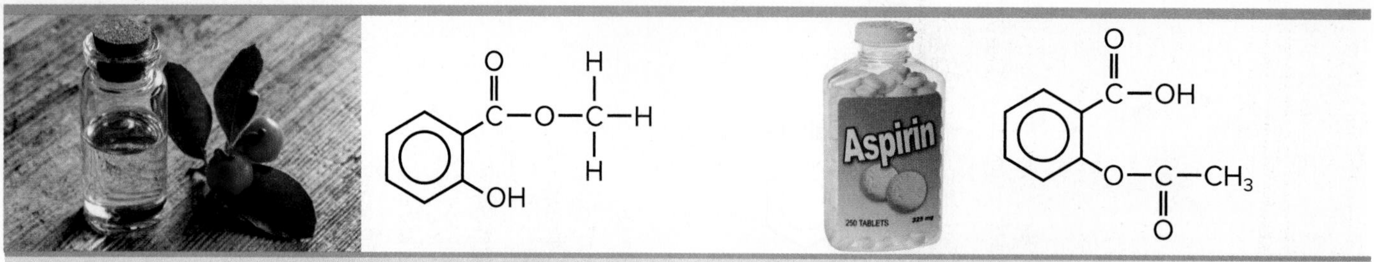

Methyl salicylate gives gum and mints a wintergreen flavor. Acetyl salicylic acid, commonly called aspirin, is a pain reliever.

Figure 15 Aromatic compounds contain a benzene ring and often have strong scents and flavors.

Aromatic Compounds

Recall that benzene is very stable and makes a good building block for compounds. All compounds that contain a benzene structure are **aromatic compounds.** Benzene's hydrogen atoms can be replaced with various functional groups to make many different aromatic compounds.

Aromatic compounds and smell

Aromatic compounds are so named because most of them have a distinctive smell. They contribute to the smell of cloves, cinnamon, and vanilla. Some aromatic compounds produce pleasant odors or tastes. For example, methyl salicylate, shown in **Figure 15,** produces a fresh wintergreen fragrance. Other aromatic compounds have less pleasant flavors and smells. Aspirin, also shown in **Figure 15,** is a sour-tasting aromatic compound. The different flavors are due to the different functional groups.

Check Your Progress

Summary

- In alcohols, the —OH group is substituted for a hydrogen atom.
- Organic acids contain the group —COOH.
- Esters are prepared by combining an alcohol and an organic acid.
- Amines contain the $-NH_2$ group, and mercaptans contain the —SH group.
- Substituted hydrocarbons containing a halogen are called halocarbons.
- Many aromatic compounds have distinct odors.

Demonstrate Understanding

8. **Classify** each of the following as a hydrocarbon or a substituted hydrocarbon: ethyne, tetrachloroethene, ethanol, benzene, propane, and acetic acid.
9. **Identify** the structure that is present in all aromatic compounds.
10. **Explain** why chemists might want to prepare substituted hydrocarbons. Give two examples of possible substitutions.
11. **Identify** possible uses for each of the following types of substituted hydrocarbons: alcohols, esters, and halocarbons.

Explain Your Thinking

12. **Explain** Chloroethane (C_2H_5Cl) can be used as a spray-on anesthetic for localized injuries. How does chloroethane fit the definition of a substituted hydrocarbon? Diagram its structure.
13. **MATH Connection** The odor of mercaptans can be detected in concentrations as low as 0.5 parts per million. Express this concentration as a percent.

LEARNSMART® Go online to follow your personalized learning path to review, practice, and reinforce your understanding.

LESSON 3

PETROLEUM—A SOURCE OF ORGANIC COMPOUNDS

FOCUS QUESTION

How do organic compounds combine to form polymers?

What is petroleum?

Do you carry a plastic comb in your pocket or purse? Do you know where that plastic originated? Chances are, it came from petroleum, a mixture of hydrocarbons and small amounts of other substances found deep within Earth. Because it is formed from the remains of fossilized material, it is a fossil fuel. The liquid part of petroleum is called crude oil. Crude oil is dark, flammable, and foul-smelling.

How can a thick, dark liquid like crude oil be transformed into a hard, brightly colored, useful object like a comb? The answer lies in the nature of crude oil. Crude oil is a mixture of thousands of organic compounds. To make items such as combs, specific compounds are refined from this mixture and then processed into consumer products.

Processing Crude Oil

The first step is to extract the crude oil from its underground source, as shown in **Figure 16.** Once the crude oil is obtained, chemists and engineers separate the crude oil into fractions containing compounds with similar boiling points. The separation process is known as fractional distillation and takes place in oil refineries.

Figure 16 To obtain crude oil, large wells must be drilled deep into Earth.

3D THINKING DCI Disciplinary Core Ideas CCC Crosscutting Concepts SEP Science & Engineering Practices

COLLECT EVIDENCE

Use your Science Journal to record the evidence you collect as you complete the readings and activities in this lesson.

INVESTIGATE

GO ONLINE to find these activities and more resources.

Review the News

Obtain information from a current news story about advantages or disadvantages of using petroleum. **Evaluate** your source and **communicate** your findings to your class.

CCC **Identify Crosscutting Concepts**

Create a table of the crosscutting concepts and fill in examples you find as you read.

The tower

If you have ever driven past a refinery, you may have seen big, metal towers called fractionating towers. These towers, like the one shown in **Figure 17,** can rise as high as 35 m and be 18 m wide. Inside the tower is a series of metal plates arranged like the floors of a building. These plates have small holes through which vapors can pass. On the outside is a maze of scaffolding and pipes at various levels.

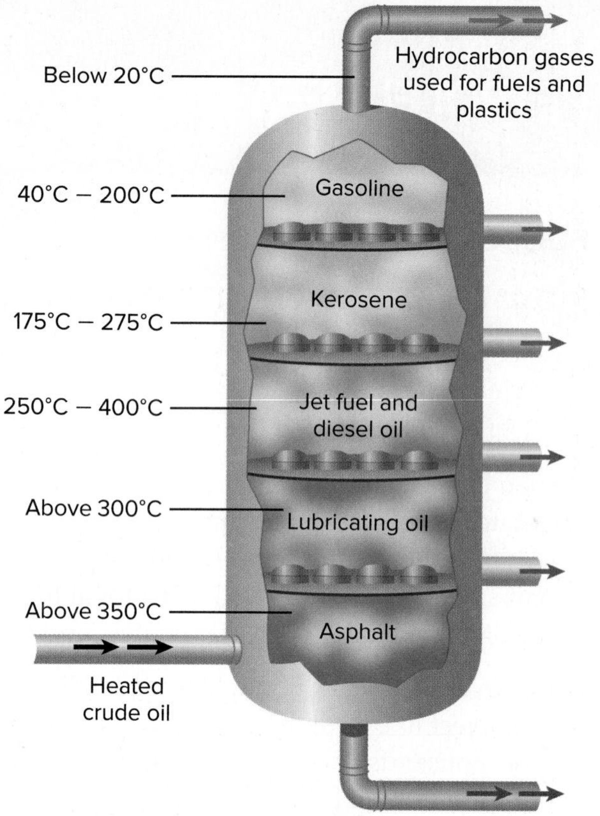

Figure 17 Fractions are separated in a fractionating tower by their boiling points. Larger hydrocarbons have high boiling points and condense on the bottom plates. The smallest hydrocarbons have much lower boiling points and are collected at the top of the tower as gases.

Infer *How might these fractions be further separated?*

Separating organic compounds

The tower separates crude oil into fractions containing compounds with a range of boiling points. Within a fraction, boiling points may range more than 100°C.

At the base of the tower, the crude oil is heated to more than 350°C. At this temperature, most hydrocarbons in the mixture become vapor and start to rise. A few compounds, such as those in asphalt, remain as liquids and are drained from the bottom of the tower. Compounds with higher boiling points condense on the lower plates, drain off through pipes on the sides of the tower, and are collected.

Fractions with lower boiling points may rise to the middle plates before condensing. Those with the lowest boiling points condense on the topmost plates. A few compounds never condense at all and are collected as gases at the top of the tower.

Figure 17 shows some typical fractions. Larger carbon chains have higher boiling points. They generally condense at the bottom of the tower, and smaller compounds condense at the top.

Why do the condensed liquids not fall back through the holes? The reason is that pressure from the rising vapors prevents this. In fact, the separation of the fractions is improved by the interaction of rising vapors with condensed liquid. The exact processes involved vary. For example, some towers add steam at the bottom to aid vaporization. The design and process used depend on the type of crude oil and on the fractions desired.

Get It?

Compare the masses of compounds collected at the top of the tower to those collected at the bottom.

CCC CROSSCUTTING CONCEPTS

Structure and Function Review **Figure 17** and the information on this page. Identify the function of a fractionating tower, describe its structure, and explain how its structure allows it to accomplish its function. Present your analysis in the form of a graphic organizer.

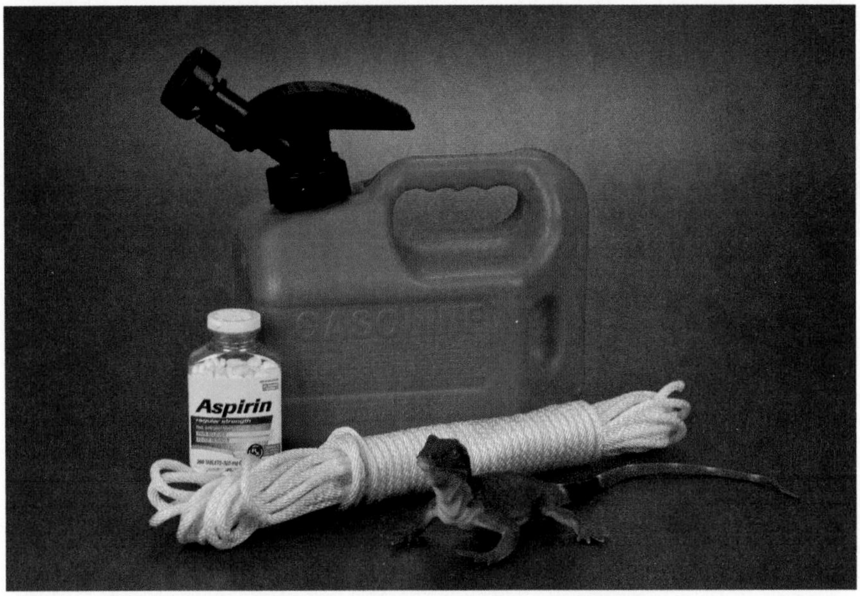

Figure 18 Crude oil provides the organic compounds used in many common products. After crude oil has been refined, the organic compounds can be used to make various types of fuel, plastics, and synthetic fibers, as well as paint, dyes, and medicines.

Uses for Petroleum Compounds

The lightest fractions from the top of the tower include butane and propane, which are used for fuel. The fractions that condense on the upper plates and contain from five to ten carbons are used for gasoline and solvents. Below these are fractions with 12 to 18 carbons that are used for kerosene and jet fuel. The bottom fractions go into lubricating oil, and the residue is used for paving asphalt.

Other petroleum products are obtained by further purifying crude oil fractions using different techniques to isolate individual compounds. After these are separated, they can be converted into substituted hydrocarbons. Chemists use these to make products ranging from medicines such as aspirin to fuel, plastic toys, and synthetic fibers, as shown in **Figure 18.** Aromatic dyes from crude oil have almost completely replaced natural dyes, such as indigo (a deep blue) and alizarin (a deep red). The first synthetic dye was a bright purple called mauve that was accidentally discovered in coal tar compounds.

Polymers

Did you ever loop together strips of paper to make a paper chain for decoration, or have you ever strung paper clips together? A paper chain can represent the structure of a polymer, as shown in **Figure 19.** Some of the smaller molecules from petroleum are monomers that act like links in a chain. *Mono–* means one. A **monomer** is a small molecule that can combine with itself repeatedly to form a long chain. When these links are hooked together, they make new, extremely large molecules known as polymers. **Polymers** are long chains of monomers. Often, two or more different monomers, known as copolymers, combine to make one polymer molecule.

Figure 19 Both paper chains and polymers are long chains made up of smaller units. Paper chains are made of loops of paper and polymers are made of monomers.

Get It?

Explain how polymers are similar to paper chains.

Polymer properties

The properties of polymers depend mostly on which monomers are used to make them. The amount of branching and the shape of the polymer also greatly affect its properties. Because of this diversity, polymers can be made light and flexible or strong enough to make plastic pipes, boats, and even some auto bodies. Other polymers can be spun into threads for use in clothing, suitcases, and backpacks.

Hydrocarbon polymers

Some polymers are made entirely of carbon and hydrogen, as shown in **Table 3**. When ethene combines with itself repeatedly, it forms the common polymer polyethylene (pah lee EH thuh leen). Polyethylene is used widely in shopping bags and plastic bottles, as shown in **Table 3**. Another common polymer, polypropylene (pah lee PRO puh leen), is made of propene monomers and has many uses.

Table 3 Hydrocarbon Polymers

Polymer	Monomer	Uses
Polyethylene	$\left[CH_2 - CH_2 \right]$	• plastic bags • plastic bottles
Polypropylene	$\left[\begin{array}{c} CH_2 - CH \\ \quad\quad\; CH_3 \end{array} \right]$	• glues • carpets • high-performance outdoor clothing
Polystyrene	$\left[CH - CH_2 \right]$ (with benzene ring)	• foam packing • disposable food containers • computer monitor casings

(t,c)Matt Meadows/McGraw-Hill Education, (b)Lawrence Manning/Fuse/Getty Images

WORD ORIGIN

polymer
from the Greek *poly–* meaning "many" and *–mer* meaning "part or segment"
Polymers are used to make plastic bags and foam cups.

STEM CAREER Connection

Prosthetic Makeup Artist
If you have a flair for the creative and enjoy working with your hands, you might be interested in a career as a prosthetic makeup artist. A prosthetic makeup artist works with various materials, including polymers such as latex and silicone, to transform actors into monsters and other characters for film, video, and stage.

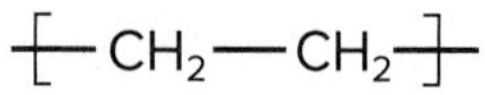

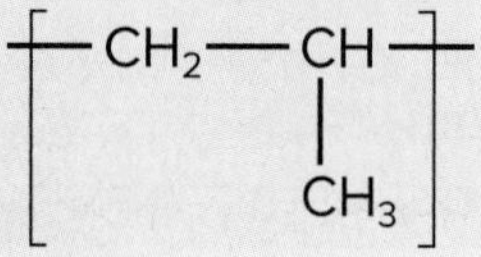

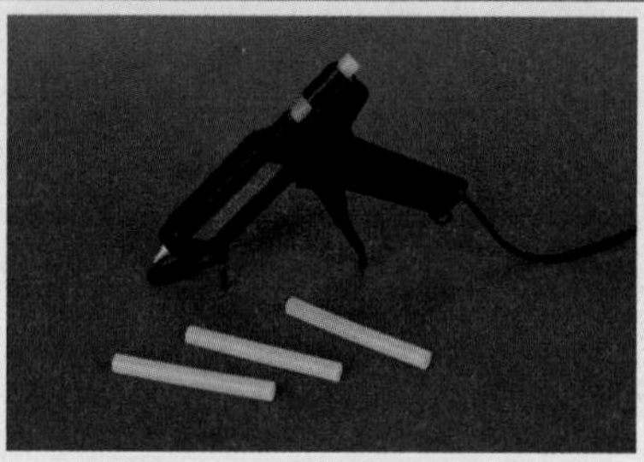

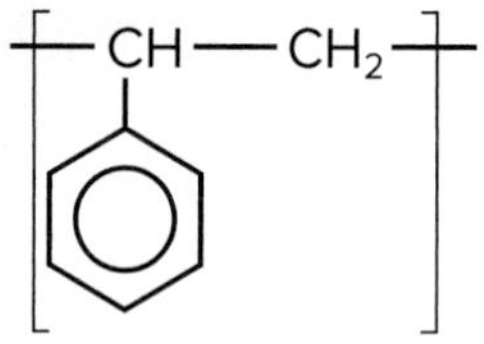

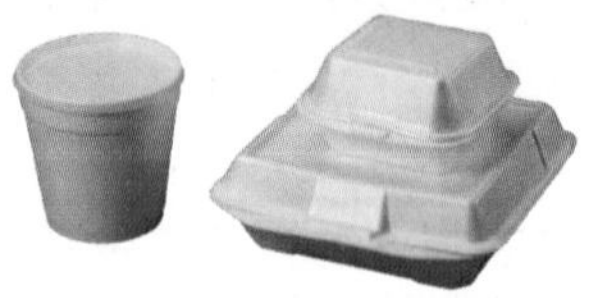

Table 4 Substituted Hydrocarbon Polymers

Polymer	Monomer	Uses
Polyurethane	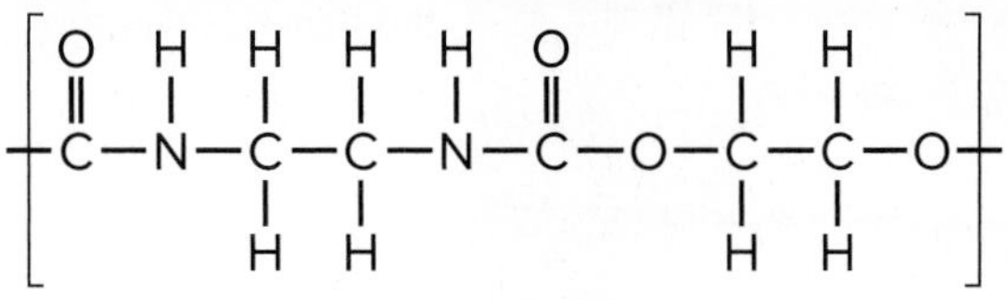	 • foam • waterproof coatings • shoe parts
Polyvinyl chloride	$+CH_2-CHCl+$	 • pipes • hoses • house siding
Polyester	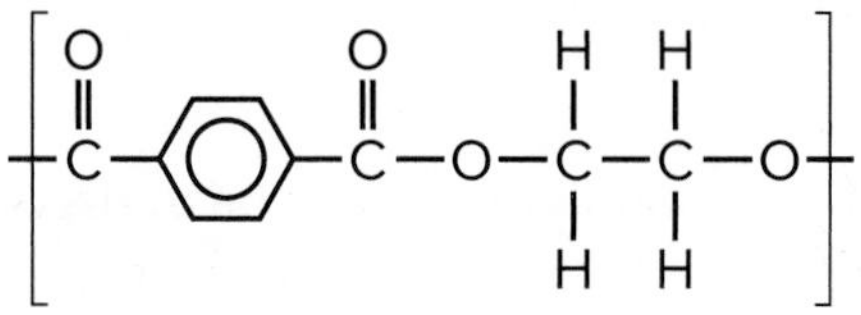	 • fabric • rope

Versatile polystyrene Some polymers can take two completely different forms. Polystyrene (pah lee STI reen) forms the brittle casings around computer monitors and other electronics. But if a gas such as carbon dioxide is blown into the melted polystyrene as it is molded, bubbles will remain within the polymer when it cools. These bubbles make polystyrene the efficient insulator used in foam cups.

Substituted hydrocarbon polymers

Polymers can contain elements other than carbon and hydrogen, as shown in **Table 4.** Polyurethane has functional groups that contain oxygen and nitrogen. The monomers of polyvinyl chloride, also known as PVC, are ethene monomers with chlorine substituted for one hydrogen atom. This substitution makes PVC harder and more heat resistant than polyethylene.

Polyesters Synthetic fibers called polyesters are made from an organic acid that has two –COOH groups and an alcohol that has two –OH groups, as shown in **Figure 20.** Polyester is strong because these chains are closely packed together. Many varieties of polyesters can be made, depending on what alcohols and acids are used. Polyesters can be woven or knitted into durable fabrics.

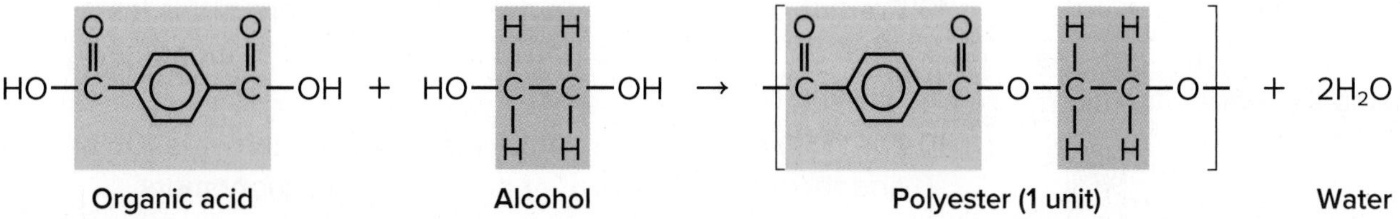

Figure 20 Polyesters are formed when an organic acid with two –COOH groups combines with an alcohol that has two –OH groups.

Figure 21 Plastics can be recycled into many structures, such as decks, gazebos, and this bench.

Depolymerization

Disposing of polymers has become a problem because polymers have been used so widely and many polymers do not decompose. One way to combat this is by recycling, which recovers clean plastics for reuse in new products, such as the bench in **Figure 21.** Many communities recycle plastics.

Another approach is depolymerization. **Depolymerization** is a process that uses heat or chemicals to break the long polymer chain into its monomer fragments. These monomers can then be reused. However, each polymer requires a different depolymerization process, and much research is needed to make this type of recycling economical.

Check Your Progress

Summary

- Crude oil is a dark, flammable liquid formed from fossilized materials.
- Organic compounds in crude oil can be separated using fractional distillation.
- Polymers are long chains of repeating chemical units called monomers.
- Polymers can be designed with specific properties.
- Depolymerization is the process of breaking a polymer into its components.

Demonstrate Understanding

14. **Identify** several items around your home that are made from organic compounds obtained from crude oil.
15. **Name** some of the fuels obtained from crude oil by fractional distillation.
16. **Describe** the process of fractional distillation.
17. **Explain** why polymers made from the same monomer can have physical properties that vary greatly.
18. **Describe** why depolymerization can be an expensive process.

Explain Your Thinking

19. **Predict** Based on the names of the polymers in this lesson, what do you think the polymer made from the monomer terpene is called?
20. **MATH Connection** If the mass of a monomer is 105 amu, find the mass of a polymer containing 122 monomers.

LEARNSMART® Go online to follow your personalized learning path to review, practice, and reinforce your understanding.

DPD ImageStock/Alamy Stock Photo

LESSON 4

BIOLOGICAL COMPOUNDS

FOCUS QUESTION

How are the structures of proteins, carbohydrates, and nucleic acids similar? How are they different?

Biological Polymers

Like all polymers, biological polymers are huge molecules made of many monomers linked together. The monomers of biological polymers are usually larger and more complex in structure than those used to make plastics. Many of the important biological compounds in your body are polymers. Among them are the proteins and starches.

Proteins

Proteins are large organic polymers formed from organic monomers called amino acids. Amino acids, like glycine and cysteine, are the monomers that combine to form proteins. As shown in **Figure 22,** the amine group ($-NH_2$) of one amino acid combines with the carboxylic acid group (–COOH) of another amino acid to form a compound called a peptide. The bond joining them is known as a peptide bond. Peptides with about 50 or more amino acids are called proteins.

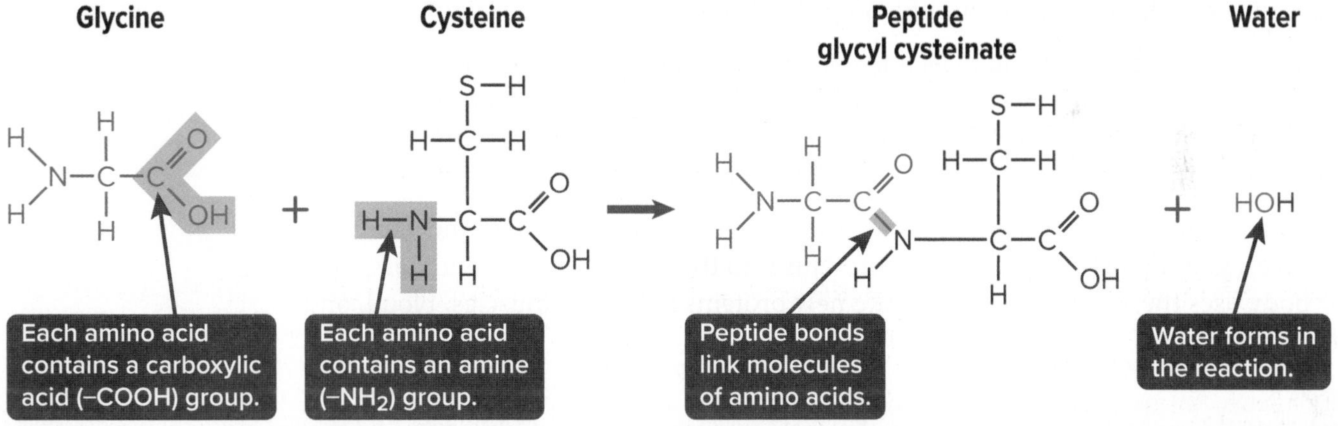

Figure 22 Amino acids link together with peptide bonds. Here, glycine and cysteine combine to form the peptide glycyl cysteinate. Long chains of peptides are called proteins.

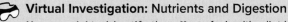

3D THINKING DCI Disciplinary Core Ideas CCC Crosscutting Concepts SEP Science & Engineering Practices

COLLECT EVIDENCE

Use your Science Journal to record the evidence you collect as you complete the readings and activities in this lesson.

INVESTIGATE

GO ONLINE to find these activities and more resources.

Virtual Investigation: Nutrients and Digestion
Use a model to identify the effect of a healthy diet by designing a menu.

Quick Investigation: Test for Starch
Carry out an investigation to determine the cause and effect of the reaction between starch and iodine solution.

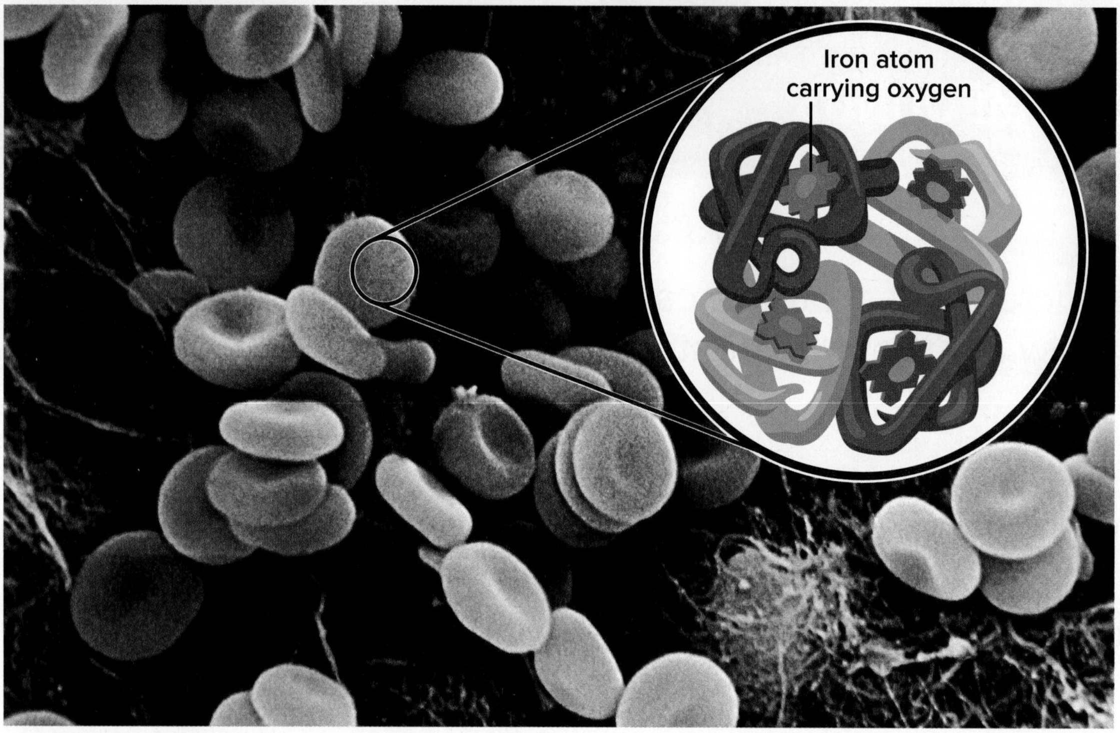

Figure 23 Human blood contains disks called red blood cells. In red blood cells, hemoglobin has four protein chains coiled around each other. Each chain has an atom of iron, which carries oxygen.

Protein structure

Even though only 20 amino acids are commonly found in nature, they can be arranged in so many ways that millions of different proteins exist. Proteins come in numerous forms and make up many of the tissues in your body, such as muscles, tendons, hair, and fingernails. In fact, proteins account for 15 percent of your total body weight.

Long protein molecules tend to twist and coil in a manner unique to each protein. For example, hemoglobin, which carries oxygen in your blood, has four chains that coil around each other, as shown in **Figure 23.** Each chain contains an iron atom that carries the oxygen. If you look closely, you can see all four iron atoms in hemoglobin.

When you eat foods that contain proteins, such as meat, dairy products, and some vegetables, your body breaks down the proteins into their amino acid monomers. Then your body uses these amino acids to make new proteins that form muscles, blood, and other body tissues.

Get It?

Explain what happens when you eat foods containing protein.

Carbohydrates

If you hear the word *carbohydrate*, you may think of bread, cookies, or pasta. You may know someone who is on a low-carbohydrate diet to control weight. Or you may have heard of carbohydrate loading by athletes. Runners, for example, often prepare for a long-distance race by eating, or loading up on, carbohydrates in foods such as vegetables and pasta. **Carbohydrates** are compounds containing carbon, hydrogen, and oxygen that have twice as many hydrogen atoms as oxygen atoms. Carbohydrates include sugars and starches.

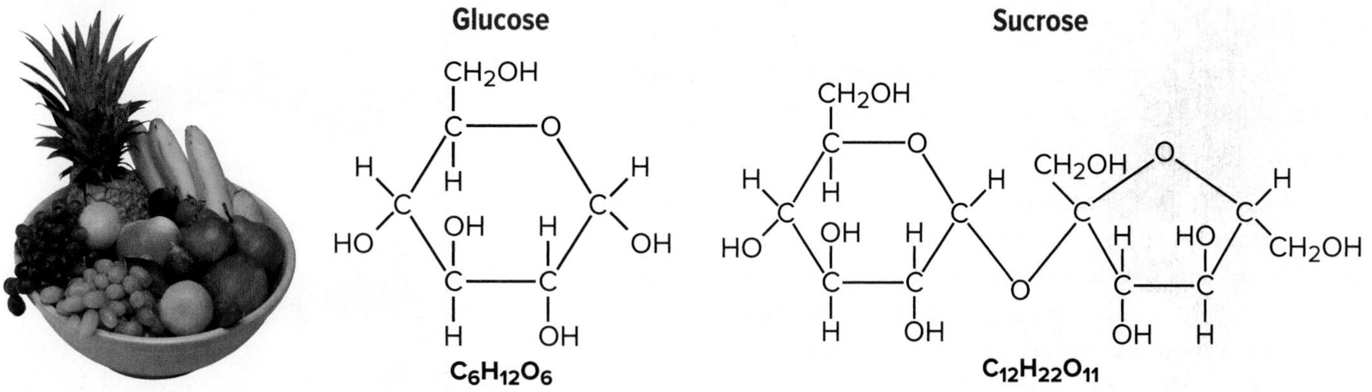

Figure 24 Fruits contain glucose and another simple sugar called fructose. Sucrose, commonly called table sugar, consists of a glucose molecule and a fructose molecule bonded together.

Explain *why sugars are carbohydrates.*

Sugars

Sugars form a major group of carbohydrates. The sugar glucose, shown in **Figure 24,** is found in your blood and also in many sweet foods, such as grapes and honey. Common table sugar, known as sucrose, is broken down by digestion into two simpler sugars—fructose, often called fruit sugar, and glucose. Unlike starches, sugars provide energy soon after eating.

Starches

Starch, shown in **Figure 25,** is a carbohydrate that is also a polymer. It is made of monomers of the sugar glucose. During digestion, the starch is broken down into similar sugars, which releases energy in your body cells. Athletes, especially long-distance runners, use starches to provide high-energy, long-lasting fuel for the body. The energy from starches can be stored in liver and muscle cells in the form of a compound called glycogen.

Get It?

Describe the difference between sugars and starches.

Lipids

Fats, oils, and related compounds make up a group of organic compounds known as **lipids.** Lipids include animal fats, such as butter, and vegetable oils, such as corn oil. Lipids contain the same elements as carbohydrates but in different proportions. For example, lipids have fewer oxygen atoms and contain carboxylic acid groups. Lipids are long hydrocarbon chains, but they are not polymers.

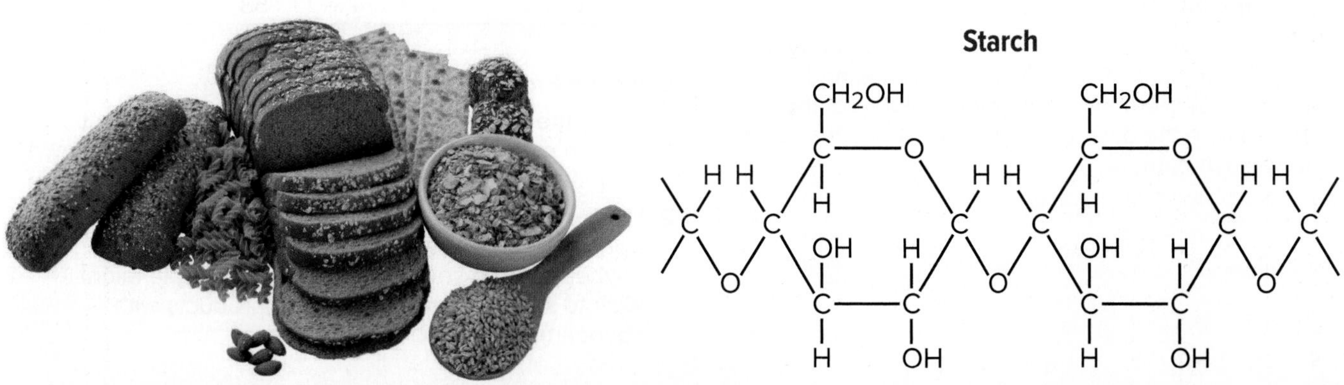

Figure 25 Starch, the major component of pasta, is a polymer made of glucose monomers.

Fats and oils

Lipids are similar in structure to hydrocarbons. They can be classified as saturated or unsaturated according to the types of bonds in their carbon chains.

Saturated fats, like saturated hydrocarbons, are saturated with hydrogen and contain only single bonds between carbon atoms. Unsaturated fats are like unsaturated hydrocarbons and have double or triple bonds. Unsaturated fats with one double bond are called monounsaturated, and those with two or more double bonds are called polyunsaturated.

Animal lipids, called fats, tend to be saturated and are solids at room temperature. Plant lipids, called oils, are unsaturated and are usually liquids, as shown in **Figure 26.** Sometimes, hydrogen is added to vegetable oils to form more saturated solid compounds called hydrogenated vegetable shortenings.

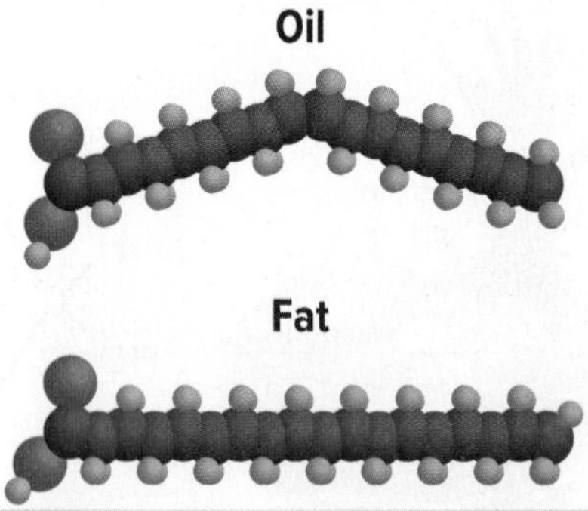

Figure 26 Fats, such as butter and lard, are saturated lipids. Oils are unsaturated lipids.

Get It?

Compare and contrast the structure and properties of unsaturated fats and saturated fats.

Have you heard that eating too much fat can be unhealthy? Evidence shows that too much saturated fat and cholesterol in the diet may contribute to heart disease and that unsaturated fats may help to prevent heart disease. It appears that saturated fats are more likely to be converted to substances that can block the arteries leading to the heart. A balanced diet includes some fats, just as it includes proteins and carbohydrates.

Get It?

Compare and contrast how saturated fats and unsaturated fats can affect human health.

APPLY SCIENCE

Which foods should you choose?

What do you like to eat? You probably choose your foods by how good they taste. A better way might be to look at their nutritional values. Your body needs nutrients, such as proteins, carbohydrates, and fats, to give it energy and to help it build cells. Almost every food has some of these nutrients in it. The trick is to choose your foods so you do not get too much of one thing and not enough of another.

Identify the Problem

The table shows some basic nutrients for a variety of foods. The amount of protein, carbohydrate, and fat is recorded as the number of grams in 100 g of the food. Examine the data, and select the foods that best provide each nutrient.

Nutritional Values for Some Common Foods

Food (100 g)	Protein (g)	Carbohydrate (g)	Fat (g)
Cheddar cheese	25	1	33
Hamburger	17	23	17
Soybeans	13	11	7
Wheat	15	68	2
Potato chips	7	53	35

Solve the Problem

1. Using the table, list the foods that supply the most protein and carbohydrates. What might be the problem with eating too many potato chips?
2. In countries where meat and dairy products are not plentiful, people eat a lot of food made from soybeans. Give several reasons why people might wish to substitute meat and dairy products with soybean-based products.

Cholesterol

Cholesterol is another lipid that is often in the news. It is found in meats, eggs, butter, cheese, and fish. Like fats and oils, cholesterol can be both helpful and harmful to your body. Some cholesterol is produced by the body to build cell membranes. It is also found in bile, a digestive fluid. However, too much cholesterol may cause serious damage to heart and blood vessels, similar to the damage caused by saturated fats.

Nucleic Acids

Nucleic acids are another important group of organic polymers that are essential for life. A **nucleic acid** is an organic polymer that controls the activities and reproduction of cells. You may have heard about DNA, which is one type of nucleic acid. **Deoxyribonucleic** (dee AHK sih ri boh noo klay ihk) **acid,** also called DNA, is an essential biological polymer found in the nuclei of cells that codes and stores genetic information. This information is known as the genetic code.

Nucleic acid monomers

The monomers that make up DNA are called nucleotides. **Nucleotides** are complex molecules that make up DNA and that contain one of four organic bases, a sugar, and a phosphate unit. **Figure 27** shows the structure of DNA, along with a closer look at a nucleotide. DNA nucleotides form chains that are unique to an organism. Two nucleotide chains twist around each other, forming what resembles a twisted ladder called a double helix. The rungs of the ladder are paired organic bases. Your genetic code gives instructions for making other nucleotides and proteins that are needed by your body. The type of protein made depends on the order of the bases in the DNA.

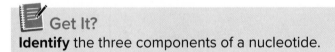

Get It?
Identify the three components of a nucleotide.

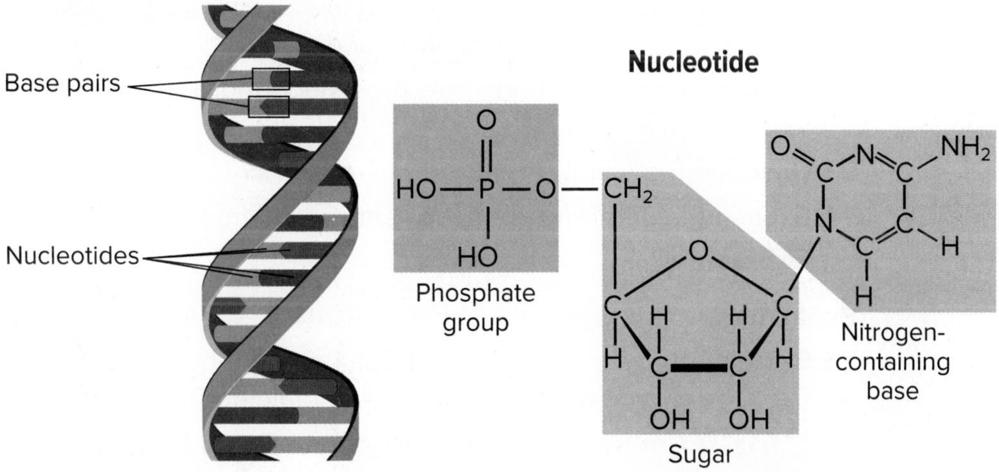

Figure 27 Each nucleotide looks like half of a ladder rung with an attached side piece. As you can see, each pair of nucleotides forms a rung on the ladder, while the side pieces cause a little twist that gives DNA its double helix shape.

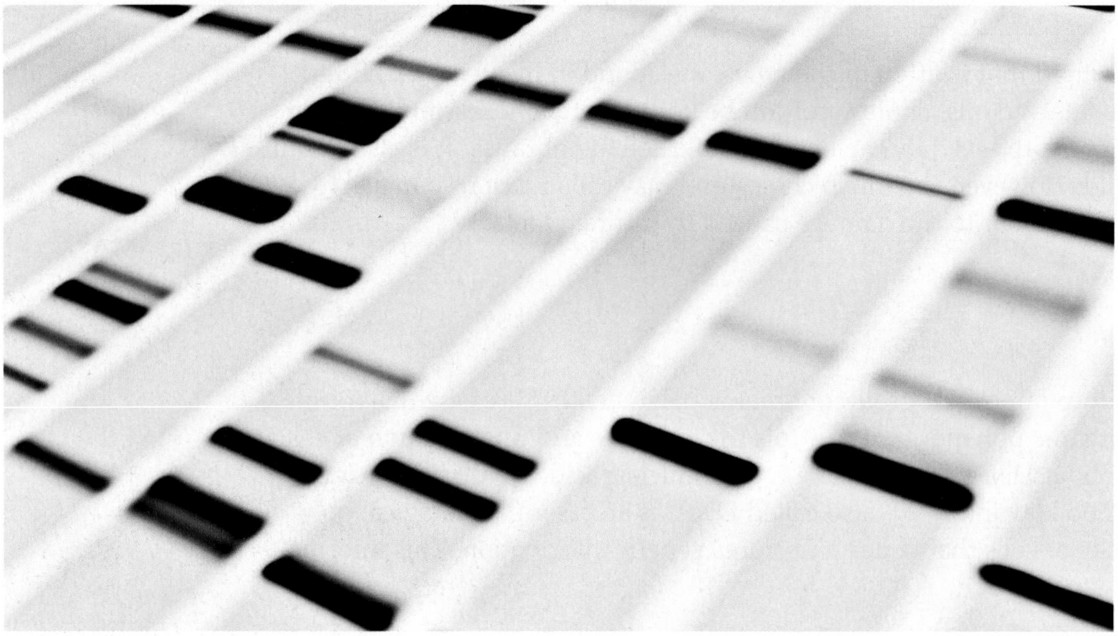

Figure 28 A DNA profile analyzes the base pairs at several specific sites along the DNA strand.

Steven Puetzer/Photographer's Choice RF/Getty Images

DNA fingerprinting

Human DNA contains more than five billion base pairs. The DNA of each person differs in some way from that of everyone else. The unique nature of DNA offers crime investigators a way to identify criminals from hair or fluids left at a crime scene. First, DNA from bloodstains or cells in saliva, such as on a soda bottle, can be extracted in the laboratory. Then, chemists can break up the DNA into its nucleotide components and use radioactive and X-ray methods to obtain a picture of the nucleotide pattern. **Figure 28** shows a DNA profile similar to those used by investigators. Comparing this pattern to one made from the DNA of a suspect can link that suspect to the crime scene.

Check Your Progress

Summary

- Proteins are large, organic polymers that form muscles, blood, and other body tissues.
- Carbohydrates contain carbon, hydrogen, and oxygen.
- Sugars and starches are carbohydrates that provide energy to your body.
- Lipids include fats and oils.
- DNA is a nucleic acid that is found in the cell nucleus.

Demonstrate Understanding

21. **Name** the monomers that make up the following biological polymers: proteins, nucleic acids, and starches.
22. **Identify** where your body gets the compounds that it needs to build proteins.
23. **Describe** the function of DNA.
24. **Explain** the difference between saturated and unsaturated fats and oils.

Explain Your Thinking

25. **Explain** Whole milk contains about 4 percent butterfat. Explain why you might choose milk containing 2 percent fat.
26. **MATH Connection** You have read that your body is about 15 percent protein. Calculate the mass of protein in your body in kilograms.

LEARNSMART® Go online to follow your personalized learning path to review, practice, and reinforce your understanding.

Molecular Clock Forensics

The case was as tragic as it was bizarre. A doctor injected his former girlfriend with a substance he claimed was vitamin B. Later, when she tested positive for HIV, the virus that causes AIDS, she was certain she had been the victim of a terrible crime. But how could it be proven?

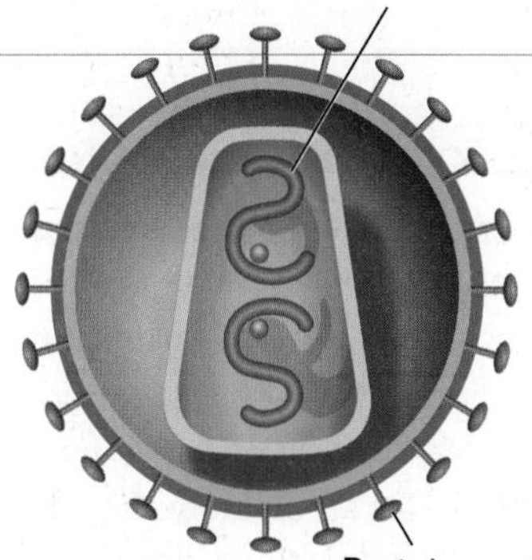

HIV is a virus that consists of a protein coat surrounding genetic material.

Evolution on the clock

Investigators used the science of the molecular clock to examine the evidence. The genetic material of HIV, shown in the diagram, mutates at a rate of around one percent per year. Measuring and tracking these mutations can tell scientists when two strains of HIV last shared a common ancestor. The genetic material itself acts as a type of clock.

The investigators examined two genes from the virus's genome. The first gene contains information for the protein coat and evolves rapidly. The second gene is involved in the genetic copying process and evolves more slowly.

The rapidly-evolving gene showed that the HIV strain from the victim was closely related to that from a blood sample found in the doctor's refrigerator. The slowly-evolving gene proved that the viruses in the victim's body had evolved from those in the sample. With this evidence, the doctor was convicted of attempted murder.

Calibrating the clock

The accuracy of molecular clocks depends on a controversial question: Just how regularly does the clock "tick"? Some parts of a genome may evolve more rapidly than others, particularly if the change increases fitness. Fortunately, there are many mutations that are neutral or nearly neutral to the survival of the organism. By locating these mutations, scientists can calibrate their molecular clocks and discover unanticipated sequences of events.

The molecular clock can also be applied to much longer intervals. It was once believed that many mammals first evolved soon after the extinction of the dinosaurs 65 million years ago (mya). Molecular clock measurements put their emergence at around 100 mya, firmly in the time of the dinosaurs. Later, fossils were found that confirmed these results.

USE A MODEL TO ILLUSTRATE

Model Molecular Clocks Have one person (the common ancestor) create two copies of a phrase (the genome). Pass each copy to a classmate, who flips a coin. On heads, make a random change to one letter. Pass copies to two other classmates. After everyone has a turn, count the mutations in each genome, and find the average rate of your molecular clock.

MODULE 23

STUDY GUIDE

GO ONLINE to study with your Science Notebook.

Lesson 1 SIMPLE ORGANIC COMPOUNDS

- Carbon can form many compounds because it has four electrons in its outer energy level.
- Hydrocarbons can be saturated or unsaturated.
- Isomers are compounds that have identical chemical formulas but different molecular structures.
- Benzene contains six carbon atoms bonded into a ring with alternating double and single bonds.

- organic compound
- hydrocarbon
- saturated hydrocarbon
- unsaturated hydrocarbon
- isomer
- benzene

Lesson 2 SUBSTITUTED HYDROCARBONS

- In alcohols, the –OH group is substituted for a hydrogen atom.
- Organic acids contain the group –COOH.
- Esters are prepared by combining an alcohol and an organic acid.
- Amines contain the $-NH_2$ group, and mercaptans contain the –SH group.
- Substituted hydrocarbons containing a halogen are called halocarbons.
- Many aromatic compounds have distinct odors.

- substituted hydrocarbon
- alcohol
- ester
- amine
- aromatic compound

Lesson 3 PETROLEUM—A SOURCE OF ORGANIC COMPOUNDS

- Crude oil is a dark, flammable liquid formed from fossilized materials.
- Organic compounds in crude oil can be separated using fractional distillation.
- Polymers are long chains of repeating chemical units called monomers.
- Polymers can be designed with specific properties.
- Depolymerization is the process of breaking a polymer into its components.

- monomer
- polymer
- depolymerization

Lesson 4 BIOLOGICAL COMPOUNDS

- Proteins are large, organic polymers that form muscles, blood, and other body tissues.
- Carbohydrates contain carbon, hydrogen, and oxygen.
- Sugars and starches are carbohydrates that provide energy to your body.
- Lipids include fats and oils.
- DNA is a nucleic acid that is found in the cell nucleus.

- protein
- carbohydrate
- lipid
- nucleic acid
- deoxyribonucleic acid (DNA)
- nucleotide

THREE-DIMENSIONAL THINKING

Module Wrap-Up

REVISIT THE PHENOMENON

Why is this natural fire always burning?

CER Claim, Evidence, Reasoning

Explain Your Reasoning Revisit the claim you made when you encountered the phenomenon. Summarize the evidence you gathered from your investigations and research and finalize your Summary Table. Does your evidence support your claim? If not, revise your claim. Explain why your evidence supports your claim.

STEM UNIT PROJECT

Now that you've completed the module, revisit your STEM unit project. You will summarize your evidence and apply it to the project.

GO FURTHER

SEP Data Analysis Lab

The Smell of Danger

Have you ever smelled leaking natural gas, propane, or liquid petroleum (LP) gas? These gases have a strong odor, even though in nature they have little or no odor. Strong-smelling substituted hydrocarbons, such as tertiary butyl mercaptan, dimethyl sulfide, and ethyl mercaptan, are added to these gases to warn people of their presence. Ethyl mercaptan is also found in the aroma of skunks and rotting meat. It is listed in the Guinness Book of World Records as one of the smelliest substances known.

Directions: Contact your local natural gas, propane, or LP gas company to find answers to the following questions.

CER Analyze and Interpret Data

1. **Claim and Reasoning** What compounds make up natural gas and LP gas?
2. **Claim, Evidence, Reasoning** What are some uses for natural gas, propane, and LP gas?
3. **Claim, Evidence, Reasoning** Draw the chemical structures and list characteristics of the following chemicals: dimethyl sulfide, ethyl mercaptan, and tertiary butyl mercaptan.

Rweisswald/iStock/Getty Images

Oleg Doroshin/Shutterstock

MODULE 24

NEW MATERIALS THROUGH CHEMISTRY

ENCOUNTER THE PHENOMENON

What makes this candy so colorful?

GO ONLINE to play a video about how scientists make synthetic food coloring.

SEP Ask Questions

Do you have other questions about the phenomenon? If so, add them to the driving question board.

CER Claim, Evidence, Reasoning

Make Your Claim Use your CER chart to make a claim about what makes this candy so colorful.

Collect Evidence Use the lessons in this module to collect evidence to support your claim. Record your evidence as you move through the module.

Explain Your Reasoning You will revisit your claim and explain your reasoning at the end of the module.

GO ONLINE to access your CER chart and explore resources that can help you collect evidence.

LESSON 1: Explore and Explain: The Metallic Properties of Alloys

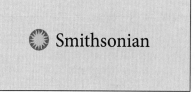

LESSON 2: Explore and Explain: Ceramics

Smithsonian

Additional Resources

(t)Video supplied by BBC Worldwide Learning, (bl)Allan Shoemake/Photographer's Choice/Getty Images, (br)Steven Puetzer/The Image Bank/Getty Images

LESSON 1
ALLOYS

FOCUS QUESTION
How did we discover that mixing metals improves them?

Alloys Through Time

For ages, people have searched for better materials to use to make their lives more comfortable and their tasks easier. Ancient cultures used stone tools until methods for processing metals became known. Today, advances in processing and blending metals still occur as scientists continue to improve the art of making alloys. Recall that an alloy is a mixture of elements that has metallic properties. Alloys can produce materials with improved properties, such as greater hardness, strength, lightness, or durability.

The first alloy

Historians think that bronze was accidentally discovered more than 5000 years ago by the ancient Sumerians. The Sumerians lived in the Tigris-Euphrates Valley (now Iraq). It is likely that Sumerians used rocks rich in copper and tin ore to keep their campfires from spreading. The hot campfire melted the copper and tin ores within the rocks, creating bronze. This first known mixture of metals became so popular and widely used that a 2000-year span of history is known as the Bronze Age. The ancients did not have a chemical language for their discovery, but their bronze tools and objects, such as those shown in **Figure 1,** helped change the history of civilization.

Figure 1 Artifacts from the Bronze Age prove that alloys of metal were used before 3000 B.C. The development of bronze led to improved tools, weapons, and building materials and encouraged development of industry and trade.

DEA/G. DAGLI ORTI/De Agostini Picture Library/Getty Images

Get It?
Describe how bronze was discovered.

3D THINKING — DCI Disciplinary Core Ideas — CCC Crosscutting Concepts — SEP Science & Engineering Practices

COLLECT EVIDENCE
Use your Science Journal to record the evidence you collect as you complete the readings and activities in this lesson.

INVESTIGATE
GO ONLINE to find these activities and more resources.

Virtual Investigation: Ratios of Elements
Use a model to determine how the ratios of elements affects an alloys' physical properties.

CCC **Identify Crosscutting Concepts**
Create a table of the crosscutting concepts and fill in examples you find as you read.

Figure 2 Copper is shiny, conducts thermal and electrical energy well, and can be easily drawn into a wire or hammered into sheets. Copper's malleability makes it ideal for use in pipes, wires, and tubing.

Explain *the difference between malleability and ductility.*

Materials change

Bronze is still used today, but it is doubtful that ancient people would recognize it. The methods of processing this alloy have undergone many changes. Other alloys, such as brass, pewter, and steel, have been developed through the ages, giving people a large selection of materials to choose from today.

The Metallic Properties of Alloys

Alloys retain the metallic properties of the metals in them. Recall that metals have four distinct properties: luster, ductility, malleability, and conductivity. **Luster,** shown in **Figure 2,** is the property of a shiny appearance due to the ability to reflect light. The shiny appearance of aluminum foil and new copper coins demonstrates the property of luster. The property that allows the material to be pulled into wires is known as ductility (duk TIH luh tee). The copper electrical wires in your home demonstrate the ductility of metals and alloys.

Recall that metals are malleable. The property that allows a material to be hammered or rolled into thin sheets is malleability (mal yuh BIH luh tee). Aluminum foil that is used in food preparation and food storage demonstrates the malleability of aluminum. Musical instruments like trombones or trumpets demonstrate the luster and malleability of brass. **Conductivity** (kahn duk TIH vuh tee) is the property of a material that allows thermal energy or electrical charges to move easily through it. Metals and alloys have high conductivities because some of their electrons are not tightly held by their atoms. Copper is used in electrical wiring because it is conductive as well as ductile.

Get It?

List five other examples of items that you know have metallic properties.

Choosing an alloy

Why are gold alloys used in jewelry, but drill bits are made from steel? The alloy chosen to make an object depends upon how the object will be used. The characteristics desired in the final product must be considered before the product is constructed. For example, the manufacturer will have to consider how hard the alloy has to be to prevent the object from breaking when it is used. He or she must also consider if the object will be exposed to chemicals that will react with the alloy and cause the alloy to fail. These questions relate to the properties of the alloy and its intended use. This represents only two of the many possible questions that must be answered while a product is being designed.

Choosing a gold alloy Look at the characteristics of familiar objects, such as the gold jewelry shown in **Figure 3.** The rings appear to be made of pure gold, but they are really made from gold-copper alloys. Gold is a bright, expensive metal that is soft and bends easily. Copper, on the other hand, is a relatively inexpensive metal that is harder than gold. When gold and copper are melted, mixed, and allowed to cool, an alloy forms with properties different from those of either pure gold or pure copper.

The properties of an alloy will vary depending upon the amount of each metal that is added. A ring made with a higher percentage of gold will bend easily due to gold's softness. This ring will be more valuable because it contains a higher percentage of gold, which is a more expensive metal. A ring with a higher percentage of copper will not bend as easily because copper is harder than gold. However, this ring will be less valuable because it contains more copper, a less-expensive metal.

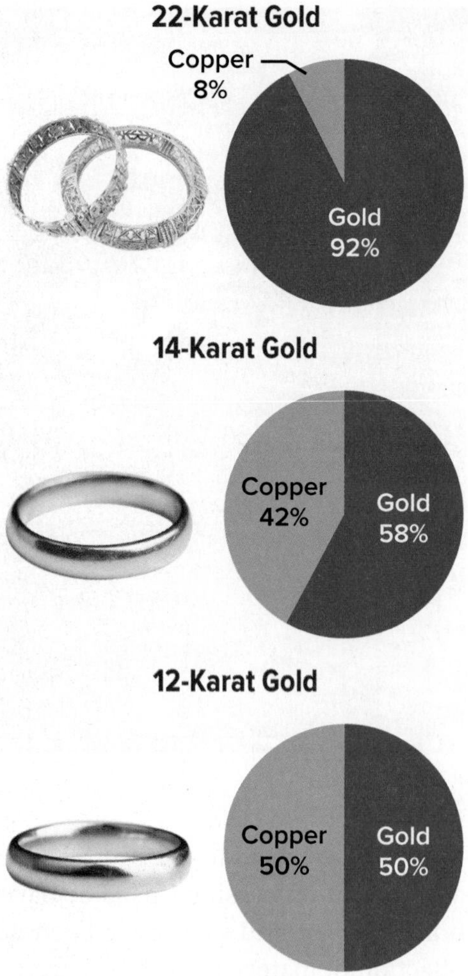

Figure 3 Not all gold jewelry is the same. These rings vary in the amount of copper that has been added to the gold. A karat is a measure of gold's purity; pure gold is 24-karat.

Infer *How do the rings compare in hardness and malleability?*

Get It?
Compare and contrast the properties of gold-copper alloys that are mostly gold with those that are mostly copper.

Uses of Alloys

If you see an object that looks metallic, it is most likely an alloy. Alloys are used in a variety of products, as shown in **Figure 4** on the next page. Alloys that are exceptionally strong are used to manufacture industrial machinery, construction beams, and railroad cars and rails. Automobile and aircraft bodies are constructed of alloys that are corrosion-resistant and lightweight but are strong and able to carry heavy loads. Other types of alloys are used in products such as food cans, carving knives, and rollerblades.

How are such a variety of alloys made? Scientists and engineers must find starting materials with some of the properties they need. Then, by adjusting the ratios of those materials, they can create a final product that suits their needs.

Figure 4 Visualizing Common Alloys

Most of the metal objects you use every day are made from alloys.

Stainless steel is an alloy of chromium, iron, and carbon. It is often used to make kitchenware.

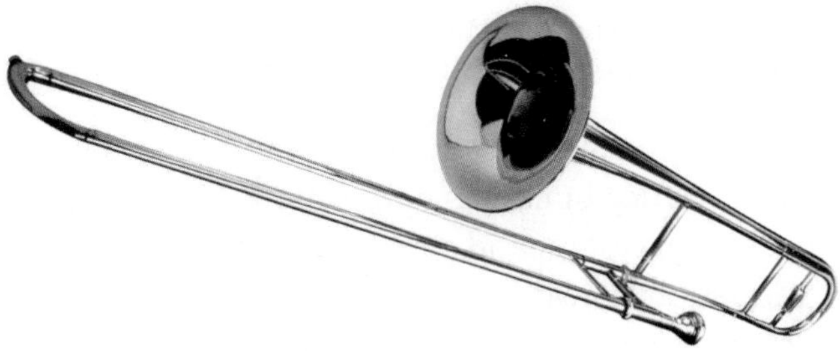

Brass is an alloy of copper and zinc. It is used in musical instruments such as this trombone.

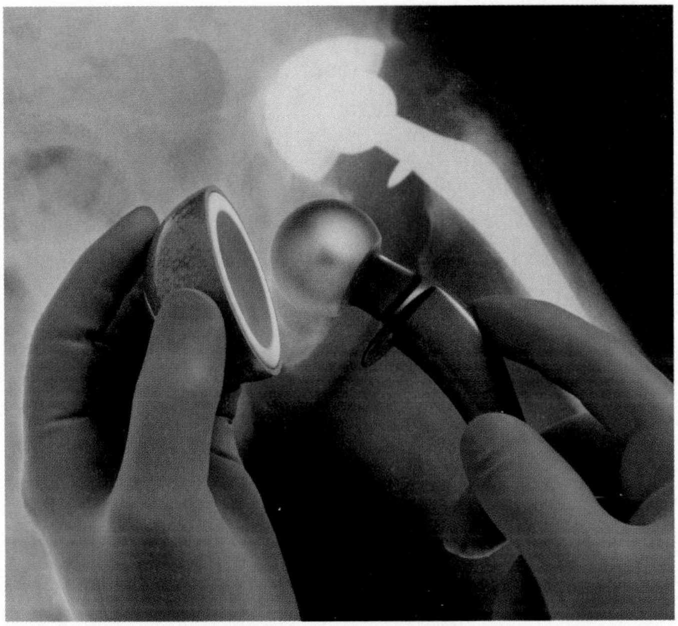

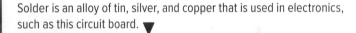

Because they are lightweight and nonreactive, titanium alloys are often used in replacement joints.

Pure silver is too soft to be very useful. Sterling silver, used in this tea service, is an alloy of silver and copper.

Solder is an alloy of tin, silver, and copper that is used in electronics, such as this circuit board. ▼

Steel is an alloy of iron and carbon. Steel is used in many buildings and bridges. ▶

(tl)Africa Studio/Shutterstock, (tr)Ingram Publishing/Age Fotostock, (cl)Don Farrall/Photodisc/Getty Images, (cr)Redfield and Rice American silversmiths 1863-1871;Tiffany & Company (American retailer jeweler and manufacturer 1853 to the present)/Yale University Art Gallery, (br)©Maskot/Corbis, (bkgd)Marty Honig/Photodisc/Getty Images

Steel—an important alloy

Steel is an alloy of iron and carbon. Sometimes other elements are added to give steel specific properties. Steel is classified by the amount of carbon and other elements present, as well as by the manufacturing process that is used to refine the iron ore. The classes of steel such as stainless, carbon, or high-strength steels, each have different properties and therefore different uses.

High-strength steel is a strong alloy and is used often if a great deal of strength is required. Office buildings have steel beams to support the weight of the structure. Bridges, overpasses, and streets also are reinforced with steel. Ship hulls, bedsprings, and automobile gears and axles are made from steel. Another class of steel, called stainless steel, is used in surgical instruments, cooking utensils, and food preparation. Stainless steel contains chromium as well as iron and carbon and gets its name from its resistance to rusting and corrosion.

Get It?

Explain why steel is an important alloy.

Alloys in flight

Steel is not the only common type of alloy. Aluminum is familiar because it is used to make soda cans and cooking foil. Did you know that engineers also are using new aluminum and titanium alloys to build large commercial aircraft? The aircraft diagram in **Figure 5** shows how extensively alloys are used in new aircraft construction. The new alloys are strong and lightweight and last longer than alloys used in the past. Some of the aluminum-zinc alloys are nearly as strong as steel, but a third of the weight. A lighter plane is less expensive to fly and produces less pollution.

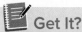

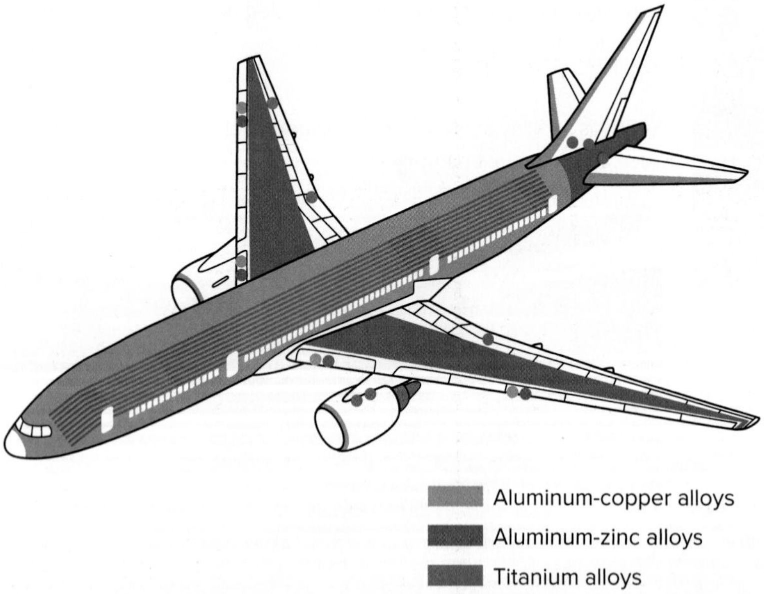

Figure 5 Airplanes feature a number of alloys in their construction. Notice that the aircraft skin is made mostly of alloys.

Infer *what characteristics make alloys useful in airplane construction.*

ACADEMIC VOCABULARY

structure

the arrangement of parts in a substance, body, or building

The house's structure was designed to withstand hurricane-force winds.

Alloys in medicine

Alloys that are resistant to tissue rejection and do not react chemically with the elements and compounds in the body can be used for a number of medical applications. One such alloy is surgical steel, a specific type of stainless steel made from iron, chromium, carbon, and nickel. Pins and screws made from surgical steel are used by doctors to connect broken bones and repair serious fractures, as shown in **Figure 6.** Surgical steel is also part of the metal plates used to repair damage to the skull. These plates protect the brain from injury and are safe to use inside the body.

If you have ever had a cavity, your dentist might have used a mercury-silver alloy to fill it, preventing further tooth decay. The alloy is malleable and durable, and it helps to prevent bacterial growth and additional cavities in the same spot.

Shape memory alloys A special type of alloy commonly used in dentistry is called a shape memory alloy. A shape memory alloy can be bent and will return to its original shape when heat is applied. The wire in braces that helps to pull teeth into the correct position—called the archwire—is often made of shape memory alloys. Orthodontists bend the wire to fit a patient's mouth and then instruct the patient to drink warm liquids to help "activate" the archwire, which will cause it to return to its original shape and reposition teeth.

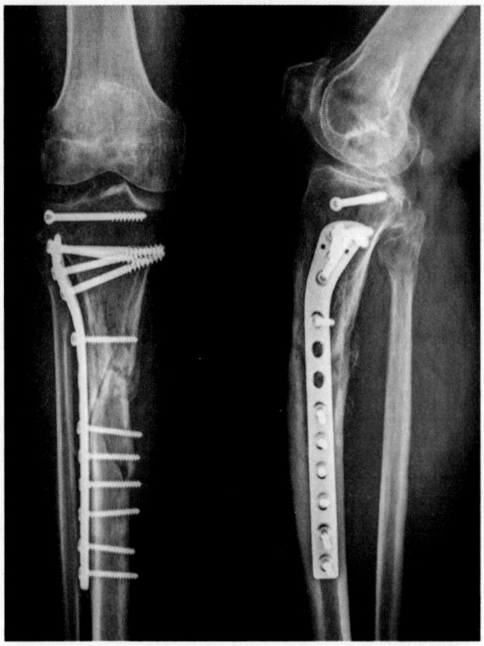

Figure 6 Surgical steel screws and plates help to hold this elbow together as it heals. Because they are made of surgical steel, they are safe for use in the human body.

Identify *the properties that make surgical steel useful for implants.*

Check Your Progress

Summary

- Alloys have metallic properties.
- An alloy has characteristics that are different from, and often improved upon, the individual elements in it.
- Scientists and engineers can control the properties of an alloy by manipulating its composition.
- Alloy materials are used in industry, food service, medicine, and aeronautics.

Demonstrate Understanding

1. **List** the metallic properties of alloys.
2. **Identify** three different alloys in your home. Describe how each alloy is used and why the properties of the alloy make it suitable for that application.
3. **Describe** the importance of steel.
4. **Identify** two medical uses of alloys. Describe the desired characteristics in medical implements made from alloys.

Explain Your Thinking

5. If you were designing a skyscraper in an earthquake zone, what properties would the structural materials need?
6. **MATH Connection** 14-karat gold is 58 percent gold and 42 percent copper. Calculate the actual amount of gold in a 65-g, 14-karat gold chain.
7. **MATH Connection** If a 7.6-g sample of copper can be hammered into a 2-cm × 2-cm sheet, calculate the number of grams necessary to hammer a 17-cm × 17-cm sheet under the same manufacturing conditions.

LEARNSMART® Go online to follow your personalized learning path to review, practice, and reinforce your understanding.

LESSON 2
VERSATILE MATERIALS

FOCUS QUESTION
How are toilets made?

Ceramics

Do you think of floor tiles, pottery, or souvenir knickknacks when you see the word *ceramic*? By definition, **ceramics** are materials that are made from dried clay or claylike mixtures. Ceramics have been around for centuries—in fact, pieces of clay pottery from 16,000 B.C. have been found. The oldest known wall, built about 8000 B.C., was around the Middle Eastern city of Jericho. The walls surrounding Jericho, as well as the homes inside the walls, were ceramic, consisting of bricks made from mud and straw that were baked in the Sun.

Around 1500 B.C., the first glass vessels were made, and kilns were used to fire and glaze pottery. By 50 B.C., the Romans had developed concrete and begun using it as a building material. Some of the structures built by the ancient Romans still stand today. About the same time that Romans were developing concrete, the Syrians were experimenting with glass-blowing techniques to make pitchers, bowls, cups, and jars. All of these items are ceramics. The china and glassware shown in **Figure 7** are also examples of ceramics.

Figure 7 Glass, china, pottery, bricks, and tile are all examples of ceramics. Because of their wide range of properties, ceramics are used in a variety of different products.

Explain *how ceramics and alloys are similar.*

Steven Puetzer/The Image Bank/Getty Images

3D THINKING DCI Disciplinary Core Ideas CCC Crosscutting Concepts SEP Science & Engineering Practices

COLLECT EVIDENCE

Use your Science Journal to record the evidence you collect as you complete the readings and activities in this lesson.

INVESTIGATE

GO ONLINE to find these activities and more resources.

Quick Investigation: Model a Ceramic
Draw a pictorial model of the structures found within the mixture.

Review the News
Obtain information from a current news story about ceramics or semiconductors. Evaluate your source and communicate your findings to your class.

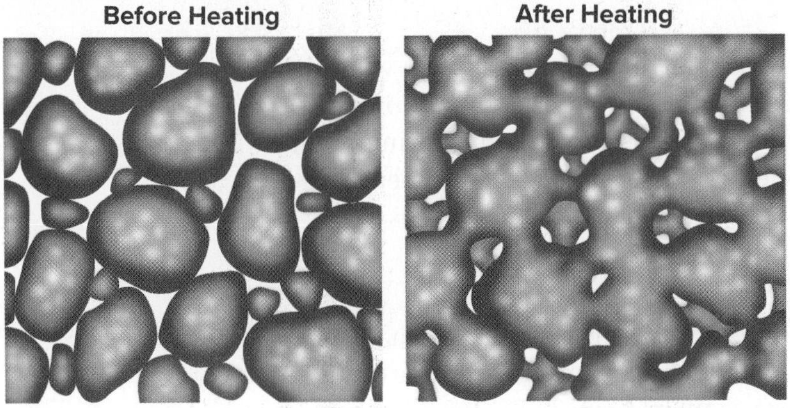

Figure 8 When a ceramic object is fired, its particles merge together due to the loss of water, called dehydration. The object shrinks, and the structure becomes more dense and stronger.

Making ceramics

Traditional ceramics are made from easily obtainable raw materials—clay, silica (sand), and feldspar (crystalline rocks). These raw materials were used by ancient civilizations to make ceramic materials and are still used today. However, some of the more recent ceramics are made from compounds of metallic and nonmetallic elements, such as carbon, nitrogen, or sulfur.

After the raw materials are processed, ceramics usually are made by molding the ceramic into the desired shape, then heating it to temperatures between 1000°C and 1700°C. The heating process, called firing, causes the spaces between the particles to shrink as a result of the loss of water, called dehydration, as shown in **Figure 8.** The entire object shrinks as the spaces become smaller. This extremely dense internal structure gives ceramics their strength. Ceramics can be very durable. However, these same ceramics can also be brittle and will break if they are dropped or if the temperature changes too quickly.

Get It?

Summarize the steps involved in making ceramics.

Properties of traditional ceramics

Ceramics are known also for their chemical resistance to oxygen, water, acids, bases, salts, and strong solvents. These qualities make ceramics useful for applications where they may encounter these substances. For instance, ceramics are used for tableware because foods contain acids, water, and salts. Ceramic tableware is not damaged by contact with foods containing these substances.

Traditional ceramics also are used as insulators because they do not conduct heat or electricity. You may have seen electric wires attached to poles or posts with ceramic insulators. These insulators keep the current flowing through the wire instead of into the ground. Ceramic tiles were also used on the heat shield of space shuttles to protect them from the extreme heat experienced when reentering Earth's atmosphere.

Get It?

Explain why ceramics make ideal tableware.

Customizing ceramics

The properties of ceramics can be customized, which makes them useful for a wide variety of applications. Changing the composition of the raw materials or altering the manufacturing process can change the properties of the ceramic. Manufacturing ceramics is similar to manufacturing alloys because scientists determine which properties are required and then attempt to create the ceramic material with those properties.

Ceramics in medicine

Ceramics also have medical uses. Replacement hip sockets, for example, are made of ceramics because they are safe for use in the human body. Ceramics are strong and durable, but what properties make them safe for use in medicine? Like the alloys used in medicine, ceramics are resistant to body fluids, which can damage other materials. They are also relatively nonreactive and are resistant to rejection by the body. In the medical field, surgeons use ceramics in conjunction with alloys for the repair and replacement of joints such as hips, knees, shoulders, elbows, fingers, and wrists. Dentists use ceramics for braces as well as tooth replacement and repair.

Get It?

Explain why ceramics are appropriate for medical applications.

APPLY SCIENCE

Can you choose the right material?

Scientists continue to learn about atoms and how they interact. With this new knowledge, chemists today are able to create substances with a wide range of properties. This is especially evident in the production of specialized ceramics. Technical ceramics include ceramic products used in engineering applications, such as tiles on the outside of spacecraft, jet engine turbine blades, and gas-burner nozzles.

Identify the Problem

As an engineer working on the design of a new car, you need to select the right ceramic materials to build parts of the car's engine and its onboard computer. The table at right shows the materials you have to choose from.

The ceramics listed in the table vary in their resistance to wear, electrical conductivity, chemical reactivity, and melting point. Using these properties, decide which materials should be used for the different components of the car.

Ceramic Properties

Material	Wear Resistant	Conducts Electricity	Reacts with Chemicals	Melting Point (°C)
A	highly	no	no	3000
B	not at all	no	yes	100
C	moderately	yes	no	1500
D	resistant	yes	no	500

Solve the Problem

1. Which of the above materials would you use when you build the engine? Explain the factors that you considered to make your decision.
2. Which of the above materials would you select when building the onboard computer? Explain your selection.
3. If you had to choose a material for building the car's bumper, what factors would you consider? Do you think that a ceramic material would be the best choice? Explain your answer.

Modern ceramics

Ceramics can be customized to have nontraditional properties. Ceramics traditionally are used as insulators, but there are exceptions. For instance, chromium dioxide conducts electricity as well as most metals, and some copper-based ceramics have superconductive properties. One application of nontraditional ceramics uses a transparent, electrically conductive ceramic in aircraft windshields to keep them free of ice and snow.

Semiconductors

Another class of versatile materials is semiconductors. Recall that semiconductors are poorer conductors of electricity than metals but better conductors than nonmetals, and their electrical conductivities can be controlled. This property makes semiconductor devices useful and makes computers and other electronic devices possible. Two common semiconducting materials are the metalloids silicon (Si) and germanium (Ge).

Get It?
Define the term *semiconductor.*

Doping

Adding other elements to some metalloids can change their electrical conductivities. The process of adding impurities, such as other elements, to a semiconductor to modify conductivity is called **doping.** For example, the conductivity of silicon can be increased by replacing some of the silicon atoms with atoms of other elements, such as arsenic (As) or gallium (Ga), as shown in **Figure 9.**

Adding even a single atom of one of these elements to a million silicon atoms significantly changes the conductivity. By controlling the type and number of atoms added, the conductivity of silicon can vary over a wide range.

Figure 9 Pure silicon is a poor conductor. Adding an element, such as gallium, as an impurity creates an area of fewer electrons, called a hole. Therefore, the impurity changes the material's conductivity.

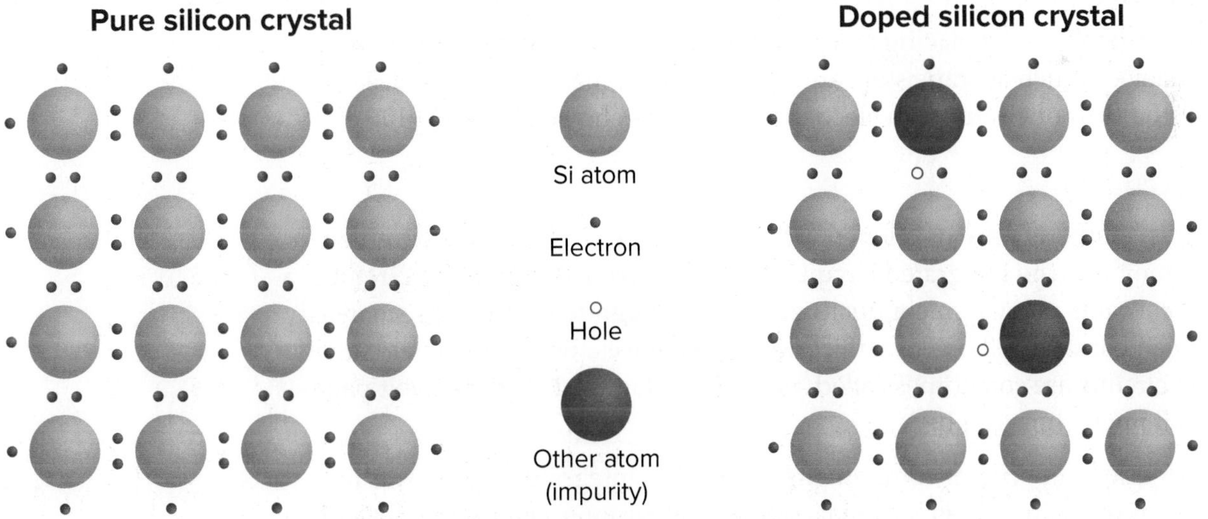

WORD ORIGIN

ceramic
comes from the Greek word *keramos,* meaning *potter's clay*
In my ceramics class, I made a beautiful vase for my mother.

STEM CAREER Connection

Biomedical Engineer
Are you interested in engineering and medical and biological sciences? If you are, you may want to explore a career as a biomedical engineer. Biomedical engineers combine their knowledge of engineering with biological science to design technology used in healthcare. Some biomedical engineers specialize in biomaterials. They study or design materials that are used to create medical technology and equipment. They might even develop materials to be used to build artificial organs or replacement body parts.

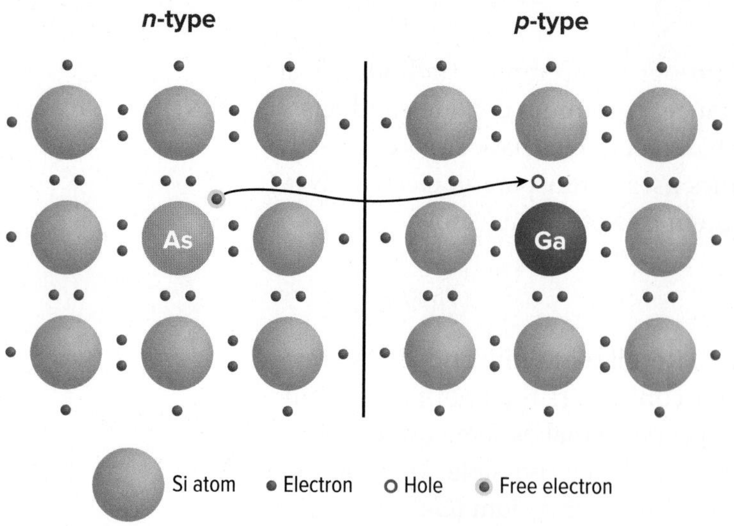

Figure 10 Arsenic as an impurity adds free electrons. These electrons flow from the arsenic-doped silicon to the germanium-doped silicon, filling the available holes created by the addition of germanium. The electron flow is controlled by the sequence of *n*-type or *p*-type semiconductors.

Compare *How are* n-type *and* p-type *semiconductors different?*

Types of doping Depending on the element added, the overall number of electrons in the semiconductor can be either increased or decreased. If the impurity causes the overall number of electrons to increase, the semiconductor is called an *n*-type semiconductor. The extra electrons are called free electrons because they are very weakly attached to the impurity and will easily flow. If doping reduces the overall number of electrons, the semiconductor is called a *p*-type semiconductor. The silicon crystals will contain holes, or areas with fewer electrons. Electrons now can move from hole to hole across the crystal, increasing conductivity.

Controlling electron flow By placing n-type and p-type semiconductors together, semiconductor devices such as transistors and diodes can be made. These devices are used to control the flow of electrons in electrical circuits, as shown in **Figure 10.**

Integrated circuits

During the 1960s, methods were developed for making these components extremely small. At the same time, the integrated circuit was developed. An **integrated circuit** is a small chip that contains many semiconducting devices. Integrated circuits as small as 1 cm on a side can contain millions of semiconducting devices. Because of their small size, integrated circuits are sometimes called microchips. **Figure 11**, on the next page, shows how small microchips can be.

Being able to pack so many circuit components onto a tiny integrated circuit was a technological breakthrough. Microchips make today's televisions, cell phones, calculators, and other devices smaller in size, cheaper to manufacture, and capable of more advanced functions than their predecessors. Also, because the circuit components are so close together, it takes less time for electric current to travel through the circuit. Therefore, computers, tablets, and other electronic devices can process electronic signals rapidly. **Figure 11,** on the next page, illustrates how integrated circuits have given us faster, smaller, and more capable computers since the 1940s.

Figure 11 The History of Computers

The earliest computers were enormous and relied on vacuum tubes to process data. Today's computers use microchips, tiny flakes of silicon engraved with millions of circuit components.

ENIAC

ENIAC's vacuum tubes

1946 One of the first computers was developed by the Army. It was called the Electronic Numerical Integrator and Computer, or ENIAC for short. ENIAC had more than 18,000 vacuum tubes and was the size of a room.

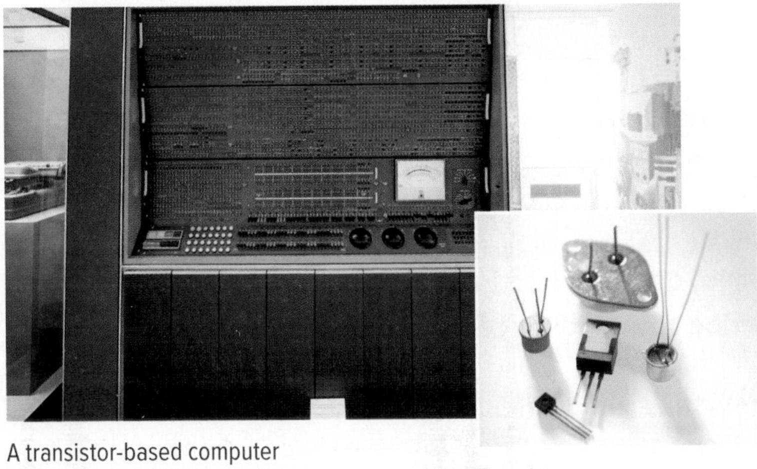

A transistor-based computer

Transistor

1953 The vacuum tubes of older computers were replaced by transistors, which used semiconductors to perform the same function in a much smaller space.

1961 The first commercially available integrated circuit (IC) combined 200,000 transistors on a single, miniature circuit called a microchip.

A microchip

1977 The first home computers included keyboards and cables to connect them to a monitor or a television.

A modern computer

An early home computer

Today Modern computers run on processors that contain millions of transistors and perform operations measured in the billions each second.

Computer components

Computers receive and store information, follow instructions to perform tasks, and communicate information to the outside world. All three of these jobs require a combination of hardware and software components.

The term *computer hardware* refers to the major physical components of a computer, such as the keyboard, monitor, mouse, and central processing unit (CPU). These components are shown in **Figure 12.** All of these components contain semiconductors in integrated circuits. Inside the CPU, a special microchip called the processor acts as a control center. It is responsible for how all the components work together.

Software refers to the instructions that tell the computer what to do. The instructions are stored briefly in the computer's memory—on a special type of semiconductor chip—before being executed by the CPU. When a computer system is functioning properly, the hardware and software work together to perform tasks, sending and receiving millions of electronic signals each second.

Figure 12 Today's computers come in all shapes and sizes. This desktop computer has a monitor, keyboard, mouse, and CPU as part of its hardware.

Predict *what computers will look like in 20 years.*

Check your Progress

Summary

- Ceramics are nonmetallic clay or claylike mixtures.
- The properties of ceramics can be customized by changing raw materials and manufacturing processes.
- Doping is the process of adding impurities to a semiconductor to change its conductivity.
- Integrated circuits are microchips that contain many semiconducting devices.

Demonstrate Understanding

8. **Describe** the electrical conductivities of traditional ceramics, modern ceramics, and semiconductors.
9. **List** five uses of ceramic materials. What properties make ceramics good choices for these applications?
10. **Describe** how modern ceramics are different from traditional ceramics.
11. **Explain** how the function of semiconductors is determined by the electrons and electrical forces between atoms.

Explain Your Thinking

12. Computers have changed the way businesses operate. If you operated a distribution center for a manufacturer, how would you use computers to assist you?
13. **MATH Connection** Ceramic A forms when heated to 1400°C and has a density of 5.3 g/cm3. Ceramic B forms at a temperature 675°C cooler and is four times as dense. What temperature is required to form Ceramic B, and what is its density?
14. **MATH Connection** A developmental ceramic is designed to be 35 percent silica and 65 percent sulfur. If a researcher needs 75 g of this material, how many grams of each component will she need?

LEARNSMART® Go online to follow your personalized learning path to review, practice, and reinforce your understanding.

LESSON 3
POLYMERS AND COMPOSITES

FOCUS QUESTION

Why is it hard for plastics to decompose?

Polymers

The term *polymer* describes a class of substances that are composed of molecules arranged in large chains of simple, repeating units called monomers. Polymer chains can be very long. Polypropylene, for example, might have 50,000 to 200,000 monomers in its chain.

Some polymers, such as proteins, cellulose, and nucleic acids, occur naturally. Humans have used natural polymers for centuries. The ancient Egyptians soaked burial wrappings in plant polymers called resins to help preserve the dead. Animal horns and turtle shells, which contain natural resins, were used to make combs and buttons for many years.

Making polymers

In the 1800s, scientists began developing processes to improve natural polymers and to create new ones in the laboratory. Materials that do not occur naturally but are manufactured in a laboratory or chemical plant are called **synthetic.** In this lesson, the focus will be on synthetic polymers. Many synthetic polymers are designed to outperform their natural counterparts. Several examples of synthetic polymers are shown in **Figure 13.**

Figure 13 Synthetic polymers are found in many materials we use every day, including computers, clothing, bedding, and wall paint.

3D THINKING DCI Disciplinary Core Ideas CCC Crosscutting Concepts SEP Science & Engineering Practices

COLLECT EVIDENCE

Use your Science Journal to record the evidence you collect as you complete the readings and activities in this lesson.

INVESTIGATE

GO ONLINE to find these activities and more resources.

Quick Investigation: Observe Drying
Carry out an investigation to determine the cause and effect fabric type has on retaining moisture.

Revisit the Encounter the Phenomenon Question
What information from this lesson can help you answer the Unit and Module questions?

History of synthetic polymers

In 1839, American inventor Charles Goodyear found that heating sulfur and natural rubber together prevented rubber from turning brittle when it became cold or soft when it became hot. In the 1850s, Alexander Parkes created a plastic synthetic that would later become known as celluloid. It was used in applications such as umbrella handles and toys. These early synthetic polymers had many drawbacks, but they were the beginning of the development of a huge class of materials. **Figure 14** shows a time line of when some of these materials were created. Today, synthetic polymers usually are made from the hydrocarbons found in fossil fuels, such as oil, coal, and natural gas. Recall that hydrocarbons are organic molecules made entirely of carbon and hydrogen.

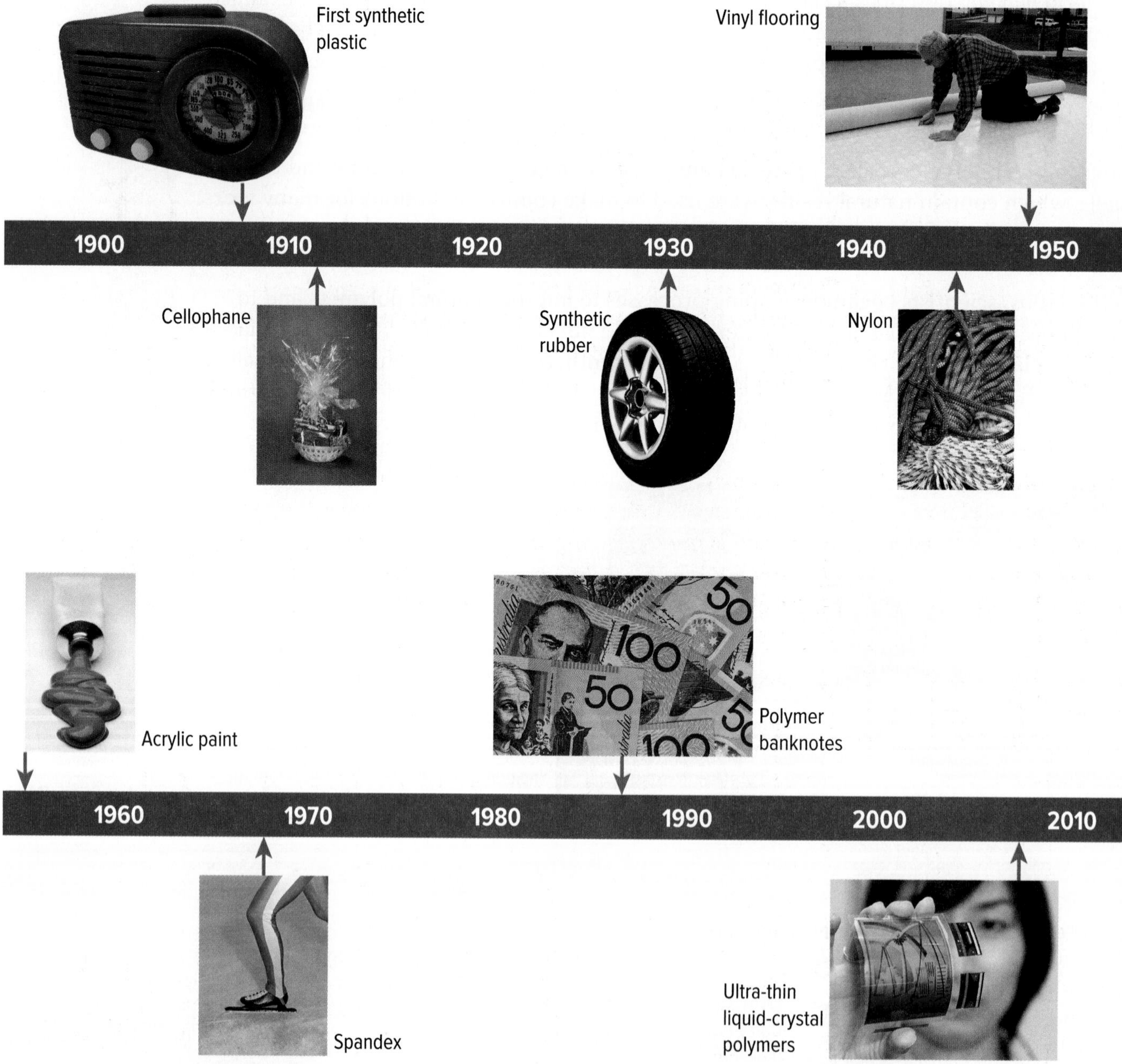

Figure 14 In the past 150 years, dozens of synthetic polymers have been developed for thousands of applications.

(l to r, t to b)©Comstock Images/Alamy Stock Photo, Ernest Prim/Getty Images, Paul Conrath/Getty Images, ©Ingram Publishing/Alamy Stock Photo, vuk8691/iStockphoto/Getty Images, Jeffery Coolidge/Photodisc/Getty Images, Sarah1810/Shutterstock, Benne Ochs/fStop/Getty Images, YOSHIKAZU TSUNO/AFP/Getty Images

Changing properties

Polymers have a wide range of uses because their properties can be easily modified. If the composition or arrangement of monomers is changed, then the material properties will change. Changing the amount of branching is one way to change a polymer's properties. Low-density polyethylene has more branching than high-density polyethylene.

Another way to change the properties of a polymer is to replace one or more of the hydrogen atoms in the monomer with another element or group. For example, if one of the hydrogen atoms in an ethylene monomer is replaced with a chlorine atom, the result is polyvinyl chloride (PVC). Because of the many ways to change polymers, the possibilities for creating new materials are almost limitless.

Using Polymers

Because so many things today are made of polymers, some people call this "The Age of Plastics." However, plastics are only one group of polymers. Other groups include synthetic fibers, adhesives, surface coatings, and synthetic rubbers.

The plastics group

Plastics are widely used for many products because they have desirable properties. They are usually lightweight, strong, impact resistant, waterproof, moldable, chemical resistant, and inexpensive. Their properties vary widely. Some are clear, some melt at high temperatures, and some are flexible. These properties are determined by the composition of the polymer from which the plastic is made.

Synthetic fibers

Nylon, polyester, acrylic, and polypropylene are examples of polymers that can be manufactured as fibers. Synthetic fibers can be mass-produced to almost any set of desired properties. For example, nylon is often used in wind and water-resistant clothing such as lightweight jackets. Polyester and polyester blended with natural fibers such as cotton are often used in clothing. Polyester fiber is also used to fill pillows and quilts. Polyurethane is the foam used in many mattresses and pillows.

Synthetic fibers called aramids are a family of nylons with special properties. Aramids are used to make fireproof clothing. Another aramid fiber is used to make bulletproof vests. Although they are lightweight, these aramids are five times stronger than steel.

SCIENCE USAGE v. COMMON USAGE

fiber

Science usage: a thread
Silk is a natural fiber created by silkworm caterpillars.

Common usage: a carbohydrate important in human diets
Whole-wheat bread is a good source of dietary fiber.

CCC CROSSCUTTING CONCEPTS

Structure and Function Look around your school environment for objects made from synthetic materials. Choose five objects, and determine what material or materials they are manufactured from. Use a graphic organizer to compare and contrast the structure and functions of the materials.

Taking a cue from nature Have you ever wondered where humans got the idea to create threads and fabrics? Spinning long fibers into threads is an idea not original to humans. Spiders spun fibers for their webs long before humans copied the idea and began spinning fibers themselves.

The idea behind nylon fiber is also borrowed from nature. Nylon was produced in the laboratory as a possible substitute for silk. **Figure 15** shows the similarities and differences between silk and nylon fibers. The silkworm is a caterpillar that produces a strong and durable fiber when creating its cocoon. The silk is a highly desirable fiber that is woven into fabric for items such as blouses and stockings. Think of how many silkworm cocoons it would take to produce enough silk for a single blouse. Why do you think natural silk fabric is more expensive than nylon fabric?

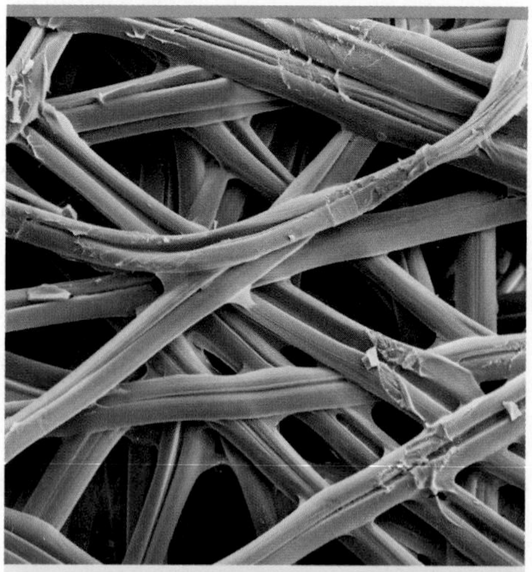

Silk fibers

Nylon fibers

Figure 15 Synthetic fibers are designed to mimic natural fibers, but they are not identical.

Describe *the similarities and differences between the nylon fiber and the silk fiber.*

Adhesives

Synthetic polymers are used to make a wide range of adhesives. Contact cements are used in the manufacture of automobile parts, furniture, leather goods, and decorative laminates. They adhere instantly, and the bond gets stronger after the adhesive dries. Structural adhesives are used in construction projects. One structural adhesive, called silicone, is used to seal windows and doors to prevent heat loss in homes and other buildings. Ultraviolet-cured adhesives are used by orthodontists to adhere braces brackets to teeth, as shown in **Figure 16.** These adhesives bond after exposure to ultraviolet light. Other types of adhesives include transparent, pressure-sensitive tape and hot-melt adhesives that are used in glue guns. Like other synthetic polymers, adhesives can be engineered to have specific properties. By altering the starting materials or the processes under which the adhesives are created, the properties of the final product can be made to suit any need.

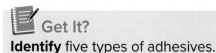

Get It?

Identify five types of adhesives.

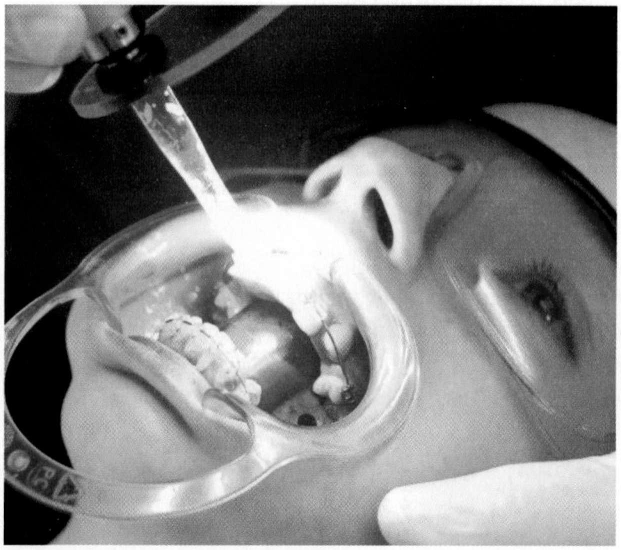

Figure 16 The use of ultraviolet light-cured adhesives allows orthodontists to ensure the correct placement of braces before they are secured in place.

Surface coatings and elastic polymers

Many surface coatings use synthetic polymers. Polyurethane is a popular polymer that is used to protect and enhance wood surfaces. Many paints use synthetic polymers in their composition, too. For example, acrylic paints are useful because they are water-soluble when wet, but water-resistant when dry. Acrylic paints are, therefore, easy to clean off brushes, but the dried paint will not wash off surfaces.

Synthetic rubber is a synthetic elastic polymer. An elastic object is one that is capable of returning to its original size and shape after it is deformed, like a rubber band. Synthetic rubber is used to manufacture tires, gaskets, belts, and hoses. The soles of some shoes also are made from this rubber.

Composites

The properties of a synthetic polymer can be altered by using more than one material. A **composite** is a mixture of two or more materials—one embedded or layered in another. Embedding one material into another can produce a composite with the desired properties. For example, if a substance is lightweight but brittle, such as some plastics, embedding flexible fibers into it can alter the brittleness. After the substance has the flexible fibers embedded, the product is less brittle and can withstand greater forces before it breaks.

Glass fibers are often used to reinforce plastics because glass is inexpensive, but other materials can be used as well. Composite materials of plastic and glass fibers are used to construct boat and car bodies, as shown in **Figure 17.** These bodies are made of a mixture of small fibers of glass embedded in a plastic. The structure of the fiberglass reinforces the plastic, making a strong, lightweight composite.

Figure 17 Many cars, boats, and other vehicles have bodies that contain or are made entirely out of fiberglass, a composite of glass and plastic.

Infer *What parts of a car's body could be made of fiberglass?*

Composites in space

Composite materials are used in the construction of satellites. Carbon fibers strengthen the plastic body, creating a material that is more rigid and stronger than aluminum. Satellites made of these composites, such as the one shown in **Figure 18,** are significantly lighter than those made of aluminum. These lighter satellites are less expensive to launch into orbit, yet still able to withstand the stress of the launch.

Aircraft made of composites also benefit from the strong yet lightweight properties of composite materials. Materials made of graphite composites are 13 percent lighter than those made with aluminum. The mass of an aircraft can be reduced by more than 2500 kg by using advanced, lightweight composite materials. The lower mass results in cost savings by reducing the amount of fuel required to operate the aircraft. Some modern commercial aircraft contain 50 percent composite materials.

Figure 18 Satellites contain durable, lightweight composite materials.

Explain *why composites are used in artificial satellites.*

Get It?

Identify the advantages of using composite materials.

Check Your Progress

Summary

- The composition and chemistry of polymers allow almost preferred limitless modifications.
- Polymers can be natural or synthetic.
- Polymers have a great variety of uses.
- Composite materials often are preferred because they can be stronger and lighter than metals or alloys.

Demonstrate Understanding

15. **Discuss** reasons why chemists create new polymers and composites.
16. **Identify** four uses of synthetic polymers.
17. **Explain** the difference between natural and synthetic polymers, and give an example of each.
18. **Explain** what a composite material is, and give three examples of items that are made from composites.

Explain Your Thinking

19. You are designing a new material for use in an airplane body. What properties should the material have?
20. **MATH Connection** A telecommunications company launches 10,000-kg satellites. A new satellite made from composites promises to reduce that mass by 25%. What is the mass of the new satellite?

LEARNSMART® Go online to follow your personalized learning path to review, practice, and reinforce your understanding.

Wonder Fiber

Chemist Stephanie Kwolek had a knack for recognizing things that other people missed. So, when her experiment yielded a thin, cloudy liquid that separated into layers when she stirred it rather than the clear, thick solution she was expecting, she didn't throw it out. Kwolek continued experimenting with the unusual solution.

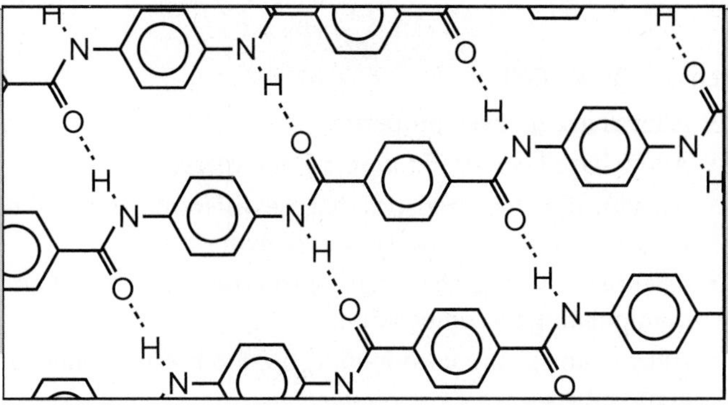

Kevlar is made of long, repeating chains that line up parallel to one another.

Kwolek studied polymers—large molecules made of repeating units of smaller molecules that string together to form chains. They can be made into long, thin fibers and incorporated into fabrics and other materials.

Creation of Kevlar

In 1964, Kwolek's research group was assigned to create a strong, rigid, heat-resistant and lightweight polymer fiber to replace the heavy steel wire used to reinforce tires. This would help cars use less gasoline. When a promising polymer would not melt during Kwolek's experiments, she searched for a solvent that could dissolve the polymer. She eventually succeeded in dissolving the polymer, but the resulting solution was very different from any polymer she had ever encountered.

Stephanie Kwolek shown here in 1967, was the fourth woman to be added to the National Inventors Hall of Fame.

Walter Oleksy/Alamy Stock Photo

When the polymer was spun into fibers, the results were astonishing. The fibers lined up parallel to each other, as shown above, creating a strong, lightweight, durable material. Her group immediately recognized possible uses for the material. In 1971, this new fiber was named Kevlar.

Kevlar is difficult to cut and does not shrink in cold temperatures. It does not rust or corrode, can be immersed in water without being affected, has low reactivity, and is an insulator. It is also flame-resistant and will not melt or soften at high temperatures.

While Kevlar is best known for its use in bulletproof vests, it is also used in hiking boots, tires, brake pads, and hundreds of other products. Kwolek's "wonder fiber" is now an important component of many products used every day.

OBTAIN, EVALUATE, AND COMMUNICATE INFORMATION

Research and compile a list of applications of Kevlar or another polymer. Create a chart showing the products or applications and the physical or chemical properties of the polymer that make it appropriate for each application.

MODULE 24

STUDY GUIDE

GO ONLINE to study with your Science Notebook.

Lesson 1 ALLOYS

- Alloys have metallic properties.
- Alloys have been used for over 5,500 years.
- An alloy has characteristics that are different from, and often improved upon, the individual elements in it.
- Scientists and engineers can control the properties of an alloy by manipulating its composition.
- Alloy materials are used in industry, food service, medicine, and aeronautics.

- luster
- conductivity

Lesson 2 VERSATILE MATERIALS

- Ceramics are nonmetallic clay or claylike mixtures.
- The properties of ceramics can be customized by changing raw materials and manufacturing processes.
- Doping is the process of adding impurities to a semiconductor to change its conductivity.
- Integrated circuits are microchips that contain many semiconducting devices.

- ceramics
- doping
- integrated circuit

Lesson 3 POLYMERS AND COMPOSITES

- The composition and chemistry of polymers allow almost limitless modifications.
- Polymers can be natural or synthetic.
- Polymers have a great variety of uses.
- Composite materials often are preferred because they can be stronger and lighter than metals or alloys.

- synthetic
- composite

THREE-DIMENSIONAL THINKING
Module Wrap-Up

REVISIT THE PHENOMENON

What makes this candy so colorful?

CER Claim, Evidence, Reasoning

Explain Your Reasoning Revisit the claim you made when you encountered the phenomenon. Summarize the evidence you gathered from your investigations and research and finalize your Summary Table. Does your evidence support your claim? If not, revise your claim. Explain why your evidence supports your claim.

STEM UNIT PROJECT

Now that you've completed the module, revisit your STEM unit project. You will apply your evidence from this module and complete your project.

GO FURTHER

SEP Data Analysis Lab: Pole Vaulting—Flexing More than Muscles

Vaulting poles need to be flexible. As a running vaulter plants the end in the box, the pole must easily flex so as not to drastically brake the vaulter's forward motion. As the vaulter flexes the pole, he or she does work on the pole, which causes the pole to bend and increases its potential energy. As the vaulter leaps, the pole flexes back to its original shape and does work on the vaulter, increasing his or her kinetic energy.

Olympic Pole Vault Record Over Time

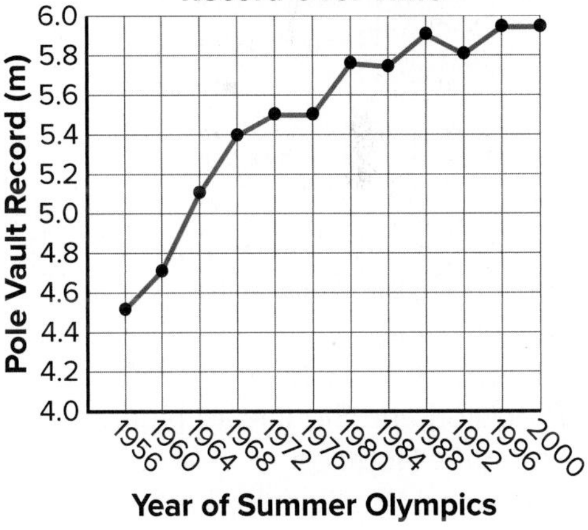

Until the early 1960s, poles were made from a variety of materials, including bamboo and metals. When poles began to be manufactured from fiberglass, vaulting records reached new heights as shown in the graph. Today, carbon-composite poles are custom-built to a vaulter's speed, weight, and grip.

CER Analyze and Interpret Data

1. **Claim and Evidence** Using the averages of the first three and last three pole-vault records shown on the graph, determine the percentage by which Olympic vaulting records have increased.
2. **Claim, Evidence, Reasoning** Explain what properties you think vaulting poles should have in addition to flexibility.
3. **Claim, Evidence, Reasoning** What other sports equipment do you think might be improved by being made from new materials that increase flexibility?
4. **Evidence and Reasoning** Select one piece of equipment from your list in question 3. Research if and how the equipment is being improved by the use of new materials.

STUDENT RESOURCES

Make Comparisons

Why learn this skill?

Suppose you want to buy a portable music player, and you must choose between three models. You would probably compare the characteristics of the three models, such as price, amount of memory, sound quality, and size in order to determine which model is best for you. In the study of chemistry, you often make comparisons between the structures of elements and compounds. You will also compare scientific discoveries or events from one time period with those from another period.

Learn the Skill

When making comparisons, you examine two or more items, groups, situations, events, or theories. You must first decide what will be compared and which characteristics you will use to compare them. Then identify any similarities and differences.

For example, comparisons can be made between the two models in the illustration on this page. The structure of the hydrogen atom can be compared to the structure of the oxygen atom. By reading the labels, you can see that both atoms have protons, neutrons, and electrons.

Practice the Skill

Create a table with the heading *Hydrogen and Oxygen Atoms*. Make three columns. Label the first column *Protons*. Label the second column *Neutrons*. Label the third column *Electrons*. Make two rows. Label the first row *Hydrogen*. Label the second row *Oxygen*. List the number of protons for each atom in the first column. Fill in the number of neutrons and electrons for each atom in the remaining columns. When you have finished the table, answer the following questions.

1. What is being compared? How are they being compared?
2. What do hydrogen and oxygen atoms have in common?
3. Describe how the differences between these two atoms affect the number of energy levels that each atom has.

Apply the Skill

Make Comparisons On page 150 you will find illustrations of a dry cell and an wet cell. Compare these two illustrations carefully. Then, identify the similarities and the differences between the two cells.

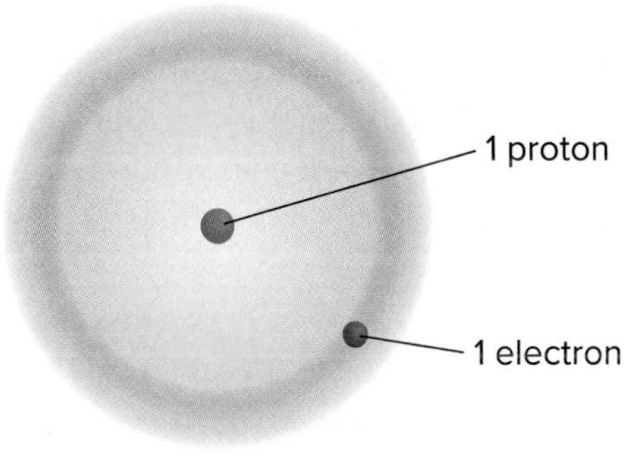

Hydrogen atom

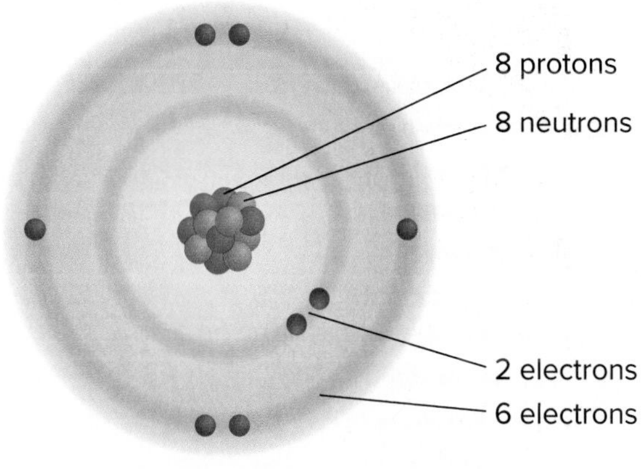

Oxygen atom

Take Notes and Outline

Why learn this skill?

One of the best ways to remember something is to write it down. Taking notes—writing down information in a brief and orderly format—not only helps you remember, but also makes studying easier.

Learn the Skill

There are several styles of note taking, but all put information in a logical order. As you read, identify and summarize the main ideas and details that support them and write them in your notes. Paraphrase—that is, state in your own words—the information rather then copying it directly from the text. Using note cards or developing a personal "shorthand"—using symbols to represent words—can help.

You might also find it helpful to create an outline when taking notes. When outlining material, first read the material to identify the main ideas. In textbooks, lesson headings provide clues to main topics. Then, identify and fill in the subheadings and subdetails.

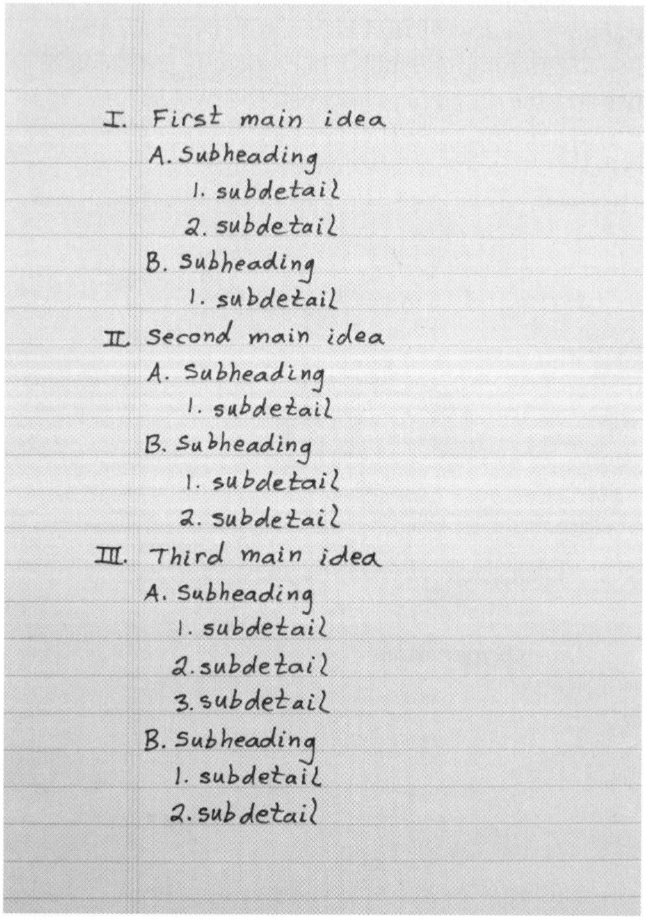

Practice the Skill

Read the following excerpt from Module 12. Use the steps you just read about to take notes or create an outline. Then answer the questions that follow.

After your eyes adjust to a dark room you will see that brightly colored objects look gray or black in the dim light. With the lights on, however, you will see all the objects in the room, including their colors. What you see depends on the amount of light in the room and the color of the objects. To see an object, it must reflect some light into to your eyes.

Objects can absorb, reflect, and transmit light. Objects that transmit light allow light to pass through them. An object's material determines the amount of light it absorbs, reflects, and transmits. The material in the third candleholder in Figure 1 is opaque. Opaque materials only absorb and reflect light; no light passes through them. As a result, you cannot see the candle.

Some materials, such as the second candleholder in Figure 1, are translucent. Translucent materials transmit light but also scatter it. You cannot see clearly through translucent materials, and objects appear blurry. The first candleholder shown in Figure 1 is transparent. Transparent materials transmit light without scattering it, so you can see objects clearly through them.

1. What is the main topic of the excerpt?
2. What are the first, second, and third ideas?
3. Name one detail for each of the ideas.

Apply the Skill

Take Notes and Outline Go to Lesson 1 of Module 15 and take notes by paraphrasing and using shorthand or by creating an outline. Use the lesson title and headings to help you create your outline. Summarize the lesson using only your notes.

Analyze Media Sources

Why learn this skill?

To stay informed, people use a variety of media sources, including print media, broadcast media, and electronic media. The Internet has become an especially valuable research tool. It is convenient to use, and the information contained on the Internet is plentiful. Whichever media source you use to gather information, it is important to analyze the source to determine its accuracy and reliability.

Learn the Skill

There are a number of issues to consider when analyzing a media source. Most important is to check the accuracy of the source and content. The author and publisher or sponsors should be credible and clearly indicated. To analyze print media or broadcast media, ask yourself the following questions:

- Is the information current?
- Are the resources revealed?
- Is more than one resource used?
- Is the information biased?
- Does the information represent both sides of an issue?
- Is the information reported firsthand or secondhand?

Practice the Skill

To analyze print media, choose two articles—one from a newspaper and the other from a news magazine—about an issue on which public opinion is divided. Then answer these questions:

1. What points are the articles trying to make? Were the articles successful? Can the facts be verified?
2. Did either article reflect a bias toward one viewpoint or another? List any unsupported statements.
3. Was the information reported firsthand or secondhand? Do the articles seem to represent both sides fairly?
4. How many resources can you identify in the articles? List them.

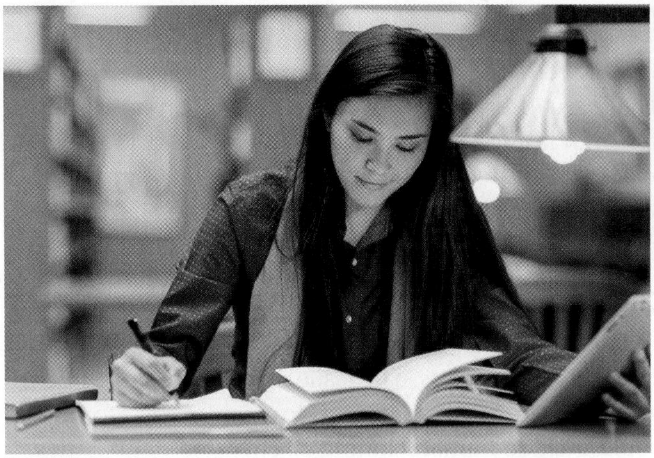

Marc Romanelli/Blend Images LLC

To analyze electronic media, visit a Web site as instructed by your teacher. Read the information on that Web site, and then answer these questions.

1. Who is the author or sponsor of the Web site?
2. What links does the Web site contain? How are they appropriate to the topic?
3. What resources were used for the information on the Web site?

Apply the Skill

Analyze Sources of Information Think of an issue on which public opinion is divided. Use a variety of media resources to read about this issue. Which news source more fairly represents the issue? Which news source has the most reliable information? Can you identify any biases?

Debate Skills

New research leads to new scientific information. There are often opposing points of view on how this research is conducted, how it is interpreted, and how it is communicated. Some of the features in your book offer a chance to debate a current controversial topic. Here is an overview on how to conduct a debate.

Choose a Position and Research

First, choose a scientific issue that has at least two opposing viewpoints. The issue can come from current events, your textbook, or your teacher. These topics could include human cloning or environmental issues. Topics are stated as affirmative declarations, such as "Cloning human beings is beneficial to society."

One speaker will argue the viewpoint that agrees with the statement, called the positive position, and another speaker will argue the viewpoint that disagrees with the statement, called the negative position. Either individually or with a group, choose the position for which you will argue. The viewpoint that you choose does not have to reflect your personal belief. The purpose of debate is to create a strong argument supported by scientific evidence.

After choosing your position, conduct research to support your viewpoint. Use resources in your media center or library to find articles, or use your textbook to gather evidence to support your argument.

Hold the Debate

You will have a specific amount of time, determined by your teacher, in which to present your argument. Organize your speech to fit within the time limit: explain the viewpoint that you will be arguing, present an analysis of your evidence, and conclude by summing up your most important points. Try to vary the elements of your argument. Your speech should not be a list of facts, a reading of a newspaper article, or a statement of your personal opinion, but an analysis of your evidence in an organized manner. It is also important to remember that you must never make personal attacks against your opponent. Argue the issue. You will be evaluated on your overall presentation, organization and development of ideas, and strength of support for your argument.

Additional Roles There are other roles that you or your classmates can play in a debate. You can act as the timekeeper. The timekeeper times the length of the debaters' speeches and gives quiet signals to the speaker when time is almost up (usually a hand signal).

You can also act as a judge. There are important elements to look for when judging a speech: an introduction that tells the audience what position the speaker will be arguing, strong evidence that supports the speaker's position, and organization. The speaker also must speak clearly and loudly enough for everyone to hear. It is helpful to take notes during the debate to summarize the main points of each side's argument.

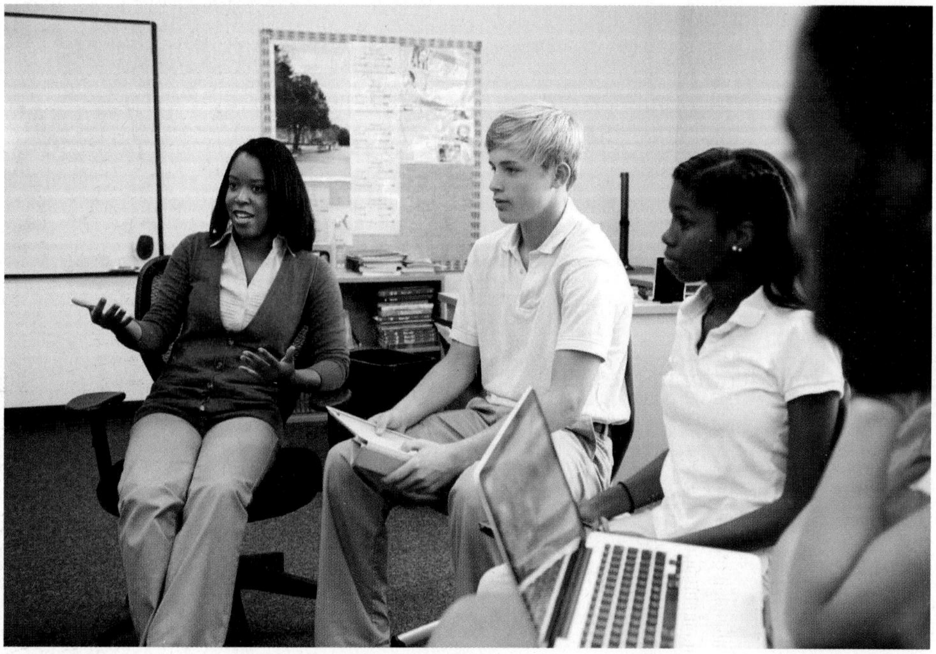

Scientific Methods

Scientists use an orderly approach called the scientific method to investigate problems. This includes organizing and recording data so others can understand them. Scientists use many variations in this method when they perform an investigation. Although there is variation, there are six common steps to the scientific methods, as shown in **Figure 1**.

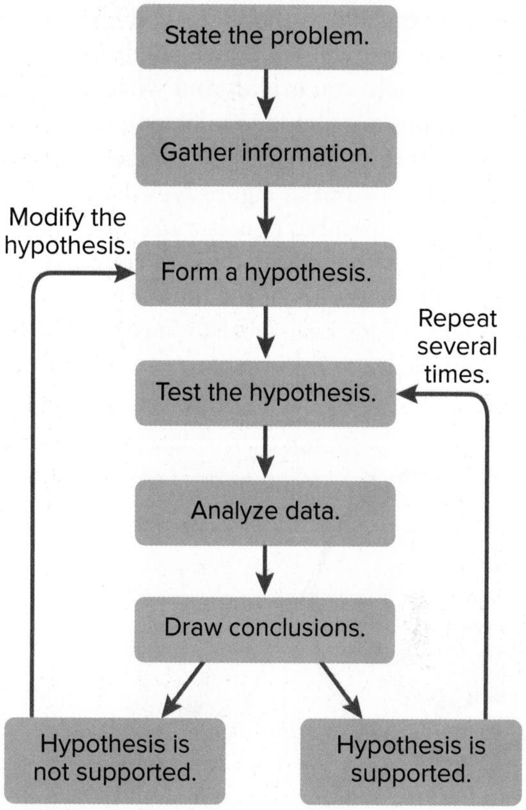

Figure 1 This flowchart can help you visualize scientific methods.

Identify a Question

The first step in a scientific investigation or experiment is to identify a question to be answered or a problem to be solved. For example, you might ask which gasoline is the most efficient.

Gather and Organize Information

After you have identified your question, begin gathering and organizing information. There are many ways to gather information, such as researching in a library, interviewing those knowledgeable about the subject, and testing and working in the laboratory and field. Fieldwork is investigations and observations done outside of a laboratory.

Researching Information Before moving in a new direction, it is important to gather the information that already is known about the subject. Start by asking yourself questions to determine exactly what you need to know. Then you will look for the information in various reference sources. Some sources may include textbooks, encyclopedias, government documents, professional journals, science magazines, and the Internet. Always list the sources of your information.

Evaluate Sources of Information Not all sources of information are reliable. You should evaluate all of your sources of information, and use only those you know to be dependable.

For example, if you are researching ways to make homes more energy efficient, a site written by the U.S. Department of Energy would be more reliable than a site written by a company that is trying to sell a new type of weatherproofing material. Also, remember that research always is changing. Consult the most current resources available to you. For example, a 1985 resource about saving energy would not reflect the most recent findings.

Sometimes scientists use data that they did not collect themselves, or conclusions drawn by other researchers. This data must be evaluated carefully. Ask questions about how the data were obtained, if the investigation was carried out properly, and if it has been duplicated exactly with the same results. Only when you have confidence in the data can you believe it is true and feel comfortable using it.

Interpret Scientific Illustrations As you research a topic in science, you will see drawings, diagrams, and photographs to help you understand what you read. Some illustrations are included to help you understand an idea that you can't see easily by yourself, like the tiny particles in an atom in **Figure 2**. A drawing helps many people to remember details more easily and provides examples that clarify difficult concepts or give additional information about the topic you are studying. Most illustrations have labels or a caption to identify or to provide more information.

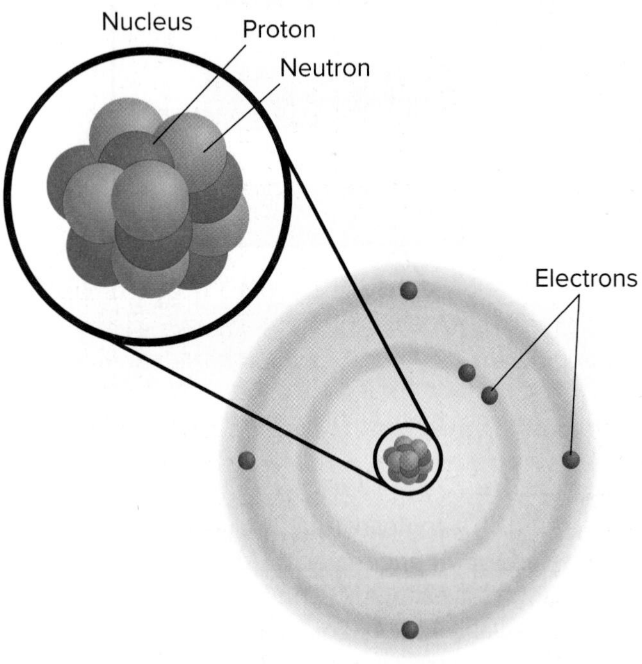

Figure 2 This drawing shows an atom of carbon with its six protons, six neutrons, and six electrons.

Concept Maps One way to organize data is to draw a diagram that shows relationships among ideas (or concepts). A concept map can help make the meanings of ideas and terms more clear, and it can help you both understand and remember what you are studying. Concept maps are useful for breaking large concepts down into smaller parts, which makes learning easier.

Network Tree A type of concept map that shows how related ideas branch out from a central concept is a network tree, shown in **Figure 3**. In a network tree, the words are written in the ovals, while the description of the type of relationship is written across the connecting lines.

When constructing a network tree, write down the topic and all major topics on separate pieces of paper or note cards. Then arrange them in order from general to specific. Branch the related concepts from the major concept and describe the relationship on the connecting line. Continue to more specific concepts until finished.

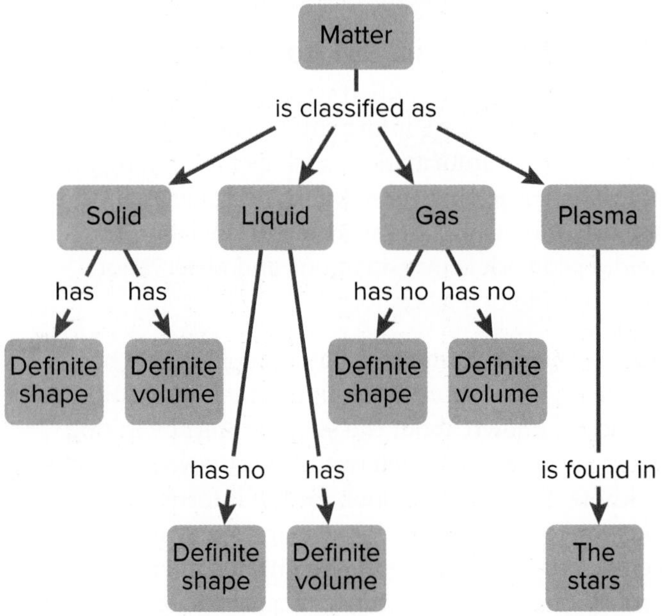

Figure 3 A network tree shows how concepts or objects are related.

Events Chain Another type of concept map is an events chain. Sometimes called a flowchart, it models the order or sequence of items. An events chain can be used to describe a sequence of events, the steps in a procedure, or the stages of a process.

When making an events chain, first find the one event that starts the chain. This event is called the initiating event. Then, find the next event and continue until the outcome is reached, as shown in **Figure 4**.

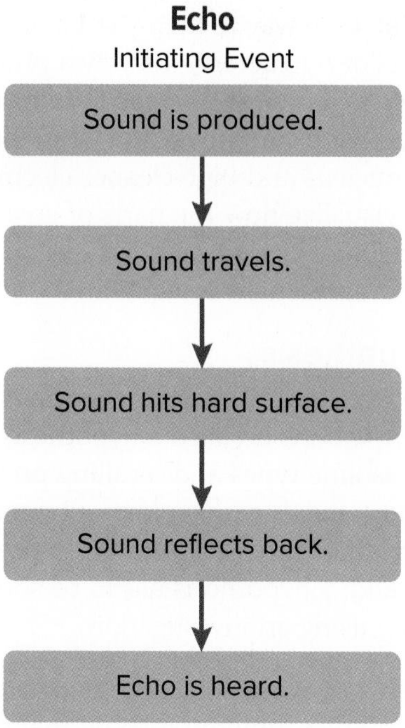

Figure 4 Events-chain concept maps show the order of steps in a process or event. This concept map shows how a sound makes an echo.

Cycle Map A cycle map is a specific type of events chain. It is used when the series of events do not produce a final outcome, but instead relate back to the beginning event, such as in **Figure 5**. Therefore, the cycle repeats itself.

To make a cycle map, first decide what event is the beginning event. This is also called the initiating event. Then list the next events in the order that they occur, with the last event relating back to the initiating event. Words can be written between the events that describe what happens from one event to the next. The number of events in a cycle map can vary, but usually contain three or more events.

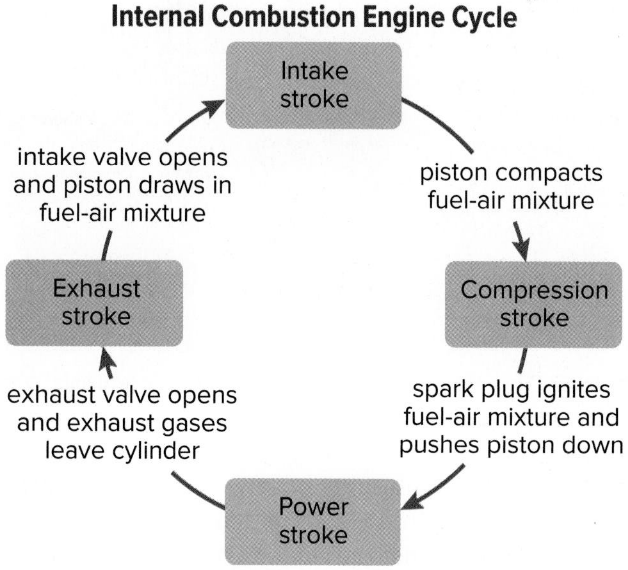

Figure 5 A cycle map shows events that occur in a cycle.

Spider Map A spider map is a type of concept map that you can use for brainstorming. When you have a central idea, you might find that you have a jumble of ideas that relate to it but are not necessarily clearly related to each other. The spider map on sound in **Figure 6** shows that if you write these ideas outside the main concept, then you can begin to separate and group unrelated terms so they become more useful.

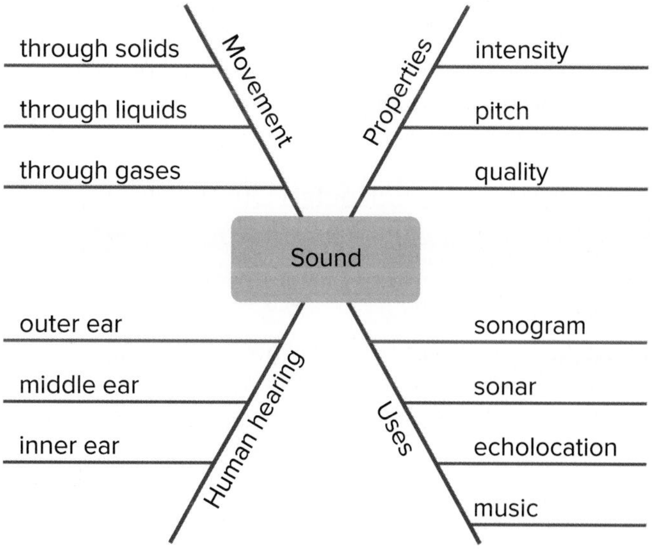

Figure 6 A spider map allows you to list ideas that relate to a central topic but not necessarily to one another.

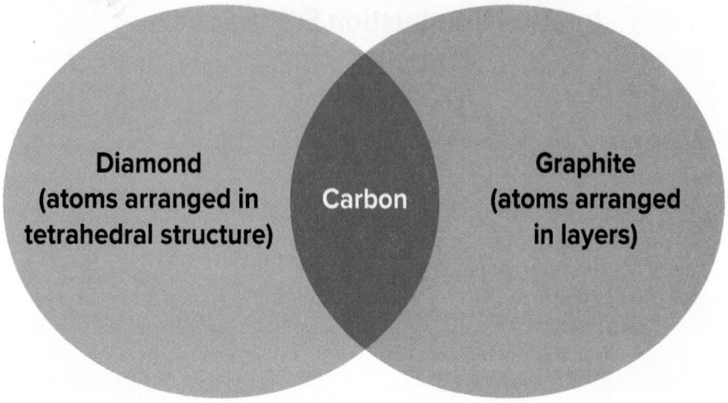

Figure 7 This Venn diagram compares and contrasts two substances made from carbon.

Venn Diagram To illustrate how two subjects compare and contrast you can use a Venn diagram. You can see the characteristics that the subjects have in common and those that they do not, as shown in **Figure 7**.

To create a Venn diagram, draw two overlapping ovals that are big enough to write in. List the characteristics unique to one subject in one oval, and the characteristics of the other subject in the other oval. The characteristics in common are listed in the overlapping section.

Make and Use Tables One way to organize information so it is easier to understand is to use a table. Tables can contain numbers, words, or both. To make a table, list the items to be compared in the first column and the characteristics to be compared in the first row. The title should clearly indicate the content of the table, and the column or row heads should be clear. Notice that in **Table 1** the units are included.

Table 1 Recyclables Collected During Week

Day of Week	Paper (kg)	Aluminum (kg)	Glass (kg)
Monday	5.0	4.0	12.0
Wednesday	4.0	1.0	10.0
Friday	2.5	2.0	10.0

Make a Model One way to help you better understand the parts of a structure, the way a process works, or to show things too large or small for viewing is to make a model. For example, an atomic model made of a plastic-ball nucleus and pipe-cleaner electron shells can help you visualize how the parts of an atom relate to each other. Other types of models can by devised on a computer or represented by equations.

Form a Hypothesis

A possible explanation based on previous knowledge and observations is called a hypothesis. After researching gasoline types and recalling previous experiences in your family's car you form a hypothesis—our car runs more efficiently because we use premium gasoline. To be valid, a hypothesis has to be something you can test by using an investigation.

Predict When you apply a hypothesis to a specific situation you predict something about that situation. A prediction makes a statement in advance, based on prior observation, experience, or scientific reasoning. Scientists test predictions by performing investigations. Based on previous observations and experiences, you might form a prediction that cars are more efficient with premium gasoline. The prediction can be tested in an investigation.

Design an Experiment A scientist needs to make many decisions before beginning an investigation. Some of these include: how to define the variables, how to carry out the investigation, what steps to follow, and how to record the data. It also is important to address any safety concerns.

Test the Hypothesis

Now that you have formed your hypothesis, you need to test it. Using an investigation, you will make observations and collect data, or information. This data might either support or not support your hypothesis. Scientists collect and organize data as numbers and descriptions.

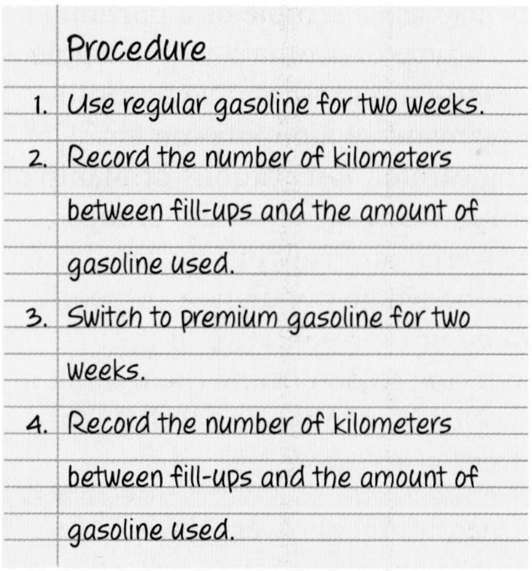
Procedure
1. Use regular gasoline for two weeks.
2. Record the number of kilometers between fill-ups and the amount of gasoline used.
3. Switch to premium gasoline for two weeks.
4. Record the number of kilometers between fill-ups and the amount of gasoline used.

Figure 8 A procedure tells you what to do step by step.

Follow a Procedure In order to know what materials to use, as well as how and in what order to use them, you must follow a procedure. **Figure 8** shows a procedure you might follow to test your hypothesis.

Identify and Manipulate Variables and Controls In any experiment, it is important to keep everything the same except for the item you are testing. The one factor you change is called the independent variable. The change that results is the dependent variable. Make sure you have only one independent variable, to assure yourself of the cause of the changes you observe in the dependent variable. For example, in your gasoline experiment the type of fuel is the independent variable. The dependent variable is the fuel economy.

Many experiments also have a control—an individual instance or experimental subject for which the independent variable is not changed. You can then compare the test results to the control results. To design a control you can have two cars of the same type. The control car uses regular gasoline for four weeks. After you are done with the test, you can compare the experimental results to the control results. All other factors in an experiment must remain constant. **Table 2** summarizes the types of variables that are used in an experiment.

Table 2 Types of Variables

Variable	Description
Dependent	Changes according to the changes of the independent variable
Independent	The variable that is changed to test the effect on the dependent variable.
Constant	A factor that does not change when other variables change.
Control	The standard by which the test results can be compared.

Collect Data

Whether you are carrying out an investigation or a short observational experiment, you will collect data, as shown in **Figure 9**. Scientists collect data as numbers and descriptions and organize it in specific ways.

Observe Scientists observe items and events, then record what they see. When they use only words to describe an observation, it is called qualitative data. Scientists' observations also can describe how much there is of something. These observations use numbers, as well as words, in the description and are called quantitative data. For example, if a sample of the element gold is described as being "shiny and very dense" the data are qualitative. Quantitative data on this sample of gold might include "a mass of 30 g and a density of 19.3 g/cm^3."

Figure 9 Collecting data is one way to gather information directly.

Digital Vision/Alamy Stock Photo

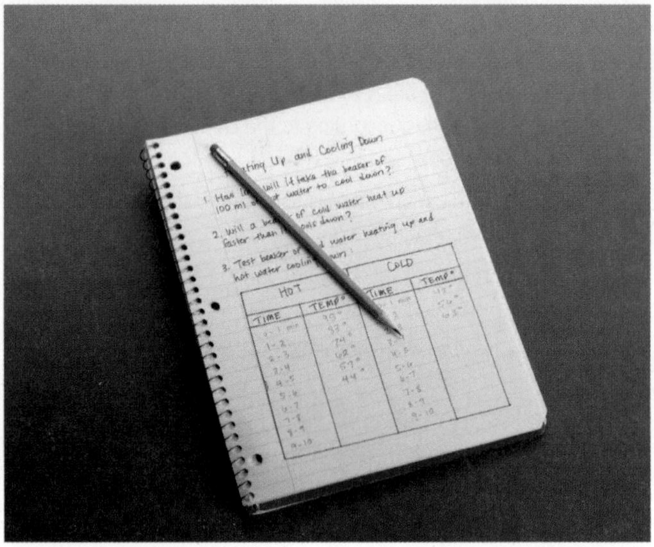

Figure 10 Record data neatly and clearly so it is easy to understand.

When you make observations you should examine the entire object or situation first, and then look carefully for details. It is important to record observations accurately and completely. Always record your notes immediately as you make them, so you do not miss details or make a mistake when recording results from memory. Never put unidentified observations on scraps of paper. Instead they should be recorded in a notebook, like the one in **Figure 10**. Write your data neatly so you can easily read it later.

At each point in the experiment, record your observations and label them. That way, you will not have to determine what the figures mean when you look at your notes later. Set up any tables that you will need to use ahead of time, so you can record any observations right away. Remember to avoid bias when collecting data by not including personal thoughts when you record observations. Record only what you observe.

Estimate Scientific work also involves estimating. To estimate is to make a judgment about the size or the number of something without measuring or counting. This is important when the number or size of an object or population is too large or too difficult to accurately count or measure.

Sample Scientists may use a sample or a portion of the total number as a type of estimation. To sample is to take a small, random representative portion of the objects or organisms of a population for research. By making careful observations or manipulating variables within that portion of the group, information is discovered and conclusions are drawn that might apply to the whole population. A poorly chosen sample can be unrepresentative of the whole. If you were trying to determine the rainfall in an area, it would not be best to take a rainfall sample from under a tree.

Measure You use measurements every day. Scientists also take measurements when collecting data. When taking measurements, it is important to know how to use measuring tools properly. Accuracy also is important.

Length To measure length, the distance between two points, scientists use meters. Smaller measurements might be measured in centimeters or millimeters.

Length is measured using a metric ruler or meter stick. When using a metric ruler, line up the 0-cm mark with the end of the object being measured and read the number of the unit where the object ends. Look at the metric ruler shown in **Figure 11**. The centimeter lines are the long, numbered lines, and the shorter lines are millimeter lines. In this instance, the length would be 4.50 cm.

Figure 11 This metric ruler has centimeter and millimeter divisions.

(t)Richard Hutchings/Digital Light Source/McGraw-Hill Education, (b)Veniamin Kraskov/Shutterstock

Mass The SI unit for mass is the kilogram (kg). Scientists can measure mass using units formed by adding metric prefixes to the unit gram (g), such as milligram (mg). To measure mass, you might use a triple-beam balance similar to the one shown in **Figure 12**. The balance has a pan on one side and a set of beams on the other side. Each beam has a rider that slides on the beam.

When using a triple-beam balance, place an object on the pan. Slide the largest rider along its beam until the pointer drops below zero. Then move it back one notch. Repeat the process for each rider proceeding from the larger to smaller until the pointer swings an equal distance above and below the zero point. Sum the masses on each beam to find the mass of the object. Move all riders back to zero when finished.

Instead of putting materials directly on the balance, scientists often take a tare of a container. A tare is the mass of a container into which objects or substances are placed for measuring their masses. To mass objects or substances, find the mass of a clean container. Remove the container from the pan, and place the object or substances in the container. Find the mass of the container with the materials in it. Subtract the mass of the empty container from the mass of the filled container to find the mass of the materials you are using.

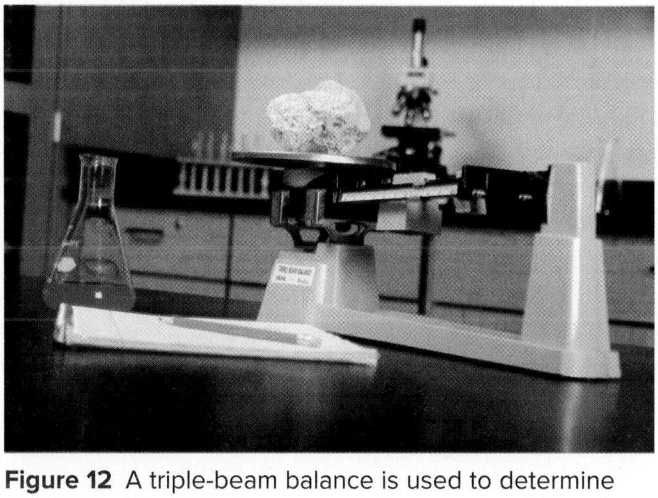

Figure 12 A triple-beam balance is used to determine the mass of an object.

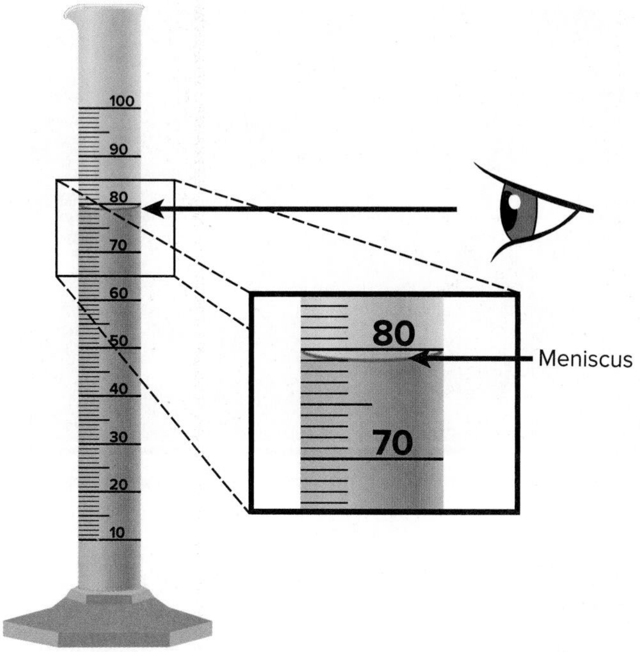

Figure 13 Graduated cylinders measure liquid volume.

Liquid Volume To measure liquids, the unit used is the liter. When a smaller unit is needed, scientists might use a milliliter. Because a milliliter takes up the volume of a cube measuring 1 cm on each side it also can be called a cubic centimeter ($cm^3 = cm \times cm \times cm$).

You can use graduated cylinders to measure liquid volume. A graduated cylinder, shown in **Figure 13**, is marked from bottom to top in milliliters. In lab, you might use a 10-mL graduated cylinder or a 100-mL graduated cylinder. When measuring liquids, notice that the liquid has a curved surface. Look at the surface at eye level, and measure the bottom of the curve. This is called the meniscus. The graduated cylinder in **Figure 13** contains 79.0 mL, or 79.0 cm^3, of a liquid.

Temperature Scientists often measure temperature using the Celsius scale. Pure water has a freezing point of 0°C and boiling point of 100°C. The unit of measurement is degrees Celsius. Two other scales often used are the Fahrenheit and Kelvin scales.

Figure 14 A thermometer measures the temperature of an object.

Scientists use a thermometer to measure temperature. Most thermometers in a laboratory are glass tubes with a bulb at the bottom end containing a liquid such as colored alcohol. The liquid rises or falls with a change in temperature. To read a glass thermometer like the thermometer in **Figure 14**, rotate it slowly until a red line appears. Read the temperature where the red line ends.

Form Operational Definitions An operational definition defines an object by how it functions, works, or behaves. For example, when you are playing hide and seek and a tree is home base, you have created an operational definition for a home base.

Objects can have more than one operational definition. For example, a ruler can be defined as a tool that measures the length of an object (how it is used). It can also be a tool with a series of marks used as a standard when measuring (how it works).

Analyze the Data

To determine the meaning of your observations and investigation results, you will need to look for patterns in the data. Then you must think critically to determine what the data mean. Scientists use several approaches when they analyze the data they have collected and recorded. Each approach is useful for identifying specific patterns.

Interpret Data The word interpret means "to explain the meaning of something." When analyzing data from an experiment, try to find out what the data show. Identify the control group and the test group to see whether or not changes in the independent variable have had an effect. Look for differences in the dependent variable between the control and test groups.

Classify Sorting objects or events into groups based on common features is called classifying. When classifying, first observe the objects or events to be classified. Then select one feature that is shared by some members in the group, but not by all. Place those members that share that feature in a subgroup.

You can classify members into smaller and smaller subgroups based on characteristics. Remember that when you classify, you are grouping objects or events for a purpose. Keep your purpose in mind as you select the features to form groups and subgroups.

Compare and Contrast Observations can be analyzed by noting the similarities and differences between two more objects or events that you observe. When you look at objects or events to see how they are similar, you are comparing them. Contrasting is looking for differences in objects or events.

Recognize Cause and Effect A cause is a reason for an action or condition. The effect is that action or condition. When two events happen together, it is not necessarily true that one event caused the other. Scientists must design a controlled experiment to recognize the exact cause and effect.

Draw Conclusions

When scientists have analyzed the data they collected, they proceed to draw conclusions about the data. These conclusions are sometimes stated in words similar to the hypothesis that you formed earlier. They may confirm a hypothesis, or lead you to a new hypothesis.

Infer Scientists often make inferences based on their observations. An inference is an attempt to explain observations or to indicate a cause. An inference is not a fact, but a logical conclusion that needs further investigation. For example, you may infer that a fire has caused smoke. Until you investigate, however, you do not know for sure.

Apply When you draw a conclusion, you must apply those conclusions to determine whether the data supports the hypothesis. If your data do not support your hypothesis, it does not mean that the hypothesis is wrong. It means only that the result of the investigation did not support the hypothesis. Maybe the experiment needs to be redesigned, or some of the initial observations on which the hypothesis was based were incomplete or biased. Perhaps more observation or research is needed to refine your hypothesis. A successful investigation does not always come out the way you originally predicted.

Avoid Bias Sometimes a scientific investigation involves making judgments. When you make a judgment, you form an opinion. It is important to be honest and not to allow any expectations of results to bias your judgments. This is important throughout the entire investigation, from researching to collecting data to drawing conclusions.

Communicate

The communication of ideas is an important part of the work of scientists. A discovery that is not reported will not advance the scientific community's understanding or knowledge. Communication among scientists also is important as a way of improving their investigations.

Scientists communicate in many ways, from writing articles in journals and magazines that explain their investigations and experiments, to announcing important discoveries on television and radio. Scientists also share ideas with colleagues on the Internet or present them as lectures, like the student is doing in **Figure 15**.

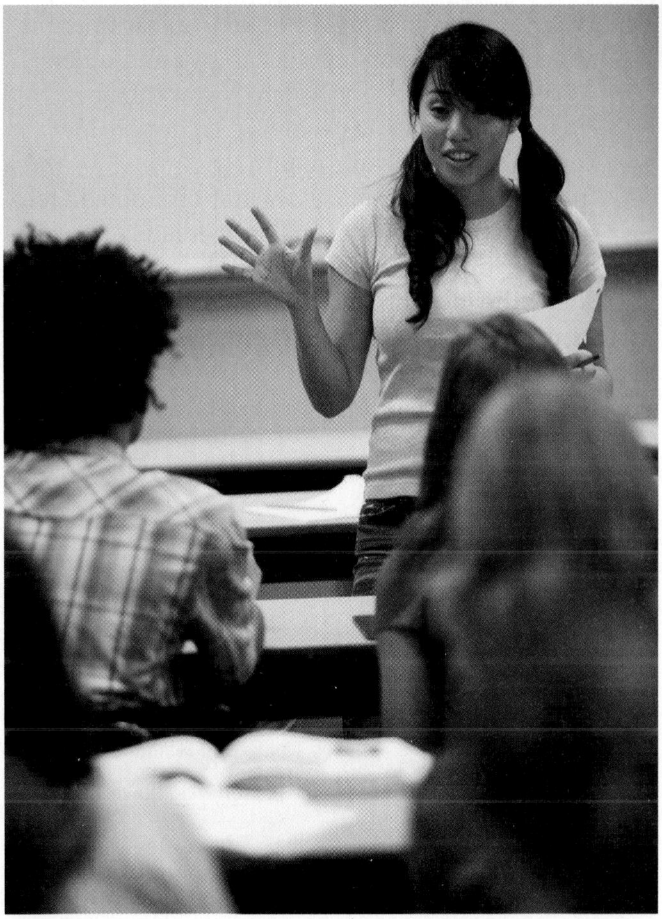

Figure 15 A student communicates to her peers about her investigation.

Rob Melnychuk/JupterImages/Brand X/Alamy Stock Photos

Safety Guidelines in the Laboratory

The laboratory is a safe place to work if you are aware of important safety rules and if you are careful. You must be responsible for your own safety and for the safety of others. The safety rules given here and the first aid instruction in **Table 3** will protect you and others from harm in the lab. While carrying out procedures in any lab, notice the safety symbols and warning statements. The safety symbols are explained in the chart on the next page.

1. Always obtain your teacher's permission to begin a lab.
2. Study the procedure. If you have questions, ask your teacher. Be sure you understand all safety symbols shown.
3. Use the safety equipment provided for you. Goggles and a safety apron should be worn when any lab calls for using chemicals.
4. When you are heating a test tube, always slant it so the mouth points away from you and others.
5. Never eat or drink in the lab. Never inhale chemicals. Do not taste any substance or draw any material into your mouth.
6. If you spill any chemical, wash it off immediately with water. Report the spill immediately to your teacher.
7. Know the location and proper use of the fire extinguisher, safety shower, fire blanket, first aid kit, and fire alarm.
8. Keep all materials away from open flames. Tie back long hair.
9. If a fire should break out in the classroom, or if your clothing should catch fire, smother it with the fire blanket or a coat, or get under a safety shower. **NEVER RUN.**
10. Report any accident or injury, no matter how small, to your teacher.

Follow these procedures as you clean up your work area.

1. Turn off the water and gas. Disconnect electrical devices.
2. Return materials to their places.
3. Dispose of chemicals and other materials as directed by your teacher. Place broken glass and solid substances in the proper containers. Never discard materials in the sink.
4. Clean your work area.
5. Wash your hands thoroughly after working in the laboratory.

Table 3 First Aid in the Laboratory

Injury	Safe Response
Burns	Apply cold water. Notify your teacher immediately.
Cuts and bruises	Stop any bleeding by applying direct pressure. Cover cuts with a clean dressing. Apply cold compresses to bruises. Notify your teacher immediately.
Fainting	Leave the person lying down. Loosen any tight clothing and keep crowds away. Notify your teacher immediately.
Foreign matter in eye	Flush with plenty of water. Use eyewash bottle or fountain. Notify your teacher immediately.
Poisoning	Note the suspected poisoning agent and call your teacher immediately.
Any spills on skin	Flush with large amounts of water or use safety shower. Notify your teacher immediately.

Safety Symbols

Safety symbols in the following table are used in the lab activities to indicate possible hazards. Learn the meaning of each symbol. **It is recommended that you wear safety goggles and apron at all times in the lab. This might be required in your school district.**

Safety Symbols	Hazard	Examples	Precaution	Remedy
Disposal	Special disposal procedures need to be followed.	certain chemicals, living organisms	Do not dispose of these materials in the sink or trash can.	Dispose of wastes as directed by your teacher.
Biological	Organisms or other biological materials that might be harmful to humans	bacteria, fungi, blood, unpreserved tissues, plant materials	Avoid skin contact with these materials. Wear mask or gloves.	Notify your teacher if you suspect contact with material. Wash hands thoroughly.
Extreme Temperature	Objects that can burn skin by being too cold or too hot	boiling liquids, hot plates, dry ice, liquid nitrogen	Use proper protection when handling.	Go to your teacher for first aid.
Sharp Object	Use of tools or glassware that can easily puncture or slice skin	razor blades, pins, scalpels, pointed tools, dissecting probes, broken glass	Practice common-sense behavior and follow guidelines for use of the tool.	Go to your teacher for first aid.
Fume	Possible danger to respiratory tract from fumes	ammonia, acetone, nail polish remover, heated sulfur, moth balls	Be sure there is good ventilation. Never smell fumes directly. Wear a mask.	Leave foul area and notify your teacher immediately.
Electrical	Possible danger from electrical shock or burn	improper grounding, liquid spills, short circuits, exposed wires	Double-check setup with teacher. Check condition of wires and apparatus. Use GFI-protected outlets.	Do not attempt to fix electrical problems. Notify your teacher immediately.
Irritant	Substances that can irritate the skin or mucous membranes of the respiratory tract	pollen, moth balls, steel wool, fiberglass, potassium permanganate	Wear dust mask and gloves. Practice extra care when handling these materials.	Go to your teacher for first aid.
Chemical	Chemicals that can react with and destroy tissue and other materials	bleaches such as hydrogen peroxide; acids such as sulfuric acid, hydrochloric acid; bases such as ammonia, sodium hydroxide	Wear goggles, gloves, and an apron.	Immediately flush the affected area with water and notify your teacher.
Toxic	Substance may be poisonous if touched, inhaled, or swallowed.	mercury, many metal compounds, iodine, poinsettia plant parts	Follow your teacher's instructions.	Always wash hands thoroughly after use. Go to your teacher for first aid.
Flammable	Flammable chemicals may be ignited by open flame, spark, or exposed heat.	alcohol, kerosene, potassium permanganate	Avoid open flames and heat when using flammable chemicals.	Notify your teacher immediately. Use fire safety equipment if applicable.
Open Flame	Open flame in use, may cause fire.	hair, clothing, paper, synthetic materials	Tie back hair and loose clothing. Follow teacher's instruction on lighting and extinguishing flames.	Notify your teacher immediately. Use fire safety equipment if applicable.

Eye Safety
Proper eye protection should be worn at all times by anyone performing or observing science activities.

Clothing Protection
This symbol appears when substances could stain or burn clothing.

Animal Safety
This symbol appears when safety of animals and students must be ensured.

Radioactivity
This symbol appears when radioactive materials are used.

Handwashing
After the lab, wash hands with soap and water before removing goggles.

Math Review

Use Fractions

A fraction compares a part to a whole. In the fraction $\frac{2}{3}$ the 2 represents the part and is the numerator. The 3 represents the whole and is the denominator.

Reduce Fractions To reduce a fraction, you must find the largest factor that is common to both the numerator and the denominator, the greatest common factor (GCF). Divide both numbers by the GCF.

Example Twelve of the 20 chemicals in the science lab are in powder form. What fraction of the chemicals are in powder form?

Step 1 Write the fraction.
$$\frac{\text{part}}{\text{whole}} = \frac{12}{20}$$

Step 2 To find the GCF of the numerator and denominator, list all of the factors of each number.

Factors of 12: 1, 2, 3, 4, 6, 12
(the numbers that divide evenly into 12)

Factors of 20: 1, 2, 4, 5, 10, 20
(the numbers that divide evenly into 20)

Step 3 List the common factors.
1, 2, 4

Step 4 Choose the greatest factor in the list.
The GCF of 12 and 20 is 4.

Step 5 Divide the numerator and denominator by the GCF.
$$\frac{12 \div 4}{20 \div 4} = \frac{3}{5}$$

$\frac{3}{5}$ of the chemicals are in powder form.

Practice Problem At an amusement park, 66 of 90 rides have a height restriction. What fraction of the rides, in its simplest form, has a height restriction?

Add and Subtract Fractions To add or subtract fractions with the same denominator, add or subtract the numerators and write the sum or difference over the denominator. After finding the sum or difference, find the simplest form for your fraction.

Example 1 In the forest outside your house, 18 of the animals are rabbits, $\frac{3}{8}$ are squirrels, and the remainder are birds and insects. How many are mammals?

Step 1 Add the numerators.
$$\frac{1}{8} + \frac{3}{8} = \frac{(1 + 3)}{8} = \frac{4}{8}$$

Step 2 Find the GCF.
$$\frac{4}{8} \text{ (GCF, 4)}$$

Step 3 Divide the numerator and denominator by the GCF.
$$\frac{4}{4} = 1, \frac{8}{4} = 2$$

$\frac{1}{2}$ of the animals are mammals.

Example 2 If $\frac{7}{16}$ of a region is covered by freshwater, and $\frac{1}{16}$ of that is in glaciers, how much freshwater is not frozen?

Step 1 Subtract the numerators.
$$\frac{7}{16} - \frac{1}{16} = \frac{(7 - 1)}{16} = \frac{6}{16}$$

Step 2 Find the GCF.
$$\frac{6}{16} \text{ (GCF, 2)}$$

Step 3 Divide the numerator and denominator by the GCF.
$$\frac{6}{2} = 3, \frac{16}{2} = 8$$

$\frac{3}{8}$ of the freshwater is not frozen.

Practice Problem A bicycle rider is riding at a rate of 15 km/h for $\frac{4}{9}$ of his ride, 10 km/h for $\frac{2}{9}$ of his ride, and 8 km/h for the remainder of the ride. How much of his ride is he riding at a rate greater than 8 km/h?

Unlike Denominators To add or subtract fractions with unlike denominators, first find the least common denominator (LCD). This is the smallest number that is a common multiple of both denominators. Rename each fraction with the LCD, and then add or subtract. Find the simplest form if necessary.

Example 1 A chemist makes a paste that is $\frac{1}{2}$ table salt (NaCl), $\frac{1}{3}$ sugar ($C_6H_{12}O_6$), and the remainder is water (H_2O). How much of the paste is a solid?

Step 1 Find the LCD of the fractions.
$\frac{1}{2} + \frac{1}{3}$ (LCD, 6)

Step 2 Rename each numerator and each denominator with the LCD.
$1 \times 3 = 3, 2 \times 3 = 6$
$1 \times 2 = 2, 3 \times 2 = 6$

Step 3 Add the numerators.
$\frac{3}{6} + \frac{2}{6} = \frac{(3+2)}{6} = \frac{5}{6}$

$\frac{5}{6}$ of the paste is a solid.

Example 2 The average precipitation in Grand Junction, CO, is $\frac{7}{10}$ inch in November and $\frac{3}{5}$ inch in December. What is the sum of these averages?

Step 1 Find the LCD of the fractions.
$\frac{7}{10} + \frac{3}{5}$(LCD, 10)

Step 2 Rename each numerator and each denominator with the LCD.
$7 \times 1 = 7, 10 \times 1 = 10$
$3 \times 2 = 6, 5 \times 2 = 10$

Step 3 Add the numerators.
$\frac{7}{10} + \frac{6}{10} = \frac{(7+6)}{10} = \frac{13}{10}$

$\frac{13}{10}$ inches total precipitation, or $1\frac{3}{10}$ inches.

Practice Problem On an electric bill, about 1/8 of the energy is from solar energy and about 1/10 is from wind power. How much of the total bill is from solar energy and wind power combined?

Example 3 In your body, $\frac{7}{10}$ of your muscle contractions are involuntary (cardiac and smooth muscle tissue). Smooth muscle makes $\frac{3}{15}$ of your muscle contractions. How many of your muscle contractions are made by cardiac muscle?

Step 1 Find the LCD of the fractions.
$\frac{7}{10} - \frac{3}{15}$(LCD, 30)

Step 2 Rename each numerator and each denominator with the LCD.
$7 \times 3 = 21, 10 \times 3 = 30$
$3 \times 2 = 6, 15 \times 2 = 30$

Step 3 Subtract the numerators.
$\frac{21}{30} - \frac{6}{30} = \frac{(21-6)}{30} = \frac{15}{30}$

Step 4 Find the GCF.
$\frac{15}{30}$ (GCF, 15)

$\frac{1}{2}$ of all muscle contractions are cardiac muscle.

Example 4 Tony wants to make cookies that call for $\frac{3}{4}$ of a cup of flour, but he only has $\frac{1}{3}$ of a cup. How much more flour does he need?

Step 1 Find the LCD of the fractions.
$\frac{3}{4} - \frac{1}{3}$(LCD, 12)

Step 2 Rename each numerator and each denominator with the LCD.
$\frac{3 \times 3}{4 \times 3} = \frac{9}{12}$
$\frac{1 \times 4}{3 \times 4} = \frac{4}{12}$

Step 3 Subtract the numerators.
$\frac{9}{12} - \frac{4}{12} = \frac{(9-4)}{12} = \frac{5}{12}$

$\frac{5}{12}$ of a cup of flour

Practice Problem Using the information provided to you in Example 3 above, determine how many muscle contractions are voluntary (skeletal muscle).

Multiply Fractions To multiply with fractions, multiply the numerators and multiply the denominators. Find the simplest form if necessary.

Example Multiply $\frac{3}{5}$ by $\frac{1}{3}$.

Step 1 Multiply the numerators and denominators.

$$\frac{3}{5} \times \frac{1}{3} = \frac{(3 \times 1)}{(5 \times 3)} = \frac{3}{15}$$

Step 2 Find the GCF.

$\frac{3}{15}$ (GCF, 3)

Step 3 Divide the numerator and denominator by the GCF.

$$\frac{3 \div 3}{15 \div 3} = \frac{1}{5}$$

$\frac{3}{5}$ multiplied by $\frac{1}{3}$ is $\frac{1}{5}$.

Practice Problem Multiply $\frac{3}{14}$ by $\frac{5}{16}$.

Find a Reciprocal Two numbers whose product is 1 are called multiplicative inverses, or reciprocals.

Example Find the reciprocal of $\frac{3}{8}$.

Step 1 Inverse the fraction by putting the denominator on top and the numerator on the bottom.

$$\frac{8}{3}$$

The reciprocal of $\frac{3}{8}$ is $\frac{8}{3}$.

Practice Problem Find the reciprocal of $\frac{4}{9}$.

Divide Fractions To divide one fraction by another fraction, multiply the dividend by the reciprocal of the divisor. Find the simplest form if necessary.

Example 1 Divide $\frac{1}{9}$ by $\frac{1}{3}$.

Step 1 Find the reciprocal of the divisor.

The reciprocal of $\frac{1}{3}$ is $\frac{3}{1}$.

Step 2 Multiply the dividend by the reciprocal of the divisor.

$$\frac{\frac{1}{9}}{\frac{1}{3}} = \frac{1}{9} \times \frac{3}{1} = \frac{(1 \times 3)}{(9 \times 1)} = \frac{3}{9}$$

Step 3 Find the GCF.

$\frac{3}{9}$ (GCF, 3)

Step 4 Divide the numerator and denominator by the GCF.

$$\frac{3 \div 3}{9 \div 3} = \frac{1}{3}$$

$\frac{1}{9}$ divided by $\frac{1}{3}$ is $\frac{1}{3}$.

Example 2 Divide $\frac{3}{5}$ by $\frac{1}{4}$.

Step 1 Find the reciprocal of the divisor.

The reciprocal of $\frac{1}{4}$ is $\frac{4}{1}$.

Step 2 Multiply the dividend by the reciprocal of the divisor.

$$\frac{\frac{3}{5}}{\frac{1}{4}} = \frac{3}{5} \times \frac{4}{1} = \frac{(3 \times 4)}{(5 \times 1)} = \frac{12}{5}$$

$\frac{3}{5}$ divided by $\frac{1}{4}$ is $\frac{12}{5}$ or $2\frac{2}{5}$.

Practice Problem Divide $\frac{3}{11}$ by $\frac{7}{10}$.

Use Ratios

A ratio compares two different numbers. For example, if you have 3 dogs and 5 cats, the ratio of dogs to cats is 3 to 5 and can be written 3:5. Notice that you have eight total animals in this example—3 dogs and 5 cats.

Ratios can represent one type of probability, called odds. This is a ratio that compares the number of ways a certain outcome occurs to the number of possible outcomes. For example, if you flip a coin 100 times, what are the odds that it will come up heads? There are two possible outcomes, heads or tails, so the most likely outcome of 100 flips is 50 heads and 50 tails, or 50:50. Like fractions, ratios can be written in simplest form—50:50 becomes 1:1. To write this as a fraction, you would say $\frac{1}{2}$ the flips are heads and $\frac{1}{2}$ are tails.

Example 1 A chemical solution contains 40 g of salt and 64 g of baking soda. What is the ratio of salt to baking soda in simplest form?

Step 1 Write the ratio.

salt:baking soda = 40:64

Step 2 Express the ratio in simplest form.
The GCF of 40 and 64 is 8.
$40 \div 8 = 5$, and $64 \div 8 = 8$

The ratio of salt to baking soda in the sample is 5:8.

Example 2 Sean rolls a 6-sided die 6 times. Predict how many times the side with a 3 will show in six rolls.

Step 1 Write the ratio as a fraction.

$$\frac{\text{number of sides with a 3}}{\text{number of total sides}} = \frac{1}{6}$$

Step 2 Multiply by the number of attempts.

$$\frac{1}{6} \times 6 \text{ attempts} = \frac{6}{6} = 1$$

In six rolls, Sean will likely roll one 3.

Practice Problem Two metal rods measure 100 cm and 144 cm in length. What is the ratio of their lengths in simplest form?

Use Decimals

A fraction with a denominator that is a power of ten can be easily written as a decimal. The decimal point separates the ones place from the tenths place. For example, $\frac{27}{100}$ means 0.27.

Any fraction can be written as a decimal using division. For example, the fraction $\frac{5}{8}$ can be written as a decimal by dividing 5 by 8. Written as a decimal, it is 0.625.

Add or Subtract Decimals When adding and subtracting decimals, line up the decimal points before carrying out the operation.

Example 1 Find the sum of 47.68 and 7.80.

Step 1 Line up the decimal places when you write the numbers.

$$\begin{array}{r} 47.68 \\ \underline{+7.80} \end{array}$$

Step 2 Add the decimals.

$$\begin{array}{r} 47.68 \\ \underline{+7.80} \\ 55.48 \end{array}$$

The sum of 47.68 and 7.80 is 55.48.

Example 2 Find the difference of 42.17 and 15.85.

Step 1 Line up the decimal places when you write the numbers.

$$\begin{array}{r} 42.17 \\ \underline{-15.85} \end{array}$$

Step 2 Subtract the decimals.

$$\begin{array}{r} 42.17 \\ \underline{-15.85} \\ 26.32 \end{array}$$

The difference of 42.17 and 15.85 is 26.32.

Practice Problem Find the sum of 1.245 and 3.842.

Multiply Decimals To multiply decimals, multiply the numbers like you multiply numbers without decimals. Count the decimal places in each factor. The product will have the same number of decimal places as the sum of the decimal places in the factors.

Example Multiply 2.4 by 5.9.

Step 1 Multiply the factors like two whole numbers.
$24 \times 59 = 1416$

Step 2 Find the sum of the number of decimal places in the factors. Each factor has one decimal place, so the sum is two decimal places.

Step 3 The product will have two decimal places.
14.16

The product of 2.4 and 5.9 is 14.16.

Practice Problem Multiply 4.6 by 2.2.

Divide Decimals When dividing decimals, change the divisor to a whole number. To do this, multiply both the divisor and the dividend by the same power of ten. Then place the decimal point in the quotient directly above the decimal point in the dividend. Then divide as you do with whole numbers.

Example Divide 8.84 by 3.4.

Step 1 Multiply both factors by 10.
$3.4 \times 10 = 34, 8.84 \times 10 = 88.4$

Step 2 Divide 88.4 by 34.

$$\begin{array}{r} 2.6 \\ 34\overline{)88.4} \\ -68 \\ \hline 204 \\ -204 \\ \hline 0 \end{array}$$

8.84 divided by 3.4 is 2.6.

Practice Problem Divide 75.6 by 3.6.

Use Proportions

An equation that shows that two ratios are equivalent is a proportion. The ratios $\frac{2}{4}$ and $\frac{5}{10}$ are equivalent, so they can be written as $\frac{2}{4} = \frac{5}{10}$. This equation is a proportion. When two ratios form a proportion, the cross products are equal. To find the cross products in the proportion $\frac{2}{4} = \frac{5}{10}$ multiply the 2 and the 10, and the 4 and the 5. Therefore $2 \times 10 = 4 \times 5$, or $20 = 20$.

Because you know that both ratios are equal, you can use cross products to find a missing term in a proportion. This is known as solving the proportion.

Example The heights of a tree and a pole are proportional to the lengths of their shadows. The tree casts a shadow of 24 m when a 6-m pole casts a shadow of 4 m. What is the height of the tree?

Step 1 Write a proportion.

$$\frac{\text{height of tree}}{\text{height of pole}} = \frac{\text{length of tree's shadow}}{\text{length of pole's shadow}}$$

Step 2 Substitute the known values into the proportion. Let h represent the unknown value, the height of the tree.

$$\frac{h}{6\text{ m}} = \frac{24\text{ m}}{4\text{ m}}$$

Step 3 Find the cross products.
$h \times 4\text{ m} = 6\text{ m} \times 24\text{ m}$

Step 4 Simplify the equation.
$(4\text{ m})\, h = 144\text{ m}^2$

Step 5 Divide each side by 4 m.

$$\frac{(4\text{ m})\, h}{4\text{ m}} = \frac{144\text{ m}^2}{4\text{ m}}$$

The height of the tree is 36 m.

Practice Problem The ratios of the weights of two objects on the Moon and on Earth are proportional. A rock weighing 3 N on the Moon weighs 18 N on Earth. How much would a rock that weighs 5 N on the Moon weigh on Earth?

Use Percentages

The word percent means "out of one hundred." It is a ratio that compares a number to 100. Suppose you read that 77 percent of the Earth's surface is covered by water. That is the same as reading that the fraction of the Earth's surface covered by water is $\frac{77}{100}$. To express a fraction as a percent, first find the equivalent decimal for the fraction. Then, multiply the decimal by 100 and add the percent symbol.

Example Express $\frac{13}{20}$ as a percent.

Step 1 Find the equivalent decimal for the fraction.

$$\begin{array}{r} 0.65 \\ 20\overline{)13.00} \\ \underline{12.0} \\ 1.00 \\ \underline{1.00} \\ 0 \end{array}$$

Step 2 Rewrite the fraction $\frac{13}{20}$ as 0.65.

Step 3 Multiply 0.65 by 100 and add the % symbol.

$0.65 \times 100 = 65 = 65\%$

So, $\frac{13}{20} = 65\%$

This also can be solved as a proportion.

Example Express $\frac{13}{20}$ as a percent.

Step 1 Write a proportion.

$\frac{13}{20} = \frac{x}{20}$

Step 2 Find the cross products.

$1300 = 20x$

Step 3 Divide each side by 20.

$\frac{1300}{20} = \frac{20x}{20}$

$65\% = x$

Practice Problem In one year, 73 of 365 days were rainy in one city. What percent of the days in that city were rainy?

Solve One-Step Equations

A statement that two expressions are equal is an equation. For example, A = B is an equation that states that A is equal to B.

An equation is solved when a variable is replaced with a value that makes both sides of the equation equal. To make both sides equal, the inverse operation is used. Addition and subtraction are inverses, and multiplication and division are inverses.

Example 1 Solve the equation $x - 10 = 35$.

Step 1 Find the solution by adding 10 to each side of the equation.

$x - 10 = 35$

$x - 10 + 10 = 35 + 10$

$x = 45$

Step 2 Check the solution.

$x - 10 = 35$

$45 - 10 = 35$

$35 = 35$

Both sides of the equation are equal, so $x = 45$.

Example 2 In the formula $a = bc$, find the value of c if $a = 20$ and $b = 2$.

Step 1 Rearrange the formula so the unknown value is by itself on one side of the equation by dividing both sides by b.

$a = bc$

$\frac{a}{b} = \frac{bc}{b}$

$\frac{a}{b} = c$

Step 2 Replace the variables a and b with the values that are given.

$\frac{a}{b} = c$

$\frac{20}{2} = c$

$10 = c$

Step 3 Check the solution.

$a = bc$

$20 = 2 \times 10$

$20 = 20$

Both sides of the equation are equal, so $c = 10$ is the solution when $a = 20$ and $b = 2$.

Practice Problem In the formula $h = gd$, find the value of d if $g = 12.3$ and $h = 17.4$.

Use Statistics

The branch of mathematics that deals with collecting, analyzing, and presenting data is statistics. In statistics, there are three common ways to summarize data with a single number—the mean, the median, and the mode.

The mean of a set of data is the arithmetic average. It is found by adding the numbers in the data set and dividing by the number of items in the set.

The median is the middle number in a set of data when the data are arranged in numerical order. If there were an even number of data points, the median would be the mean of the two middle numbers.

The mode of a set of data is the number or item that appears most often.

Another number that often is used to describe a set of data is the range. The range is the difference between the largest number and the smallest number in a set of data.

A frequency table shows how many times each piece of data occurs, usually in a survey. **Table 1** below shows the results of a student survey on favorite color.

Table 1 Student Color Choice

Color	Tally	Frequency
Red	IIII	4
Blue	~~IIII~~	5
Black	II	2
Green	III	3
Purple	~~IIII~~ II	7
Yellow	~~IIII~~ I	6

Based on the frequency table data, which color is the favorite?

Example The speeds (in m/s) for a race car during five different time trials are 39, 37, 44, 36, and 44.

Find the mean:

Step 1 Find the sum of the numbers.
$39 + 37 + 44 + 36 + 44 = 200$

Step 2 Divide the sum by the number of items, which is 5.
$200 \div 5 = 40$

The mean is 40 m/s.

Find the median:

Step 1 Arrange the measures from least to greatest.
36, 37, 39, 44, 44

Step 2 Determine the middle measure.
36, 37, <u>39</u>, 44, 44

The median is 39 m/s.

Find the mode:

Step 1 Group the numbers that are the same together.
44, 44, 36, 37, 39

Step 2 Determine the number that occurs most in the set.
<u>44</u>, <u>44</u>, 36, 37, 39

The mode is 44 m/s.

Find the range:

Step 1 Arrange the measures from greatest to least.
44, 44, 39, 37, 36

Step 2 Determine the greatest and least measures in the set.
<u>44</u>, 44, 39, 37, <u>36</u>

Step 3 Find the difference between the greatest and least measures.
$44 - 36 = 8$

The range is 8 m/s.

Practice Problem Find the mean, median, mode, and range for the data set 8, 4, 12, 8, 11, 14, 16.

Use Geometry

The branch of mathematics that deals with the measurement, properties, and relationships of points, lines, angles, surfaces, and solids is called geometry.

Perimeter The perimeter (P) is the distance around a geometric figure. To find the perimeter of a rectangle, add the length and width and multiply that sum by two, or $2(l + w)$. To find perimeters of irregular figures, add the lengths of the sides.

Example 1 Find the perimeter of a rectangle that is 3 m long and 5 m wide.

Step 1 You know that the perimeter is 2 times the sum of the width and length.

$P = 2(3\text{ m} + 5\text{ m})$

Step 2 Find the sum of the width and length.

$P = 2(8\text{ m})$

Step 3 Multiply by 2.

$P = 16\text{ m}$

The perimeter is 16 m.

Example 2 Find the perimeter of a shape with sides measuring 2 cm, 5 cm, 6 cm, 3 cm.

Step 1 You know that the perimeter is the sum of all the sides.

$P = 2\text{ cm} + 5\text{ cm} + 6\text{ cm} + 3\text{ cm}$

Step 2 Find the sum of the sides.

$P = 2\text{ cm} + 5\text{ cm} + 6\text{ cm} + 3\text{ cm}$

$P = 16\text{ cm}$

The perimeter is 16 cm.

Practice Problem Find the perimeter of a rectangle with a length of 18 m and a width of 7 m.

Practice Problem Find the perimeter of a triangle measuring 1.6 cm by 2.4 cm by 2.4 cm.

Area of a Rectangle The area (A) is the number of square units needed to cover a surface. To find the area of a rectangle, multiply the length by the width, or $l \times w$. When finding area, the units also are multiplied. Area is given in square units.

Example Find the area of a rectangle with a length of 1 cm and a width of 10 cm.

Step 1 You know that the area is the length multiplied by the width.

$A = (1\text{ cm} \times 10\text{ cm})$

Step 2 Multiply the length by the width. Also multiply the units.

$A = 10\text{ cm}^2$

The area is 10 cm^2.

Practice Problem Find the area of a square whose sides measure 4 m.

Area of a Triangle To find the area of a triangle, use the formula:

$A = \frac{1}{2}(\text{base} \times \text{height})$

The base of a triangle can be any of its sides. The height is the perpendicular distance from a base to the opposite endpoint, or vertex.

Example Find the area of a triangle with a base of 18 m and a height of 7 m.

Step 1 You know that the area is $\frac{1}{2}$ the base times the height.

$A = \frac{1}{2}(18\text{ m} \times 7\text{ m})$

Step 2 Multiply $\frac{1}{2}$ by the product of 18×7. Multiply the units.

$A = \frac{1}{2}(126\text{ m}^2)$

$A = 63\text{ m}^2$

The area is 63 m^2.

Practice Problem Find the area of a triangle with a base of 27 cm and a height of 17 cm.

Math Skill Handbook

Circumference of a Circle The diameter (d) of a circle is the distance across the circle through its center, and the radius (r) is the distance from the center to any point on the circle. The radius is half of the diameter. The distance around the circle is called the circumference (C). The formula for finding the circumference is:

$C = 2\pi r$ or $C = \pi d$

The circumference divided by the diameter is always equal to 3.1415926... This nonterminating and nonrepeating number is represented by the Greek letter π (pi). An approximation often used for π is 3.14.

Example 1 Find the circumference of a circle with a radius of 3 m.

Step 1 You know the formula for the circumference is 2 times the radius times π.
$C = 2\pi$ (3 m)

Step 2 Multiply 2 times the radius.
$C = \pi$ (6 m)

Step 3 Multiply by π.
$C = 19$ m

The circumference is 19 m.

Example 2 Find the circumference of a circle with a diameter of 24.0 cm.

Step 1 You know the formula for the circumference is the diameter times π.
$C = \pi$ (24.0 cm)

Step 2 Multiply the diameter by π.
$C = 75.4$ cm

The circumference is 75.4 cm.

Practice Problem Find the circumference of a circle with a radius of 19 cm.

Area of a Circle The formula for the area of a circle is: $A = \pi r^2$

Example 1 Find the area of a circle with a radius of 4.0 cm.

Step 1 $A = \pi$ (4.0 cm)^2

Step 2 Find the square of the radius.
$A = 16\pi$ cm^2

Step 3 Multiply the square of the radius by π.
$A = 50$ cm^2

The area of the circle is 50 cm^2

Example 2 Find the area of a circle with a radius of 225 m.

Step 1 $A = \pi$ (225 m)^2

Step 2 Find the square of the radius.
$A = 50{,}625\pi$ m^2

Step 3 Multiply the square of the radius by π.
$A = 158{,}962.5$m^2

The area of the circle is 158,962 m^2

Example 3 Find the area of a circle whose diameter is 20.0 mm.

Step 1 You know the formula for the area of a circle is the square of the radius times π and that the radius is half of the diameter.

$$A = \pi \left(\frac{20.0 \text{ mm}}{2}\right)^2$$

Step 2 Find the radius.
$A = \pi$ (10.0 mm)^2

Step 3 Find the square of the radius.
$A = 100\pi$ mm^2

Step 4 Multiply the square of the radius by π.
$A = 314$ mm^2
The area is 314 mm^2.

Practice Problem Find the area of a circle with a radius of 16 m.

Volume The measure of space occupied by a solid is the volume (V). To find the volume of a rectangular solid, multiply the length by the width by the height, or $V = l \times w \times h$. It is measured in cubic units, such as cubic centimeters (cm^3).

Example Find the volume of a rectangular solid with a length of 2.0 m, a width of 4.0 m, and a height of 3.0 m.

Step 1 You know the formula for volume is the length times the width times the height.
$V = 2.0 \text{ m} \times 4.0 \text{ m} \times 3.0 \text{ m}$

Step 2 Multiply the length by the width by the height.
$V = 24 \text{ m}^3$

The volume is 24 m^3.

Practice Problem Find the volume of a rectangular solid that is 8 m long, 4 m wide, and 4 m high.

To find the volume of other solids, multiply the area of the base by the height.

Example 1 Find the volume of a prism that has two triangular bases with lengths of 8.0 m and heights of 7.0 m. The height of the entire solid is 15.0 m.

Step 1 You know that the base is a triangle, and the area of a triangle is $\frac{1}{2}$ the base times the height, and the volume is the area of the base times the height.

$V = \left[\frac{1}{2}(b \times h)\right] \times 15 \text{ m}$

Step 2 Find the area of the base.

$V = \left[\frac{1}{2}(8 \text{ m} \times 7 \text{ m})\right] \times 15 \text{ m}$

$V = \left(\frac{1}{2} \times 56 \text{ m}^2\right) \times 15 \text{ m}$

Step 3 Multiply the area of the base by the height of the solid.

$\text{V} = 28 \text{ m}^2 \times 15 \text{ m}$

$\text{V} = 420 \text{ m}^3$

The volume is 420 m^3.

Example 2 Find the volume of a cylinder that has a base with a radius of 12.0 cm and a height of 21.0 cm.

Step 1 You know that the base is a circle, and the area of a circle is the square of the radius times π, and the volume is the area of the base times the height.
$V = (\pi r^2) \times 21 \text{ cm}$
$V = \pi\,(12 \text{ cm})^2 \times 21 \text{ cm}$

Step 2 Find the area of the base.
$V = 144\pi \text{ cm}^2 \times 21 \text{ cm}$
$V = 452 \text{ cm}^2 \times 21 \text{ cm}$

Step 3 Multiply the area of the base by the height of the solid.
$V = 9{,}490 \text{ cm}^3$

The volume is $9{,}500 \text{ cm}^3$

Example 3 Find the volume of a cylinder that has a diameter of 15 mm and a height of 4.8 mm.

Step 1 You know that the base is a circle with an area equal to the square of the radius times π. The radius is one-half the diameter. The volume is the area of the base times the height.

$V = (\pi r^2) \times 4.8 \text{ mm}$

$V = \left[\pi\left(\frac{1}{2} \times 15 \text{ mm}\right)^2\right] \times 4.8 \text{ mm}$

$V = \pi\,(7.5 \text{ mm})^2 \times 4.8 \text{ mm}$

Step 2 Find the area of the base.

$V = 56.25\pi \text{ mm}^2 \times 4.8 \text{ mm}$

$V = 176.63 \text{ mm}^2 \times 4.8 \text{ mm}$

Step 3 Multiply the area of the base by the height of the solid.
$V = 847.8 \text{ mm}^3$

The volume is 847.8 mm^3.

Practice Problem Find the volume of a cylinder with a diameter of 7 cm in the base and a height of 16 cm.

Science Applications

Measure in SI

The metric system of measurement was developed in 1795. A modern form of the metric system, called the International System (SI), was adopted in 1960 and provides the standard measurements that all scientists around the world can understand.

The SI system is convenient because unit sizes vary by powers of 10. Prefixes are used to name units. Look at **Table 2** for some common SI prefixes and their meanings.

Table 2 Common SI Prefixes

Prefix	Symbol	Meaning	
kilo–	k	1,000	thousand
hecto–	h	100	hundred
deka–	da	10	ten
deci–	d	0.1	tenth
centi–	c	0.01	hundredth
milli–	m	0.001	thousandth

Example How many grams equal one kilogram?

Step 1 Find the prefix *kilo–* in **Table 2.**

Step 2 Using **Table 2,** determine the meaning of *kilo–*. According to the table, it means 1,000. When the prefix kilo– is added to a unit, it means that there are 1,000 of the units in a "kilounit."

Step 3 Apply the prefix to the units in the question. The units in the question are grams. There are 1,000 grams in a kilogram.

Practice Problem Is a milligram larger or smaller than a gram? How many of the smaller units equal one larger unit? What fraction of the larger unit does one smaller unit represent?

Dimensional Analysis

Convert SI Units In science, quantities such as length, mass, and time sometimes are measured using different units. A process called dimensional analysis can be used to change one unit of measure to another. This process involves multiplying your starting quantity and units by one or more conversion factors. A conversion factor is a ratio equal to one and can be made from any two equal quantities with different units. If 1,000 mL equal 1 L, then two ratios can be made.

$$\frac{1{,}000\text{ mL}}{1\text{ L}} = \frac{1\text{ L}}{1{,}000\text{ mL}} = 1$$

One can convert between units in the SI system by using the equivalents in **Table 2** to make conversion factors.

Example 1 How many cm are in 4 m?

Step 1 Write conversion factors for the units given. From **Table 2,** you know that 100 cm = 1 m. The conversion factors are

$$\frac{100\text{ cm}}{1\text{ m}} \text{ and } \frac{1\text{ m}}{100\text{ cm}}$$

Step 2 Decide which conversion factor to use. Select the factor that has the units you are converting from (m) in the denominator and the units you are converting to (cm) in the numerator.

$$\frac{100\text{ cm}}{1\text{ m}}$$

Step 3 Multiply the starting quantity and units by the conversion factor. Cancel the starting units with the units in the denominator. There are 400 cm in 4 m.

$$4\,\cancel{\text{m}} \times \frac{100\text{ cm}}{1\,\cancel{\text{m}}} = 400\text{ cm}$$

Practice Problem How many milligrams are in one kilogram? (Hint: You will need to use two conversion factors from **Table 2.**)

Table 3 Unit System Equivalents

Type of Measurement	Equivalent
Length	1 in = 2.54 cm 1 yd = 0.91 m 1 mi = 1.61 km
Mass and weight (measured in standard Earth gravity)	1 oz = 28.35 g 1 lb = 0.45 kg 1 ton (short) = 0.91 tonnes (metric tons) 1 lb = 4.45 N
Volume	1 in^3 = 16.39 cm^3 1 qt = 0.95 L 1 gal = 3.78 L
Area	1 in^2 = 6.45 cm^2 1 yd^2 = 0.83 m^2 1 mi^2 = 2.59 km^2 1 acre = 0.40 hectares
Temperature	$°C = \frac{(°F - 32)}{1.8}$ K = °C + 273

Convert Between Unit Systems **Table 3** gives a list of equivalents that can be used to convert between English and SI units.

Example If a meterstick has a length of 100 cm, how long is the meterstick in inches?

Step 1 Write the conversion factors for the units given. From **Table 3**, 1 in = 2.54 cm.

$$\frac{1 \text{ in}}{2.54 \text{ cm}} \text{ and } \frac{2.54 \text{ cm}}{1 \text{ in}}$$

Step 2 Determine which conversion factor to use. You are converting from cm to in. Use the conversion factor with cm on the bottom.

$$\frac{1 \text{ in}}{2.54 \text{ cm}}$$

Step 3 Multiply the starting quantity and units by the conversion factor. Cancel the starting units with the units in the denominator. Round your answer to the nearest tenth.

$$100 \text{ cm} \times \frac{1 \text{ in}}{2.54 \text{ cm}} = 39.37 \text{ in}$$

The meterstick is about 39.4 in long.

Practice Problem A book has a mass of 5 lbs. What is the mass of the book in kg?

Practice Problem Use the relationship 1 in = 2.54 cm to show how 1 in^3 = 16.39 cm^3.

Precision and Significant Figures

When you make a measurement, the value you record depends on the precision of the measuring instrument. This precision is represented by the number of significant figures recorded in the measurement. When counting the number of significant figures, all figures are counted except zeros at the end of a number with no decimal point such as 2,050, and zeros at the beginning of a decimal such as 0.03020. When adding or subtracting numbers with different precision, round the answer to the smallest number of decimal places of any number in the sum or difference. When multiplying or dividing, the answer is rounded to the smallest number of significant figures of any number being multiplied or divided.

Example The lengths 5.28 and 5.2 are measured in meters. Find the sum of these lengths and record your answer using the correct number of significant figures.

Step 1 Find the sum. [align the decimal points]

5.28 m	2 digits after the decimal
+ 5.2 m	1 digit after the decimal
10.48 m	

Step 2 Round to one digit after the decimal because the least number of digits after the decimal of the numbers being added is 1.

The sum is 10.5 m.

Practice Problem How many significant figures are in the measurement 7,071,301 m? How many significant figures are in the measurement 0.003010 g?

Practice Problem Multiply 5.28 and 5.2 using the rule for multiplying and dividing. Record the answer using the correct number of significant figures.

Scientific Notation

Often times the numbers used in science are very small or very large. Because these numbers are difficult to work with, scientists use scientific notation. To write numbers in scientific notation, move the decimal point until only one non-zero digit remains on the left. Then count the number of places you moved the decimal point and use that number as a power of ten. For example, the average distance from the Sun to Mars is 227,800,000,000 m. In scientific notation, this distance is 2.278×10^{11} m. Because you moved the decimal point to the left, the number is a positive power of ten.

The mass of an electron is about 0.000 000 000 000 000 000 000 000 000 000 911 kg. Expressed in scientific notation, this mass is 9.11×10^{-31} kg. Because the decimal point was moved to the right, the number is a negative power of ten.

Example Earth is 149,600,000 km from the Sun. Express this in scientific notation.

Step 1 Move the decimal point until one non-zero digit remains on the left. 1.496 000 00

Step 2 Count the number of decimal places you have moved. In this case, eight.

Step 3 Show that number as a power of ten, 10^8. Earth is 1.496×10^8 km from the Sun.

Practice Problem Express each of the following in scientific notation: 0.005835 g, 300,000 m/s, 15,000,000 K, 0.00020 cm.

Make and Use Graphs

Data in tables can be displayed in a graph—a visual representation of data. Common graph types include line graphs, bar graphs, and circle graphs.

Line Graph A line graph shows a relationship between two variables that change continuously. The independent variable is changed and is plotted on the *x*-axis. The dependent variable is observed and is plotted on the *y*-axis.

Example Draw a line graph of the data in **Table 4** from a cyclist in a long-distance race.

Table 4 Bicycle Race Data

Time (h)	Distance (km)
0	0
1	8
2	16
3	24
4	32
5	40

Step 1 Determine the *x*-axis and *y*-axis variables. Time varies independently of distance and is plotted on the *x*-axis. Distance is dependent on time and is plotted on the *y*-axis.

Step 2 Determine the scale of each axis. The *x*-axis data ranges from 0 to 5. The *y*-axis data ranges from 0 to 50.

Step 3 Using graph paper, draw and label the axes. Include units in the labels.

Step 4 Draw a point at the intersection of the time value on the *x*-axis and corresponding distance value on the *y*-axis. Connect the points and label the graph with a title, as shown in **Figure 1**.

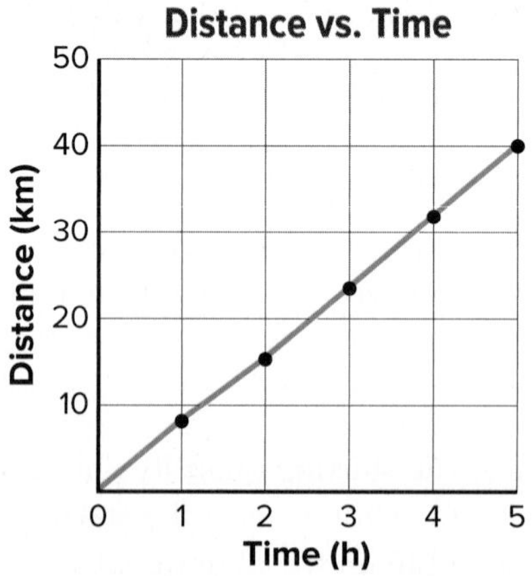

Figure 1 This line graph shows the relationship between distance and time during a bicycle ride.

Practice Problem A puppy's shoulder height is measured during the first year of her life. The following measurements were collected: (3 mo, 52 cm), (6 mo, 72 cm), (9 mo, 83 cm), (12 mo, 86 cm). Graph this data.

Find a Slope The slope of a straight line is the ratio of the vertical change, rise, to the horizontal change, run.

$$\text{slope} = \frac{\text{vertical change (rise)}}{\text{horizontal change (run)}} = \frac{\text{change in } y}{\text{change in } x}$$

Example Find the slope of the graph in Figure 1.

Step 1 You know that the slope is the change in y divided by the change in x.

$$\text{slope} = \frac{\text{change in } y}{\text{change in } x}$$

Step 2 Determine the data points you will be using. For a straight line, choose the two sets of points that are the farthest apart.

$$\text{slope} = \frac{(40 - 0)\text{ km}}{(5 - 0)\text{ h}}$$

Step 3 Find the change in y and x.

$$\text{slope} = \frac{40\text{ km}}{5\text{ h}}$$

Step 4 Divide the change in y by the change in x.

$\text{slope} = 8\text{ km/h}$

The slope of the graph is 8 km/h.

Bar Graph To compare data that does not change continuously, you might use a bar graph. A bar graph uses bars to show the relationships between variables. The x-axis variable is divided into parts. The parts can be numbers, such as years, or a category, such as a type of animal. The y-axis is a number and increases continuously along the axis.

Example A recycling center collects 4.0 kg of aluminum on Monday, 1.0 kg on Wednesday, and 2.0 kg on Friday. Create a bar graph of this data.

Step 1 Select the x-axis and y-axis variables. The measured numbers (the masses of aluminum) should be placed on the y-axis. The variable divided into parts (collection days) is placed on the x-axis.

Step 2 Create a graph grid like you would for a line graph. Include labels and units.

Step 3 For each measured number, draw a vertical bar above the x-axis value up to the y-axis value. For the first data point, draw a vertical bar above Monday up to 4.0 kg.

Practice Problem Draw a bar graph of the gases in air: 78% nitrogen, 21% oxygen, 1% other gases.

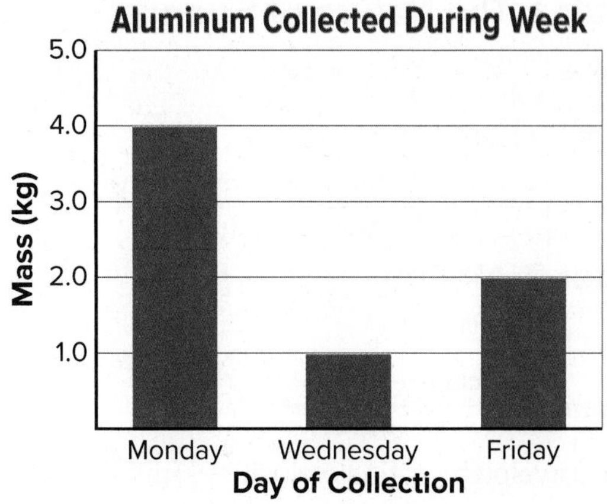

Circle Graph A circle graph is a circle divided into sections that represent the relative size of each piece of data. The entire circle represents 100%, half represents 50%, and so on.

Example Air is made up of 78% nitrogen, 21% oxygen, and 1% other gases. Display the composition of air in a circle graph.

Step 1 Multiply each percent by 360° and divide by 100 to find the angle of each section in the circle.

$$78 \times \frac{360^\circ}{100} = 280.8^\circ$$

$$21 \times \frac{360^\circ}{100} = 75.6^\circ$$

$$1 \times \frac{360^\circ}{100} = 3.6^\circ$$

Step 2 Use a compass to draw a circle and to mark the center of the circle. Draw a straight line from the center to the edge of the circle.

Step 3 Use a protractor and the angles you calculated to divide the circle into parts.

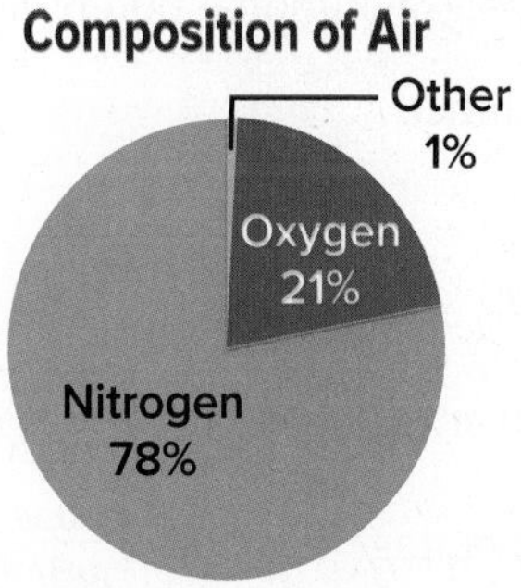

Practice Problem Draw a circle graph to represent the amount of aluminum collected during the week shown in the bar graph above.

Math Skill Handbook

Formulas

Chapter 1 The Nature of Science

Density = $\frac{\text{mass}}{\text{volume}}$

Kelvin = °Celsius + 273

% Error = $\left|\frac{(\text{Accepted value} - \text{Experimental value})}{\text{Accepted value}}\right| \times 100$

Chapter 2 Motion

Speed = $\frac{\text{distance}}{\text{time}}$

Acceleration = $\frac{\text{change in velocity}}{\text{time}}$

Change in velocity = final velocity − initial velocity

Chapter 3 Forces and Newton's Laws

Acceleration = $\frac{\text{net force}}{\text{mass}}$

Force = mass × acceleration

Gravitational force = mass × (acceleration due to gravity)

Weight = mass × gravity

Momentum (p) = mass × velocity

Force = $\frac{(mv_f - mv_i)}{\text{time}}$

Change in position = initial velocity (change in time) + $\frac{1}{2}$ acceleration (change in time)^2

Average velocity = $\frac{\text{change in position}}{\text{change in time}}$

Average acceleration = $\frac{\text{change in velocity}}{\text{change in time}}$

Chapter 4 Work and Energy

Kinetic energy = $\frac{1}{2}$ (mass) × (velocity)^2

Gravitational potential energy (GPE) = mass × gravity × height

Mechanical energy = gravitational potential energy + kinetic energy

Work = force × distance

Power = $\frac{\text{work}}{\text{time}}$

Efficiency = $(\frac{\text{work}_{\text{out}}}{\text{work}_{\text{in}}}) \times 100\%$

Ideal mechanical advantage (IMA) = $\frac{\text{length of effort arm}}{\text{length of resistance arm}} = \frac{L_e}{L_r}$

Ideal mechanical advantage (IMA) = $\frac{\text{radius of wheel}}{\text{radius of axle}} = \frac{r_w}{r_a}$

(IMA) = $\frac{\text{effort distance}}{\text{resistance distance}} = \frac{\text{length of slope}}{\text{height of slope}} = \frac{l}{h}$

Chapter 5 Thermal Energy

Change in thermal energy = mass × change in temperature × specific heat   or

$Q = m \times (T_{final} - T_{initial}) \times C_p$

Chapter 6 Electricity

Electric current = $\frac{\text{voltage difference}}{\text{resistance}}$ or $I = \frac{V}{R}$

Electric power = current × voltage difference   or $P = I \times V$

Electric energy = power × time   or $E = P \times t$

Series Circuits	**Parallel Circuits**
$I_t = I_1 = I_2 = I_3 = \ldots$	$I_t = I_1 + I_2 + I_3 + \ldots$
$V_t = V_1 + V_2 + V_3 + \ldots$	$V_t = V_1 = V_2 = V_3 = \ldots$
$R_t = R_1 + R_2 + R_3 + \ldots$	$\frac{1}{R_t} = \frac{1}{R_1} + \frac{1}{R_2} + \frac{1}{R_3} + \ldots$

Chapter 9 Introduction to Waves

Wave velocity = wavelength × frequency

or $v_w = \lambda \times f$

Chapter 12 Light

Index of refraction = $\frac{\text{speed of light in a vacuum}}{\text{speed of light in a substance}}$ or $n = \frac{c}{v}$

Chapter 14 Solids, Liquids, and Gases

Pressure = $\frac{\text{force}}{\text{area}}$ or $P = \frac{F}{A}$

Boyle's law $P_1 \times V_1 = P_2 \times V_2$

Charles's law $\frac{V_1}{T_1} = \frac{V_2}{T_2}$

Chapter 21 Solutions

Surface area of a rectangular solid = $2(h \times w) + 2(h \times l) + 2(w \times l)$

ADDITIONAL PRACTICE PROBLEMS

MODULE 1
THE NATURE OF SCIENCE

1. How many centimeters are in four meters?

2. How many deciliters are in 500 mL?

3. How many liters are in 2,540 cm^3?

4. A young child has a mass of 40 kg. What is the mass of the child in grams?

5. Iron has a density of 7.9 g/cm^3. What is the mass in kg of an iron statue that has a volume of 5.4 L?

6. A 2-L bottle of soda has a volume of 2,000 cm^3. What is the volume of the bottle in cubic meters?

7. A big summer movie has a running time of 96 minutes. What is the movie's running time in seconds?

8. The temperature in space is approximately 3 K. What is this temperature in degrees Celsius?

9. The x-axis of a certain graph is distance traveled in meters and the y-axis is time in seconds. Two points are plotted on this graph with coordinates (2, 43) and (5, 68). What is the elapsed time between the two points?

10. A circle graph has labeled segments of 57%, 21%, 13%, and 6%. What percentage does the unlabeled segment have?

11. A car can travel 14 km on 1 L of gasoline. What percent of its fuel efficiency does another car have if it travels 10 km on 1 L of gasoline?

12. What is the fuel efficiency of a car if it gets 45 percent of the fuel efficiency of another car that travel 15 km on 1 L of gasoline?

13. Data from a new Web site takes 17 min to download. How many seconds does it take?

14. A circle graph shows the comparative effects of five new technologies. The circle graph has segments labeled 45 percent, 32 percent, 12 percent, and 3 percent. What is the correct label of the fifth segment?

15. If a new Internet connection reduces a download time of 17 min by 85 percent, how much time (in minutes and seconds) will it take now?

16. Because of a new health center in a town, the cases of influenza are 25 percent fewer. How many cases should you expect during the next month if the town usually has 300 cases per month?

17. If a person can sell 10 computers per day and a new technology allows her to triple her sales, how many computers can she expect to sell the next day?

18. A technologically advanced engine runs at twice the efficiency of an older engine. If the older engine allows a car to travel at a fuel efficiency of 8 km/L, what is the fuel efficiency of the advanced engine?

MODULE 2
MOTION

1. John rides his bike 2.3 km to school. After school, he rides an additional 1.4 km to the mall in the opposite direction. What is his total distance traveled?

2. A squirrel runs 4.8 m across a lawn, stops, then runs 2.3 m back in the opposite direction. What is the squirrel's displacement from its starting point?

3. An ant travels 75 cm in 5 s. What was the ant's speed?

4. It takes a car one minute to go from rest to 30 m/s east. What is the acceleration of this car?

5. It took you 6.5 h to drive 550 km. What was your average speed?

6. A bus leaves at 9 A.M. with a group of tourists. They travel 350 km before they stop for lunch. Then they travel an additional 250 km until the end of their trip at 3 p.m. What was the average speed of the bus?

7. Halfway through a cross-country meet, a runner's speed is 4 m/s. In the last stretch, she increases her speed to 7 m/s. What is her change in speed?

8. You are in a car traveling an average speed of 60 km/h. The total trip is 240 km. How long does the trip take?

9. You are riding in a train that is traveling at a speed of 120 km/h. How long will it take to travel 950 km?

10. A car goes from rest to a velocity of 108 km/h north in 10 s. What is the car's acceleration in m/s^2?

11. A cart rolling south at a speed of 10 m/s comes to a stop in 2 s. What is the cart's acceleration?

12. A car with a mass of 1,200 kg has a velocity of 30 m/s west. What is the car's momentum?

13. If a 5,000-kg mass is moving at a velocity of 40 m/s south, what is its momentum?

14. How fast must a 50-kg mass travel to have a momentum of 1,500 kg·m/s east?

MODULE 3

FORCES AND NEWTON'S LAWS

1. If you are pushing on a box with a force of 20 N and there is a force of 7 N on the box due to sliding friction, what is the net force on the box?

2. A weight lifter is trying to lift a 1,500-N weight but can apply a force of only 1,200 N on the weight. One of his friends helps him lift it at a constant velocity. What force was applied to the weight by the weight lifter's friend?

3. During a tug-of-war, Team A pulls with a force of 5,000 N while Team B pulls with a force of 8,000 N. What is the net force applied to the rope?

4. A 80-kg mass has an acceleration of 5.5 m/s^2 north. What is the net force applied?

5. A force of 3200 N west is applied to a 160-kg mass. What is the acceleration of the mass?

6. A 2.5-kg object is dropped from a height of 1000 m. What is the force of air resistance on the object when it reaches terminal velocity?

7. How much force is needed to lift a 25-kg mass at a constant velocity?

8. A person is on an elevator that moves downward with an acceleration of 1.8 m/s^2.
If the person weighs 686 N, what is the net force on the person?

9. What is the net force on a 4,000-kg car that doubles its velocity from 15 m/s west to 30 m/s west over 10 seconds?

10. A book with a mass of 1 kg is sliding to the left on a table. If the frictional force on the book is 5 N, calculate the book's acceleration. Is it speeding up or slowing down?

MODULE 4

WORK AND ENERGY

1. When moving a couch, you exert a force of 400 N and push it 4.0 m. How much work have you done on the couch?

2. How much work is needed to lift a 50-kg weight 3.0 m?

3. By applying a force of 50 N, a pulley system can lift a box with a mass of 20 kg. What is the mechanical advantage of the pulley system?

4. How much energy do you save per hour if you replace a 60-watt lightbulb with a 55-watt lightbulb?

5. What is the efficiency of a machine if your work on the machine is 1,200 J and the machine's output work is 300 J?

6. What power is used by a machine to perform 800 J of work in 25 s?

7. A person pushes a box up a ramp that is 3 m long and 1 m high. If the box has a mass of 20 kg and the person pushes with a force of 80 N, what is the efficiency of the ramp?

8. A lever has a mechanical advantage of 5. How large would a force need to be to lift a rock with a mass of 100 kg?

9. What is the kinetic energy from the motion of a 5.0-kg object moving at 7.0 m/s?

10. An object has 600 J of kinetic energy from its motion and a speed of 10 m/s. What is its mass?

11. If you throw a 0.4-kg ball at a speed of 20 m/s, what is the kinetic energy from the ball's motion?

12. If you have a mass of 80 kg and you are standing on a platform 3.0 m above the ground, what is the gravitational potential energy between you and Earth relative to the ground?

13. A 2.0-kg book is moved from a shelf that is 2.0 m off the ground to a shelf that is 1.5 m off the ground. What is the change in GPE?

14. A car moving at 30 m/s has 900 kJ of kinetic energy from its motion. What is the car's mass?

15. A car with a mass of 900 kg is traveling at a speed of 25 m/s. What is the kinetic energy from the car's motion?

16. If your weight is 500 N, and you are standing on a floor that is 20 m above the ground, what is the gravitational potential energy between you and Earth, relative to the ground?

MODULE 5

THERMAL ENERGY

1. Water has a specific heat of 4,184 J/(kg·°C). How much energy is needed to increase the temperature of a kilogram of water 5.0°C?

2. The temperature of a block of iron, which has a specific heat of 450 J/(kg·°C), increases by 3°C when 2,700 J of energy are added to it. What is the mass of this block of iron?

3. How much energy is needed to heat 1.0 kg of sand, which has a specific heat of 664 J/(kg·°C), from 30°C to 50°C?

4. 1 kg of liquid water (specific heat = 4,184 J/(kg·°C)) is heated from freezing (0°C) to boiling (100°C). What is the water's change in thermal energy?

5. A concrete statue (specific heat = 600 J/(kg·°C)) sits in sunlight and warms up to 40°C. Overnight, it cools to 15°C and transfers 90,000 J of thermal energy to its surroundings. What is its mass?

6. A substance with a mass of 10.0 kg transfers 106.5 kJ of thermal energy to its surroundings when its temperature drops 15°C. What is this substance's specific heat?

7. How much heat is needed to raise the temperature of 100 g of water by 50°C, if the specific heat of water is 4,184 J/kg·°C?

8. A calorimeter contains 1.0 kg of water (specific heat = 4,184 J/(kg·°C)). An object with a mass of 4.23 kg is added to the water. If the water temperature increases by 3.0°C and the temperature of the object decreases by 1.0°C, what is the specific heat of the object?

9. A sample of an unknown metal has a mass of 0.5 kg. Adding 1,985 J of thermal energy to the metal raises its temperature by 10°C. What is the specific heat of the metal?

MODULE 6

ELECTRICITY

1. A circuit has a resistance of 4.0 Ω. What voltage difference will produce a current of 1.4 A in the circuit?

2. How many amperes of current will there be in a circuit if the voltage difference is 9.0 V and the resistance in the circuit is 3.0 Ω?

3. If a voltage difference of 3.0 V causes a 1.5 A current in a circuit, what is the resistance in the circuit?

4. The current in an appliance is 3.0 A and the voltage difference is 120 V. How much power is being supplied to the appliance?

5. What is the current into a microwave oven that requires 700 W of power if the voltage difference is 120 V?

6. What is the voltage difference in a circuit that uses 2,420 W of power when the current through the circuit is 11 A?

7. How much energy is converted when a 110 kW appliance is used for 3.0 hours?

8. How much does it cost to light six 100-W lightbulbs for six hours if the price of electrical energy is $0.09/kWh?

9. An electric clothes dryer uses 4 kW of electric power. How long did it take to dry a load of clothes if electric power costs $0.09/kWh and the cost of using the dryer was $0.27?

10. What is the resistance of a lightbulb that draws 0.50 amp of current when plugged into a 120-V outlet?

11. How large is the current through a 100-W lightbulb that is plugged into a 120-V outlet?

12. The current through a hair dryer connected to a 120-V outlet is 8 A. How much electrical power does the hair dryer use?

MODULE 7

MAGNETISM AND ITS USES

1. How many turns are in the secondary coil of a step-down transformer that reduces a voltage from 900 V to 300 V and has 15 turns in the primary coil?
2. A step-down transformer reduces voltage from 2,400 V to 120 V. What is the ratio of the number of turns in the primary coil to the number of turns in the secondary coil of the transformer?
3. The current produced by an AC generator switches direction twice for each revolution of the coil. How many times does a 110-Hz alternating current switch direction each second?
4. What is the output voltage from a step-down transformer with 200 turns in the primary coil and 100 turns in the secondary coil if the input voltage is 800 V?
5. The coil of a 60-Hz generator makes 60 revolutions each second. How many revolutions does the coil make in five minutes?
6. What is the output voltage from a step-up transformer with 25 turns in the primary coil and 75 turns in the secondary coil if the input voltage was 120 V?
7. How many turns are in the primary coil of a step-down transformer that reduces a voltage from 400 V to 100 V and has 80 turns in the secondary coil?
8. How many turns are in the secondary coil of a step-up transformer that increases voltage from 30 V to 150 V and has seven turns in the primary coil?
9. If a generator coil makes 6,000 revolutions in two minutes, how many revolutions does it make each second?

MODULE 8

ENERGY SOURCES AND THE ENVIRONMENT

1. A gallon of gasoline contains about 2,800 g of gasoline. If burning one gram of gasoline releases about 48 kJ of energy, how much energy is released when a gallon of gasoline is burned? (1 kJ = 1,000 J)
2. An automobile engine converts the energy released by burning gasoline into mechanical energy with an efficiency of about 25%. If burning 1 kg of gasoline releases about 48,000 kJ of energy, how much mechanical energy is produced by the engine when 1 kg of gasoline is burned?
3. Refer to the pie chart *Energy Sources* in **Figure 2.** According to the pie chart, petroleum represents 38 percent of the consumable energy sources in the United States. If 39.7 quadrillion BTUs of petroleum are consumed each year, what amount of energy (in quadrillion BTUs) does coal use account for?
4. Refer to **Figure 8,** a graph of carbon dioxide concentration (ppm) vs. time (years). If the average carbon dioxide concentration in 1960 was 315 parts per million and the average carbon dioxide concentration in 2010 is 385 parts per million, what is the percentage change in the concentration of carbon dioxide over the past 50 years?
5. A nuclear reactor contains 100,000 kg of enriched uranium. About 4% of the enriched uranium is the isotope uranium-235. What is the mass of uranium-235 in the reactor core?
6. Suppose the number of uranium-235 nuclei that are split doubles at each stage of a chain reaction. If the chain reaction starts with one nucleus split in the first stage, how many nuclei will have been split after six stages?
7. From 1970 to 2010, the carbon dioxide concentration in Earth's atmosphere increased from about 325 parts per million to about 385 parts per million. What is the percentage change in the concentration of carbon dioxide over the past 40 years?
8. About 85% of the energy used in the U.S. comes from fossil fuels. How many times greater is the amount of energy used from fossil fuel than the amount used from all other energy sources?

MODULE 9

INTRODUCTION TO WAVES

1. What is the wavelength of a wave with a frequency of 0.4 kHz traveling at 16 m/s?
2. What is the wavelength of a wave with a frequency of 5 Hz traveling at 15 m/s?

3. A wave has a wavelength of 250 cm and a frequency of 4 Hz. What is its speed?

4. Two waves are traveling in the same medium with a speed of 340 m/s. What is the difference in frequency of the waves if the one has a wavelength of 5.0 m and the other has a wavelength of 0.2 m?

5. What is the speed of a wave that has a wavelength of 6.0 m and a frequency of 3.0 Hz?

6. What is the frequency of a wave with a wavelength of 7 m traveling at 21 m/s?

7. A light ray strikes a plane mirror. The angle between the incident light ray and the normal to the mirror is 55°. What is the angle between the reflected ray and the normal?

MODULE 10
SOUND

1. What is the wavelength of a 440-Hz sound wave traveling with a speed of 347 m/s?

2. A sound wave with a frequency of 440 Hz travels in steel with a speed of 5,200 m/s. What is the wavelength of the sound wave?

3. A wave traveling in water has a wavelength of 750 m and a frequency of 2 Hz. How fast is this wave moving?

4. At 0°C sound travels through air with a speed of about 331 m/s and through aluminum with a speed of 4,877 m/s. How many times longer is the wavelength of a sound wave in aluminum compared to the wavelength of a sound wave in air if both waves have the same frequency?

5. The speed of sound in air at 0°C is 331 m/s, and at 20°C is 344 m/s. What is the percentage change in the speed of sound at 20°C compared to 0°C?

6. What is the frequency of the first overtone of a 440-Hz wave?

7. The wreck of the *Titanic* is at a depth of about 3,800 m. A sonar unit on a ship above the *Titanic* emits a sound wave that travels at a speed of 1,500 m/s. How long does it take a sound wave reflected from the *Titanic* to return to the ocean surface?

8. A sonar unit on a ship emits a sound wave. The echo from the ocean floor is detected two seconds later. If the speed of sound in water is 1,500 m/s, how deep is the ocean beneath the ship?

9. One flute plays a note with a frequency of 443 Hz, and another flute plays a note with a frequency of 440 Hz. What is the frequency of the beats that the flute players hear?

10. A sound wave has a wavelength of 50 m and a frequency of 22 Hz. What is the speed of the sound wave?

MODULE 11
ELECTROMAGNETIC WAVES

1. Express the number 20,000 in scientific notation.

2. An electromagnetic wave has a wavelength of 0.054 m. What is the wavelength in scientific notation?

3. Earth is about 4,500,000,000 years old. Express this number in scientific notation.

4. The speed of electromagnetic waves in air is 300,000 km/s. What is the frequency of electromagnetic waves that have a wavelength of 5×10^{-3} km?

5. Radio waves with a frequency of 125,000 Hz have a wavelength of 1.84 km when traveling in ice. What is the speed of the radio waves in ice?

6. The speed of radio waves in water is about 2.26×10^5 km/s. What is the frequency of radio waves that have a wavelength of 3.0 km?

7. Some infrared waves have a frequency of 10,000,000,000,000 Hz. Express this frequency in scientific notation.

8. An infrared wave has a frequency of 1×10^{13} Hz and a wavelength of 3×10^{-5} m. Express the wave's speed as a decimal number.

9. An AM radio station broadcasts at a frequency of 620 kHz. Express this frequency in Hz using scientific notation.

10. An FM radio station broadcasts at a frequency of 101 MHz. Express this frequency in Hz using scientific notation.

MODULE 12

LIGHT

1. A ray of light hits a plane mirror at 35° from the normal. What angle does the reflected ray make with the normal?

2. A light ray strikes a plane mirror. The angle between the light ray and the surface of the mirror is 25°. What angle does the reflected ray make with the normal?

3. A light ray is reflected from a plane mirror. If the angle between the incident ray and the reflected ray is 104°, what is the angle of incidence?

4. What will happen to a ray of light leaving water and entering air if it hits the boundary at an angle of 49° to the normal? (The critical angle for water and air is 49°.)

5. A ray of light hits a plane mirror at 60° from the normal. What is the angle between the reflected ray and the surface of the mirror?

6. About 8% of men and 0.5% of women have some form of color blindness. The percentage of men who experience color blindness is how many times larger than the percentage of women who experience color blindness?

7. The index of refraction of a material is the speed of light in a vacuum divided by the speed of light in the material. If the index of refraction of the mineral rock salt is 1.52, and the speed of light in a vacuum is 300,000 km/s, what is the speed of light in rock salt?

8. A laser is used to measure the distance from Earth to the Moon. The laser beam is reflected from a mirror on the Moon's surface. If the time needed for the laser to reach the Moon and be reflected back is 2.56 s, and the laser beam travels at 300,000 km/s, what is the distance to the Moon?

9. In the human eye, there are about 7,000,000 cone cells distributed over an area of 5 cm^2. If cone cells are evenly distributed, how many cone cells are distributed over an area of 2 cm^2? Express your answer in scientific notation.

10. When a light beam is reflected from a glass surface, only 4% of the energy carried by the beam is reflected. If a light beam is reflected from one glass surface and then another, what is the ratio of the energy carried by the beam after the second reflection, compared to the energy carried by the beam before the first reflection?

MODULE 13

MIRRORS AND LENSES

1. A convex lens in a magnifying glass has a focal length of 5 cm. How far should the lens be from an object if the image formed is virtual, enlarged, and upright?

2. The magnification of a mirror or lens equals the image size divided by the object size. If a plant cell with a diameter of 0.0035 mm is magnified so that the diameter of the image is 0.028 cm, what is the magnification?

3. Magnification equals the image size divided by the object size. Magnification also equals the distance of the image from the lens divided by the distance of the object from the lens. A penny has a diameter of 2.0 cm. A convex lens forms an image with a diameter of 5.2 cm and is 6.0 cm from the lens. What is the distance between the penny and the lens?

4. Light enters the human eye through the pupil. In the dark, the pupil is dilated and has a diameter of about 1 cm. The Keck telescope has a mirror with a diameter of 10 m. If both the pupil and the Keck mirror are circles, what is the ratio of the area of the Keck telescope mirror to the area of a dilated human pupil?

5. A small insect is viewed in a compound microscope. The objective lens of the microscope forms a real image 20 times larger than the insect. The eyepiece lens then magnifies this real image by 10 times. What is the magnification of the microscope?

6. A light source is placed a distance of 1.2 m from a concave mirror on the optical axis. The reflected light rays are parallel and form a light beam. What is the focal length of the mirror?

7. Astronomers have proposed building the Thirty Meter Telescope with a mirror 30.0 m in diameter. The diameter of the *Hubble Space Telescope* mirror is 2.4 m. What percent of the surface area of the telescope's mirror would be covered by the surface area of the *Hubble* mirror?

8. In some types of reflecting telescopes the eyepiece is located behind the concave mirror. A small curved mirror in front of the concave mirror reflects light through a hole in the concave mirror to the eyepiece. Suppose a circular concave mirror with a diameter of 50 cm has a hole with a diameter of 10 cm. What is the ratio of the reflecting area of the mirror with the 10-cm hole to the reflecting area of the same mirror without the hole?

MODULE 14

SOLIDS, LIQUIDS, AND GASES

1. A book is sitting on a desk. The area of contact between the book and the desk is 0.06 m^2. If the book's weight is 30 N, what is the pressure the book exerts on the desk?

2. A skater has a weight of 500 N. The skate blades are in contact with the ice over an area of 0.001 m^2. What is the pressure exerted on the ice by the skater?

3. The weight of the water displaced by a person floating in the water is 686 N. What is the person's mass?

4. The pressure on a balloon that has a volume of 7 L is 100 kPa. If the temperature stays the same and the pressure on the balloon is increased to 250 kPa, what is the new volume of the balloon?

5. The air in a tire pump has a volume of 1.50 L at a temperature of 5°C. If the temperature is increased to 30°C and the pressure remains constant, what is the new volume?

6. A block of wood with a mass of 1.2 kg is floating in a container of water. If the density of water is 1.0 g/cm^3, what is the volume of water displaced by the floating wood?

7. Two cylinders contain pistons that are connected by fluid in a hydraulic system. A force of 1,300 N is exerted on one piston with an area of 0.05 m^2. What is the force exerted on the other piston that has an area of 0.08 m^2?

8. A gas-filled weather balloon floating in the atmosphere has an initial volume of 850 L. The weather balloon rises to a region where the pressure is 56 kPa, and its volume expands to 1,700 L. If the temperature remains the same, what was the initial pressure on the weather balloon?

9. In a hydraulic system, a force of 7,500 N is exerted on a piston with an area of 0.05 m^2. If the force exerted on a second piston in the hydraulic system is 1,500 N, what is the area of this second piston?

10. A gold bar weighs 17.0 N. If the density of gold is 19.3 g/cm^3, what is the volume of the gold bar?

11. A book is sitting on a desk. If the surface area of the book's cover is 0.05 m^2 and atmospheric pressure is 100.0 kPa, what is the downward force of the atmosphere on the book?

MODULE 15

CLASSIFICATION OF MATTER

1. The size of particles in a solution is about 1 nm (1 nm = 0.000000001 m). Write 0.000000001 m in scientific notation.

2. A chemical reaction produces two new substances, one with a mass of 34 g and the other with a mass of 39 g. What is the total mass of the reactants?

3. The human body is about 65% oxygen by mass. If a person has a mass of 75.0 kg, what is the mass of oxygen in his body?

4. Two solutions, one with a mass of 450 g and the other with a mass of 350 g, are mixed. A chemical reaction occurs and 125 g of solid crystals are produced that settle on the bottom of the container. What is the mass of the remaining solution?

5. Carbon reacts with oxygen to form carbon dioxide according to the following equation: $C + O_2 \rightarrow CO_2$. When 120 g of carbon reacts with oxygen, 440 g of carbon dioxide are formed. How much oxygen reacted with the carbon?

6. Salt water is distilled by boiling it and condensing the vapor. After distillation, 1,164 g of water have been collected and 12 g of salt are left behind in the original container. What was the original mass of the salt water?

7. Calcium carbonate, $CaCO_3$, decomposes according to the following reaction: $CaCO_3 \rightarrow CaO + CO_2$. When 250 g of $CaCO_3$ decompose completely, the mass of CaO is 56% of the mass of the products of this reaction. What is the mass of CO_2 produced?

8. Water breaks down into hydrogen gas and oxygen gas according to the reaction $2H_2O \rightarrow 2H_2 + O_2$. In this reaction, the mass of oxygen produced is eight times greater than the mass of hydrogen produced. If 36 g of water form hydrogen and oxygen gas, what is the mass of hydrogen gas produced?

9. A 112-g serving of ice cream contains 19 g of fat. What percentage of the serving is fat?

10. The mass of the products produced by a chemical reaction is measured. The reaction is repeated five times, with the same mass of reactants used each time. The measured product masses are 50.17 g, 50.12 g, 50.17 g, 50.10 g, and 50.14 g. What is the average of these measurements?

MODULE 16

PROPERTIES OF ATOMS AND THE PERIODIC TABLE

1. A boron atom has a mass number of 11 and an atomic number of 5. How many neutrons are in the boron atom?

2. A magnesium atom has 12 protons and12 neutrons. What is its mass number?

3. Iodine-127 has a mass number of 127 and 74 neutrons. What percentage of the particles in an iodine-127 nucleus are protons?

4. How many neutrons are in an atom of phosphorus-31?

5. What is the ratio of neutrons to protons in the isotope radium-234?

6. About 80% of all magnesium atoms are magnesium-24, about 10% are magnesium-25, and about 10% are magnesium-26. What is the average atomic mass of magnesium?

7. The half-life of the radioactive isotope rubidium-87 is 48,800,000,000 years. Express this half-life in scientific notation.

8. The radioactive isotope nickel-63 has a half-life of 100 years. How much of a 10.0-g sample of nickel-63 is left after 300 years?

9. A sample of the radioactive isotope cobalt-62 is prepared. The sample has a mass of 1.00 g. After three minutes, the mass of cobalt-62 remaining is 0.25 g. What is the half-life of cobalt-62?

10. A neutral phosphorus atom has 15 electrons. How many electrons are in the third energy level?

MODULE 17

ELEMENTS AND THEIR PROPERTIES

1. In seawater the concentration of fluoride ions, F^-, is 1.3×10^{-3} g/L. How many liters of seawater would contain 1.0 g of F^-?

2. There are three isotopes of hydrogen. The isotope deuterium, with one proton and one neutron in the nucleus, makes up 0.015% of all hydrogen atoms. Of every million hydrogen atoms, how many are deuterium?

3. A vitamin and mineral supplement pill contains 1.0×10^{-5} g of selenium. According to the label on the bottle, this amount is 18% of the recommended daily value. What is the recommended daily value of selenium in g?

4. The density of silver is 10.5 g/cm^3 and the density of copper is 8.9 g/cm^3. What is the difference in mass between a piece of silver with a volume of 5 cm^3 and a piece of copper with a volume of 5 cm^3?

5. A person has a mass of 68.3 kg. If 18% of the mass of a human body is carbon, what is the mass of carbon in this person's body?

6. A gold ore produces about 5 g of gold for every 1,000 kg of ore that is mined. If one ounce = 28.3 g, how many kg of ore must be mined to produce an ounce of gold?

7. A metal bolt with a mass of 26.6 g is placed in a 50-mL graduated cylinder containing water. The water level in the cylinder rises from 27.0 mL to 30.5 mL. What is the density of the bolt in g/cm^3?

8. On a circle graph showing the percentage of elements in the human body, the wedge representing nitrogen takes up 10.8°. What is the percentage of nitrogen in the human body?

9. The melting point of aluminum is 660.0°C. What is the melting point of aluminum on the Fahrenheit temperature scale?

10. The synthetic element hassium-269 has a half-life of 9.3 s. The synthetic element fermium-255 has a half-life of 20.1 h. How many times longer is the half-life of fermium-255 than the half-life of hassium-261?

MODULE 18

CHEMICAL BONDS

1. What is the formula of the compound formed when ammonium ions (NH_4^+) and phosphate ions (PO_4^{3-}) combine?

2. Show that the sum of positive and negative charges in a unit of calcium chloride ($CaCl_2$) equals zero.

3. What is the formula for iron(III) oxide?

4. How many hydrogen atoms are in three molecules of ammonium sulfate, $(NH_4)_2SO_4$?

5. The overall charge on the polyatomic phosphate ion (PO_4^{3-}) is 3−. What is the oxidation number of phosphorus in the phosphate ion?

6. The overall charge on the polyatomic dichromate ion ($Cr_2O_7^{2-}$) is 2−. What is the oxidation number of chromium in this polyatomic ion?

7. What is the formula for lead(IV) oxide?

8. What is the formula for potassium chlorate?

9. What is the formula for carbon tetrachloride?

10. What is the name of NaF?

11. What is the name of Al_2O_3?

12. What percentage of the mass of a sulfuric acid molecule (H_2SO_4) is sulfur?

MODULE 19

CHEMICAL REACTIONS

1. Lithium reacts with oxygen to form lithium oxide according to the following equation: $4Li + O_2 \rightarrow 2Li_2O$. If 27.8 g of Li react completely with 32.0 g of O_2, how many grams of Li_2O are formed?

2. What coefficients balance the following equation: $\_Zn(OH)_2 + \_H_3PO_4 \rightarrow \_Zn_3(PO_4)_2 + \_H_2O$?

3. Aluminum hydroxide, $Al(OH)_3$, decomposes to form aluminum oxide, Al_2O_3, and water according to the reaction $2Al(OH)_3 \rightarrow Al_2O_3 + 3H_2O$. If 156.0 g of $Al(OH)_3$ decompose to from 102.0 g of Al_2O_3, how many grams of H_2O are formed?

4. In the following balanced chemical reaction one of the products is represented by the symbol X:$BaCO_3 + C + H_2O \rightarrow Ba(OH)_2 + 2X$.What is the formula for the compound represented by X?

5. When propane (C_3H_8) is burned, carbon dioxide and water vapor are produced according to the following reaction: $C_3H_8 + 5O_2 \rightarrow 3CO_2 + 4H_2O$.How much propane is burned if 160.0 g of O_2 are used and 132.0 g of CO_2 and 72.0 g of H_2O are produced?

6. Increasing the temperature usually causes the rate of a chemical reaction to increase. If the rate of a chemical reaction doubles when the temperature increases by 10°C, by what factor does the rate of reaction increase if the temperature increases by 30°C?

7. When acetylene gas (C_2H_2) is burned, carbon dioxide and water are produced. Find the coefficients that balance the chemical equation for the combustion of acetylene:$\_C_2H + \_O_2 \rightarrow \_CO_2 + \_H_2O$.

8. What coefficients balance the following equation: $\_CS_2 + \_O_2 \rightarrow \_CO_2 + \_SO_2$?

9. When methane (CH_4) is burned, 50.1 kJ of energy per gram are released. When propane (C_3H_8) is burned, 45.8 kJ of energy are released. If a mixture of 1.0 g of methane and 1.0 g of propane is burned, how much energy is released per gram of mixture?

10. A chemical reaction produces 0.050 g of a product in 0.18 s. In the presence of a catalyst, the reaction produces 0.050 g of the same product in 0.0070 s. How much faster is the rate of reaction in the presence of the enzyme?

MODULE 20

RADIOACTIVITY AND NUCLEAR REACTIONS

1. How many protons are in the nucleus $^{81}_{36}Kr$?

2. How many neutrons are in the nucleus $^{56}_{26}Fe$?

3. What is the ratio of neutrons to protons in the nucleus $^{241}_{95}Am$?

4. How many alpha particles are emitted when the nucleus $^{222}_{86}Rn$ decays $^{218}_{84}Po$?

5. An alpha particle is the same as the helium nucleus $^{4}_{2}He$. What nucleus is produced when the nucleus $^{226}_{88}Ra$ decays by emitting an alpha particle?

6. A sample of $^{38}_{71}Cl$ is observed to decay to 25% of the original amount in 74.4 minutes. What is the half-life of $^{38}_{71}Cl$?

7. How many beta particles are emitted when the nucleus $^{40}_{19}K$ decays to the nucleus $^{40}_{20}Ca$?

8. How long will it take a sample of $^{194}_{84}Po$ to decay to $\frac{1}{8}$ of its original amount if $^{194}_{84}Po$ has a half-life of 0.70 s?

9. The half-life of $^{131}_{53}I$ is 8.04 days. How much time would be needed to reduce 1.00 g of $^{131}_{53}I$ to 0.25 g?

10. A sample of radioactive carbon-14 has decayed to 12.50% of its original amount. If the half-life of carbon-14 is 5,730 years, how old is this sample?

11. Recall that objects in motion have kinetic energy. How fast would a 1,500-kg car need to travel in order to increase its mass by 1.0 kg?

12. The Kashiwazaki-Kariwa nuclear power plant is capable of converting approximately 8 billion J of nuclear energy into electrical energy every second. How much energy does the Kashiwazaki-Kariwa nuclear power plant convert to electrical energy in 1 year? How much mass is this equivalent to?

MODULE 21

SOLUTIONS

1. A cup of orange juice contains 126 mg of vitamin C and $\frac{1}{2}$ cup of strawberries contain 42 mg of vitamin C. How many cups of strawberries contain as much vitamin C as one cup of orange juice?

2. A Sacagawea dollar coin is made of manganese brass alloy that is $\frac{1}{25}$ nickel. Express this number as a percentage.

3. What is the total surface area of a 2-cm cube?

4. A cube has 2-cm sides. If it is split in half, what is the total surface area of the two pieces?

5. What is the increase in surface area when a cube with 2-cm sides is divided into eight equal parts?

6. How much surface area is lost if two 4-cm cubes are attached at one face?

7. At 20°C, the solubility in water of potassium bromide (KBr) is 65.3 g/100 mL. What is the maximum amount of potassium bromide that will dissolve in 237 mL of water?

8. At 20°C, the solubility of sodium chloride (NaCl) in water is 35.9 g/100 mL. If the maximum amount of sodium chloride is dissolved in 500 mL of water at 20°C, the mass of the dissolved sodium chloride is what percentage of the mass of the solution?

9. At 60°C, the solubility of sucrose (sugar) in water is 287.3 g/100 mL. At this temperature, what is the minimum amount of water needed to dissolve 50.0 g of sucrose?

10. A fruit drink contains 90% water and 10% fruit juice. How much fruit juice does 500 mL of fruit drink contain?

ADDITIONAL PRACTICE PROBLEMS

MODULE 22

ACIDS, BASES, AND SALTS

1. The difference between the pH of an acidic solution and the pH of pure water is 3. What is the pH of the solution?

2. The pH of rain that fell over a region had measured values of 4.6, 5.1, 4.8, 4.5, 4.5, 4.9, 4.7, and 4.8. What was the mean value of the measured pH?

3. A molecule of acetylsalicylic acid, or aspirin, has the chemical formula $COOHC_6H_4COOCH_3$. What is the mass of a molecule of acetylsalicylic acid in amu?

4. If 5.5% of 473.0 mL of vinegar is acetic acid, how many milliliters of acetic acid are there?

5. The difference between the pH of a basic solution and the pH of pure water is 2. What is the pH of the solution?

6. On the pH scale, a decrease of one unit means that the concentration of H^+ ions increases 10 times. If the pH of a solution changes from 6.5 to 4.5, how has the concentration of H^+ ions changed?

7. Write the balanced chemical equation for the neutralization of H_2SO_4, sulfuric acid, by KOH, potassium hydroxide.

8. Write the balanced chemical equation for the neutralization of hydrobromic acid (HBr) by aluminum hydroxide ($Al(OH)_3$).

9. Write the equation for the reaction when nitric acid (HNO_3) ionizes in water.

10. When sodium oxide (Na_2O) reacts with water, the base sodium hydroxide (NaOH) is formed. Write the balanced equation for this reaction.

MODULE 23

ORGANIC COMPOUNDS

1. Fats supply 9 Calories per gram; carbohydrates and proteins each supply 4 Calories per gram. If 100 g of potato chips contain 7 g of protein, 53 g of carbohydrates, and 35 g of fats, how many Calories are in 100 g of potato chips?

2. The basal metabolic rate (BMR) is the amount of energy required to maintain basic body functions. The BMR is approximately 1.0 Calories/hr per kilogram of body mass. For a person with a mass of 65 kg, how many Calories are needed each day to maintain basic body functions?

3. The hydrocarbon octane (C_8H_{18}) has a boiling point of 259°F. What is its boiling point on the Celsius temperature scale?

4. Four molecules of a saturated hydrocarbon contain carbon atoms and 56 hydrogen atoms. What is the formula for a molecule of this hydrocarbon?

5. For saturated hydrocarbons, the number of hydrogen atoms in a molecule can be calculated by the formula $N_H = 2N_C + 2$, where N_H is the number of hydrogen atoms and N_C is the number of carbon atoms in the molecule. If a molecule of a saturated hydrocarbon has 22 hydrogen atoms, how many carbon atoms does that molecule contain?

6. A food Calorie is an energy unit equal to 4,184 joules. If a person uses 2,070 Calories in one day, what is the power being used? Express your answer in watts.

7. In each 100 g of cheddar cheese, there are 33 g of fat. Calculate how many grams of fat are in 250 g of cheddar cheese.

8. A car gets 25 miles per gallon of gas. If the car is driven 12,000 miles in one year and gasoline costs $2.55 per gallon, what was the cost of the gasoline used in one year?

MODULE 24

NEW MATERIALS THROUGH CHEMISTRY

1. A 14-karat gold earring has a mass of 10 g. What is the mass of gold in the earring?

2. In 1997, about 6,400,000,000 kg of polyvinyl chloride were used in the United States. About 6% of the PVC used was for packaging. Express in scientific notation how many kilograms of PVC were used for packaging in 1997.

3. A stainless steel spoon contains 30.0 g of iron, 6.8 g of chromium, and 3.2 g of nickel. What percentage of the stainless steel is chromium?

4. The molecules in a sample of polypropylene have an average length of 60,000 monomers. The monomer of polypropylene has the formula C_3H_6. Express in scientific notation the mass, in amu, of a polypropylene molecule made of 60,000 monomers.

5. A certain process for manufacturing integrated circuits packs 47,600,000 transistors into an area of 340 mm^2. If this process is used to produce an integrated circuit with an area of 1 cm^2, express in scientific notation the number of transistors in this integrated circuit.

6. The melting points of five different samples of a new aluminum alloy have measured values of 631.5°C, 632.3°C, 636.1°C, 637.4°C, and 630.2°C. What is the mean of these measurements?

7. The measured values of the copper content of seven bronze buttons found at an archaeological site are 83%, 90%, 91%, 72%, 79%, 87%, and 89%. What is the median of these measurements?

8. The number of transistors and other components per mm^2 on an integrated circuit has doubled, on average, every two years. If integrated circuits contained 100,000 transistors in 1992, estimate how many transistors an integrated circuit of the same size contained in 2008.

9. A car contains 200 kg of plastic parts that replaced steel parts. The density of steel is twice the density of plastic. If the volume of the plastic parts equals the volume of the same parts made of steel, how much less is the mass (kg) of the car made by using plastic parts instead of steel?

Physical Science Reference Tables

Standard Units

Symbol	Name	Quantity
m	meter	length
kg	kilogram	mass
Pa	pascal	pressure
K	kelvin	temperature
mol	mole	amount of a substance
J	joule	energy, work, quantity of heat
s	second	time
C	coulomb	electric charge
V	volt	electric potential
A	ampere	electric current
Ω	ohm	resistance

Physical Constants and Conversion Factors

Acceleration due to gravity	g	9.8 m/s/s or m/s^2
Avogadro's Number	N_A	6.02×10^{23} particles per mole
Electron charge	e	1.6×10^{-19} C
Electron rest mass	m_e	9.11×10^{-31} kg
Gravitation constant	G	6.67×10^{-11} N $\times$ m^2/kg^2
Mass-energy relationship		1 u (amu) = 9.3×10^2 MeV
Speed of light in a vacuum	c	3.00×10^8 m/s
Speed of sound at STP	331 m/s	
Standard Pressure	1 atmosphere 101.3 kPa 760 Torr or mmHg 14.7 lb/in.^{2}	

Wavelengths of Light in a Vacuum

Violet	$4.0 - 4.2 \times 10^{-7}$ m
Blue	$4.2 - 4.9 \times 10^{-7}$ m
Green	$4.9 - 5.7 \times 10^{-7}$ m
Yellow	$5.7 - 5.9 \times 10^{-7}$ m
Orange	$5.9 - 6.5 \times 10^{-7}$ m
Red	$6.5 - 7.0 \times 10^{-7}$ m

The Index of Refraction for Common Substances

($\lambda = 5.9 \times 10^{-7}$ m)

Air	1.00
Alcohol	1.36
Canada Balsam	1.53
Corn Oil	1.47
Diamond	2.42
Glass, Crown	1.52
Glass, Flint	1.61
Glycerol	1.47
Lucite	1.50
Quartz, Fused	1.46
Water	1.33

Heat Constants

	Specific Heat (average) (kJ/kg × °C) (J/g × °C)	Melting Point (°C)	Boiling Point (°C)	Heat of Fusion (kJ/kg) (J/g)	Heat of Vaporization (kJ/kg) (J/g)
Alcohol (ethyl)	2.43 (liq.)	–117	79	109	855
Aluminum	0.90 (sol.)	660	2467	396	10500
Ammonia	4.71 (liq.)	–78	–33	332	1370
Copper	0.39 (sol.)	1083	2567	205	4790
Iron	0.45 (sol.)	1535	2750	267	6290
Lead	0.13 (sol.)	328	1740	25	866
Mercury	0.14 (liq.)	–39	357	11	295
Platinum	0.13 (sol.)	1772	3827	101	229
Silver	0.24 (sol.)	962	2212	105	2370
Tungsten	0.13 (sol.)	3410	5660	192	4350
Water (solid)	2.05 (sol.)	0	–	334	–
Water (liquid)	4.18 (liq.)	–	100	–	–
Water (vapor)	2.01 (gas)	–	–	–	2260
Zinc	0.39 (sol.)	420	907	113	1770

Electromagnetic Spectrum

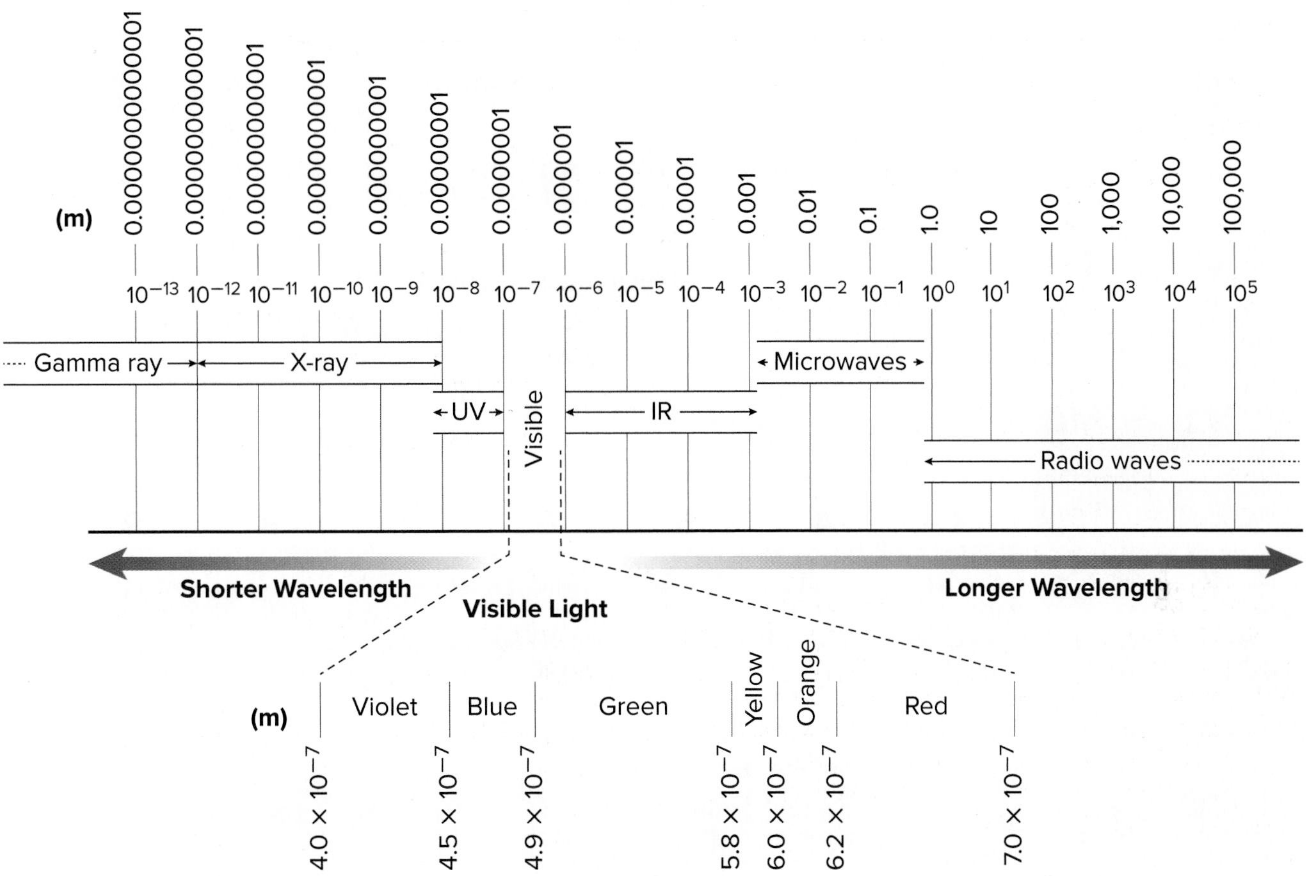

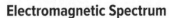

Uranium Decay Series

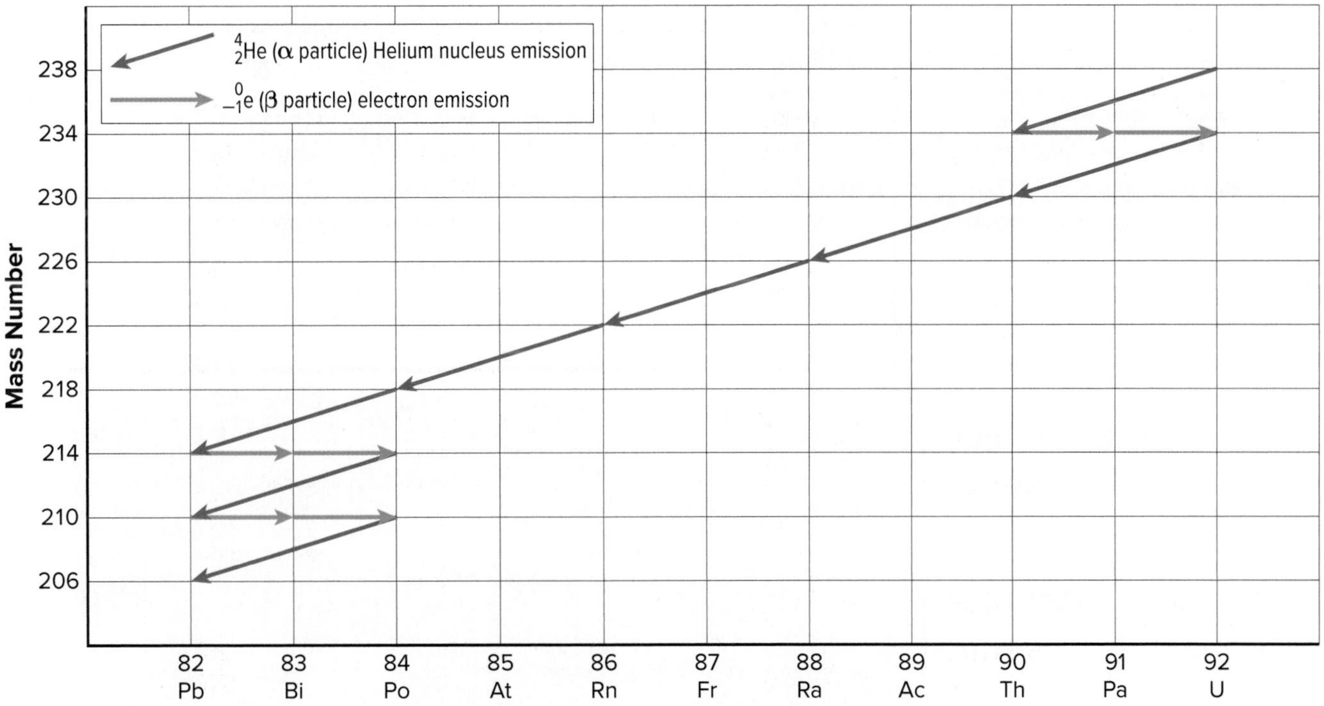

Reference Handbook

PERIODIC TABLE OF THE ELEMENTS

Atomic number — 1 | Symbol — H | Element — Hydrogen | Atomic mass — 1.008

Period	1	2	3	4	5	6	7	8	9
1	1 H Hydrogen 1.008								
2	3 Li Lithium 6.941	4 Be Beryllium 9.012							
3	11 Na Sodium 22.990	12 Mg Magnesium 24.305							
4	19 K Potassium 39.098	20 Ca Calcium 40.078	21 Sc Scandium 44.956	22 Ti Titanium 47.867	23 V Vanadium 50.942	24 Cr Chromium 51.996	25 Mn Manganese 54.938	26 Fe Iron 55.847	27 Co Cobalt 58.933
5	37 Rb Rubidium 85.468	38 Sr Strontium 87.62	39 Y Yttrium 88.906	40 Zr Zirconium 91.224	41 Nb Niobium 92.906	42 Mo Molybdenum 95.95	43 Tc Technetium (98)	44 Ru Ruthenium 101.07	45 Rh Rhodium 102.906
6	55 Cs Cesium 132.905	56 Ba Barium 137.327	57 La Lanthanum 138.905	72 Hf Hafnium 178.49	73 Ta Tantalum 180.948	74 W Tungsten 183.84	75 Re Rhenium 186.207	76 Os Osmium 190.23	77 Ir Iridium 192.217
7	87 Fr Francium (223)	88 Ra Radium (226)	89 Ac Actinium (227)	104 Rf Rutherfordium * (267)	105 Db Dubnium * (270)	106 Sg Seaborgium * (269)	107 Bh Bohrium * (270)	108 Hs Hassium * (277)	109 Mt Meitnerium * (278)

The number in parentheses is the mass number of the longest-lived isotope for that element.

Lanthanide series	58 Ce Cerium 140.115	59 Pr Praseodymium 140.908	60 Nd Neodymium 144.242	61 Pm Promethium (145)	62 Sm Samarium 150.36	63 Eu Europium 151.965
Actinide series	90 Th Thorium 232.038	91 Pa Protactinium 231.036	92 U Uranium 238.029	93 Np Neptunium (237)	94 Pu Plutonium (244)	95 Am Americium (243)

Metal

Metalloid

Nonmetal

⊙ Synthetic

10	11	12	13	14	15	16	17	18
								2 He Helium 4.003
			5 B Boron 10.811	6 C Carbon 12.011	7 N Nitrogen 14.007	8 O Oxygen 15.999	9 F Fluorine 18.998	10 Ne Neon 20.180
			13 Al Aluminum 26.982	14 Si Silicon 28.086	15 P Phosphorus 30.974	16 S Sulfur 32.066	17 Cl Chlorine 35.453	18 Ar Argon 39.948
28 Ni Nickel 58.693	29 Cu Copper 63.546	30 Zn Zinc 65.39	31 Ga Gallium 69.723	32 Ge Germanium 72.61	33 As Arsenic 74.922	34 Se Selenium 78.971	35 Br Bromine 79.904	36 Kr Krypton 83.80
46 Pd Palladium 106.42	47 Ag Silver 107.868	48 Cd Cadmium 112.411	49 In Indium 114.82	50 Sn Tin 118.710	51 Sb Antimony 121.757	52 Te Tellurium 127.60	53 I Iodine 126.904	54 Xe Xenon 131.290
78 Pt Platinum 195.08	79 Au Gold 196.967	80 Hg Mercury 200.59	81 Tl Thallium 204.383	82 Pb Lead 207.2	83 Bi Bismuth 208.980	84 Po Polonium 208.982	85 At Astatine 209.987	86 Rn Radon 222.018
110 Ds Darmstadtium * (281) ⊙	111 Rg Roentgenium * (281) ⊙	112 Cn Copernicium * (285) ⊙	113 Nh Nihonium * (286) ⊙	114 Fl Flerovium * (289) ⊙	115 Mc Moscovium * (289) ⊙	116 Lv Livermorium * (293) ⊙	117 Ts Tennessine * (294) ⊙	118 Og Oganesson * (294) ⊙

* Properties are largely predicted.

64 Gd Gadolinium 157.25	65 Tb Terbium 158.925	66 Dy Dysprosium 162.50	67 Ho Holmium 164.930	68 Er Erbium 167.259	69 Tm Thulium 168.934	70 Yb Ytterbium 173.04	71 Lu Lutetium 174.967
96 Cm Curium (247) ⊙	97 Bk Berkelium (247) ⊙	98 Cf Californium (251) ⊙	99 Es Einsteinium * (252) ⊙	100 Fm Fermium * (257) ⊙	101 Md Mendelevium * (258) ⊙	102 No Nobelium * (259) ⊙	103 Lr Lawrencium * (262) ⊙

GLOSSARY/GLOSARIO

Multilingual eGlossary The multilingual science glossary includes Arabic, Bengali, Chinese, English, Haitian Creole, Hmong, Korean, Portuguese, Russian, Tagalog, Urdu, and Vietnamese.

Pronunciation Key

Use the following key to help you sound out words in the glossary.

a	back (BAK)	**ew**	food (FEWD)
ay	day (DAY)	**yoo**	pure (PYOOR)
ah	father (FAH thur)	**yew**	few (FYEW)
ow	flower (FLOW ur)	**uh**	comma (CAHM uh)
ar	car (CAR)	**u** (+con)	rub (RUB)
e	less (LES)	**sh**	shelf (SHELF)
ee	leaf (LEEF)	**ch**	nature (NAY chur)
ih	trip (TRIHP)	**g**	gift (GIHFT)
i (i+con+e)	idea, life (i DEE uh, life)	**j**	gem (JEM)
oh	go (GOH)	**ing**	sing (SING)
aw	soft (SAWFT)	**zh**	vision (VIHZH un)
or	orbit (OR but)	**k**	cake (KAYK)
oy	coin (COYN)	**s**	seed, cent (SEED, SENT)
oo	foot (FOOT)	**z**	zone, raise (ZOHN, RAYZ)

Como usar el glosario en español:

1. Busca el termino en ingles que desees encontrar.
2. El termino en espanol, junto con la definicion, se encuentran en la columna de la derecha.

ENGLISH — A — ESPAÑOL

acceleration: rate of change of velocity; can be calculated by dividing the change in the velocity by the time it takes the change to occur.

aceleración: tasa de cambio de la velocidad; se calcula dividiendo el cambio en la velocidad por el tiempo que toma para que ocurra el cambio.

acid: substance that produces hydrogen ions (H^+) in a water solution.

ácido: sustancia que produce iones de hidrógeno (H^+) en una solución de agua.

acid precipitation: water with a pH below 5.6 that falls to Earth as rain or snow and can harm plants and animals and corrode buildings.

precipitación ácida: el agua con un pH inferior a 5,6 que cae a la Tierra como lluvia o nieve y puede dañar a las plantas y los animales, y oxida a los edificios.

acoustics: the study of sound.

acústica: el estudio del sonido.

air resistance: force that opposes the motion of objects that move through the air.

resistencia del aire: fuerza que se opone al movimiento de los objetos que se mueven por el aire.

alcohol: substituted hydrocarbon, such as ethanol, that is formed when –OH groups replace one or more hydrogen atoms in a hydrocarbon.

alcohol: hidrocarbonados sustituidos, como el etanol, que se forma cuando grupos –OH reemplazan a uno o más átomos de hidrógeno en un hidrocarburo.

allotropes: different molecular structures of the same element.

alótropos: estructuras moleculares diferentes de un mismo elemento.

alloy: a mixture of elements that has metallic properties.

aleación: una mezcla de elementos que tiene propiedades metálicas.

alpha particle: particle consisting of two protons and two neutrons that is emitted from a decaying atomic nucleus.

partícula alfa: partícula compuesta por dos protones y dos neutrones y que es emitida por un núcleo atómico en descomposición.

alternating current (AC): electric current that reverses its direction of flow in a regular pattern.

corriente alterna (CA): corriente eléctrica que invierte su dirección de flujo en un patrón regular.

amine: a substituted hydrocarbon with an amine group ($-NH_2$).

amina: un hidrocarburo sustituido con un grupo amino ($-NH_2$).

amplitude: a measure of the size of the disturbance of a wave, related to the energy that it carries.

amplitud: medida del tamaño del desplazamiento por una onda, indica la cantidad de la energía que transporta.

analog signal: an electric signal whose value changes smoothly over time.

aromatic compound: an organic compound that contains the benzene ring structure, most have a distinctive smell.

atom: the smallest particle of an element that still retains the properties of the element.

atomic number: number of protons in an atom's nucleus.

average atomic mass: weighted-average mass of an element's isotopes according to their natural abundance.

average speed: total distance an object travels divided by the total time it takes to travel that distance.

señal analógica: una señal eléctrica cuya value cambia fluidamente a lo largo del tiempo.

compuesto aromático: compuesto orgánico que contiene la estructura del anillo bencénico, la mayoría tiene una fragancia característica.

átomo: la partícula más pequeña de un elemento que mantiene las propiedades del elemento.

número atómico: número de protones en el núcleo de un átomo.

masa atómica promedio: masa media ponderada de los isótopos de un elemento en función de su abundancia natural.

velocidad promedio: distancia que recorre un objeto dividida por el tiempo que dura en recorrer dicha distancia.

B

balanced chemical equation: chemical equation with the same number of atoms of each element on both sides of the equation.

base: a substance that produces hydroxide ions **(OH^-)** in a water solution.

benzene: (C_6H_6) cyclic hydrocarbon whose carbon atoms are joined with alternating single and double bonds.

beta particle: high-energy electron that is emitted when a neutron decays into a proton.

bias: occurs when a scientist's expectations change how the results of an experiment are viewed.

binary compound: compound that is composed of two elements.

biomass: renewable organic matter from plants and animals, such as wood and animal manure, that can be burned to provide thermal energy.

boiling point: the temperature at which the pressure of the vapor of a liquid is equal to the external pressure acting on the surface of the liquid.

Boyle's law: states that the volume and pressure of a gas are related, such that if the temperature of a gas remains constant, an increase in volume causes a proportional decrease in the pressure.

buffer: solution that resists changes in pH when limited amounts of acid or base are added.

buoyancy: ability of a fluid, which include liquids and gases, to exert an upward force on an object immersed in it.

ecuación química equilibrada: ecuación química con el mismo número de átomos de cada elemento en ambos lados de la ecuación.

base: sustancia que forma iones de hidróxido (OH^-) en una solución de agua.

benceno: **(C_6H_6)** hidrocarburos cíclico cuyos átomos de carbono están unidos con una alternancia de enlaces simples y dobles.

partícula beta: electrón de alta energía que se emite cuando un neutrón se convierte en un protón.

predisposición: ocurre cuando las expectativas de un científico cambian la forma en que son vistos los resultados de un experimento.

compuesto binario: compuesto conformado por dos elementos.

biomasa: materia orgánica renovable que proviene de plantas y animales, tales como madera y estiércol animal, que puede ser incinerada para proveer energía térmica.

punto de ebullición: temperatura a la cual la presión del vapor de un líquido es igual a la presión externa que actúa sobre la superficie del líquido.

ley de Boyle: establece que el volumen y la presión de un gas son relacionados de manera que si la temperatura se mantiene constante, un aumento el el volumen causaría una disminución proporcional en la presión.

buffer: solución que se resiste a los cambios en el pH al añadir cantidades limitadas de ácido o base.

fuerza de flotación: capacidad de un fluido, líquido o gas, para ejercer una fuerza ascendente sobre un objeto inmerso en el fluido.

C

carbohydrate: group of biological compounds containing carbon, hydrogen, and oxygen with twice as many hydrogen atoms as oxygen atoms.

carrier wave: specific frequency that a radio station is assigned and uses to broadcast signals.

carrying capacity: maximum number of individuals of a given species that the environment can support.

catalyst: substance that speeds up a chemical reaction without being permanently changed itself.

carbohidrato: grupo de compuestos biológicos que contienen carbono, hidrógeno y oxígeno que contienen el doble de átomos de hidrógeno que de oxígeno.

onda transportadora: frecuencia específica que se le asigna a una estación de radio y que la usa para emitir señales.

capacidad de carga: máximo número de individuos de una especie que el medio ambiente puede apoyar.

catalizador: sustancia que acelera una reacción química sin cambiar el mismo permanentemente.

GLOSSARY/GLOSARIO

centripetal acceleration: acceleration of an object toward the center of a curved or circular path.

centripetal force: a force that is directed toward the center of a curved or circular path.

ceramics: versatile materials made from dried clay or clay-like mixtures with customizable properties; produced by a process in which an object is molded and then heated to high temperatures, increasing its density.

chain reaction: series of fission reactions caused by the release of additional neutrons in every step.

charging by contact: the transferring of electrical charge between objects by touching or rubbing.

charging by induction: the rearranging of electrons on a neutral object caused by bringing a charged object close to it.

Charles's law: states that the temperature and volume of a gas are related such that, if the pressure is constant, an increase in temperature will produce a proportionate increase in the volume.

chemical bond: force that holds atoms together in a compound.

chemical change: change of one substance into a new substance.

chemical equation: shorthand method used to describe chemical reactions using chemical formulas and other symbols.

chemical formula: chemical shorthand that uses symbols to tell what elements are in a compound and their ratios.

chemical potential energy: energy that is due to chemical bonds.

chemical property: any characteristic of a substance, such as flammability, that can be observed that produces a new substance.

chemical reaction: process in which one or more substances are changed into new substances.

circuit: closed conducting loop through which an electric current can flow.

cochlea: spiral-shaped, fluid-filled structure in the inner ear that converts sound waves to nerve impulses.

coefficient: number in a chemical equation that represents the number of units of each substance taking part in a chemical reaction.

coherent light: light of one wavelength that travels in one direction with a constant distance between the corresponding crests of the waves.

collision model: explains why certain factors affect reaction rates, states that particles must collide in order to react.

colloid: heterogeneous mixture whose particles never settle.

combustion reaction: a type of chemical reaction that occurs when a substance reacts with oxygen to produce energy in the form of heat and light.

aceleración centrípeta: aceleración de un objeto dirigida hacia el centro de un trayecto curvo o circular.

fuerza centrípeta: fuerza dirigida hacia el centro de un trayecto curvo o circular.

cerámicas: materiales versátiles hechos con arcilla seca o mezclas parecidas a la arcilla con propiedades adaptables, producidos mediante un proceso en el cual un objeto es moldeado y luego sujeto a altas temperaturas, aumentando su densidad.

reacción en cadena: serie continua de reacciones de fisión causado por la liberación de neutrones en cada paso.

cargar por contacto: transferir carga eléctrica entre objetos por contacto o frotaciòn.

cargar por inducción: redistribuir los electrones de un objeto neutro debido al acercarle a un objeto con carga.

ley de Charles: establece que la temperatura y el volumen de un gas son relacionados de tal manera que sin un cambio de la presión, un aumento de temperatura poducirá un aumento proporcional en el volumen.

enlace químico: fuerza que mantiene a los átomos juntos en un compuesto.

cambio químico: transformación de una sustancia en una nueva sustancia.

ecuación química: método simplificado utilizado para describir reacciones químicas por medio de fórmulas químicas y otros símbolos.

fórmula química: nomenclatura química que usa símbolos para expresar cuales elementos están en un compuesto y en cuales proporciones.

energía química potencial: energía debido a los enlaces químicos.

propiedad química: cualquier característica de una sustancia, como por ejemplo la combustibilidad, que puede ser observado y produce un nuevo sustancia.

reacción química: proceso en el cual una o más sustancias son cambiadas por nuevas sustancias.

circuito: circuito conductor cerrado a través del cual puede fluir una corriente eléctrica.

cóclea: estructura en el oído interno, con forma de espiral y llena de un fluido, la cual convierte las ondas sonoras en impulsos nerviosos.

coeficiente: número en una ecuación química que representa el número de unidades de cada una de las sustancias que participan en una reacción química.

luz coherente: luz de una sola longitud de onda que viaja en una sola dirección con una distancia constante ente las crestas de las olas.

modelo de colisión: explica por qué ciertos factores afectan las velocidades de reacción, afirma que las partículas deben chocar para que reaccionen.

coloide: mezcla heterogénea cuyas partículas nunca se sedimentan.

reacción de combustión: un tipo de reacción química que ocurre cuando una sustancia reacciona con oxígeno y produce energía en forma de calor y luz.

composite: mixture of two materials, one of which is embedded or layered in the other.

compound: substance in which the atoms of two or more elements are combined in a fixed proportion.

compound machine: machine that is a combination of two or more simple machines.

compression: denser region of a longitudinal wave.

concave lens: a lens that is thicker at the edges than in the middle; causes light rays to diverge and forms reduced, upright, virtual images; often used in combination with other lenses.

concave mirror: a reflective surface that curves inward and can magnify objects or create real images.

concentration: the amount of solute actually dissolved in a given amount of solvent.

conduction: transfer of thermal energy by collisions between the particles that make up matter.

conductivity: property of metals and alloys that allows heat or electrical charges to pass through the material easily.

conductor: material, such as copper wire, through which electrons can move easily.

constant: in an experiment, a variable that does not change.

control: standard used for comparison of test results in an experiment.

convection: transfer of thermal energy in a fluid by the movement of warmer and cooler fluid from one place to another.

convex lens: a lens that is thicker in the middle than at the edges and can form real or virtual images.

convex mirror: a reflective surface that curves outward, away from the viewer, and forms a reduced, upright, virtual image.

cornea: transparent covering on the eyeball through which light enters the eye.

covalent bond: attraction formed between atoms when they share electrons.

crest: the highest point on a transverse wave.

material compuesto: mezcla de dos materiales, uno de los cuales está integrado por capas o fundido en el otro.

compuesto: sustancia en la que los átomos de dos o más elementos se combinan en una proporción fija.

máquina compuesta: máquina compuesta por dos o más máquinas simples.

compresión: la región densa de una onda longitudinal.

lente cóncavo: lente que es más grueso en los bordes que en el centro; hace que los rayos de luz se desvíen y forma imágenes reducidas, verticales y virtuales, frecuentemente se utiliza en combinación con otros lentes.

espejo cóncavo: superficie reflexiva que se curva hacia el interior puede aumentar imágenes o crear imágenes reales.

concentración: la cantidad de soluto que se disuelve en una cantidad dada de disolvente.

conducción: transferencia de energía térmica por colisiones entre las partículas que componenuna materia.

conductividad: propiedad de los metales y aleaciones que permite fácilmente el paso de calor o cargas eléctricas a través del material.

conductor: material, como el alambre de cobre, a través del cual los electrones se pueden pasar con facilidad.

constante: en un experimento, una variable que no cambia.

control: estándar usado para la comparación de resultados de pruebas en un experimento.

convección: transferencia de energía térmica en un fluido por el movimiento de fluidos con mayores y menores temperaturas de un lugar a otro.

lente convexo: lente que es más grueso en el centro que en los bordes y que puede formar imágenes reales o virtuales.

espejo convexo: una superficie reflectante que se curva hacia afuera, lejos del espectador, y forma una imagen virtual, reducida, en posición vertical.

córnea: cubierta transparente del globo ocular a través de la cual entra la luz al ojo.

enlace covalente: atracción formada entre átomos que comparten electrones.

cresta: la punta más alta en una onda transversal.

D

decibel: unit for sound intensity; abbreviated dB.

decomposition reaction: chemical reaction in which one substance breaks down into two or more substances.

density: mass per unit volume of a material.

deoxyribonucleic acid: also called DNA, a type of essential biological compound found in the nuclei of cells that codes and stores genetic information.

dependent variable: factor that changes as a result of changes in the other variables.

decibel: unidad que mide la intensidad del sonido; se abrevia dB.

reacción de descomposición: reacción química en la cual una sustancia se descompone en dos o más sustancias.

densidad: masa por unidad de volumen de un material.

ácido desoxirribonucleico: también conocido como ADN, compuesto biológico esencial encontrado en el núcleo de las células, codifica y almacena información genética.

variable dependiente: factor que varía como resultado de los cambios en las otras variables.

depolymerization: process using heat or chemicals to break a polymer chain into its monomer fragmentss.

despolimerización: proceso en el que se utilizan calor o químicos para descomponer una cadena de polímeros en sus fragmentos de monómeros.

diatomic molecule: a molecule that consists of two atoms of the same element, joined by a covalent bond.

molécula diatómica: molécula que consta de dos átomos del mismo elemento, unidas por un enlace covalente.

diffraction: the bending of waves around an obstacle; can also occur when waves pass through a narrow opening.

difracción: un cambio de dirección de las ondas al rozar el borde de un obstáculo, la cual también puede ocurrir cuando éstas pasan a través de una abertura angosta.

diffusion: spreading of particles throughout a given volume until they are uniformly distributed.

difusión: propagación de partículas en la totalidad de un volumen determinado hasta que se distribuyen de manera uniforme.

digital signal: an electric signal with only two possible values: ON and OFF.

señal digital: una señal eléctrica, con sólo dos valores posibles: ON y OFF.

direct current (DC): electric current that flows in only one direction.

corriente directa (CD): también conocido como corriente continua (CC), corriente eléctrica que fluye en una sola dirección.

displacement: distance and direction of an object's change in position from the starting point.

desplazamiento: distancia y dirección del cambio de posición de un objeto desde el punto inicial.

dissociation: process in which an ionic compound separates into its positive and negative ions in solution.

disociación: proceso en el cual un compuesto iónico se separa en sus iones positivos y negativos en una solución.

distance: the length an object travels, measured in SI units of meters.

distancia: la longitud de un objeto de viajes, medido en unidades SI de metros.

distillation: process that can separate two substances in a mixture by evaporating a liquid and recondensing its vapor.

destilación: proceso que puede separar dos sustancias de una mezcla por medio de la evaporación de un líquido y la recondensación de su vapor.

doping: process of adding impurities to a semiconductor to modify its conductivity.

dopaje: proceso que consiste en añadir impurezas a un semiconductor para modificar a su conductividad.

Doppler effect: change in frequency that occurs when a source is moving relative to an observer.

efecto Doppler: cambio en la frecuencia que se produce cuando una fuente se mueve respecto a un observador.

double-displacement reaction: reaction in which two ionic compounds in solution are combined, can produce a precipitate, water, or a gas.

reacción de doble desplazamiento: reacción química en cual se combinan dos compuestos iónicos en una solución, puede producir un precipitado, agua o gas.

ductile: property of metals and alloys that allows them to be drawn into wires.

ductibilidad: propiedad de los metales y aleaciones que les permiten ser convertidos en alambres.

E

eardrum: tough membrane in the outer ear that is about 0.1 mm thick and transmits sound vibrations into the middle ear.

tímpano: membrana fuerte del oído externo que tiene aproximamente 0.1 mm de grueso y transmite las vibraciones del sonido al oído medio.

echolocation: process by which objects are located by emitting sounds and interpreting the sound waves that are reflected from those objects.

ecolocalización: proceso por el cual los objetos son localizados por emitir sonidos e interpretando las ondas de sonido que se reflejan.

efficiency: ratio of the output work done by the machine to the input work done on the machine, expressed as a percentage.

eficiencia: relación del trabajo efectuado por una máquina y el trabajo hecho en ésta, expresada en porcentaje.

elastic potential energy: energy that is stored by compressing or stretching an object.

energía elástica potencial: energía almacenada por compresionar o estrechar un objeto.

electrical power: rate at which electrical energy is converted to another form of energy; expressed in watts (W).

potencia eléctrica: velocidad a la cual la energía eléctrica se convierte en otra forma de energía; se expresa en vatios (W).

electric circuit: a closed path that electric current follows.

circuito eléctrica: un camino cerrado que sigue la corriente electrica.

electric current: the net movement of electric charges in a single direction, measured in amperes (A).

corriente eléctrica: movimiento neto de cargas eléctricas en una sola dirección, medido en amperios (A).

electric field: a region surrounding every electric charge in which a force of attraction or repulsion is exerted on other electric charges.

campo eléctrico: una región en torno a toda carga eléctrica en la que una fuerza de de atración o repulsión se ejerce hacia otras cargas eléctricas.

electric motor: device that converts electrical energy to mechanical energy by using the magnetic forces between an electromagnet and a permanent magnet to make a shaft rotate.

motor eléctrico: dispositivo que convierte la energía eléctrica en energía mecánica por las fuerzas magnéticas entre un electroimán y un imán permanente, y hace que un eje gire.

electrolyte: compound that breaks apart in water, producing charged particles (ions) that can conduct electricity.

electrolito: compuesto que se descompone en agua y así produce partículas cargadas (iones) que pueden conducir electricidad.

electromagnet: temporary magnet created when there is a current in a wire coil.

electroimán: imán temporal crea cuando hay una corriente en una bobina de alambre.

electromagnetic force: the attractive or repulsive force between electric charges and magnets.

fuerza electromagnética: fuerza de atracción o repulsión entre cargas eléctricas y imánes.

electromagnetic induction: process by which electric current is produced in a wire loop by a changing magnetic field.

inducción electromagnética: proceso mediante el cual se produce la corriente eléctrica en un aro de alambre por un campo magnético variable.

electromagnetic wave: waves created by vibrating electric charges; consists of vibrating electric and magnetic fields, and can travel through a vacuum or through matter.

ondas electromagnéticas: olas creadas por la vibración de cargas eléctricas; consta de vibrar los campos eléctricos y magnéticos, y puede viajar a través del vacío o través de la materia.

electromagnetism: the interaction between electric charges and magnets.

electromagnetismo: la interacción entre las cargas eléctricas y los imánes.

electron: particle with an electric charge of 1–, surrounds the nucleus of an atom.

electrón: partícula que rodea con carga eléctrica de 1-, rodea el núcleo de un átomo.

electron cloud: area around the nucleus of an atom where the atom's electrons are most likely to be found.

nube de electrones: área alrededor del núcleo de un átomo en donde hay más probabilidad de encontrar los electrones de los átomos.

electron dot diagram: uses the symbol for an element and dots representing the number of electrons in the element's outer energy level.

diagrama de punto de electrones: usa el símbolo de un elemento y puntos que representan el número de electrones en el nivel de energía externo del elemento.

electroscope: a device, sometimes consisting of two leaves of metallic foil, used to detect electric charge.

electroscopio: un aparato, a veces con hojas metales, que detecta a carga eléctrica.

element: substance with atoms that are all alike.

elemento: sustancia en la cual todos los átomos son iguales.

endergonic reaction: chemical reaction that requires energy input in the form of light, thermal energy or electricity in order to proceed.

reacción endergónica: reacción química que requiere entrada de energía en la forma de luz, energía térmica o electricidad para proceder.

endothermic reaction: chemical reaction that requires thermal energy in order to proceed.

reacción endotérmica: reacción química que requiere energía de energía térmica para proceder.

energy: the ability to cause change, measured in joules.

energía: la habilidad para efectuar un cambio, medida en julio

equilibrium: state in which forward and reverse reactions or processes occur at equal rates.

equilibrio: estado en que reacciones o procesos avancen y retrocesan a tasas iguales.

ester: a substituted hydrocarbon with a –COOC– group.

ester: un hidrocarburo sustituido con un –COOC–.

exergonic reaction: chemical reaction that releases some form of energy, such as light or thermal energy.

reacción exergónica: reacción química que libera una forma de energía, tal como luz o energía térmica.

exothermic reaction: chemical reaction in which energy is primarily given off in the form of thermal energy.

reacción exotérmica: reacción química en la cual la energía es inicialmente emitida en forma de energía térmica.

experiment: organized procedure for testing a hypothesis; tests the effect of one thing on another under controlled conditions.

experimento: procedimiento organizado para probar una hipótesis; prueba el efecto de una cosa sobre otra bajo condiciones controladas.

F

field: a region of space in which every point has a physical quantity, such as a force.

campo: una región en cual cada punto se puede expresar por una cantidad, por ejemplo, una fuerza.

filter: a transparent material that selectively transmits light.

filtro: un material transparente que transmite la luz de forma selectiva.

first law of thermodynamics: states that if the mechanical energy of a system is constant, the increase in the thermal energy of the system equals the sum of the thermal energy transferred into the system and the work done on the system.

primera ley de la termodinámica: establece que si la energía mecánica del sistema es constante, el aumento de la energía térmica del sistema es igual a la suma de la energía térmica transferida al sistema y el trabajo realizado en el sistema.

fission: process of splitting an atomic nucleus into two or more nuclei with smaller masses, releasing large amounts of energy.

fisión: proceso de división en el cual un núcleo atómico de divide entre dos o más núcleos con masas más pequeñas, produciendo grandes cantidades de energía.

fluorescent light: light generated by using phosphors to convert ultraviolet radiation to visible light.

luz fluorescente: luz generada mediante el uso de fósforo para convertir la radiación ultravioleta a la luz visible.

focal length: distance from the center of a lens or mirror to the focal point.

longitud focal: distancia desde el centro de un lente o espejo al punto focal.

focal point: the point on the optical axis of a curved mirror or lens where light rays that are initially parallel to the optical axis converge after striking the mirror or lens.

punto focal: el punto en el eje óptico de un espejo o lente curvo en el cual los rayos de luz, que inicialmente son paralelos al eje óptico, convergen después de chocar al espejo o lente.

force: a push or pull exerted on an object.

fuerza: impulso o tracción sobre un objeto.

fossil fuel: oil, natural gas, and coal; formed from the remains of ancient plants and animals that were buried and altered over millions of years.

combustibles fósiles: petróleo, gas natural y carbón; formado a partir de los restos de antiguas plantas y animales que fueron enterrados y alterados durante millones de años.

free fall: describes the fall of an object on which only the force of gravity is acting.

caida libre: describe la caída de un objeto sobre cual la única fuerza que le actua es la de la gravedad.

frequency: the number of wavelengths that pass a fixed point each second; is expressed in hertz (Hz).

frecuencia: el número de longitudes de onda que pasan por un punto fijo en un segundo; se expresa en hercios (Hz).

friction: force that opposes the sliding motion between two touching surfaces.

fricción: fuerza que se opone al movimiento deslizante entre dos superficies en contacto.

fusion: reaction in which two or more atomic nuclei form a nucleus with a larger mass, releasing large amounts of energy.

fusión: proceso en la cual dos o más núcleos atómicos forman un núcleo con mayor masa, produciendo grandes cantidades de energía.

G

galvanometer: a device that uses an electromagnet to measure electric current.

galvanómetro: un dispositivo que utiliza y un electroimán para medir la corriente eléctrica.

gamma ray: electromagnetic wave with a wavelength less than about 100 trillionths of a meter; usually emitted from a decaying atomic nucleus.

rayo gama: onda electromagnética con longitud de onda menor a cien trillonésimas de un metro, generalmente emitido por un núcleo atómico en descomposición.

Geiger counter: radiation detector that produces a click or a flash of light when a charged particle is detected.

contador Geiger: detector de radiación que produce un sonido seco o un destello de luz al detectar una partícula cargada.

generator: device that uses electromagnetic induction to convert mechanical energy to electrical energy.

generador: dispositivo que usa inducción electromagnética para convertir energía mecánica en energía eléctrica.

geothermal energy: thermal energy contained in and around magma; can be converted by a power plant into electrical energy.

energía geotérmica: energía térmica en y alrededor del magma, la cual se puede convertir mediante una planta industrial en energía eléctrica.

Global Positioning System (GPS): a system of satellites and ground monitoring stations that enable a receiver to determine its location at or above Earth's surface.

Sistema de Posicionamiento Global (GPS): sistema de satélites y estaciones de monitoreo en tierra que permiten que un receptor determine su ubicación en o sobre la superficie terrestre.

graph: visual display of information or data.

gráfica: presentación visual de información que puede suministrar una forma rápida de comunicar gran cantidad de información.

gravitational potential energy (GPE): energy that is due to the gravitational force between objects.

energía gravitacional potencial: energía debida a la fuerza gravitoria entre objetos

gravity: attractive force between two objects that depends on the masses of the objects and the distance between them.

gravedad: fuerza de atracción entre dos objetos que depende de las masas de los objetos y de la distancia entre ellos.

group: vertical column in the periodic table.

grupo: columna vertical en la tabla periódica.

H

half-life: amount of time it takes for half the nuclei in a sample of a radioactive isotope to decay.

vida media: tiempo requerido para que se descomponga la mitad de los núcleos de una muestra de isótopo radiactivo.

hazardous waste: wastes that are poisonous, cause cancer, or can catch fire.

residuo peligroso: residuo tóxico, provoca cáncer, o puede incendiarse.

heat: energy that is transferred between objects due to a temperature difference between those objects.

calor: la energía que se transfiere entre los objetos debido a una diferencia de temperatura entre esos objetos.

heat engine: device that converts some thermal energy into mechanical energy.

motor de calor: dispositivo que convierte térmica en energía mecánica.

heat of fusion: amount of energy required to change a substance from the solid phase to the liquid phase.

calor de fusión: cantidad de energía necesaria para cambiar una sustancia del estado sólido al líquido.

heat of vaporization: the amount of energy required for a liquid at its boiling point to become a gas.

calor de vaporización: cantidad de energía necesaria para que un líquido en su punto de ebullición se convierta en gas.

heterogeneous mixture: substance in which its different components are easily distinguished.

mezcla heterogénea: sustancia en la que los diferentes componentes se distinguen fácilmente.

holography: technique that produces a complete three-dimensional photographic image of an object.

holografía: técnica que produce una imagen fotográfica tridimensional completa de un objeto.

homogeneous mixture: substance containing two or more components that are blended uniformly so that individual components are indistinguishable with a microscope.

mezcla homogénea: sustancia que contiene dos o más componentes que se combina de manera uniforme a fin de que los componentes individuales no se pueden distinguir con un microscopio.

hydrate: compound that has water chemically attached to its atoms and written into its chemical formula.

hidrato: compuesto que contiene agua químicamente apegado a sus átomos y escrita en su fórmula química.

hydrocarbon: compound containing only carbon and hydrogen atoms.

hidrocarburo: compuesto que contiene únicamente átomos de carbono e hidrógeno.

hydroelectricity: electricity produced from the energy of moving water.

hidroelectricidad: electricidad producida por la energía de movimiento de agua.

hydronium ion: H_3O^+ ion, forms when an acid dissolves in water and H^+ ions interact with water.

ion de hidronio: ion H_3O^+, se forma cuando un ácido se disuelve en agua y el ion H^+ interactúa con el agua.

hydroxide ion: OH^- ion, forms when a base dissolves in water.

hydroxide ion: ion OH^-, forma cuando una base disuelve en agua.

hypothesis: possible explanation for a problem using what is known and what is observed.

hipótesis: explicación posible para resolver un problema al utilizar lo que conoce y lo que observa.

I

incandescent light: light produced by heating a piece of metal, usually tungsten, until it glows.

luz incandescente: luz que se produce al calentar una pieza de metal, generalmente tungsteno, hasta que brille.

incoherent light: light that can contain more than one wavelength, travel in more than one direction, and has varying distances between the corresponding crests of the waves.

luz incoherente: luz que puede contener más de una longitud de onda, viaja en más de una dirección, y tiene distancias variables entre las crestas correspondientes de las olas.

independent variable: factor that, as it changes, affects the measure of another variable.

variable independiente: factor que, a medida que cambia, afecta la medida de otra variable.

index of refraction: property of a material indicating how much the speed of light is reduced in the material compared to the speed of light in a vacuum.

índice de refracción: propiedad de un material que indica cuanto la velocidad de la luz se reduce en el material en comparación con la velocidad de la luz en el vacío.

indicator: organic compound that changes color in acids and bases.

indicador: compuesto orgánico que cambia de color en presencia de ácidos y bases.

inertia: tendency of an object to resist any change in its motion.

inercia: tendencia de un objeto a resistir cualquier cambio en su movimiento.

infrared wave: electromagnetic wave with a wavelength between about 1 mm and 700 billionths of a meter.

inhibitor: substance that slows down a chemical reaction or prevents it from occurring by combining with a reactant.

instantaneous speed: speed of an object at a given point in time; is constant for an object moving with constant speed, and changes with time for an object that is slowing down or speeding up.

insulator: material in which electrons and thermal energy are not able to move easily.

integrated circuit: tiny chip that can contain millions of transistors, diodes, and other components.

intensity: amount of energy that flows through a certain area in a specific amount of time.

interference: the process of two or more waves overlapping and combining to form a new wave.

internal combustion engine: heat engine that burns fuel inside the engine in chambers or cylinders.

ion: charged particle that has either more or fewer electrons than protons.

ionic bond: the force of attraction between the opposite charges of the ions in an ionic compound.

ionization: process in which electrolytes dissolve in water and separate into charged particles.

isomers: compounds with identical chemical formulas but different molecular structures and shapes.

isotope: atom of an element that has a specific number of neutrons.

onda infrarroja: onda electromagnética que tiene una longitud de onda entre aproximadamente 1 mm y 700 billonésimas de metro.

inhibidor: sustancia que reduce una reacción química o previene que ocurra por combinar con un reactivo.

velocidad instantánea: velocidad de un objeto en un punto dado en el tiempo; es constante para un objeto que se mueve a una velocidad constante y cambia con el tiempo en un objeto que está reduciendo o aumentando su velocidad.

aislador: material a través del cual los electrones y la energía térmica no se pueden transferir con facilidad.

circuito integrado: pedazo minúsculo que puede contener millones de transistores, diodos y otros componentes.

intensidad: cantidad de energía que fluye a través de cierta área en un tiempo específico.

interferencia: el proceso de dos o más ondas se superponen y se combinan para formar una nueva ola.

motor de combustión interna: motor de calor que quema combustible en su interior en cámaras o cilindros.

ion: partícula cargada que tiene ya sea más o menos electrones que protones.

enlace iónico: fuerza de atracción entre las cargas opuestas de los iones en un compuesto iónico.

ionización: proceso en el cual los electrolitos se disuelven en agua y se separan en partículas cargadas.

isómeros: compuestos con fórmulas químicas idénticas pero con estructuras moleculares diferentes.

isótopo: átomo de un elemento que tiene un cierto número de neutrones.

J

joule: SI unit of work and energy.

julio: unidad del SI de trabajo y energía.

K

kinetic energy: energy a moving object has because of its motion; determined by the mass and speed of the object.

kinetic theory: explanation of the behavior of particles in gases; states that matter is made of constantly moving particles that collide without losing energy.

energía cinética: energía que tiene un cuerpo debido a su movimiento, se determina pore la masa y velocidad del objeto.

teoría cinética: explicación del comportamiento de las partículas en un gas, la cual establece que las sustancias son compuestas de partículas en constante movimiento que se chocan sin perder energía.

L

law of conservation of charge: states that charge can be transferred from one object to another but it cannot be created or destroyed.

law of conservation of energy: states that energy cannot be created or destroyed.

ley de la conservación de carga: estados de que se puede ser transferida de un objeto a otro, pero no puede ser creada o destruida.

ley de la conservación de energía: establece que la energía no puede crearse ni destruirse.

law of conservation of mass: states that the mass of all substances present before a chemical change equals the mass of all the substances remaining after the change.

law of conservation of momentum: states that if no external forces act on a group of objects, their total momentum does not change.

Le Châtelier's principle: states that if a stress is applied to a reaction at equilibrium, the reaction shifts in the direction opposite of the stress.

lever: simple machine consisting of a bar free to pivot about a fixed point called the fulcrum.

linearly polarized light: light whose magnetic field vibrates in only one direction.

lipid: group of biological compounds that contains the same elements as carbohydrates but in different proportions, includes saturated and unsaturated fats and oils.

longitudinal wave: a wave in which the matter in the medium moves back and forth along the direction that the wave travels.

loudness: human perception of sound volume, depends primarily on intensity.

luster: property of metals and alloys that describes having a shiny appearance.

ley de conservación de la masa: establece que la masa de todas las sustancias presente antes de un cambio químico es igual a la masa de todas las sustancias resultantes después del cambio.

ley de conservación del momentum: establece que si no actúan fuerzas externas sobre un grupo de objetos, su cantidad de movimiento total no cambia.

principio de Le Châtelier: establece que si una tensión se aplica a una reacción en el equilibrio, la reacción se acelera en la dirección opuesta de la tensión.

palanca: máquina simple que consiste de una barra que puede girar sobre un punto fijo llamado pivote.

luz polarizada lineal: luz cuya campo magnético vibra en una sola dirección.

lípido: grupo de compuestos biológicos que contiene los mismos elementos que los hidratos de carbono, pero en diferentes proporciones, incluye grasas saturadas e insaturadas y aceites.

onda longitudinal: onda por la cual la materia en el medio se mueve para adelante y para atrás en la dirección en que viaja la onda.

volumen de sonido: percepción humana de la fuerza del sonido, depende primariamenta en la intensidad.

lustre: propiedad de los metales y aleaciones que describe la manera come brilla.

M

machine: device that makes doing work easier by increasing the force applied to an object, changing the direction of an applied force, or increasing the distance over which a force can be applied.

magnetic domain: group of atoms in a magnetic material in which the magnetic poles of the atoms are aligned in the same direction.

magnetic field: region surrounding a magnet that exerts a force on other magnets and objects made of magnetic materials.

magnetic pole: region on a magnet where the magnetic force exerted by a magnet is strongest; like poles repel and opposite poles attract.

magnetism: the properties and interactions of magnets.

malleable: property of metals and alloys that allows them to be hammered or rolled into thin sheets.

mass: amount of matter in an object.

mass number: sum of the number of protons and neutrons in an atom's nucleus.

matter: anything that takes up space and has mass.

mechanical advantage (MA): ratio of the output force exerted by a machine to the input force applied to the machine.

mechanical energy: sum of the potential energy and kinetic energy of the objects in a system.

mechanical wave: a wave that can only travel through matter.

medium: matter through which a wave travels.

máquina: artefacto que facilita la ejecución del trabajo por aumentar la fuerza que se aplica a un objeto, cambiar la dirección de una fuerza aplicada o aumentar la distancia sobre la cual se puede aplicar una fuerza.

dominio magnético: grupo de átomos en un material magnético en el cual los polos magnéticos de los átomos están alineados en la misma dirección.

campo magnético: región que rodea a un imán que ejerce una fuerza sobre otros imanes y objetos hechos de materiales magnéticos.

polo magnético: zona en un imán en donde la fuerza magnética ejercida por un imán es la más fuerte; los polos-iguales se repelen y los polos-opuestos se atraen.

magnetismo: propiedades e interacciones de los imanes.

maleable: la propiedad de los metales y aleaciones que les permite ser martillados o enrollados en láminas delgadas.

masa: cantidad de materia en un objeto.

número de masa: suma del número de protones y neutrones en el núcleo de un átomo.

materia: todo lo que ocupa espacio y tiene masa.

ventaja mecánica (MA): relación de la fuerza ejercida por una máquina y la fuerza aplicada a dicha máquina.

energía mecánica: suma de la energía potencial y energía cinética de los objetos en un sistema.

onda mecánica: una ola que sólo puede viajar a través de la materia.

medio: materia a través de la cual viaja una onda.

melting point: temperature at which a solid begins to liquefy.

metal: element that is shiny, malleable, ductile, and a good conductor of heat and electricity.

metallic bonding: occurs because some electrons move freely among a metal's positively charged ions, explains properties such as ductility and the ability to conduct electricity.

metalloid: element that shares some properties with metals and some with nonmetals.

microscope: instrument that uses two convex lenses to magnify small, close objects.

microwave: electromagnetic wave with wavelength between about 0.1 mm and 30 cm.

mirage: image of a distant object produced by the refraction of light through air layers of different densities.

model: can be used to represent an idea, object, or event that is too big, too small, too complex, or too dangerous to observe or test directly.

modulation: process of adding a signal to a carrier wave by altering the carrier wave's amplitude, frequency, or other properties.

molar mass: the mass in grams of one mole of a substance.

mole: SI unit for quantity equal to 6.022×10^{23} units of that substance.

molecule: a neutral particle that forms as a result of electron sharing among atoms.

momentum: property of a moving object that equals its mass times its velocity.

monomer: small molecule that can combine with itself repeatedly to form a long chain.

motion: a change in an object's position relative to a reference point.

music: collection of sounds deliberately used in a regular pattern.

punto de fusión: temperatura a la cual un sólido comienza a licuarse.

metal: elemento que por lo general es brillante, maleable, dúctil, y un buen conductor del calor y la electricidad.

enlace metálico: ocurre debido a que algunos electrones se mueven libremente entre los iones de cargados positiva de un metal y explica propiedades tales como la ductibilidad y la capacidad para conducir electricidad.

metaloide: elemento que tiene algunas propiedades de los metales y algunas de los no metales.

microscopio: instrumento que usa dos lentes convexos para amplificar objetos pequeños y cercanos.

microonda: onda electromagnética con longitud de onda entre aproximadamente 0.1 mm y 30 cm.

espejismo: imagen de un objeto distante producida por la refracción de la luz a través de capas de aire de diferentes densidades.

modelo: puede ser usado para representar una idea, objeto o evento que es demasiado grande, demasiado pequeño, demasiado complejo o demasiado peligroso para ser observado o probado directamente.

modulación: proceso de agregar una señal a una onda portadora mediante la alteración de la amplitud de la onda portadora, la frecuencia, u otra propiedad.

masa molar: la masa en gramos de un mol de una sustancia.

mol: unidad SI de cantidad igual a $6{,}022 \times 10^{23}$ unidades de dicha sustancia.

molécula: partícula neutra que se forma al compartir electrones entre átomos.

momentum: propiedad de un objeto en movimiento que es igual a su masa por su velocidad.

monómero: pequeña molécula que se puede formar una cadena por combinar consigo misma repetidamente.

movimiento: un cambio de puesto en relación con un punto de referencia.

música: colección de sonidos que se usan deliberadamente en un patrón regular.

N

net force: sum of all of the forces that are acting on an object.

neutralization: chemical reaction that occurs when the H_3O^+ ions from an acid react with the OH^- ions from a base to produce water molecules and a salt.

neutron: electrically neutral particle inside the nucleus of an atom.

Newton's first law of motion: states that an object moving at a constant velocity keeps moving at that velocity unless an unbalanced force acts on it.

Newton's second law of motion: states that the acceleration of an object is in the same direction as the net force on the object, and that the acceleration equals the net force divided by its mass.

fuerza neta: suma de las fuerzas que actúan sobre un objeto.

neutralización: reacción química que ocurre cuando los iones H_3O^+ de un ácido reaccionan con los iones OH^- de una base para producir moléculas de agua.

neutrón: partícula con carga eléctrica neutral del núcleo de un átomo.

primera ley del movimiento de Newton: establece que un objeto con una velocidad constante continuará a menos que una fuerza neta se interpone.

segunda ley de movimiento de Newton: establece que la aceleración de un objeto es en la misma dirección que la fuerza neta del objeto y que la aceleración es igual a la fuerza neta dividida por su masa.

Newton's third law of motion: states that when one object exerts a force on a second object, the second object exerts a force on the first object that is equal in strength and in the opposite direction.

tercera ley de movimiento de Newton: establece que cuando un objeto ejerce una fuerza sobre un segundo objeto, el segundo objeto ejerce una fuerza igual de fuerte sobre el primer objeto y en dirección opuesta.

node: a point in a standing wave at which the interfering waves always cancel.

nodo: un punto en una onda estacionaria en la que las olas siempre interfiriendo cancelar.

nonelectrolyte: substance that does not ionize in water and cannot conduct electricity.

no electrolito: sustancia que no se ioniza en el agua y no puede conducir electricidad.

nonmetal: element that usually is a gas or brittle solid at room temperature, is not malleable or ductile, is a poor conductor of heat and electricity, and typically is not shiny.

no metal: elemento que por lo general es un gas o un sólido frágil a temperatura ambiente, no es maleable o dúctil, es mal conductor del calor y la electricidad, y por lo general no es brillante.

nonpolar bond: a covalent bond in which electrons are shared equally by both atoms.

enlace no polar: enlace covalente en el cual los electrones son compartidos por igual por ambos átomos.

nonpolar molecule: molecule that shares electrons equally and does not have oppositely charged ends.

molécula no polar: molécula que comparte equitativamente los electrones y que no tiene extremos con cargas opuestas.

nonrenewable resources: natural resources, such as fossil fuels, that cannot be replaced by natural processes as quickly as they are used.

recursos no renovables: recursos naturales, tales como combustibles fósiles, que no son reemplazados por procesos naturales tan pronto como son usados.

nuclear reactor: an apparatus in which controlled nuclear chain reactions generate electricity.

reactor nuclear: aparato en lo que las reacciones nucleares controladas generan la electricidad.

nuclear waste: radioactive by-product that results when radioactive materials are used.

desperdicio nuclear: subproducto radioactivo que resulta del uso de materiales radiactivos.

nucleic acid: essential organic polymer that controls the activities and reproduction of cells.

ácido nucleico: polímero orgánico esencial que controla las actividades y la reproducción de las células.

nucleotide: complex, organic molecule that makes up DNA; contains an organic base, a phosphoric acid unit, and a sugar.

nucleótido: molécula orgánica compleja que compone el ADN, contiene una base orgánica, una unidad de ácido fosfórico y un azúcar.

nucleus: the small, positively charged center of an atom, contains protons and neutrons.

núcleo: el pequeño centro de carga positiva de un átomo, contiene protones y neutrones.

O

Ohm's law: states that the current in a circuit equals the voltage difference divided by the resistance.

ley de Ohm: establece que la corriente en un circuito es igual a la diferencia de voltaje dividida entre la resistencia.

opaque: material that absorbs or reflects all light and does not transmit any light.

opaco: material que absorbe o refleja toda la luz pero no la transmite.

optical axis: imaginary straight line that is perpendicular to the surface at the center of a mirror or lens.

eje óptico: línea recta imaginaria que es perpendicular a la superficie en el centro de un espejo o lente.

optical scanner: device that reads intensities of reflected light and converts the information to digital signals.

escáner óptico: dispositivo que lee la intensidad de la luz reflejada y convierte la información en señales digitales.

organic compound: one of a large number of compounds that contain the element carbon.

compuesto orgánicos: una de un gran número de compuestos que contiene el elemento carbono.

overtone: vibration whose frequency is a multiple of the fundamental frequency.

sobretono: vibración cuya frecuencia es un múltiplo de la frecuencia fundamental.

oxidation: the loss of electrons from the atoms of a substance in a chemical reaction.

oxidación: la pérdida de electrones de los átomos de una sustancia en una reacción quimica.

oxidation number: positive or negative number that indicates how many electrons an atom has gained, lost, or shared to become stable.

número de oxidación: número positivo o negativo que indica cuántos electrones ha ganado, perdido o compartido un átomo para alcanzar la estabilidad.

P

parallel circuit: circuit in which electric current has more than one path to follow.

circuito paralelo: circuito en el cual la corriente eléctrica tiene más de una trayectoria para seguir.

pascal: SI unit of pressure.

pascal: unidad SI de presión.

period: horizontal row in the periodic table. the amount of time it takes one wavelength to pass a fixed point; is expressed in seconds.

período: fila horizontal en la tabla periódica. El tiempo que requiere para que una longitud de onda pase un punto fijo; se expresa en segundos.

periodic table: organized list of all known elements that are arranged by increasing atomic number and by changes in chemical and physical properties.

tabla periódica: lista organizada de todos los elementos conocidos y ordenados de manera ascendente por número atómico y por cambios en sus propiedades químicas y físicas.

petroleum: liquid fossil fuel formed from decayed remains of ancient organisms; can be refined into fuels and used to make plastics.

petróleo: combustible fósil líquido que se forma a partir de residuos en descomposición de organismos ancestrales y que puede ser refinado para producir combustibles y fabricar plásticos.

pH: a measure of the concentration of hydronium ions in a solution using a scale ranging from 0 to 14, with 0 being the most acidic and 14 being the most basic.

pH: medida de la concentración de iones de hidronio en una solución, usando una escala de 0 a 14, en la cual 0 es la más ácida y 14 la más básica.

photochemical smog: the ozone-containing pollution that results from the reaction between sunlight and vehicular or industrial exhaust.

smog fotoquímico: la contaminación que incluye ozono y que resulta de la reacción entre la luz solar y el escape vehicular or industriales.

photon: massless energy-containing particle that electromagnetic waves sometimes behave like; the frequency of the electromagnetic wave increases with the energy of the particle.

fotón: partícula que contiene energía como la cual algunas veces se comportan las ondas electromagnéticas; la frecuencia de la onda electromagnética aumenta con la energía del fotón.

photovoltaic cell: device that converts solar energy into electricity; also called a solar cell.

célula fotovoltaica: dispositivo que convierte la energía solar en electricidad; también llamada celda solar.

physical change: any change in size, shape, or state of matter in which the identity of the substance remains the same.

cambio físico: cualquier cambio en tamaño, forma o estado de una sustancia en la cual la identidad de la sustancia sigue siendo la misma.

physical property: any characteristic of a material, such as size or shape, that can be observed without changing the identity of the material.

propiedad física: cualquier característica de un material, tal como tamaño o forma, que se puede observar sin cambiar la identidad del material.

pigment: colored material that is used to change the color of other substances.

pigmento: material de color que se usa para cambiar el color de otras sustancias.

pitch: perception of how high or low a sound is; related to the frequency of the sound waves.

tono: percepción de qué tan alto o bajo es un sonido; relacionado a la frecuencia de las ondas sonoras.

plane mirror: flat, smooth mirror that reflects light to form upright, virtual images.

espejo plano: espejo plano y liso que refleja la luz y forma imágenes verticales y virtuales.

plasma: matter with enough energy to overcome the attractive forces within its atoms, composed of positively and negatively charged particles.

plasma: materia con la energía suficiente para superar las fuerzas de atracción entre sus átomos, consiste de partículas con cargas positivas y negativas.

polar bond: a covalent bond in which the electrons are not shared equally, resulting in a slightly positive end and a slightly negative end.

enlace polar: enlace en el que los electrones no se comparten por igual, resultando en un lado ligeramente positivo y un lado ligeramente negativo.

polar molecule: a neutral molecule in which unequal electron sharing results in a slightly positive end and a slightly negative end.

molécula polar: molécula con un extremo ligeramente positivo y otro ligeramente negativo como resultado de un compartir desigual de los electrones.

pollutant: any substance that contaminates the environment.

contaminante: cualquier sustancia que ensucia al medioambiente.

polyatomic ion: positively or negatively charged, covalently bonded group of atoms.

ion poliatómico: grupo de átomos enlazados covalentemente, con carga positiva o negativa.

polyethylene: polymer formed from a chain containing many ethylene units; often used in plastic bags and plastic bottles.

polietileno: polímero formado por una cadena que contiene varias unidades de etileno; es comúnmente usado en la fabricación de bolsas y envases plásticos.

polymer: class of natural or synthetic substances made up of many smaller, simpler molecules, called monomers, arranged in large chains.

population: the total number of individuals of one species occupying the same area.

potential energy: energy that is stored due to the interactions between objects.

power: rate at which energy is converted; measured in watts (W).

precipitate: insoluble compound that is formed in a solution during a double-displacement reaction.

pressure: amount of force exerted per unit area; SI unit is the pascal (Pa).

product: in a chemical reaction, the new substance or substances formed.

protein: large, complex, biological polymer formed from amino acid units; make up many body tissues such as muscles, tendons, hair, and fingernails.

proton: particle in the nucleus with an electric charge of 1+.

polímero: clase de sustancias naturales o sintéticas compuestas por muchas moléculas más simples y pequeñas, llamadas monómeros, ordenadas en largas cadenas.

población: el número total de individuos de una especie que ocupa la misma zona.

energía potencial: energía almacenada que un objeto tiene debido a las interaciónes entre objetos.

potencia: tasa de cambio en que la energía se convierte; medida en vatios (W).

precipitado: compuesto insoluble que resulta en una solución por medio de una reacción de doble desplazamiento.

presión: cantidad de fuerza ejercida por unidad de área; la unidad SI es el pascal (Pa).

producto: en una reacción química, la sustancia formada or las sustancias formados.

proteína: polímero biológico extenso y complejo formado por unidades de aminoácidos; conforma muchos tejidos del cuerpo como los músculos, los tendones, el pelo y las uñas.

protón: partícula en el núcleo con una carga eléctrica de 1+.

Q

quark: particle of matter that makes up protons and neutrons.

quark: partícula de materia que constituye los protones y neutrones.

R

radiant energy: energy carried by an electromagnetic wave.

radiation: transfer of energy by electromagnetic waves.

radioactive element: element, such as radium, whose nucleus breaks down and emits particles and energy.

radioactivity: process that occurs when a nucleus decays and emits matter and energy.

radio wave: electromagnetic wave with wavelength longer than about 10 cm, used for communications.

rarefaction: the less-dense region of a longitudinal wave.

reactant: in a chemical reaction, the substance that reacts.

reaction rate: the rate at which reactants change into products in a chemical reaction.

real image: an image that appears at a certain location as a result of rays of light converging at that location.

reduction: the gain of electrons by the atoms of a substance in a chemical reaction.

reflecting telescope: uses mirrors and lenses to collect and focus light from distant objects.

refracting telescope: uses lenses to gather and focus light from distant objects.

refraction: the bending of a wave caused by a change in its speed as it travels from one medium to another.

energía radiante: energía transportada por una onda electromagnética.

radiación: transferencia de energía mediante ondas electromagnéticas.

elemento radiactivo: elemento, como el radio, cuyo núcleo se divide y emite partículas y energía.

radiactividad: proceso que ocurre cuando un núcleo se descompone y emite materia y energía.

onda de radio: onda electromagnética con longitud de onda más larga de aproximadamente 10 cm y que se usa en las comunicaciones.

rarefacción: la región menos densa de una onda longitudinal.

reactante: la sustancia que reacciona en una reacción química.

velocidad de reacción: velocidad en que se transforman los reactivos en productos en una reacción química.

imagen real: imagen que aparece en un cierto lugar como consecuencia de rayos de luz que convergen en ese lugar.

reducción: la obtención de electrones por los átomos de una sustancia en una reacción química.

telescopio reflexivo: usa espejos y lentes para recolectar y enfocar la luz proveniente de objetos distantes.

telescopio refractivo: usa lentes para reunir y enfocar la luz proveniente de objetos distantes.

refracción: el cambio en dirección de una ola debido a un cambio en su velocidad por viajar de un medio a otro.

renewable resource: energy source that is replaced by natural processes faster than it is used.

recursos renovables: fuente de energía que es se reemplaza mas rápido de que se consume.

resistance: tendency for a material to oppose electron flow and to convert electrical energy into other forms of energy, such as thermal energy and light; measured in ohms (Ω).

resistencia: tendencia de un material de oponerse al fluido de los electrones y convertir la energía eléctrica en energía térmica y luz; se mide en ohmios (Ω).

resonance: the process by which an object is made to vibrate by absorbing energy at its natural frequencies.

resonancia: el proceso por el cual un objeto vibra al absorber energía en sus frecuencias naturales.

resonator: hollow, air-filled chamber that amplifies sound when the air inside it vibrates.

resonador: cámara hueca, llena de aire, que amplifica el sonido cuando vibra el aire en su interior.

retina: inner lining of the eye that has cells which convert light images into electric signals for interpretation by the brain.

retina: capa interna del ojo que posee células que convierten imágenes iluminadas en señales eléctricas para que el cerebro las interprete.

reversible reaction: a reaction that can proceed in both the forward and the reverse directions.

reacción reversible: una reacción que puede proceder tanto en el avance que en la dirección retroceso.

S

salt: compound formed when negative ions from an acid combine with positive ions from a base.

sal: compuesto iónico que se forma cuando un halógeno adquiere un electrón de un metal.

saturated hydrocarbon: hydrocarbon, such as propane or methane, in which all the carbon atoms are connected by single covalent bonds.

hidrocarburo saturado: hidrocaburo, como el propano y el metano, en la que todos los átomos de carbono están unidos por enlaces covalentes simples.

saturated solution: any solution that contains all the solute it can hold at a given temperature.

solución saturada: cualquier solución que contiene todo el soluto que puede retener a una temperatura determinada.

scientific law: statement about what happens in nature that seems to be true all the time; does not explain why or how something happens.

ley científica: enunciado acerca de lo que ocurre en la naturaleza, lo cual parece ser cierto en todo momento, no explica cómo o por qué algo ocurre.

scientific methods: pattern of investigation procedures that can include stating a problem, forming a hypothesis, researching and gathering information, testing a hypothesis, analyzing data, and drawing conclusions.

método científico: patrón organizado de procedimientos de investigación que puede incluir el planteamiento de un problema, formulación de una hipótesis, investigación y recopilación de información, comprobación de la hipótesis, análisis de datos y elaboración de conclusiones.

second law of thermodynamics: states that energy spontaneously spreads from regions of higher concentration to regions of lower concentration.

segunda ley de la termodinámica: afirma que la energía de forma espontánea se extiende desde las regiones de mayor concentración a regiones de menor concentración.

semiconductor: material that conducts an electric current under certain conditions.

semiconductor: los material que conduce la corriente eléctrica bajo ciertas condiciones.

series circuit: circuit in which electric current has only one path to follow.

circuito en serie: circuito en el cual la corriente eléctrica tiene una sola trayectoria para seguir.

SI: International System of Units—the improved, universally accepted version of the metric system that is based on multiples of ten and includes the meter (m), liter (L), and kilogram (kg).

SI: Sistema Internacional de Unidades: la versión del sistema métrico mejorada y aprobada universalmente por científicos que se basa en múltiples de diez e incluye el metro (m), el litro (L) y el kilogramo (Kg).

simple machine: machine that does work with only one movement; examples include lever, pulley, wheel and axle, inclined plane, screw, and wedge.

máquina simple: máquina que realiza el trabajo con un solo movimiento; ejemplos incluye palanca, polea, rueda y eje, plano inclinado, tornillo y cuña.

single-displacement reaction: chemical reaction in which one element replaces another element in a compound.

reacción de un solo desplazamiento: reacción química en la cual un elemento reemplaza a otro elemento en un compuesto.

sliding friction: frictional force that opposes the motion of two surfaces sliding past each other.

fricción deslizante: fuerza de fricción que se opone al movimiento de dos superficies que se deslizan entre sí.

soap: organic salt with a nonpolar, hydrocarbon end that interacts with oils and dirt and a polar end that causes it to dissolve in water.

jabón: sal orgánica con un extremos de hidrocarburo no polar que interactúa con aceites y suciedad, y un extremo polar que ayuda a disolverlo en agua.

society: group of people that share similar values and beliefs.

sociedad: grupo de personas que comparten valores y creencias similares.

solar collector: device used in an active solar heating system that transforms radiant energy from the Sun into thermal energy.

recolector solar: dispositivo usado en un sistema de calefacción solar activa que transforma la energía radiante del Sol en energía térmica.

solenoid: a cylindrical coil of wire, used to produce a magnetic field when an electrical current passes through the wire.

solenoide: bonina cilíndrica de alambre, utilizado para producir un campo eléctrico cuando pasa una corriente eléctrica.

solubility: maximum amount of a solute that can be dissolved in a given amount of solvent at a given temperature.

solubilidad: máxima cantidad de soluto que puede ser disuelto en una cantidad dada de solvente a una temperatura determinada.

solute: in a solution, the substance being dissolved.

soluto: en una solución, la sustancia que está disuelta.

solution: homogenous mixture, remains constantly and uniformly mixed and has particles that are so small they cannot be seen with a microscope.

solución: mezcla homogénea, permanece constante y uniformemente mezclada y tiene partículas tan pequeñas que no pueden ser vistas en un microscopio.

solvent: in a solution, the substance in which the solute is dissolved.

solvente: en una solución, la sustancia en la cual se disuelve el soluto.

sonar: system that uses the reflection of sound waves to detect objects underwater.

sonar: sistema que usa la reflexión de las ondas sonoras para detectar objetos bajo el agua.

sound quality: result of the differences between sounds having the same pitch and loudness.

timbre: resulto de las diferencias entre sonidos del mismo tono e intensidad sonora.

specific heat: amount of heat needed to raise the temperature of 1 kg of a material 1°C.

calor específico: cantidad de calor necesaria para aumentar la temperatura un grado centígrado en un kilogramo de material.

speed: distance an object travels per unit of time.

velocidad: distancia que recorre un objeto por unidad de tiempo.

standard: exact, agreed-upon quantity used for comparison.

estándar: cantidad exacta y acordada, usada para hacer comparaciones.

standing wave: a wave pattern that forms when waves of equal wavelength and amplitude, but traveling in opposite directions, continuously interfere with each other; does not appear to be travelling.

onda estacionaria: patrón de una onda que se forma cuando ondas con la misma longitud de onda y amplitud, pero que viajan en direcciones opuestas, interfieren continuamente entre sí; parece que no se mueve.

static electricity: the accumulation of excess electric charge on an object.

electricidad estática: la acumulación del exceso de carga eléctrica en un objeto.

static friction: frictional force that prevents two surfaces from sliding past each other.

fricción estática: fuerza que evita que dos superficies en contacto se deslicen una sobre otra.

strong acid: any acid that dissociates almost completely in solution.

ácido fuerte: cualquier ácido que se disocie casi por completo en una solución.

strong base: any base that dissociates completely in solution.

base fuerte: cualquier base que se disocie completamente en una solución.

strong force: attractive force that acts between protons and neutrons in an atomic nucleus.

interacción nuclear fuerte: fuerza de atracción que mantiene juntos los protones y neutrones en un núcleo atómico.

sublimation: the process of a solid changing directly to a vapor without forming a liquid.

sublimación: proceso mediante el cual un sólido se convierte directamente en vapor sin pasar por el estado líquido.

substance: element or compound that cannot be broken down into simpler components without losing the properties of the original substance.

sustancia: elemento o compuesto que no se puede descomponer en componentes más simples sin perder las propiedades de la sustancia original.

substituted hydrocarbon: hydrocarbon with one or more of its hydrogen atoms replaced by atoms, or groups of atoms, of other elements.

hidrocarburo sustituido: un hidrocarburo en el cual uno o más de sus átomos de hidrógeno son reemplazados por un átomo, o grupo de átomos, de otros elementos.

supersaturated solution: any solution that contains more solute than a saturated solution at the same temperature.

solución sobresaturada: cualquier solución que contenga más soluto que una solución saturada a la misma temperatura.

suspension: heterogeneous mixture containing a liquid, and in which visible particles slowly settle due to gravity.

synthesis reaction: chemical reaction in which two or more substances combine to form a different substance.

synthetic: a material that is made in a laboratory or chemical plant and does not occur naturally.

system: a region or set of regions around which a boundary can be defined.

suspensión: mezcla heterogénea que contiene un líquido, y en el cual las partículas visibles lentamente se sedimentan.

reacción síntesis: reacción química en la cual se combinan dos o más sustancias y forman una sustancia diferente.

sintético: un material que se realiza en un laboratorio of fábrica de productos químicos y no se produce de forma natural.

sistema: una región o conjunto de regiones alrededor de lo cual se puede distinguir con un límite.

T

technology: application of science to benefit people.

temperature: measure of the average kinetic energy of all the particles that make up an object.

terminal velocity: the maximum speed an object will reach when falling through a substance, such as air.

theory: explanation of things or events based on knowledge gained from many observations and investigations.

thermal energy: sum of the kinetic and potential energy of the particles that make up an object.

thermal expansion: increase in the volume of a substance when the temperature is increased.

thermal insulator: a material through which thermal energy moves slowly.

thermodynamics: study of the relationship between thermal energy, heat, and work.

titration: process in which a solution of known concentration is used to determine the concentration of another solution.

total internal reflection: the complete reflection of light at a boundary that occurs when light strikes at an angle greater than the critical angle, and light travels faster in the second medium than in the first medium.

tracer: radioactive isotope, such as iodine-131, that can be detected by the radiation it emits after it is absorbed by a living organism.

transceiver: device that transmits radio signals at one frequency and receives radio signals at a different frequency, allowing a user to talk and listen at the same time.

transformer: device that uses electromagnetic induction to increase or decrease the voltage of an alternating current.

transition elements: elements in groups 3 through 12 of the periodic table; occur in nature as uncombined elements and include the iron triad and coinage metals.

translucent: material that transmits and scatters light so that objects viewed through it appear blurry.

transmutation: process of changing one element to another through radioactive decay.

tecnología: aplicación de la ciencia para el beneficio de la población.

temperatura: medida de la energía cinética promedio de todas las partículas que componen un objeto.

velocidad límite: la velocidad máxima que un objeto puede alcanzar cuando pasa en caída libre por una sustancia como aire.

teoría: explicación de las objetos o eventos que se basa en el conocimiento obtenido a partir de numerosas observaciones e investigaciones.

energía térmica: suma de la energía cinética y potencial de las partículas que componen un objeto.

expansión térmica: aumento del volumen de una sustancia al aumentar la temperatura.

aislante térmico: un material a través del cual la energia térmica se mueve lentamente.

termodinámica: estudio de la relación entre la energía térmica, el calor y el trabajo.

titulación: proceso mediante el cual una solución con una concentración conocida es usada parea determinar la concentración de otra solución.

reflexión interna total: la reflexión completa de la luz en un límite que se produce cuando la luz incide en un ángulo mayor que el ángulo crítico, y la luz viaja más rápido en el segundo medio que en el primer medio.

indicador radiactivo: isótopo radioactivo, tal como el yodo-131, que se detecta por la radiación que emite después de ser absorbido por un organismo vivo.

radio transmisor-receptor: dispositivo que transmite señales de radio a una frecuencia y recibe señales de radio en una frecuencia diferente, lo que permite al usuario hablar y escuchar al mismo tiempo.

transformador: dispositivo que usa inducción electromagnética para aumentar o disminuir el voltaje de una corriente alterna.

elementos de transición: los elementos de los grupos 3 al 12 de la tabla periódica que se encuentran en la naturaleza como elementos sin combinar e incluyen la tríada de hierro y los metales con los que se fabrican las monedas.

translúcido: material que transmite y dispersa la luz para que los objetos vistos a través de ella se vean borrosos.

transmutación: proceso de cambio de un elemento a otro mediante la descomposición radioactiva.

transparent: material that transmits light without scattering so that objects are clearly visible through it.

transuranium elements: elements having more than 92 protons, all of which are synthetic and unstable.

transverse wave: wave in which the matter in the medium moves at right angles to the direction of the wave, has crests and troughs.

trough: the lowest point on a transverse wave.

turbine: large wheel that rotates when pushed by steam, wind, or water and provides mechanical energy to a generator.

Tyndall effect: tendency for a beam of light to scatter as it passes through a colloid.

transparente: material que transmite luz sin dispersarse de manera que los objetos son claramente visibles a través de ella.

elementos transuránicos: elementos con más de 92 protones, que son sintéticos e inestables.

onda transversal: onda por la cual la materia en el medio se mueve en ángulos rectos con respecto a la dirección en que viaja la onda; tiene crestas y depresiones.

valle: el punto más bajo de una onda transversal.

turbina: rueda grande que gira al ser impulsada por vapor, viento o agua y que suministra energía mecánica a un generador.

efecto Tyndall: tendencia de un rayo de luz para dispersar al pasar a través de un coloide.

U

ultrasound: sound waves with frequency above 20,000 Hz; cannot be heard by humans.

ultraviolet wave: electromagnetic wave with wavelength between about 400 billionths and 10 billionths of a meter.

unsaturated hydrocarbon: hydrocarbon, such as ethene or ethyne, that contains at least one double or triple bond between carbon atoms.

unsaturated solution: any solution that can dissolve more solute at a given temperature.

ultrasonido: onda de sonido con frecuencia superiore a 20,000 Hz, no percebido por seres humanos.

onda ultravioleta: ondas electromagnética con longitud de onda entre aproximadamente 10 y 400 billonésimas de metro.

hidrocarburo no saturado: hidrocarburo, como el etileno, que contiene al menos un enlace doble o triple entre los átomos de carbono.

solución no saturada: cualquier solución que puede disolver más soluto a una temperatura determinada.

V

variable: quantity that can have more than a single value, can cause a change in the results of an experiment.

velocity: the speed and direction of a moving object.

virtual image: an image formed by diverging light rays that is perceived by the brain, even though the light rays do not actually originate from the place where the image appears to be located.

viscosity: a fluid's resistance to flowing.

visible light: electromagnetic waves with wavelengths of 700 to 400 billionths of a meter that can be detected by human eyes.

voltage difference: related to the force that causes electric charges to flow; measured in volts (V).

volume: amount of space occupied by an object.

variable: una cantidad que puede tener más de un valor, puede causar un cambio en los resultados de un experimento.

velocidad direccional: la rapidez y dirección de un objeto en movimiento.

imagen virtual: una imagen formada por las distintas rayos de luz que son percibidas por el cerebro, a pesar de que los rayos de luz en realidad no originan donde la imagen parece que se encuentra.

viscosidad: resistencia de un fluido al flujo.

luz visible: ondas electromagnéticas con longitudes de onda entre 400 y 700 billonésimas de metro y que pueden ser detectadas por el ojo humano.

diferencia de voltaje: relacionados con la fuerza que hace que las cargas eléctricas a fluir; se mide en voltios (V).

volumen: cantidad de espacio ocupado por un objeto.

W

wave: a repeating disturbance that transfers energy as it travels through matter or space.

wavelength: distance between one point on a wave and the nearest point just like it.

weak acid: any acid that only partly dissociates in solution.

onda: alteración repetitiva que transfiere energía a través de la materia o el espacio.

longitud de onda: distancia entre un punto en una onda y el semejante punto más cercano.

ácido débil: cualquier ácido que solamente se disocie parcialmente en una solución.

weak base: any base that does not dissociate completely in solution.

weight: gravitational force exerted on an object.

work: transfer of energy when a force is applied over a distance; measured in joules.

base débil: cualquier base que no se disocie completamente en una solución.

peso: fuerza gravitacional ejercida sobre un objeto.

trabajo: transferencia de energía cuando una fuerza actúa por una distancia y que se mide en julios.

X

X-ray: electromagnetic wave with wavelength between about 10 billionths of a meter and 10 trillionths of a meter, often used for medical imaging.

rayo X: onda electromagnética con longitud de onda entre 10 billonésimas de metro y 10 trillonésimas de metro, la cual se utiliza con frecuencia para producir imágenes de uso médico.

INDEX

Index Key

Bolded page numbers refer to vocabulary terms.
The following abbreviations appear after page numbers.
Activities and labs = *act.* Tables = *table*
Illustrations/photographs = *illus.* Problems = *prob.*

A

B

INDEX

C

D

E

F

G

H

INDEX

M

INDEX

Q

R

Index

S

T

INDEX